Advanced Electrical Machines and Drives Technologies—Volume I

Advanced Electrical Machines and Drives Technologies—Volume I

Editors

Loránd Szabó
Feng Chai

MDPI
Basel • Beijing • Wuhan • Barcelona • Belgrade • Novi Sad • Cluj • Manchester

Editors
Loránd Szabó
Technical University of Cluj-Napoca
Cluj-Napoca
Romania

Feng Chai
Harbin Institute of Technology
Harbin
China

Editorial Office
MDPI AG
Grosspeteranlage 5
4052 Basel, Switzerland

This is a reprint of articles from the Topic published online in the open access journals *Applied Sciences* (ISSN 2076-3417), *Designs* (ISSN 2411-9660), *Energies* (ISSN 1996-1073), *Machines* (ISSN 2075-1702), and *Vehicles* (ISSN 2624-8921) (available at: https://www.mdpi.com/topics/electrical_machines_drives).

For citation purposes, cite each article independently as indicated on the article page online and as indicated below:

Lastname, A.A.; Lastname, B.B. Article Title. *Journal Name* **Year**, *Volume Number*, Page Range.

Volume I
ISBN 978-3-7258-2307-9 (Hbk)
ISBN 978-3-7258-2308-6 (PDF)
doi.org/10.3390/books978-3-7258-2308-6

Set
ISBN 978-3-7258-2305-5 (Hbk)
ISBN 978-3-7258-2306-2 (PDF)

Contents

machines

MDPI

Review

The Key Role of 3D Printing Technologies in the Further Development of Electrical Machines

Loránd Szabó [1,*] and Dénes Fodor [2]

1 Electrical Machines and Drives Department, Technical University of Cluj-Napoca, RO-400114 Cluj-Napoca, Romania
2 Vehicle Industry Research Center, Széchenyi István University, H-9026 Győr, Hungary; fodor.denes@ga.sze.hu
* Correspondence: lorand.szabo@emd.utcluj.ro

Abstract: There is a strong general demand for the permanent improvement of electrical machines. Nowadays, these are at their near maximum potential, and even small further improvements can only be achieved with great effort and high cost. The single solution should be a paradigm shift in their development, by using radically new approaches to topology, materials, and fabrication. Therefore, the application of diverse 3D printing techniques for advanced fabrication in this field is inevitable. Therefore, these new approaches are receiving a great deal of attention among electrical machines designers. In the paper, the possible applications of these new fabrication technologies in the field of electrical machines are surveyed. The focus is set on emphasizing the advancement over the traditional manufacturing approaches.

Keywords: additive manufacturing process; advanced materials; applications of additive manufacturing; electrical machines

Citation: Szabó, L.; Fodor, D. The Key Role of 3D Printing Technologies in the Further Development of Electrical Machines. *Machines* **2022**, *10*, 330. https://doi.org/10.3390/machines10050330

Academic Editor: Francisco J. G. Silva

Received: 3 April 2022
Accepted: 26 April 2022
Published: 1 May 2022

1. Introduction

Electrical machines (EM) are an essential part of both advanced industry and the electrification of transportation. The global electrical machine market is expected to increase to 195.1 billion USD by 2030, growing at a compound annual growth rate (CAGR) of 6.3% during this period [1].

The applications of EMs come up against several frequently contradictory requirements, e.g., great power/torque density, high efficiency, wide speed range, good dynamics, high degree of reliability and robustness, low level of acoustic noise, and, of course, reduced cost [2]. However, EMs seem to have almost reached their maximum potential. Even small further improvements can only be made by investing an enormous amount of effort at a significant expense. It is sufficient to mention the hard work carried out by designers to ensure induction machines (IMs) comply with drastic energy efficiency regulations [3]. It can be assumed that advancing to the next technological level in this field can only be achieved by a paradigm shift, namely by employing entirely new approaches with regard to topologies, the materials used, and manufacturing [4]. These novel and innovative advances are interdependent, and only when applied together will they bring the needed breakthrough.

The classical manufacturing technologies available strongly limit how industrially widespread both the new conductive and magnetic materials are, as well as innovative EM constructions. Therefore, the possible application of novel, revolutionary fabrication technologies, such as additive manufacturing (AM), must be very seriously considered, even more so since this is a constituent component of so-called "smart factories", which maximize benefits brought about by the Fourth Industrial Revolution (named Industry 4.0). This technology is considerably different from traditional manufacturing processes such as subtractive (drilling, milling), formative (casting), and joining processes (welding, fastening) [5].

This paper focuses on the possible improvements brought about by the implementation of AM with regard to the fabrication of iron cores, permanent magnets (PMs), windings, and cooling systems for EMs and usable materials. Furthermore, some typical applications of using 3D printing in this field are surveyed.

2. Additive Manufacturing Technologies

AM, also referred to as 3D printing, involves the layer-by-layer processing of parts. It is an advanced technology that combines digitally controlled high-precision mechanical processing equipment, and the use of advanced materials fitted to this kind of processing. Huge research, development, and investment efforts over the past few years have made this technology essential in almost all industries.

By applying AM methods, components of essentially any complex shape of a great variety of materials can be precisely produced without scratches. The obtained parts can have lower volume and mass than if they were traditionally manufactured. Outputs from advanced topology optimization (including artificial intelligence-based ones) can be closely achieved. Most of the AM technologies enable the easy integration (simultaneously processing) of components even made of different materials, which previously had to be separately fabricated. A notable advantage of this method is that the processing costs do not depend on the component complexity. By using AM, the fabrication of all the basic components of an EM (magnetic core, PMs, conductors/windings, cooling system, and housing) is possible [6].

The many AM technologies have been classified into seven categories by the American Society for Testing and Materials (ASTM) [7,8]. A large proportion of them is useful for the fabrication of EMs. Next, only these are mentioned.

Powder bed fusion (PBF) is one of the most widely used AM methods [9]. Its most common variants are selective laser melting (SLM), also called laser beam melting (LBM) [10,11], and selective laser sintering (SLS) [12]. It consists of the successive melting of thin layers of metal powders by a laser beam. Even parts compounded by two different materials can be together fabricated [13]. Direct metal laser sintering (DMLS) is another similar technique, developed especially for the 3D printing of alloys composed of materials with different melting points that are fused on a molecular level at high temperatures [14].

Directed energy deposition (DED) is also a popular 3D printing process, which uses lasers, electron beams, or electric arcs to melt and deposit metal powders or wires in successive layers [9,15]. Within this technology, the laser-engineered net shaping (LENS) method seems the most suitable for manufacturing the iron cores of EMs from a diverse range of metal powders. As this device can inject the metal powder anywhere in the heated workspace, the metal layer can be deposited on any complicated curved surfaces.

Fused deposition modeling (FDM), also known as fused filament fabrication (FFF), involves placing a thin continuous melted filament composed of a thermoplastic material layer-by-layer to form the desired component [16]. This is adequate for the fabrication of bonded PMs.

The binder jetting (BJ) AM technology consists of a printhead selectively depositing drops of binding liquid onto a thin layer of powder, thereby forming a layer of the structure to be fabricated [17]. Cold spraying (CS) is based upon the amalgamation of microparticles accelerated faster than the speed of sound and sufficiently directed onto an appropriate substrate. It is a cheap technology that uses inexpensive powders, but the precision of the final products is low [18]. By utilizing CS manufacturing, iron cores, windings, and PMs can be produced. Laminated Object Manufacturing (LOM) is also a sheet lamination 3D printing technique that is widely applied. It uses temperature and pressure to fuse or laminate layered materials [9]. For a great variety of materials, stereolithography (SLA) technology can also be used [19,20]. Upon this, a liquid photosensitive prime material is turned into a 3D solid object layer-by-layer by using a low-power laser beam and photopolymerization.

Despite rapid advances in various AM technologies, in some high-tech applications, standalone metal-powder-based 3D printing cannot achieve the very tight geometric tolerances and surface integrity required. In such cases, the so-called hybrid AM techniques must be applied, where the 3D printed parts are further post-processed to precisely meet all the design constraints. There are already hybrid AM systems that are commercially available that combine laser-based 3D printing with advanced post-processing techniques (such as machining, heat treatment, surface treatment, etc.) that can also produce complex parts of the EMs [21].

3. Materials for 3D Printing of Electrical Machines

3.1. Soft Magnetic Materials

High permeability and saturation, besides low iron core losses, are the basic technical requirements for materials the magnetic core of EMs are manufactured of. Traditionally, they are made of punched non-oriented electric steel laminations stacked together [22]. Even if the soft magnetic materials used are continuously enhanced (see, for example, the improvements of iron-silicon alloys in [23]), significant progress could be made concerning the iron losses by simply reducing the thickness of the laminations. However, due to the low mechanical strength of the thin sheet, the laminations are bent during stamping and cannot be packed at a high stacking coefficient. Electrical steel lamination technology also limits the flexibility in EM design since complex-shaped 3D iron cores make it difficult to assemble laminations [22].

AM can bring advancements in this field by enabling the 3D printing of excellent-quality powdered soft and hard magnetic materials. By using this, complex-shaped iron cores that allow true 3D magnetic flux paths can be made, and good control over electromagnetic and mechanical properties of the cores can be achieved [24].

A great variety of soft magnetic materials to be used in AM of EMs is at the disposal of design engineers. Most of the traditional soft magnetic alloys, such as Fe-Si, Fe-Ni, Fe-Co, and soft magnetic composites (SMCs) can be 3D printed [25]. Besides these, there is intensive research to improve the quality of other magnetic materials (such as amorphous and nanocrystalline ones) to shape their properties to the requirement laid down for EM iron cores [26].

Nowadays, iron alloyed with a low quantity of silicon is the most-used material in the construction of iron cores (in both grain-oriented and non-oriented states). Usually, up to 1% aluminum and less than 0.5% manganese are also added to the alloy to improve its performance. Finding the ideal recipe is complicated since in all cases the change in the amount of a material has a contradictory effect on the overall performance of the alloy [23].

Recent studies have demonstrated that iron-silicon alloys with a high 6.5% silicon composition (called Fe-6.5Si) are most adequate for the AM of the iron cores of EMs [27]. Parts made of Fe-6.5Si employing DMLS technology have better magnetization properties than Fe-Si alloys with a lower silicon content (Fe-3Si having only 3% Si), and the eddy current losses at 50 Hz are reduced by more than 50% [28].

Moreover, Fe-Ni alloys with up to 80% nickel composition (the *Permalloy* or the *Supermalloy*) can be candidates for 3D printing the iron cores of EMs. These have excellent magnetic properties concerning their losses and magnetic permeability, but relatively low saturation magnetic flux density. These alloys can be easily 3D printed due to their exceptional mechanical permeability. For fabricating EMs' iron cores, Fe-Ni alloys can be processed by diverse AM technologies, such as DED [29], SLM [30], and FDM [31].

Fe-Co alloys have the greatest saturation magnetic flux density among the soft magnetic materials considered here, with 2.45 T for the 50-50% Fe-Co alloy, trade named *Permendur* [32]. Supplementarily, this alloy has low coercivity and losses and high magnetic permeability and Curie temperature.

Adding nickel, *Perminvar* (30Fe-25Co-45Ni) can be obtained, which has constant magnetic permeability across a wide field range. By enhancing it with 2% vanadium (*Hiperco*—49Fe-49Co-2V), its ductility (machinability) and electrical resistivity are radically improved, without significantly losing its other good magnetic properties [33,34].

All the Fe-Co alloys are suitable for AM by both SLM [32] and LENS [34] methods.

In the past, SMCs were the single potential alternatives to electrical steel laminations in the EMs iron core construction, especially when complex 3D-shaped iron cores had to be manufactured [35]. These are iron powder particles bounded by an insulating coating. Their magnetic characteristics (saturation flux density, magnetic permeability, and iron core losses) compared with electric steel laminations at low frequencies are slightly worse. This is mainly due to the non-magnetic space between the iron powder particles. However, due to its great electrical resistance, the eddy current losses are very small. This advantage becomes essential as the frequency increases (for example, in the case of high-speed EMs) [22,23,36].

Unfortunately, in the past, SMC-based iron cores could only be fabricated by employing costly, time-consuming, and sophisticated techniques. This drawback can be now easily solved by applying diverse 3D printing technologies [37–39].

Amorphous and nanocrystalline irons are newcomer candidates for the fabrication of high-quality EM iron cores.

Amorphous (glass) iron is a non-crystalline solid that has its atoms randomly arranged. This unique property is twofold. It has excellent magnetic properties (such as low coercivity, magnetic anisotropy, hysteresis losses, high magnetic permeability, resistivity, etc.) since it responds more rapidly to the changes in magnetic fields than the ordinary crystalline iron and its alloys [40]. Due to its ten times lower iron losses compared to the usual laminations, they were generally considered crucial in achieving IE5 motor efficiency standards [41,42]. On the other hand, its manufacturing is complicated since, to obtain the glass-like material, a very fast (about 1 million °C/s) cooling rate must be ensured. AM technologies also brought about a breakthrough in the practical use of glass iron alloys in EMs. SLM and LENS were proven to also be adequate for processing parts made of amorphous iron [43–45].

The very complex microstructure of the nanocrystalline iron alloys contains very small grains (<50 nm) of iron, boron, phosphorus, silicon, and noble metals serving as nucleating agents for the ferromagnetic nanocrystalline phase. These alloys also have very good magnetic properties, comparable with those of amorphous iron alloys, but due to their more stable structure, their magnetic and mechanical properties can be more easily tailored for specific requirements. Unfortunately, they are fragile, and thus difficult to handle during EMs' fabrication [23]. *FINEMET* (73.5Fe-1Cu-3Nb-15.5Si-7B) is the on-shelf available flagship of such soft magnetic materials. Current research activities are targeting the AM of nanocrystalline iron alloys by LENS [29].

A comparison of the soft iron materials that could be used for the iron cores of EMs to be produced via AM is of interest. The above-detailed materials are compared upon four critical characteristics in designing EMs: Specific iron core losses, saturation flux density, maximum magnetic permeability, and electric resistivity. The data from Table 1 were collected from diverse data sheets [46–49] and the reported measurements [22,23,26,50–54]. The scores range from 1 (poor) to 5 (excellent).

Table 1. Comparison of soft magnetic materials to be used for iron cores processed by AM.

Soft Magnetic Material	Iron Core Losses@50 Hz, 1 T	Saturation Flux Density	Magnetic Permeability	Resistivity
Non-oriented electrical steel (*M330-50A*)	2	4	2	3
Fe-Si alloy (*6.5% Si*)	3	4	4	3
Fe-Ni alloy (*Permalloy*)	4	2	4	3
Fe-Co alloy (*Hiperco*)	3	5	3	3
SMC (*Somaloy 700*)	1	3	1	5
Amorphous iron (*MetalGlass*)	4	4	4	4
Nanocrystalline iron (*FINEMET*)	5	3	5	4

Note: Poor, Fair, Good, Very good, Excellent

As it can be seen from the table, amorphous and nanocrystalline soft iron materials are the most promising. The former has good properties for all the selected criteria, while the latter has excellent, small iron core losses at low frequencies and great magnetic permeability. It must be mentioned that when selecting the material for the iron core of an EM, not only must these issues be considered but also other important features related to mechanical and thermal behavior. Last but not least, economical concerns (price, availability, etc.) must also be evaluated.

3.2. Hard Magnetic Materials

PM EMs are potential candidates for a wide range of industrial and automotive applications, mainly due to their high efficiency and specific power density. For such EMs, both cheap low- and more expensive high-energy density PMs are required to be integrated into the iron cores. Commonly, PMs are produced by means of sintering or bonding. Both technologies are complicated and time-consuming. Therefore, in this case, AM can also bring a significant advance in EM manufacturing [24].

In the literature, no special PM materials are proposed to be produced via AM. Most of the papers deal with the fabrication of the strongest available PM, Neodymium-Iron-Boron (NdFeB). Besides its very good magnetic properties (great residual magnetic flux density, coercive magnetic field strength, and energy density), this also has favorable mechanical features. On the other hand, it has limitations concerning temperature, corrosion, and availability.

There are some AM processes experimented with for the fabrication of both bonded and purely metallic NdFeB PMs. Bonded PMs can be manufactured by many AM technologies, for example, by using FDM, BJ, or SLA. Through these, the constituent metallic particles can be adequately mixed with the polymer binder material during the fabrication of each layer [22,55]. Moreover, other technologies such as SLM can be used. In this case, a stratum of a mixture of two powder materials (NdFeB-based hard magnetic material and a low melting eutectic infiltration alloy) is laid, and the laser fuses the required surface [56].

Regardless of the 3D techniques used to manufacture PMs for EMs, they offer a huge advantage as they allow the fabrication of PMs with arbitrary, even very complex, PM shapes [57].

3.3. Conducting Materials

Nowadays, almost all the EMs have windings made of copper, except the squirrel cages of low power/voltage IMs, which are manufactured from aluminum [58]. Both materials have very good electrical and heat conductivity. The first property is crucial in the case of EMs since it enables low resistance to the windings resulting in low winding losses.

Furthermore, in the case of 3D printing the windings, these two materials are the principal candidates. AM of pure copper is difficult because, due to its very good thermal conductivity, it cools prematurely during any 3D printing process that heats the material to be processed. Furthermore, freshly fabricated pure copper parts oxidize very quickly [59]. To prevent all of this, the *C18150* CuCrZr alloy is applied, which contains 98.9% copper, 1% chrome, and 0.1% zirconium. It has 80 ÷ 90% of the electric conductivity of the reference International Annealed Copper Standard (IACS) [60]. It can be 3D printed by using both DMLS and PBF technology [61].

Despite the near 40% less electric and thermal conductivity of aluminum as compared with copper, it has the advantage of much lower density and price. AlSi10Mg is the most customary aluminum alloy, used inclusively in the EMs industry due to its suitable casting properties, strong corrosion resistance, and good strength and rigidity, even if it has less than 70% of the electric conductivity of IACS [62]. It can be 3D printed employing DMLS or SLM methods [63].

When looking for the possibilities of improving the efficiency of the EMs, obligatorily graphene, the "wonder material" of the 21st century, must also be considered [64,65]. It has numerous outstanding characteristics, among them the extreme electric and thermal conductivity [66]. Its one-dimensional (1D) allotrope, the carbon nanotube (CNT), is also very promising for improving the conductibility of 3D printed windings.

Graphene is also a superior nanofiller, hence it can be well-compounded with copper to obtain very low resistivity conductors [51,67]. In [68], a CNT-Cu composite was reported that has 2.6 times higher conductivity than pure copper. Moreover, its temperature coefficient of resistivity is one order of magnitude lower than that of copper. By increasing the CNT content of the composite, a 42% density reduction was obtained, which is a very important issue for mobile applications, as was already stated. Windings of such composites can be manufactured via FDM technology [69]. Through the same process, flexible graphene-polylactic acid (PLA) composite wires can also be prepared for manufacturing the windings of EMs [70].

3.4. Insulator Materials

The classical enamel coatings of the conductors (magnet wires) can be replaced by better temperature-resistant insulator materials, such as ceramics. Due to the excellent thermal stability of the ceramic, such insulated wires can endure over 300 °C, a temperature much greater than the maximum allowed within the H insulating class (180 °C) of the EMs [50]. Due to its high melting point, the processing of these ceramics was inappropriate using classical manufacturing methods. However, by using diverse AM technologies, for example, LOM or FDM, ceramic insulators of any shape can be much more straightforwardly fabricated [50,71,72].

Even if they are hard to 3D print, polyether ether ketone (PEEK)-based polymers must be also considered here. They can be used in EMs for insulating the slots, but also the bearings, preventing the dangerous bearing currents [73]. Such polymers can be 3D printed by diverse methods, such as FDM, SLA, or SLS [12].

3.5. Materials for Thermal Management

For the thermal management of EMs, several devices, such as heatsinks, heat exchangers, pipes, cooling packets, etc., are used. By using a variety of AM techniques, very fine cooling structures can be made precisely.

A common requirement for the materials used in this scope is to have a high thermal conductivity coefficient (TCC) for easy heat evacuation and obviously good electrical insulation [74]. For these, a very low coefficient of thermal expansion (CTE) is also a basic requirement [75].

Due to their favorable thermophysical properties, aluminum and copper are generally the most widely used materials in thermal management devices such as radiators, pipes, and heat exchangers. They can be easily processed by a variety of AM techniques. However, aluminum is preferred because copper will rapidly (during the printing process) interact with oxygen in the air. In practical applications, instead of pure aluminum, alloys with magnesium and silicon are used, such as the already-mentioned AlSi10Mg or the Al 6061 [76,77].

Ceramics are one of the most commonly emerging materials today, and they are also well-suited for AM of thermal management systems. Furthermore, these low-density materials are not corroded by any coolant agent [78]. Aluminum nitride, beryllium oxide (BeO), or silicon carbide (SiC) ceramics seem to be the best choice for such applications as they have high TCC.

Graphene must also be considered for such applications [65,79] since it has a very small density and its TCC is at least 7 times and 4 times higher than that of aluminum and copper, respectively [80]. It also has other outstanding properties, such as a very high specific surface area, elasticity, and strength, which make it well-suited for a wide range of thermal management applications [81].

A great diversity of composites for cooling applications are proposed in the literature. One option is to use metal-matrix composites, such as *Invar* (64Fe-36Ni) or *Kovar* (54Fe-29Ni-17Co), which have a very low CTE but higher TCC [82,83]. Furthermore, carbon-based composites can be considered for such applications. Besides relatively high TCC and low CTE, carbon also has corrosion resistance and a low density [84,85]. It is fully compatible with a variety of carbon fibers, so its thermal conductivity can be improved by applying graphitized carbon fibers [75].

Metal foams are highly porous structures composed of solid metal with closed or open pores [86]. Aluminum or copper-based metal foams with open cells (so-called metal sponges) are ideal for the AM of fluid-cooled heat exchangers to be placed inside the EM constructions since they have high strength and bending rigidity, good TCC, and are lightweight.

4. 3D Printing of Windings

The optimal design of the windings for the EMs can significantly contribute to the improvement of their performances. By using low electrical resistivity materials, the Joule (copper) losses can be reduced. By achieving a high slot filling factor, the iron core volume can be diminished [87], and thus also the iron core losses [24]. An adequate winding design can decrease the air-gap magnetomotive force (MMF) harmonics, too.

By applying 3D printing for the purpose of manufacturing, several advancements can be achieved, e.g., special forms of windings, better heat resistant insulations, and particular wire profiles that can also be equipped with cooling channels. A disadvantage of AM windings is that since they cannot be made of pure copper or aluminum, their resistance is somewhat greater.

EMs either have distributed or concentrated windings. Unfortunately, both have drawbacks. Those that are distributed have longer end-windings (resulting in higher Joule losses) and smaller slot fill factors. By using concentrated windings, higher harmonics of air-gap MMF result, and the heat generated in the windings is not evenly distributed over the circumference of the machine [88].

For low-power EMs, the windings are fabricated from round copper conductors. Even though present-day winding technologies (including automated ones) have been well-established for tens of years, they all have a basic drawback, namely only a low fill factor is achievable due to spaces between the wires. By using flat conductors with a rectangular cross-section, theoretically, the space required for the slot can be more efficiently filled [89].

However, the use of commercially available pre-insulated flat conductors has some limitations concerning the accessible cross-sections and the minimum bend radii, which both restrict the reduction in the winding volumes. By using AM, such conductors can fill the slots with a very high factor. Therefore, smaller slots and teeth cross-sections can be used, leading to a reduction in the iron core volume and, as a result, in the iron losses as well as an increase in the specific power/torque of the EMs [88]. The AM of windings comprised of flat conductors enables EM designers to implement windings of any optimal shape. As a result of these advanced manufacturing technologies, the feedstock of a wide range of powdered conductive and insulating materials are selectively and successively bonded together.

Concentrated windings seem to be the most appropriate in terms of AM. They can be manufactured of pure copper, having air between the flat layered turns, as can be seen in Figure 1a [90]. For better insulation of the layers, the researchers from the Chemnitz University of Technology (Germany) used a 3D-printed coating of ceramic material and a suitable binding agent [91]. The coils designed for an SRM have an optimized shape, as well as exhibiting good conductive insulation and superior mechanical properties. By using AM technologies such as DMLS, concentrated coils of more complex shapes can be made, which closely follow the form and location of the slot where it is to be placed [92]. A salient pole stator with such coils (shown in Figure 1b) was developed for a racing car by Additive Drives (Germany).

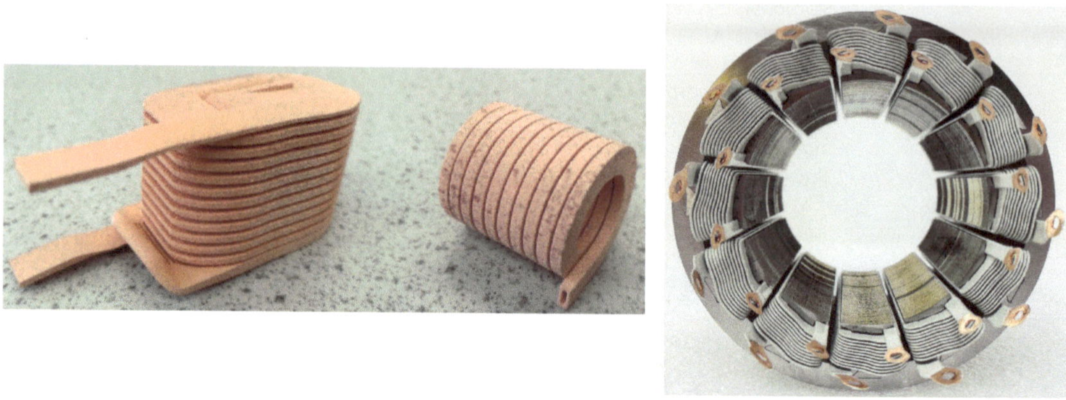

(a) (b)

Figure 1. Concentrated coils made by AM technologies. (**a**) Samples made by laser PBF, reproduced with permission from the author of [90]; (**b**) coils used on a salient pole stator, reproduced with permission from the owner, Additive | Drives GmbH (Dresden) [93].

AM can also bring about advancements regarding distributed windings in EMs. Basically, optimized windings of any shape can be fabricated utilizing this technology.

One so-called hairpin winding can be constructed from rectangular conductors, which are curved into a U-shape (as can be seen in Figure 2a) and inserted into the iron core slots. The ends of the windings are joined together (usually welded) on the other side of the iron core. For very high current windings, multiple insulated copper conductors can be printed in parallel upon the so-called Roebel transposition (see Figure 2b), thus improving the efficiency and other performances of high-power EMs [94,95].

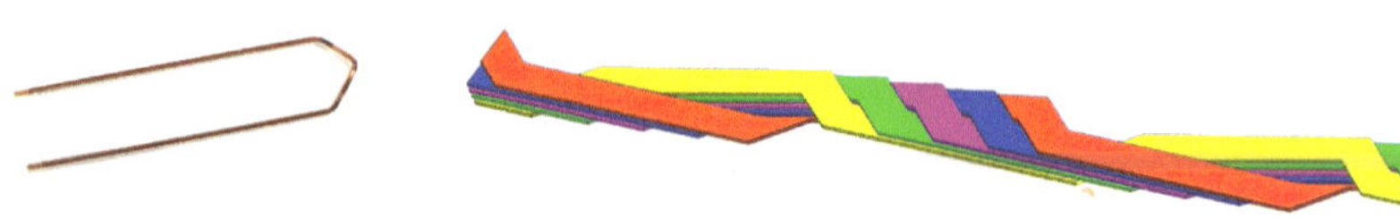

(a) (b)

Figure 2. Hairpin stator windings. (**a**) A hairpin winding branch for low currents, reproduced with permission from the owner, Otto Bihler Maschinenfabrik GmbH & Co. KG (Germany) [96]; (**b**) half-bar comprising of five strands assembled upon the Roebel-transposition, which can be processed by AM [97].

Such windings have several advantages, e.g., a significantly increased fill factor (up to 0.7÷0.8), the feasibility of automated processing as well as AM, and good thermal performances due to the small amount of air between the conductors. Such distributed windings have shorter end-windings (and an inherently lower winding resistance) and do not come into contact with each other.

Therefore, more effective cooling systems (for example, in-slot water jackets) can be applied [98]. On the other hand, they have drawbacks, namely the low degree of flexibility with regard to winding designs limited by the reduced number of conductors per slot, substantial AC winding losses when operating at high-speed, and high manufacturing costs [88,99,100].

Hairpin windings are typically used in pretentious EMs. A typical example of a 3D printed hairpin stator winding manufactured at Additive Drives (Germany) is given in Figure 3 [101].

(a) (b)

Figure 3. 3D-printed hairpin stator winding, reproduced with permission from the owner, Additive | Drives GmbH (Germany) [102]. (**a**) Wound stator; (**b**) end-windings.

Upon such windings, a similar technology, namely plug-in windings, was proposed. Basically, a pair of hairpin windings are combined via male–female plug-in features. Although these windings are more easily automated manufactured, special measures have to be taken to join the two branches, and the resulting contact exhibits a high contact resistance [103,104].

AM enables the fabrication of non-conventional windings proposed for EMs of any level of complexity [105,106]. As a potential example, Figure 4 presents a special, radially displaced segmented laminated winding proposed by Reinap, A. in [105].

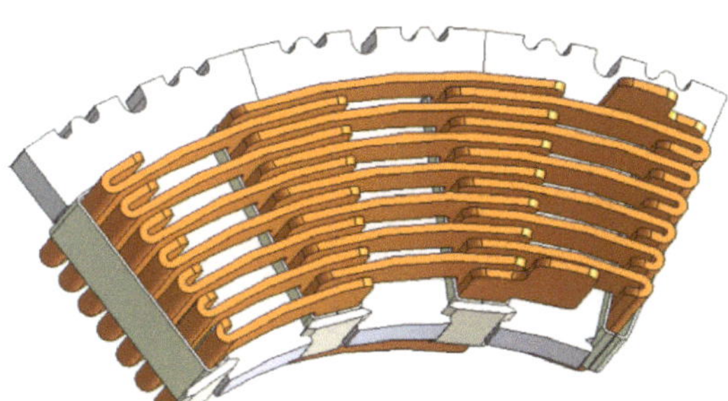

Figure 4. Radially displaced segmented laminated winding, reproduced with permission from the owner, Lund University (Sweden) [105].

5. Additive Manufacturing of the Iron Cores

As noted, the possible use of AM technologies has had the lowest impact on the fabrication of the iron cores of the EMs, despite the great variety of utilizable ferromagnetic materials. Even if the above-mentioned prospective soft magnetic materials can be 3D printed in very thin layers (for example, via LOM technology), an insulating coating among them is needed to reduce the eddy current losses. This makes the use of AM technologies less attractive since it offers only two minor advantages: The possibility to construct rarely used, radially unsymmetrical iron cores and the elimination of the negative influence of punching, welding, and clamping over the traditional iron core stacks [50,107]. As a consequence, most of the practical realizations in the field of 3D printing of these iron cores are related to very small scale (moreover, demonstrative) EMs [108,109], and to axially flux PM ones [110].

As stated already, FeSi alloys with a high silicon content (up to 7%) are excellent materials for manufacturing iron cores of EMs. However, as the silicon content is increased by over 4.5% and the electric and magnetic properties (such as electrical resistivity, magnetic saturation, magneto-crystalline anisotropy, magnetostriction, etc.) are improved, the steel begins to brittle. Thus, punching technology cannot be applied, but the DMLS approach can solve this mechanical issue [111].

In [112], iron composite filaments were proposed for the fabrication of EMs' iron cores. The very dense material must be sequentially 3D printed and sintered. It was demonstrated in the paper that the obtained material has high magnetic permeability. Unfortunately, the paper did not deal with the important iron losses.

In [27], an innovative hollow-core transformer iron core made of silicon electrical steel alloys (Fe-3Si or Fe-6Si) was proposed. It was designed upon a certain Hilbert curve geometry, a continuous fractal space-filling curve (see Figure 5), and it can be exclusively fabricated by AM means, for example, SLM.

Figure 5. 3D-printed hollow iron core designed upon a Hilbert curve geometry [27].

This construction idea can be extended to salient pole EMs, too. Even if the iron core losses were not reduced radically and the core volume was increased, the proposed structure has the potential of internal cooling the iron core legs.

By using the same manufacturing technology and Fe6Si powder, the iron core of an SRM was produced [113]. The obtained results showed greater iron core losses as compared with traditional construction made of laminations. These results are not relevant since the iron core was built up without insulating thin layers of the material and thus reducing eddy current losses. The paper only demonstrated the possibility of manufacturing iron cores using 3D printing.

Here, coreless PM rotor EMs should also be mentioned. Their rotor is comprised only of PMs and the potentially 3D-printed nonferrous (usually plastic) PM holders of any complexity. Such EMs have the advantage of not having iron losses (even at very high speeds) in the rotor, and not producing cogging torque, which increases the torque ripple. On the other hand, they have low torque/power density [114].

6. Additive Manufacturing of the Cooling Systems

Thanks to the optimized design of EMs, their speed and power density are pushed to their maximum limits. Therefore, a significant amount of heat is generated within them. The resulting thermal stress not only influences the performance of the EMs but also their lifespan. As a consequence, complicated cooling systems are required to maintain the internal temperature below its maximum allowable value [115]. Effective cooling is a challenging engineering task hampered by the closed waterproof housing, variable ambient temperatures, and humidity.

Conventional EMs are all air-cooled via a fan attached to the shaft and are equipped with housing fins to increase the convective cooling surface. This approach is inadequate for EMs applied in variable speed drives since, at low velocities, the fan does not guarantee the airflow required to cool the machine. In such cases, an independently driven radial fan is used. This method must also be applied for high-speed EMs since, at great velocities, the shaft-mounted fans are very noisy and account for a significant proportion of the developed torque.

This last method is not adequate for transportation applications since, due to the very limited amount of space under the hood of a vehicle, there is insufficient space for cooling fins on the housing and external cooling fans. For such applications, liquid cooling systems are the most efficient and widespread approaches. They consist of a pump, radiator, and assorted jackets or pipes within the housing of the EM. The heat generated is removed via the cooling agent that circulates through the jacket or pipes and passes through the radiator.

In terms of AM, cooling systems can bring about the most significant breakthrough in the field of EMs. Several older but very good designs of thermal management systems were infeasible due to their inability to be manufactured. 3D printing can enable the fabrication of specially shaped frames, cooling channels, and heat exchangers optimized for better thermal management of the EMs.

6.1. Hollow Winding Conductors

Conductor windings with a heat pipe or cooling channel in the middle of them, which can transport coolant fluids or gases, can ensure a good degree of direct local cooling of the windings. At higher currents, as in the case of high-power EMs, large cross-sections of hollow conductors are hard to shape as optimally required. However, by using AM, such conductors can be formed in any shape [51,116–118]. Obviously, although hollow conductors with the same active cross-section as their solid counterparts have a larger effective area, the current density in the windings can be multiplied as many as 3.5 times [119,120].

6.2. Direct Cooling of the Windings

AM enabled the global thermal performance of EMs to be improved by directly cooling the windings instead of optimizing the airflow inside the machine or of the active convective cooling surface of the frame [121].

Designers can make use of two of the basic benefits of diverse AM technologies. On the one hand, tubes with walls as thin as 60 μm can be produced, which enables the fabrication of coolers of low masses and volumes but with huge convective cooling surfaces, and hollow structures of any degree of complexity through which the cooling agent can flow [10]. On the other hand, AM technologies support the manufacturing of the coolers implemented in special thermally conductive electrical insulators, which are hard to process using conventional methods [122]. These materials must not only exhibit high thermal conductivities but also still be good electrical insulators with high dielectric strength. Moreover, they must be non-magnetic and corrosion-resistant [123–125]. The selection of the most suitable material in each cooling system has a significant influence on the thermal management of EMs.

Both the end-windings and the coil branches inside the slots can be cooled directly by applying new, possibly 3D-printed components made of such materials. These cooling methods are very effective since the cooling surfaces are very close to the heat sources (the windings).

In [126], a specially shaped duct is presented, which is placed through the end-windings and circulates a coolant agent (as can be seen in Figure 6a). In another approach, a micro-structured heat pipe surrounds the end-windings of an experimental permanent magnet synchronous machine (PMSM), as shown in Figure 6b [127]. Both technical solutions have the advantage of increasing the cooling efficiency without increasing the volume of the EM.

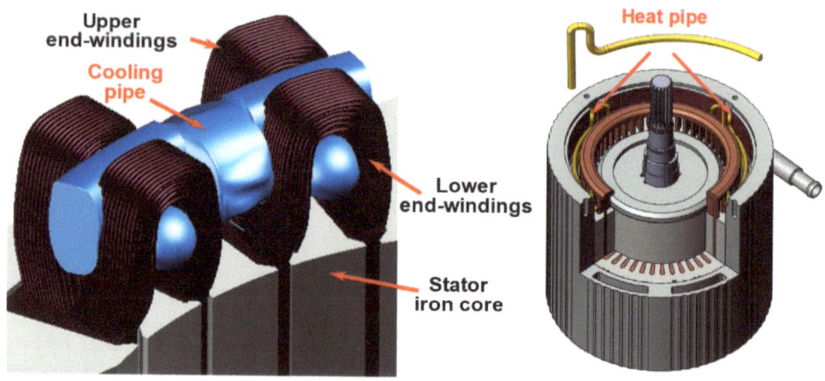

(a) (b)

Figure 6. Direct cooling of the end-windings. (**a**) Cooling duct inserted in the middle of the end-windings, reproduced with permission from the owner, Madonna, V. [126]; (**b**) heat pipe placed near the end-windings, reproduced with permission from [127]. Copyright Elsevier 2019.

By using 3D-printing techniques, cooling ducts effectively surround the windings, and even more complicated shapes with very thin walls can be manufactured [127–129].

By applying another approach, particularly formed thermally conductive parts, which effectively serve as heat exchangers with a large active cooling surface, can be inserted in the EMs to surround the concentrated windings, especially the ends of the coils. The extra parts ensure a larger contact surface among the coils of the housing. This concept is suitable in terms of both cooling by traditional airflows and liquid cooling jackets [130]. The optimized shape of the cooling insert can only be effectively manufactured by employing 3D printing.

A great variety of technical solutions have been proposed in the literature that adopts the so-called in-slot cooling approach. Most typically, this can be applied to EMs with concentrated windings, where a space between the coils is found in their slots, mainly for insulation purposes (see the blue areas in Figure 7).

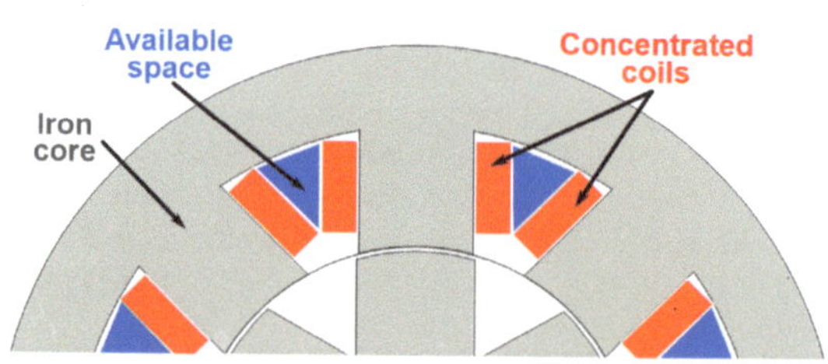

Figure 7. Free space available for placing coolers in the slots with concentrated windings.

By applying the simplest approach, these gaps can be filled with specifically shaped thermally conductive insulating materials, which can exhibit much greater thermal conductivity than air [131]. Improved cooling can be achieved if a heat exchanger is placed in the available space, even if the EM is only cooled by the fan and the fins on the housing.

In the slots, among the turns, 3D-printed cooling channels with very thin walls also can be placed, as can be seen in Figure 8a [132–134]. In [135], H. Yin proposed to place a two-way thin copper pipeline in the slots of a PMSM, as shown in Figure 8b. By applying this relatively simple pipe arrangement, called an internal water cooling (IWC) system, effective thermal management of the PMSM was ensured.

Wrobel and Hussein [136] placed small, optimally designed, 3D-printed so-called heat guides in these free spaces inside the slots. Both solutions profoundly contributed to increasing the extent to which the coils are cooled.

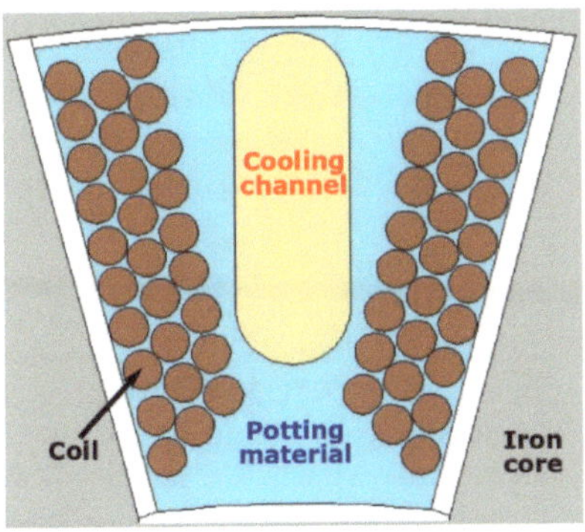

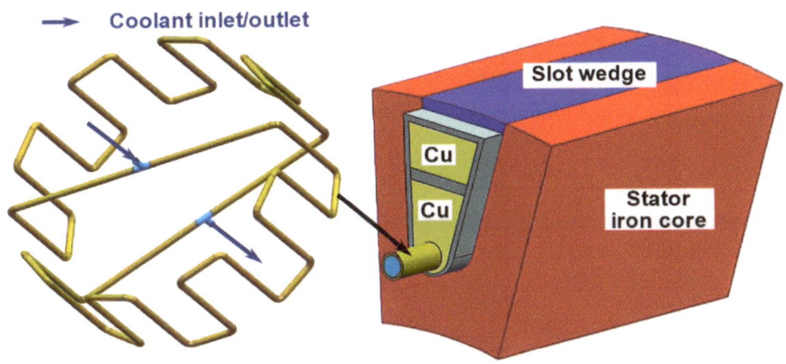

Figure 8. In-slot cooling approaches. (**a**) Liquid cooling channel in the slot of the EM [137]; (**b**) two-way copper pipeline placed in the slots of a PMSM, reproduced with permission from [135]. Copyright the authors 2018.

Another achievement of AM (as well PBF and SLM in particular) is enabling the fabrication of small-scale heat exchangers consisting of complex internal geometries with a large surface area density (that is, having a very large active surface but a small volume) with regard to low-density materials, which cannot be manufactured using conventional technologies [138]. By integrating heat exchangers into the EMs, their cooling capability can be significantly improved.

W. Sixel et al. proposed an innovative thermal management system consisting of polycarbonate-aluminum flake composite pieces placed between the concentrated coils of an EM. Each component contains three cooling channels. All the ends of the cooling pieces are connected to a common, 3D-printed liquid-cooled heat exchanger [139].

A more demanding direct winding cooling solution was proposed in [140]. Rectangular microfeature-enhanced flat heat exchangers (with the fine structure shown in Figure 9) can be inserted inside the windings. These are washed by the cooling fluid in the axial direction of the machine. Due to their innovative large surface structure, they exhibit a huge heat transfer coefficient, approximately 30 kW/(m^2·K). The special non-conductive material produced, which can be the basis for the manufacturing of microchannels, can be 3D printed by applying DMLS [141]. However, this method also has some drawbacks, as a high pumping pressure is required and only the windings are cooled [88].

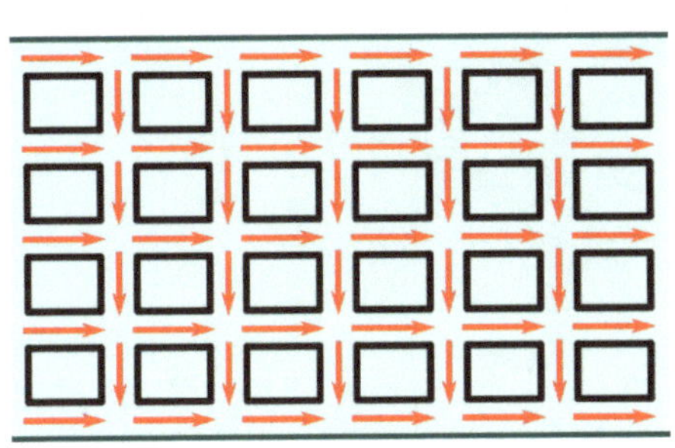

Figure 9. Rectangular microfeatures of a flat heat exchanger used in direct winding cooling.

As coils can be 3D printed layer-by-layer, the concept of integrating the cooling channels inside the windings can be achieved [63]. The increase in the coil volume due to the inside channels is compensated for by the application of very high current densities (over 100 A/mm^2). The diameter and position of the channels must be optimized in a way that ensures the current density distribution over the entire cross-section of the coil is uniform.

6.3. Cooling the Iron Core

The iron cores within EMs can also be directly cooled by heat exchangers and cooling jackets.

A large proportion of EMs is cooled by a closed-loop liquid cooling system, which usually consists of a cooling jacket around the machine housing in direct contact with the iron core, arranged in numerous ways, and using various coolants [138]. Such cooling jackets can consist of squirrel-cage or serpentine duct structures [115,142]. Further optimization of such cooling systems is limited by the classical manufacturing techniques and traditional thermally conductive materials available. However, AM can also bring about significant advancements in this case since it enables structures of practically any shape to be fabricated (toroidal, axial, helical, circumferential, serpentine pipes) with new, advanced materials.

Materials with extremely high thermal conductivities can be applied in such thermal management systems. These can contribute to enhancing the cooling performance and reducing the mass of the cooling system, which are both important issues in automotive applications.

Besides these benefits, another significant advantage of 3D-printed cooling jackets is the reduced likelihood of them leaking due to their mechanical integrity since, rather than being assembled from different components, they are manufactured as a single unit [50].

The cooling jacket surrounding the outer side of the iron core of the EM can similarly be extended to penetrate the iron core, and also serve as a flux barrier in the yoke of the stator iron core, as it was proposed for a segmented stator PMSM by A. Nollau and D. Gerling in [143]. In this case, the cooling jacket not only cools the iron core, but also magnetically separates the stator modules of the machine.

In [144], an axial heat exchanger fabricated by employing a metal AM has been proposed for an outer rotor EM with concentrated windings, which does not have end-windings. The heat exchanger, as can be seen in Figure 10, is placed around the shaft and is in thermal contact with the yoke of the stator iron core. The air flowing axially through the machine cools the heat exchanger as well as the iron core indirectly.

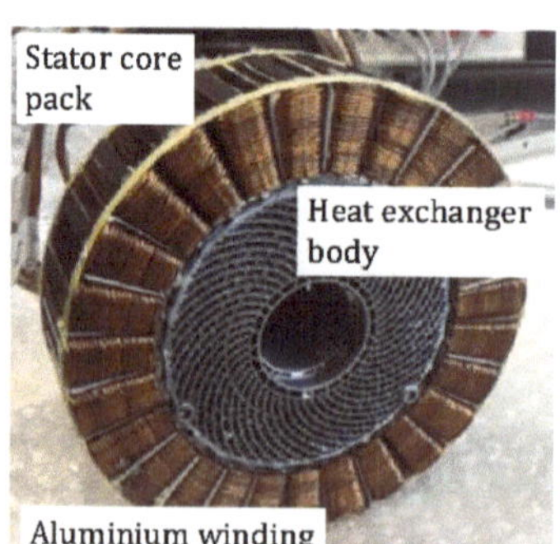

Figure 10. 3D-printed axial heat exchanger of an outer rotor EM with concentrated coils, reproduced with permission from [144]. Copyright Elsevier 2020.

6.4. The Housing and the Cooling Fan

Although the housing of EMs seems to be the simplest part of an EM, it plays a crucial role in terms of cooling and protecting the interior of the EM and assuring its required rigidity. All of these are achieved by the smallest mass and at the minimum cost. Furthermore, in the case of housing, AM can bring about significant improvements.

In Figure 11, a relevant example of the possible complexity of a 3D-printed housing of an electric traction system is presented. The topologically optimized design, as well as the use of the most advanced materials and AM technologies, enabled the mass of the housing to be reduced by 40%. The innovative honeycomb structure of the surface not only increased the active cooling surface but also the rigidity of the housing at a minimal mass [145].

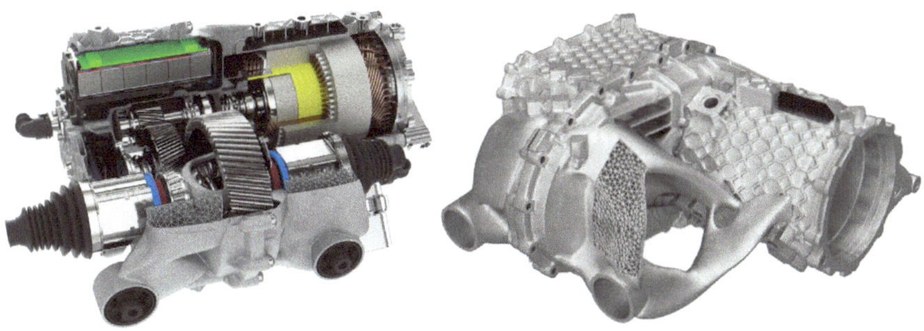

Figure 11. Porsche's electric traction drive system and its housing realized via AM technology, reproduced with permission from the owner, tuningblog.eu (Germany) [146].

The cooling fan of an EM is also a simple, albeit essential, component. It must be optimized to achieve maximum airflow efficiency as well as minimum mass and noise. Until recently, its optimization was significantly limited by the materials available and manufacturing technologies.

As is the case with other components of an EM, further improvement of the fan and blades, in particular, was catalyzed by the multitude of potentials of AM technologies. The reduction in the mass of this part helps to reduce the total weight of the EM and increase its overall performance.

Diverse fan blades enhanced by laser sintering can be fabricated, as is shown in Figure 12. The excellent thermomechanical properties of the low-density polyamide powder from which the blades are produced are the key to their increased performance.

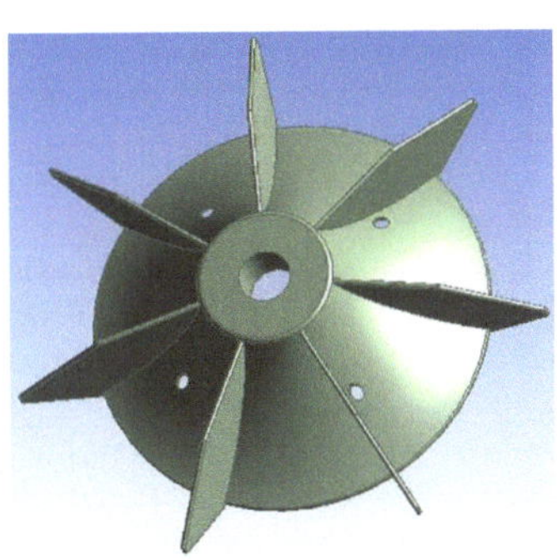

Figure 12. 3D-printed radial cooling fan with optimized blades, reproduced with permission from [147]. Copyright Elsevier 2018.

7. Representative and Potential Examples of Using 3D Printing in the Fabrication of Electric Machines

For EMs, the power/torque density and efficiency are the two key performance indicators that can be enhanced by improving any of the components of EMs. In Figure 13, the influence of a diverse range of possible advancements with regard to increasing the efficiency of EMs is presented [67].

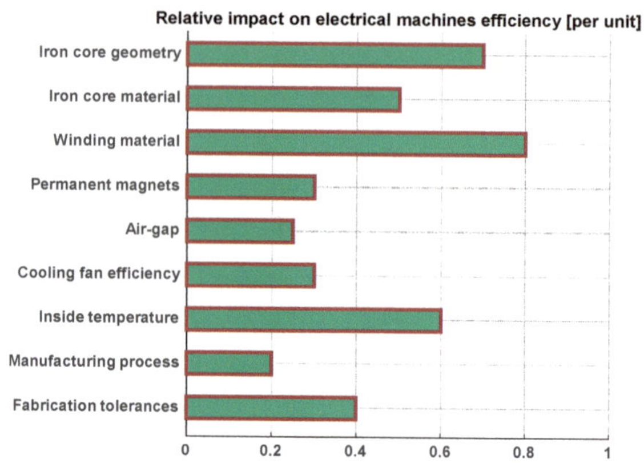

Figure 13. Impact of various advancements concerning the efficiency of EMs.

Over recent years, leading scientific journals have been overwhelmed with papers concerning the use of diverse AM technologies concerning essentially all the matters mentioned in Figure 13. In the following section, the most significant achievements will be surveyed.

7.1. Permanent Magnet Synchronous Machines

PMSMs, being highly efficient and as a result of their great specific power and torque, are preferred in numerous industrial and automotive applications [148–150].

AM can contribute to reductions in the mass, volume, and inertia of the rotors of such machines. As was reported in [151], by introducing optimized gaps, holes, smaller cavities, and lattice structures into a PMSM rotor, the moment of inertia was reduced by 23% without sacrificing its mechanical strength.

Furthermore, by applying this manufacturing technology, the PMs can be excellently fitted in special places, as can be seen in Figure 14 [152]. Therefore, the likelihood of the PM detaching is low, even in the case of high-speed EMs.

In [153], a spoke-type PM was reported. Among EMs, this is considered to exhibit one of the greatest power densities as its maximum power is 125 kW at 12,000 r/min with a mass of just 14 kg. This was achieved mainly due to the 3D-printed direct oil cooling of the stator, combined with the forced air cooling of the rotor. The single-piece cooling system is comprised of very thin walls and an optimized fine active surface.

Furthermore, axial flux PMSMs are proposed for light traction applications. Their high torque density is counterbalanced by more complicated structures, which also require fabrication using AM technologies [154].

AM empowers specialists to design PMs of practically any shape for PMSMs. In [155], PMs with a sinusoidal-shaped surface fabricated by CS technology are proposed. It was demonstrated that both the harmonic content in the induced electromotive force (EMF) and the cogging torque were appreciably reduced.

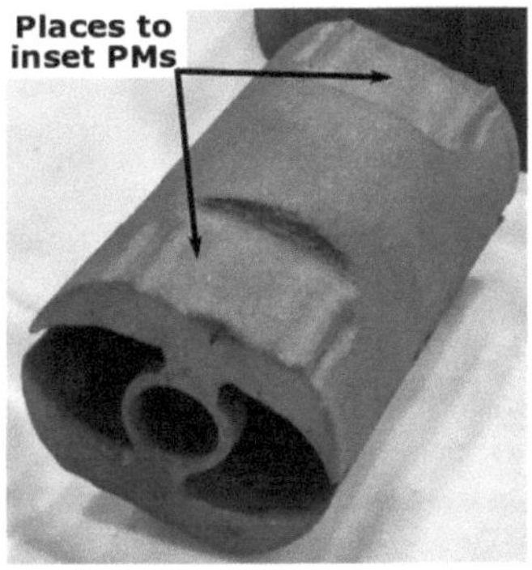

Figure 14. 3D-printed low-mass PMSM rotor with special locations for placing the PMs, reproduced with permission from the owner, Garibaldi, M. [152].

7.2. Induction Machines

Without a doubt, IMs dominate the EM market, mainly due to their simple and robust construction, low maintenance requirements, and affordable cost [156]. Even though their manufacturing technology has been well-established for many decades, AM can also lead to new approaches in this field.

Advanced AM technologies enable the 3D printing of multiple diverse materials simultaneously. By taking into consideration this advantage, Chinthavali [157] demonstrated that the entire squirrel-cage rotor can be fabricated within a single processing stage.

The CS AM method can also bring about improvements in the field of high-speed IMs, where the construction of robust rotors is the main bottleneck. If this technology is used for the fabrication of solid rotors, very good diffusion bonding between the iron core and the copper layer that covers it can be achieved, which prevents the covering from detaching [61].

7.3. Synchronous Reluctance Machines

Over recent years, SyncRels have become very serious competitors of IMs due to their passive rotor and high efficiency [158]. To achieve significant levels of performance, their rotor must be optimized, resulting in complicated structures with diverse forms of flux guides and barriers [159,160]. Due to the technical limitations of laminated punching, in numerous cases, the rotor iron core could not be fabricated exactly as determined from the optimization. Moreover, these rotors cannot be used in high-speed applications due to the mechanical limitations of the low-rigidity rotor structures.

Although PM-assisted SyncRels, which have frequently been investigated over recent years, have PMs placed in the flux barriers of complex shapes, it is difficult to acquire PMs of any specific shapes.

These drawbacks of SyncRels can be overcome by applying diverse AM technologies [161]. The 3D printing of both the iron cores and PMs of any shape by SLM and FDM has been thoroughly investigated [162,163].

A significant drawback of SyncRels that operate at high speeds is the necessity to place so-called ribs to rigidize the flux bridges, as can be seen in Figure 15a [160,164]. These small zones diminish the magnetic properties of these EMs. This disadvantage can be eliminated by applying dual-phase magnetic components for the construction of the iron core, which can be processed utilizing AM technologies. These magnetic materials have the remarkable property that certain regions (in this case, those corresponding to the ribs) can be locally treated via a nitriding process to transform them into non-magnetic zones [165].

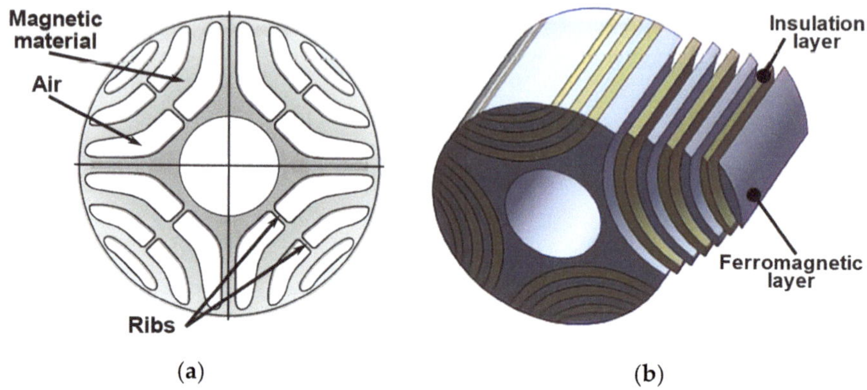

Figure 15. SyncRel rotor constructions. (**a**) With fluid-shaped flux barriers [166]; (**b**) ALA rotor made of bent laminations [167].

A higher saliency ratio (L_d/L_q) can be obtained in the case of the SyncRels with axially laminated anisotropic (ALA) rotors, as shown in Figure 15b [168,169]. However, both structures are difficult and expensive to manufacture when employing traditional punched lamination technology since numerous, differently sized laminations are required. Furthermore, in the case of the sandwiched rotor, the designer must deal with the different thermal expansion coefficients of aluminum and iron, while in the case of topology, extra bending of the laminations is required. Concerning the manufacture of both structures, AM can be a source of significant stimuli.

7.4. Switched Reluctance Machines

SRMs are also competitive candidates for the EMs used in variable speed applications, mainly due to their simple and robust construction [170,171]. Having two basic drawbacks (high torque ripples and noise), their structure was frequently optimized to address these disadvantages. In several cases, the optimized structure could not be processed via traditional methods [172].

AM has also brought about important breakthroughs in the case of SRMs, too. Laser PBF-fabricated concentrated coils (similar to those seen in Figure 1) can be used on the stator of the SRMs, as is shown in Figure 16a. 3D printing can also bring advancements in the case of the SRM rotors. Given that the complex optimized rotor structure proposed in [172] and presented in Figure 16b has several notches and holes, both low mass and inertia were achieved. Additionally, the rotor can be easily cooled by the air passing through the machine. It was fabricated by employing an SLM of an Fe-Co alloy [32].

By using any AM technology, continuous rotor skewing can be easily solved for any type of EM. This issue is very important in the case of SRMs, which exhibit high torque ripples.

These new manufacturing technologies simply enable even small components to be added to the traditional machine structure. In [173], small ribs were placed between the 3D-printed rotor poles of an SRM to collect the leakage fluxes and thus reduce losses. The ribs must be sufficiently small so as to not influence the inertia of the rotor, but still mechanically hard to withstand the great forces they are subjected to at high speeds. Therefore, a rigid honeycomb structure was selected.

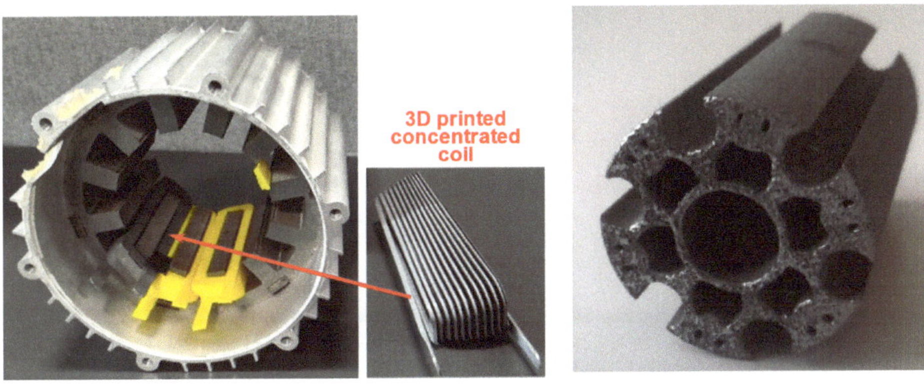

(a) (b)

Figure 16. SRM components fabricated using AM technologies. (**a**) 3D-printed concentrated coil placed on the stator poles [90], reproduced with permission from the owner, Silbernagel, C.; (**b**) optimized 3D-printed SRM rotor [32], reproduced with permission from the owner, VTT Technical Research Centre of Finland.

7.5. Construction of Special Electrical Machines

Several unconventional EMs are in use (customarily with complex structures and magnetic flux paths), e.g., flux modulation, flux switching, transverse flux, claw-pole machines, etc., all of which are difficult and expensive to manufacture using traditional methods but are suitable for use with AM [50,174].

Perhaps the most complex among them is transverse flux machines (TFMs). Given their performance, they could be an excellent solution for direct-drive electric traction applications since their torque density is very high and their rated speed is low since they can be constructed with a great number of poles. The elimination of the gearbox increases both their efficiency and reliability while lowering the overall cost of the entire drivetrain. In turn, since their power factor is low, converters of increased power are required [175,176]. However, the main drawback of TFMs is their very complex structure, as can be seen in Figure 17 [177].

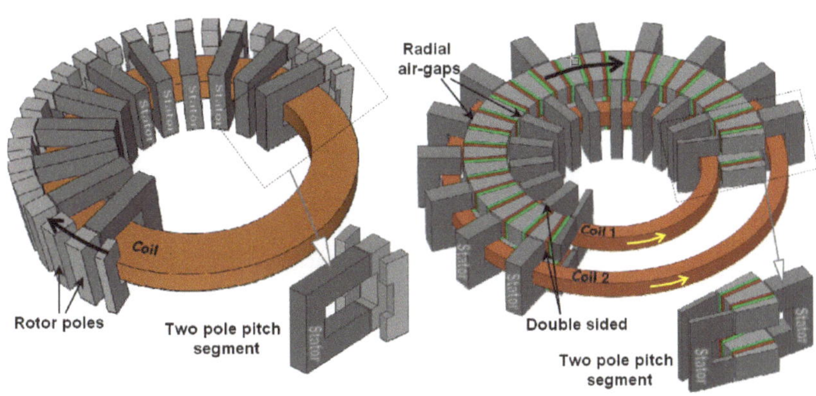

(a) (b)

Figure 17. Two typical TFM constructions [178], reproduced with permission from the owner, Baserrah, S. (**a**) Reluctance TFM; (**b**) flux concentrating PM-TFM.

As has already been stated, new AM technologies have been envisaged by design engineers. Specialists eliminating the limitations of traditional fabrication technologies and materials have proposed several EMs of intriguing topologies offering superior levels of performance.

In [179], a dual-conical (hourglass-shaped) air-gap PMSM was proposed and investigated (see Figure 18). The machine, which weighs 40 kg, can produce 112 kW of power at 3750 r/min. Due to liquid cooling, its admitted current density is 23 A/mm^2. The uncommon structure consists of a large air-gap surface, and the magnetic flux passes through it along true 3D paths. This is the "secret" of both its high-power density (twice that of traditional machines) and efficiency.

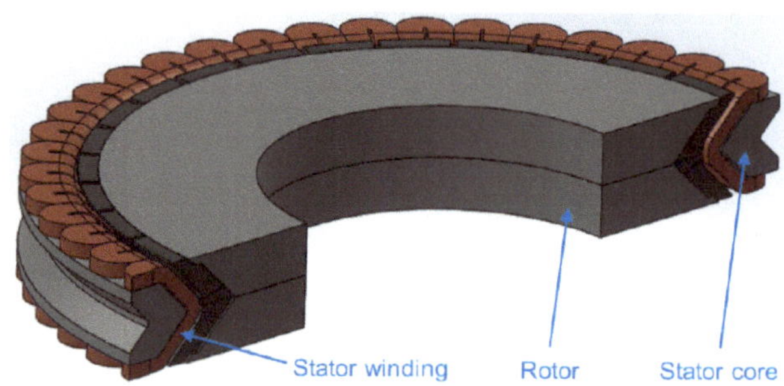

Figure 18. Dual-conical air-gap PMSM with components fabricated via AM, reproduced with permission from [179]. Copyright ASME 2016.

It is obvious that this EM of an unusual shape cannot be fabricated by employing traditional technologies. Its iron cores were 3D printed using CS technology [180]. The stator windings were made of rectangular cross-sectional conductors to achieve a high slot fill factor. Such windings could also be produced via AM technologies.

The aforementioned cases are just few examples of the multitude of EM proposals that could be exclusively fabricated using AM technologies.

8. Conclusions

As can be seen, a great variety of AM technologies can be used for the fabrication of a diverse range of (partially novel) components and variants of EMs. Finally, the advantages, drawbacks, and future of implementing AM technologies in the field of EMs should be summarized.

Designers of EMs can make use of several advantages concerning the 3D printing of diverse components of EMs. The complexity of newly developed EM structures is practically unlimited. It may not be possible, or is at least extremely hard, to fabricate the shapes of the designed components by conventional, subtractive, or powder metallurgical manufacturing methods. 3D printing enables different EM parts to be built simultaneously, avoiding the need for them to be assembled later.

Some EM components can be fabricated with mixed materials, even of different types (conductors with insulators, iron cores with heating pipes, structural components with thermal management components, etc.). By using AM technologies, it is possible to vary the composition of iron cores in such a way that their properties can be tuned to suit the requirements of a certain type of EM. Continuous rotor skewing can be achieved by 3D printing, which is the key to reducing the torque ripple of most EMs.

The impact of AM on the development of diverse types of EMs is different, mainly due to the complexity of the materials and structures used. In Table 2, the impact scores for the different parts of the most typical EM fabricated by using AM are provided. As in the case of Table 1, the scores here also range from 1 (poor) to 5 (excellent).

Table 2. Impact score of applying AM to the fabrication of diverse parts of the most typical EMs.

EM	Iron Core	PM	Winding	Cooling System	Mean Score
Low-power IM	2	–	1	2	1.66
High-power IM	1	–	2	3	2
Low-Power PMSM	2	3	1	2	2
High-power PMSM	1	2	2	3	2
SyncRel	3	3	2	2	2.5
SRM	2	–	3	4	3
TFM	5	3	3	2	3.25
Special EMs	5	4	4	4	4.25

Note: Poor, Fair, Good, Very good, Excellent.

As can be seen from the table, for the most widely used low-power IM with a well-established structure and manufacturing technology, the new manufacturing technologies and materials are the least likely to be applied. As the complexity of EMs increases, so does the need for additive manufacturing of their key components. 3D printing has the greatest potential in the case of special EMs with outstanding topologies.

Most of the 3D printing processes do not result in considerable scratches, so material losses of close to zero can be achieved. It is also possible to reduce the usage of raw materials by designing EMs without excessive amounts of material. By only placing ferrous material where the magnetic flux must pass (for example, by modularizing the iron cores [181]), the other parts of the stator or rotor can be manufactured from non-magnetic materials, which are cheaper and of lower density. Therefore, both the price of the EM and iron core losses can be decreased, while its power/torque density increases. This approach can be predominantly beneficial in the case of rotors since their mass and inertia can be reduced, while the dynamic characteristics of EMs improved [182]. Moreover, the segmented parts can more easily be decommissioned and recycled [183].

These material savings are in line with the Circular Economy Action Plan (CEAP) adopted by the European Commission in March 2020 [184]. This is one of the main constituents of the European Green Deal, Europe's new agenda for sustainable growth, but it conforms with other regulations around the world concerning the reduction in the footprint of EMs [185].

All the above aspects facilitate the development of EMs being better suited to the given applications with increased manufacturability, efficiency, and power/torque density [50].

On the other hand, some limitations of these new technologies can be mentioned. Despite the intensive level of research conducted, AM technology has not yet become fully mature. Consistently, countless improvements are reported concerning the technologies themselves (manufacturing speed, finishing, etc.), the applied materials, and, of course, costs. These all surely will contribute to the widespread growth of 3D printing technologies in the manufacturing of different EM parts.

Currently, 3D printing is only revolutionizing the prototyping and very-small-series production of EMs since no expensive special-purpose tools are needed. The main advances in terms of the fabrication of cooling systems were achieved, which has a significant impact on the performance of all EMs.

The complete manufacturing of EMs exclusively by AM technologies without any need for assembly or post-processing during or after 3D printing seems futuristic, although some attempts have been reported albeit only for demonstrative purposes [186].

Finally, it can be concluded that given the forecasted and expected enhancements in AM, this new manufacturing technology will gain more and more ground, and the performances of EMs could be considerably enhanced. New designs of EMs of a high degree of complexity and customization are also expected soon [187].

Funding: This work was supported in part by the TKP2020-NKA-10 Project through the 2020-4.1.1-TKP2020 Thematic Excellence Program by the National Research, Development and Innovation Fund of Hungary; in part by ZalaZONE Automotive Proving Ground Zala Ltd.; and in part by the Ministry of Innovation and Technology National Research, Development and Innovation (NRDI) Office within the framework of the Autonomous Systems National Laboratory Program.

Institutional Review Board Statement: Not applicable.

Informed Consent Statement: Not applicable.

Data Availability Statement: Not applicable.

Conflicts of Interest: The authors declare no conflict of interest.

References

1. Global Electric Motor Market Report 2021–2030: Development of High Power-to-Weight Ratio Electric Motors Gaining Momentum. Available online: https://www.globenewswire.com/news-release/2021/12/20/2355052/28124/en/Global-Electric-Motor-Market-Report-2021-2030-Development-of-High-Power-to-Weight-Ratio-Electric-Motors-Gaining-Momentum.html (accessed on 27 January 2022).
2. Cheng, M.; Chan, C. General Requirement of Traction Motor Drives. In *Encyclopedia of Automotive Engineering*; Crolla, D., Foster, D.E., Kobayashi, T., Vaughan, N.D., Eds.; Wiley: Chichester, UK, 2014; pp. 1–18.
3. Dorrell, D.G. A review of the methods for improving the efficiency of drive motors to meet IE4 efficiency standards. *J. Power Electron.* **2014**, *14*, 842–851. [CrossRef]
4. Cantwell, J.; Hayashi, T. *Paradigm Shift in Technologies and Innovation Systems*; Springer Nature: Singapore, 2019.
5. Conner, B.P.; Manogharan, G.P.; Martof, A.N.; Rodomsky, L.M.; Rodomsky, C.M.; Jordan, D.C.; Limperos, J.W. Making sense of 3D printing: Creating a map of additive manufacturing products and services. *Addit. Manuf.* **2014**, *1–4*, 64–76. [CrossRef]
6. Tiismus, H.; Kallaste, A.; Belahcen, A.; Tarraste, M.; Vaimann, T.; Rassõlkin, A.; Asad, B.; Shams Ghahfarokhi, P. AC magnetic loss reduction of SLM processed Fe-Si for additive manufacturing of electrical machines. *Energies* **2021**, *14*, 1241. [CrossRef]
7. *ISO/ASTM 52900-15*; Additive manufacturing—General principles—Terminology. ASTM International: West Conshohockeas, PA, USA, 2015.
8. MacDonald, E.; Wicker, R. Multiprocess 3D printing for increasing component functionality. *Science* **2016**, *353*, aaf2093. [CrossRef] [PubMed]
9. Razavykia, A.; Brusa, E.; Delprete, C.; Yavari, R. An overview of additive manufacturing technologies—A review to technical synthesis in numerical study of selective laser melting. *Materials* **2020**, *13*, 3895. [CrossRef]
10. Tiismus, H.; Kallaste, A.; Vaimann, T.; Rassõlkin, A.; Belahcen, A. Electrical resistivity of additively manufactured silicon steel for electrical machine fabrication. In Proceedings of the Electric Power Quality and Supply Reliability Conference (PQ 2019) & Symposium on Electrical Engineering and Mechatronics (SEEM 2019), Kärdla, Estonia, 12–15 June 2019. [CrossRef]
11. Pfaff, A.; Jäcklein, M.; Hoschke, K.; Wickert, M. Designed materials by additive manufacturing—Impact of exposure strategies and parameters on material characteristics of AlSi10Mg processed by laser beam melting. *Metals* **2018**, *8*, 491. [CrossRef]
12. Kafle, A.; Luis, E.; Silwal, R.; Pan, H.M.; Shrestha, P.L.; Bastola, A.K. 3D/4D Printing of polymers: Fused deposition modelling (FDM), selective laser sintering (SLS), and stereolithography (SLA). *Polymers* **2021**, *13*, 3101. [CrossRef]
13. Demir, A.G.; Previtali, B. Multi-material selective laser melting of Fe/Al-12Si components. *Manuf. Lett.* **2017**, *11*, 8–11. [CrossRef]
14. Kocsis, B.; Fekete, I.; Varga, L. Metallographic and magnetic analysis of direct laser sintered soft magnetic composites. *J. Magn. Magn. Mater.* **2020**, *501*, 166425. [CrossRef]
15. Dass, A.; Moridi, A. State of the art in directed energy deposition: From additive manufacturing to materials design. *Coatings* **2019**, *9*, 418. [CrossRef]
16. Wong, K.V.; Hernandez, A. A review of additive manufacturing. *ISRN Mech. Eng.* **2012**, *2012*, 208760. [CrossRef]
17. Gokuldoss, P.K.; Kolla, S.; Eckert, J. Additive manufacturing processes: Selective laser melting, electron beam melting and binder jetting—Selection guidelines. *Materials* **2017**, *10*, 672. [CrossRef] [PubMed]
18. Najmon, J.C.; Raeisi, S.; Tovar, A. Review of additive manufacturing technologies and applications in the aerospace industry. In *Additive Manufacturing for the Aerospace Industry*; Froes, F., Boyer, R., Eds.; Elsevier: Amsterdam, The Netherlands, 2019; pp. 7–31.
19. Zakeri, S.; Vippola, M.; Levänen, E. A comprehensive review of the photopolymerization of ceramic resins used in stereolithography. *Addit. Manuf.* **2020**, *35*, 101177. [CrossRef]
20. Huang, J.; Qin, Q.; Wang, J. A review of stereolithography: Processes and systems. *Processes* **2020**, *8*, 1138. [CrossRef]
21. Jiménez, A.; Bidare, P.; Hassanin, H.; Tarlochan, F.; Dimov, S.; Essa, K. Powder-based laser hybrid additive manufacturing of metals: A review. *Int. J. Adv. Manuf. Technol.* **2021**, *114*, 63–96. [CrossRef]
22. Pham, T.; Kwon, P.; Foster, S. Additive manufacturing and topology optimization of magnetic materials for electrical machines—A review. *Energies* **2021**, *14*, 283. [CrossRef]
23. Krings, A.; Boglietti, A.; Cavagnino, A.; Sprague, S. Soft magnetic material status and trends in electric machines. *IEEE Trans. Ind. Electron.* **2017**, *64*, 2405–2414. [CrossRef]

24. Naseer, M.U.; Kallaste, A.; Asad, B.; Vaimann, T.; Rassãlkin, A. A review on additive manufacturing possibilities of electrical machines. *Energies* **2021**, *14*, 1940. [CrossRef]
25. Sarap, M.; Kallaste, A.; Ghahfarokhi, P.S.; Tiismus, H.; Vaimann, T. Determining the thermal conductivity of additively manufactured metal specimens. In Proceedings of the 29th International Workshop on Electric Drives: Advances in Power Electronics for Electric Drives (IWED 2022), Moscow, Russia, 26–29 January 2022.
26. Lamichhane, T.N.; Sethuraman, L.; Dalagan, A.; Wang, H.; Keller, J.; Paranthaman, M.P. Additive manufacturing of soft magnets for electrical machines—A review. *Mater. Today Phys.* **2020**, *15*, 100255. [CrossRef]
27. Plotkowski, A.; Carver, K.; List, F.; Pries, J.; Li, Z.; Rossy, A.M.; Leonard, D. Design and performance of an additively manufactured high-Si transformer core. *Mater. Des.* **2020**, *194*, 108894. [CrossRef]
28. Stornelli, G.; Faba, A.; Di Schino, A.; Folgarait, P.; Ridolfi, M.R.; Cardelli, E.; Montanari, R. Properties of additively manufactured electric steel powder cores with increased Si content. *Materials* **2021**, *14*, 1489. [CrossRef] [PubMed]
29. Mikler, C.; Chaudhary, V.; Borkar, T.; Soni, V.; Jaeger, D.; Chen, X.; Contieri, R.; Ramanujan, R.; Banerjee, R. Laser additive manufacturing of magnetic materials. *JOM* **2017**, *69*, 532–543. [CrossRef]
30. Schönrath, H.; Spasova, M.; Kilian, S.; Meckenstock, R.; Witt, G.; Sehrt, J.; Farle, M. Additive manufacturing of soft magnetic permalloy from Fe and Ni powders: Control of magnetic anisotropy. *J. Magn. Magn. Mater.* **2019**, *478*, 274–278. [CrossRef]
31. Ding, C.; Liu, L.; Mei, Y.; Ngo, K.D.; Lu, G.-Q. Magnetic paste as feedstock for additive manufacturing of power magnetics. In Proceedings of the 2018 IEEE Applied Power Electronics Conference and Exposition (APEC 2018), San Antonio, TX, USA, 4–8 March 2018; pp. 615–618. [CrossRef]
32. Metsä-Kortelainen, S.; Lindroos, T.; Savolainen, M.; Jokinen, A.; Revuelta, A.; Pasanen, A.; Ruusuvuori, K.; Pippuri, J. Manufacturing of topology optimized soft magnetic core through 3D printing. In Proceedings of the NAFEMS Exploring the Design Freedom of Additive Manufacturing through Simulation, Helsinki, Finland, 22–23 November 2016.
33. Schmool, D.; Markó, D. Magnetism in Solids: Hysteresis. In *Reference Module in Materials Science and Materials Engineering*; Hashmi, S., Ed.; Elsevier: Amsterdam, The Netherlands, 2018.
34. Chaudhary, V.; Mantri, S.A.; Ramanujan, R.V.; Banerjee, R. Additive manufacturing of magnetic materials. *Prog. Mater. Sci.* **2020**, *114*, 100688. [CrossRef]
35. Szabó, L.; Viorel, I.A.; Iancu, V.; Popa, D.C. Soft magnetic composites used in transverse flux machines. *Oradea Univ. Ann. Electrotech. Fascicle* **2004**, 134–141.
36. Pham, T.Q.; Do, T.T.; Kwon, P.; Foster, S.N. Additive manufacturing of high performance ferromagnetic materials. In Proceedings of the Energy Conversion Congress and Exposition (ECCE 2018), Portland, OR, USA, 23–27 September 2018; pp. 4303–4308. [CrossRef]
37. Pyo, H.-J.; Jeong, J.W.; Yu, J.; Lee, S.G.; Kim, W.-H. Design of 3D-printed hybrid axial-flux motor using 3D-printed SMC core. *IEEE Trans. Appl. Supercond.* **2020**, *30*, 5202004. [CrossRef]
38. Goll, D.; Schuller, D.; Martinek, G.; Kunert, T.; Schurr, J.; Sinz, C.; Schubert, T.; Bernthaler, T.; Riegel, H.; Schneider, G. Additive manufacturing of soft magnetic materials and components. *Addit. Manuf.* **2019**, *27*, 428–439. [CrossRef]
39. Benack, N.C.; Wang, T.; Matthews, K.; Taheri, M.L. Additive manufacturing methods for soft magnetic composites (SMCs). *Microsc. Microanal.* **2018**, *24*, 1066–1067. [CrossRef]
40. Kong, F.; Chang, C.; Inoue, A.; Shalaan, E.; Al-Marzouki, F. Fe-based amorphous soft magnetic alloys with high saturation magnetization and good bending ductility. *J. Alloy. Compd.* **2014**, *615*, 163–166. [CrossRef]
41. Enomoto, Y.; Tokoi, H.; Imagawa, T.; Suzuki, T.; Obata, T.; Souma, K. Amorphous motor with IE5 efficiency class. *Hitachi Rev.* **2015**, *64*, 480–487.
42. Cepoi, R.D.; Jaşcău, F.F.; Szabó, L. Current trends in energy efficient electrical machines. *J. Electr. Electron. Eng.* **2017**, *10*, 13–18.
43. Nam, Y.G.; Koo, B.; Chang, M.S.; Yang, S.; Yu, J.; Park, Y.H.; Jeong, J.W. Selective laser melting vitrification of amorphous soft magnetic alloys with help of double-scanning-induced compositional homogeneity. *Mater. Lett.* **2020**, *261*, 127068. [CrossRef]
44. Zou, Y.; Qiu, Z.; Tan, C.; Wu, Y.; Li, K.; Zeng, D. Microstructure and mechanical properties of Fe-based bulk metallic glass composites fabricated by selective laser melting. *J. Non-Cryst. Solids* **2020**, *538*, 120046. [CrossRef]
45. Ozden, M.G.; Morley, N.A. Laser additive manufacturing of Fe-based magnetic amorphous alloys. *Magnetochemistry* **2021**, *7*, 20. [CrossRef]
46. Treadgold, T. Nickel Soars and Could Keep Flying as Demand Rises and Supply Falls. Available online: https://www.forbes.com/sites/timtreadgold/2021/01/14/nickel-soars-and-could-keep-flying-as-demand-rises-and-supply-falls/?sh=7fe1e74675ba (accessed on 1 March 2022).
47. Permalloy 80. Available online: https://www.espimetals.com/index.php/technical-data/175-Permalloy%2080 (accessed on 22 April 2021).
48. Somaloy®3P Material Data; Höganäs: 2018. Available online: https://www.hoganas.com/globalassets/download-media/sharepoint/brochures-and-datasheets---all-documents/somaloy-3p_material-data_june_2018_2273hog.pdf (accessed on 1 March 2022).
49. Metals, H. Nanocrystalline Soft Magnetic Material FINEMET®; Hitachi Metals: 2018. Available online: https://elnamagnetics.com/wp-content/uploads/catalogs/Finemet/FINEMET%20Materials%20(HL-FM10-D).pdf (accessed on 1 March 2022).
50. Wu, F.; El-Refaie, A.M. Towards fully additively-manufactured permanent magnet synchronous machines: Opportunities and challenges. In Proceedings of the International Electric Machines & Drives Conference (IEMDC 2019), San Diego, CA, USA, 12–15 May 2019; pp. 2225–2232. [CrossRef]

51. El-Refaie, A. Role of advanced materials in electrical machines. *CES Trans. Electr. Mach. Syst.* **2019**, *3*, 124–132. [CrossRef]
52. Yan, Y. Design Methodology and Materials for Additive Manufacturing of Magnetic Components. Ph.D. Thesis, Virginia Tech, Blacksburg, VA, USA, 2017.
53. Li, Z.; Yao, K.; Li, D.; Ni, X.; Lu, Z. Core loss analysis of Finemet type nanocrystalline alloy ribbon with different thickness. *Prog. Nat. Sci. Mater. Int.* **2017**, *27*, 588–592. [CrossRef]
54. Nonaka, T.; Zeze, S.; Makino, S.; Ohto, M. Research on motor with nanocrystalline soft magnetic alloy stator cores. *Electr. Eng. Jpn.* **2020**, *211*, 55–62. [CrossRef]
55. Compton, B.G.; Kemp, J.W.; Novikov, T.V.; Pack, R.C.; Nlebedim, C.I.; Duty, C.E.; Rios, O.; Paranthaman, M.P. Direct-write 3D printing of NdFeB bonded magnets. *Mater. Manuf. Processes* **2018**, *33*, 109–113. [CrossRef]
56. Volegov, A.; Andreev, S.; Selezneva, N.; Ryzhikhin, I.; Kudrevatykh, N.; Mädler, L.; Okulov, I. Additive manufacturing of heavy rare earth free high-coercivity permanent magnets. *Acta Mater.* **2020**, *188*, 733–739. [CrossRef]
57. Wu, J.; Aboulkhair, N.T.; Degano, M.; Ashcroft, I.; Hague, R.J. Process-structure-property relationships in laser powder bed fusion of permanent magnetic Nd-Fe-B. *Mater. Des.* **2021**, *209*, 109992. [CrossRef]
58. Marfoli, A.; Dinardo, M.; Degano, M.; Gerada, C.; Jara, W. Squirrel cage induction motor: A design-based comparison between aluminium and copper cages. *IEEE Open J. Ind. Appl.* **2021**, *2*, 110–120. [CrossRef]
59. Jiang, Q.; Zhang, P.; Yu, Z.; Shi, H.; Wu, D.; Yan, H.; Ye, X.; Lu, Q.; Tian, Y. A review on additive manufacturing of pure copper. *Coatings* **2021**, *11*, 740. [CrossRef]
60. *Copper CuCrZr Datasheet*; 3T Additive Manufacturing: Newbury, UK, 2021.
61. Wrobel, R.; Mecrow, B. A comprehensive review of additive manufacturing in construction of electrical machines. *IEEE Trans. Energy Convers.* **2020**, *35*, 1054–1064. [CrossRef]
62. Cabrini, M.; Calignano, F.; Fino, P.; Lorenzi, S.; Lorusso, M.; Manfredi, D.; Testa, C.; Pastore, T. Corrosion behavior of heat-treated AlSi10Mg manufactured by laser powder bed fusion. *Materials* **2018**, *11*, 1051. [CrossRef] [PubMed]
63. Wohlers, C.; Juris, P.; Kabelac, S.; Ponick, B. Design and direct liquid cooling of tooth-coil windings. *Electr. Eng.* **2018**, *100*, 2299–2308. [CrossRef]
64. Bonaccorso, F.; Colombo, L.; Yu, G.; Stoller, M.; Tozzini, V.; Ferrari, A.C.; Ruoff, R.S.; Pellegrini, V. Graphene, related two-dimensional crystals, and hybrid systems for energy conversion and storage. *Science* **2015**, *347*, 1246501. [CrossRef] [PubMed]
65. Szabó, L.; Szabó, G.S.; Szabó, R. Usage of Graphene in Power Systems. A Survey. In Proceedings of the 11th International Conference on Electrical Power Drive Systems (ICEPDS 2020), Saint-Petersburg, Russia, 4–7 October 2020. [CrossRef]
66. Rallabandi, V.; Taran, N.; Ionel, D.M.; Eastham, J.F. On the feasibility of carbon nanotube windings for electrical machines—Case study for a coreless axial flux motor. In Proceedings of the IEEE Energy Conversion Congress and Exposition (ECCE 2016), Milwaukee, WI, USA, 18–22 September 2016. [CrossRef]
67. Ramesh, P.; Lenin, N. High power density electrical machines for electric vehicles—Comprehensive review based on material technology. *IEEE Trans. Magn.* **2019**, *55*, 0900121. [CrossRef]
68. Subramaniam, C.; Yamada, T.; Kobashi, K.; Sekiguchi, A.; Futaba, D.N.; Yumura, M.; Hata, K. One hundred fold increase in current carrying capacity in a carbon nanotube–copper composite. *Nat. Commun.* **2013**, *4*, 2202. [CrossRef]
69. Podsiadły, B.; Matuszewski, P.; Skalski, A.; Słoma, M. Carbon Nanotube-Based Composite Filaments for 3D Printing of Structural and Conductive Elements. *Appl. Sci.* **2021**, *11*, 1272. [CrossRef]
70. Zhang, D.; Chi, B.; Li, B.; Gao, Z.; Du, Y.; Guo, J.; Wei, J. Fabrication of highly conductive graphene flexible circuits by 3D printing. *Synth. Met.* **2016**, *217*, 79–86. [CrossRef]
71. Zocca, A.; Colombo, P.; Gomes, C.M.; Günster, J. Additive manufacturing of ceramics: Issues, potentialities, and opportunities. *J. Am. Ceram. Soc.* **2015**, *98*, 1983–2001. [CrossRef]
72. Chen, Z.; Li, Z.; Li, J.; Liu, C.; Lao, C.; Fu, Y.; Liu, C.; Li, Y.; Wang, P.; He, Y. 3D printing of ceramics: A review. *J. Eur. Ceram. Soc.* **2019**, *39*, 661–687. [CrossRef]
73. *Advanced APTIV™ Film Electrical Insulation for Cost-Effective, High Power Density E-Motors*; Victrex Plc: Thornton-Cleveleys, UK, 2017.
74. Szabó, L. Additive manufacturing of cooling systems used in power electronics. A brief survey. In Proceedings of the 29th International Workshop on Electric Drives: Advances in Power Electronics for Electric Drives (IWED 2022), Moscow, Russia, 26–29 January 2022; p. 32.
75. Chung, D.D.L. Materials for thermal conduction. *Appl. Therm. Eng.* **2001**, *21*, 1593–1605. [CrossRef]
76. Tommasi, A.; Maillol, N.; Bertinetti, A.; Penchev, P.; Bajolet, J.; Gili, F.; Pullini, D.; Mataix, D.B. Influence of surface preparation and heat treatment on mechanical behavior of hybrid aluminum parts manufactured by a combination of laser powder bed fusion and conventional manufacturing processes. *Metals* **2021**, *11*, 522. [CrossRef]
77. Nafis, B.M.; Whitt, R.; Iradukunda, A.-C.; Huitink, D. Additive manufacturing for enhancing thermal dissipation in heat sink implementation: A review. *Heat Transf. Eng.* **2021**, *42*, 967–984. [CrossRef]
78. Nawrot, W.; Malecha, K. Additive manufacturing revolution in ceramic microsystems. *Microelectron. Int.* **2020**, *37*, 79–85. [CrossRef]
79. Silva, C.K.A.; Alves, A.K. Graphene Application. In *Technological Applications of Nanomaterials*; Kopp Alves, A., Ed.; Springer: Cham, Switzerland, 2022; pp. 195–206.

80. Wang, N.; Liu, Y.; Ye, L.; Li, J. Highly thermal conductive and light-weight graphene-based heatsink. In Proceedings of the 22nd European Microelectronics and Packaging Conference & Exhibition (EMPC 2019), Pisa, Italy, 16–19 September 2019. [CrossRef]
81. Fu, Y.; Hansson, J.; Liu, Y.; Chen, S.; Zehri, A.; Samani, M.K.; Wang, N.; Ni, Y.; Zhang, Y.; Zhang, Z.-B. Graphene related materials for thermal management. *2D Materials* **2019**, *7*, 12001. [CrossRef]
82. Sidhu, S.S.; Kumar, S.; Batish, A. Metal matrix composites for thermal management: A review. *Crit. Rev. Solid State Mater. Sci.* **2016**, *41*, 132–157. [CrossRef]
83. Baig, M.M.A.; Hassan, S.F.; Saheb, N.; Patel, F. Metal matrix composite in heat sink application: Reinforcement, processing, and properties. *Materials* **2021**, *14*, 6257. [CrossRef]
84. Chawla, K.K. Carbon Fiber/Carbon Matrix Composites. In *Composite Materials*; Chawla, K.K., Ed.; Springer: Cham, Switzerland, 2019; pp. 297–311.
85. Windhorst, T.; Blount, G. Carbon-carbon composites: A summary of recent developments and applications. *Mater. Des.* **1997**, *18*, 11–15. [CrossRef]
86. Meenashisundaram, G.K.; Xu, Z.; Nai, M.L.S.; Lu, S.; Ten, J.S.; Wei, J. Binder jetting additive manufacturing of high porosity 316L stainless steel metal foams. *Materials* **2020**, *13*, 3744. [CrossRef] [PubMed]
87. Wawrzyniak, B.I.; Tangudu, J. *Design Analysis of High Power Density Additively Manufactured Induction Motor*; 0148-7191; SAE Technical Paper no. 2016-01-2063; SAE International: Warrendale, PA, USA, 2016.
88. Liu, M.; Li, Y.; Ding, H.; Sarlioglu, B. Thermal management and cooling of windings in electrical machines for electric vehicle and traction application. In Proceedings of the Transportation Electrification Conference and Expo (ITEC 2017), Chicago, IL, USA, 22–24 June 2017; pp. 668–673. [CrossRef]
89. Zhao, K.; Luo, J. Performance comparison and analysis of different rotor structures of vehicle permanent magnet synchronous flat wire motor. *Machines* **2022**, *10*, 212. [CrossRef]
90. Silbernagel, C. Investigation of the Design, Manufacture and Testing of Additively Manufactured Coils for Electric Motor Applications. Ph.D. Thesis, University of Nottingham, Nottingham, UK, 2019.
91. Lorenz, F.; Rudolph, J.; Wemer, R. Design of 3D printed high performance windings for switched reluctance machines. In Proceedings of the 23rd International Conference on Electrical Machines (ICEM 2018), Alexandroupoli, Greece, 3–6 September 2018; pp. 2451–2457. [CrossRef]
92. Simpson, N.; North, D.J.; Collins, S.M.; Mellor, P.H. Additive manufacturing of shaped profile windings for minimal AC loss in electrical machines. *IEEE Trans. Ind. Appl.* **2020**, *56*, 2510–2519. [CrossRef]
93. Use-Case: Racing Engine. Available online: https://www.additive-drives.de/en/project/racing-engine/ (accessed on 19 December 2021).
94. Jung, J.; Helm, A.; Liebold, J. Improved efficiency of electric drives with additively manufactured roebel bar windings. *MTZ Worldw.* **2021**, *82*, 54–58. [CrossRef]
95. Ghahfarokhi, P.S.; Podgornovs, A.; Cardoso, A.J.M.; Kallaste, A.; Belahcen, A.; Vaimann, T. AC losses analysis approaches for electric vehicle motors with hairpin winding configuration. In Proceedings of the 47th Annual Conference of the IEEE Industrial Electronics Society, Toronto, ON, Canada, 13–16 October 2021. [CrossRef]
96. Hairpin Winding. Available online: https://www.bihler.de/files/bihler/images/branchen/e-mobilitaet/bihler-hairpin-elektromobilitaet.png (accessed on 11 February 2022).
97. Vervecken, L. Assebled Roebel Cable. Available online: https://commons.wikimedia.org/wiki/Category:Roebel_bars#/media/File:Roebel_assembled_coated_conductor_cable_(RACC).svg (accessed on 21 February 2022).
98. Venturini, G.; Volpe, G.; Villani, M.; Popescu, M. Investigation of cooling solutions for hairpin winding in traction application. In Proceedings of the 24th International Conference on Electrical Machines (ICEM 2020), Online, 23–26 August 2020; pp. 1573–1578. [CrossRef]
99. Rivière, N.; Volpe, G.; Villani, M.; Fabri, G.; Di Leonardo, L.; Popescu, M. Design analysis of a high speed copper rotor induction motor for a traction application. In Proceedings of the International Electric Machines & Drives Conference (IEMDC 2019), San Diego, CA, USA, 12–15 May 2019; pp. 1024–1031. [CrossRef]
100. Xue, S. *Maximising E-Machine Efficiency with Hairpin Windings*; Motor Design: Wrexham, UK, 2021.
101. Sher, D. *Additive Drives Introduces Highly Efficient 3D Printed Windings for Electric Motors*; 3D Printing Media Network (3dpbm): Farnham, UK, 2020.
102. Use-Case: E-Traction Motor. Available online: https://www.additive-drives.de/en/project/e-traction-motor/ (accessed on 5 December 2021).
103. Al-Ani, M. *Plug-In Winding for Electrical Machine and Electromagnetic Devices*; Fluxmakers: Nottinghamshire, UK, 2019.
104. Fleischer, J.; Haag, S.; Hofmann, J. *Quo Vadis Winding Technology? A study on the State of the Art and Research on Future Trends in Automotive Engineering*; Institute of Production Science, Karlsruhe Institute of Technology: Karlsruhe, Germany, 2017.
105. Reinap, A. *Direct Cooled Laminated Windings—Radially Displaced Laminated Winding Segments*; Technical report; LUTEDX/(TEIE-7261); Lund University: Lund, Sweden, 2015.
106. Högmark, C.; Andersson, R.; Reinap, A.; Alaküla, M. Electrical machines with laminated winding for hybrid vehicle applications. In Proceedings of the International Electric Drives Production Conference (EDPC 2012), Nuremberg, Germany, 15–18 October 2012. [CrossRef]

107. Gmyrek, Z.; Cavagnino, A. Influence of punching, welding, and clamping on magnetic cores of fractional kilowatt motors. *IEEE Trans. Ind. Appl.* **2018**, *54*, 4123–4132. [CrossRef]
108. Peng, H.; Guimbretière, F.; McCann, J.; Hudson, S. A 3D printer for interactive electromagnetic devices. In Proceedings of the 29th Annual Symposium on User Interface Software and Technology (UIST 2016), Tokyo, Japan, 16–19 October 2016; pp. 553–562. [CrossRef]
109. Elaskri, A.; Ellery, A. 3D printed electric motors as a step towards self-replicating machines. In Proceedings of the International Symposium Artificial Intelligence Robotics & Automation in Space (i-SAIRAS 2020), Online, 19–23 October 2020.
110. Ferrara, M.; Rinaldi, M.; Pigliaru, L.; Cecchini, F.; Nanni, F. Investigating the use of 3D printed soft magnetic PEEK-based composite for space compliant electrical motors. *J. Appl. Polym. Sci.* **2022**, *139*, 52150. [CrossRef]
111. Stornelli, G.; Folgarait, P.; Ridolfi, M.R.; Corapi, D.; Repitsch, C.; Pietro, O.D.; Schino, A.D. Feasibility study of ferromagnetic cores fabrication by additive manufacturing process. *Mater. Proc.* **2021**, *3*, 28. [CrossRef]
112. Zhang, Y.; Poli, L.; Garratt, E.; Foster, S.; Roch, A. Utilizing fused filament fabrication for printing iron cores for electrical devices. *3D Print. Addit. Manuf.* **2020**, *7*, 279–287. [CrossRef]
113. Yakout, M.; Elbestawi, M.; Wang, L.; Muizelaar, R. Selective laser melting of soft magnetic alloys for automotive applications. In Proceedings of the Joint Special Interest Group meeting between euspen and ASPE Advancing Precision in Additive Manufacturing, Nantes, France, 12 September 2019.
114. Gnanaraj, S.D.; Gundabattini, E.; Singh, R.R. Materials for lightweight electric motors–a review. *IOP Conf. Ser. Mater. Sci. Eng.* **2020**, *906*, 12020. [CrossRef]
115. Gai, Y.; Kimiabeigi, M.; Chong, Y.C.; Widmer, J.D.; Deng, X.; Popescu, M.; Goss, J.; Staton, D.A.; Steven, A. Cooling of automotive traction motors: Schemes, examples, and computation methods—A review. *IEEE Trans. Ind. Electron.* **2018**, *66*, 1681–1692. [CrossRef]
116. Vialva, T. *TRUMPF Introduces Precious Metal and Copper 3D Printing Powered by Green Laser*; 3D Printing Industry: Hong Kong, China, 2018.
117. Lindh, P.M.; Petrov, I.; Semken, R.S.; Niemelä, M.; Pyrhönen, J.J.; Aarniovuori, L.; Vaimann, T.; Kallaste, A. Direct liquid cooling in low-power electrical machines: Proof-of-concept. *IEEE Trans. Energy Convers.* **2016**, *31*, 1257–1266. [CrossRef]
118. Reinap, A.; Andersson, M.; Márquez-Fernández, F.J.; Abrahamsson, P.; Alaküla, M. Performance estimation of a traction machine with direct cooled hairpin winding. In Proceedings of the Transportation Electrification Conference and Expo (ITEC 2019), Detroit, MI, USA, 19–21 June 2019. [CrossRef]
119. Bräuer, P.; Lindner, M.; Studnitzky, T.; Kieback, B.; Rudolph, J.; Werner, R.; Krause, G. 3D Screen Printing technology-Opportunities to use revolutionary materials and machine designs. In Proceedings of the 2nd International Electric Drives Production Conference (EDPC 2012), Nuremberg, Germany, 15–18 October 2012. [CrossRef]
120. Lindh, P.; Petrov, I.; Jaatinen-Värri, A.; Grönman, A.; Martinez-Iturralde, M.; Satrústegui, M.; Pyrhönen, J. Direct liquid cooling method verified with an axial-flux permanent-magnet traction machine prototype. *IEEE Trans. Ind. Electron.* **2017**, *64*, 6086–6095. [CrossRef]
121. Orosz, T.; Rassõlkin, A.; Kallaste, A.; Arsénio, P.; Pánek, D.; Kaska, J.; Karban, P. Robust design optimization and emerging technologies for electrical machines: Challenges and open problems. *Appl. Sci.* **2020**, *10*, 6653. [CrossRef]
122. Thermally Conductive Polymer Materials for 3D Printing. Available online: https://3dprinting.com/3d-printing-use-cases/thermally-conductive-polymer-materials/ (accessed on 29 November 2021).
123. Diana, M.; Colussi, J.; La Ganga, A.; Guglielmi, P. An innovative slot cooling for integrated electric drives. In Proceedings of the Workshop on Electrical Machines Design, Control and Diagnosis (WEMDCD 2019), Athens, Greece, 22–23 April 2019; pp. 191–196. [CrossRef]
124. *High Thermal Conductivity 3D Printing Materials and Parts*; TCPoly: Atlanta, GA, USA, 2019.
125. Xu, Y.; Wang, X.; Zhou, J.; Song, B.; Jiang, Z.; Lee, E.M.; Huberman, S.; Gleason, K.K.; Chen, G. Molecular engineered conjugated polymer with high thermal conductivity. *Sci. Adv.* **2018**, *4*, eaar3031. [CrossRef] [PubMed]
126. Madonna, V.; Giangrande, P.; Walker, A.; Galea, M. On the effects of advanced end-winding cooling on the design and performance of electrical machines. In Proceedings of the 23rd International Conference on Electrical Machines (ICEM 2018), Alexandroupoli, Greece, 3–6 September 2018; pp. 311–317. [CrossRef]
127. Fang, G.; Yuan, W.; Yan, Z.; Sun, Y.; Tang, Y. Thermal management integrated with three-dimensional heat pipes for air-cooled permanent magnet synchronous motor. *Appl. Therm. Eng.* **2019**, *152*, 594–604. [CrossRef]
128. Schiefer, M.; Doppelbauer, M. Indirect slot cooling for high-power-density machines with concentrated winding. In Proceedings of the International Electric Machines & Drives Conference (IEMDC 2015), Coeur d'Alene, ID, USA, 10–13 May 2015; pp. 1820–1825. [CrossRef]
129. Huang, J.; Naini, S.S.; Miller, R.; Rizzo, D.; Sebeck, K.; Shurin, S.; Wagner, J. A hybrid electric vehicle motor cooling system—design, model, and control. *IEEE Trans. Veh. Technol.* **2019**, *68*, 4467–4478. [CrossRef]
130. Vansompel, H.; Sergeant, P. Extended end-winding cooling insert for high power density electric machines with concentrated windings. *IEEE Trans. Energy Convers.* **2019**, *35*, 948–955. [CrossRef]
131. Galea, M.; Gerada, C.; Raminosoa, T.; Wheeler, P. A thermal improvement technique for the phase windings of electrical machines. *IEEE Trans. Ind. Appl.* **2011**, *48*, 79–87. [CrossRef]

132. Acquaviva, A.; Skoog, S.; Thiringer, T. Design and verification of in-slot oil-cooled tooth coil winding PM machine for traction application. *IEEE Trans. Ind. Electron.* **2020**, *68*, 3719–3727. [CrossRef]
133. Gundabattini, E.; Mystkowski, A.; Idzkowski, A.; Solomon, D.G. Thermal mapping of a high-speed electric motor used for traction applications and analysis of various cooling methods—A review. *Energies* **2021**, *14*, 1472. [CrossRef]
134. Geng, W.; Zhu, T.; Li, Q.; Zhang, Z. Windings indirect liquid cooling method for a compact outer-rotor PM starter/generator with concentrated windings. *IEEE Trans. Energy Convers.* **2021**, *36*, 3282–3293. [CrossRef]
135. Yin, H.; Huang, S.; Li, H. Study on the influence of internal water cooling system on the loss of permanent magnet synchronous motor. *AIP Conf. Proc.* **2018**, *1971*, 040043. [CrossRef]
136. Wrobel, R.; Hussein, A. Design considerations of heat guides fabricated using additive manufacturing for enhanced heat transfer in electrical machines. In Proceedings of the Energy Conversion Congress and Exposition (ECCE 2018), Portland, OR, USA, 23–27 September 2018; pp. 6506–6513. [CrossRef]
137. Acquaviva, A.; Skoog, S.; Grunditz, E.; Thiringer, T. Electromagnetic and calorimetric validation of a direct oil cooled tooth coil winding PM machine for traction application. *Energies* **2020**, *13*, 3339. [CrossRef]
138. Szabó, L. Survey on applying 3D printing in manufacturing the cooling systems of electrical machines. In Proceedings of the International Conference on Automation, Quality and Testing, Robotics (AQTR 2022), Cluj-Napoca, Romania, 21–22 May 2022; in print.
139. Sixel, W.; Liu, M.; Nellis, G.; Sarlioglu, B. Cooling of windings in electric machines via 3-D printed heat exchanger. *IEEE Trans. Ind. Appl.* **2020**, *56*, 4718–4726. [CrossRef]
140. Semidey, S.A.; Mayor, J.R. Experimentation of an electric machine technology demonstrator incorporating direct winding heat exchangers. *IEEE Trans. Ind. Electron.* **2014**, *61*, 5771–5778. [CrossRef]
141. Panse, S.S.; Ekkad, S.V. Forced convection cooling of additively manufactured single and double layer enhanced microchannels. *Int. J. Heat Mass Transf.* **2021**, *168*, 120881. [CrossRef]
142. Carriero, A.; Locatelli, M.; Ramakrishnan, K.; Mastinu, G.; Gobbi, M. *A Review of the State of the Art of Electric Traction Motors Cooling Techniques*; 0148-7191; SAE Technical Paper no. 2018-01-0057; SAE International: Warrendale, PA, USA, 2018.
143. Nollau, A.; Gerling, D. A flux barrier cooling for traction motors in hybrid drives. In Proceedings of the International Electric Machines & Drives Conference (IEMDC 2015), Coeur d'Alene, ID, USA, 10–13 May 2015; pp. 1103–1108. [CrossRef]
144. Wrobel, R.; Scholes, B.; Hussein, A.; Law, R.; Mustaffar, A.; Reay, D. A metal additively manufactured (MAM) heat exchanger for electric motor thermal control on a high-altitude solar aircraft—Experimental characterisation. *Therm. Sci. Eng. Prog.* **2020**, *19*, 100629. [CrossRef]
145. Amelia, H. Porsche Present 40% Lighter 3D Printed Electric Drive Housing. Available online: https://www.3dnatives.com/en/porsche-present-3d-printed-electric-drive-housing-261220204/ (accessed on 30 September 2021).
146. Porsche Electric Traction Drive System Housing. Available online: https://www.tuningblog.eu/en/wp-content/uploads/2020/12/E-Antrieb-Gehaeuse-3D-Drucker-Porsche-2.jpg; https://www.tuningblog.eu/en/wp-content/uploads/2020/12/E-Antrieb-Gehaeuse-3D-Drucker-Porsche-4.jpg (accessed on 30 September 2021).
147. Galloni, E.; Parisi, P.; Marignetti, F.; Volpe, G. CFD analyses of a radial fan for electric motor cooling. *Therm. Sci. Eng. Prog.* **2018**, *8*, 470–476. [CrossRef]
148. De Santiago, J.; Bernhoff, H.; Ekergård, B.; Eriksson, S.; Ferhatovic, S.; Waters, R.; Leijon, M. Electrical motor drivelines in commercial all-electric vehicles: A review. *IEEE Trans. Veh. Technol.* **2011**, *61*, 475–484. [CrossRef]
149. Candelo-Zuluaga, C.; Espinosa, A.G.; Riba, J.-R.; Tubert, P. PMSM design for achieving a target torque-speed-efficiency map. *IEEE Trans. Veh. Technol.* **2020**, *69*, 14448–14457. [CrossRef]
150. Zhao, X.; Kou, B.; Huang, C.; Zhang, L. Optimization design and performance analysis of a reverse-salient permanent magnet synchronous motor. *Machines* **2022**, *10*, 204. [CrossRef]
151. Lammers, S.; Adam, G.; Schmid, H.J.; Mrozek, R.; Oberacker, R.; Hoffmann, M.J.; Quattrone, F.; Ponick, B. Additive manufacturing of a lightweight rotor for a permanent magnet synchronous machine. In Proceedings of the International Electric Drives Production Conference (EDPC 2016), Nuremberg, Germany, 30 November–1 December 2016; pp. 41–45. [CrossRef]
152. Garibaldi, M. Laser Additive Manufacturing of Soft Magnetic Cores for Rotating Electrical Machinery: Materials Development and Part Design. Ph.D. thesis, University of Nottingham, Nottingham, UK, 2018.
153. Bailey, P. Light work. *Electr. Hybrid Veh. Technol. Int.* **2020**, 24–26.
154. Młot, A.; Łukaniszyn, M. Analysis of axial flux motor performance for traction motor applications. *COMPEL-Int. J. Comput. Math. Electr. Electron. Eng.* **2019**, *38*, 1306–1322. [CrossRef]
155. Singh, S.; Pillay, P. Sinusoidal shaped surface permanent magnet motor using cold spray additive manufacturing. In Proceedings of the Energy Conversion Congress and Exposition (ECCE 2020), Detroit, MI, USA, 11–15 October 2020; pp. 2089–2094. [CrossRef]
156. Demir, U. Improvement of the power to weight ratio for an induction traction motor using design of experiment on neural network. *Electr. Eng.* **2021**, *103*, 2267–2284. [CrossRef]
157. Chinthavali, M. Additive manufacturing technology for power electronics applications. In Proceedings of the Applied Power Electronics Conference and Exposition (APEC 2016), Long Beach, CA, USA, 20–24 March 2016.
158. Bianchi, N.; Bolognani, S.; Carraro, E.; Castiello, M.; Fornasiero, E. Electric vehicle traction based on synchronous reluctance motors. *IEEE Trans. Ind. Appl.* **2016**, *52*, 4762–4769. [CrossRef]

159. Uberti, F.; Frosini, L.; Szabó, L. An optimization procedure for a synchronous reluctance machine with fluid shaped flux barriers. In Proceedings of the 24th International Conference on Electrical Machines (ICEM 2020), Online, 23–26 August 2020; pp. 389–395. [CrossRef]
160. Uberti, F.; Frosini, L.; Szabó, L. A new design procedure for rotor laminations of synchronous reluctance machines with fluid shaped barriers. *Electronics* **2022**, *11*, 134. [CrossRef]
161. Ibrahim, M.; Bernier, F.; Lamarre, J.-M. Design of a PM-assisted synchronous reluctance motor utilizing additive manufacturing of magnetic materials. In Proceedings of the Conversion Congress and Exposition (ECCE 2019), Baltimore, MD, USA, 29 September–3 October 2019; pp. 1663–1668. [CrossRef]
162. Freeman, F.S.; Lincoln, A.; Sharp, J.; Lambourne, A.; Todd, I. Exploiting thermal strain to achieve an in-situ magnetically graded material. *Mater. Des.* **2019**, *161*, 14–21. [CrossRef]
163. Zhang, Z.-Y.; Jhong, K.J.; Cheng, C.-W.; Huang, P.-W.; Tsai, M.-C.; Lee, W.-H. Metal 3D printing of synchronous reluctance motor. In Proceedings of the International Conference on Industrial Technology (ICIT 2016), Taipei, Taiwan, 14–17 March 2016; pp. 1125–1128. [CrossRef]
164. Ban, B.; Stipetic, S. Absolutely feasible synchronous reluctance machine rotor barrier topologies with minimal parametric complexity. *Machines* **2022**, *10*, 206. [CrossRef]
165. Reddy, P.B.; El-Refaie, A.M.; Zou, M.; Pan, D.; Alexander, J.P.; Tapadia, N.; Grace, K.; Huh, K.-K.; Johnson, F. Performance testing and analysis of synchronous reluctance motor utilizing dual-phase magnetic material. *IEEE Trans. Ind. Appl.* **2018**, *54*, 2193–2201. [CrossRef]
166. RotorVagatiSynRM. Available online: https://upload.wikimedia.org/wikipedia/commons/c/c2/RotorVagatiSynRM.png (accessed on 10 February 2022).
167. Cai, S.; Shen, J.; Hao, H.; Jin, M. Design methods of transversally laminated synchronous reluctance machines. *CES Trans. Electr. Mach. Syst.* **2017**, *1*, 164–173. [CrossRef]
168. Pyrhönen, J.; Jokinen, T.; Hrabovcová, V.; Niemelä, H. *Design of Rotating Electrical Machines*; John Wiley and Sons: Chichester, UK, 2008.
169. Abramenko, V.; Petrov, I.; Nerg, J.; Pyrhönen, J. Synchronous reluctance motors with an axially laminated anisotropic rotor as an alternative in high-speed applications. *IEEE Access* **2020**, *8*, 29149–29158. [CrossRef]
170. Son, I.; Lukman, G.F.; Shah, M.H.; Jeong, K.-I.; Ahn, J.-W. Design considerations and selection of cost-effective switched reluctance drive for radiator cooling fans. *Electronics* **2021**, *10*, 917. [CrossRef]
171. Dúbravka, P.; Rafajdus, P.; Makyš, P.; Szabó, L. Control of switched reluctance motor by current profiling under normal and open phase operating condition. *IET Electr. Power Appl.* **2017**, *11*, 548–556. [CrossRef]
172. Lee, J.; Seo, J.H.; Kikuchi, N. Topology optimization of switched reluctance motors for the desired torque profile. *Struct. Multidiscip. Optim.* **2010**, *42*, 783–796. [CrossRef]
173. Tseng, G.-M.; Jhong, K.-J.; Tsai, M.-C.; Huang, P.-W.; Lee, W.-H. Application of additive manufacturing for low torque ripple of 6/4 switched reluctance motor. In Proceedings of the 19th International Conference on Electrical Machines and Systems (ICEMS 2016), Chiba, Japan, 13–16 November 2016.
174. Szabó, L. A survey on modular variable reluctance generators for small wind turbines. *IEEE Trans. Ind. Appl.* **2019**, *55*, 2548–2557. [CrossRef]
175. Wan, Z.; Ahmed, A.; Husain, I.; Muljadi, E. A novel transverse flux machine for vehicle traction aplications. In Proceedings of the Power & Energy Society General Meeting, Denver, CO, USA, 26–30 July 2015. [CrossRef]
176. Drabek, T.; Kapustka, P.; Lerch, T.; Skwarczyński, J. A novel approach to transverse flux machine construction. *Energies* **2021**, *14*, 7690. [CrossRef]
177. Szabó, L. Novel variable reluctance generators used in small wind turbines. The modular approach. In Proceedings of the 19th International Carpathian Control Conference (ICCC 2018), Szilvásvárad, Hungary, 28–31 May 2018. [CrossRef]
178. Baserrah, S. Theoretical and Experimental Investigations of a Permanent Magnet Excited Transverse Flux Machine with a Segmented Stator for In-Wheel Motor Applications. Ph.D. Thesis, University of Bremen, Bremen, Germany, 2014.
179. Hosek, M.; Krishnasamy, J.; Sah, S.; Bashaw, T. Spray-formed hybrid-field electric motor. In Proceedings of the International Design Engineering Technical Conferences and Computers and Information in Engineering Conference, Charlotte, NC, USA, 21–24 August 2016. [CrossRef]
180. Lamarre, J.-M.; Bernier, F. Permanent magnets produced by cold spray additive manufacturing for electric engines. *J. Therm. Spray Technol.* **2019**, *28*, 1709–1717. [CrossRef]
181. Szabó, L. Modular switched reluctance machines to be used in automotive applications. In Proceedings of the SRM Drives an Alternative for E-traction Workshop, Vilanova i la Geltrú, Spain, 2 February 2018.
182. Kallaste, A.; Vaimann, T.; Rassãlkin, A. Additive design possibilities of electrical machines. In Proceedings of the 59th International Scientific Conference on Power and Electrical Engineering of Riga Technical University (RTUCON 2018), Riga, Latvia, 12–13 November 2018. [CrossRef]
183. Garcia Gonzalez, A.; Wang, D.; Dubus, J.-M.; Rasmussen, P.O. Design and experimental investigation of a hybrid rotor permanent magnet modular machine with 3D flux paths accounting for recyclability of permanent magnet material. *Energies* **2020**, *13*, 1342. [CrossRef]

184. European Commission. *A New Circular Economy Action Plan—For a Cleaner and More Competitive Europe*; Commission to the European Parliament: Bruegel, Belgium, 2020.
185. Claeys, G.; Tagliapietra, S.; Zachmann, G. *How to Make the European Green Deal Work*; Bruegel: Brussels, Belgium, 2019.
186. Aguilera, E.; Ramos, J.; Espalin, D.; Cedillos, F.; Muse, D.; Wicker, R.; MacDonald, E. 3D printing of electro mechanical systems. In Proceedings of the Solid Freeform Fabrication Symposium, Austin, TX, USA, 12–14 August 2013; pp. 950–961.
187. Petrick, I.J.; Simpson, T.W. 3D printing disrupts manufacturing: How economies of one create new rules of competition. *Res.-Technol. Manag.* **2013**, *56*, 12–16. [CrossRef]

energies

MDPI

Article

Design of High-Dynamic PMSM Servo Drive Using Nonlinear Predictive Controller with Harmonic Disturbance Observer

Zhanqing Zhou [1], Shuaijiang Yao [1], Chaolei Ma [2], Guozheng Zhang [1,*] and Qiang Geng [1,*]

[1] School of Electrical Engineering, Tiangong University, Tianjin 300387, China; zhzhq@tju.edu.cn (Z.Z.); 20170021@tiangong.edu.cn (S.Y.)

[2] School of Electrical and Information Engineering, Tianjin University, Tianjin 300387, China; machaolei@tju.edu.cn

* Correspondence: zhanggz@tju.edu.cn (G.Z.); gengqiang@tju.edu.cn (Q.G.)

Abstract: The high-dynamic permanent magnet (PM) motor servo system with high-bandwidth is the core equipment for industrial production, and the control bandwidth is also an important indexes to evaluate the performance of the servo system. The non-cascaded direct predictive speed control is an appropriate scheme to optimize the dynamic performance of the PM motor servo system. However, the high bandwidth of the non-cascaded control structure results in poor anti-interference ability, and it cannot effectively deal with the coupling relationship between current and speed, leading to poor control performance in the current limit region. Regarding the above problems, a nonlinear predictive speed control strategy combined with harmonic disturbance observer is proposed. In the proposed strategy, the disturbances of the servo system are separated from the mathematical model according to the nonlinear modeling theory, and the traditional disturbance observer is modified to estimate the harmonics. A nonlinear control law with strong disturbance suppression ability was designed. Furthermore, a complete current and speed prediction mechanism was introduced into the algorithm, in which the proportional differential (PD) controller is employed as the connection medium between the reference current and speed to solve the coupling problem of the non-cascaded control structure.

Keywords: model predictive control; nonlinear observer; disturbances attenuation; permanent magnet synchronous motor (PMSM); observer-based control

Citation: Zhou, Z.; Yao, S.; Ma, C.; Zhang, G.; Geng, Q. Design of High-Dynamic PMSM Servo Drive Using Nonlinear Predictive Controller with Harmonic Disturbance Observer. *Energies* **2022**, *15*, 4107. https://doi.org/10.3390/en15114107

Academic Editors: Lorand Szabo and Feng Chai

Received: 8 May 2022
Accepted: 1 June 2022
Published: 2 June 2022

1. Introduction

High precision servo control has been widely used in computer numerical control (CNC) machines, artificial intelligence, etc. Because permanent magnet (PM) motors have the advantages of simple structure, high specific power, and wide speed range [1,2], they are usually employed as the main motor system of servo control. The speed control algorithm of the PM servo system usually adopts the cascade control structure with the current inner loop and the speed outer loop. This control structure requires that the control bandwidth of the current loop is much larger than that of the speed loop, to meet the stability condition of system, resulting in the limitation of the dynamic response speed of the motor [3–5]. To match the growing demand for highly dynamic and high-precision modern servo systems, the non-cascaded direct predictive speed control (DPSC) has become an important research direction in servo system control strategy [6,7]. Because of its excellent dynamic performance and multivariable online optimization abilities, DSPC is an appropriate scheme to construct the direct speed control structure of a PM servo motor [8,9].

In recent years, the model predictive control algorithm has achieved great success in the permanent magnet synchronous motor (PMSM) drive, which improves the dynamic performance of the motor drive system [10,11]. DPSC is a feasible means of further optimizing the dynamic performance of the PM servo system. Currently, scholars are trying to improve DPSC in terms of the two aspects of disturbance suppression and efficient operation, so as to make it possible for industrial application.

The excellent dynamic performance of DPSC is due to its high-bandwidth characteristics; however, the anti-interference capability of this high-bandwidth control structure is relatively weak. There are a large number of harmonics in the servo system, causing the problems of parameter mismatch, machining error, and control misalignment. A high-bandwidth structure cannot effectively suppress these medium- and high-frequency harmonics, resulting in the deterioration of the steady-state performance [12,13]. To solve this problem, a repetitive controller is embedded into the model predictive control (MPC) algorithm according to the Fourier series, which is equivalent to adding periodic motion modes to the controller. This improves the suppression ability of the MPC to the corresponding periodic disturbance [14]. Similarly, an iterative learning MPC controller was proposed in [15]. Compared with the repetitive MPC or iterative learning MPC, the frequency characteristic of the resonant controller is more stable and has less impact on the stability of the control system. On this basis, the predictive-resonant controller is formed by integrating multiple resonant units into the MPC, which enhances the control ability of the controller to the signal at a specific frequency [16,17]. However, DPSC is a typical multiple input and multiple output (MIMO) controller. Both repetitive MPC and resonant MPC algorithms adopt the design idea of a single input and single output (SISO) controller, which is not suitable for the design of DPSC. The robust MPC controller is a kind of MIMO MPC strategy, and is suitable for the transformation to DPSC [18–20]. Instead of the accurate frequency information of disturbances, the key control information of the robust MPC is only the extreme range of disturbances. For this reason, the robust MPC can only satisfy the minimum control objectives (such as tracking and stability) in the presence of disturbances. Due to its simple structure, intuitive principle, and suitability for MIMO design, the observer-based MPC (OB-MPC) is a sophisticated methods to suppress disturbances in a control system that adopts a predictive algorithm [21–23]. According to the estimated value of the observer, this kind of algorithm will generate supplementary values that can be directly superimposed in the control quantity, so as to improve the control performance. However, these algorithms pay more attention to the estimation and suppression of constant disturbances [24], and fewer reports have been made about the application of the algorithms to harmonic disturbances.

The cascade control structure is abandoned in DPSC, resulting in the disappearance of the direct relationship between speed and current. Thus, it is difficult to utilize the maximum operation current of the motor for DPSC, due to considerations of safe operation and stability. Specifically, assuming that the motor boundary constraint is too strict, the motor cannot reach its maximum efficiency; otherwise, the transient current of the motor will be large, which means that it is easy to trigger an overcurrent fault. To solve the above problem, an additional current limiting loop is introduced into the output [25–28]. However, this method is equivalent to re-imposing a constraint boundary on the solution of an unconstrained optimization problem, which deviates from the actual optimal solution. Hence, the dynamic performance of the servo system will be degraded when tracing the reference, such as a slope signal or acceleration signal. Thus, the current limiting loop is rejected in favor of a PI compensation link, which provides a new idea to deal with the operation problem of DPSC on the current boundary [29]. Due to the integral saturation phenomenon of PI controller, this algorithm needs to be improved further. The explicit model predictive algorithm can suitably deal with the above problems, using different control laws under different control partitions [30,31]. However, the huge number of control partitions increases the difficulty of the algorithm implementation. Therefore, alternative direct speed optimization algorithms without a current limiting link should be researched, under the condition of meeting the computing power of existing microprocessors [32,33]. This would be of great value to the promotion of DPSC.

Aiming at addressing the problems of DPSC in terms of disturbance suppression and efficient operation, a nonlinear predictive speed control strategy combined with harmonic disturbance observer (HDO-NPSC) is proposed for the design of a PM servo system with high dynamic response and high operation accuracy. In the proposed strategy, the disturbances in the servo system are separated from its classical model according to the nonlinear modeling theory, and the corresponding predictive model including harmonic and constant disturbances is constructed. Then, the nonlinear predictive control law suitable for PM servo system is established by solving the given cost function. In addition, the observer, which can effectively observe the harmonic disturbances of the system, was designed, and the observation values were introduced into the control law to improve the suppression ability of the proposed algorithm, so as to eliminate the control problems caused by non-ideal modeling factors, such as parameter mismatch and control inaccuracy. In contrast to the existing DPSC strategies, the output variables of HDO-NPSC completely include the motor current and speed variables, and a PD controller is used to connect the speed errors and current reference. Hence, HDO-NPSC can employ the maximum working current of the motor without the help of a current limiting link, by appropriately distributing the weight coefficient.

2. Harmonics Problem of PMSM Servo Drive

The speed fluctuations of a PM servo motor mainly come from the non-ideal modeling factors in the system, including the parameter mismatch of the motor, the dead-time effect of the converter, and the current sampling errors of the controller.

2.1. Parameter Mismatch

The value of the stator resistance is affected by the external environment temperature, and the actual resistance of motor at a given temperature can be expressed with the resistance coefficient α, as [34,35]:

$$R_s = R_{s0}[1 + \alpha(Temp. - 20)], \tag{1}$$

Considering that the value of α is small, and usually selected as 0.00393 ppm/°C, it can be approximately supposed that there is only a constant disturbance in R_s, that is:

$$R_s = R_{s0} + \Delta R, \tag{2}$$

The stator inductance of the motor can be calculated by the finite element method, and its calculation expression is [36]:

$$L_s = \frac{2L_{ef}\sum_{l=1}^{\infty} H_l B_l \Delta S_l}{3i_s^2}, \tag{3}$$

When the magnetic circuit of the motor is unsaturated, the energy in the magnetic field is stable, and the inductance value will not change greatly. In other words, the nominal value of inductance will only change significantly after the saturation of the magnetic circuit. Therefore, it can still be considered that there are only constant disturbances for the inductance parameter, i.e.,

$$\begin{cases} L_d = L_{d0} + \Delta L_d \\ L_q = L_{q0} + \Delta L_q \end{cases}, \tag{4}$$

The moment of inertia of the PM servo system is composed of the motor and the load; its calculation expression is [37]:

$$J_m = \underbrace{\frac{1}{2}\left[M_{iron}c\left(R_{iron}^2 + r_{iron}^2\right) + M_{shaft}R_{shaft}^2\right]}_{Motor} + J_{Load}, \tag{5}$$

From (5), the moment of inertia of the motor can be calculated according to the design parameters. However, it is usually difficult to obtain the load inertia; as a result, the estimated value of the servo system deviates from its actual value. However, the moment of inertia only depends on the shape of the object, the mass distribution, and the position of the rotation axis, so it can be considered that there is only a constant estimation deviation for J_{m}, that is:

$$J_{\mathrm{m}} = J_{\mathrm{m0}} + \Delta J, \tag{6}$$

In the process of system modeling, the permanent flux ψ_{f} is usually considered to be a constant value, that is, the rotor flux is assumed to be an ideal sinusoidal distribution in space. However, due to the influence of stator slotting, core material saturation, manufacturing process constraints, etc., it is difficult to achieve an ideal sinusoidal distribution in the permanent flux, which leads to the existence of harmonics in the stator winding. The harmonics of the permanent magnet flux linkage in the d-q synchronous coordinates can be expressed as [12]:

$$\begin{cases} \psi_{\mathrm{fd}} = \psi_{\mathrm{d0}} + \sum_{l=1}^{\infty} \psi_{\mathrm{d6},l} \cos(6l\omega_{\mathrm{e}}t) \\ \psi_{\mathrm{fq}} = \sum_{l=1}^{\infty} \psi_{\mathrm{q6},l} \cos(6l\omega_{\mathrm{e}}t) \end{cases}, \tag{7}$$

It can be seen from (7) that subharmonics with $6l\omega_{\mathrm{e}}$ can be found in the permanent magnet, which are represented as $\psi_{\mathrm{d6},l}$ and $\psi_{\mathrm{q6},l}$.

2.2. Dead-Time Effects and Current Sampling Errors

Avoiding the operation faults of a voltage sourced inverter, a dead time is usually added to the IGBT control signal, which leads to the deviation in the output voltage. This deviation voltage can be expressed as [38]:

$$\Delta u_{\mathrm{x}} = -\mathrm{sign}(i_{\mathrm{x}}) \frac{T_{\mathrm{dead}}}{T_{\mathrm{s}}} u_{\mathrm{dc}}, \tag{8}$$

According to the Park transformation, the deviation voltages of A, B, and C phases can be transformed into the d-q frame, which can be expressed as:

$$\begin{cases} \Delta u_{\mathrm{d,dead}} = \frac{4T_{\mathrm{d}}u_{\mathrm{dc}}}{\pi T_{\mathrm{s}}} \sum_{l=1}^{\infty} \frac{12l\sin(6l\omega_{\mathrm{e}}t)}{36l^2-1} \\ \Delta u_{\mathrm{q,dead}} = \frac{4T_{\mathrm{d}}u_{\mathrm{dc}}}{\pi T_{\mathrm{s}}} \left\{ -1 + \sum_{l=1}^{\infty} \frac{2\cos(6l\omega_{\mathrm{e}}t)}{36l^2-1} \right\} \end{cases}, \tag{9}$$

From (9), when the motor operates stably with the frequency of ω_{e}, the deviation voltage contains the harmonic disturbances with the frequency of $6l\omega_{\mathrm{e}}$ (l = 1, 2, 3, ...).

The stator current is the necessary information for the feedback control, and its measurement process includes sensor detection, low-pass filtering, polarity conversion, and analog-to-digital conversion. However, offset errors exist in this measurement because of the imbalanced power supply of the current sensors. In addition, proportional errors will also be detected, as the value of the sampling resistance changes with temperature. To summarize, the current measurement results can be expressed with an undetermined coefficient $k_{1\sim4}$ as [12]:

$$\Delta i_{\mathrm{d}} = k_1 \sin(\omega_{\mathrm{e}}t) + k_2 \sin(2\omega_{\mathrm{e}}t) + o_{\mathrm{d}}, \tag{10}$$

$$\Delta i_{\mathrm{q}} = k_3 \cos(\omega_{\mathrm{e}}t) + k_4 \cos(2\omega_{\mathrm{e}}t) + o_{\mathrm{q}}, \tag{11}$$

From the above analysis, it can be seen that the main disturbance components in the PM servo system can be divided as follows: (1) the constant disturbances caused by the parameter mismatch of resistance, inductance, and moment of inertia; (2) disturbances with a frequency of $6\omega_{\mathrm{e}}$ caused by the flux harmonics and dead-time effect; and (3) disturbances with a frequency of ω_{e} and $2\omega_{\mathrm{e}}$ caused by current sampling errors.

3. Design of Nonlinear Predictive Speed Controller

3.1. Continuous-Time Predictive Model

The dynamic model of the PM servo system in the d-q frame can be expressed as [39]:

$$u_{\mathrm{sd}} = R_{\mathrm{s}} i_{\mathrm{sd}} + L_{\mathrm{d}} \frac{\mathrm{d}i_{\mathrm{sd}}}{\mathrm{d}t} - L_{\mathrm{q}} n_{\mathrm{p}} \omega_{\mathrm{m}} i_{\mathrm{sq}}, \tag{12}$$

$$u_{\mathrm{sq}} = R_{\mathrm{s}} i_{\mathrm{sq}} + L_{\mathrm{q}} \frac{\mathrm{d}i_{\mathrm{sq}}}{\mathrm{d}t} + L_{\mathrm{d}} n_{\mathrm{p}} \omega_{\mathrm{m}} i_{\mathrm{sd}} + n_{\mathrm{p}} \omega_{\mathrm{m}} \psi_{\mathrm{f}}, \tag{13}$$

$$J_{\mathrm{m}} \frac{\mathrm{d}\omega_{\mathrm{m}}}{\mathrm{d}t} = T_{\mathrm{e}} - T_L - B\omega_{\mathrm{m}}, \tag{14}$$

Considering the disturbance factors in (12)–(14), then we can obtain:

$$\dot{x}(t) = f(t) + g_1 u(t) + g_2(\chi_{\mathrm{c}}(t) + \chi_{\mathrm{h}}(t)), \tag{15}$$

where $x = [i_{\mathrm{sd}}\ i_{\mathrm{sq}}\ \omega_{\mathrm{m}}]^{\mathrm{T}}$, $u = [u_{\mathrm{sd}}\ u_{\mathrm{sq}}]^{\mathrm{T}}$,

$$\chi_{\mathrm{c}}(t) = \begin{bmatrix} \Delta R i_{\mathrm{sd}} + \Delta L_{\mathrm{d}}(\mathrm{d}i_{\mathrm{sd}}/\mathrm{d}t) - n_{\mathrm{p}}\omega_{\mathrm{m}}\Delta L_{\mathrm{q}} i_{\mathrm{sq}} \\ \Delta R i_{\mathrm{sq}} + \Delta L_{\mathrm{q}}(\mathrm{d}i_{\mathrm{sq}}/\mathrm{d}t) + n_{\mathrm{p}}\omega_{\mathrm{m}}\Delta L_{\mathrm{d}} i_{\mathrm{sd}} \\ T_L + B\omega_{\mathrm{m}} + \Delta J(\mathrm{d}\omega_{\mathrm{m}}/\mathrm{d}t) \end{bmatrix},$$

$$\chi_{\mathrm{h}}(t) = \begin{bmatrix} -\Delta u_{\mathrm{d}} + \Delta i_{\mathrm{sd}}(R_{\mathrm{s}} + \Delta R) + (L_{\mathrm{d}} + \Delta L_{\mathrm{d}})(\mathrm{d}\Delta i_{\mathrm{sd}}/\mathrm{d}t) \\ -n_{\mathrm{p}}\omega_{\mathrm{m}}(L_{\mathrm{q}} + \Delta L_{\mathrm{q}})\Delta i_{\mathrm{sq}} \\ \hline -\Delta u_{\mathrm{q}} + \Delta i_{\mathrm{sq}}(R_{\mathrm{s}} + \Delta R) + (L_{\mathrm{q}} + \Delta L_{\mathrm{q}})(\mathrm{d}\Delta i_{\mathrm{sq}}/\mathrm{d}t) \\ +n_{\mathrm{p}}\omega_{\mathrm{m}}(L_{\mathrm{d}} + \Delta L_{\mathrm{d}})\Delta i_{\mathrm{sd}} + p_{\mathrm{r}}\omega_{\mathrm{m}}\psi_{\mathrm{fd}} \\ \hline 1.5 p_{\mathrm{r}} \begin{bmatrix} -i_{\mathrm{sq}} \sum_{l=1}^{\infty} \psi_{\mathrm{d}6,l} \cos(6l\omega_{\mathrm{e}} t) + \psi_{\mathrm{d}0}\Delta i_{\mathrm{q}} - \\ (L_{\mathrm{d}} - L_{\mathrm{q}}) i_{\mathrm{sd}}\Delta i_{\mathrm{q}} - (L_{\mathrm{d}} - L_{\mathrm{q}}) i_{\mathrm{sq}}\Delta i_{\mathrm{d}} \end{bmatrix} \end{bmatrix}.$$

The other coefficient matrices are:

$$f(t) = \begin{bmatrix} f_{\mathrm{d}} \\ f_{\mathrm{q}} \\ f_{\mathrm{w}} \end{bmatrix} = \begin{bmatrix} (-R_{\mathrm{s}} i_{\mathrm{sd}} + n_{\mathrm{p}}\omega_{\mathrm{m}} L_{\mathrm{q}} i_{\mathrm{sq}})/L_{\mathrm{d}} \\ (-R_{\mathrm{s}} i_{\mathrm{sq}} - n_{\mathrm{p}}\omega_{\mathrm{m}} L_{\mathrm{d}} i_{\mathrm{sd}} - p_{\mathrm{r}}\omega_{\mathrm{m}}\psi_{\mathrm{f}})/L_{\mathrm{q}} \\ \{1.5 n_{\mathrm{p}}[\psi_{\mathrm{d}0} i_{\mathrm{sq}} + (L_{\mathrm{d}} - L_{\mathrm{q}}) i_{\mathrm{sd}} i_{\mathrm{sq}}] - B\omega_{\mathrm{m}}\}/J_{\mathrm{m}} \end{bmatrix}, g_1 = \begin{bmatrix} 1/L_{\mathrm{d}} & 0 \\ 0 & 1/L_{\mathrm{q}} \\ 0 & 0 \end{bmatrix},$$

$$g_2 = \begin{bmatrix} -1/L_{\mathrm{d}} & 0 & 0 \\ 0 & -1/L_{\mathrm{q}} & 0 \\ 0 & 0 & -1/J_{\mathrm{m}} \end{bmatrix}.$$

It can be seen from (20) that $\chi_{\mathrm{c}}(t)$ is mainly introduced by parameter mismatch and external load disturbance, and $\chi_{\mathrm{h}}(t)$ is mainly introduced by the dead-time effects and current sampling errors. Taking the disturbance frequency as the modeling characteristic, the nonlinear model of PM servo system can be constructed as:

$$\begin{cases} \dot{x}(t) = f(t) + g_1 u(t) + g_2 \chi(t) \\ y(t) = h(x) \end{cases}, \tag{16}$$

where $\chi(t) = \chi_{\mathrm{c}}(t) + \chi_{\mathrm{h}}(t)$, $h(x) = [h_1\ h_2\ h_3]^{\mathrm{T}} = [i_{\mathrm{sd}}\ i_{\mathrm{sq}}\ \omega_{\mathrm{m}}]^{\mathrm{T}}$

3.2. Long-Horizon Optimization and Its Control Law

From (17)–(19), the relative order of i_{sd} and i_{sq} is equal to 1, and the relative order of ω_{m} is equal to 2. Then, we can obtain:

$$i_{\mathrm{sd}}(t+\tau) = i_{\mathrm{sd}}(t) + \tau \dot{i}_{\mathrm{sd}}(t), \tag{17}$$

$$i_{\mathrm{sq}}(t+\tau) = i_{\mathrm{sq}}(t) + \tau \dot{i}_{\mathrm{sq}}(t), \tag{18}$$

$$\omega_{\mathrm{m}}(t+\tau) = \omega_{\mathrm{m}}(t) + \tau\dot{\omega}_{\mathrm{m}}(t) + \frac{\tau^2}{2}\ddot{\omega}_{\mathrm{m}}(t), \tag{19}$$

where:

$$\begin{cases} \dot{i}_{\mathrm{sd}}(t) = L_f i_{\mathrm{sd}} + L_{g_1} i_{\mathrm{sd}} \boldsymbol{u}(t) + L_{g_2} i_{\mathrm{sd}} \boldsymbol{\chi}(t) \\ \dot{i}_{\mathrm{sd/q}}(t) = L_f i_{\mathrm{sq}} + L_{g_1} i_{\mathrm{sq}} \boldsymbol{u}(t) + L_{g_2} i_{\mathrm{sq}} \boldsymbol{\chi}(t) \end{cases}, \tag{20}$$

$$\dot{\omega}_{\mathrm{m}}(t) = L_f \omega_{\mathrm{m}} + L_{g_2} \omega_{\mathrm{m}} \boldsymbol{\chi}(t), \tag{21}$$

$$\ddot{\omega}_{\mathrm{m}}(t) = L_f^2 \omega_{\mathrm{m}} + L_{g_1} L_f \omega_{\mathrm{m}} \boldsymbol{u}(t) + L_{g_2} L_f \omega_{\mathrm{m}} \boldsymbol{\chi}(t), \tag{22}$$

To evaluate the tracking performance of output variables in the future time scale, the cost function is defined as:

$$J = \frac{1}{2} \int_0^{T_{\mathrm{i}}} q_{\mathrm{i}} \left\{ \begin{array}{l} \left[i_{\mathrm{sd}}^{\mathrm{ref}}(t+\tau) - i_{\mathrm{sd}}(t+\tau) \right]^2 \\ + \left[i_{\mathrm{sq}}^{\mathrm{ref}}(t+\tau) - i_{\mathrm{sq}}(t+\tau) \right]^2 \end{array} \right\} \mathrm{d}\tau + \frac{1}{2} \int_0^{T_{\mathrm{w}}} q_{\mathrm{w}} \left[\omega_{\mathrm{m}}^{\mathrm{ref}}(t+\tau) - \omega_{\mathrm{m}}(t+\tau) \right]^2 \mathrm{d}\tau, \tag{23}$$

Substituting (20)–(22) into (23), then we obtain:

$$J = \frac{1}{2} \left[\overline{\boldsymbol{Y}}(t) - \overline{\boldsymbol{W}}(t) \right]^{\mathrm{T}} \overline{\boldsymbol{H}}(\tau) \left[\overline{\boldsymbol{Y}}(t) - \overline{\boldsymbol{W}}(t) \right], \tag{24}$$

where:

$$\begin{aligned} &\overline{\boldsymbol{Y}}(t) = \begin{bmatrix} i_{\mathrm{sd}} & \dot{i}_{\mathrm{sd}} & i_{\mathrm{sq}} & \dot{i}_{\mathrm{sq}} & \omega_{\mathrm{m}} & \dot{\omega}_{\mathrm{m}} & \ddot{\omega}_{\mathrm{m}} \end{bmatrix}^{\mathrm{T}}, \\ &\overline{\boldsymbol{W}}(t) = \begin{bmatrix} i_{\mathrm{sd}}^{\mathrm{ref}} & \dot{i}_{\mathrm{sd}}^{\mathrm{ref}} & i_{\mathrm{sq}}^{\mathrm{ref}} & \dot{i}_{\mathrm{sq}}^{\mathrm{ref}} & \omega_{\mathrm{m}}^{\mathrm{ref}} & \dot{\omega}_{\mathrm{m}}^{\mathrm{ref}} & \ddot{\omega}_{\mathrm{m}}^{\mathrm{ref}} \end{bmatrix}^{\mathrm{T}}, \\ &\overline{\boldsymbol{H}}(\tau) = \mathrm{diag}\begin{pmatrix} q_{\mathrm{i}} \int_0^{T_{\mathrm{i}}} \boldsymbol{H}_1^{\mathrm{T}} \boldsymbol{H}_1 \mathrm{d}\tau, q_{\mathrm{i}} \int_0^{T_{\mathrm{i}}} \boldsymbol{H}_2^{\mathrm{T}} \boldsymbol{H}_2 \mathrm{d}\tau, \\ q_{\mathrm{w}} \int_0^{T_{\mathrm{i}}} \boldsymbol{H}_3^{\mathrm{T}} \boldsymbol{H}_3 \mathrm{d}\tau \end{pmatrix}, \boldsymbol{H}_1 = \boldsymbol{H}_2 = \begin{bmatrix} 1 & \tau \end{bmatrix}, \boldsymbol{H}_3 = \begin{bmatrix} 1 & \tau & \tau^2/2 \end{bmatrix}. \end{aligned}$$

According to the unconstrained quadratic optimization theory [36], the optimal solution of the optimization problem corresponding to cost function (24) is equivalent to $\partial J/\partial \boldsymbol{u} = 0$; then, we can obtain the corresponding law of the nonlinear predictive speed controller (NPSC), i.e.,

$$\boldsymbol{u}(t) = -\boldsymbol{G}^{-1} \boldsymbol{M} [\boldsymbol{Y} - \boldsymbol{W} + \boldsymbol{N} \boldsymbol{\chi}(t)], \tag{25}$$

where:

$$\begin{aligned} &\boldsymbol{G} = \frac{q_{\mathrm{i}} T_{\mathrm{i}}^3}{3} \left[L_{g_1}^{\mathrm{T}} i_{\mathrm{d}} L_{g_1} i_{\mathrm{d}} + L_{g_1}^{\mathrm{T}} i_{\mathrm{q}} L_{g_1} i_{\mathrm{q}} \right] + \frac{q_{\mathrm{w}} T_{\mathrm{w}}^3}{20} L_{g_1} L_f^{\mathrm{T}} \omega_{\mathrm{m}} L_{g_1} L_f \omega_{\mathrm{m}}, \\ &\boldsymbol{M} = \begin{bmatrix} q_{\mathrm{i}} T_{\mathrm{i}}^2 L_{g_1}^{\mathrm{T}} i_{\mathrm{d}}/2 & q_{\mathrm{i}} T_{\mathrm{i}}^3 L_{g_1}^{\mathrm{T}} i_{\mathrm{d}}/3 & q_{\mathrm{i}} T_{\mathrm{i}}^2 L_{g_1}^{\mathrm{T}} i_{\mathrm{q}}/2 & q_{\mathrm{i}} T_{\mathrm{i}}^3 L_{g_1}^{\mathrm{T}} i_{\mathrm{q}}/3 \\ q_{\mathrm{w}} T_{\mathrm{i}}^3 L_{g_1} L_f^{\mathrm{T}} \omega_{\mathrm{m}}/6 & q_{\mathrm{w}} T_{\mathrm{i}}^4 L_{g_1} L_f^{\mathrm{T}} \omega_{\mathrm{m}}/8 & q_{\mathrm{w}} T_{\mathrm{i}}^5 L_{g_1} L_f^{\mathrm{T}} \omega_{\mathrm{m}}/20 & \end{bmatrix} \\ &\boldsymbol{Y} = \begin{bmatrix} L_f^0 i_{\mathrm{sd}} & L_f^1 i_{\mathrm{sd}} & L_f^0 i_{\mathrm{sq}} & L_f^1 i_{\mathrm{sd}} & L_f^0 \omega_{\mathrm{m}} & L_f^1 \omega_{\mathrm{m}} & L_f^2 \omega_{\mathrm{m}} \end{bmatrix}^{\mathrm{T}}, \\ &\boldsymbol{W} = \begin{bmatrix} i_{\mathrm{sd}}^{\mathrm{ref}} & \dot{i}_{\mathrm{sd}}^{\mathrm{ref}} & i_{\mathrm{sq}}^{\mathrm{ref}} & \dot{i}_{\mathrm{sq}}^{\mathrm{ref}} & \omega_{\mathrm{m}}^{\mathrm{ref}} & \dot{\omega}_{\mathrm{m}}^{\mathrm{ref}} & \ddot{\omega}_{\mathrm{m}}^{\mathrm{ref}} \end{bmatrix}^{\mathrm{T}}, \\ &\boldsymbol{N} = \begin{bmatrix} \boldsymbol{O} & L_{g2} i_{\mathrm{sd}} & \boldsymbol{O} & L_{g2} i_{\mathrm{sq}} & \boldsymbol{O} & L_{g2} \omega_{\mathrm{m}} & L_{g2} L_f \omega_{\mathrm{m}} \end{bmatrix}^{\mathrm{T}}. \end{aligned}$$

where $\boldsymbol{O}$ is the zero matrix.

3.3. Asymptotic Stability Analysis

Assuming that $\chi(t)$ and its derivatives are continuous and bounded, $\chi(t)$ can be estimated by constructing a disturbance observer. Therefore, on the premise of ignoring the influence of disturbance, by substituting (25) into (15), we can obtain:

$$\dot{\boldsymbol{e}} = \begin{bmatrix} -\frac{3}{2T_{\mathrm{i}}} & 0 & 0 & 0 \\ 0 & -\frac{3}{2T_{\mathrm{i}}} & -\frac{q_{\mathrm{w}} n_{\mathrm{p}} \psi_{\mathrm{f}}}{2J_{\mathrm{m}}} & -\frac{3q_{\mathrm{w}} n_{\mathrm{p}} \psi_{\mathrm{f}}}{8J_{\mathrm{m}}} \\ 0 & 0 & 0 & 1 \\ 0 & -\frac{10 q_{\mathrm{i}} J_{\mathrm{m}}}{n_{\mathrm{p}} \psi_{\mathrm{f}} T_{\mathrm{i}}^3} & -\frac{10}{3T_{\mathrm{w}}^2} & -\frac{5}{2T_{\mathrm{w}}} \end{bmatrix} \boldsymbol{e}, \tag{26}$$

where $\boldsymbol{e} = \begin{bmatrix} i_{\mathrm{sd}} - i_{\mathrm{sd}}^{\mathrm{ref}} & i_{\mathrm{sq}} - i_{\mathrm{sq}}^{\mathrm{ref}} & \omega_{\mathrm{m}} - \omega_{\mathrm{m}}^{\mathrm{ref}} & \dot{\omega}_{\mathrm{m}} - \dot{\omega}_{\mathrm{m}}^{\mathrm{ref}} \end{bmatrix}^{\mathrm{T}}$.

From (26), the characteristic polynomial of system is:

$$\begin{cases} \left(s + \frac{3}{2T_{\mathrm{i}}}\right)^2 = 0 \\ s^2 + \frac{5}{2T_{\mathrm{w}}}s + \frac{10}{3T_{\mathrm{w}}^2} = 0 \end{cases}, \tag{27}$$

From (27), the solutions of the characteristic polynomial are $s_{1,2} = -3/2T_{\mathrm{i}}$, $s_{3,4} = (-1.25 \pm \mathrm{j}1.307)/T_{\mathrm{w}}$. Because the predictive horizon T_{i} and T_{w} are positive, the eigenvalues are located in the left plane of the complex plane, which means the designed system is stable.

It can also be seen from Equation (27) that NPSC has four system characteristic roots, which correspond to the current and speed control performance of the d-axis and q-axis, respectively. Among these, the current control performance is approximately equal to the first-order controller, and the speed control performance is approximately equal to the under-damped second-order controller. In addition, the dynamic regulation time of speed is usually longer than the control period; as a result, it is necessary to implement multi-step prediction for predictive speed control to ensure the stability of the algorithm, and the predictive horizon is required to cover the dynamic process of speed. Thus, $T_{\mathrm{w}} > T_{\mathrm{i}}$, which shows that the current dynamic performance of NPSC is better than that of speed. Furthermore, with the increase in T_{i} and T_{w}, the eigenvalues of the system are closer to the imaginary axis of the complex plane, and the dynamic process of the system slows accordingly. To summarize, the values of T_{i} and T_{w} should not be selected as being too large or too small to ensure the stability of system.

The high bandwidth is the key element for high-performance control of NPSC. Hence, it is necessary to undertake quantitative analysis of the actual bandwidth of NPSC. Employing the transformation results of $i_{\mathrm{d}}^{\mathrm{ref}}$, i_{d}, $\omega_{\mathrm{e}}^{\mathrm{ref}}$, and ω_{e} in the Laplace domain, $I_{\mathrm{d}}^{\mathrm{ref}}(s)$, $I_{\mathrm{d}}(s)$, $W^{\mathrm{ref}}(s)$, and $W(s)$, the control bandwidth of the proposed strategy is analyzed as follows:

$$\begin{cases} \frac{I_{\mathrm{d}}^{\mathrm{ref}}(s)}{I_{\mathrm{d}}(s)} = 1 + \frac{1}{6}T_{\mathrm{i}}s \\ \frac{W^{\mathrm{ref}}(s)}{W(s)} = \frac{K^*\left[2q_{\mathrm{w}}T_{\mathrm{i}}T_{\mathrm{w}}^5s^2 + \left(12q_{\mathrm{w}}T_{\mathrm{w}}^5 + 5q_{\mathrm{w}}T_{\mathrm{i}}T_{\mathrm{w}}^4\right)s + 30q_{\mathrm{w}}T_{\mathrm{w}}^4\right]}{2q_{\mathrm{w}}T_{\mathrm{i}}T_{\mathrm{w}}^5s^2 + \left(12q_{\mathrm{w}}T_{\mathrm{w}}^5 + 5q_{\mathrm{w}}T_{\mathrm{i}}T_{\mathrm{w}}^4\right)s + 30q_{\mathrm{w}}T_{\mathrm{w}}^4 + 240J} \end{cases}, \tag{28}$$

It can be seen from (28) that NPSC is essentially a type of PD controller for the d-axis current control. The control characteristics are externally manifested as an all-pass characteristic, which has high-control bandwidth and poor noise suppression. However, the application of the following harmonic disturbance observer (HDO) can effectively compensate for this defect. For the speed control part, HDO-NPSC is approximately equivalent to a class of P-type controller. Similarly, HDO is used to suppress harmonics in the control system.

4. Harmonic Disturbance Observer

4.1. Modified Observer Targeting at Harmonics Estimation

There are still unknown disturbances in the optimal control law shown in (25), with the exceptions of the given nominal parameters, predetermined control parameters, and predictive horizon. According to the construction principle of the nonlinear observer [40], the disturbance observer with (16) can be constructed as:

$$\begin{cases} \dot{\hat{\chi}}(t) = \boldsymbol{L}(\chi(t) - \hat{\chi}(t)) \\ \chi(t) = g_2^{-1}(-\boldsymbol{f}(t) - g_1\boldsymbol{u}(t)) \end{cases}, \tag{29}$$

From (34), the nonlinear observer converges when $\dot{\hat{\chi}}(t) = 0$, which means $\hat{\chi}(t) \to \chi(t)$. That is, the classical nonlinear observer can only observe the constant or slowly varying disturbances. However, not only are constant disturbances $\chi_c(t)$ present in the PM servo system, but also harmonic disturbances $\chi_h(t)$.

To observe the harmonic disturbances with frequencies of ω_e, $2\omega_e$, and $6\omega_e$, which are denoted as $\chi_h(t)$, $\chi_{2h}(t)$, and $\chi_{6h}(t)$, $\chi(t)$ can be reconstructed according to the different disturbance frequencies, as:

$$\dot{\xi} = X\xi = \begin{bmatrix} O & & & \\ & X(\omega_e) & & \\ & & X(2\omega_e) & \\ & & & X(6\omega_e) \end{bmatrix} \xi, \quad (30)$$

where: $X(\omega_e) = \left[\begin{array}{ccc|c} & & & I_3 \\ \hline -\omega_e^2 & & & \\ & -\omega_e^2 & & \\ & & -\omega_e^2 & \end{array}\right]$,

$\xi = \left[\chi_c \mid \chi_h(t) \;\; \dot{\chi}_h(t) \mid \chi_{2h}(t) \;\; \dot{\chi}_{2h}(t) \mid \chi_{6h}(t) \;\; \dot{\chi}_{6h}(t)\right]^T$.

Combined with (16) and (30), the modified nonlinear disturbance observer can be contrasted as:

$$\begin{cases} \dot{\hat{\xi}} = X\hat{\xi} + Lg_2\left(\chi(t) - \hat{\chi}(t)\right) \\ \chi(t) = g_2^{-1}(\dot{x} - f(t) - g_1u(t)) \\ \hat{\chi}(t) = C\hat{\xi} \end{cases}, \quad (31)$$

where $C = [1_{1\times3}\ 1_{1\times3}\ 0\ 1_{1\times3}\ 0\ 1\ 0]$, $1_{1\times3}$ is the ones matrix, and $L = [\,L_c\ L_h\ L_{2h}\ L_{6h}]^T$ is the coefficient matrix of the observer.

4.2. Stability of Modified Observer

The observation errors are defined as $\varepsilon = \xi - \hat{\xi}$; then, the following equation can be obtained according to (31):

$$\dot{\varepsilon} = (X - Lg_2C)\varepsilon, \quad (32)$$

and the Lyapunov function is defined as follows:

$$V = \varepsilon^T P\varepsilon, \quad (33)$$

where P is an arbitrary symmetric positive definite matrix. The derivative of the Lyapunov function is:

$$\dot{V} = \varepsilon^T P\dot{\varepsilon} + \dot{\varepsilon}^T P\varepsilon = \varepsilon^T\left[P(X - Lg_2C) + (X - Lg_2C)^T P\right]\varepsilon, \quad (34)$$

For the nonlinear disturbance observer in (37), it can be seen from (36) and (37) that there is always an observer coefficient matrix L which makes $P(X - Lg_2C)$ negative definite. That is, the observation errors $\varepsilon \to 0$, $\hat{\xi} \to \xi$ when $t \to \infty$.

Since the derivative information of the state variables is difficult to obtain, we can solve this problem by introducing the relationship of $\hat{\xi} = z + p(x)$ into (37). Finally, the mathematical expression of HDO is constructed as:

$$\begin{cases} \dot{z} = (X - Lg_2C)z + Xp(x) - Lf(t) - \\ \quad Lg_1u(t) - Lg_2Cp(x) \\ \hat{\chi}(t) = C[z + p(x)] \\ p(x) = Lx \end{cases}, \quad (35)$$

5. Proposed Scheme and Its Parameter Tuning

From substituting the estimated value in (35) into (25), the principal block diagram of HDO-NPSC can be shown as in Figure 1.

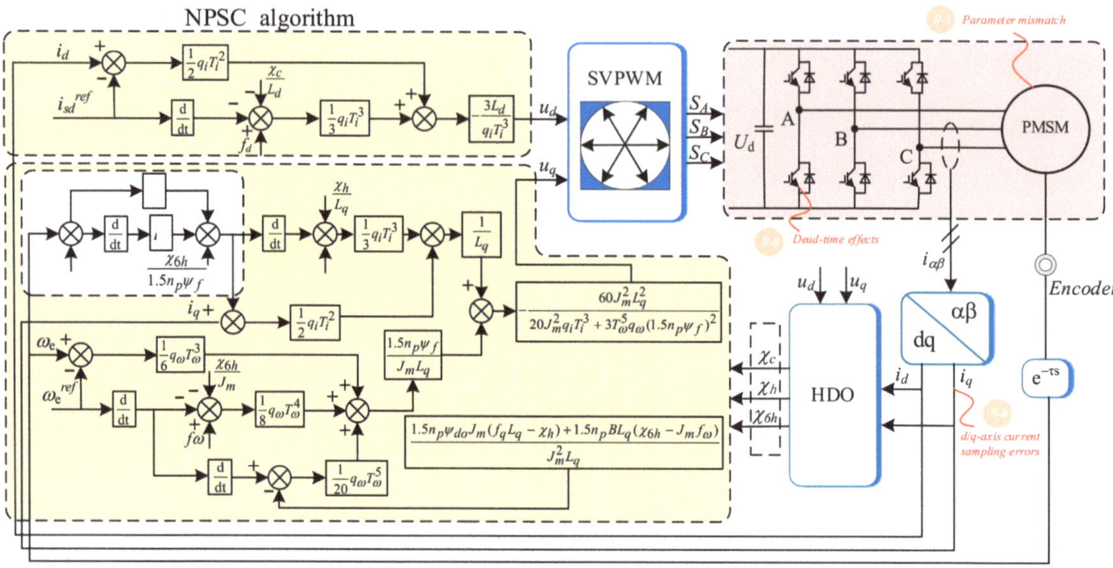

Figure 1. Control block diagram of the proposed HDO-NPSC.

In the proposed algorithm, $i_{sd}{}^{ref} = 0$. Furthermore, an additional PD controller is introduced to adjust the relationship between the q-axis reference current and the speed; its expression is:

$$i_{sq}^{ref} = \begin{bmatrix} 0 & 0 & 1/1.5n_p\psi_f \end{bmatrix} \hat{\chi}(t) + K_p\left(\omega_m^{ref} - \omega_m\right) + K_d\left(\dot{\omega}_m^{ref} - \dot{\omega}_m\right), \tag{36}$$

where K_p and K_d are the proportional and differential coefficients, respectively. The PD controller is employed to modify the reference value of the q-axis current in advance to ensure the dynamic performance, but it does not change the control structure and stability of HDO-NPSC.

From the system model in (15), state errors in (26), and observation errors in (32), it can be concluded that:

$$\begin{bmatrix} \dot{e} \\ \dot{\varepsilon} \end{bmatrix} = \left[\begin{array}{cccc|c} -\frac{3}{2T_i} & 0 & 0 & 0 & \\ 0 & -\frac{3}{2T_i} & -\frac{q_w n_p \psi_f}{2J_m} & -\frac{3q_w n_p \psi_f}{8J_m} & B \\ 0 & 0 & 0 & 1 & \\ 0 & -\frac{10 q_i J_m}{n_p \psi_f T_i^3} & -\frac{10}{3T_w^2} & -\frac{5}{2T_w} & \\ \hline & & & & X - L_\xi g_2 C \end{array}\right] \begin{bmatrix} e \\ \varepsilon \end{bmatrix}, \tag{37}$$

In (37), $\boldsymbol{B}$ is a parameter matrix independent of $\boldsymbol{L}$. The control parameters of NPSC are the current weight q_i and predictive horizon T_i, and speed weight q_w and predictive horizon T_w. The control parameters of HDO are contained in the coefficient matrix $\boldsymbol{L}$. From (37), it can be found that HDO and NPSC can be designed separately, and there is no direct parameter coupling relationship between them. Therefore, the parameter design of

NPSC can be completed under the assumption that $\chi(t) = 0$, and then $\boldsymbol{L}$ can be adjusted based on the pole assignment of the NPSC-based system. The specific parameter tuning process is shown in Figure 2.

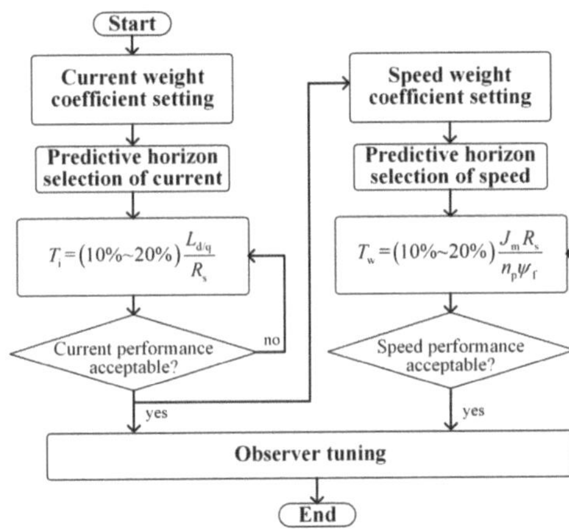

Figure 2. Parameter tuning flowchart of the proposed HDO-NPSC.

According to the flowchart of Figure 2, the detailed parameter setting procedure is listed as follows:

Step I: Current control parameters. The parameters of the current control part should be set first in the case of disconnecting the speed control part (i.e., set ${i_{sq}}^{ref}$ to a fixed value and block the output of the PD controller). The specific parameter tuning process is as follows. Firstly, the current horizon T_i within the allowable range should be selected to set the current weight coefficient (e.g., $T_i = T_s$ is also allowed). The setting process of the current weight coefficient is completed until the motor system can operate normally and the q-axis current can track the given value. Then, the current horizon should be optimized under the condition of keeping the current weight coefficient unchanged. It is worth mentioning that the speed predictive horizon should be selected to be sufficiently long to cover the dynamic process of the motor system as much as possible. However, it can be seen from Section 3.3 that a predictive horizon having a too-large value will lead to dynamic performance degradation. For this reason, the current predictive horizon is generally selected as 10%–20% of the transient time constant of servo system; then:

$$T_{id} = (10\% \sim 20\%)\frac{L_d}{R_s}, T_{iq} = (10\% \sim 20\%)\frac{L_q}{R_s}, \tag{38}$$

Step II: Speed Control Parameters. Similarly, the speed predictive horizon is generally selected as:

$$T_w = (10\% \sim 20\%)\frac{J_m R_s}{n_p \psi_f}, \tag{39}$$

The weight coefficient determines the importance of each variable in the control process; however, a too-large value will destroy the stability of the system. Therefore, it is necessary to gradually increase the value of q_w from a smaller value until a satisfactory control effect is achieved.

Step III: Observer parameters. To ensure the observation speed and interference suppression performance, the observer poles are usually assigned to be 2~5 times faster than the system poles.

6. Experimental Results

To verify the feasibility and effectiveness of NDO-NPSC, the experiments were carried out on a 2.3 kW surface-mounted PMSM. The parameters of the motor are given in Table 1. In the testbench, the load torque is produced by a PM servo motor of the same model. In the following experiments, the algorithms were implemented with a TMS320F28377D micro controller unit (MCU) produced by TI and a Cyclone V FPGA produced by Intel. DSP was used to execute the code of the main algorithm, and FPGA was used to implement the high-precision analog-to-digital conversion (ADC), digital-to-analog conversion (DAC), and pulse generation.

Table 1. PMSM parameters.

Parameter	Symbol	Value
Rated power	P_N	2.3 kW
Number of pole-pairs	p	2
Rated speed	n_N	1500 r/min
Rated torque	T_N	15 Nm
Rated current	I_N	10 A
Permanent magnet flux	ψ_r	0.33 Wb
Stator resistance	R_s	0.63 Ω
Stator inductance	L_s	4.0 mH
Moment of inertia	J_m	0.00272 kg·m^2
Control period	T_s	100 µs

Figure 3 shows the steady-state experimental waveforms of the motor speed, current, and rotor position with the proposed HDO-NPSC. In the experiments, the operating performance of PMSM adopting HDO-NPSC under the condition of high/low speed with heavy/light load were verified. It can be seen from Figure 3 that the servo motor system can work stably at the given speed with load when HDO-NPSC is adopted. The proposed HDO-NPSC algorithm can achieve favorable speed tracking performance regardless of high/low speed and light/heavy load conditions. At the low speed and light load condition, the low-frequency harmonics introduced by non-ideal flux harmonics are more significant and the three-phase current has slight distortion. However, the overall steady-state control performance is still excellent.

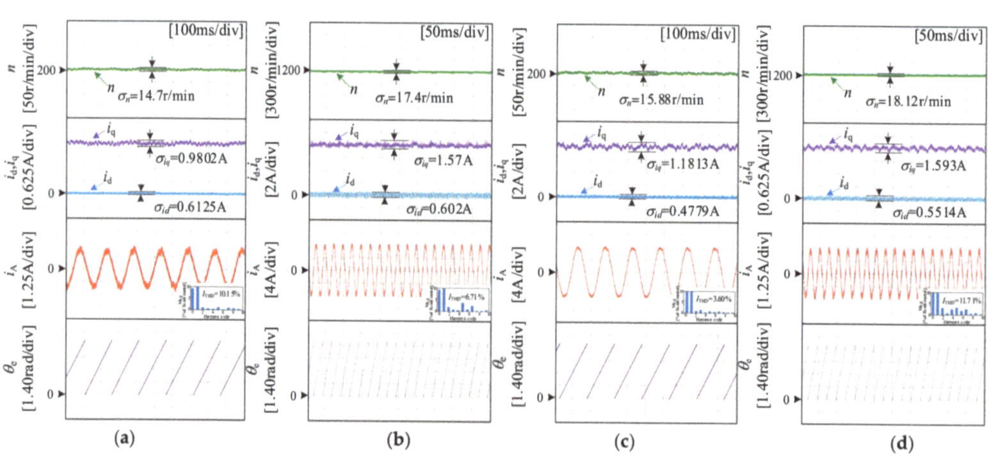

Figure 3. Experimental results of the proposed algorithm under the steady-state operation. (**a**) Low speed (200 r/min) with light-load condition (3.75 Nm). (**b**) High speed (1200 r/min) with heavy-load condition (12 Nm). (**c**) Low speed (200 r/min) with heavy-load condition (12 Nm). (**d**) High speed (1200 r/min) with light-load condition (3.75 Nm).

Figure 4 shows the dynamic experimental waveforms of speed, stator current, and rotor position of FOC with the Luenberger observer (DOB-FOC), predictive speed control with the Luenberger observer [41], and HDO-NPSC, respectively. In Figure 4, the rotor speed is accelerated abruptly from 0 to 800 r/min with no load; the given speed is then unchanged and the load torque is abruptly increased to 8.2 Nm. As shown in Figure 4, when the speed reference is a step signal, all three strategies can track the speed reference value without bias. The speed response time of DOB-FOC is 697.1 ms, whereas that of DOB-PSC is 43.5 ms and that of HDO-NPSC is 204 ms. More importantly, when load disturbance occurs, the speed recovery time of DOB-FOC is 65 ms and that of DOB-PSC is 319.2 ms, whereas that of HDO-NPSC is 40.2 ms. Furthermore, comparing the q-axis current waveforms of the three methods, it can be seen that the q-axis current waveform of HDO-NPSC is relatively stable.

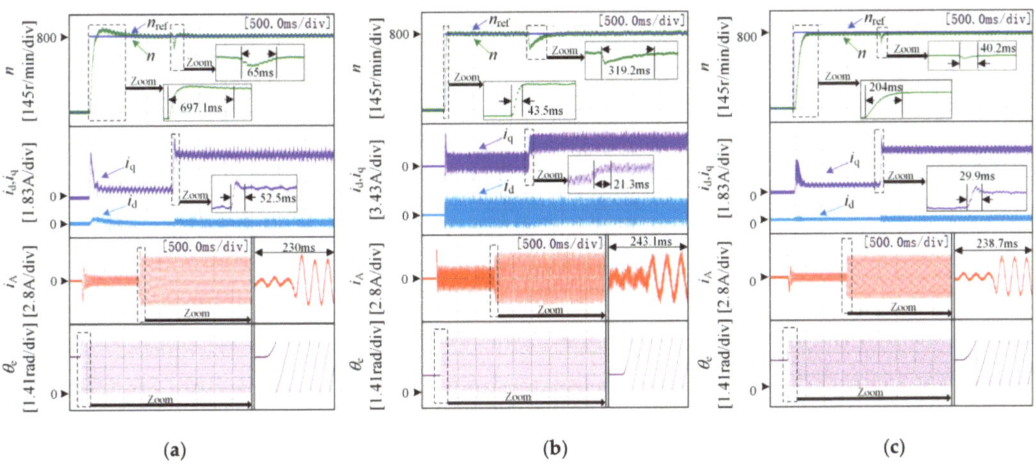

Figure 4. Comparison results of the proposed algorithm with DOB-FOC and DOB-PSC. (**a**) DOB-FOC. (**b**) DOB-PSC. (**c**) HDO-NPSC.

Furthermore, in order to comprehensively verify the performance of the proposed algorithm, the above three strategies without auxiliary observers were been compared. By comparing Figures 4 and 5, it can be found that, after adding the observer, the dynamic performance of FOC is indeed improved, but it is still inferior to HDO-NPSC. When the observer is removed, the prediction algorithms (PSC and NPSC) have some small steady-state errors. However, the dynamic performance of the proposed algorithm is satisfactory in the absence of an observer. In general, compared with the existing strategies, the proposed algorithm can significantly improve the dynamic performance of the motor system without increasing the cost of the control hardware and deteriorating the steady-state performance of the motor system.

Figure 6 shows the dynamic experimental waveforms of speed, stator current, and rotor position of HDO-NPSC, FOC (with the feedforward and anti-windup link), and PI-PCC, respectively. In Figure 6, the given speed of the servo motor system is set to a triangular wave with a peak value of 800 r/min with no external load. After a period of time, the load torque is abruptly increased to 8.2 Nm. From Figure 6, it can be seen that the proposed HDO-NPSC algorithm has better dynamic performance and faster response to the slope reference signal compared with the FOC and PI-PCC strategies. More importantly, it can also effectively avoid the tracking errors caused by the saturation effect of the PI controller. However, HDO-NPSC has a slight tracking error due to the lack of an integral link. This error is relatively small, and within the acceptable range for applications. From Figure 6, for the given signal of triangular wave speed, it can be seen more intuitively that

HDO-NPSC has smaller tracking error and faster response speed when tracking the given signal of a high type, which has obvious advantages.

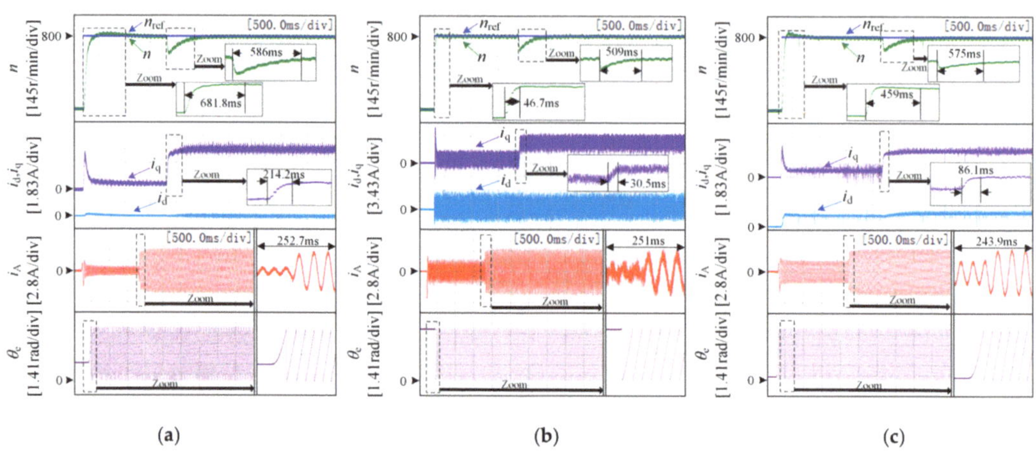

Figure 5. Comparison results of the proposed algorithm with FOC and PSC; all strategies are not equipped with observers. (**a**) FOC. (**b**) PSC. (**c**) NPSC.

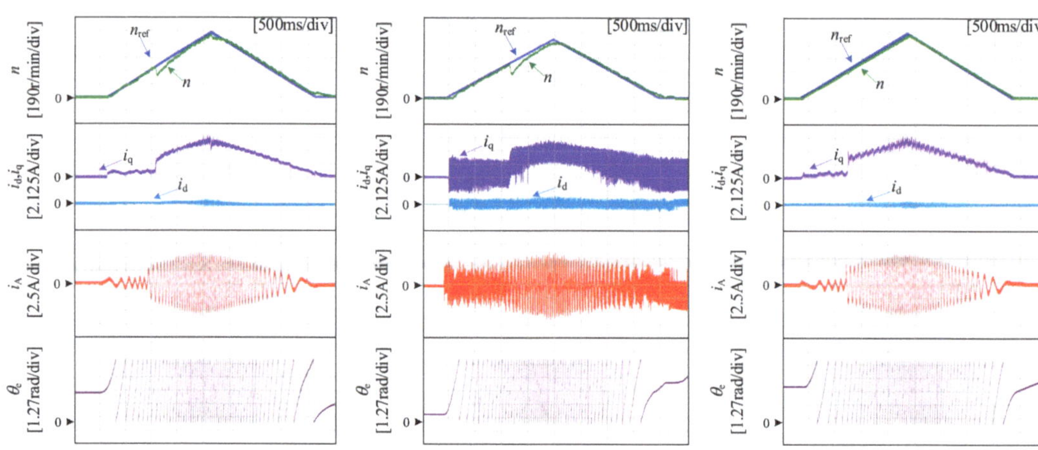

Figure 6. Tracking performance validation with triangular reference signals. (**a**) FOC (with the feedforward and anti-windup link). (**b**) PI-PCC. (**c**) HDO-NPSC.

7. Conclusions

An HDO-NPSC strategy for the PM servo system is presented in this paper. By separating the disturbance terms from the mathematical model of the system, a high-bandwidth nonlinear speed controller with non-cascaded structure was constructed, and the matching harmonic disturbance observer was designed to improve the anti-interference ability of the controller. Furthermore, the PD controller was employed to eliminate the traditional limitation loop, which further improves the tracking ability of the predictive speed algorithm for the high-type reference signals, e.g., the ramping signals and acceleration signal. To summarize, HDO-NPSC effectively solves the problems of classical DPSC in terms of disturbance suppression and efficient operation. Specifically, compared with FOC, the dynamic performance of HDO-NPSC is significantly improved; and, compared with

PI-PCC, it can achieve more stable steady-state performance, which means it has great value for the promotion and application of the predictive algorithm in servo systems.

The algorithm in this paper has the following disadvantages, and thus needs to be further improved. The essence of the proposed algorithm is to observe the spectrum of the disturbance distribution in the system by applying HDO. On this basis, it can inject the corresponding compensation into the control quantity (d-axis and q-axis voltages), so as to improve the control performance of motor speed. In conclusion, the compensation values that are opposite to the system disturbance in phase will be injected into the d-axis and q-axis currents, which will lead to a certain amount of compensation harmonics in the three-phase current.

Author Contributions: Conceptualization, Z.Z. and S.Y.; methodology, Z.Z.; software, C.M.; validation, Z.Z., S.Y. and C.M.; formal analysis, Z.Z.; investigation, Z.Z.; resources, Z.Z.; data curation, G.Z.; writing—original draft preparation, Z.Z.; visualization, G.Z.; supervision, Q.G.; project administration, Q.G.; funding acquisition, Z.Z. and Q.G. All authors have read and agreed to the published version of the manuscript.

Funding: This research was funded by the financial support from the National Natural Science Foundation of China under Grant (51907142, 52077154), and in part by the Natural Science Foundation of Tianjin City under Grant 20JCQNJC00030.

Institutional Review Board Statement: Not applicable.

Informed Consent Statement: Not applicable.

Data Availability Statement: Not applicable.

Conflicts of Interest: The authors declare no conflict of interest.

Nomenclature

i, u, ψ_f, T_s	Current, voltage, flux linkage of rotor, control period
ω_e, ω_m	Electric angular frequency, mechenical angular frequency
R, L, J_m, J_{load}	Resistance, inductance, the moment of inertia, the moment of inertia of load
p_r, B, T_L	Pole pairs, friction coefficient, load torque
L_{ef}, H, B, S	Core length of stator winding, magnetic field intensity, magnetic flux density, calculation area
M_{iron}, M_{shaft}, R_{iron}, r_{iron}, R_{shaft}	Core mass, the shaft mass, the inner radius of core, the outer radius of the core, the inner radius of shaft
x, u, χ, h	State vector, input vector, disturbance vector, output vector
q, T	Weight coefficient, predictive horizon
τ, L	Future time scale, Lie derivative
ξ, L	State variable of observer, coefficient matrix of observer
K_p, K_d	Proportional and differential coefficien of PD controller
Δ, o	Incremental quantity of variables, high-order infinitesimal variable
A, B, C, d, q (subscript)	Variables under A-, B-, C-, d-, q-axis
dc, s, 0 (subscript)	DC bus variables, stator variables, nominal value
c, h (subscript)	Constant disturbances variables, harmonics variables
dead (subscript)	Variables related to dead-time effects
i, w (subscript)	Current control loop variables, speed control loop variables
ref, $\wedge$ (superscript)	Reference values, estimated values

References

1. Wu, G.; Huang, S.; Wu, Q.; Rong, F.; Zhang, C.; Liao, W. Robust Predictive Torque Control of N*3-Phase PMSM for High-Power Traction Application. *IEEE Trans. Power Electron.* **2020**, *35*, 10799–10809. [CrossRef]
2. Ping, Z.; Wang, T.; Huang, Y.; Wang, H.; Lu, J.G.; Li, Y. Internal Model Control of PMSM Position Servo System: Theory and Experimental Results. *IEEE Trans. Ind. Inform.* **2020**, *16*, 2202–2211. [CrossRef]
3. Tang, Z.; Akin, B. A New LMS Algorithm Based Deadtime Compensation Method for PMSM FOC Drives. *IEEE Trans. Ind. Appl.* **2018**, *54*, 6472–6484. [CrossRef]

4. ZWang, Z.; Chen, J.; Cheng, M.; Chau, K.T. Field-Oriented Control and Direct Torque Control for Paralleled VSIs Fed PMSM Drives with Variable Switching Frequencies. *IEEE Trans. Power Electron.* **2016**, *31*, 2417–2428.
5. Mendoza-Mondragón, F.; Hernández-Guzmán, V.M.; Rodríguez-Reséndiz, J. Robust Speed Control of Permanent Magnet Synchronous Motors Using Two-Degrees-of-Freedom Control. *IEEE Trans. Ind. Electron.* **2018**, *65*, 6099–6108. [CrossRef]
6. Garcia, C.; Rodriguez, J.; Silva, C.; Rojas, C.; Zanchetta, P.; Abu-Rub, H. Full Predictive Cascaded Speed and Current Control of an Induction Machine. *IEEE Trans. Energy Convers.* **2016**, *31*, 1059–1067. [CrossRef]
7. Cisneros, P.S.; Werner, H. A Velocity Algorithm for Nonlinear Model Predictive Control. *IEEE Trans. Control. Syst. Technol.* **2020**, *29*, 1310–1315. [CrossRef]
8. Karamanakos, P.; Geyer, T.; Kennel, R. Reformulation of the long-horizon direct model predictive control problem to reduce the computational effort. In Proceedings of the 2014 IEEE Energy Conversion Congress and Exposition (ECCE), Pittsburgh, PA, USA, 14–18 September 2014; pp. 3512–3519.
9. Hang, J.; Wu, H.; Zhang, J.; Ding, S.; Huang, Y.; Hua, W. Cost Function-Based Open-Phase Fault Diagnosis for PMSM Drive System with Model Predictive Current Control. *IEEE Trans. Power Electron.* **2021**, *36*, 2574–2583. [CrossRef]
10. Errouissi, R.; Ouhrouche, M.; Chen, W.H.; Trzynadlowski, A.M. Robust Cascaded Nonlinear Predictive Control of a Permanent Magnet Synchronous Motor with Antiwindup Compensator. *IEEE Trans. Ind. Electron.* **2012**, *59*, 3078–3088. [CrossRef]
11. Chen, W.H.; Ballance, D.J.; Gawthrop, P.J.; O'Reilly, J. A Nonlinear Disturbance Observer for Robotic Manipulators. *IEEE Trans. Ind. Electron.* **2000**, *47*, 932–938. [CrossRef]
12. Xia, C.; Liu, N.; Zhou, Z.; Yan, Y.; Shi, T. Steady-State Performance Improvement for LQR-Based PMSM Drives. *IEEE Trans. Power Electron.* **2018**, *33*, 10622–10632. [CrossRef]
13. Yang, J.; Chen, W.H.; Li, S.; Guo, L.; Yan, Y. Disturbance/Uncertainty Estimation and Attenuation Techniques in PMSM Drives—A Survey. *IEEE Trans. Ind. Electron.* **2017**, *64*, 3273–3285. [CrossRef]
14. Wang, L.; Freeman, C.T.; Chai, S.; Rogers, E. Predictive-repetitive control with constraints: From design to implementation. *J. Process. Control* **2013**, *23*, 956–967. [CrossRef]
15. Gupta, M.; Lee, J.H. Period-robust repetitive model predictive control. *J. Process. Control* **2006**, *16*, 545–555. [CrossRef]
16. Zhou, Z.; Xia, C.; Yan, Y.; Wang, Z.; Shi, T. Disturbances Attenuation of Permanent Magnet Synchronous Motor Drives Using Cascaded Predictive-Integral-Resonant Controllers. *IEEE Trans. Power Electron.* **2018**, *33*, 1514–1527. [CrossRef]
17. Song, Z.; Zhou, F. Observer-Based Predictive Vector-Resonant Current Control of Permanent Magnet Synchronous Machines. *IEEE Trans. Power Electron.* **2019**, *34*, 5969–5980. [CrossRef]
18. Gao, Y.; Chong, K.T. The Explicit Constrained Min-Max Model Predictive Control of a Discrete-Time Linear System with Uncertain Disturbances. *IEEE Trans. Autom. Control* **2012**, *57*, 2373–2378. [CrossRef]
19. Köhler, J.; Soloperto, R.; Müller, M.A.; Allgöwer, F. A Computationally Efficient Robust Model Predictive Control Framework for Uncertain Nonlinear Systems. *IEEE Trans. Autom. Control* **2021**, *66*, 794–801. [CrossRef]
20. Türker, T.; Buyukkeles, U.; Bakan, A.F. A Robust Predictive Current Controller for PMSM Drives. *IEEE Trans. Ind. Electron.* **2016**, *63*, 3906–3914. [CrossRef]
21. Zhou, P.; Chai, T.Y.; Zhao, J.H. DOB Design for Nonminimum-Phase Delay Systems and Its Application in Multivariable MPC Control. *IEEE Trans. Circuits Syst. II Exp. Briefs* **2012**, *59*, 525–529. [CrossRef]
22. Liu, H.; Li, S. Speed Control for PMSM Servo System Using Predictive Functional Control and Extended State Observer. *IEEE Trans. Ind. Electron.* **2012**, *59*, 1171–1183. [CrossRef]
23. Chen, M.; Mei, R. Robust tracking control of uncertain nonlinear systems using disturbance observer. In Proceedings of the 2011 International Conference on System Science and Engineering, Macau, China, 8–10 June 2011; pp. 431–436.
24. Chen, W.-H. Harmonic Disturbance Observer for Nonlinear Systems. *J. Dyn. Syst. Meas. Control* **2003**, *125*, 114–117. [CrossRef]
25. Zhang, X.; He, Y. Robust Model Predictive Direct Speed Control for SPMSM Drives Based on Full Parameters and Load Observer. In Proceedings of the 2019 IEEE International Symposium on Predictive Control of Electrical Drives and Power Electronics (PRECEDE), Quanzhou, China, 31 May–2 June 2019; pp. 1–5.
26. Zhang, X.; He, Y. Direct Voltage-Selection Based Model Predictive Direct Speed Control for PMSM Drives without Weighting Factor. *IEEE Trans. Power Electron.* **2019**, *34*, 7838–7851. [CrossRef]
27. Yuan, X.; Zhang, S.; Zhang, C. Improved Model Predictive Current Control for SPMSM Drives with Parameter Mismatch. *IEEE Trans. Ind. Electron.* **2020**, *67*, 852–862. [CrossRef]
28. Wei, Y.; Wei, Y.; Sun, Y.; Qi, H.; Guo, X. Prediction Horizons Optimized Nonlinear Predictive Control for Permanent Magnet Synchronous Motor Position System. *IEEE Trans. Ind. Electron.* **2020**, *67*, 9153–9163. [CrossRef]
29. Tarczewski, T.; Grzesiak, L.M. Constrained State Feedback Speed Control of PMSM Based on Model Predictive Approach. *IEEE Trans. Ind. Electron.* **2016**, *63*, 3867–3875. [CrossRef]
30. Feng, W.; O'reilly, J.; Ballance, D.J. MIMO nonlinear PID predictive controller. *IEE Proc. Control Theory Appl.* **2002**, *149*, 203–208. [CrossRef]
31. Tavernini, D.; Vacca, F.; Metzler, M.; Savitski, D.; Ivanov, V.; Gruber, P.; Hartavi, A.E.; Dhaens, M.; Sorniotti, A. An Explicit Nonlinear Model Predictive ABS Controller for Electro-Hydraulic Braking Systems. *IEEE Trans. Ind. Electron.* **2020**, *67*, 3990–4001. [CrossRef]
32. Errouissi, R.; Ouhrouche, M.; Chen, W.H.; Trzynadlowski, A.M. Robust Nonlinear Predictive Controller for Permanent-Magnet Synchronous Motors with an Optimized Cost Function. *IEEE Trans. Ind. Electron.* **2012**, *59*, 2849–2858. [CrossRef]

33. Liu, X.D.; Li, K.; Zhang, Q.; Zhang, C. Single-loop Predictive Control of PMSM Based on Nonlinear Disturbance Observers. *Proc. CSEE* **2018**, *38*, 2153–2162.
34. ZSong, Z.; Zhang, Z.; Komurcugil, H.; Lee, C.H. Controller-Based Periodic Disturbance Mitigation Techniques for Three-Phase Two-Level Voltage-Source Converters. *IEEE Trans. Ind. Inform.* **2021**, *17*, 6553–6568.
35. Hao, Z.; Zhou, W.; Ji, T.; Huang, X.; Zhang, C. Multi-Objective Optimization of Double Primary Tubular Permanent Magnet Synchronous Linear Motor in Wide Temperature Range Environment Based on Pareto Front Method. *IEEE Access* **2020**, *8*, 207193–207203. [CrossRef]
36. Wang, S.J.; Lin, S. K Analytical prediction of the incremental inductance of the permanent magnet synchronous motors. *IEEE Trans. Magn.* **2004**, *40*, 2044–2046. [CrossRef]
37. Lian, C.; Xiao, F.; Gao, S.; Liu, J. Load Torque and Moment of Inertia Identification for Permanent Magnet Synchronous Motor Drives Based on Sliding Mode Observer. *IEEE Trans. Power Electron.* **2018**, *34*, 5675–5683. [CrossRef]
38. Hwang, S.H.; Kim, J.M. Dead Time Compensation Method for Voltage-Fed PWM Inverter. *IEEE Trans. Energy Convers.* **2010**, *25*, 1–10. [CrossRef]
39. Linder, A.; Kanchan, R.; Stolze, P.; Kennel, R. *Model-Based Predictive Control of Electric Drives*; Cuvillier Verlag: Gottingen, Germany, 2010.
40. Li, S.; Yang, J.; Chen, W.H.; Chen, X. *Disturbance Observer-Based Control: Methods and Applications*; CRC Press: Boca Raton, FL, USA, 2014.
41. Wang, H.; Zhang, S.; Liu, W.; Geng, Q.; Zhou, Z. Finite control-set model predictive direct speed control of a PMSM drive based on the Taylor series model. *IET Electr. Power Appl.* **2021**, *15*, 1452–1465. [CrossRef]

energies

MDPI

Article

Virtual Signal Injection Maximum Torque per Ampere Control Based on Inductor Identification

Ning-Zhi Jin, Hong-Chao Chen, Dong-Yang Sun *, Zhi-Qiang Wu, Kai Zhou and Long Zhang

Engineering Research Center of Automotive Electronics Drive Control and System Integration, Ministry of Education, Harbin University of Science and Technology, Harbin 150080, China; sharon0716@126.com (N.-Z.J.); hongchaochen311@163.com (H.-C.C.); zhiqiangwu1120@163.com (Z.-Q.W.); zhoukai@hrbust.edu.cn (K.Z.); longz919@163.com (L.Z.)

* Correspondence: dongyang0521@126.com

Abstract: The high-frequency signal injection-type maximum torque per ampere (MTPA) algorithm is usually employed to control the operation of interior permanent magnet synchronous motors (IPMSMs). The MTPA algorithm exhibits good dynamic performance and anti-interference ability. However, due to the injection of a high-frequency current signal, problems such as torque ripple and additional loss are encountered. Therefore, in this paper, a virtual signal injection control (VSIC) method that does not require actual injection is proposed for solving the aforementioned problems while yielding good performance. However, in the control process of the proposed method, the d-axis inductance parameter affects the accuracy of the torque information, resulting in errors in the system. To solve this problem, an online identification algorithm of model reference adaptive systems (MRAS) based on the Popov super stability theory as the basis for the design of the adaptive law is proposed in this paper. The d-axis inductance parameter of the motor is obtained in real-time and then introduced into the control system by using the VSIC method. Finally, VSIC-type MTPA control based on inductance identification is realized. The proposed algorithm does not depend on the design parameters of the motor and exhibits good dynamic response and anti-interference performance.

Keywords: permanent magnet synchronous motor; model reference adaptive system; virtual signal injection control; maximum torque per ampere

Citation: Jin, N.-Z.; Chen, H.-C.; Sun, D.-Y.; Wu, Z.-Q.; Zhou, K.; Zhang, L. Virtual Signal Injection Maximum Torque per Ampere Control Based on Inductor Identification. *Energies* **2022**, *15*, 4851. https://doi.org/10.3390/en15134851

Academic Editors: Lorand Szabo and Feng Chai

Received: 11 May 2022
Accepted: 30 June 2022
Published: 1 July 2022

1. Introduction

With the advancements in high-performance permanent magnet materials, especially the emergence of rare-earth permanent magnet materials, the performance of interior permanent magnet synchronous motors (IPMSMs) has been greatly improved [1]. Compared with the traditional electric excitation motor, the IPMSM offers the advantages of simple structure, reliable operation, small size, low weight, low loss, high efficiency, and superior performance [2]. The maximum torque per ampere (MTPA) algorithm is widely employed as an efficient control method for IPMSMs. It can output the maximum torque under a certain stator current, thereby improving the operating efficiency of the entire system [3].

The mainstream MTPA control methods include direct formula calculation, parameter identification, table look-up, automatic search, and high-frequency signal injection.

In the direct formula calculation method, the minimum value of the electromagnetic torque equation is obtained by derivation, and then the value of the stator current on the alternating and direct axis components is directly obtained [4]. Although the direct formula calculation method is relatively simple, the parameters change during the operation. The parameter identification method can obtain accurate motor parameters in real-time and improve the running accuracy of the motor by utilizing the direct formula method combined with online parameter identification. However, this method involves tedious parameter monitoring and estimation; thus, it requires the controller to have high computing power and additional hardware [5]. In view of the problems encountered in the formula calculation method, scholars proposed the table look-up method [6]. The table look-up method simplifies the computational complexity and can take into account the parameter change factors, but it takes up a lot of storage space and requires a large offline test workload. In the automatic search method, the vector angle of the stator current is adjusted by continuously giving a small step angle under the steady-state operation of the system to achieve the optimal control strategy of MTPA [7]. The automatic search method can automatically and gradually approach the MTPA trajectory when the parameters change, but the convergence speed and torque control accuracy are low. The high-frequency signal injection method involves observing the feedback amount of the high-frequency signal injected into the system and calculating and analyzing it to obtain the optimal working state [8,9]. The high-frequency signal injection method adjusts the optimal operating point of the MTPA control in real-time according to the motor torque pulsation generated by the injected signal, but the injected high-frequency current increases the system torque pulsation and results in increased power loss. To solve this problem, a virtual signal injection control (VSIC) method is proposed. This method is basically the same as the high-frequency signal injection method. The sine wave signal is virtually injected into the stator current angle so that the torque change information of the motor is included in the motor torque model, and the torque information containing high-frequency signals is extracted to determine the degree of deviation, which definitely avoids the high-frequency pulsation caused by the actual signal injection [10,11]. However, the VSIC method is related to the inductance parameters of the motor. When the motor parameters are inaccurate, it will deviate from the optimal operating point. Therefore, introducing an appropriate parameter identification method into the VSIC system can effectively improve the operation accuracy of the control system.

Online identification has been widely used due to its good real-time performance. Online identification methods mainly include recursive least squares, extended Kalman filter, model reference adaptive system (MRAS), and neural networks.

The recursive least squares method is the most widely used online identification method in engineering. In this method, the data obtained in the previous step is corrected according to the data obtained in the current step based on the estimated value of the model parameters in the previous step to obtain the model parameters at the current step [12,13]. However, data saturation occurs during the recursive operation process, thus requiring high system hardware and software programming. The extended Kalman filter method integrates the discrete space model into the filtering algorithm and realizes the optimal estimation of the system state by making the estimated covariance reach the minimum value [14,15]. When the extended Kalman filter method identifies multiple parameters at the same time, the operation process becomes more complicated, which increases the difficulty. The core idea of the MRAS method is to gradually converge to the actual parameters in the reference model by adjusting the parameters of the adjustable model and through the pre-designed adaptation law [16]. The MRAS method is simple in principle, accurate in identification, and fast in convergence. The neural network method can obtain better convergence characteristics, but its algorithm is very complex; thus, it is subject to many limitations in practical applications [17]. With further advances in the field of parameter identification in recent years, many new identification methods have been established. A previous study proposed the identification of PMSM parameters by using the particle swarm algorithm combined with cloud model theory, which has fast convergence speed, high precision, and efficient local evolution and mutation capabilities, avoiding the problem of easily falling into the local maximum that is encountered in the traditional particle swarm algorithm [18]. A previous study employed a genetic algorithm in a Markov chain model for parameter identification, taking into account the identification speed and accuracy [19].

To sum up, in this paper, the MTPA control of the IPMSM is realized using the virtual signal injection control (VSIC) method to improve the operation efficiency of the motor. Moreover, according to the influence of d-axis inductance parameters on the accuracy of torque information, the MRAS is proposed to identify the motor parameters online. Finally, the MRAS and MTPA are combined to form the overall control system, and the MTPA control based on real-time online identification of inductance is realized.

2. IPMSM Mathematical Model and MTPA Control

The voltage of the IPMSM in the d-q coordinate system can be expressed as follows:

$$\begin{cases} u_{\mathrm{d}} = Ri_{\mathrm{d}} - \omega L_{\mathrm{q}} i_{\mathrm{q}} + L_{\mathrm{d}} \frac{\mathrm{d}i_{\mathrm{d}}}{\mathrm{d}t} \\ u_{\mathrm{q}} = Ri_{\mathrm{q}} + \omega(\psi_{\mathrm{f}} + L_{\mathrm{d}} i_{\mathrm{d}}) + L_{\mathrm{q}} \frac{\mathrm{d}i_{\mathrm{q}}}{\mathrm{d}t} \end{cases}, \tag{1}$$

The electromagnetic torque can be expressed as follows:

$$T_{\mathrm{e}} = 1.5p(\psi_{\mathrm{f}} i_{\mathrm{q}} + (L_{\mathrm{d}} - L_{\mathrm{q}}) i_{\mathrm{d}} i_{\mathrm{q}}), \tag{2}$$

where i_{d}, i_{q}, and u_{d}, u_{q} are the current and voltage of the stator d-q axis, respectively. L_{d} and L_{q} are the stator d-q-axis inductance, respectively; R is the stator resistance; ψ_{f} is the permanent magnet flux linkage; ω is the electrical angular velocity; T_{e} is the electromagnetic torque, and p is the pairs of poles.

From Equation (2), it can be seen that in addition to the excitation torque generated by the permanent magnet, there is the reluctance torque generated by the d-q axis inductances in the electromagnetic torque, and the magnitude of the electromagnetic torque outputted by the motor depends on i_{d} and i_{q}. As shown in Figure 1, the angle between the stator current I_{s} and the d-axis is called the current vector angle β. The calculation relationship between I_{s}, i_{d}, i_{q}, and β is shown in Equation (3).

$$\begin{cases} i_{\mathrm{d}} = I_{\mathrm{s}} \cos\beta \\ i_{\mathrm{q}} = I_{\mathrm{s}} \sin\beta \end{cases}, \tag{3}$$

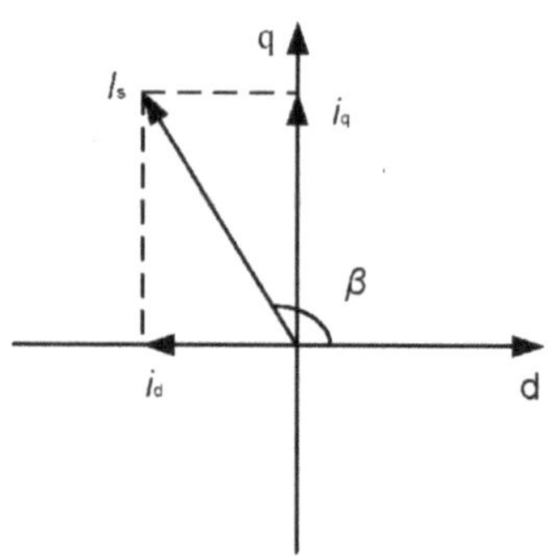

Figure 1. The current vector relationship in the d-q coordinate system.

By substituting Equation (3) into the torque formula (Equation (2)), we obtain the electromagnetic torque calculation formula expressed by the stator current and the current angle as shown in Equation (4).

$$T_{\mathrm{e}} = 1.5p(\psi_{\mathrm{f}} I_{\mathrm{s}} \sin\beta + (L_{\mathrm{d}} - L_{\mathrm{q}}) I_{\mathrm{s}}^2 \sin\beta\cos\beta), \tag{4}$$

The MTPA current angle can be obtained using Equation (5).

$$\beta = \arccos\frac{-\psi_{\mathrm{f}} + \sqrt{\psi_{\mathrm{f}}^2 + 8(L_{\mathrm{d}} - L_{\mathrm{q}})^2 I_{\mathrm{s}}^2}}{4(L_{\mathrm{d}} - L_{\mathrm{q}}) I_{\mathrm{s}}}, \tag{5}$$

To realize MTPA, the proportion relationship between i_{d} and i_{q} must be reasonably distributed to achieve the maximum electromagnetic torque output in the IPMSM.

3. MTPA Control Based on Virtual Signal Injection

3.1. Determining the Torque by Using a High-Frequency Signal

When the motor is in a stable operation state, the differential term in the d-q axis voltage equation (Equation (1) can be regarded as zero, and the d-q axis voltage can be expressed as follows:

$$\begin{cases} L_q = -\frac{u_d - Ri_d}{i_q \omega} \\ \psi_f = \frac{u_q - Ri_q}{\omega} - L_d i_d \end{cases}, \tag{6}$$

Substituting Equation (6) into Equation (2), we obtain

$$T_e = \frac{3}{2}p(\frac{u_q - Ri_q}{\omega} - L_d i_d + (\frac{u_d - Ri_d}{i_q \omega} + L_d)i_d)i_q, \tag{7}$$

After calculating the re-expressed electromagnetic torque, a high-frequency sinusoidal small-signal $\Delta\beta$ (Equation (8) is mathematically injected into the current angle β in Equation (3); the corresponding d-q axis currents i_d and i_q can be expressed as Equation (9), where ω_h is the injection frequency of the high-frequency signal.

$$\Delta\beta = A\sin(\omega_h t), \tag{8}$$

$$\begin{cases} i_d^h = I_s \cos(\beta + \Delta\beta) \\ i_q^h = I_s \sin(\beta + \Delta\beta) \end{cases}, \tag{9}$$

The injection frequency ω_h in the high-frequency sinusoidal small-signal $\Delta\beta$ must be greater than the bandwidth of the outer speed loop, and the frequency ω_h must be much smaller than the inverter switching frequency to ensure the integrity of the injected signal. Moreover, the amplitude A must be sufficiently small to ensure that the injected high-frequency signal is not affected by the speed change.

Substituting Equations (8) and (9) into Equation (7), the torque expression containing high-frequency information can be obtained as shown in Equation (10).

$$T_e^h(\beta + A\sin(\omega_h t)) = \frac{3}{2}p(\frac{u_q - Ri_q}{\omega} - L_d(i_d - i_d^h) + \frac{u_d - Ri_d}{i_q \omega} i_d^h)i_q^h, \tag{10}$$

From Equation (10), it can be seen that to achieve accurate torque information estimation, only the L_d information is needed. Although (ψ_f + $L_d i_d$) varies with temperature and d-q axis current, it can be considered constant during signal injection because $(u_q - Ri_q)/\omega$ and $(Ri_d - u_d)/i_q\omega$ represent parameter information (ψ_f + $L_d i_d$) and L_q. Thus, (ψ_f + $L_d i_d$) can be considered constant during signal injection, and no high-frequency signal is injected into the stator current angle contained in i_d and i_q.

3.2. Processing of High-Frequency Torque Signals

The expression of the torque model containing high-frequency information expanded according to Taylor's formula is shown in Equation (11).

$$T_e^h(\beta + A\sin(\omega_h t)) = T_e(\beta) + \frac{\partial T_e}{\partial \beta}A\sin(\omega_h t) + \frac{\partial}{\partial \beta}(\frac{\partial T_e}{\partial \beta})A^2\sin^2(\omega_h t) + \cdots, \tag{11}$$

Because the injected high-frequency signal amplitude A is sufficiently small, the first-order term in Equation (11) is the main part of the torque change information, and the higher-order terms (above the second-order term) have little influence on the torque change and can be ignored. Further, the quadratic term in the Taylor expansion can be expanded into the sum of the constant term and the double frequency term of the injected signal by using trigonometric functions as follows:

$$\frac{\partial}{\partial \beta}(\frac{\partial T_e}{\partial \beta})A^2 \sin^2(\omega_h t) = \frac{1}{2}\frac{\partial}{\partial \beta}(\frac{\partial T_e}{\partial \beta})A^2(1 - \cos(2\omega_h t)), \tag{12}$$

As previously mentioned, the core principle of MTPA implementation is to make the first-order partial derivative of torque to current angle $\partial T_e/\partial \beta = 0$. Next, we need some methods and filters to perform the signal processing of the clutter in Equation (11), as shown in Figure 2. Finally, the part containing only $\partial T_e/\partial \beta$ term is obtained to realize MTPA control.

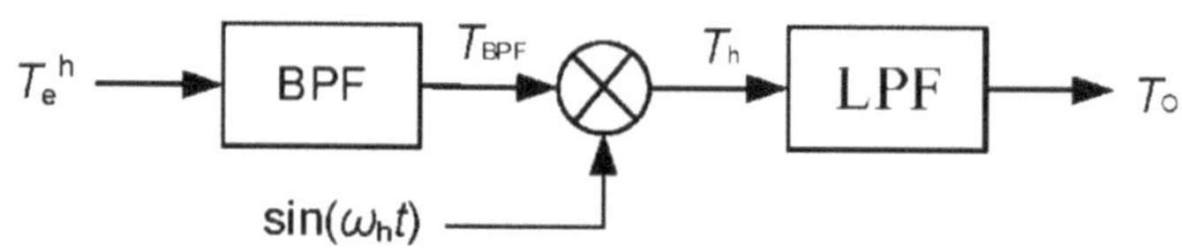

Figure 2. Extraction of virtual signal injection MTPA criterion.

In Figure 2, the torque signal containing high-frequency components is passed through the band-pass filter (BPF) to obtain only the fundamental frequency component of the injected signal. The BPF is a device that allows certain frequency bands of waves to pass while blocking other frequency bands. The center frequency of the BPF is ω_h, and all the signal components of other frequencies are filtered out. Next, after multiplying with $\sin(\omega_h t)$, the quadratic component of $\sin(\omega_h t)$ is obtained, and Equation (13) is obtained by expanding Equation (12).

$$\frac{\partial T_e}{\partial \beta}A \sin^2(\omega_h t) = \frac{1}{2}\frac{\partial T_e}{\partial \beta}A(1 - \cos(2\omega_h t)), \tag{13}$$

Equation (13) consists of a constant term containing only $\partial T_e/\partial \beta$ and the double frequency term of the injected signal. Finally, the first-order deviation of the electromagnetic torque is obtained after the double frequency term is filtered out by the low-pass filter (LPF) (Figure 2). The action of the LPF is to suppress the high-frequency of waves and allow the low-frequency of waves to pass.

To sum up, in the proposed method, the calculation formula of electromagnetic torque is re-expressed as follows: by injecting a high-frequency, small-amplitude current angle signal, the high-frequency signal of the current is included in the expression of the electromagnetic torque. Next, the torque model containing high-frequency information is expanded according to the Taylor formula, $\partial T_e/\partial \beta$ is obtained after multiple filtering, and the integral result is taken as the given value of the reference current angle. If it does not work at the MTPA point at this time and $\partial T_e/\partial \beta \neq 0$, the integrator will continue to integrate and adjust the current angle until it works at the MTPA point; at this time, $\partial T_e/\partial \beta = 0$, and the reference current angle remains unchanged from the previous state due to the zero input to the integrator, and the motor continues to run at the MTPA point.

3.3. Error Analysis and Application Occasions

Compared with the formula calculation method, the VSIC method involves fewer motor parameters; thus, the influence of parameter changes on the MTPA control accuracy is greatly reduced, high-frequency signals do not need to be injected into the motor, and the torque to the current angle change rate can be obtained through signal processing technology to achieve complete MTPA control. However, there is still a certain amount of error in the MTPA angle obtained using this method. From Equation (10), it can be seen that to accurately obtain the change rate $\partial T_e/\partial \beta$ of the electromagnetic torque to the stator current angle, it is necessary to replace the permanent magnet flux linkage ψ_f, the direct-axis inductance L_d, and the quadrature-axis inductance L_q. However, ψ_f and $L_d i_d$ are coupled in the motor parameter estimation formula (Equation (6) and cannot be extracted separately. Therefore, the error of the direct-axis inductance L_d is ignored.

When the reluctance torque ratio of IPMSM is large, $L_d(i_d - i_d{}^h)$ in Equation (10) can be ignored. The approximate calculation formula (Equation (14) further ignores the influence of L_d change so that the MTPA control strategy based on VSIC is not affected by any parameters; thus, it is simpler and more convenient.

$$T_e^h(\beta + A\sin(\omega_h t)) = \frac{3}{2}p(\frac{u_q - Ri_q}{\omega} + \frac{u_d - Ri_d}{i_q\omega}i_d^h)i_q^h, \tag{14}$$

However, in the IPMSM with a small proportion of reluctance torque, $L_d(i_d - i_d{}^h)$ cannot be ignored; otherwise, the error between the automatically found MTPA point and the correct MTPA point in the algorithm will be too large, resulting in the failure of optimization. In the simulation, by setting an IPMSM with a small reluctance torque, the optimization results obtained using the exact torque calculation formula (Equation (10) are compared with those obtained using the approximate torque calculation formula (Equation (14)), and the simulation data are recorded (Figure 3). The optimization effect is better in the IPMSM with a small reluctance torque using the accurate calculation formula.

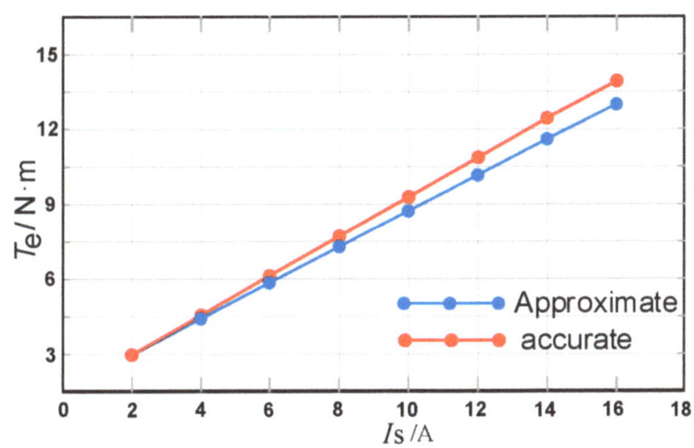

Figure 3. Comparison of simulation results obtained using different torque calculation formulas.

Based on the above analysis and simulation verification, for the IPMSM with a large proportion of reluctance torque, the approximate torque calculation formula can be used to ignore the influence of the d-axis inductance and thus simplify the control system. When the reluctance torque is small, the change in the inductance term cannot be ignored; otherwise, the control accuracy will be reduced, and the optimization will fail. Therefore, the value of the d-axis inductance must be obtained in real-time.

4. MTPA Control Based on Inductance Identification

4.1. Model Reference Adaptive System

As previously discussed, for realizing high-precision control by using the MTPA virtual signal injection method in an IPMSM with a small reluctance torque, the influence of the inductance term cannot be ignored, and the d-axis inductance of the motor must be known. Therefore, the online identification of the inductance must be realized using the MRAS system, which consists of a reference model, an adjustable model, and a parameter adaptive law. The reference model specifies the performance of the system, and its output represents the desired ideal output curve of the closed-loop control system, which shows the main features of MRAS. The actual motor is set as the reference model and is inputted at the same voltage as the adjustable model; the error signal e is generated by comparing the current output in real-time. The parameters of the adjustable model are modified using appropriate adaptive laws. When the error signal decays to 0, the parameters of the adjustable model converge to the actual value. The structure of MRAS is shown in Figure 4.

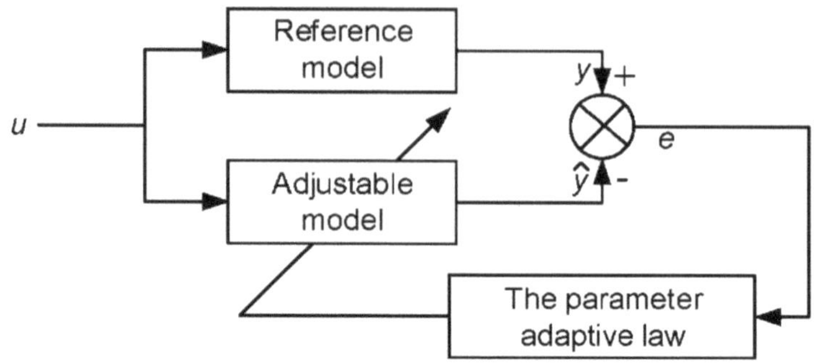

Figure 4. Block diagram of the model reference adaptive system.

For the IPMSM system, the input is $u = [u_d u_q]^T$, the output of the reference model is $y = i = [i_d i_q]^T$, the output of the adjustable model is $\hat{y} = \hat{i} = [\hat{i}_d \hat{i}_q]^T$, and the output quantity error is $e = i - \hat{i}$.

The adaptive law of the model reference adaptive system is a vital part of the system. Currently, the following three common methods are available for determining the appropriate adaptive law:

1. Design method based on local parameter optimization theory (MIT method);
2. Design method based on Lyapunov function;
3. Design method based on ultra-stable and positive dynamic theory (Popov super stability theory).

Among the three methods, the MIT method causes significant defects, and the stability of the adaptive system is not known. In contrast, although the Lyapunov function-based design method is feasible for designing a stable model reference adaptive system, the Lyapunov function has no definite form and includes different types. The Popov super stability theory involves using the reverse solution of the Popov integral inequality to construct the parameter adaptation rate. Because the solution process is concise, it is widely used in the construction of the parameter adaptation rate of the MRAS system.

To sum up, in this study, the design method based on the Popov super stability theory was adopted to construct the parameter adaptation rate of the MRAS system. For the MRAS system to reach an asymptotically stable state, the following two conditions of the Popov law of super stability must be satisfied:

1. The conditions of the Popov integral inequality must be satisfied according to the nonlinear time-varying feedback loop;
2. An appropriate gain matrix M must be selected to ensure that the transfer function matrix of the linear steady forward loop is strictly positive and real.

4.2. Reference Model of the IPMSM System

According to Equation (1), the state space equation of the IPMSM system can be described as follows:

$$\begin{bmatrix} \frac{di_d}{dt} \\ \frac{di_q}{dt} \end{bmatrix} = \begin{bmatrix} -\frac{R_s}{L_d} & \frac{L_q}{L_d}\omega \\ -\frac{L_d}{L_q}\omega & -\frac{R_s}{L_q} \end{bmatrix} \begin{bmatrix} i_d \\ i_q \end{bmatrix} + \begin{bmatrix} \frac{1}{L_d} & 0 \\ 0 & \frac{1}{L_q} \end{bmatrix} \begin{bmatrix} u_d \\ u_q \end{bmatrix} + \begin{bmatrix} 0 \\ -\frac{\psi_f}{L_q}\omega \end{bmatrix}, \tag{15}$$

Equation (15) is considered as the reference model of IPMSM in the MRAS and is simplified as follows:

$$\dot{i} = Ai + Bu + C, \tag{16}$$

where $A = \begin{bmatrix} -\frac{R_s}{L_d} & \frac{L_q}{L_d}\omega \\ -\frac{L_d}{L_q}\omega & -\frac{R_s}{L_q} \end{bmatrix}$, $B = \begin{bmatrix} \frac{1}{L_d} & 0 \\ 0 & \frac{1}{L_q} \end{bmatrix}$, and $C = \begin{bmatrix} 0 \\ -\frac{\psi_f}{L_q}\omega \end{bmatrix}$.

4.3. Adjustable Model of the IPMSM System

In the identification process, the motor speed ω and rotor flux linkage ψ_f are assumed to be constant; thus, the parameter adjustable model can be constructed from the following reference model:

$$\dot{\hat{i}} = \hat{A}\hat{i} + \hat{B}u + \hat{C} - Me, \tag{17}$$

where $\hat{A} = \begin{bmatrix} -\frac{R_s}{\hat{L}_d} & \frac{\hat{L}_q}{\hat{L}_d}\omega \\ -\frac{\hat{L}_d}{\hat{L}_q}\omega & -\frac{R_s}{\hat{L}_q} \end{bmatrix}$, $\hat{B} = \begin{bmatrix} \frac{1}{\hat{L}_d} & 0 \\ 0 & \frac{1}{\hat{L}_q} \end{bmatrix}$, $\hat{C} = \begin{bmatrix} 0 \\ -\frac{\psi_f}{\hat{L}_q}\omega \end{bmatrix}$, M is the gain matrix, and $M = \begin{bmatrix} -m_{11} & -m_{12} \\ m_{21} & -m_{22} \end{bmatrix}$.

The error equation of the model-referenced adaptive system is obtained by subtracting the adjustable model from the following reference model:

$$\dot{e} = (A + M)e + \Delta A\hat{i} + \Delta Bu + \Delta C, \tag{18}$$

where $\Delta A = A - \hat{A}$, $\Delta B = B - \hat{B}$, and $\Delta C = C - \hat{C}$.

4.4. Model Reference Adaptive System Parameter Identifier Design

The error equation of MRAS is transformed into an equivalent nonlinear time-varying feedback system that consists of a linear steady-state forward loop and a nonlinear time-varying feedback loop.

Let $w = -(\Delta A\hat{i} + \Delta Bu + \Delta C)$. Accordingly, the parameter identification model formula (Equation (18)) can be rewritten as follows:

$$\dot{e} = (A + M)e - w, \tag{19}$$

The equivalent nonlinear time-varying feedback system can be derived from Equation (19) and is shown in Figure 5, where $\varphi(e)$ is the adaptive law with respect to the parameter matrices A, B, and C.

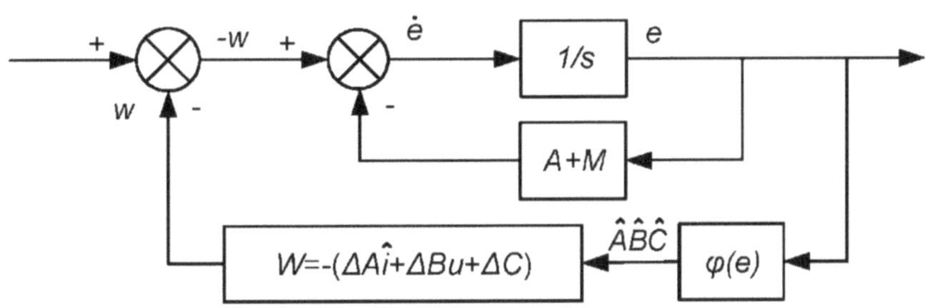

Figure 5. Equivalent nonlinear time−varying feedback system.

The nonlinear time−varying feedback system constructed above must satisfy the conditions of the Popov integral inequality. The Popov integral inequality is given as follows:

$$\eta(0, t_1) = \int_0^{t_1} e^T w dt \geq -\gamma^2, \tag{20}$$

In Equation (20), for any $t \geq 0$, γ is a finite constant independent of t.

Taking L_d and L_q as identification objects and substituting w into Equation (20), we obtain the follows:

$$\eta(0,t_1) = \int_0^{t_1} e^T w dt = \int_0^{t_1} (-e^T \Delta A\hat{i} - e^T \Delta Bu - e^T \Delta C)dt \geq -\gamma^2, \tag{21}$$

Equation (21) can be decomposed into the following three formulas:

$$\eta_1(0,t_1) = -\int_0^{t_1} e^T \Delta A\hat{i} dt \geq -{\gamma_1}^2, \tag{22}$$

$$\eta_2(0,t_1) = -\int_0^{t_1} e^T \Delta Budt \geq -{\gamma_2}^2, \tag{23}$$

$$\eta_3(0,t_1) = -\int_0^{t_1} e^T \Delta Cdt \geq -{\gamma_3}^2, \tag{24}$$

According to the Popov super stability theory, as long as Equations (22)–(24) are satisfied, the nonlinear feedback system remains stable. Then, the adaptive law can be derived from Equations (22)–(24). Because Equation (23) does not contain the inductive coupling term of the quadrature axis, the adaptive law of the identification of the quadrature axis inductance can be deduced using Equation (23). Thus, Equation (23) can be transformed into Equation (25) as follows:

$$\eta_2(0,t_1) = -\int_0^{t_1} (i_d - \hat{i}_d)(\frac{1}{L_d} - \frac{1}{\hat{L}_d})u_d + (i_q - \hat{i}_q)(\frac{1}{L_q} - \frac{1}{\hat{L}_q})u_q dt \geq -{\gamma_2}^2, \tag{25}$$

Next, Equation (25) can be decomposed into Equations (26) and (27) as follows:

$$\eta_{21}(0,t_1) = \int_0^{t_1} (i_d - \hat{i}_d)(\frac{1}{\hat{L}_d} - \frac{1}{L_d})u_d dt \geq -{\gamma_{21}}^2, \tag{26}$$

$$\eta_{22}(0,t_1) = \int_0^{t_1} (i_q - \hat{i}_q)(\frac{1}{\hat{L}_q} - \frac{1}{L_q})u_q dt \geq -{\gamma_{22}}^2, \tag{27}$$

The PI adaptive law about L_d is usually expressed as follows:

$$\frac{1}{\hat{L}_d} = \frac{1}{L_d} + f_1(t) + \int_0^t f_1(\tau)d\tau, \tag{28}$$

Only the adaptation law of the direct-axis inductance identification is analyzed; the derivation process of the quadrature-axis adaptation law is similar. Substituting Equation (28) into Equation (26), $\eta_{21}(0,t_1)$ can be further decomposed into the following two sub-inequalities:

$$\eta_{211}(0,t_1) = \int_0^{t_1} u_d(i_d - \hat{i}_d)f_1(t)dt \geq -{\gamma_{211}}^2, \tag{29}$$

$$\eta_{212}(0,t_1) = \int_0^{t_1} u_d(i_d - \hat{i}_d)\int_0^t f_1(\tau)d\tau dt \geq -{\gamma_{212}}^2, \tag{30}$$

From Equations (29)–(30), the PI adaptive law of parameter L_d can be obtained as follows:

$$\frac{1}{\hat{L}_d} = \frac{1}{L_d} + (k_1 + \frac{1}{\tau_1 s})u_d(i_d - \hat{i}_d), \tag{31}$$

where k_1 and τ_1 are the proportional and integral coefficients of the L_d adaptive law, respectively.

Similarly, the adaptive law of L_q can be obtained as follows:

$$\frac{1}{\hat{L}_q} = \frac{1}{L_q} + (k_2 + \frac{1}{\tau_2 s})u_q(i_q - \hat{i}_q), \tag{32}$$

where k_2 and τ_2 are the proportional and integral coefficients of the L_q adaptive law, respectively.

The overall structure of the IPMSM MTPA drive control system is illustrated in Figure 6.

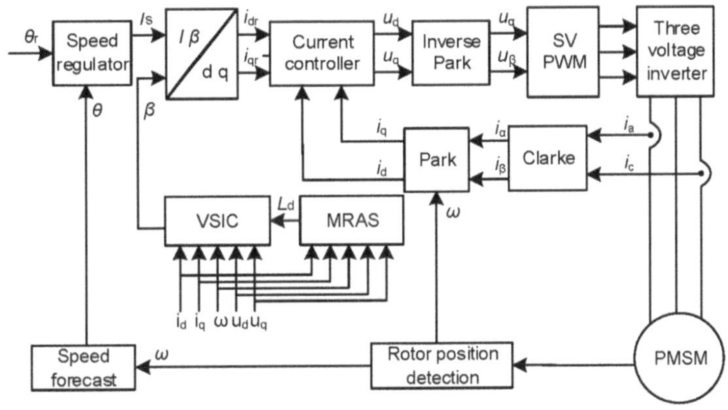

Figure 6. Interior permanent magnet synchronous motor MTPA drive control system.

5. Simulation Results and Analysis

In this study, the simulation model of the IPMSM drive control system was established in the MATLAB/Simulink software environment, and the simulation tests and analysis were performed. The main parameters of the motor are presented in Table 1. The main parameters of the VSIC-type MTPA control method based on inductance identification are presented in Table 2.

Table 1. Main parameters of the IPMSM in the simulation.

Main Parameters	Rated Value
Rated speed	1000 r/min
Rated torque	10 N·m
q-axis inductance	12 mH
d-axis inductance	5.5 mH
The rotor flux	0.1827 Wb
Pairs of poles	4
The moment of inertia	0.003 kg·m^2

Table 2. Main parameters of the VSIC-type MTPA control method based on inductance identification in the simulation.

Main Parameters	Numerical Value
Injected signal amplitude A	0.05 A
Injected signal frequency ω_h	300 Hz
MRAS L_d PI proportional coefficient k_1	0.01
MRAS L_d PI integral coefficient τ_1	10.25
MRAS L_q PI proportional coefficient k_2	0.01
MRAS L_q PI integral coefficient τ_2	1.02

5.1. MTPA Operation Effectiveness Analysis

The motor was run at 1000 r/min, 10 N·m. When the motor was in a stable operation state, the stator current angle gradually increased from 1.57 rad to 2.14 rad. Figure 7 expresses the corresponding waveform changes of the stator current amplitude and $\partial T_e/\partial\beta$ when the stator current angle gradually increases with time.

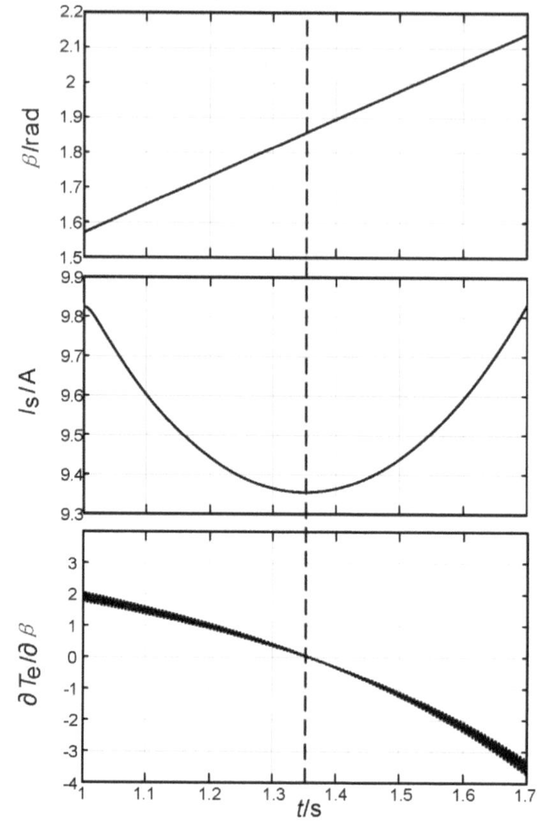

Figure 7. Stator current angle, current amplitude, and $\partial T_e/\partial\beta$ waveform.

The stator current amplitude first decreased and then increased, reaching the minimum at t = 1.35 s. The minimum stator current RMS was 9.37 A, and the corresponding MTPA angle was approximately 1.85 rad, which is in line with the expected actual value. The MTPA points were the same. In addition, the corresponding torque versus current angular change rate revealed that when $\beta \leq \beta_{\rm MTPA}$, $\partial T_e/\partial\beta \geq 0$; when $\beta \geq \beta_{\rm MTPA}$, $\partial T_e/\partial\beta \leq 0$; when $\beta = \beta_{\rm MTPA}$, $\partial T_e/\partial\beta = 0$. This result is consistent with the theoretical analysis, thus, proving that the proposed method can effectively track the MTPA current angle.

Under the condition of no load, the motor reached the given speed value of 1000 r/min after t = 0.06 s. When the motor reached a stable running state, a load torque of 10 N·m was added at t = 1.5 s. The command and feedback tracking performance of the motor speed and load torque, the motor torque tracking performance of the current angle change rate, and the command and feedback of the stator vector angle were compared to demonstrate the improved algorithm accuracy.

After a sudden load was applied, the speed decreased, and the system quickly returned to the steady state (Figure 8), thus, showing that the proposed method has a good speed tracking effect. After the system started, the speed tracked the command very well. The overshoot was only 8 r/min (0.8%), the maximum speed drop after the addition of load was only 12 r/min (1.2%), and it quickly returned to the steady state. The simulation results show that the speed loop controlled by MTPA has good dynamic and steady-state performance and strong anti-load disturbance capability.

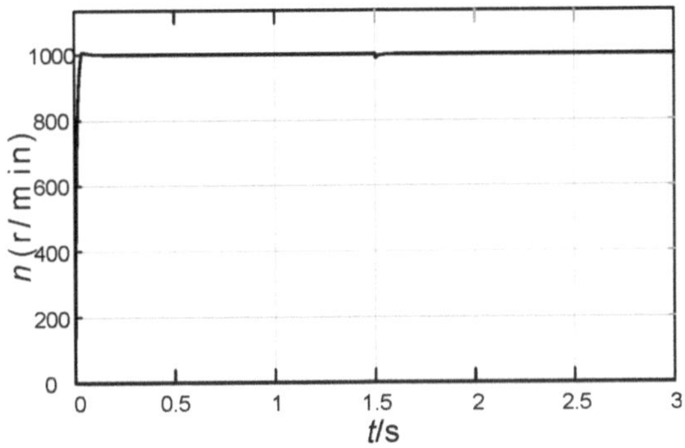

(a)

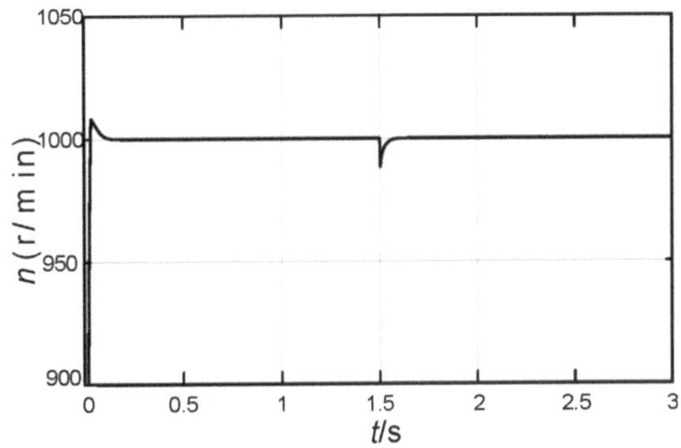

(b)

Figure 8. Waveform of rotational speed. (**a**) Waveform of rotational speed; (**b**) Zoomed waveform of rotational speed.

Figure 9 shows that IPMSM can provide a large torque when starting. The motor load torque was 10 N·m, and the electromagnetic torque was 10.32 N·m, indicating that the motor has good accuracy. When the rated speed and rated load were used, the minimum stator current value was 9.37 A, and the optimal stator current value was almost the same, thereby demonstrating very good accuracy.

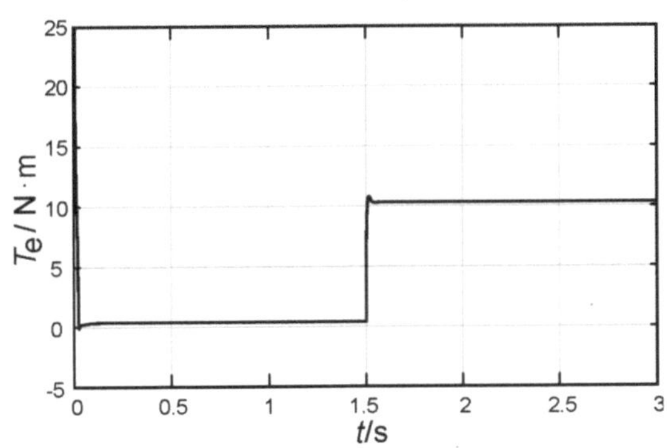

(a)

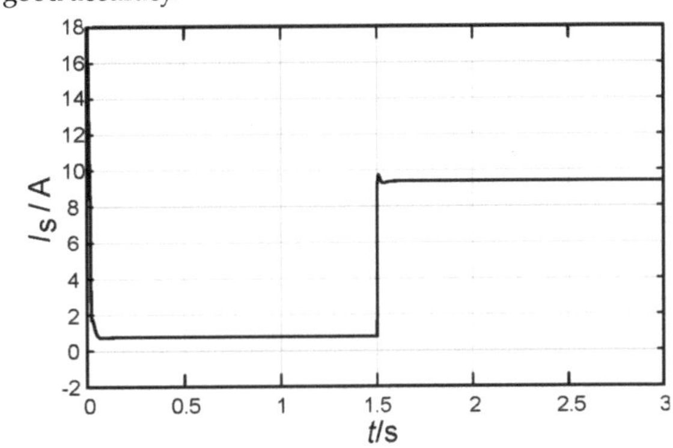

(b)

Figure 9. Torque and current waveforms. (**a**) Torque waveform; (**b**) Stator current waveform.

Figure 10 shows that at the moment of sudden torque change, the method responded immediately to make the MTPA angle converge, and the corresponding MTPA angle after stabilization was approximately 1.85 rad, which is the same as the expected actual MTPA point. In addition, the PI regulator quickly adjusted it to a steady state of 0 to achieve the purpose of automatically tracking the MTPA angle. However, due to the influence of the filter and the integrator in the virtual signal injection method, it required 1 s to reach the steady state.

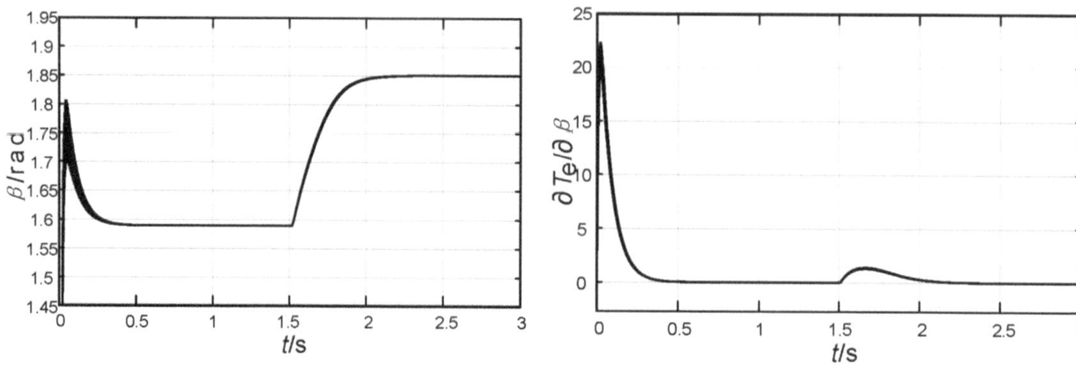

(a) (b)

Figure 10. Stator vector angle and $\partial T_e/\partial\beta$ waveform. (**a**) Stator vector angle waveform; (**b**) $\partial T_e/\partial\beta$ waveform.

5.2. MTPA Operation Accuracy Analysis

To verify that the improved virtual signal injection control (IVSIC) method MTPA control maintains strong robustness and accuracy under the parameter mismatch condition, the rated parameters of the motor model in the simulation were changed. L_d and L_q were set at 7 and 15 mH, respectively, which is an increase of approximately 1.25 times the rated value. ψ_f was set as 0.14 Wb, which is a reduction of approximately 77% of the rated value. Then, the formula control (FC) method, virtual signal injection control (VSIC) method, and IVSIC method were simulated and compared. For the IPMSM running at 1000 r/min under a load of 10 N·m, when the motor was in a stable operation state, the stator current angle gradually increased from 1.57 rad to 2.14 rad. The relational waveform of the angular change rate is shown in Figure 11.

Figure 11 shows that under this working condition (1000-r/min running speed and 10 N·m load torque), the current vector amplitude changed with the current vector angle; the minimum value was 11.12 A, and the corresponding radian was 1.99 rad. The $\partial T_e/\partial\beta$ zero-crossing point obtained using the formula calculation method was far away from the actual operating point of the MTPA, and the corresponding angle was 1.90 rad; thus, demonstrating that the control effect of the FC method is greatly affected by changes in the motor parameters and the obtained MTPA angle is not accurate. Further, the position of $\partial T_e/\partial\beta$ zero−crossing point obtained using the VSIC method was close to the actual operating point of MTPA, and the corresponding radian was 2.02 rad, but it was still inaccurate. In contrast, the $\partial T_e/\partial\beta$ zero-point obtained using the IVSIC method was very close to the actual operating point of MTPA, which shows that the proposed method has high accuracy. That is, the proposed method can more accurately track the MTPA operating point and has high accuracy.

Figures 12–14 illustrate the tracking response curves and d- and q-axis current error curves of the parameters to be identified. The value of L_d was increased to 7 mH (Figure 12), that is, approximately 1.27 times the rated value, and the value of L_q was increased to 15 mH (Figure 13), that is, approximately 1.25 times the rated value. Both parameters changed simultaneously (Figure 14); the IPMSM was run under the conditions of 1000 r/min and 10 N·m. From Figures 12–14, it can be seen that when the parameters changed, the dynamic response of the parameters to be identified had a small overshoot and converged quickly to the reference value, thereby demonstrating that the designed parameter identifier has good parameter change tracking characteristics.

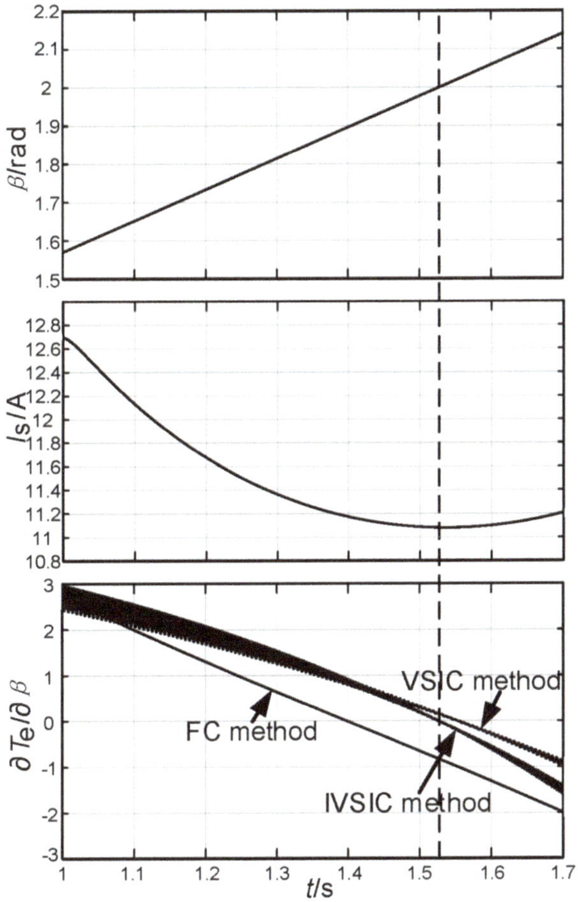

Figure 11. Stator current angle, current amplitude, and $\partial T_e/\partial \beta$ waveform.

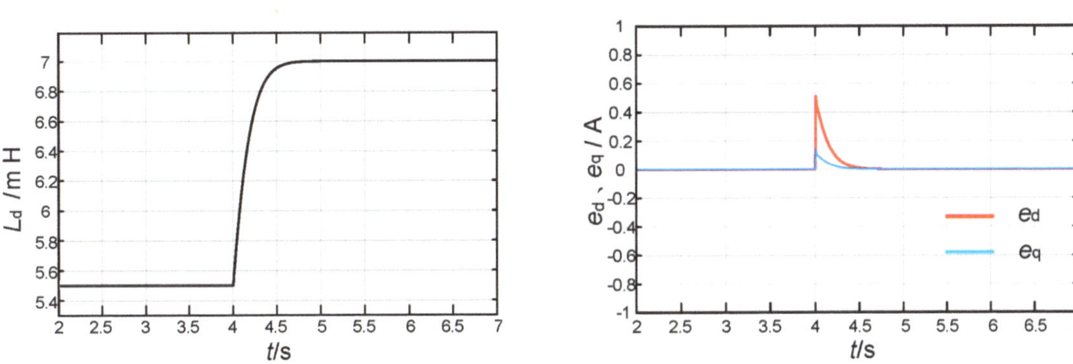

(a) (b)

Figure 12. Identification of sudden drop in d-axis inductance. (**a**) d-axis inductance identification; (**b**) d-axis and q-axis current errors.

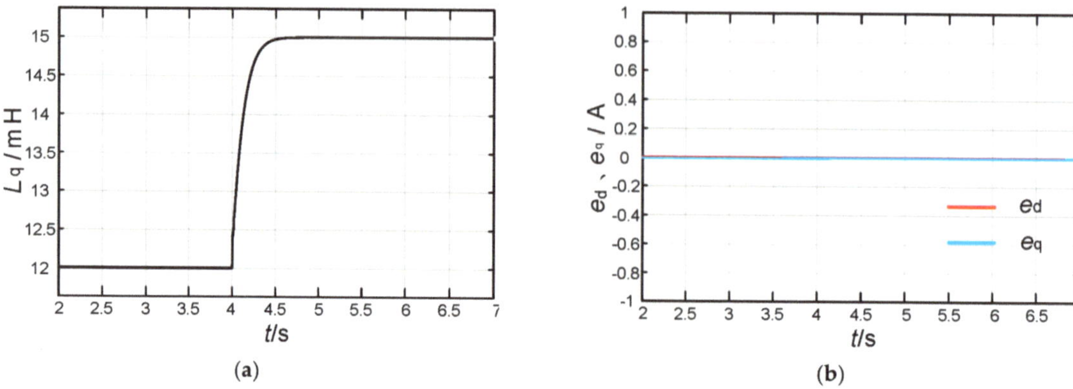

Figure 13. Identification of sudden drop in q-axis inductance. (**a**) q-axis inductance identification; (**b**) d-axis and q-axis current errors.

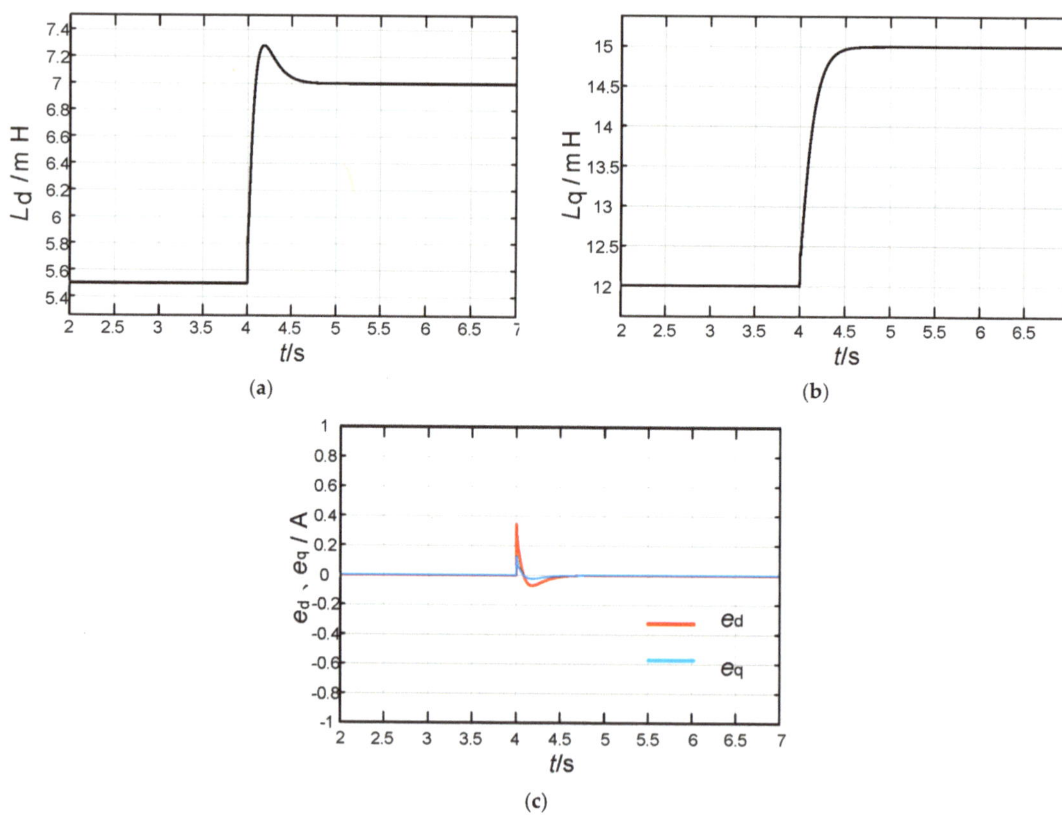

Figure 14. Identification of simultaneous changes in L_d and L_q. (**a**) d-axis inductance identification; (**b**) q-axis inductance identification; (**c**) d-axis and q-axis current errors.

Figures 15–23, respectively, show the comparison charts of the tracking curves of the d-axis current, q-axis current, and stator current when the FC method, VSIC method, and IVSIC method were used. To highlight the influence of the change in the d-axis inductance parameter on the working efficiency of the IPMSM, three parameter changes were employed in this study. In the first case, the d-axis inductance parameter was increased from the initial state inductance value to 7 mH, the q-axis inductance parameter was increased from the initial state inductance value to 15 mH, and the other parameters were left unchanged. In the second case, the d-axis inductance parameter was decreased from the initial state inductance value to 4 mH, the q-axis inductance parameter was decreased from the initial state inductance value to 7 mH, and the other parameters were left unchanged. In the third case, the d-axis inductance parameter was increased from the initial state inductance value to 7 mH, and the q-axis inductance parameter was decreased from the initial state inductance value to 9 mH, and the other parameters were left unchanged. For these parameter changes, the calculated expected d-axis current, q-axis current, and the operating point of the stator current changed accordingly. When the motor was run at 1000 r/min under a load torque of 10 N·m, the waveform obtained when the motor was in a stable operation state is shown in Figures 15–17.

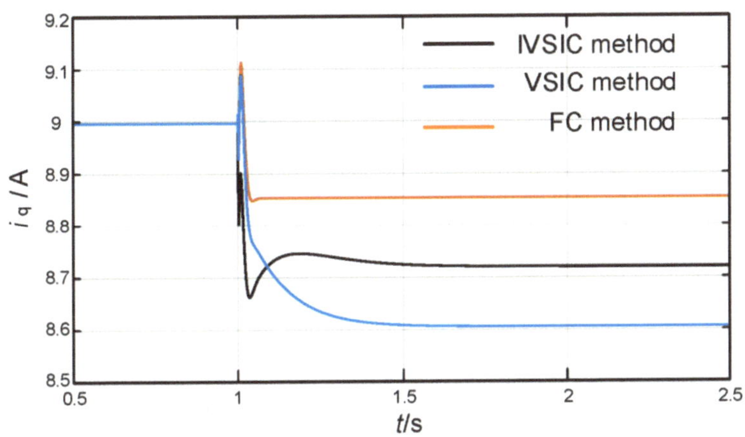

Figure 15. Comparison of tracking curves of q-axis current.

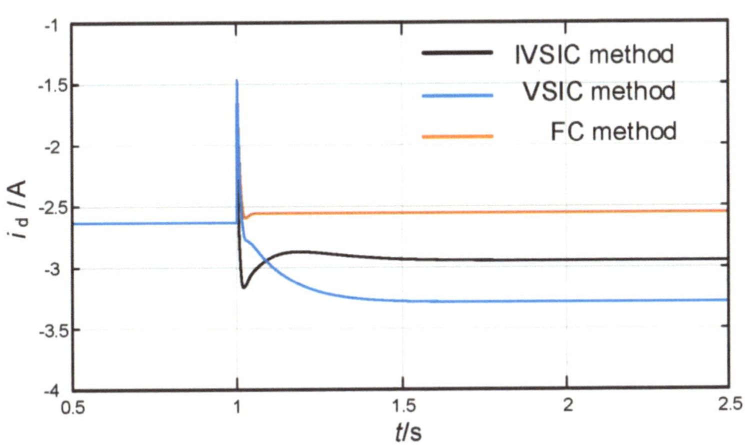

Figure 16. Comparison of tracking curves of d-axis current.

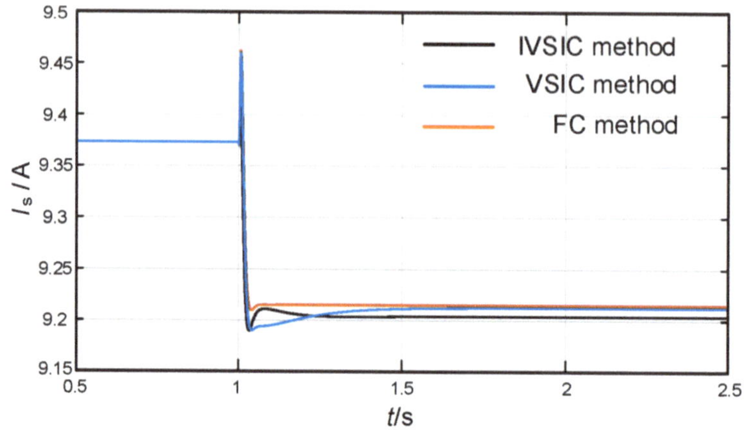

Figure 17. Comparison of tracking curves of stator current.

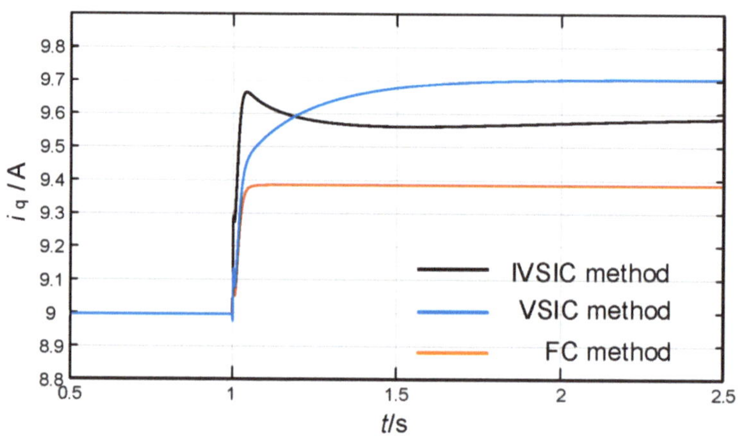

Figure 18. Comparison of tracking curves of q-axis current.

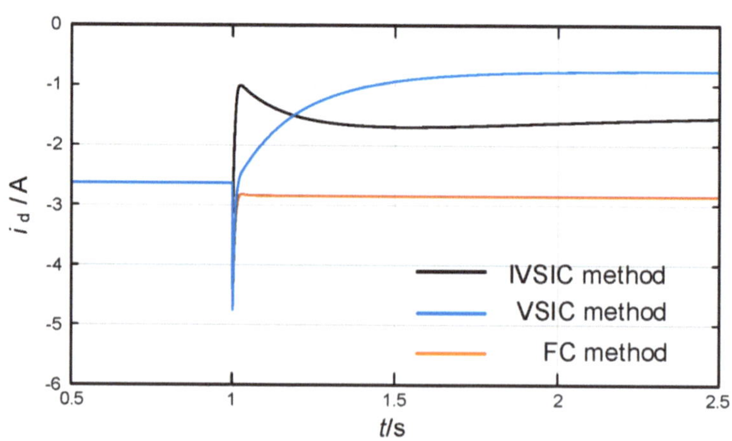

Figure 19. Comparison of tracking curves of d-axis current.

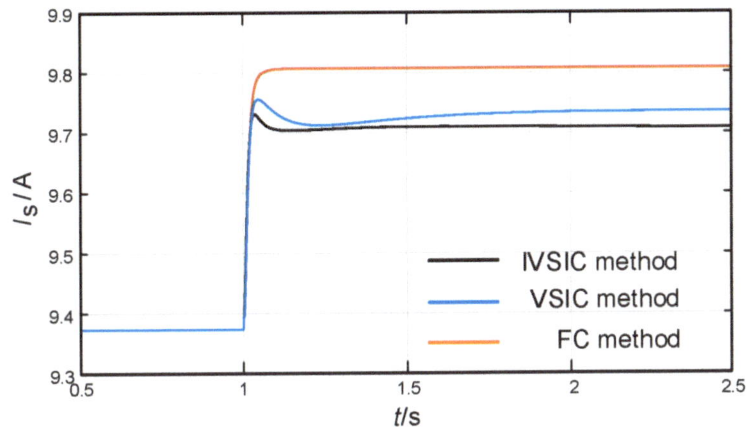

Figure 20. Comparison of tracking curves of stator current.

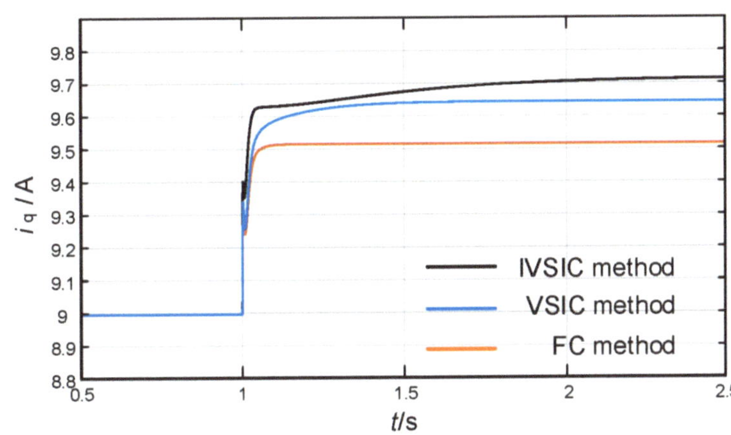

Figure 21. Comparison of tracking curves of q-axis current.

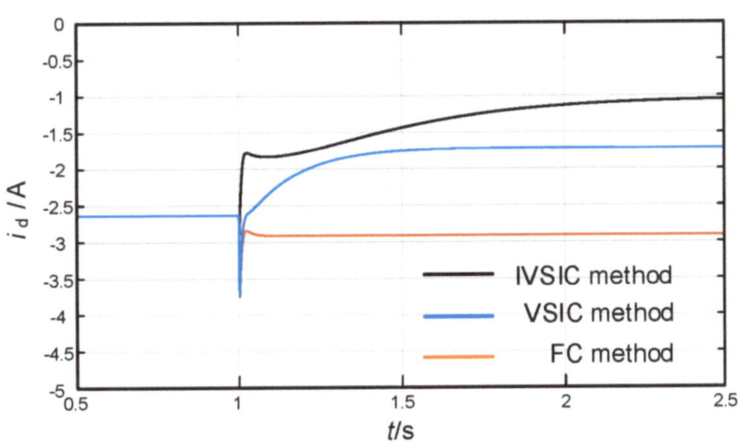

Figure 22. Comparison of tracking curves of d-axis current.

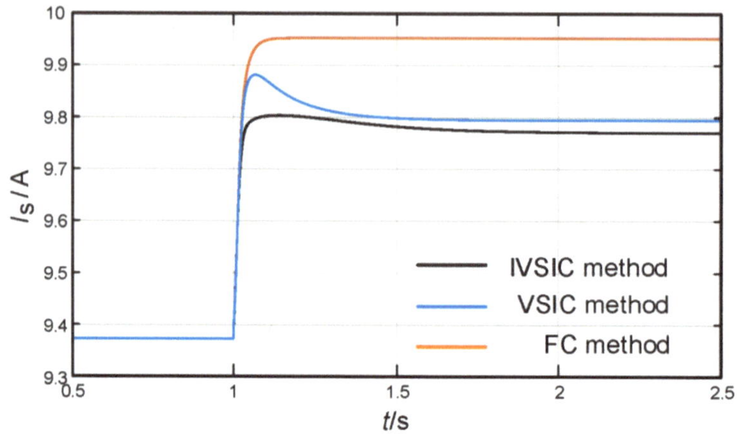

Figure 23. Comparison of tracking curves of stator current.

From Figures 15–17, it can be seen that when the inductance value was increased to 15 mH while the other parameters were left unchanged, the magnitude of the current vector changed. The largest stator current amplitude was obtained using the FC method (9.22 A), followed by the VSIC method (9.21 A) and the IVSIC method (9.20 A). The value obtained using the VSIC method was inaccurate, whereas that obtained using the IVSIC method was the same as the expected value, thus showing that the proposed control algorithm has high accuracy. Additionally, when the parameter changes, the longest current transition time t_s was obtained using the IVSIC method (0.7 s), followed by the VSIC method (0.6 s) and the FC method (0.1 s). It can be seen that the control speed drops when using the IVSIC method.

Figures 18–20 show that when the d-axis inductance parameter was decreased from the initial state inductance value to 4 mH and the q-axis inductance parameter was decreased from the initial state inductance value to 7 mH, the magnitude of the current vector changed. The largest stator current amplitude was obtained using the FC method (9.80 A), followed by the VSIC method (9.73 A) and the IVSIC method (9.71 A). The stator current amplitude obtained using the VSIC method was inaccurate, whereas that obtained using the IVSIC was the same as the expected value, thus showing that the proposed control algorithm has high accuracy. Additionally, when the parameter changes, the longest current transition time t_s was obtained using the IVSIC method (1.5 s), followed by the VSIC method (1.1 s) and the FC method (0.1 s). It can be seen that the control speed drops when using the IVSIC method.

Figures 21–23 show that when the d-axis inductance parameter was increased from the initial state inductance value to 7 mH and the q-axis inductance parameter was decreased from the initial state inductance value to 9 mH, the magnitude of the current vector changed. The largest stator current amplitude was obtained using the FC method (9.95 A), followed by the VSIC method (9.80 A) and the IVSIC method (9.77 A). The stator current amplitude obtained using the VSIC method was inaccurate, whereas that obtained using the IVSIC method was the same as the expected value, thus showing that the proposed control algorithm has high accuracy. In summary, the simulation results prove that the IVSIC method has good trajectory tracking characteristics when the parameters change. Additionally, when the parameter changes, the longest current transition time t_s was obtained using the IVSIC method (1.1 s), followed by the VSIC method (0.7 s) and the FC method (0.1 s). It can be seen that the control speed drops when using the IVSIC method.

In order to make the obtained results more significant and easier to analyze, we have organized the main data from Figures 15–23. in Table 3.

Table 3. Arrangement of the main data in Figures 15–23.

Inductor Parameter Value		FC Method				VSIC Method				IVSIC Method			
L_d (mH)	L_q (mH)	i_d (A)	i_q (A)	I_s (A)	t_s (s)	i_d (A)	i_q (A)	I_s (A)	t_s (s)	i_d (A)	i_q (A)	I_s (A)	t_s (s)
7	15	−2.6	8.85	9.22	0.1	−3.3	8.60	9.21	0.6	−3.0	8.73	9.20	0.7
4	7	−2.8	9.39	9.80	0.1	−0.7	9.70	9.73	1.1	−1.5	9.58	9.71	1.5
7	9	−3.0	9.50	9.95	0.1	−1.7	9.64	9.80	0.7	−1.1	9.71	9.77	1.1

In Table 3, we can find that when the d-axis inductance value and the q-axis inductance value increase or decrease at the same time, or one increases and the other decreases, the stator current value required by the IVSIC method (9.20 A, 9.71 A, 9.77 A) is the smallest. It can also be seen that although the IVSIC method increases efficiency compared to the VSIC method, the increased efficiency is less than that between the VSIC method and the FC method. Taking the third parameter change case as an example, the IVSIC method (9.77 A) reduces the stator current by 0.03 A compared to the VSIC method (9.80 A), and the VSIC method (9.80 A) reduces the stator current by 0.15 A compared to the FC method (9.95 A). This can also be explained by the side that the influence of the q-axis inductance parameters on the motor control efficiency is greater than that of the d-axis inductance parameters. At the same time, it can also be found that the mathematical relationship between the d-axis current value, the q-axis current value, and the stator current value satisfies Equation (3), which verifies the validity of the data. Additionally, when the parameter changes, the longest current transition time was obtained using the IVSIC method (0.7 s, 1.5 s, 1.1 s). It can be seen that the control speed drops when using the IVSIC method.

6. Conclusions

Aiming at the problem that the MTPA trajectory deviates due to changes in the motor parameters when the IPMSM is in MTPA control mode when using VSIC, in this paper, we proposed the online identification of the inductance by using MRAS and realize MTPA trajectory adjustment with changes in motor inductance parameters. The identifier model designed using this method is easy to implement, has strong practicability, and has a good parameter identification effect. The simulation results showed that the control strategy can inherit the performance of VSIC control well and has good robustness to changes in shaft inductance, thereby greatly improving the anti-interference ability of the system. However, this control method is relatively complex, and the application cost is high, so it is suitable for applications with relatively high-performance requirements.

Author Contributions: Conceptualization, N.-Z.J. and K.Z.; Methodology, N.-Z.J. and H.-C.C.; software, H.-C.C., Z.-Q.W. and L.Z.; validation, H.-C.C., Z.-Q.W. and L.Z.; supervision, N.-Z.J., K.Z. and D.-Y.S.; writing-original draft, H.-C.C.; writing-review and editing, N.-Z.J., H.-C.C. and D.-Y.S. All authors have read and agreed to the published version of the manuscript.

Funding: This research was funded by The Natural Science Foundation of Heilongjiang Province Joint Guide Project of China, grant number LH2021E086.

Conflicts of Interest: The authors declare no conflict of interest. The funders had no role in the design of the study; in the collection, analyses, or interpretation of data; in the writing of the manuscript, or in the decision to publish the results.

References

1. Zhu, J. Application present situation & development trend of rare earths permanent magnet motor. *Heavy Mach. Sci. Technol.* **2008**, *4*, 38–42.
2. Dutta, R.; Rahman, M.F. Design and analysis of an interior permanent magnet (IPM) machine with very wide constant power operation range. *IEEE Trans. Energy Convers.* **2008**, *23*, 25–33. [CrossRef]
3. Zhu, Z.Q.; Howe, D. Electrical machines and drives for electric, hybrid, and fuel cell vehicles. *Proc. IEEE* **2007**, *95*, 746. [CrossRef]
4. Li, C.H.; Chen, M.J. Study of a maximum ratio of torque to current control method for PMSM. *Proc. CSEE* **2005**, *25*, 169–174.

5. Kim, S.; Yoon, Y.D.; Sul, S.K. Parameter independent maximum torque per ampere (MTPA) control of IPM machine based on signal injection. In Proceedings of the IEEE Applied Power Electronics Conference and Exposition, Palm Springs, CA, USA, 21–25 February 2010.
6. Cheng, B.; Tesch, T.R. Torque feedforward control technique for permanent-magnet synchronous motors. *IEEE Trans. Ind. Electro.* **2010**, *57*, 969–974. [CrossRef]
7. Liao, Y.; Wu, Z.D.; Liu, R. Improved maximum torque per ampere control strategy study for PMSM used in electric vehicles. *Electr. Mach. Control* **2012**, *16*, 12–17.
8. Tang, Q.P.; Shen, A.W.; Luo, P. IPMSMs sensorless MTPA control based on virtual q-axis inductance by using virtual high-frequency signal injection. *IEEE Trans. Ind. Electron.* **2020**, *67*, 136–146. [CrossRef]
9. Liu, G.H.; Wang, J.; Zhao, W.X. A novel MTPA control strategy for IPMSM drives by space vector signal injection. *IEEE Trans. Ind. Electron.* **2017**, *64*, 9243–9252. [CrossRef]
10. Sun, T.F.; Wang, J.B.; Chen, X. Maximum Torque Per Ampere (MTPA) Control for Interior Permanent Magnet Synchronous Machine Drives Based on Virtual Signal Injection. *IEEE Trans. Power Electron.* **2015**, *30*, 5036–5045. [CrossRef]
11. Wang, J.; Huang, X.Y.; Yu, T. An Accurate Virtual Signal Injection Control of MTPA for an IPMSM with Fast Dynamic Response. *IEEE Trans. Power Electron.* **2018**, *33*, 7916–7926. [CrossRef]
12. Yang, Z.J.; Wang, L.N. Online multi-parameter identification for surface-mounted permanent magnet synchronous motors. *Trans. China Electrotech. Soc.* **2014**, *29*, 111–118.
13. Xun, Q.; Wang, P.L.; Li, Z.X. PMSM parameters identification based on recursive least square method. *Trans. China Electrotech. Soc.* **2016**, *31*, 161–169.
14. Wen, C.B.; Qi, L. On-line detection method of magnet flux linkage for permanent magnet synchronous motor. *Proc. CSU-EPSA* **2010**, *22*, 22–26.
15. Salvatore, N.; Caponio, A.; Neri, F. Optimization of delayed-state Kalman-filter-based algorithm via differential evolution for sensorless control of induction motors. *IEEE Trans. Ind. Electro.* **2010**, *57*, 385–394. [CrossRef]
16. An, Q.T.; Sun, L.; Zhao, K. An adaptive on-line identification method for the parameters of permanent magnet synchronous motor. *Trans. China Electrotech. Soc.* **2008**, *23*, 31–35.
17. Wang, L.H.; Zhang, X.; Zhang, W.F. Parameter identification of permanent magnet synchronous motor based on neural network. *Power Electron.* **2020**, *54*, 47–49.
18. Shi, W.G.; Yan, X.Y. Improved particle swarm algorithm of parameter identification for PMSM based on cloud model. *J. Dalian Jiaotong Univ.* **2019**, *40*, 113–119.
19. Zhong, W.P. Parameter Identification of Permanent Magnet Synchronous Motor Based on Genetic Algorithm. Master's Thesis, Changsha University of Science and Technology, Changsha, China, 2019.

machines

MDPI

Review

Design and Analysis of the High-Speed Permanent Magnet Motors: A Review on the State of the Art

Qiping Shen *, Ziyao Zhou, Shan Li, Xinglin Liao, Tao Wang, Xiaorong He and Jingshan Zhang

School of Electrical and Electronic Engineering, Chongqing University of Technology, Chongqing 400054, China; 52200712112@stu.cqut.edu.cn (Z.Z.); lishan@cqut.edu.cn (S.L.); lxl023040@cqut.edu.cn (X.L.); wangtao20190031@cqut.edu.cn (T.W.); hexiaorong@cqut.edu.cn (X.H.); 52200712123@stu.cqut.edu.cn (J.Z.)
* Correspondence: shenqp1983@cqut.edu.cn

Abstract: This paper provides an overview of the design and analysis of high-speed PM motors by focusing on prominent issues such as motor losses, temperature rise, rotor strength and vibration. The design challenges of high-speed PM motors are briefly described, and the application of various stator and rotor structures and materials is presented in electromagnetic design. Combined with the temperature distribution of the motor, various heat dissipation measures to suppress the temperature rise are summarized. Strength and dynamics analysis of the rotor are outlined with respect to the safety of rotor operation. The current status of coupled multi-physics domain design used to improve the comprehensive design capability of high-speed PM motors is reviewed. Future directions of technologies related to the design of high-speed PM motors are pointed out.

Keywords: high-speed PM motor; electromagnetic design; temperature rise; strength and dynamics analysis; multi-physics domain

Citation: Shen, Q.; Zhou, Z.; Li, S.; Liao, X.; Wang, T.; He, X.; Zhang, J. Design and Analysis of the High-Speed Permanent Magnet Motors: A Review on the State of the Art. *Machines* **2022**, *10*, 549. https://doi.org/10.3390/machines10070549

Academic Editors: Lorand Szabo and Feng Chai

Received: 14 June 2022
Accepted: 4 July 2022
Published: 7 July 2022

1. Introduction

High-speed motors have been developed over a long period of time and have been widely used in machine tools, turbochargers, mechanical turbo-charging systems, flywheel energy storage systems, gas compressors, blowers, vacuum pumps, shipborne power supply systems and aero-engines, etc. They typically operate at speeds in excess of 10,000 r/min, rotor circumferential speeds (v_c) in excess of 50 m/s and r/min$\sqrt{kW}$ (the product of speed and power square root) in excess of 1×10^5 [1–4]. In terms of structural principle, both DC motors and AC motors can realize high-speed operation. The main structural types include IM, PMSM and SRM. Referring to [3–60], the power and speed distributions of the current high-speed motor studies are given in Figure 1. IMs operate by the interaction of the induced current generated by the rotor winding in the stator magnetic field with the stator magnetic field. SRMs rely on the principle of minimum reluctance to generate torque, that is, the magnetic flux always closes along the path of minimum reluctance, thus creating a magnetic pull to rotate the rotor. PM motors are driven by the interaction of the magnetic field generated by the permanent magnets and the rotating magnetic field generated by the stator winding. The characteristics of these high-speed motors are shown in Table 1 [61]. As can be seen from Table 1, the high-speed PM motor has outstanding advantages at the efficiency and power density levels. Among them, the power density distribution of high-speed PM motors at different speeds and powers is shown in Figure 2. Moreover, compared with other motor structures, PMSM has the advantages of diverse stator and rotor structures, as well as characteristics of good control. Therefore, PM motors are more widely used in high-speed fields.

Figure 3a shows the status of the speed-power study for a variety of structures of high-speed PM motors. Detailed raw data are presented in Appendix A. The distribution of v_c and r/min$\sqrt{kW}$ of high-speed PM motors is given in Figure 3b. As we can see from the figure, the circumferential speed v_c of the SPM structure design is generally higher than that of the IPM structures and cylindrical PM structures. Moreover, the SPM structure is easier to achieve higher power of 100 kW and above, as well as higher speed of 100 krpm and above. However, there are more challenges in electromagnetic design, thermal management, mechanical structure and dynamics analysis for high-speed PM motors compared with traditional low and medium-speed PM motors.

Table 1. Characteristics of various types of high-speed motors.

	High-Speed IM	High-Speed SRM	High-Speed PM Motor
Advantages	easily starting low cost rotor can withstand high temperature	simple rotor structure low rotor loss short end of winding	high efficiency high power density high power factor
Disadvantages	high rotor loss low power factor laminated rotor end rings are easily damaged	low efficiency high noise large wind friction	low rotor strength permanent magnets are easy to demagnetize higher cost

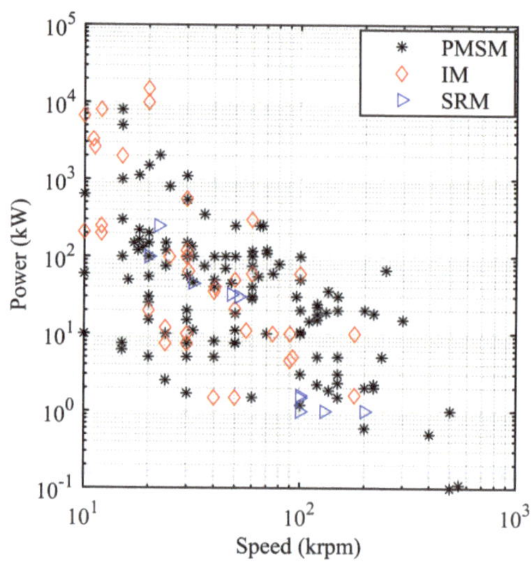

Figure 1. Speed and power distribution of high-speed motors of different structure types.

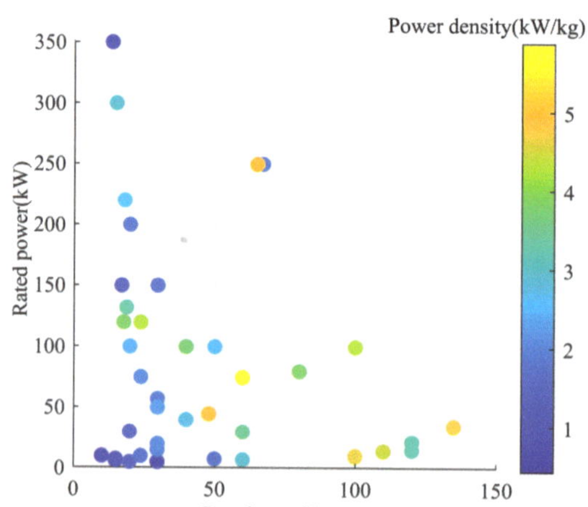

Figure 2. Distribution of power density of high-speed PM motors at different speed and power.

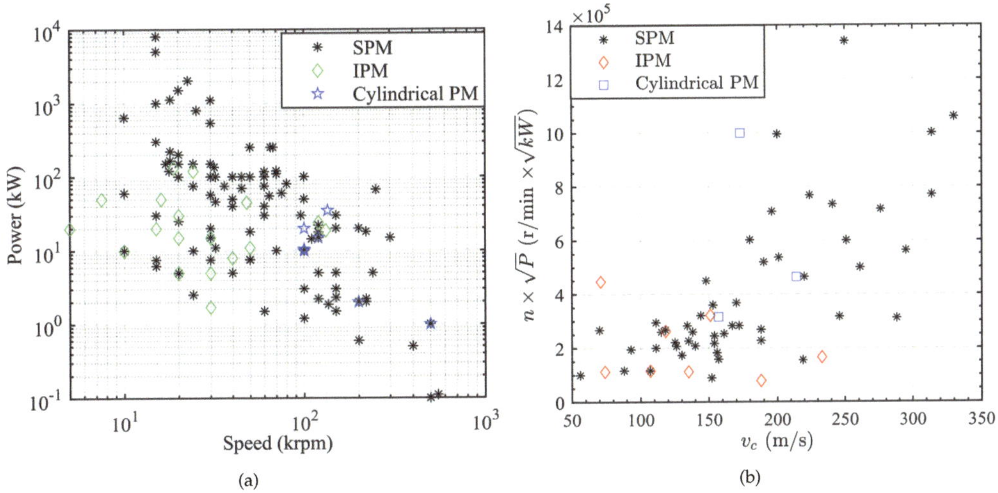

Figure 3. Power and speed distribution of high-speed motor: (**a**) Speed and power distribution of high-speed PM motors with different structures; (**b**) Distribution of v_c and $\text{r/min}\sqrt{kW}$ of high-speed PM motors.

Firstly, the frequency of high-speed PM motors is generally up to thousands of hertz, so the loss density and thermal effect of the motor is much higher than that of low and medium speed motors. To overcome these problems, it is necessary to innovate motor structures and to be more careful in the choice of materials compared to conventional motors [54].

Secondly, the high-speed PM motor is compact, while its wind friction loss and rotor eddy current loss are much higher than those of conventional motors, so the highest temperature region in the motor is shifted from the stator to the rotor [60]. Seeking thermal management measures to solve the efficient heat dissipation of the rotor is extremely important for the efficient operation of the motor.

Thirdly, the rotor circumferential speed (v_c) of high-speed PM motors can reach up to 200 m/s and above. The problem of rotor strength is prominent due to high-speed centrifugal force and thermal stress [62]. Finally, compared with conventional motors, the rotor of high-speed PM motors is more prone to bending vibration. To ensure the safe and stable operation of the motor, rotor strength analysis and calculation of critical speed are indispensable parts of the design of high-speed PM motors.

In summary, there are more factors to be considered at each design level for high-speed PM motors than for conventional low and medium speed motors. This paper first outlines the selection of stator and rotor structures and materials for high-speed PM motors from the perspective of improving electromagnetic performance, and then reviews the current status and key issues of research in four areas, namely thermal management, rotor strength, dynamics analysis and multi-physics domain coupling in turn. Finally, issues related to the further efficient design of high-speed PM motors are discussed.

2. Electromagnetic Design

The electromagnetic performance of high-speed PM motors is closely related to the structure and material of the motor. In order to avoid the harm caused by high loss, the structure and material of the stator and rotor need to be designed reasonably.

2.1. Stator Design

Stator design includes the selection of major dimensions and calculation of stator losses, as well as the structure and material selection of cores and windings. Stator design is one of the important links to reduce motor losses and improve motor stiffness.

The major dimensions of the motor can usually be estimated by (1).

$$D_{i1}^2 L_{ef} = \frac{6.1 \times 10^4}{K_\phi K_{dp} A B_\delta \alpha_p'} \frac{P'}{n_N} \quad (1)$$

where $D_{i1}^2 L_{ef}$ is the armature volume, D_{i1} is the stator inner diameter, L_{ef} is the armature calculation length, α_p' is the calculating pole arc factor, K_{dp} is the winding factor, K_ϕ is the waveform coefficient of the air gap magnetic field, A is the electrical load, B_δ is the air gap magnetic density, AB_δ is the Electromagnetic load, P' is the calculated power, and n_N is the rated speed.

Determining the major dimension ratio λ is also a common method for estimating the overall size of a motor, which is defined as the ratio of the L_{ef} to the pole pitch τ, as shown in (2).

$$\lambda = \frac{L_{ef}}{\tau} \quad (2)$$

The selection of $D_{i1}^2 L_a$ and AB_δ for high-speed PM motors at different speeds and powers are given in Figure 4a,b, respectively. It can be seen that in the listed speed and power range, the $D_{i1}^2 L_a$ selection is mainly in the range of 2000 cm^3 and below, and there is a tendency to decrease with the increase in speed. The selection of AB_δ is mainly concentrated around $(0.5 \sim 3) \times 10^4$ $A \cdot T/m$. Figure 5a,b show the selection of λ and D_{i1}/D_1 for different speed and power, respectively. The larger λ is, the slimmer the motor is, which helps to reduce the centrifugal force caused by high speed. However, a motor core that is too long will affect its operational stability. Choosing a relatively small value of λ facilitates the placement of more permanent magnets to improve motor performance. D_1 is the outer diameter of stator. From Figure 5, we can see that the λ is often taken in the range of 0.5~2.5, and the D_{i1}/D_1 is mostly taken in the range of 0.3~0.8.

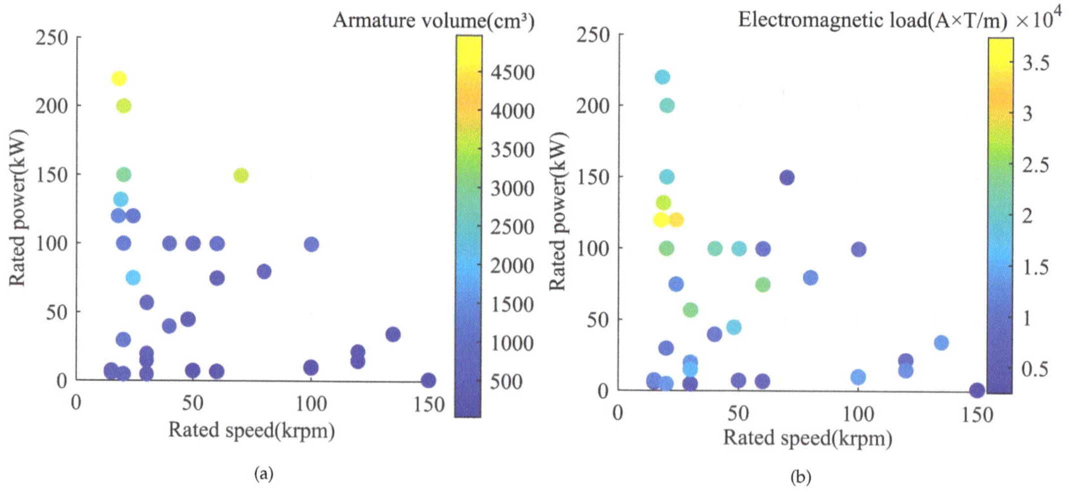

Figure 4. $D_{i1}^2 L_a$ and AB_δ distribution: (**a**) Distribution of $D_{i1}^2 L_a$ of high-speed PM motors; (**b**) Distribution of AB_δ of high-speed PM motors.

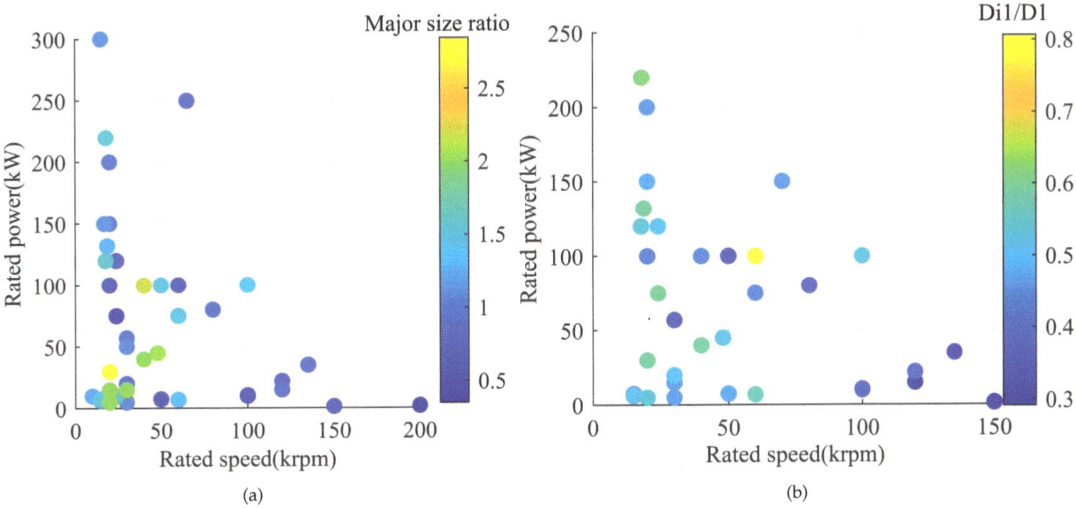

Figure 5. λ and D_{i1}/D_1 distribution: (**a**) Distribution of λ of high-speed PM motors; (**b**) Distribution of D_{i1}/D_1 of high-speed PM motors.

For the calculation of stator core losses. Generally speaking, the alternating frequencies of winding currents and core magnetic fields of high-speed PM motors are as high as thousands of hertz, which is likely to cause additional losses and serious temperature rise in motors. Figure 6 shows the loss curve of 35W270 silicon steel at different frequencies. It can be seen from Figure 6 that the iron loss at high frequencies is higher than that at low frequencies. Therefore, the accuracy of the early Bertotti iron loss separation model is greatly reduced. To accommodate high frequency operating conditions, Shanlin et al. [63] of the Harbin Institute of Technology propose a variable loss coefficient orthogonal decomposition core loss model, which can simultaneously consider the rotational magnetization in the core and the high-frequency skin effect, and is practical for high-speed motors with high fundamental frequency. The loss factor of the core has been shown to be related to the temperature rise and stress variation of the material as well [64]. In response, an improved model of iron loss in high-speed PM motors that can take various physical factors such as magnetic field, temperature and stress into account has been studied and applied [14,65].

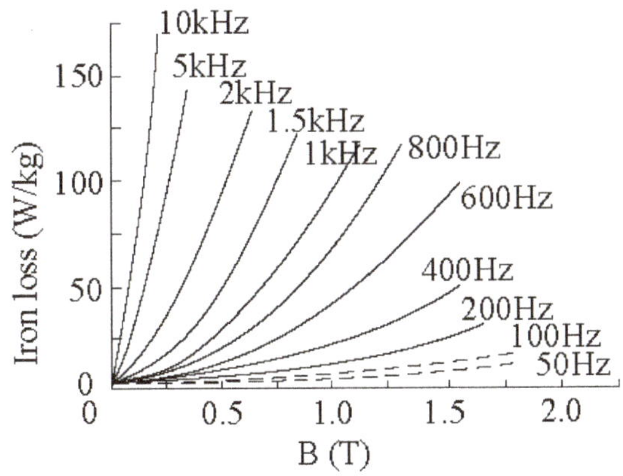

Figure 6. The loss curve of 35W270 at different frequencies.

High-speed PM motors are usually selected in 2 or 4-pole configurations, the purpose of which is to reduce the operating frequency, improve the sinusoidality of the current waveform and, in particular, reduce the eddy current losses in the permanent magnets. The common stator structures are few-slot, multi-slot, no-slot and fictitious slot, as shown in Figure 7.

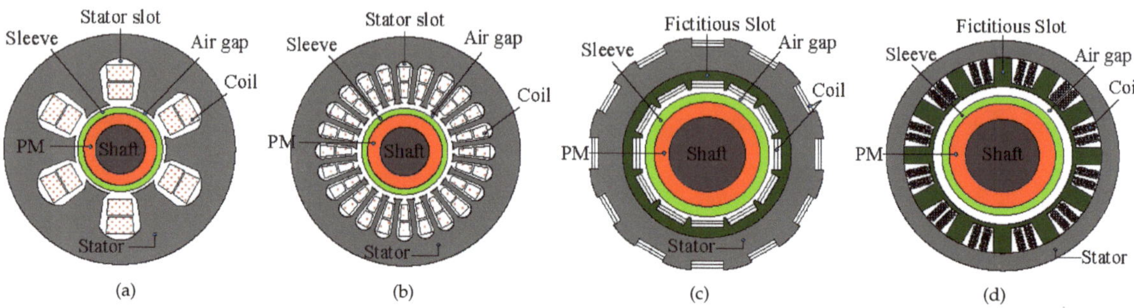

Figure 7. Stator configurations with different number of slots: (**a**) Few-slot structure; (**b**) Multi-slot structure; (**c**,**d**) is fictitious slots.

The slotted structure on the inner surface of the stator can achieve higher air-gap flux density and power density. However, stator slotting will cause changes in air-gap permeability and bring tooth harmonics, especially for the less-slotted structure shown in Figure 7a. However, due to the high torque density of this structure, it has some applications in low-power and lightweight high-speed PM motors [66]. The stator multi-slot structure shown in Figure 7b can reduce the cogging torque and weaken the harmonic air gap magnetic density. Meanwhile, measures such as adjusting the slot opening width of the stator [20], optimizing the pole-slot fit [53], and using the stator slant slot and magnetic slot wedge [67] can contribute to the weakening of motor eddy current loss and the improvement of material utilization.

On the contrary, the slotless structure has a stable air gap permeability, but the air gap magnetic density and torque of this structure is weak [68]. To make full use of the advantages of slotless and slotted structures, the virtual stator slot structures, as shown in Figure 7c,d, have gradually emerged in recent years to avoid the slot effect [69].

For the diversity of each stator structure, silicon steel with high silicon content and thickness between 0.2 mm and 0.5 mm is usually chosen as the core material, but the eddy current losses of silicon steel is obvious under high frequency conditions. With the improvement of process level, amorphous alloy material with lower hysteresis loss and thin strip thickness of 0.02$\sim$0.03 mm is gradually becoming the alternative material for traditional silicon steel in the high-speed field. However, the saturation magnetic density of this material is lower, while the more stress-sensitive nature of this material is not conducive to the low-vibration and low-noise operation of the motor. [70,71]. Another alternative material is the SMC, which has an insulating coating for each pure iron powder nanoparticle as illustrated in Figure 8a. Due to the special physical structure, the permeability of SMC is significantly lower than the first two materials, as shown in Figure 8b. Meanwhile, with the presence of the inter-particle insulation layer, the eddy current loss of this material is much lower than that of the silicon steel material in high frequency situations at upper kilohertz, and the material is easier to process to accommodate more complex core shapes [72–74]. Nanocrystalline materials with good magnetic properties at high frequencies are currently in the research and development stage and are expected to be widely used in the future [75].

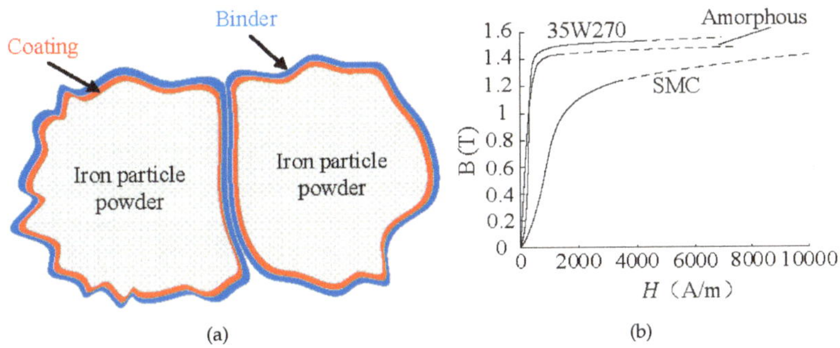

Figure 8. Material properties: (**a**) The basic components of soft magnetic composite [76]; (**b**) Magnetization curves of amorphous alloys, soft magnetic composite and silicon steel sheets [77,78].

At the winding configuration level, the concentrated winding structure is simple and compact with short end windings, which is usually used in some slotless motors or fields requiring compact structure. However, the Back-EMF waveform of this winding structure is not sinusoidal, and the rotor loss is large caused by high harmonics [79]. Compared to centralized windings, double-layer short-distance distribution winding can improve the Back-EMF waveform and magnetomotive force waveforms, thus improving electromagnetic performance, but the end winding of this structure is long [80]. The distributed double-layer toroidal winding is evenly distributed around the stator to produce a sinusoidal Back-EMF waveform. Meanwhile, the toroidal winding minimizes the length of the end winding as shown in Figure 9, which is conducive to the reduction in copper consumption and the improvement of motor stiffness [69]. Therefore, the toroidal winding structure is currently superior to other winding structures in high-speed, high-performance applications.

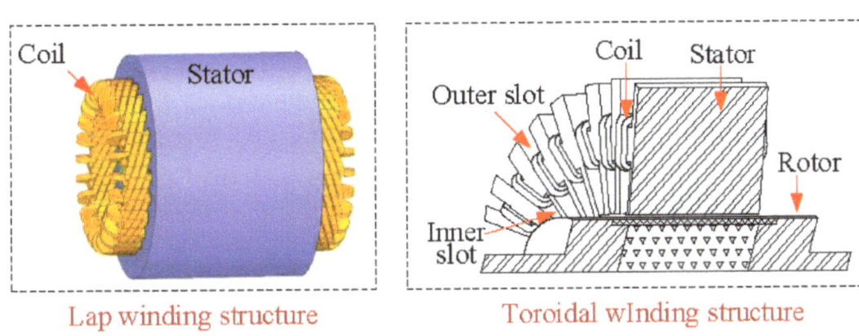

Figure 9. End length comparison between toroidal winding and conventional winding.

The winding losses in high-speed PM motors include basic copper consumption and additional losses due to skin and proximity effects at high frequency currents [68]. Figure 10 shows the winding losses at different frequencies and different wire diameters. Usually, the high-speed PM motors mostly use multi-stranded Litz wire as the stator winding to minimize the impact of skin effect [81]. Yet, the use of round Litzs wire generally requires a special process of dipping or lacquering, which increases the complexity of the process [82]. Another type of rectangular cross-section wire has a simple process, along with a better fill factor and lower Joule losses, but its eddy current loss is higher [83].

Copper is the most commonly used material for most windings. In pursuit of low loss and light weight of the windings, Volpe et al. [84] propose the use of aluminum instead of copper, but it was found in the experiments that the AC winding loss of the aluminum winding could not be reduced effectively. Currently, a team from Lappeenranta University of Technology has designed a carbon nanotube braided yarn based on nanotechnology. This material is extremely lightweight, has twice the electrical conductivity of copper-wound wire, and is thermally stable. It also has half the heat loss of copper winding , which is expected to replace copper as winding wire in the future [76]. In addition, the composite technology of copper-CNT developed by Oak Ridge Laboratory provides more options for winding materials.

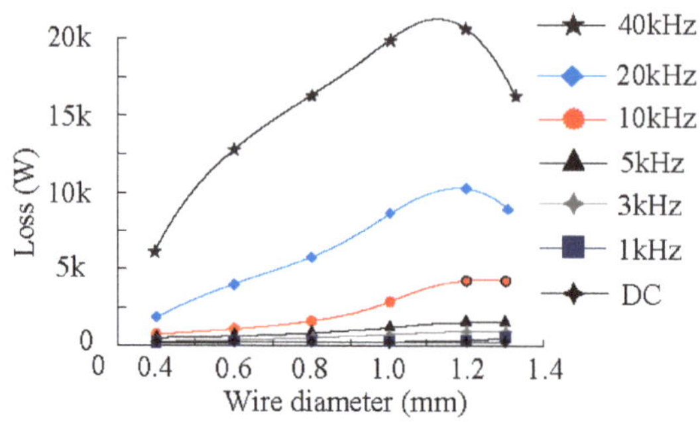

Figure 10. Winding losses at different frequencies and different wire diameters.

2.2. Rotor Design

Eddy current loss and wind friction loss in high-speed PM motors are prominent. The rotor eddy current loss is mainly caused by the time and space harmonics of the armature magneto-dynamic potential and the stator slotting effect [85]. For eddy current loss, there are usually two ways to calculate. The analytical method is suitable for eddy current loss calculations of simple structured rotors. The FEM, on the other hand, can take into account many nonlinearities at the same time and obtain more accurate calculation results, but the calculation speed is slower [86].

The protection of permanent magnets must also be taken into account at high speed. PM has the characteristic of pressure resistance but not tensile resistance, so the rotor of the high-speed PM motor often adopts the SPM structure with sleeve or the IPM structure.

For the SPM structure, fiber or alloy materials are mostly used to protect the magnets. The alloy material has better heat dissipation and stronger stiffness, but the harmonics of armature magneto dynamic potential will generate large eddy current loss in the sleeve. On the contrary, although the fiber composite material is less effective in heat dissipation, it does not produce eddy current loss itself, so it is more likely to fulfill the low loss requirement of the rotor [87]. Both of these materials have non-permeable properties, which can be detrimental to the increase in magnetic load. Yon et al. [88], at the University of Bristol, designed a novel PM rotor sealing technique using semi-conducting materials as a protective sleeve to increase the motor magnetic load and counter potential fundamental wave amplitude by 20%.

In fact, the eddy current losses in SPM rotors are mainly concentrated on the surface of the PMs. The reduction in rotor surface loss can be achieved by adding shielding [89], adjusting the shield size [80,90] or by taking measures such as magnet segmentation, sleeve segmentation [67,87] and sleeve slotting [43]. When the permanent magnets are segmented, the air gap magnetic field is difficult to meet the design requirements of the sine wave. In response, reference [91] proposes a structure in which samarium cobalt and ferrite are placed in blocks and mixed along the circumference, as shown in Figure 11a, and indicates that this structure can improve the sinusoidal distribution of the air gap magnetic density waveform.

However, for high-power and high-speed PM motors, the segmentation of the magnet and sleeve will reduce the rotor stiffness to a certain extent, which is not conducive to high-speed operation. Additionally, the shield will no longer have a significant effect on eddy current suppression in high power and high speed motors. Reference [41] combined the advantages of low conductivity of ferrite and high remanence of rare-earth PMs, and proposed a composite excitation structure, as shown in Figure 11b. SmCo is used to improve the magneto-dynamic potential, while ferrite is mainly used to suppress rotor eddy currents instead of shielding. The eddy currents on the surface of this composite excitation structure are significantly reduced compared with that of a single-excited rotor, as shown in Figure 12.

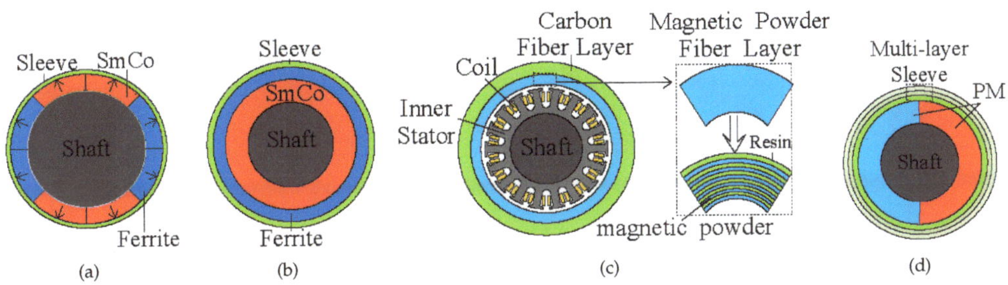

Figure 11. Rotor structures of high-speed PM motor with sleeves: (**a**) Circumferential composite excitation structure; (**b**) Radial composite excitation structure; (**c**) Magnetic powder fiber composite rotor structure; (**d**) Multi-layer sleeve mating structure.

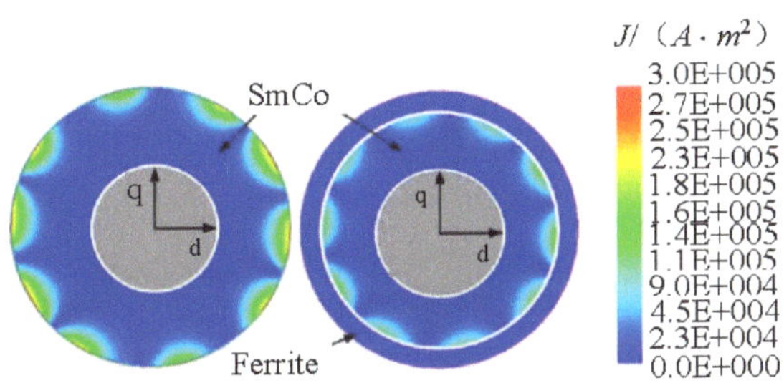

Figure 12. Comparison of eddy current distribution between single excitation and combined excitation [41].

To further reduce the rotor eddy current loss, Jingyue et al. [92] mixed NdFeB magnetic powder with epoxy resin and cured magnetization with reinforcing fiber based on the outer rotor structure, as shown in Figure 11c. The sleeves' layered coordination structure as shown in Figure 11d has also been further researched and applied in recent years. Its function is to reduce the thickness of the original sleeve by the interference fit of the multi-layer thin sleeve, which increases the physical air gap and greatly reduces the eddy current loss of the sleeve without losing rotor strength [93].

Compared with the SPM strucure, the poles of the IPM strucure are less influenced by the armature magnetic field, and its overload capacity and the capacity of PM resistance to demagnetization are better. However, the problem of stress concentration in isolated magnetic bridges has to some extent made them less used in high-speed applications [7,12]. The PM of the IPM structure is mostly segmented and circumferentially layered to improve rotor strength and suppress eddy currents inside the permanent magnet, as shown in Figure 13a. The interior tangential rotor structure shown in Figure 13b is also usable in high-speed applications, which is able to reduce the leakage of the IPM structure while providing a larger reluctance torque and reducing the number of magnets [94]. Figure 13c shows a non-uniform air-gap IPM (NUA-IPM) structure. The reverse "V-shaped" magnet improves the pole arc coefficient of the magnet, and by adjusting the non-uniform air gap, the harmonic amplitude in the air gap magnetic density waveform can be reduced, but the air friction loss of this structure is high [30]. Markus et al. of Karlsruhe Institute of Technology [95] also proposed a special hollow shaft structure made of amorphous alloy, as shown in Figure 13d. This special hollow shaft structure has obvious advantages of low core loss and good weak magnetic capability, but the eddy current loss of the hollow shaft is high and the slotting process is complex.

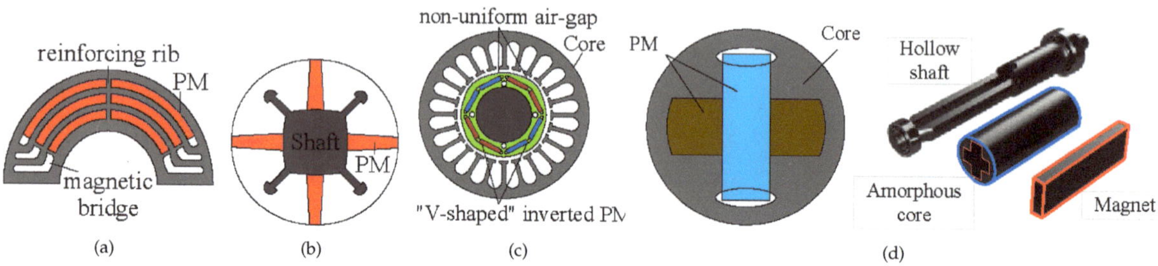

Figure 13. Interior permanent magnet structures of high-speed PM motor: (**a**) Interior permanent magnet segmented structure; (**b**) Interior permanent magnet tangential structure; (**c**) Non-uniform air-gap interior permanent magnet structure; (**d**) Hollow shaft permanent magnet special structure.

The rotor wind friction loss of the high-speed PM motor mainly includes air gap loss p_{air}, rotor end and thrust collar loss p_{end}, axial cooling air loss p_{axial}, which can be expressed as [70,96]:

$$\begin{cases} p_{\text{air}} = C_f \pi \rho_a \omega^3 r^4 L_a \\ p_{\text{end}} = 0.5 C_f \rho_a \omega^3 (r_2^5 - r_1^5) \\ p_{\text{axial}} = 0.67 \pi \rho \left((r+\delta)^3 - r^3 \right) v_m u_m \omega \end{cases} \tag{3}$$

where C_f is the friction coefficient, which is related to the rotor surface roughness; ρ_a is the air density; ω is the rotor angular velocity; r is the rotor radius; L_a is the rotor axial length; r_1 and r_2 are the inner and outer radii of the rotor end and thrust ring; v_m is the axial cooling air velocity and u_m is the tangential circumferential velocity of the cooling gas at the air gap exit; δ is the length of the air gap.

Referring to (3), it can be seen that the wind friction loss is mainly affected by rotor speed, thrust disc size, air gap thickness, wind speed and contact surface roughness. Among them, Figure 14 shows the common values of the air gap length for different linear speeds and different rotor outer diameters (v_c). From Figure 14, we can see that the rotor outer diameter of high-speed PM motor is mostly taken between 40~170 mm, and the rotor outer diameter of individual ultra-high-speed motor can be taken below 40 mm. The choice of air gap length is mostly concentrated in the range of 1~4 mm.

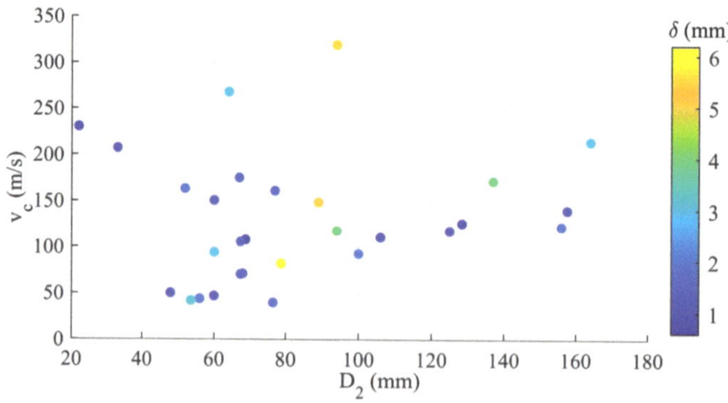

Figure 14. Common values of air gap length for different rotor outside diameters and different linear speeds.

3. Thermal Management

The goal of thermal management is to accurately solve the temperature rise distribution of the motor and design an efficient heat dissipation system to ensure the efficient and continuous operation of the motor.

Currently, the commonly used calculation methods of motor temperature rise include LPTN, FEM and CFD. LPTN is to solve the temperature distribution by constructing the thermal network of the motor. The calculation speed of this method is fast, but the accuracy is low [13,97]. The FEM can directly profile the motor entity in 2D or 3D modeling and load each loss density and heat transfer condition obtained by empirical methods to obtain a more detailed and intuitive temperature rise distribution than LPTN [98]. The CFD obtains an accurate distribution of temperature rise by jointly modeling the heat transfer entity and the cooling fluid of the motor, which has greater advantages for analyzing and optimizing the heat dissipation structure of the motor, but at the same time, CFD also requires very high computer performance [99]. The comprehensive use of the above schemes to solve the temperature is also widely used in practice.

Efficient heat dissipation is the core of thermal management. The temperature rise of high-speed PM motors mainly comes from the core and windings of the stator and inside the rotor, as shown in Figure 15. In this regard, the cooling system of high-speed motors mostly uses a combination of air-cooled and liquid-cooled structures. As shown in Figure 16a, the air-cooled system is mainly used to achieve heat dissipation between the stator and rotor through air flow, while the liquid-cooled system is mostly used to absorb heat within the stator core section or to exchange heat with the ventilation system to achieve effective heat dissipation [100]. Its inner air channel can also use a closed oil cooling system to replace air to achieve a better cooling effect, as shown in Figure 16b. However, isolation devices are required between the cooling channels and the rotor when oil is passed, which makes the system more complicated, and it is prone to contamination problems caused by oil leakage over a long period of operation [101].

In the ventilation cooling part, the traditional double-end inlet air cooling can achieve better end winding heat dissipation, but the heat dissipation along the stator and rotor axial air gap is poor. Considering that heat dissipation between the stator and rotor is mainly through heat exchange of convective air, a ventilation structure with only one air inlet and one air outlet is applied to force the cooling air to flow along a narrow air gap, and thus the heat dissipation capacity of the air gap is improved [16,60]. Fengge et al. [102] of Shenyang University of Technology designed a hybrid radial and axial cooling structure shown in Figure 17a, which achieves better effect than axial air cooling with double ventilation capacity at the same air speed. Baojun et al. [56] of Harbin Institute of Technology adopted a "hammer slot type" structure based on the semi-closed slot structure, as shown in Figure 17b, which further improves the axial heat dissipation effect of the motor by increasing the axial heat dissipation area with a larger inner slot design. In addition, measures such as arranging cooling fins, creating shallow slots in the rotor sheath and adding appropriate amount of wind spurs on the rotor end face to improve the air flow in the end cavity have also been used to improve the rotor cooling capacity [103,104].

The housing water cooling system has no wind-motion losses, and its structure generally has "Z" axial cooling and circumferential spiral cooling water channel structure. The spiral water-cooling structure has a relatively uniform overall water velocity distribution and is more suitable as a water-cooling structure for high-speed PM motors [105]. Reference [100] pointed out that the cooling efficiency can be improved to some extent by increasing the number of water channels and cooling water flow rate, but this measure also increases the pressure loss in the water channels. In order to circumvent the pressure loss in water pipes and to solve the problem that the water-cooling system is prone to scale, the spray cooling and evaporative cooling methods based on phase change cooling technology have been further applied in recent years [100,106,107]. In addition, the motor cooling solution with additional heat path enhancement also provides ideas for efficient heat dissipation in high-speed PM motors, as shown in Figure 18a,b.

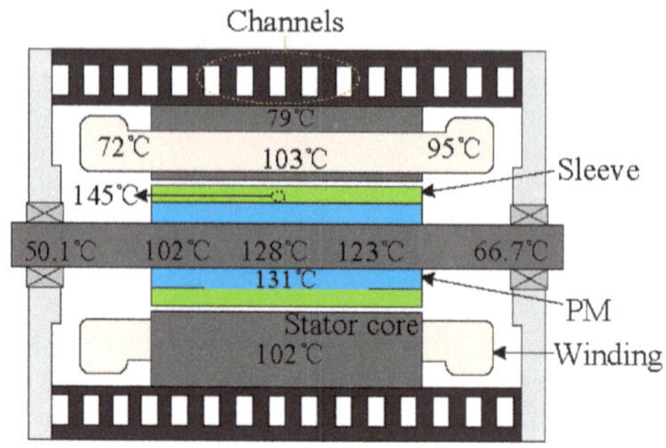

Figure 15. Example of temperature distribution of 100 kW 100 kr/min high-speed PM motor.

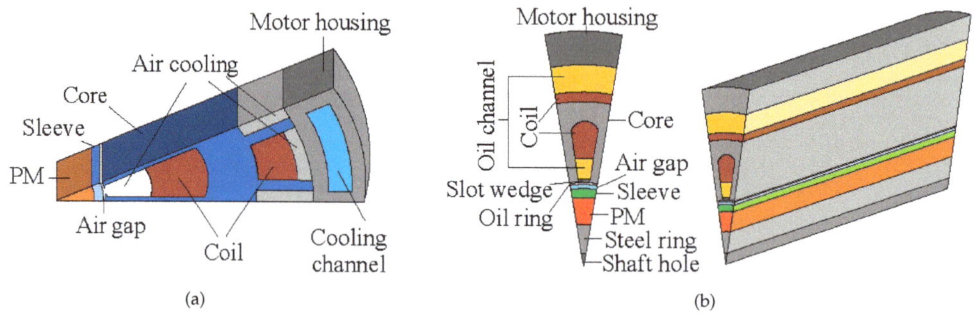

Figure 16. Structure of stator opening internal and external cooling channels: (**a**) Hybrid cooling model for toroidal windings; (**b**) Oil cooling model.

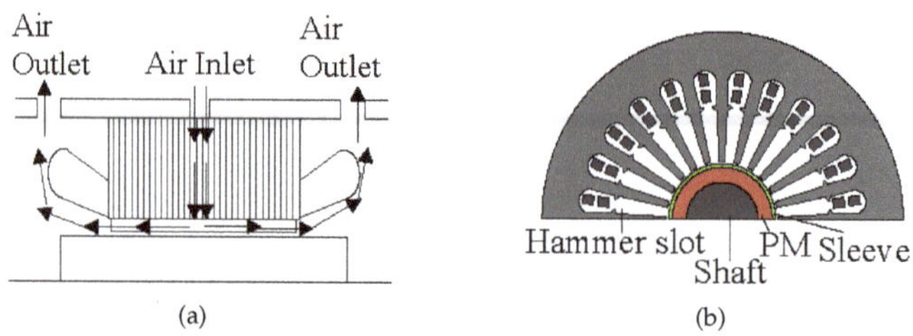

Figure 17. Cooling structure: (**a**) The structure of mixed ventilation [102]; (**b**) The structure of "hammer slot type" [56].

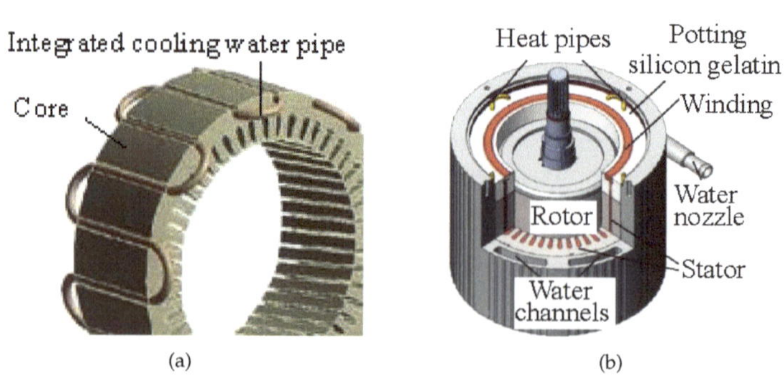

Figure 18. Motor cooling solution with additional heat path enhancement: (**a**) Stator integrated water pipe [108]; (**b**) Heat pipe and thermal adhesive to enhance heat dissipation [109].

4. Rotor Strength and Dynamics

The rotors of the high-speed PM motor are usually designed as slender structures to reduce the centrifugal force on the rotor surface, but the slender structures pose additional challenges to the rotor stiffness and critical speed.

4.1. Rotor Strength Analysis

The rotor circumferential speed (v_c) of the high-speed PM motor can reach 200 m/s and above, at that time, the high-speed centrifugal force and thermal stress can seriously affect the rotor safety performance. When the rotor is unable to withstand the stresses, it will fail in strength, which will seriously cause the "chamber sweeping" of the stator and rotor and damage the motor [110], as shown in Figure 19.

(a) (b)

Figure 19. Destruction of brittle magnets underneath the broken bandage [110].

To protect the rotor, the SPM structure needs to provide a certain amount of preload through the protective material to counteract the tensile stresses generated by the high speed centrifugal force. The pre-stress of the sleeve should be sufficient to counteract the centrifugal force generated by high-speed operation, but should not exceed the compressive degree of PM. The value of pre-stress depends on the temperature, speed, sleeve thickness and the value of interference, where the value of interference is usually less than 0.2 mm [111,112]. Usually, the pre-stress is formed by the interference fit of the alloy sleeve to the PM or by the winding of the fiber composite, as shown in Figure 20a,b. Table 2 shows several rotor bandage materials for high-speed PM motors [94,113–115]. As seen in Table 2, the composite materials have very low density and high strength and resistivity. Alloy materials have significant advantages in terms of heat dissipation capability, temperature stability and stiffness.

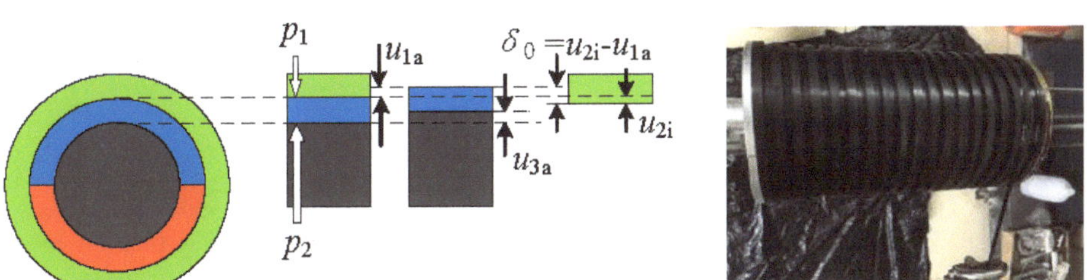

(a) (b)

Figure 20. Two common permanent magnet protection measures: (**a**) Three-layer interference fit. The meanings of the characters in the figure are detailed in [116]; (**b**) Winding process of carbon fiber composites [114].

Table 2. Mechanical properties of several rotor bandage materials.

Parameters	KEVLAR	Carbon Fiber	Fiberglass	Inconel718
$\rho/(\text{g/cm}^3)$	1.14	1.76	2.54	8.2
$\sigma_{t,\max}/(\text{MPa})$	2920	3750	3447	1030
$k/(\text{W}/(m \times K))$	0.04	5.0	1.0	11.4
$\rho_t/(\Omega \cdot \text{m})$	Insulation	1.5×10^{-5}	4×10^{12}	1.25×10^{-7}

During high-speed operation of the motor, the rotor stresses, strains and displacements will change, and the accurate calculation of each change is critical to the durability of the PM and sleeve. According to the theory related to elasticity mechanics, the equilibrium differential equation of the internal micro-element body of the high-speed rotor subjected to centrifugal force is [117]:

$$\frac{d\sigma_r}{dr} + \frac{\sigma_r - \sigma_\theta}{r} + \rho\omega^2 r = 0 \tag{4}$$

where σ_θ and σ_r are the rotor radial and tangential forces, r is the rotor radius, ω is the rotor angular velocity, and ρ is the material density.

Refer to (4), the rotor strength analysis of cylindrical PM and ring PM structures with alloy sleeves for different fit models is derived in detail in [116,117]. In engineering practice, metal sleeves are considered to be all isotropic materials, and carbon fibers are considered to be anisotropic materials, hence the strength analysis of the two is significantly different [118]. The detailed derivation of the rotor stress analysis for the carbon fiber tied structure is shown in [114]. When magnets are divided into pieces and produce pole gaps, the gaps need to be filled with non-magnetic material of densities and thermal expansion coefficients closer to those of the magnets to avoid stress concentrations in the sleeves at the edges of the magnetic poles. For this type of structure, C. Liangliang et al. of Zhejiang University [119] derived the strength analytical formula that can be applied under the protection of non-permeable metal sleeve and under the protection of carbon fiber sleeve, respectively. In addition, in order to seek a more accurate theoretical analysis, a generalized plane strain theory based on plane strain was proposed in [120] and verified by FEM, and the results showed that the theory is more suitable for rotors with long shafts.

The equivalent force of the rotor under normal temperature is mainly caused by centrifugal force, and when temperature rise is considered, the rotor will be subjected to uneven thermal stress. The stress distribution of the sleeve for both cases is shown in Figure 21. Among them, Figure 21b take the characteristic of each material having a temperature gradient at high speed into account, which makes the stress distribution closer to the actual situation. To improve the rotor strength, literature [121] indicates that an appropriate increase in the static interference between the PM and the sleeve or increase in the sleeve thickness can improve the preload pressure on the PM and facilitate rotor protection. In addition, measures combining different sleeve materials or mixing magnetic powder with carbon fiber have also been used to improve the bending resistance of the sleeve [92,93].

The IPM structure can simplify the sleeve assembly link, and the magnets are not easily damaged by the direct protection of the rotor core, but the isolated magnetic bridge of the rotor in the IPM structure is subjected to the greatest stress under high-speed operation and is easily damaged, as shown in Figure 22. A moderate increase in the number and thickness of the isolated magnetic bridge is beneficial to improving the mechanical strength of the rotor. For the strength analysis of the IPM structure, due to the complexity of their structure, it is generally necessary to equate the magnets and the protective bridges into equal mass rings and then combine them with finite element modeling for analysis.

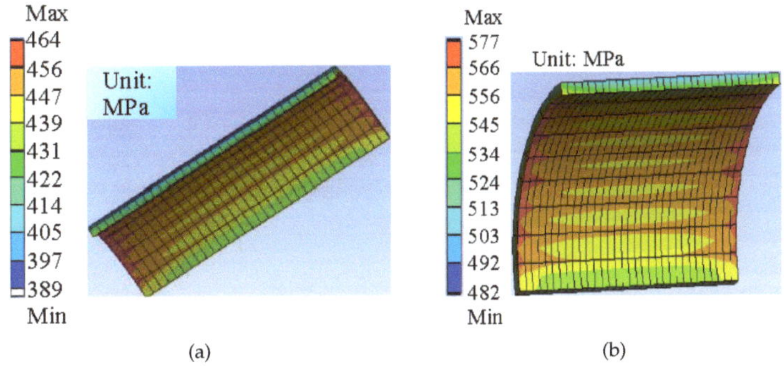

Figure 21. The equivalent stress distribution of sleeve at different temperature conditions [122]: (**a**) 20 °C; (**b**) 150 °C.

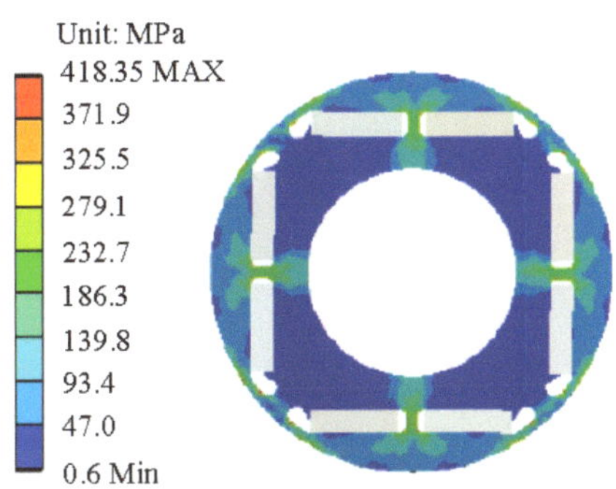

Figure 22. Equivalent force distribution of IPM [122].

4.2. Rotor Dynamics Analysis

In addition to fulfilling the requirements of rotor strength, stable dynamic characteristics are also pursued for high-speed PM motors, so it is particularly important to perform dynamics analysis. The study of dynamics analysis mainly includes stability analysis, critical speed calculation and unbalanced response.

Stability analysis is closely related to the rotor bearing system, and the main bearings used in high-speed PM motors are ball bearings, oil film bearings, air bearings, and magnetic levitation bearings [82]. Figure 23 shows the distribution of the various bearing schemes reported in [6,7,13,40,50,53,82,93,123–127] relative to the power and speed of the high-speed PM motor.

Table 3 summarizes the application characteristics of several types of bearings. Among them, ball bearings dominate the early stages of high-speed motor bearing supply due to their low cost, high stiffness and ease of assembly. Ball bearings are characterized by the DN number (the product of bearing inner diameter D and rotational speed N) as an indication of high-speed operating capability. By optimizing the topology and improving the lubricating material, etc., Nippon Seiko Co. has developed a high-speed ball bearing with a DN number of more than 1.8 million in 2021. Oil-filled bearings, which support the rotor by the pressure of the oil film formed by the lubricating oil, can achieve lower loss operation than ball bearings, but the additional equipment is complex and the oil leakage problem is prominent [102]. For the demand of high-speed, low loss and high operating life of motors, air bearings and magnetic levitation bearings have gradually become popular research objects. The support stiffness of these non-contact bearings is generally smaller than that of contact bearings, and because of their larger clearances, non-contact bearings have become the main object of stability analysis. The specific study of rotor stability analysis has been discussed in detail in [40,127,128].

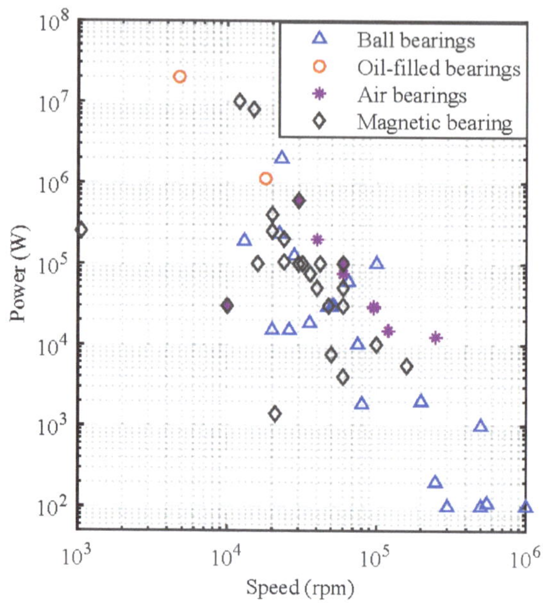

Figure 23. Bearing distribution of high-speed PM motor.

Table 3. Application characteristics of several types of bearings.

Bearing Type	Bearing Stiffness (N/m)	Advantages	Disadvantages
Ball bearing	$10^6 \sim 10^9$	High robustness, small size, low cost and high stiffness.	High bearing loss and short application life at high speed.
Oil-filled bearing	$10^7 \sim 10^9$	Friction coefficient is lower than ball bearings, and it has high impact resistance.	The cooling system is complex and has oil leakage problems.
Air bearing	$10^5 \sim 10^8$	High damping, low friction loss, long service life, compact system, good adaptability.	The load capacity is limited, the dynamic stability is poor, and the performance and processing accuracy of the bearing material are extremely demanding.
Magnetic bearing	$10^5 \sim 10^7$	Great load bearing capacity, no friction loss, good stability.	Complex control system and high cost.

Due to the presence of an imbalance in the rotor system, when the rotor is running at critical speed, it will excite the natural frequency of the order and cause resonance. To avoid this phenomenon, the motor speed should be kept away from the critical speed. For rigid rotors, the gyroscopic effect on the critical rotor speed can be ignored, and usually, its operating speed is required to run below 80% of the first-order bending critical speed to ensure the safe operation of the rotor. For flexible rotors, the operating speed should be controlled between 1.3 times the first-order critical speed and 0.7 times the second-order critical speed [129]. Typically, non-contact bearing rotors are mostly flexible and more likely to produce dynamic imbalance at high-speed, which in turn generates vibration and affects the stability of the rotor [130].

To facilitate the theoretical resolution of the critical speed, the differential equation for the rotor dynamics can be expressed as [126]:

$$M\ddot{u} + (\Omega G + C)\dot{u} + Ku = F \quad (5)$$

where M is the mass matrix of the rotor system; K is the stiffness matrix; C is the damping matrix; G is the gyroscopic matrix; F is the excitation force acting on the rotor system; u is the nodal displacement response vector; $\dot{u}$ is the nodal velocity response vector; $\ddot{u}$ is the nodal acceleration response vector; and Ω is the resonance speed.

Equation (5) states that support stiffness, damping, acceleration and support location all affect the vibration of the system. Among them, the increase in the support stiffness can raise the critical rotor speed, but this method is only suitable for adjustment in a specific range, and beyond this range it will no longer work [120]. In addition, the rotor critical speed also has an inverse relationship with the shaft length and rotor diameter, which also provides a design reference to increase the rotor critical speed and prevent the rotor from developing into a flexible rotor.

The unbalanced response of high-speed PM motors is mainly caused by the unbalanced magnetic pull effect due to the asymmetry of the rotor magnetic circuit. Domestic universities such as Southeast University [131], and Harbin Institute of Technology [132,133] have analyzed the vibration caused by unbalanced rotor response in terms of rotor eccentricity, different magnet thickness and magnetization effects, slot-pole combinations, and air pressure fluctuations to provide a theoretical basis for balancing techniques.

5. Multi-Physics Domain Coupling Design

There is a complex coupling relationship between the physical fields of the high-speed PM motor. The interactions between the parameters related to different physical domains are shown in Figure 24. Therefore, the design method of studying a certain part alone can no longer meet the requirements for efficient design of high-speed motors, and the coupled design considering multiple physical domains is more suitable for the design of high-speed PM motors [98].

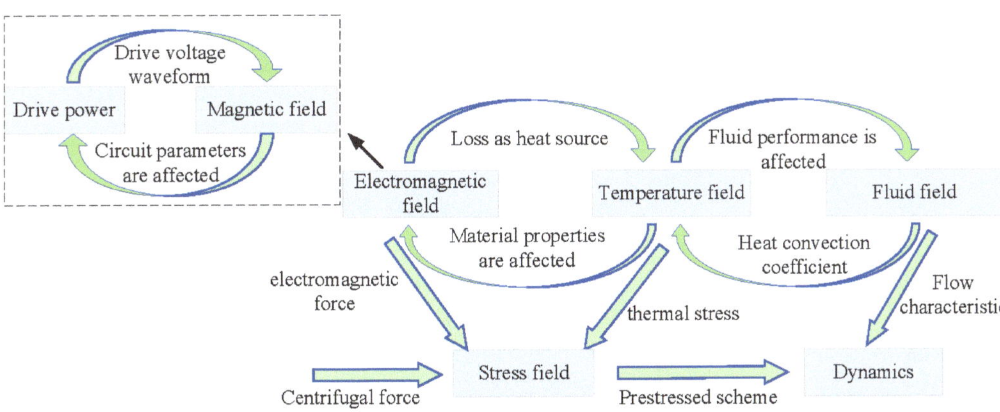

Figure 24. Mutual coupling relationship between multiple physical domains.

Coupled design combined with FEM is a common approach in the design of high-speed PM motors today. The study of field-circuit coupling calculation in [134] shows that the field-circuit coupling in the design of high-speed PM motors can ensure that the maximum error of the electromagnetic loss calculations is within 2%. The loss will directly affect the temperature rise. Magnetic-thermal coupling is necessary for the accurate distribution of temperature rise, which mainly including unidirectional coupling and bidirectional coupling. Compared with unidirectional coupling, bidirectional coupling can further take into account the change of relevant electromagnetic parameters with temperature rise and obtain more accurate results, but the calculation of bidirectional coupling is time-consuming [13,25,89,103]. In addition, the coupling between the mechanical stress and the temperature field of the motor under high speed conditions is also of concern, which is discussed in detail and in depth in [122,135].

6. Development Trend of High-Speed PM Motor Design Technology

With the increase in demand for high-speed PM motor applications, the efficient design of high-speed PM motors has put forward higher and higher requirements. Based on the analysis of the key technologies of high-speed PM motor design, the research direction of high-speed PM motor design technology is summarized. The main aspects are as follows.

1. Achieving low loss, high efficiency, high strength, long life, low cost and stable operation of motors is the goal of high-speed PM motor design. Motor losses and rotor strength are related to material properties and structure selection. Therefore, the development of new core materials and innovative structures will still be worth focusing on in the future;
2. PM material is the most important material for high speed permanent magnet motor. The tensile strength and temperature resistance level of permanent magnet materials are the two major factors that limit the speed and power increase in high-speed permanent magnet motors. Improving these two properties will be a long-term work in the development of permanent magnet materials.
3. Currently, high-speed PM motors are mostly cooled by a mixture of air-cooled and water-cooled cooling methods, which have a more complex structure and limited cooling effect. With the goal of improving the reliable and efficient operation of motors, innovative motor cooling solutions are also needed in the design phase;
4. For rotor supports. On the one hand, with the progress of material science and lubrication technology, the engineering problems such as the life of ball bearings in various high-speed occasions need to be further tested and summarized; on the other hand, in domestic, the application and performance evaluation of air bearings and magnetic levitation bearings in practice is still relatively few.
5. The multi-domain coupling design of high-speed PM motors is extremely important. In terms of design methods, scholars now commonly use FEM or CFD for multi-physics domain coupling design, which can obtain relatively accurate loss, temperature rise and stress distribution, but the calculation process is exceptionally time-consuming. The method of combining FEM and analysis method for integrated and rapid coupling solution of electromagnetic, electrical, mechanical and thermal will be a better choice for the design and analysis of high-speed permanent magnet motors in the future.

7. Conclusions

With the application and promotion of permanent magnet motors in the high-speed field, the demand for the efficient design of high-speed PM motors has been pushed to a new height. The design process of high-speed PM motors needs to face problems of high losses, high temperature rise and more complex mechanical characteristics, which must be achieved by closely considering the interaction of the electromagnetic, temperature and mechanical fields. This article briefly describes the design strategies of high-speed PM motors, and the research and application of new materials and structures in high-speed PM motors. The selection basis of the major parameters of high-speed PM motors is outlined, and the characteristics and applications of different structures and materials are summarized. The calculation method of temperature rise and the efficient management measures of temperature at the present stage are outlined. Common protection methods for permanent magnets and strength analysis methods for different rotor structures are discussed. The development and applications of different bearings are overviewed, and solutions for reducing the resonance and unbalance response of the rotor are discussed. The mechanism of the interaction between the relevant parameters in each physical domain is summarized. Finally, the future direction of the efficient design of high-speed PM motors is foreseen. Permanent magnet motors have great potential for development in the high-speed field, so the material development, structural innovation and design strategies of high-speed PM motors require further in-depth research.

Author Contributions: Conceptualization, Q.S., Z.Z., S.L., X.L., T.W., X.H. and J.Z.; methodology, Q.S., Z.Z. and S.L.; software, Q.S. and Z.Z.; validation, Q.S., Z.Z., S.L., X.L., T.W., X.H. and J.Z.; formal analysis, Q.S., Z.Z. and S.L.; writing—original draft preparation, Q.S., Z.Z. and S.L.; writing—review and editing, Q.S., Z.Z. and S.L.; supervision, Q.S.; project administration, Q.S.; funding acquisition, Q.S. All authors have read and agreed to the published version of the manuscript.

Funding: This research was funded by National Natural Science Foundation of China, grant number 51607017 and Scientific Research Foundation of Chongqing University of Technology, grant number 2019ZD78.

Institutional Review Board Statement: Not applicable.

Informed Consent Statement: Not applicable.

Data Availability Statement: Not applicable.

Conflicts of Interest: The authors declare no conflict of interest.

Abbreviations

The following abbreviations are used in this manuscript:

DC	Direct Current
AC	Alternating Current
SPM	Surface-mounted Permanent Magnet
IPM	Interior Permanent Magnet
IM	Induction Motor
PM	Permanent Magnet
PMSM	Permanent Magnet Synchronous Motor
SRM	Switched Reluctance Motor
SMC	Soft Magnetic Composite
Back-EMF	Back Electromotive Force
LPTN	Lumped-Parameter Thermal-Network
FEM	finite element method
CFD	Computational Fluid Dynamics
Litz	Litzendraht

Appendix A

Table A1. Data resources for different dimensional parameters of high-speed permanent magnet motors with different structures.

Rated Power (kW)	Rated Speed (krpm)	Efficiency	Poles	Stator Slots	Rotor Structure	D_{i1}/D_1 (mm)	L_{ef} (mm)	Iron Loss (W)	Copper Loss (W)	Air Gap Height (mm)	Reference
300	15	-	2	36	SPM	160/350	280	2225	1137	2	[8]
132	18.75	0.98	4	24	IPM	130/220	130	-	-	0.8	[11]
120	24	0.95	2	24	IPM	102/200	112	-	-	4	[12]
75	60	0.98	2	24	SPM	56/126	128	230	294	2	[13]
200	20	-	2	24	SPM	135/295	-	2500	1200	-	[16]
100	100	-	2	24	SPM	79/150	165	-	-	1	[19]
15	30	-	2	18	SPM	60/130	70	-	-	-	[20]
100	50	-	2	36	SPM	70/200	160	814	458.2	1.5	[22]
150	30	-	2	24	SPM	125/250	-	1291.8	-	-	[14]
150	20	-	4	36	SPM	158/336	120	1684	165.5	-	[25]
250	67	-	2	24	SPM	-/368	155	11990	-	5.5	[26]
220	18	0.97	2	24	SPM	127/205	308	434.4	121.6	1	[27]
200	20	-	2	24	SPM	135/295	200	2500	1200	-	[60]
120	17.75	-	4	24	SPM	104/195	130	963	280	2	[15]
100	40	-	4	36	SPM	80/180	138	-	-	-	[18]
10.5	100	-	2	24	cylindrical	38/104	33	-	-	-	[21]
35	135	-	2	12	cylindrical	43.4/138	62	-	-	-	[24]
100	32	-	4	24	SPM	99/230	-	-	-	5	[28]
45	48	-	4	24	IPM	62.4/120	100	642.7	200.5	1.2	[30]
220	18	-	4	24	SPM	75/120	-	-	-	-	[33]
20	30	-	2	36	SPM	66/130	90	-	-	3	[34]
10	100	-	2	24	SPM	40/104	36	-	-	-	[35]
15	120	-	2	12	cylindrical	35/120	50	124	520	1	[36]

Table A1. *Cont.*

Rated Power (kW)	Rated Speed (krpm)	Efficiency	Poles	Stator Slots	Rotor Structure	D_{i1}/D_1 (mm)	L_{ef} (mm)	Iron Loss (W)	Copper Loss (W)	Air Gap Height (mm)	Reference
800	25	-	4	24	SPM	170/360	-	-	-	3	[37]
6.2	15	0.96	2	24	SPM	62/120	110	70	47.6	1	[39]
80	80	-	2	24	SPM	70/182	100	617	325	3	[41]
100	60	-	2	12	SPM	100/124	102	-	-	-	[42]
100	20	-	2	18	SPM	108/250	104	700.3	668	1	[43]
150	17	-	4	36	SPM	160/350	140	1135.3	432.6	1.2	[90]
7.5	50	-	2	18	SPM	55/116	50	-	-	-	[46]
20	100	-	2	12	cylindrical	34.8/-	-	-	-	-	[48]
75	24	0.98	2	24	SPM	145/240	100	660	545	4	[49]
40	40	-	2	24	SPM	-/-	-	-	-	-	[52]
7.5	15	0.94	2	18	SPM	60/130	70	156.6	64.2	3.2	[97]
15	30	-	4	18	SPM	70/130	110	274	102	1.3	[89]
10	24	-	4	24	SPM	60/115	70	-	-	-	[77]
10	10	-	4	12	SPM	80/160	78	133.4	87.3	1.75	[58]
1120	18	-	4	27	SPM	-/550	400	-	-	3	[102]
5	20	0.948	4	18	IPM	70/130	110	215.7	158.8	1.3	[136]
7.5	15	0.95	4	18	SPM	60/-	70	-	-	2	[137]
15	20	0.96	4	18	IPM	70/-	110	-	-	1	[137]
5	20	0.94	4	24	SPM	50/100	55	32	-	1	[138]
100	50	-	2	36	SPM	-/200	160	-	-	1.5	[111]
30	20	-	6	36	IPM	91.2/154.2	135.9	-	-	6.2	[139]
2	200	-	2	Slotless	cylindrical	23.6/35	13	9.85	20.3	0.8	[140]
40	40	-	4	36	SPM	80/135	125	-	-	1.5	[141]
5	30	-	4	24	IPM	70/150	59	-	-	0.6	[142]
100	32	-	4	24	SPM	-/245	-	-	-	-	[143]
250	65	-	2	36	SPM	105/240	140	5009	1025.7	5.5	[144]

References

1. Mallin, L.R.; Barrans, S.M. A Review of the High-speed Permanent Magnet Rotor Stress Analysis used for Automotive Air-handling Machines. *Eur. J. Eng. Res. Sci.* **2020**, *5*, 448–456. [CrossRef]
2. Du, G.; Huang, N.; Zhao, Y.; Lei, G.; Zhu, J. Comprehensive sensitivity analysis and multiphysics optimization of the rotor for a high speed permanent magnet machine. *IEEE Trans. Energy Convers.* **2021**, *36*, 358–367. [CrossRef]
3. Tenconi, A.; Vaschetto, S.; Vigliani, A. Electrical Machines for High-Speed Applications: Design Considerations and Tradeoffs. *IEEE Trans. Ind. Electron.* **2013**, *61*, 3022–3029. [CrossRef]
4. Gerada, D.; Mebarki, A.; Brown, N.L.; Gerada, C.; Cavagnino, A.; Boglietti, A. High-speed electrical machines: Technologies, trends, and developments. *IEEE Trans. Ind. Electron.* **2014**, *61*, 2946–2959. [CrossRef]
5. Moghaddam, R.R. High speed operation of electrical machines, a review on technology, benefits and challenges. In Proceedings of the 2014 IEEE Energy Conversion Congress and Exposition (ECCE), Pittsburgh, PA, USA, 14–18 September 2014; pp. 5539–5546.
6. Pfister, P.D.; Perriard, Y. Very-high-speed slotless permanent-magnet motors: Analytical modeling, optimization, design, and torque measurement methods. *IEEE Trans. Ind. Electron.* **2010**, *57*, 296–303. [CrossRef]
7. Zhang, F.; Du, G.; Wang, T.; Liu, G. Development and design of high speed motor. *Trans. Chin. Electrotech. Soc.* **2016**, *31*, 5–22.
8. Zhao, G. Design and Analysis of High Speed Permanent Magnet Synchronous Motor. Ph.D. Thesis, Shenyang University of Technology, Shenyang, China, 2018.
9. Noguchi, T.; Takata, Y.; Yamashita, Y.; Komatsu, Y.; Ibaraki, S. 220,000-r/min, 2-kw pm motor drive for turbocharger. *IEEJ Trans. Ind. Appl.* **2005**, *125*, 854–861. [CrossRef]
10. Gao, Q.; Wang, X.; Ding, Q.; Shao, Y. Strength analysis and structure design of ultra high speed micro permanent magnet motor rotor. *Proc. CSEE* **2021**, *41*, 11.
11. Zhang, W. Design and Application of High Speed Permanent Magnet Synchronous Motor. Ph.D. Thesis, Huazhong University of Science and Technology, Wuhan, China, 2017.
12. Dou, Q. Electromagnetic Design and Rotor Structure Research of High-Speed Permanent Magnet Motor. Ph.D. Thesis, Harbin University of Science and Technology, Harbin, China, 2019.
13. Huang, X. Multi-Objective Optimization of High Speed Permanent Magnet Synchronous Motor Based on Multi-Physical Field. Ph.D. Thesis, Harbin Institute of Technology, Harbin, China, 2019.
14. Liu, G.; Liu, M.; Zhang, Y.; Wang, H.; Chris, G. High-speed permanent magnet synchronous motor iron loss calculation method considering multiphysics factors. *IEEE Trans. Ind. Electron.* **2020**, *67*, 5360–5368. [CrossRef]
15. Duan, C.; Guo, H.; Xing, W.; Tian, W.; Xu, J. Design and analysis of a 120 kw high-speed permanent magnet motor with a novel evaporative cooling configuration for centrifugal compressor. In Proceedings of the 2018 21st International Conference on Electrical Machines and Systems (ICEMS), Jeju, Korea, 7–10 October 2018; pp. 393–397.
16. Ding, S.; Shen, S.; Yang, Z.; Chen, S.; Dai, Y. Fluid-solid coupling simulation and performance analysis of high speed permanent magnet synchronous motor. *J. Mot. Control* **2021**, *25*, 112–121.
17. Tong, W.; Sun, R.; Zhang, C.; Wu, S.; Tang, R. Loss and thermal analysis of a high-speed surface-mounted pmsm with amorphous metal stator core and titanium alloy rotor sleeve. *IEEE Trans. Magn.* **2019**, *55*, 8102104. [CrossRef]

18. Qu, W.; Guo, Z.; Mu, Z.; Xing, Y.; Liang, R. Research on mechanical characteristics of high-speed permanent magnet motor. In Proceedings of the 2020 IEEE International Conference on Applied Superconductivity and Electromagnetic Devices (ASEMD), Tianjin, China, 16–18 October 2020; pp. 1–2.
19. Cheng, X.; Zeng, G.; Du, G. Rotor structure for ultra-high speed permanent magnet motor. *J. Mech. Electr. Eng.* **2019**, *36*, 1226–1230.
20. Xu, G.; Wan, D.; Zhang, C.; Wang, X. Analytical calculation of eddy current loss in high-speed permanent magnet motor. *Micromotors* **2020**, *53*, 6.
21. Xu, H.; Geng, H.; Lin, H.; Qi, Y.; Yin, X. Rotor design and analysis of a high speed permanent magnet synchronous motor for cryogenic centrifugal pump. In Proceedings of the 2019 IEEE International Conference on Mechatronics and Automation (ICMA), Tianjin, China, 4–7 August 2019; pp. 1756–1760.
22. Hao, Y. Design and Analysis of 100 kW, 50,000 r/min High Speed Permanent Magnet Synchronous Motor. Ph.D. Thesis, Shenyang University of Technology, Shenyang, China, 2017.
23. Shao, Y.; Wang, X.; Gao, Q.; Li, Y. Rotor strength analysis of ultra-high speed permanent magnet synchronous motor. In Proceedings of the 2019 22nd International Conference on Electrical Machines and Systems (ICEMS), Harbin, China, 11–14 August 2019; pp. 1–4.
24. Girard, N.; Olmedo, L.E.; Schiffmann, J.; Chevailler, S. Design of a very high speed, high power pmsynchronous motor. In Proceedings of the 2019 19th International Symposium on Electromagnetic Fields in Mechatronics, Electrical and Electronic Engineering (ISEF), Nancy, France, 29–31 August 2019; pp. 1–2.
25. Guo, E. Design of High Speed Permanent Magnet Motor and Thermal Analysis Based on Magnetic—Thermal Coupling. Ph.D. Thesis, Shenyang University of Technology, Shenyang, China, 2020.
26. Luo, Q. Analysis of Stator Core Loss of High Speed Permanent Magnet Synchronous Motor. Ph.D. Thesis, Harbin University of Science and Technology, Harbin, China, 2018.
27. Zhang, D. Study on Electromagnetic Characteristics and Efficiency Optimization of High Speed and High Power Permanent Magnet Synchronous Motor. Ph.D. Thesis, Hunan University, Changsha, China, 2016.
28. Huang, Z.; Fang, J.; Liu, X.; Han, B. Loss calculation and thermal analysis of rotors supported by active magnetic bearings for high-speed permanent-magnet electrical machines. *IEEE Trans. Ind. Electron.* **2016**, *63*, 2027–2035. [CrossRef]
29. Chen, D.; Feng, M. The influence of magnetic field on losses of high-speed permanent magnet motor. In Proceedings of the 2016 IEEE International Conference on Mechatronics and Automation, Harbin, China, 7–10 August 2016; pp. 27–31.
30. Xu, H.; Li, J. A sleeve-free interior permanent magnet high speed motor with non-uniform airgap. In Proceedings of the 2019 IEEE International Electric Machines Drives Conference (IEMDC), San Diego, CA, USA, 12–15 May 2019; pp. 733–738.
31. Dong, B.; Wang, K.; Han, B.; Zheng, S. High-speed permanent magnet motor with magnetic bearings: Multi-physics analysis, cooling design and experiment. In Proceedings of the 2016 19th International Conference on Electrical Machines and Systems (ICEMS), Chiba, Japan, 13–16 November 2016; pp. 1–4.
32. Qin, X.F.; Shen, J.X. Multi-physics design of high-speed large-power permanent magnet synchronous motor. In Proceedings of the 2020 Fifteenth International Conference on Ecological Vehicles and Renewable Energies (EVER), Monte-Carlo, Monaco, 10–12 September 2020; pp. 1–5.
33. Yi, L.; Yin, X. Optimization of magnetic pole structure of high-speed permanent magnet synchronous motor weakens cogging torque. In Proceedings of the 2020 IEEE 1st China International Youth Conference on Electrical Engineering (CIYCEE), Wuhan, China, 1–4 November 2020; pp. 1–5.
34. Hong, Z.F.; Cheng, Z.Y.; Wang, Y.C. Large-airgap high-speed permanent magnet machines for flywheel system applications. In Proceedings of the 2020 23rd International Conference on Electrical Machines and Systems (ICEMS), Hamamatsu, Japan, 24–27 November 2020; pp. 1936–1939.
35. Yin, X.; Geng, H.; Lv, H.; Xu, H.; Qi, Y. Analysis of thermal characteristics of 10 kw high speed permanent magnet synchronous motor. In Proceedings of the 2019 IEEE International Conference on Mechatronics and Automation (ICMA), Tianjin, China, 4–7 August 2019; pp. 192–197.
36. Hong, D.K.; Woo, B.C.; Lee, J.Y.; Koo, D.H. Ultra high speed motor supported by air foil bearings for air blower cooling fuel cells. *IEEE Trans. Magn.* **2012**, *48*, 871–874. [CrossRef]
37. Du, G.; Xu, W.; Huang, N.; Cheng, X.; Xiao, X. Rotor design of high power high speed permanent magnet machine considering multiphysics constraints. In Proceedings of the 2019 22nd International Conference on Electrical Machines and Systems (ICEMS), Harbin, China, 11–14 August 2019; pp. 1–5.
38. Jastrzebski, R.P.; Jaatinen, P.; Pyrhönen, O.; Chiba, A. Design of 6-slot inset pm bearingless motor for high-speed and higher than 100 kw applications. In Proceedings of the 2017 IEEE International Electric Machines and Drives Conference (IEMDC), Miami, FL, USA, 21–24 May 2017; pp. 1–6.
39. Fukushima, T.; Ooshima, M. Combination of pole number of motor and suspension windings and the influence on suspension force in bearingless motors. In Proceedings of the 2020 23rd International Conference on Electrical Machines and Systems (ICEMS), Hamamatsu, Japan, 24–27 November 2020; pp. 44–49.
40. Farhan, A.; Johnson, M.; Hanson, K.; Severson, E.L. Design of an ultra-high speed bearingless motor for significant rated power. In Proceedings of the 2020 IEEE Energy Conversion Congress and Exposition (ECCE), Detroit, MI, USA, 11–15 October 2020; pp. 246–253.

41. Wang, D.; Shen, Q.; Chen, X.; Han, L. Electromagnetic analysis of compound excitation high speed permanent magnet synchronous motor. In Proceedings of the 2019 22nd International Conference on Electrical Machines and Systems (ICEMS), Harbin, China, 11–14 August 2019; pp. 1–6.
42. Lv, Y.; Xia, D.; Wang, Q.; Li, L. Post-assembly magnetization of high speed pm rotor with metallic sleeve. In Proceedings of the 2017 20th International Conference on Electrical Machines and Systems (ICEMS), Sydney, NSW, Australia, 11–14 August 2017; pp. 1–5.
43. Jun, H.-W.; Lee, J.; Lee, H.-W.; Kim, W.-H. Study on the optimal rotor retaining sleeve structure for the reduction of eddy-current loss in high-speed spmsm. *IEEE Trans. Magn.* **2015**, *51*, 8103004.
44. Fang, J.; Liu, X.; Han, B.; Wang, K. Analysis of circulating current loss for high-speed permanent magnet motor. *IEEE Trans. Magn.* **2015**, *51*, 1–13. [CrossRef]
45. Boztas, G.; Aydogmus, O. Design of a high-speed pmsm for flywheel systems. In Proceedings of the 2019 4th International Conference on Power Electronics and their Applications (ICPEA), Elazig, Turkey, 25–27 September 2019; pp. 1–5.
46. Zhang, Z.; Deng, Z.; Gu, C.; Sun, Q.; Peng, C.; Pang, G. Reduction of rotor harmonic eddy-current loss of high-speed pm bldc motors by using a split-phase winding method. *IEEE Trans. Energy Convers.* **2019**, *34*, 1593–1602. [CrossRef]
47. Jikumaru, T.; Kuwata, G. 1.2 kw 100,000 rpm high speed motor for aircraft. In Proceedings of the 2018 International Power Electronics Conference (IPEC-Niigata 2018 -ECCE Asia), Niigata, Japan, 20–24 May 2018; pp. 177–180.
48. Zhang, Y.; Geng, H.; Zhou, J.; Yu, L. Analysis of electromagnetic force waves of solid cylindrical permanent magnet synchronous motors. In Proceedings of the 2018 IEEE International Conference on Mechatronics and Automation (ICMA), Changchun, China, 5–8 August 2018; pp. 689–694.
49. Li, Y.; Zhu, C.; Wu, L.; Zheng, Y. Multi-objective optimal design of high-speed surface-mounted permanent magnet synchronous motor for magnetically levitated flywheel energy storage system. *IEEE Trans. Magn.* **2019**, *55*, 1–8. [CrossRef]
50. Dong, B.; Wang, K.; Han, B.; Zheng, S. Thermal analysis and experimental validation of a 30 kw 60,000 r/min high-speed permanent magnet motor with magnetic bearings. *IEEE Access* **2019**, *7*, 92184–92192. [CrossRef]
51. Kim, J.H.; Kim, D.M.; Jung, Y.H.; Lim, M.S. Design of ultra-high-speed motor for fcev air compressor considering mechanical properties of rotor materials. *IEEE Trans. Energy Convers.* **2021**, *36*, 2850–2860. [CrossRef]
52. Chen, P.; Chen, J.; Liao, Y.; Zhang, C.; Du, J. Vibration and noise of high speed amorphous alloy permanent magnet synchronous motor with different slot width. In Proceedings of the 2019 IEEE 4th International Future Energy Electronics Conference (IFEEC), Singapore, 25–28 November 2019; pp. 1–5.
53. Zhang, C.; Tong, W.; Ma, X.; Tang, R.; Zhu, J. Development of a high-speed permanent magnet machine using amorphous alloy cores. In Proceedings of the 2016 IEEE Vehicle Power and Propulsion Conference (VPPC), Hangzhou, China, 17–20 October 2016; pp. 1–5.
54. Han, T.; Wang, Y.; Shen, J.X. Analysis and experiment method of influence of retaining sleeve structures and materials on rotor eddy current loss in high-speed pm motors. *IEEE Trans. Ind. Appl.* **2020**, *56*, 4889–4895. [CrossRef]
55. Jastrzebski, R.P.; Jaatinen, P.; Sugimoto, H.; Pyrhönen, O.; Chiba, A. Design of a bearingless 100 kw electric motor for high-speed applications. In Proceedings of the 2015 18th International Conference on Electrical Machines and Systems (ICEMS), Pattaya, Thailand, 25–28 October 2015; pp. 2008–2014.
56. Baojun, G.; Qiantong, L.; Likun, W. Analysis of iron losses of high-speed permanent magnet synchronous motor. *Electr. Mach. Control* **2020**, *24*, 36–43.
57. Wang, D.; Zhao, G. Analysis of high speed permanent magnet motor for different rotor sheath. *J. Electr. Eng.* **2018**, *13*, 17–21.
58. Wang, D.; Ma, T. Loss and temperature field analysis of permanent magnet brushless dc high speed motor with two rotor structures. *J. Electr. Eng.* **2018**, *13*, 16–21.
59. Zhang, Z.; Deng, Z.; Sun, Q.; Xu, Z.; Li, K. Influences of copper shield on eddy-current loss and stress for a rotor of high-speed pm bldc motor. *Proc. CSEE* **2018**, *38*, 275–285+356.
60. Wu, C. Research on Flow-Thermal Coupling of 200 kW High-Speed Permanent Magnet Synchronous Motor. Ph.D. Thesis, Harbin University of Science and Technology, Harbin, China, 2019.
61. Fodorean, D.; Popa, D.C.; Minciunescu, P.; Irimia, C.; Szabó, L. Study of a high-speed motorization for electric vehicle based on pmsm, im and vrsm. In Proceedings of the 2014 International Conference on Electrical Machines (ICEM), Berlin, Germany, 2–5 September 2014; pp. 2577–2582.
62. Zhang, C.; Zhu, J.; Han, X. Rotor strength analysis of high speed surface mounted permanent magnet motor. *Proc. CSEE* **2016**, *36*, 4719–4727.
63. Jiang, S.; Zou, J.; Xu, Y.; Liang, W. Variable coefficient iron loss calculating model considering rotational flux and skin effect. *Proc. CSEE* **2011**, *3*, 104–110.
64. Krings, A.; Mousavi, S.A.; Wallmark, O.; Soulard, J. Temperature influence of nife steel laminations on the characteristics of small slotless permanent magnet machines. *IEEE Trans. Magn.* **2013**, *49*, 4064–4067. [CrossRef]
65. Liu, M. Study on Iron Loss Model of High Speed Permanent Magnet Motor Considering Multiple Physical Fields. Ph.D. Thesis, Shenyang University of Technology, Shenyang, China, 2019.
66. He, T.R.; Zhu, Z.Q.; Xu, F.; Bin, H.; Wu, D.; Gong, L.M.; ; Chen, J.T. Comparative study of 6-slot/2-pole high-speed permanent magnet motors with different winding configurations. *IEEE Trans. Ind. Appl.* **2021**, *57*, 5864–5875. [CrossRef]

67. Chen, C.; Wang, Y.; Wei, J.; Wen, X. Research on method for reducing eddy current loss of magnet in high-speed permanent magnet synchronous motor. In Proceedings of the 2019 22nd International Conference on Electrical Machines and Systems (ICEMS), Harbin, China, 11–14 August 2019; pp. 1–4.
68. Burnand, G.; Perriard, Y. Very-high-speed miniaturized permanent magnet motors: Modeling and experimental validation. In Proceedings of the 2019 IEEE Energy Conversion Congress and Exposition (ECCE), Baltimore, MD, USA, 29 September–3 October 2019; pp. 5251–5257.
69. Neethu, S.; Nikam, S.P.; Pal, S.; Wankhede, A.K.; Fernandes, B.G. Toroidal winding with fictitious slots for high speed permanent magnet synchronous motors. In Proceedings of the 2018 IEEE International Conference on Industrial Technology (ICIT), Lyon, France, 20–22 February 2018; pp. 546–550.
70. Zhang, C.; Chen, L.; Wang, X.; Tang, R. Loss calculation and thermal analysis for high-speed permanent magnet synchronous machines. *IEEE Access* **2020**, *8*, 92627–92636. [CrossRef]
71. Zhu, G.; Liu, W.; Liu, X.; Li, L.; Tong, W.; Han, X. Analytical analysis and cooling system design of a high-speed permanent magnet motor utilizing an amorphous metal core. In Proceedings of the 2018 IEEE International Conference on Applied Superconductivity and Electromagnetic Devices (ASEMD), Tianjin, China, 15–18 April 2018; pp. 1–2.
72. Agravat, J.; Chavda, N. Different design considerations for designing of high speed permanent magnet motors. In Proceedings of the 2020 21st National Power Systems Conference (NPSC), Gandhinagar, India, 17–19 December 2020; pp. 1–5.
73. Wang, X.; Zhou, S.; Wu, L.; Zhao, M.; Hu, C. Iron loss and thermal analysis of high speed pm motor using soft magnetic composite material. In Proceedings of the 2019 22nd International Conference on Electrical Machines and Systems (ICEMS), Harbin, China, 11–14 August 2019; pp. 1–4.
74. Wei, F.; Wang, J.; Liu, P. Electromagnetic design and analysis of ultra-high-speed motorbased on soft magnetic composite material. *Electr. Mach. Control. Appl.* **2021**, *48*, 5.
75. Kauder, T.; Hameyer, K. Performance factor comparison of nanocrystalline, amorphous, and crystalline soft magnetic materials for medium-frequency applications. *IEEE Trans. Magn.* **2017**, *53*, 8401504. [CrossRef]
76. O'Donnell, D.; Bartos, S.; Tjong, J.; Kar, N.C. Utilization of innovative materials toward permanent magnet synchronous e-motors for traction application: A review. In Proceedings of the 2020 2nd International Conference on Smart Power Internet Energy Systems (SPIES), Bangkok, Thailand, 15–18 September 2020; pp. 380–385.
77. Liu, M.; Sun, Z.; Zhan, Y.; Wu, J.; Zhao, H.; Fang, Y. Analysis and calculation of no-load loss of stator core of amorphous alloy high speed permanent magnet motor. *J. North China Electr. Power Univ.* **2019**, *5*, 77–82.
78. Zhao, G.; Kong, D.; Gao, X. Study on performance difference between soft magnetic composite and silicon steel sheet permanent magnet motor. *Trans. China Electrotech. Soc.* **2018**, *33*, 75–81.
79. Neethu , S.; Pal, S.; Wankhede, A.K.; Fernandes, B.G. High performance axial flux permanent magnet synchronous motor for high speed applications. In Proceedings of the IECON 2017—43rd Annual Conference of the IEEE Industrial Electronics Society, Beijing, China, 29 October–1 November 2017; pp. 5093–5098.
80. Zhang, F.; Du, G.; Wang, T.; Wang, F.; Cao, W.; Kirtley, J.L. Electromagnetic design and loss calculations of a 1.12-mw high-speed permanent-magnet motor for compressor applications. *IEEE Trans. Energy Convers.* **2016**, *31*, 132–140. [CrossRef]
81. Raminosoa, T.; Wiles, R.; Cousineau, J.E.; Bennion, K.; Wilkins, J. A high-speed high-power-density non-heavy rare-earth permanent magnet traction motor. In Proceedings of the 2020 IEEE Energy Conversion Congress and Exposition (ECCE), Detroit, MI, USA, 11–15 October 2020; pp. 61–67.
82. Shen, J.; Qin, X.; Wang, Y. High-speed permanent magnet electrical machines-applications, key issues and challenges. *CES Trans. Electr. Mach. Syst.* **2018**, *2*, 23–33. [CrossRef]
83. Burn, G.; Araujo, D.M.; Perriard, Y. Optimization of shape and topology for slotless windings in bldc machines. In Proceedings of the 2018 21st International Conference on Electrical Machines and Systems (ICEMS), Jeju, Korea, 7–10 October 2018; pp. 31–36.
84. Volpe, G.; Popescu, M.; Foley, I.; Goss, J. Winding material effect on high speed brushless permanent magnet machines. In Proceedings of the 2019 IEEE Energy Conversion Congress and Exposition (ECCE), Baltimore, MD, USA, 29 September–3 October 2019; pp. 3144–3149.
85. Woo, J.H.; Bang, T.K.; Lee, H.K.; Kim, K.H.; Shin, S.H.; Choi, J.Y. Electromagnetic characteristic analysis of high-speed motors with rare-earth and ferrite permanent magnets considering current harmonics. *IEEE Trans. Magn.* **2021**, *57*, 8201805. [CrossRef]
86. Sun, Z. *Study on Loss and Thermal Characteristics of High Speed Permanent Magnet Synchronous Motor*; Harbin Institute of Technology: Harbin, China, 2011.
87. Kawanishi, K.; Matsuo, K.; Mizuno, T.; Yamada, K.; Okitsu, T.; Matsuse, K. Development and performance of high-speed spm synchronous machine. In Proceedings of the 2018 International Power Electronics Conference (IPEC-Niigata 2018–ECCE Asia), Niigata, Japan, 20–24 May 2018; pp. 169–176.
88. Yon, J.M.; Mellor, P.H.; Wrobel, R.; Booker, J.D.; Burrow, S.G. Analysis of semipermeable containment sleeve technology for high-speed permanent magnet machines. *IEEE Trans. Energy Convers.* **2012**, *27*, 646–653. [CrossRef]
89. Han, X.; Li, C.; Song, C.; Wang, S.; Zhu, T. Temperature rise calculation and application of high speed permanent magnet motor based on magneto-thermal coupling method. *Micromotors* **2020**, *53*, 13–19.
90. Zhang, Y.; McLoone, S.; Cao, W. High speed permanent magnet motor design and power loss analysis. In Proceedings of the 2017 IEEE Transportation Electrification Conference and Expo, Asia-Pacific (ITEC Asia-Pacific), Harbin, China, 7–10 August 2017; pp. 1–6.

91. Zhang, J.; Liu, C.; Wang, G.; Yang, G. The rotor structure and performance analysis of a two magnet-hybrid excitation high speed permanent magnet machine. *Large Electr. Mach. Hydraul. Turbine* **2020**, *273*, 24–29+35.
92. Su, J.; Xu, W.; Zhang, Y.; Liu, Y. Design and analysis of high-speed permanent magnet machine with low rotor loss for flywheel energy storage system. In Proceedings of the 2020 23rd International Conference on Electrical Machines and Systems (ICEMS), Hamamatsu, Japan, 24–27 November 2020; pp. 851–856.
93. Fang, H.; Qu, R.; Li, J.; Zheng, P.; Fan, X. Rotor design for high-speed high-power permanent-magnet synchronous machines. *IEEE Trans. Ind. Appl.* **2017**, *53*, 3411–3419. [CrossRef]
94. Dong, J.; Huang, Y.; Jin, L.; Lin, H. Review on high speed permanent magnet machines including design and analysis technologies. *Proceedings Chin. Soc. Electr. Eng.* **2014**, *34*, 4640–4653.
95. Liu, Y.; Ou, J.; Schiefer, M.; Breining, P.; Grilli, F.; Doppelbauer, M. Application of an amorphous core to an ultra-high-speed sleeve-free interior permanent-magnet rotor. *IEEE Trans. Ind. Electron.* **2018**, *65*, 8498–8509. [CrossRef]
96. Huang, Z.; Fang, J. Multiphysics design and optimization of high-speed permanent-magnet electrical machines for air blower applications. *IEEE Trans. Ind. Electron.* **2016**, *63*, 2766–2774. [CrossRef]
97. Sun, Y. Research on Loss of High Speed Permanent Magnet Motor Based on Field-Circuit Coupling Analysis. Ph.D. Thesis, Shenyang University of Technology, Shenyang, China, 2020.
98. Lu, W. Multi-Physics Coupling Analysis and Optimization of a High-Speed Permanent Magnet Motor. Ph.D. Thesis, Guizhou University, Guiyang, China, 2019.
99. Zhang, D. *Fluid Field Analysis and Temperature Rise Calculation of High Speed Permanent Magnet Motor*; Shenyang University of Technology: Shenyang, China, 2009.
100. Tang, Y.; Sun, Y.; Guo, Z.; Zhang, S.; Yuan, W.; Tang, H.; Liang, F. Development status and perspective trend of motor cooling systems. *China Mech. Eng.* **2021**, *32*, 16.
101. Kong, X.; Wang, F.; Xing, J. Losses calculation and temperature field analysis of high speed permanent magnet machines. *Trans. China Electrotech. Soc.* **2012**, *27*, 166–173.
102. Zhang, F.; Du, G.; Wang, T.; Wang, F.; Cao, W. Temperature field analysis of 1.12 mw high speed permanent magnet machine with different cooling schemes. *Trans. China Electrotech. Soc.* **2014**, *29*, 66–72.
103. Du, G.; Zhou, Q.; Liu, S.; Huang, N.; Chen, X. Multiphysics design and multiobjective optimization for high-speed permanent magnet machines. *IEEE Trans. Transp. Electrif.* **2020**, *6*, 1084–1092. [CrossRef]
104. Tong, W.; Cheng, X.; Sun, J.; Chen, P.; Tang, R. Inhibitory effect of rotor winding on temperature rise of high speed permanent magnet motor. *Proc. CSEE* **2017**, *37*, 1526–1534.
105. Tong, W.; Cheng, X. Cooling system analysis of high-speed water cooling permanent magnet motor. *Electr. Mach. Control. Appl.* **2016**, *43*, 16–21.
106. Zhang, Q.; Li, Z.; Dong, X.; Liu, Y.; Wang, Y. Analysis method and experimental research on cooling system of high power submersible motor. *China Mech. Eng.* **2021**, *32*, 10. [CrossRef]
107. Gai, Y.; Kimiabeigi, M.; Chong, Y.C.; Widmer, J.D.; Deng, X.; Popescu, M.; Goss, J.; Staton, D.A.; Steven, A. Cooling of automotive traction motors: Schemes, examples, and computation methods. *IEEE Trans. Ind. Electron.* **2019**, *66*, 1681–1692. [CrossRef]
108. Wei, X. Temperature Field Analysis and Cooling System of Permanent Magnet Synchronous Motor. Ph.D. Thesis, Xiangtan University, Xiangtan, China, 2017.
109. Sun, Y.; Zhang, S.; Chen, G.; Tang, Y.; Liang, F. Experimental and numerical investigation on a novel heat pipe based cooling strategy for permanent magnet synchronous motors. *Appl. Therm. Eng.* **2020**, *170*, 114970. [CrossRef]
110. Binder, A.; Schneider, T.; Klohr, M. Fixation of buried and surface-mounted magnets in high-speed permanent-magnet synchronous machines. *IEEE Trans. Ind. Appl.* **2006**, *42*, 1031–1037. [CrossRef]
111. Zhang, F.; Hao, Y.; Liu, G.; Qiu, F. Design and analysis of 100kW high speed permanent magnet synchronous motor. In Proceedings of the 2016 IEEE Transportation Electrification Conference and Expo, Asia-Pacific (ITEC Asia-Pacific), Busan, Korea, 1–4 June 2016; pp. 843–847.
112. Cui, J. Research on Rotor Structure Strength and Dynamics of High-Speed Permanent Magnet Motor. Ph.D. Thesis, Harbin University of Science and Technology, Harbin, China, 2020.
113. Hu, J. Research on Rotor with Carbon Fiber Sleeve in High-Speed Permanent Magnet Motors. Ph.D. Thesis, Nanjing University of Aeronautics and Astronautics, Nanjing, China, 2013.
114. Zhou, F. Design of Carbon Fiber Composite Sheath for High Speed Permanent Magnet Motor Rotor. Ph.D. Thesis, Wuhan University of Technology, Wuhan, China, 2018.
115. Hui, P. Study on the Structural Design of Carbon Fiber Composite Flywheel Rotor in Large Tension Winding. Ph.D. Thesis, Wuhan University of Technology, Wuhan, China, 2018.
116. Cheng, W.; Geng, H.; Feng, S.; Yu, L.; Sun, Y.; Yang, L. Rotor strength analysis of high-speed permanent magnet synchronous motors. *Proc. Chin. Soc. Electr. Eng.* **2012**, *32*, 87–94.
117. Chen, L.; Zhu, C. Rotor strength analysis for high speed permanent magnet machines. In Proceedings of the 2014 17th International Conference on Electrical Machines and Systems (ICEMS), Hangzhou, China, 22–25 October 2014; pp. 65–69.
118. Zhang, F.; Du, G.; Wang, T.; Huang, N. Strength analysis of dprotection measures for high-speed permanent magnet motor rotor. *Proc. CSEE* **2013**, *33*, 195–202.

119. Chen, L.; Zhu, C.; Qiao, X.; Wu, J.; Li, Q.; Zhou, D. Rotor strength analysis for the high speed segmented surface mounted permanent magnet synchronous machine with a carbon fiber sleeve. *Electr. Mach. Control* **2019**, *23*, 97–107.
120. Mallin, L.; Barrans, S. Comparison of theoretical approaches to determine the stresses in surface mounted permanent magnet rotors for high speed electric machines. *J. Strain Anal. Eng. Des.* **2022**, *57*, 177–192. [CrossRef]
121. Chen, L.; Zhu, C.; Wang, M. Thermal rotor strength of carbon fiber sheathed high speed permanent magnet motor. *J. Zhejiang Univ.* **2015**, *49*, 162–172.
122. Yang, Y. Thermal Stress Calculation and Mechanical Properties Analysis of High Speed Permanent Magnet Motor. Ph.D. Thesis, Shenyang University of Technology, Shenyang, China, 2018.
123. Gao, Q.; Wang, X.; Gu, C.; Liu, S.; Li, D. Design of ultra high speed micro permanent magnet motor with integrated support type based on multi coupling characteristics. *Trans. China Electrotech. Soc.* **2021**, *36*, 11.
124. Bartolo, J.B.; Zhang, H.; Gerada, D.; De Lillo, L.; Gerada, C. High speed electrical generators, application, materials and design. In Proceedings of the 2013 IEEE Workshop on Electrical Machines Design, Control and Diagnosis (WEMDCD), Paris, France, 11–12 March 2013; pp. 47–59.
125. Lin, H.; Geng, H.; Du, T.; Xu, X.; Zhang, Y.; Cheng, W.; Yu, L. Analysis of high speed permanent magnet synchronous motor rotor supported by air foil bearings. *Int. J. Appl. Electromagn. Mech.* **2018**, *59*, 755–764. [CrossRef]
126. Liu, G.; Mao, K. A novel power failure compensation control method for active magnetic bearings used in high-speed permanent magnet motor. *IEEE Trans. Power Electron.* **2016**, *31*, 4565–4575. [CrossRef]
127. Zhang, Y.; Niu, J. Design method of magnetic bearing system applied in high-speed centrifugal blower. *Chin. J. Turbomach.* **2021**, *63*, 6.
128. Looser, A.; Kolar, J.W. An active magnetic damper concept for stabilization of gas bearings in high-speed permanent-magnet machines. *IEEE Trans. Ind. Electron.* **2014**, *61*, 3089–3098. [CrossRef]
129. Dajun, T.; Ruifeng, M.; Likun, W.; Baojun, G. Analysis of stiffness and critical speed of high speed permanent magnet motor rotor system. In Proceedings of the 2019 22nd International Conference on Electrical Machines and Systems (ICEMS), Harbin, China, 11–14 August 2019; pp. 1–5.
130. Liu, C.; Zhan, J.; Yang, Y.; Liu, Z. Review of research status and development of flexible rotor-magnetic bearing. *Proc. CSEE* **2020**, *40*, 13.
131. Dong, J. Research on Comprehensive Design Method of High Speed Permanent Magnet Motor. Ph.D. Thesis, Southeast University, Nanjing, China, 2015.
132. Wu, Q.; Sun, Y.; Chen, W.; Wang, Q.; Chen, G. Theoretical prediction and experimental verification of the unbalanced magnetic force in air bearing motor spindles. *Proc. Inst. Mech. Eng. Part J. Eng. Manuf.* **2019**, *233*, 2330–2344. [CrossRef]
133. Wu, Q. Influence of unbalanced electromagnetic force and air supply pressure fluctuation in air bearing spindles on machining surface topography. *Int. J. Precis. Eng. Manuf.* **2020**, *22*, 1–22.
134. Li, W.; Zhang, X.; Cheng, S.; Cao, J. Thermal optimization for a hspmg used for distributed generation systems. *IEEE Trans. Ind. Electron.* **2013**, *60*, 474–482. [CrossRef]
135. Li, Z.; Zhan, Y. Rotor strength analysis of high speed permanent magnet machine. *Mech. Electr. Eng. Mag.* **2016**, *33*, 900–903.
136. He, X. Mechanical Strength Analysis of Amorphous Alloy High Speed Permanent Magnet Motor. Ph.D. Thesis, Shenyang University of Technology, Shenyang, China, 2017.
137. Zhang, C. Research on Design Technology of High Speed Amorphous Alloy Permanent Magnet Motor. Ph.D. Thesis, Shenyang University of Technology, Shenyang, China, 2017.
138. Qiu, R. Research on Electromagnetic Design and Optimization of Permanent Magnet Synchronous Motor. Ph.D. Thesis, Qingdao University, Qingdao, China, 2020.
139. Jannot, X.; Vannier, J.C.; March, C.; Gabsi, M.; Saint-Michel, J.; Sadarnac, D. Multiphysic modeling of a high-speed interior permanent-magnet synchronous machine for a multiobjective optimal design. *IEEE Trans. Energy Convers.* **2011**, *26*, 457–467. [CrossRef]
140. Zheng, L.; Wu, T.X.; Acharya, D.; Sundaram, K.B.; Vaidya, J.; Zhao, L.; Zhou, L.; Murty, K.; Ham, C.H.; Arakere, N.; et al. Design of a super-high speed permanent magnet synchronous motor for cryogenic applications. In Proceedings of the IEEE International Conference on Electric Machines and Drives, San Antonio, TX, USA, 15 May 2005; pp. 874–881.
141. Munteanu, G.; Binder, A.; Schneider, T.; Funieru, B. No-load tests of a 40 kw high-speed bearingless permanent magnet synchronous motor. In Proceedings of the SPEEDAM 2010, Pisa, Italy, 14–16 June 2010; pp. 1460–1465.
142. Jaatinen, P.; Jastrzebski, R.P.; Pyrhönen, O. Comparison of winding arrangements of a high-speed interior permanent magnet bearingless machine. In Proceedings of the 2016 19th International Conference on Electrical Machines and Systems (ICEMS), Chiba, Japan, 13–16 November 2016; pp. 1–6.
143. Gao, P.; Fang, J.; Han, B.; Sun, J. Analysis of rotor eddy-current loss in high-speed permanent magnet motors. *Micromotors* **2013**, *46*, 5–11.
144. Dai, N. Multi-Physics Coupling Simulation Calculation of High-Power Ultra-High-Speed Permanent Magnet Synchronous Motor. Ph.D. Thesis, North China Electric Power University, Beijing, China, 2019.

applied sciences

MDPI

Article

A Dynamic Load Simulation Algorithm Based on an Inertia Simulation Predictive Model

Kuizhi Lin [1,2], Wei Wang [1,2,*], Zhiqiang Wang [1,3], Xinmin Li [1,3] and Huang Zhang [1,4]

1 National Joint Local Engineering Research Center of Electrical Machine System Intelligent Design and Manufacturing, Tianjin 300387, China; linkuizhi@tiangong.edu.cn (K.L.); wangzhiqiang@tiangong.edu.cn (Z.W.); lixinmin@tju.edu.cn (X.L.); zhanghuang0415@163.com (H.Z.)
2 School of Electronics and Information Engineering, Tiangong University, Tianjin 300387, China
3 School of Electrical Engineering, Tiangong University, Tianjin 300387, China
4 School of Mechanical Engineering, Tiangong University, Tianjin 300387, China
* Correspondence: wangweibit@163.com

Abstract: In this study, an electric dynamic load simulation system (EDLSS) algorithm was proposed based on an inertia simulation predictive model to mitigate strong-coupling torque disturbance and load torque fluctuation caused by the change in the motion state of a bearing system. First, the inertia simulation model was proposed by combining the dynamic equations of both the EDLSS and the target system. The aforementioned inertia simulation model converted the conventional realisation method of the inertia simulation into the tracking of the motion characteristics for the target system under the same working conditions. Next, based on the aforementioned inertia simulation model while considering the strong-coupling torque effect and motor braking state disturbance as two influential factors, an inertia simulation predictive model and a load simulation algorithm were proposed. The predicted speed calculated by the predictive model was consistent with the dynamic characteristics of the target system under the same working conditions and input into the control loop. Based on the analysis of the braking state and power model of the permanent magnet synchronous motor, an energy feedback control method was proposed to improve EDLSS stability caused by the braking state of the loading motor. Finally, the experimental data revealed that the maximum speed fluctuation range of the loading motor was approximately 7.5, which was 84% lower than the range before the application of the aforementioned algorithm, which was about 46.8. Furthermore, the maximum range of the torque ripple was close to 1.5, which was 75% lower than before, which was roughly 6. All experimental data were consistent with simulation data.

Keywords: electric dynamic load simulation system (EDLSS); inertia simulation; load simulation; regenerative braking; permanent magnet synchronous motor (PMSM)

Citation: Lin, K.; Wang, W.; Wang, Z.; Li, X.; Zhang, H. A Dynamic Load Simulation Algorithm Based on an Inertia Simulation Predictive Model. *Appl. Sci.* **2022**, *12*, 7142. https://doi.org/10.3390/app12147142

Academic Editors: Lorand Szabo and Feng Chai

Received: 25 June 2022
Accepted: 13 July 2022
Published: 15 July 2022

1. Introduction

Permanent magnet synchronous motors (PMSMs) are widely used in electric dynamic load simulation systems (EDLSSs) owing to their high power factor, high power density, and high operating efficiency. EDLSSs are used to perform semi-physical simulations to test the performance indicators of servo systems. When EDLSSs are used for the servo system load test, the loading motor is dragged by the bearing system to follow its synchronous motion and simulate the load and inertia torques. During this period, the loading motor is affected by the motion state of the bearing system to produce load torque fluctuation [1–3]. Therefore, suppressing the strong-coupling torque disturbance of the servo system to the EDLSS is crucial.

In current research, interference torque suppression is typically realised using the following aspects. On the loading motor control side, the loading motor controller can be tuned, the control parameters can be optimised, and the control error can be decreased. Furthermore, the disturbance torque generated by the loading motor side can be eliminated.

Some model predictive control methods have been proposed in [4–7]. In [4], an improved three-vector model predictive control strategy was proposed. In this model, the principle of deadbeat synchronisation between torque and flux linkage was adopted to reduce six candidate vectors in conventional torque prediction to two, and a value function was designed to select the optimal voltage vector to obtain small torque ripple and current harmonics. In [5], a model predictive current control strategy was proposed. In this model, the stator current predictive model was derived, and the first and second optimal voltage vectors and their respective action times were calculated using the value function. These values were then output directly to the inverter so that the torque ripple was reduced by 9.40% and 4.80%, respectively. In [6], a modified model predictive control method with the virtual model was proposed to reduce the dynamic tracking error of the position loop and the dynamic tracking error and stability time by generating a virtual reference. Adaptive control methods were adopted in [8–12]. In [8], a feedforward control scheme based on an adaptive observer was proposed for tracking problems with unknown mechanical parameters and unknown load torque. The adaptive law of inertia was deduced and Popov's theory was introduced into the stability analysis of the adaptive observer to improve the dynamic response performance of the controller. In [9], a control method for the PMSM based on adaptive dynamic programming was proposed to establish two neural networks to control torque and realise low torque ripple under an unknown load. In [10], an adaptive optimal control method was proposed for the dynamic disturbance of velocity and current. The conventional velocity and current tracking control problem was transformed into an optimal control problem, and the optimal solution was obtained by the Hamilton–Jacobi–Isaacs equation so that the d- and q-axes current disturbance tends to be 0 under the effect of external disturbance torque.

From the perspective of the position error between the servo and the loading systems, a compensation signal was introduced to mitigate the disturbance of the position signal of the servo system caused by the disturbance torque for realising a synchronous movement of the loading motor and the bearing system [13–18]. In [13], an adaptive linear active disturbance rejection control controller was proposed. The suppression performance of the system for disturbance torque and position noise was improved by changing the series structure of the position and velocity in the conventional adaptive linear active disturbance rejection control loop into a parallel structure and combining the linear full-order extended state observer with the control loop. In [14], a fractional-order extended state observer was combined with the active disturbance rejection controller and subsequently applied to the position loop of the servo system. The disturbance was eliminated before the output of the system was affected by internal and external disturbances. Compared with integer-order proportional-integral-differential and linear-active disturbance rejection controllers, the method has high bandwidth, fast response speed, and excellent anti-interference performance.

When the EDLSS loads the servo system, the loading motor functions in the power generation state. By reversing current i_q, magnetic resistance is applied to the loading motor under the action of reverse drag, and the process of simulating the load torque and inertia torque can be approximated as braking of the loading motor. At this stage, braking is classified into two categories, namely energy feedback and energy consumption braking [19–22]. The experimental results revealed that when the loading motor enters the energy consumption braking state, the additional energy consumption causes the driver that is coupled directly with the loading motor to produce undervoltage because of the braking state. This phenomenon affects EDLSS stability.

The aforementioned schemes introduce feedforward compensation to eliminate disturbance in the control loop according to the sampled position signal, realise synchronous motion, and suppress interference. However, when the position signal passes through the high-order differential term of the observer, the error is amplified to form interference. We strove to design a dynamic load simulation algorithm that can suppress the strong-coupling torque and load torque fluctuation during the loading process of EDLSSs.

In this study, a load simulation algorithm was proposed based on the inertia simulation predictive model. First, a novel inertia simulation model was proposed. In this model, the dynamic equations of the EDLSS and the target system were combined so that the conventional realisation method for inertia simulation was converted into the tracking of the motion characteristics of the target system under the same driving torque. Next, based on the aforementioned inertia simulation model, a novel inertia simulation predictive model and a load simulation algorithm were proposed. By considering the strong-coupling torque disturbance and the motor braking state disturbance as the two influencing factors in the model and using the non-catastrophic property of the inertia system, the predicted speed that satisfies the dynamic characteristics of the target system under the same working conditions was obtained. The braking state and power model of PMSM were analysed, and an energy feedback control method was proposed to improve EDLSS stability caused by the braking state of the loading motor. Finally, the simulation analysis and experimental verification were performed in MATLAB/Simulink on a self-developed experimental platform.

2. Nonsalient-Pole PMSM Model

A d-q synchronous rotating coordinate system was established on the rotor, and Clarke and Park transforms of three-phase current vector $i_a i_b i_c$ were performed to obtain vector projection $i_d i_q$ in a d-q synchronous rotating coordinate system.

Considering the magnetic coupling effect and ignoring the field-weakening control, the flux model, voltage model, and torque model of PMSM were established in a d-q synchronous rotating coordinate system.

The flux linkage equation of a PMSM in a d-q synchronous rotating coordinate system can be expressed as follows:

$$\begin{bmatrix} \psi_d \\ \psi_q \end{bmatrix} = \begin{bmatrix} L_d & 0 \\ 0 & L_q \end{bmatrix} \begin{bmatrix} i_d \\ i_q \end{bmatrix} + \begin{bmatrix} \psi_f \\ 0 \end{bmatrix}, \tag{1}$$

where ψ_d is the stator flux linkage component in the d-axis of a synchronous rotating coordinate system; ψ_q is the stator flux component in the q-axis of a synchronous rotating coordinate system; ψ_f is the flux linkage generated by the permanent magnet; L_d is the inductance in the d-axis of a PMSM; L_q is the inductance in the q-axis of a PMSM; i_d and i_q are the vector projections of three-phase current vector $i_a i_b i_c$ in a d-q synchronous rotating coordinate system obtained by Clarke and Park transforms.

The vector relationship of each magnetic chain component in the flux linkage model of a PMSM is displayed in Figure 1, where ψ_s is the vector resultant of each magnetic chain component in a d-q synchronous rotating coordinate system.

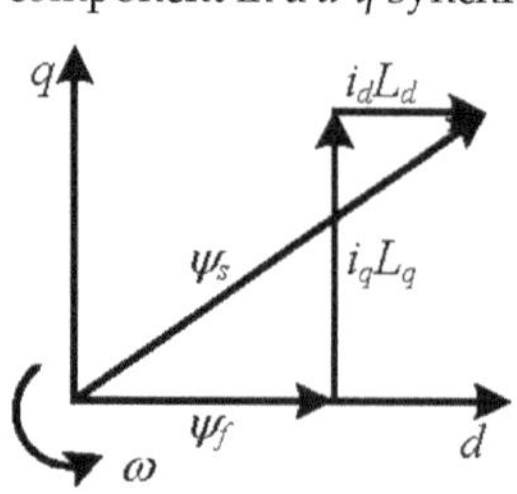

Figure 1. The flux linkage model of a PMSM.

The voltage equation of the PMSM in a d-q synchronous rotating coordinate system can be expressed as follows:

$$\begin{bmatrix} u_d \\ u_q \end{bmatrix} = R \begin{bmatrix} i_d \\ i_q \end{bmatrix} + P \begin{bmatrix} \psi_d \\ \psi_q \end{bmatrix} + \omega \begin{bmatrix} -\psi_q \\ \psi_d \end{bmatrix}, \tag{2}$$

where u_d and u_q are the vector projections of three-phase voltage vector $u_a u_b u_c$ in a d-q synchronous rotating coordinate system after Clarke and Park transforms; R is the stator resistance; P is a differential operator; and ω is the rotor angular velocity.

The simultaneous Equations (1) and (2) can be solved as follows:

$$u_d = Ri_d + L_d \frac{di_d}{dt} - \omega L_q i_q, \tag{3}$$

$$u_q = Ri_q + L_q \frac{di_q}{dt} + \omega\left(L_d i_d + \psi_f\right). \tag{4}$$

The voltage–current vector relationship is displayed in Figure 2, where u_s is the vector resultant of voltage components in a d-q synchronous rotating coordinate system, i_s is the vector resultant of current components in a d-q synchronous rotating coordinate system.

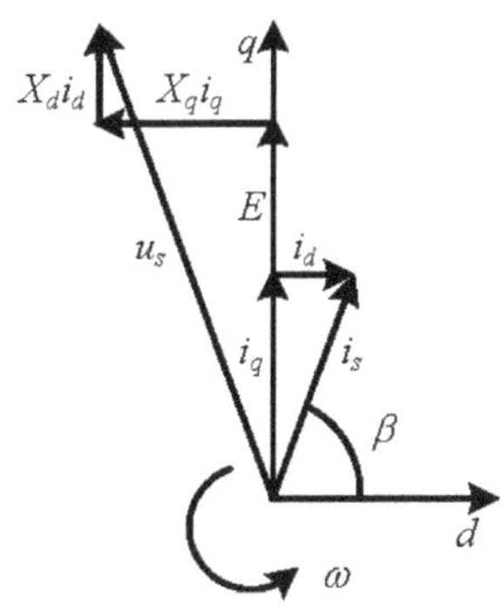

Figure 2. The voltage–current model of a PMSM.

The torque equation of the three-phase PMSM can be expressed as follows:

$$\begin{aligned} T_e &= p_n\left(\psi_d i_q - \psi_q i_d\right) \\ &= p_n\left[\psi_f i_q + \left(L_d - L_q\right) i_d i_q\right] \end{aligned}, \tag{5}$$

where p_n is the pole pairs.

Equation (5) reveals that the electromagnetic torque consists of the following two parts:

$$T_{e1} = p_n \psi_f i_q \tag{6}$$

$$T_{e2} = p_n\left(L_d - L_q\right) i_d i_q. \tag{7}$$

Similarly, Equation (5) can be expressed as follows:

$$T_e = T_{e1} + T_{e2}, \tag{8}$$

where Equation (6) is the electromagnetic torque generated by the interaction between the electromagnetic field and the permanent magnet, namely the excitation torque; Equation (7) is the torque caused by the uneven magnetic resistance of the magnetic circuit caused by the saliency effect, namely the magnetic resistance torque.

For a nonsalient-pole PMSM, no salient pole exists in its rotor structure, that is, $L_d = L_q$. Therefore, the magnetic resistance torque does not exist in Equation (7), that is, $T_{e2} = 0$. In this case, only the single effect of the excitation torque should be considered. Thus, the torque equation of a nonsalient-pole PMSM can be obtained as follows:

$$T_e = T_{e1} = p_n \psi_f i_q = p_n \psi_f i_s \sin\beta, \tag{9}$$

where β is the phase angle between i_s and the d-axis in a d-q synchronous rotating coordinate system.

For a PMSM, ψ_f generated by the permanent magnet is fixed. Therefore, according to Equation (9), the excitation torque is only related to i_q, the vector projection of the three-phase current vector in the q-axis, and when the phase angle β is $\pi/2$, the excitation torque has a maximum value. In this case, i_s coincides with the q-axis, that is, $i_d = i_s cos\beta = 0$.

3. Inertia Simulation Model

In the EDLSS, inertia simulation is typically realised using the load torque. The load and inertia torque simulated by the loading motor are tracked by controlling the current i_q, which causes the simulation system to be consistent with the dynamic characteristics of the target system.

Compared with the load torque, the friction torque of the mechanical structure is too small to be ignored. Assuming that the dynamic equation of the target system in theory is as follows:

$$T_D - T_{basic} = J_s \frac{d\omega}{dt}, \tag{10}$$

where T_D is the driving torque; T_{basic} is the basic load torque; J_s is the moment of inertia of the target system; and $d\omega/dt$ is the angular acceleration.

For the same reason, the dynamic equation of the EDLSS can be assumed as follows:

$$T_D - T_m = J_m \frac{d\omega}{dt}, \tag{11}$$

where T_m is the load torque of motor output; and J_m is the moment of inertia of the simulation system.

At this stage, if the aforementioned two systems are under the same driving torque, the dynamic equation of the inertia simulation system can be expressed as follows:

$$T_m - T_{basic} = (J_s - J_m) \frac{d\omega}{dt}. \tag{12}$$

If the simulation system and the target system have the same ω and $d\omega/dt$ under the same driving torque, the simulation system satisfies the same dynamic characteristics as the target system by simulating the load torque and inertia torque of the target system.

According to the aforementioned analysis, by establishing the inertia simulation model of the simulation system, the realisation method of inertia simulation can be transformed from 'tracking the load torque and inertia torque required for the simulation of the loading motor' to 'tracking the motion characteristics of the target system under the same driving torque', such as the angular velocity and angular acceleration.

The inertia simulation structure is displayed in Figure 3.

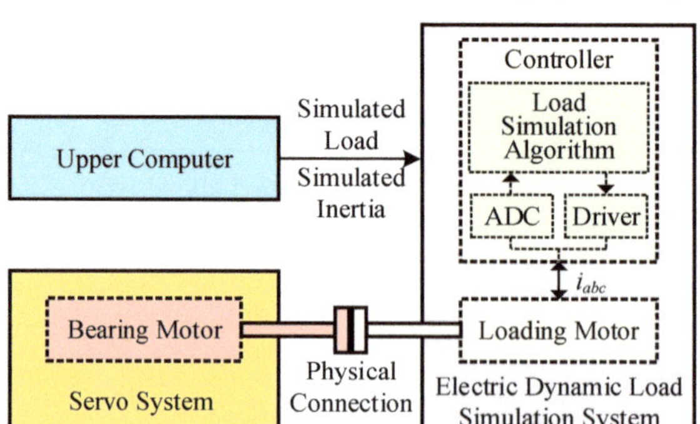

Figure 3. The inertia simulation structure.

In the figure, the bearing motor and the loading motor are connected by the mechanical structure, and the bearing motor drives the synchronous movement of the loading motor. The bearing motor, loading motor, and controller constitute a fast-response closed-loop torque control system. The dynamic compensation torque is generated based on the acceleration variation of the loading motor, which is used to compensate for the torque error caused by the difference between the inertia of the simulation system and the inertia of the target system under the same driving torque. The dynamic compensation torque is converted into the control signal of the inverter so that the error between the output torque and the target torque tends to be 0.

4. Regenerative Braking

When the EDLSS loads the servo system, the loading motor is in the power generation state. The process of simulating the load and inertia torques can be approximated as the braking of the loading motor through reversing current i_q and applying the magnetic resistance to the loading motor under the action of back drag. At this stage, the braking is classified into two categories, namely energy feedback braking and energy consumption braking. When the back electromotive force is sufficient to provide the braking current, the loading motor is in the energy feedback braking state; when the back electromotive force is insufficient to provide the required braking current, the loading motor requires an additional braking current, which is in the braking state of energy consumption.

Experimental results revealed that when the loading motor is in the energy consumption braking state, the additional energy consumption renders the driver that is coupled directly with the loading motor to undervoltage because of the effect of the braking state. Thus, the stability of the EDLSS is affected.

Based on a synchronous rotating coordinate system, the input power of the PMSM can be expressed as follows:

$$P_{in} = u_d i_d + u_q i_q. \tag{13}$$

Equations (3), (4) and (13) can be solved simultaneously as follows:

$$P_{in} = \left(Ri_d - \omega L_q i_q\right) i_d + \left(Ri_q + \omega L_d i_d + \omega \psi_f\right) i_q. \tag{14}$$

If the field-weakening control and the iron loss current of the d-axis are ignored, the maximum torque control is adopted, that is, $i_d = 0$.

Assuming that ω, the electric speed of the rotor, is a constant, then the input power $P_{in}(i_q)$ is a quadratic function of the current i_q, from which the power model of PMSM in the steady state is obtained as follows:

$$P_{in} = \left(\omega \psi_f + Ri_q\right) i_q \tag{15}$$

The input power function diagram of PMSM at various ω values is displayed in Figure 4.

In the figure, the positive direction of i_q is its direction in the motor drive state. Let $P_{in} = 0$ to obtain two zeros of $P_{in}(i_q)$ at $i_q = -\omega\psi_f/R$ and $i_q = 0$. When $i_q < -\omega\psi_f/R$, $P_{in} > 0$. At this stage, the braking requires a positive input power, which indicates that the motor functions in the state of energy consumption braking. When $-\omega\psi_f/R < i_q < 0$, $P_{in} < 0$. At this stage, the input power is negative, which indicates that the motor operates in the energy feedback braking state. When $i_q = -\omega\psi_f/2R$, a minimum value exists in $P_{in}(i_q)$, indicating that the energy feedback of motor braking has a maximum value.

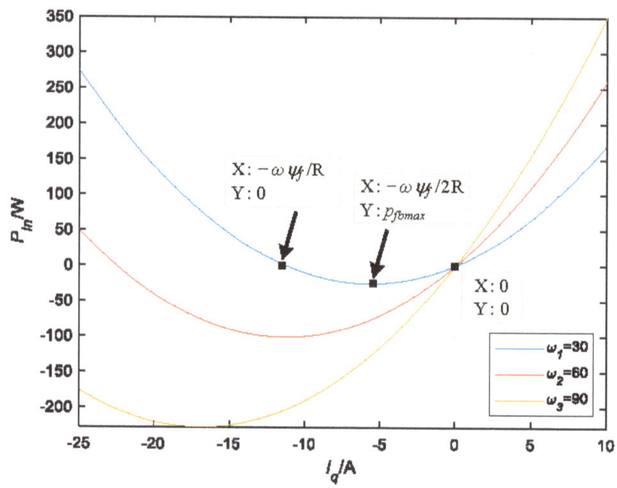

Figure 4. The input power function diagram of PMSM at various ω values (ψ_f = 0.3 Wb, R = 0.8 Ω).

Based on the aforementioned analysis of the braking state and the power model, an energy feedback control method was proposed to improve EDLSS stability caused by the braking state of the loading motor, as displayed in Figure 5. In the figure, dcK_p and dcK_i are the proportional and the integral coefficients of the energy feedback proportional integral (PI) controller; ω_r is the angular velocity of the motor rotor; τ_{dc} is the time constant of the filter; K_ω is the regenerative braking factor, which can be expressed as follows:

$$K_\omega = \frac{C_1 \omega_r \psi_f}{R}, \tag{16}$$

where C_1 is a fixed coefficient.

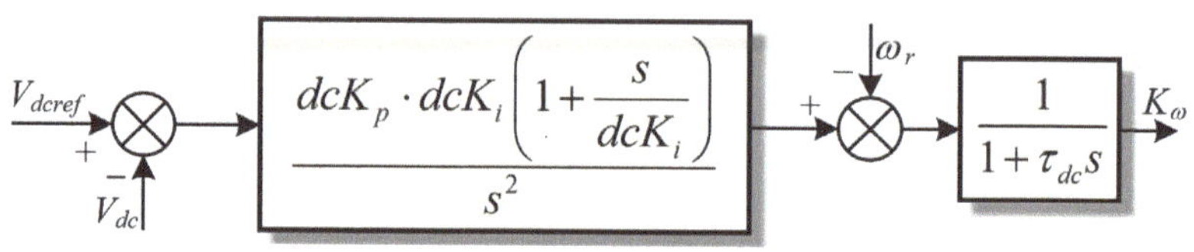

Figure 5. The energy feedback control method.

5. Inertia Simulation Predictive Model

In addition to the effect of the braking state of the loading motor, when the EDLSS is used for an on-load test of a servo system, the disturbance torque introduced by the loading motor in response to the change in the motion state of the bearing system renders achieving synchronous motion with the bearing system difficult, which results in torque fluctuation. Therefore, the smooth load simulation cannot be output.

Such load fluctuations caused by position disturbance can be suppressed by compensating the position signal of the servo system to realise the synchronous motion of the loading motor and the bearing system. However, external sensors are required and other interference can be easily introduced because of its complex calculation. Thus, the load simulation system loses robustness.

Based on the inertia simulation model of the aforementioned simulation system, an inertia simulation predictive model and a load simulation algorithm were proposed. The speed of the loading motor that conforms to the dynamic characteristics at a continuous time is predicted by analysing the dynamic characteristics of the loading motor under the aforementioned two main disturbances when simulating the inertia torque and the load torque and using the non-catastrophic property of the inertia system to suppress the load torque fluctuation caused by position disturbance.

The dynamic equation of the aforementioned inertia simulation model can be rewritten considering strong-coupling torque effect and motor braking state disturbance as two influencing factors of the model to obtain the following expression:

$$T_m - K^2{}_{accbasic}(T_{basic} + K_\omega) = (J_s - J_m)\frac{\omega_{k+1} - \omega_k}{T_s}, \tag{17}$$

where ω_{k+1} is the k + 1 moment predicted rotor angular velocity; ω_k is the k moment rotor angular velocity; T_s is the system control cycle; K_ω is the aforementioned regenerative braking factor; $K_{accbasic}$ is the load matching factor, which can be expressed as follows:

$$K_{accbasic} = \begin{cases} 0 & \omega_k < 0 \\ \frac{\omega_k}{C_2\sqrt{T_{basic}}} & 0 < \omega_k < C_2\sqrt{T_{basic}}, \\ 1 & C_2\sqrt{T_{basic}} < \omega_k \end{cases} \tag{18}$$

where C_2 is a fixed coefficient.

On trimming Equation (17), we obtain the expression of the predicted speed:

$$\omega_{k+1} = \omega_k + T_s\left(\frac{T_m - K^2{}_{accbasic}(T_{basic} + K_\omega)}{(J_s - J_m)}\right). \tag{19}$$

Based on the inertia simulation predictive model, the predicted speed is parsed and input into the control loop of the load simulation algorithm as displayed in Figure 6. The PI controller parameters of the inner current loop and the outer speed loop are tuned by calculating the bandwidth of the current loop and the damping coefficient of the speed loop.

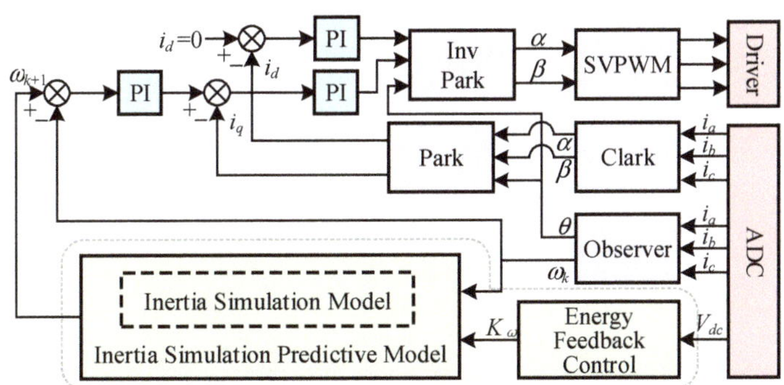

Figure 6. The complete block diagram of the load simulation algorithm.

Assuming that the current bandwidth is 20 times lower than the sampling frequency of the system, we have the following equation:

$$BW_i = \frac{2\pi F_s}{20}, \tag{20}$$

where F_s is the sampling frequency of the system.

Combining Equations (20) and (21), the proportional coefficient iK_p and the integral coefficient iK_i in the current loop can be calculated as follows:

$$\begin{cases} iK_p = BW_i \cdot L_s \\ iK_i = \frac{R}{L_s} \end{cases}. \tag{21}$$

Because the speed controller is in the mechanical domain, which has considerably slower time constants where phase delays can be tighter, system stability is considerably affected. When tuning the speed loop, the moment of inertia is also a crucial parameter to be considered in the controller.

Combining Equation (22), the proportional coefficient $spdK_p$ and the integral coefficient $spdK_i$ in the speed loop are obtained as follows:

$$\begin{cases} spdK_p = \dfrac{4(J_s - J_m)}{3p_n\lambda_r\delta\tau} \\ spdK_i = \dfrac{1}{\delta^2\tau} \end{cases}, \tag{22}$$

where τ is the time constant of velocity filter; λ_r is the back-emf coefficient; δ is the damping coefficient. If δ decreases, the bandwidth of the speed loop increases, which results in considerable overshoot. This phenomenon results in a longer time for speed rise and stability. When $\delta > 1$, the system phase margin is typically greater than 0, which renders the system stable. When δ is close to 1, severe underdamping performance typically results in oscillations.

6. Simulation and Experimental Verification

6.1. Simulation and Analysis

The inertia simulation model, the inertia simulation predictive model, and the load simulation algorithm based on the two models were simulated in MATLAB/Simulink to verify their effectiveness. A nonsalient-pole PMSM model was considered as the loading motor model. The parameters of this model are presented in Table 1. In the model of the bearing motor, an additional PMSM equipped with an independent speed regulation system is required to output the periodic driving torque in the simulation. The start and end moments of load simulation are t = 1 s and t = 10 s, respectively. When t = 6 s, the bearing motor removes the driving torque, and the loading motor runs continuously under the simulated inertial load. Other simulation parameters are as follows: the simulated load inertia is 5.06 kg·m^2; the preset speed is 145 r/min.

Table 1. The parameters of the motor.

Parameter	Value
Number of pole pairs p_n	16
Stator resistance R (Ω)	0.38
d-axis inductance L_d (H)	0.001315
q-axis inductance L_q (H)	0.001315
Permanent magnet flux linkage ψ_f (Wb)	0.4425

The simulation results are displayed in Figure 7. Figure 7a,b respectively detail the speed curve and load torque waveform of the loading motor before and after the application of this method; Figure 7c displays the speed fluctuation of the loading motor at a steady state from t = 3 s to t = 5 s, before and after the application of this method; Figure 7d details the torque ripple of the loading motor at a steady state from t = 3 s to t = 5 s, before and after the application of this method.

Figure 7 reveals that before using the aforementioned algorithm, the maximum speed fluctuation range of the loading motor is approximately 30, and the RMS of the speed fluctuation waveform is 6.953 at a steady state in the t = 3–5 s period. Furthermore, the maximum range of the torque ripple is close to 4, and the RMS of the torque ripple waveform is 0.781 at a steady state in the t = 3–5 s period. After utilising the aforementioned algorithm, the maximum speed fluctuation range of the loading motor is approximately 5, and the RMS of the speed fluctuation waveform is 1.106, which is 84% lower than before using the aforementioned algorithm when it is at a steady state in the t = 3–5 s period. The maximum range of the torque ripple is approximately 0.8, and the RMS of the torque ripple waveform is 0.281, which is 64% lower than before using the aforementioned algorithm when it is at a steady state in the t = 3–5 s period.

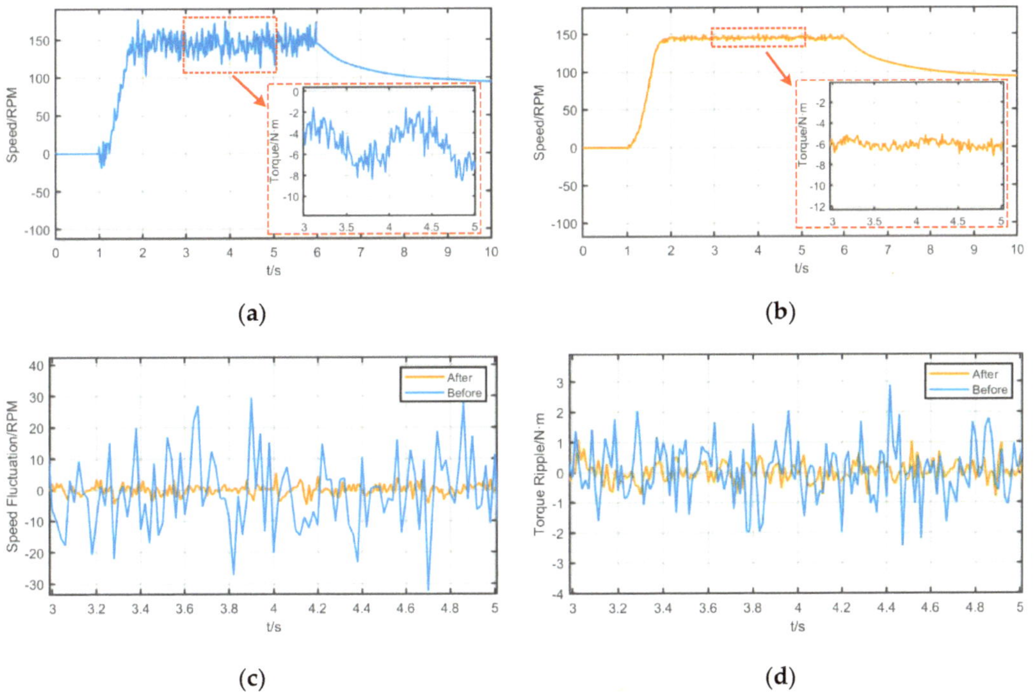

Figure 7. The simulation results. (**a**) The speed curve and load torque waveform of the loading motor before the application of this method. (**b**) The speed curve and load torque waveform of the loading motor after the application of this method. (**c**) The speed fluctuation of the loading motor at a steady state from t = 3 s to t = 5 s, before and after the application of this method. (**d**) The torque ripple of the loading motor at a steady state from t = 3 s to t = 5 s, before and after the application of this method.

Under the same simulation conditions, the simulation results revealed that the predictive model described in this study can calculate the predicted speed of the loading motor during the loading process based on the aforementioned inertia simulation model so that the loading process is not susceptible to strong-coupling torque interference, which is consistent with the dynamic characteristics of the target system under the same working conditions, and the inhibitory effect on the load torque ripple is considerable.

6.2. Experiments and Analysis

To verify the described method, a comparative experiment was performed on a self-developed experimental platform. TMS320F28069, a DSP of Texas Instruments, was used as the controller. A PMSM equipped with an independent speed regulation system was used as the bearing system. The bearing motor outputs periodic driving torque to drive the loading motor. The torque transfer can occur between the bearing motor and loading motor only through mechanical connection. The parameters of the loading motor are the same as in Table 1. The simulated load inertia is 5.06 kg·m^2. The experimental platform is displayed in Figure 8.

Figure 8. The experimental platform.

Figure 9a,b display the preset speed of the bearing motor and the actual speed of the loading motor before and after the application of this method, respectively. The aforementioned speed curves detail the degree of 'synchronous motion' between the loading motor and the bearing motor when simulating the load inertia.

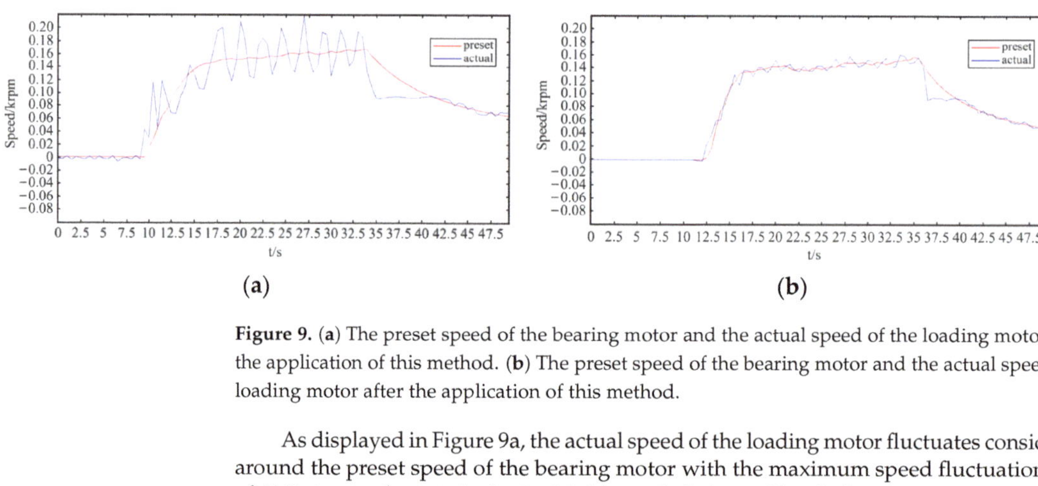

Figure 9. (**a**) The preset speed of the bearing motor and the actual speed of the loading motor before the application of this method. (**b**) The preset speed of the bearing motor and the actual speed of the loading motor after the application of this method.

As displayed in Figure 9a, the actual speed of the loading motor fluctuates considerably around the preset speed of the bearing motor with the maximum speed fluctuation range of 46.8 at a steady state in the t = 16–34 s period; At t = 34 s, the bearing motor removes the drive torque, and the loading motor runs continuously under the simulated inertial load. At this point, the actual speed of the loading motor drops to 93 r/min with a decrease of 74 r/min, which then converges to the preset speed of the bearing motor after 7 s.

Figure 9b displays the result under the same working conditions. The maximum fluctuation range of the actual speed of the loading motor is approximately 7.5, which is 84% lower than before application of the aforementioned algorithm when it is at a steady state in the t = 16–35.5 s period. The bearing motor removes the drive torque at t = 35.5 s, which causes the actual speed of the loading motor to decrease to 92 r/min with a decrease of 63 r/min, which is 14.86% lower than before using the aforementioned algorithm. The speed then converges to the preset speed of the bearing motor after 4 s, which is 42.86% less than before.

Figure 10a,b display the load torque waveform of the loading motor at a steady state from t = 20 s to t = 30 s, before and after the application of this method, respectively. The figures reveal that after using the aforementioned algorithm, the maximum range of the torque ripple is approximately 1.5, which is 75% lower than its value before the application of the aforementioned algorithm, which is approximately 6.

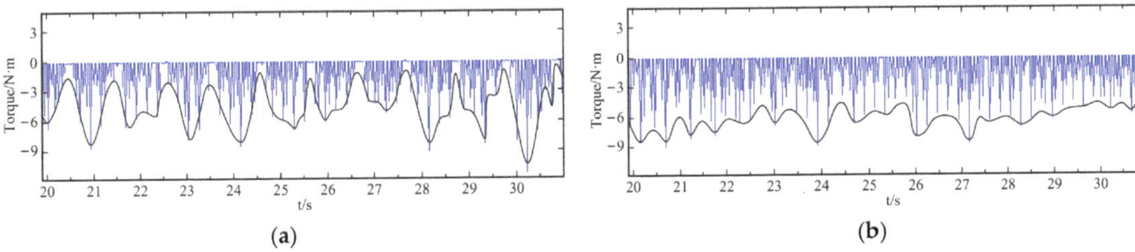

Figure 10. (**a**) The load torque waveform of the loading motor at a steady state from t = 20 s to t = 30 s, before the application of this method. (**b**) The load torque waveform of the loading motor at a steady state from t = 20 s to t = 30 s, after the application of this method.

Combined with the analysis in Figures 9a and 10a, the strong-coupling torque disturbance that is introduced by the loading motor responds to the periodic driving torque, which renders the 'synchronous motion' of the loading motor with the bearing system difficult, resulting in the fluctuation of the load torque output. Therefore, the smooth load torque cannot be output.

Combined with the results of the analysis in Figures 9b and 10b, the predicted speed based on the predictive model satisfies the dynamic characteristics of the target system under the same working conditions to realise the synchronous motion. The strong-coupling torque and load torque fluctuation caused by the position disturbance between the servo system and the loading system were considerably suppressed. The experimental results reveal that after using the aforementioned algorithm, the maximum speed fluctuation range of the loading motor decreases by 84% at a steady state, which is consistent with the simulation results. Furthermore, the maximum range of the torque ripple decreases by 75%, which conforms to the simulation results.

Figure 11a,b display the bus voltage of the system before and after this method is applied. Combining Equation (15) with the previous analysis revealed that because of the effect of the load inertia value, when $i_q < -\omega\psi_f/R$, $P_{in} > 0$, which indicates that the loading motor operates in the state of energy consumption braking and requires additional positive input power. When the motor operates in the state of energy consumption braking, an additional input power results in a decrease in the bus voltage, which affects EDLSS stability. In this experiment, when the bus voltage decreases to less than 36 V, the EDLSS is in a critical state and cannot maintain the normal operating voltage of the processor and the peripheral circuit.

As displayed in Figure 11a, during the t_1–t_2 period, both the loading motor and the bearing motor were in a standby state, and the bus voltage of the system was 24 V; during the t_2–t_5 period, when the loading motor was dragged by the bearing motor and simulated the aforementioned load inertia, the bus voltage should have increased and maintained its value at 83.2 V. However, because of the influence of the energy consumption braking state, the bus voltage decreased 12 times in 7 s. At t_4 and t_5 moments, the bus voltage decreased sharply to 36 V. At the t_6 moment, because of the braking state of energy consumption, the bus voltage could not maintain the normal operating voltage.

Utilising the proposed method to ensure the motor runs under the same working conditions, an oscilloscope was used to measure the bus voltage (Figure 11b). In the t_3–t_5 period, the jitter of the bus voltage decreased 7 times in 9 s. At the t_4 moment, through the effective control of the loading motor braking state, the bus voltage drop was limited to 52 V, which negatively affected the energy consumption braking state on EDLSS suppression.

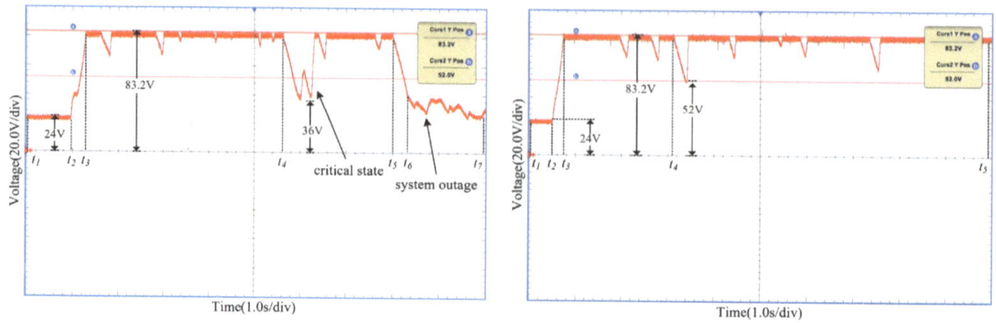

(a) (b)

Figure 11. (**a**) The bus voltage of the system before this method is applied. (**b**) The bus voltage of the system after this method is applied.

7. Conclusions

In this study, a load simulation algorithm based on an inertia simulation predictive model was proposed to mitigate strong-coupling torque disturbance and load torque fluctuation resulting from the motion state change of the bearing system in EDLSS. First, dynamic equations of both the EDLSS and the target system were combined to propose an inertia simulation model. Next, based on the inertia simulation model, an inertia simulation predictive model and a load simulation algorithm were proposed. Both the braking state and the power model of PMSM were analysed, and an energy feedback control method was proposed. The proposed method was simulated and experimentally verified. The following results were obtained:

(1) The inertia simulation model in this paper converted the conventional realisation method into the tracking of the motion characteristics of the target system under the same working conditions so that the loading process was not limited to the load torque and was not susceptible to the interference of strong-coupling torque. During the process of simulating the load and inertia torque, the proposed model paid more attention to the dynamic characteristics of the target system rather than the machine parameters, which made the dynamic characteristics of the simulation system consistent with the target system without a specific motor.

(2) The load simulation algorithm in this study analysed the predicted speed of the loading motor based on the predictive model so that the simulation system could satisfy the dynamic characteristics of the target system under the same working conditions. By realising synchronous motion, the strong-coupling torque and the load torque fluctuation caused by position disturbance of the servo system and the loading system were suppressed considerably.

(3) The energy feedback control method effectively controlled the braking state of the loading motor so that the adverse effects of its energy consumption braking state on the EDLSS were suppressed.

In conclusion, the proposed method can suppress the strong-coupling torque and load torque fluctuation considerably, as well as the adverse effects of the energy consumption braking state of the loading motor in the loading process of the EDLSS. With the exception of EDLSSs, this method has good reference value in servo system dynamic stiffness tests, mechanical back-to-back tests, and other applications.

Author Contributions: Conceptualization, W.W.; methodology, W.W., K.L., Z.W., X.L. and H.Z.; software, formal analysis, validation, K.L. and W.W.; writing—original draft preparation, K.L.; writing—review and editing, K.L. and W.W.; supervision, W.W.; project administration, W.W. All authors have read and agreed to the published version of the manuscript.

Funding: This research was funded by the National Natural Science Foundation of China, grant number 51977150.

Institutional Review Board Statement: Not applicable.

Informed Consent Statement: Not applicable.

Data Availability Statement: Not applicable.

Conflicts of Interest: The authors declare no conflict of interest.

References

1. Junejo, A.K.; Xu, W.; Mu, C.; Ismail, M.M.; Liu, Y. Adaptive speed control of PMSM drive system based anew sliding-mode Reaching Law. *IEEE Trans. Power Electron.* **2020**, *35*, 12110–12121. [CrossRef]
2. Li, L.; Pei, G.; Liu, J.; Du, P.; Pei, L.; Zhong, C. 2-DOF Robust H_∞ Control for Permanent Magnet Synchronous Motor with Disturbance Observer. *IEEE Trans. Power Electron.* **2021**, *36*, 3462–3472. [CrossRef]
3. Xu, W.; Junejo, A.K.; Liu, Y.; Hussien, M.G.; Zhu, J. An Efficient Antidisturbance Sliding-Mode Speed Control Method for PMSM Drive Systems. *IEEE Trans. Power Electron.* **2021**, *36*, 6879–6891. [CrossRef]
4. Xiao, Q.; Li, Z.; Luo, B.; Wang, T.; Wen, D.; Zhang, Y. Improved Three Vector Model Predictive Torque Control of PMSM. *Prog. Electromagn. Res. M* **2022**, *109*, 217–229. [CrossRef]
5. Zhao, X.; Zhu, J.; Wang, Z. Model Predictive Current Control of Permanent Magnet Synchronous Motor for Marine Electric Propulsion. *Prog. Electromagn. Res. M* **2022**, *109*, 241–251. [CrossRef]
6. Liu, T.; Tong, Q.; Zhang, Q.; Liu, K. A Low Dynamic Deviation and High Robustness Position Control Strategy Based on Virtual Reference. *IEEE Access* **2022**, *10*, 32290–32300. [CrossRef]
7. Zhu, H.; Wu, M. Direct Control of Bearingless Permanent Magnet Synchronous Motor Based on Prediction Model. *Prog. Electromagn. Res. M* **2021**, *101*, 127–138. [CrossRef]
8. Zhao, S.; Chen, Y.; Mao, Y. Adaptive load observer-based feed-forward control in PMSM drive system. *Trans. Inst. Meas. Control.* **2015**, *37*, 414–424. [CrossRef]
9. Khiabani, A.G.; Heydari, A. Optimal torque control of permanent magnet synchronous motors using adaptive dynamic programming. *IET Power Electron.* **2020**, *13*, 2442–2449. [CrossRef]
10. Tan, L.N.; Pham, T.C. Optimal Tracking Control for PMSM With Partially Unknown Dynamics, Saturation Voltages, Torque, and Voltage Disturbances. *IEEE Trans. Ind. Electron.* **2022**, *69*, 3481–3491. [CrossRef]
11. Pan, Y.; Liu, X.; Zhu, Y.; Liu, B.; Li, Z. Feedforward Decoupling Control of Interior Permanent Magnet Synchronous Motor with Genetic Algorithm Parameter Identification. *Prog. Electromagn. Res. M* **2021**, *102*, 117–126. [CrossRef]
12. Sun, Y.; Cui, Q.; Yuan, Y. Research on Control of Permanent Magnet Synchronous Motor Based on Second-Order Sliding Mode. *Prog. Electromagn. Res. M* **2019**, *85*, 11–20. [CrossRef]
13. Liu, C.; Luo, G.; Duan, X.; Chen, Z.; Zhang, Z.; Qiu, C. Adaptive LADRC-Based Disturbance Rejection Method for Electromechanical Servo System. *IEEE Trans. Ind. Appl.* **2020**, *56*, 876–889. [CrossRef]
14. Huang, J.; Ma, P.; Bao, G.; Gao, F.; Shi, X. Research on Position Servo System Based on Fractional-Order Extended State Observer. *IEEE Access* **2020**, *8*, 102748–102756. [CrossRef]
15. Huang, Y.; Tao, T.; Liu, Y.; Chen, K.; Yang, F. DSC-FLL Based Sensorless Control for Permanent Magnet Synchronous Motor. *Prog. Electromagn. Res. M* **2020**, *98*, 171–181. [CrossRef]
16. Feng, L.; Yan, H. Nonlinear Adaptive Robust Control of the Electro-Hydraulic Servo System. *Appl. Sci.* **2020**, *10*, 4494. [CrossRef]
17. Tian, Y.; Chai, Y.; Feng, L. Simultaneous Load Disturbance Estimation and Speed Control for Permanent Magnet Synchronous Motors in Full Speed Range. *Appl. Sci.* **2020**, *10*, 9006. [CrossRef]
18. Pietrusewicz, K.; Waszczuk, P.; Kubicki, M. MFC/IMC Control Algorithm for Reduction of Load Torque Disturbance in PMSM Servo Drive Systems. *Appl. Sci.* **2019**, *9*, 86. [CrossRef]
19. Choo, K.-M.; Won, C.-Y. Design and Analysis of Electrical Braking Torque Limit Trajectory for Regenerative Braking in Electric Vehicles with PMSM Drive Systems. *IEEE Trans. Power Electron.* **2020**, *35*, 13308–13321. [CrossRef]
20. Heydari, S.; Fajri, P.; Rasheduzzaman, M.; Sabzehgar, R. Maximizing Regenerative Braking Energy Recovery of Electric Vehicles Through Dynamic Low-Speed Cutoff Point Detection. *IEEE Trans. Transp. Electrif.* **2019**, *5*, 262–270. [CrossRef]
21. Choo, K.-M.; Jeong, W.-S.; Kim, B.-J.; Park, J.-M.; Lee, J.-H.; Won, C.-Y. Maximum Regenerative Power Braking for Interior Permanent Magnet Synchronous Motor. In Proceedings of the 2019 IEEE Transportation Electrification Conference and Expo, Asia-Pacific (ITEC Asia-Pacific), Seogwipo, Korea, 8–10 May 2019.
22. Lu, D.; Ouyang, M.; Gu, J.; Li, J. Optimal Regenerative Braking Control for Permanent Magnet Synchronous Motors in Electric Vehicles. *Proc. CSEE* **2013**, 33, 83–91+12. [CrossRef]

machines

Article

Effect of Flux Barriers on Short-Circuit Current and Braking Torque in Dual Three-Phase PM Machine

Lin Liu, Kai Wang *, Lingling Guo and Jian Li

College of Automation, Nanjing University of Aeronautics and Astronautics, 29 Jiang Jun Road, Nanjing 211106, China; linliu@nuaa.edu.cn (L.L.); llingguo@nuaa.edu.cn (L.G.); jli@nuaa.edu.cn (J.L.)
* Correspondence: k.wang@nuaa.edu.cn

Abstract: This paper investigates the influence of stator flux barriers on the short-circuit current (SCC) and braking torque of a dual three-phase permanent magnet (PM) synchronous machine. By optimizing the position and width of stator flux barriers, the machine has a lower amplitude of short-circuit current and brake torque when the short-circuit fault occurs. First, the SCC and braking torque are analytically derived. The amplitude of SCC is proportional to the PM flux linkage and inversely proportional to the inductance. The braking torque is proportional to the square of the PM flux linkage and inversely proportional to inductance. Then, the equivalent magnetic circuit model of flux barriers is established. Its influence on flux linkage and inductance is analyzed, and the improvement mechanism of output torque and fault tolerance is revealed. Furthermore, the flux barriers' width is optimized by finite element analysis and the theoretical analysis is verified. Finally, experiments on the prototype machine are carried out for the validation.

Keywords: permanent magnet machine; dual three-phase winding; stator flux barriers; short circuit current; braking torque

Citation: Liu, L.; Wang, K.; Guo, L.; Li, J. Effect of Flux Barriers on Short-Circuit Current and Braking Torque in Dual Three-Phase PM Machine. *Machines* **2022**, *10*, 611. https://doi.org/10.3390/machines10080611

Academic Editor: Gyftakis Konstantinos

Received: 22 June 2022
Accepted: 23 July 2022
Published: 26 July 2022

1. Introduction

Multi-phase machines have been widely used in aviation, aerospace, and electric vehicles due to their high torque density, low torque ripple, and high reliability [1–3]. As a specific multi-phase machine, dual three-phase machines have received extensive attentions in recent years [4–6]. A complete fault-tolerant system needs not only fault-tolerant control, but also the corresponding fault-tolerant machines. Higher fault-tolerance capability requires that the fault phase does not affect the healthy phase, and that it has a smaller short-circuit current (SCC) and braking torque [6–10].

It is well known that higher self-inductance can suppress the SCC and braking torque, and lower mutual inductance can reduce the influence of the fault phase relative to the healthy phase. In [6–8], the influence of slot/pole mating on inductance and output characteristics was studied. However, the SCC, braking torque, and fault tolerant operation were not analyzed. In [9,10], based on the particularity of two sets of three-phase windings of a dual three-phase machine, the influence of the deviation angle of two sets three-phase windings on the stator magneto motive force (MMF) harmonics, torque capability, torque ripple, and eddy-current losses of the machine were analyzed. Then in [11], a special 15° phase shift was proposed to suppress the SCC of a 24-slot/10-pole dual three-phase machine. However, in [12], compared with other phase shifts, the 15° phase shift had the largest SCC amplitude. This phenomenon is due to the fact that 1st-order stator MMF harmonic suppression effect is different [13]. In [13], for a 48-slot/22-pole dual three-phase machine, the phase shift of 30° and 7.5° had a smaller SCC. It is worth noting that the machines above all adopted the overlapping winding structure with a coil span greater than 1. The winding structure not only had a long end, but the two sets of three-phase winding also had an electrical and magnetic coupling. Therefore, concentrated windings are more suitable for fault-tolerant machines. In [14–16], various concentrated winding configurations of 12-slot/10-pole dual three-phase windings were taken into account. The results showed that the single-layer (SL) concentrated winding had the maximum self-inductance and the minimum mutual inductance, which can improve the fault tolerance of the machine. However, the MMF harmonics of SL concentrated winding is greater than that of double-layer (DL) concentrated winding and distributed winding, which increases the losses, vibration, and noise of the machine. Therefore, one of the difficulties in designing a fault-tolerant PM machine is how to improve the fault-tolerant capacity of the machine while ensuring that the other electromagnetic characteristics are not weakened.

Furthermore, the stator flux barrier structure was proposed in [17–23]. This design not only effectively suppressed the sub-harmonics of the stator MMF, but also realized the electrical, magnetic, thermal, and physical isolation between phases, thus effectively improving the fault tolerance of the machine. Then, in [22,23], it was found that the stator flux barriers can effectively reduce the SCC of the machine. However, it is worth noting that the flux barriers studied in the above papers were all located on the auxiliary teeth, and this will severely reduce the torque density of the machine with the slot number greater than the pole number. The effects of different position of flux barriers on SCC, braking torque, and fault-tolerant operation are rarely investigated. Therefore, this paper mainly focuses on the evaluation and comparison of the influence of different stator flux barrier positions and width on the SCC and braking torque for the 12-slot/10-pole dual three-phase machine, and carries out deep analysis of fault-tolerant operation.

The purpose of this paper is to reveal the principle of SCC and braking torque suppression of a dual three-phase machine with stator flux barrier structure, so as to improve the fault-tolerant operation ability of the machine. In Section 2, different structures of stator flux barriers are proposed. Then, the analytical expressions of SCC and braking torque for single-phase short-circuit fault are derived, and the effect of the stator flux barriers on flux linkage and inductance is revealed by simplified equivalent magnetic circuit model in Section 3. In Section 4, the 12-slot/10-pole PM machine with different stator flux barriers is designed, and the electromagnetic and short-circuit performances are comprehensively calculated by using the finite-element method (FEM) to verify the theoretical analysis. In Section 5, experiments on the prototype machine are carried out for validation. Finally, conclusions are drawn in Section 6.

2. Machine with Different Flux Barriers and Winding Distribution

In this section, the cross-section models of a traditional stator machine and stator flux barrier machine are first presented, as shown in Figure 1. The main parameters of the machine are shown in Table 1. There are two kinds of winding structures for the traditional 12-slot/10-pole dual three-phase machine, which are single-layer (SL) concentrated winding and double-layer (DL) concentrated winding respectively. Figure 1a shows the machine with the traditional stator with DL winding (DL-TSM), and two sets of three-phase winding deviations of 30°. Figure 1b shows the machine with the traditional stator with SL winding (SL-TSM), and two sets of three-phase winding deviations of 60°. Compared with the traditional dual three-phase machine, the stator of a flux barrier machine is composed of six E-cores. There is a flux barrier between each E-core and the adjacent core. Two types of flux barriers are used in the stator: the armature tooth flux barriers and the auxiliary tooth flux barriers. Figure 1c shows the stator with armature tooth flux barriers and SL windings. Figure 1d shows the machine with auxiliary tooth flux barriers and SL windings.

The above four kinds of machines adopt the same rotor, and the permanent magnet material is NdFeB 35SH. The DL winding structure has two coils for each phase, and the SL winding structure has one coil for each phase. However, to highlight the effect of stator flux barriers, the number of turns per phase is the same for all investigated machines.

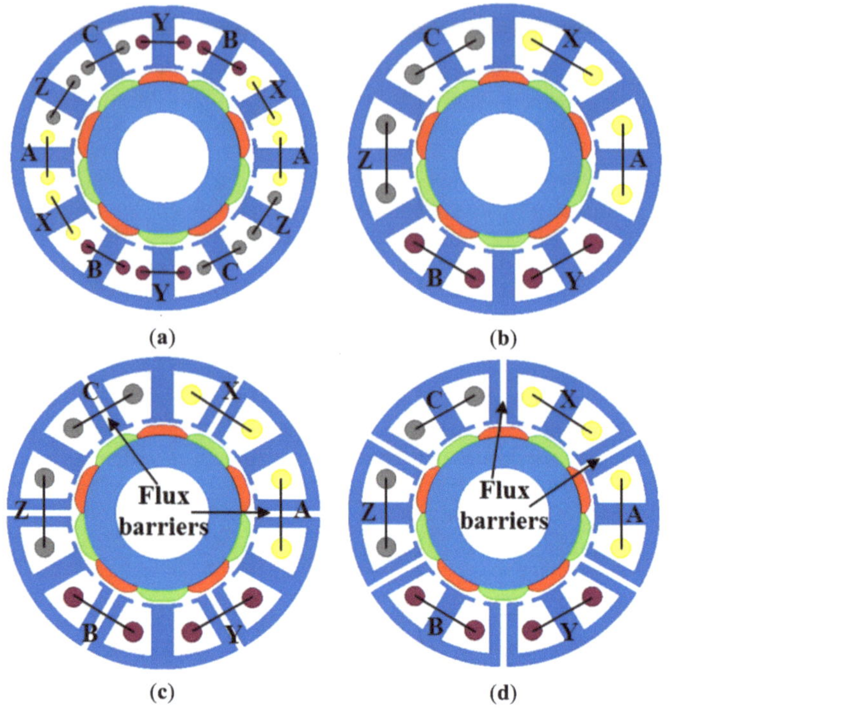

Figure 1. Cross sections of 12-slot/10-pole dual three-phase machines. (**a**) DL-TSM. (**b**) SL-TSM. (**c**) SL-ATM. (**d**) SL-ABM.

Table 1. Main Machine Parameters.

Parameter	Value	Parameter	Value
Stator slot number	12	Stack length (mm)	50
Rotor pole number	10	Air-gap length (mm)	1
Rated speed (r/min)	1000	Rotor outer radius (mm)	25
Rated torque (N·m)	4	Rotor inner radius (mm)	15
Stator outer radius (mm)	45	Magnet thickness (mm)	3
Stator inner radius (mm)	26	Flux barrier width (mm)	2.5
Tooth width (mm)	6	NdFeB 35 remanence (T)	1.2
Yoke thickness (mm)	3	Number of turns/phase	80

3. Influence of Stator Flux Barriers on SCC and Braking Torque

3.1. SCC and Braking Torque Calculation

The winding short circuit fault is one of the common fault types in the machine. Short circuit positions can be divided into winding end short-circuits and inter-turn short-circuits. Compared with inter-turn short-circuit, the possibility of short-circuit at winding ends is less. Usually, in order to alleviate the inter-turn short-circuit fault, the winding ends are artificially short-circuited [24]. The equivalent circuit of the healthy condition of A-phase winding and the short-circuit at the end is shown in Figure 2.

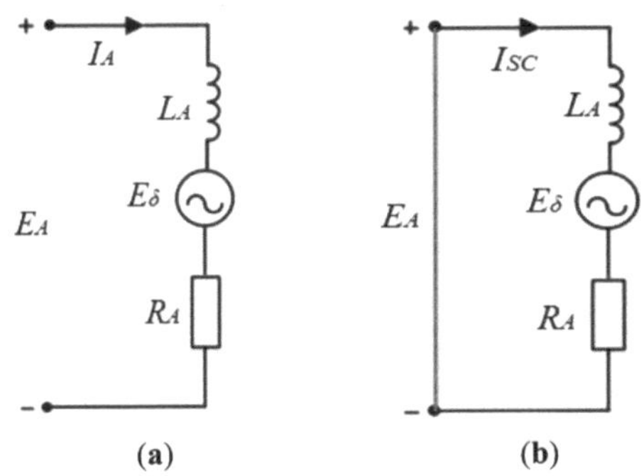

Figure 2. Equivalent circuit. (**a**) Healthy winding. (**b**) Single phase short circuit fault.

When a short circuit occurs at the winding end, the terminal voltage is 0, and the voltage equation when a fault occurs is shown in Equation (1).

$$E_{\mathrm{A}} = L_{\mathrm{A}}\frac{di_{\mathrm{SCC}}}{dt} + E_{\delta} + R_{\mathrm{A}} i_{\mathrm{SCC}} = 0 \tag{1}$$

where L_{A} is the inductance of phase A, R_{A} is the resistance of phase A, i_{SCC} is the short-circuit current, E_{δ} is the induced voltage, t is the time. The expression of SCC can be obtained to solve the above differential by using Laplace transform. The SCC can be expressed as [25,26]

$$i_{\mathrm{SCC}}(t) = N\psi_m\omega\frac{R_{\mathrm{A}}\sin(\theta_0+\omega t) - \omega L_{\mathrm{A}}\cos(\omega t)}{R_{\mathrm{A}}^2 + (\omega L_{\mathrm{A}})^2} + \left[i_{\mathrm{A}}\cos(\theta_0+\delta) + \frac{N\psi_m\omega(L_{\mathrm{A}}\omega\cos\theta_0 - R_{\mathrm{A}}\sin\theta_0)}{L_{\mathrm{A}}^2\omega^2 + R_{\mathrm{A}}^2}\right]e^{-\frac{R_{\mathrm{A}}}{L_{\mathrm{A}}}t} \tag{2}$$

where N is the number of turns per phase, ψ_m is the PM flux linkage, ω is electric angular frequency, and θ is the electrical angle between the axis of the short-circuited coil and d-axis. θ_0 is the initial value of θ, $\theta = \theta_0 + \omega t$, the electrical angle $\delta = \arctan(-L_A\omega/R_A)$. Therefore, the steady-state SCC can be expressed as

$$i_{\mathrm{SCC\_steady}}(t) = \frac{N\psi_m\omega\cos(\omega t+\theta_0+\delta)}{\sqrt{L_{\mathrm{A}}^2\omega^2 + R_{\mathrm{A}}^2}} \tag{3}$$

According to (3), ωL_{A} is much greater than R_{A} ($\omega L_{\mathrm{A}} >> R_{\mathrm{A}}$) for the machine, so the amplitude of steady SCC is given by

$$i_{\mathrm{SCC\_steady}} = \frac{N\psi_m}{L_{\mathrm{A}}} \tag{4}$$

From Equation (4), the amplitude of steady SCC is proportional to PM flux linkage and inversely proportional to inductance. When a short circuit fault occurs in the A phase, the output torque of the machine can be represented as [27]

$$\begin{aligned} T_{\mathrm{torque}} &= \frac{i_{\mathrm{SCC}}E_{\mathrm{A}} + i_{\mathrm{B}}E_{\mathrm{B}} + i_{\mathrm{C}}E_{\mathrm{C}} + i_{\mathrm{X}}E_{\mathrm{X}} + i_{\mathrm{Y}}E_{\mathrm{Y}} + i_{\mathrm{Z}}E_{\mathrm{Z}}}{\omega} \\ &= \frac{\left(i_{\mathrm{A}}\cos(\theta_0+\delta) - \omega N\psi_m\frac{R_{\mathrm{A}}\sin\theta_0 - \omega L_{\mathrm{A}}\cos\theta_0}{R_{\mathrm{A}}^2 + \omega L_{\mathrm{A}}^2}\right)e^{-\frac{R_{\mathrm{A}}}{L_{\mathrm{A}}}t}\sin(\omega t+\theta_0)}{\omega} \\ &+ N\psi_m\frac{R_{\mathrm{A}}\sin(\omega t+\theta_0) - \omega L_{\mathrm{A}}\cos(\omega t+\theta_0)}{R_{\mathrm{A}}^2 + \omega L_{\mathrm{A}}^2}\sin(\omega t+\theta_0) + \frac{i_{\mathrm{B}}E_{\mathrm{B}} + i_{\mathrm{C}}E_{\mathrm{C}} + i_{\mathrm{X}}E_{\mathrm{X}} + i_{\mathrm{Y}}E_{\mathrm{Y}} + i_{\mathrm{Z}}E_{\mathrm{Z}}}{\omega} \end{aligned} \tag{5}$$

From (5), the steady-state braking torque generated by the SCC is

$$T_{\mathrm{Braking\ torque}}(t) = -\frac{(N\psi_m\omega)^2\cos(\omega t+\theta_0)\sin(\omega t+\theta_0)}{\sqrt{L_{\mathrm{A}}^2\omega^2 + R_{\mathrm{A}}^2}} \tag{6}$$

It can be seen from (6) that braking torque is not only related to self-inductance and flux linkage, but also to the speed and resistance. When the machine runs in the high-speed area, the ωL_A is much greater than R_A (ωL_A>>R_A) for the machine, so the amplitude of steady braking torque is given by [26]

$$T_{\text{Braking torque}} = -\frac{\omega N^2 \psi_m^2 \sin(2\theta)}{2L_A} \tag{7}$$

Obviously, the amplitude of braking torque is proportional to the square of PM linkage and inversely proportional to the inductance. However, it is worth noting that when a short-circuit fault occurs, it brings a significant torque ripple of twice the electrical frequency. Two-phase short circuits and three-phase symmetric short circuits can be analyzed similarly and will not be described here. Overall, both short-circuit current and braking torque are ultimately related to the flux linkage and inductance. The fault-tolerant capacity of machine can be improved effectively by optimizing flux linkage and inductance.

3.2. Magnetic Flux Analysis with Flux Barriers

According to the above analysis, the permanent magnet flux linkage is related to the SCC amplitude and the maximum braking torque. Therefore, it is necessary to accurately calculate the influence of different stator structures on the PM flux linkage. For the dual three phase permanent magnet machine, the PM flux linkage can be expressed as

$$\psi_m = k_w N R_o L_{ef} \int B_g(\alpha, t) d\alpha \tag{8}$$

where k_w is the winding coefficient, B_g is the no-load air-gap flux density, α is the mechanical angle, R_o is the outer radius of rotor, and L_{ef} is the lamination length. For a 12-slot/10-pole dual three-phase machine, the distribution coefficients of SL winding structure and DL winding structure are 1, so the winding coefficient is determined by the short distance coefficient. However, the existence of stator flux barriers changes the machine short-distance coefficient, and different positions and widths have different influences on the short-distance coefficient, as shown in Table 2 and Figure 3.

Table 2. Winding Coefficient for Dual Three-Phase Machine with Different Winding Structure and Stator Structure.

Machine Type	k_d	k_p	$k_w = k_d \cdot k_p$
DL-TSM	1	$\sin\left(\frac{\tau_s}{\tau_p}\cdot\frac{\pi}{2}\right)$	0.966
SL-TSM	1	$\sin\left(\frac{\tau_s}{\tau_p}\cdot\frac{\pi}{2}\right)$	0.966
SL-ATM	1	$\sin\left(\frac{(\tau_s+\Delta)}{\tau_p}\cdot\frac{\pi}{2}\right)$	1
SL-ABM	1	$\sin\left(\frac{(\tau_s-\Delta)}{\tau_p}\cdot\frac{\pi}{2}\right)$	0.928

Where $\tau_s = 2\pi/N_s$ is the slot pitch, $\tau_p = 2\pi/(2p)$ is the pole pitch, Δ is the radius of the stator flux barrier, β is the width of the stator magnetic barrier, and R_i is the stator inner diameter, $\Delta = \text{asin}(\beta/2R_i)$. Table 2 shows the winding coefficients of the machines when the width of the stator flux barrier is 2.5 mm. Figure 3 shows the variation of the winding coefficient of the stator flux barrier machine with the width of the flux barriers.

The no-load air-gap flux density can be calculated by the following expression:

$$B_g(\alpha, t) = F(\alpha, t) \times \Lambda(\alpha) \tag{9}$$

where F is PM MMF and Λ is the air-gap permeance. Figure 4 shows the stator flux line paths of different machines. It can be seen that the presence of stator flux barriers changes the flux line paths.

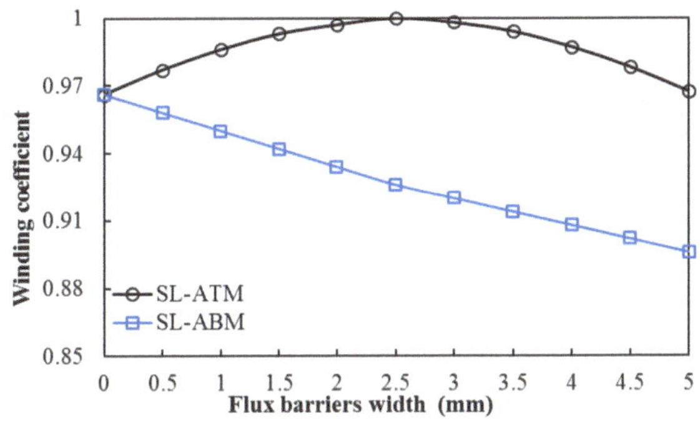

Figure 3. Winding coefficient versus flux barrier width for machines.

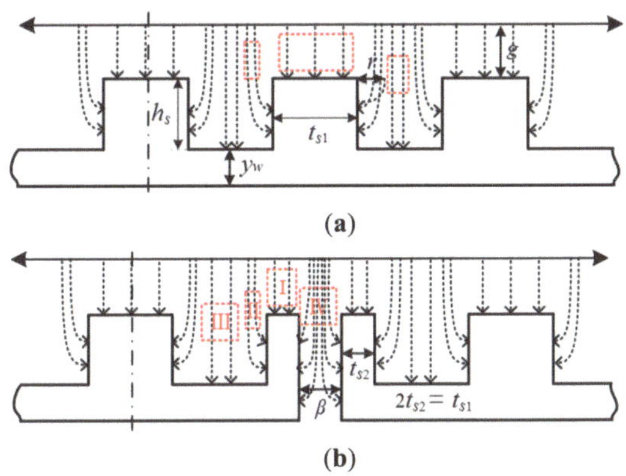

Figure 4. Path of flux lines in air-gap for different machines. (**a**) Traditional stator machine. (**b**) The machine with stator flux barriers.

According to the path of flux lines shown in Figure 4, when the flux lines go directly to the stator tooth (part I), the air-gap permeability can be expressed as

$$\Lambda_1 = \frac{\mu_0}{g + (\mu_r/h_m)} \tag{10}$$

When the flux lines go through the slot (part II and part III), Λ can be expressed as

$$\Lambda_{2,3} = \begin{cases} \frac{2\mu_0}{2(g+(\mu_r/h_m))g+\pi r_1}, 0 < r_1 < \frac{2h_s}{\pi} \\ \frac{\mu_0}{g+(\mu_r/h_m)+h_s}, r_1 \geq \frac{2h_s}{\pi} \end{cases} \tag{11}$$

The stator barrier can be thought of as a slot; the slot depth is infinite, and the width of the slot is β. When the flux lines go through stator flux barriers (part IV), Λ can be expressed as

$$\Lambda_4 = \begin{cases} \frac{2\mu_0}{2(g+(\mu_r/h_m))g+\pi r_2}, \beta \leq r_2 \\ 0, \beta > r_2 \end{cases} \tag{12}$$

where μ_0 is the air-gap permeability, μ_r is the PM permeability, g is the air-gap length, h_m is the magnet thickness, h_s is the slot depth, β is the width of flux barriers, and r_1 and r_2 are the flux path lengths in slot openings and flux barriers. In addition, the air-gap permeance $\Lambda(\alpha)$ due to the slotting can be given by ignoring the high-order harmonics, i.e., [26]

$$\Lambda(\alpha) = \sum_{i=0,1,2...}^{\infty} \Lambda_i \cos\left(i\frac{N_s}{2}\alpha\right) \tag{13}$$

where N_s is the stator slot number and Λ_i is the amplitude of ith order harmonic of air-gap permeance. The rotors of the four different machines are consistent, so the rotor magnetomotive force can be expressed as

$$F(\alpha,t) = \sum_{j=1,3,5...}^{\infty} \frac{gh_m + \mu_r}{\mu_0 h_m} B_r \sin(jp_r\alpha - j\omega t) \tag{14}$$

where B_r is the PM remanence and p_r is the pole pair of the machine.

By substituting (13) and (14) into (8), the PM flux linkage can be expressed as

$$\psi_m = \sum_{i=0,1,2...}^{\infty} \sum_{j=1,3,5...}^{\infty} \frac{F_j \Lambda_i N k_w L_{ef} R_o}{i\frac{N_s}{2} \pm jP_r} \times \sin\left[\left(i\frac{N_s}{2} \pm jp_r\right)\frac{2\pi}{N_s}\right] \cos(\pm jp_r\omega t) \tag{15}$$

An analytical model similar to the above air-gap permeance can be used regardless of the presence of flux barriers on the stator, and the calculation method of the PM flux linkage is basically the same, so it will not be described again. There is a slight deviation between the analytical results and the finite element analysis (FEA) because the saturation of the stator core is ignored in the analytical process. Figure 5 shows the variation of the PM flux linkage of different machines with the width of the flux barriers, which the trend of analytical calculation is consistent with that of FEA. It can be seen that the PM flux linkage of the SL-ATM will first increase and then decrease with the increase of the flux barrier width. However, the PM flux linkage of the SL-ABM will decrease with the increase of the flux barrier width.

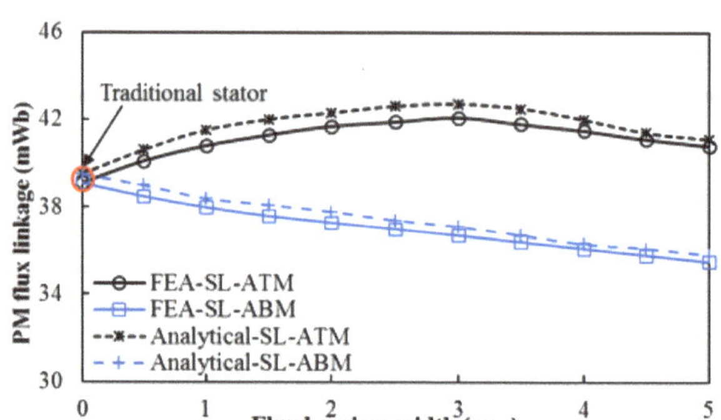

Figure 5. Variation of the PM flux linkage of different machines with the width of the flux barriers.

3.3. Inductance Analysis with Flux Barriers

As mentioned above, a larger self-induction can effectively suppress the short-circuit current and prevent damage to the machine windings. Due to the fact that the machine adopts a surface-mounted permanent magnet, the effective air gap of the machine is large, and the stator slot thickness of the permanent magnet fault-tolerant machine is high; the machine is not easily saturated, so the influence of the saturation of the ferromagnetic material of the machine can be ignored when solving the inductance analytical formula. According to the definition, the calculation formula of inductance is expressed as [28]:

$$L_A = \frac{\psi}{I} = \frac{N\Phi}{I} = \frac{NF_m}{IR_m} = \frac{N^2}{R_m} \tag{16}$$

where I is the excitation current of the winding, ψ is the magnetic flux linkage by the winding current, Φ is the magnetic flux, N is coil turns per phase, R_m is the total reluctance of the magnetic circuit, and F_m is the stator magnetomotive force. According to (16), when the number of coil turns is fixed, the inductance is related to the reluctance of the machine and its distribution. According to the distribution of magnetic flux, the magnetic circuit model of one branch of A-phase winding of the armature tooth flux barrier machine is established, as shown in Figure 6. The magnetic circuit model includes the iron reluctance, the air-gap reluctance, the slot opening reluctance, the inter-slot reluctance, and the flux barrier reluctance. The pressure drop of magnetic flux on the iron material is ignored, so the magnetic circuit model can be simplified as shown in Figure 6c.

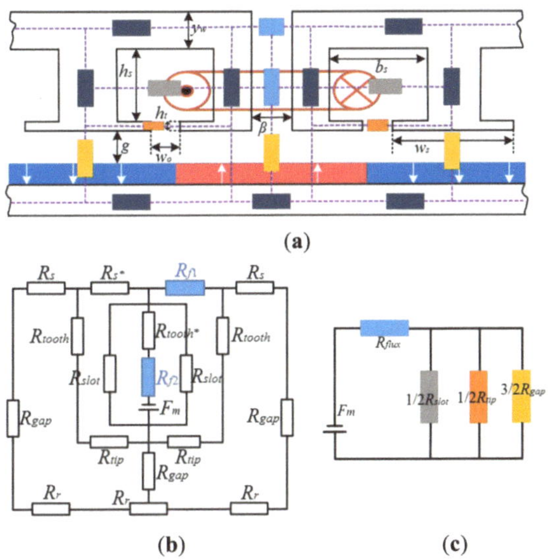

Figure 6. The equivalent reluctance network model of armature tooth flux barrier machine. (**a**) The cross section model of phase A. (**b**) Equivalent reluctance network model of phase A. (**c**) Simplified reluctance network model of phase A.

Where R_{flux} is the reluctance of flux barriers, R_{gap} is the air-gap reluctance, and R_{slot} and R_{tip} are the reluctance of inter-slot and slot opening. R_{flux}, R_{slot}, R_{tip}, and R_{gap} can be expressed as

$$\begin{cases} R_{flux} = \frac{\beta}{\mu_0 (h_s + h_t + y_w) L_{ef}}, R_{slot} = \frac{b_s}{\mu_0 h_s L_{ef}} \\ R_{tip} = \frac{w_o}{\mu_0 h_t L_{ef}}, R_{gap} = \frac{g h_m + \mu_r}{\mu_0 w_s h_m L_{ef}} \end{cases} \tag{17}$$

By substituting (17) into (16), the A-phase inductance of the armature tooth flux barrier machine can be obtained. Similarly, the A-phase inductance of DL winding with traditional stator machine, SL Winding with traditional stator machine, and auxiliary tooth flux barrier machine can be obtained by the method. Table 3 shows A-phase self-induction of the four machines.

Table 3. Self-inductance for Different machines.

Machine Type	L_A (mH)	Machine Type	L_A (mH)
DL-TSM	1.08	SL-TSM	2.14
SL-ATM	2.19	SL-ABM	2.12

However, when the width of the flux barrier increases, the slot width decreases and the leakage inductance in the slot increases. Meanwhile, the width of the tooth tip also decreases with the increase of the width of the flux barrier, and the slot opening leakage inductance will also increase. Figure 7 shows the variation of A-phase self-inductance with the flux barrier width of the machine.

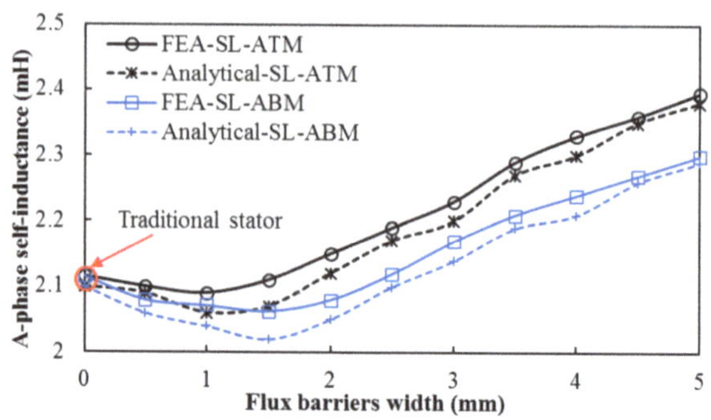

Figure 7. Variation of the A-phase self-inductance with the width of the flux barriers.

It can be seen from Figure 7 that the A-phase self-inductance of both SL-ATM and SL-ABM decreases first and then increases with the increase of the width of the flux barriers. When the decrease of air-gap inductance is greater than the increase of slot leakage inductance and tooth tip leakage inductance, the overall inductance of the machine shows a downward trend. When the decrease of air-gap inductance is less than the increase of slot leakage inductance and tooth tip leakage inductance, the overall inductance of the machine shows an upward trend.

4. Analysis and Comparison

To further explain the effect of stator flux barriers on machine fault-tolerant performance, four 12-slot/10-pole machines with different flux barriers and winding distribution were designed, as shown in Figure 1 and the main parameters are shown in Table 1. In this section, the electromagnetic characteristics of four kinds of machines under short circuit conditions are compared.

4.1. Selection of Flux Barrier Width

An analytical model and FEA were used to analyze the variation of the peak SCC and the steady SCC with the width of flux barriers of the machines, as shown in Figure 8. It can be seen that the peak SCC and the steady SCC have the same trend with the change of the flux barrier width. The peak SCC and steady SCC of SL-ATM will increase first and then decrease with the increase of the magnetic barrier width. This is because when the width of the flux barrier is less than 1.5 mm ($0 < \beta < 1.5$), the PM flux linkage is increasing, while the self-induction is decreasing, so the SCC is increasing. When the width of the flux barrier is greater than 1.5 mm ($1.5 < \beta$), although the PM flux linkage of the machine is still increasing, the self-inductance of the machine is increasing more, so the SCC shows a decreasing trend. The peak SCC and steady SCC of the SL-ABM decrease with the increase of the width of the flux barriers. Figure 9 shows the variation of the braking torque with the width of flux barriers. Compared with the SL-TSM, the braking torque of SL-ATM first increases and then decreases with the increase of the flux barrier width, while the braking torque of SL-ABM decreases with the increase of the flux barrier width. After comprehensive consideration, we determined that the width of the stator flux barrier is 2.5 mm.

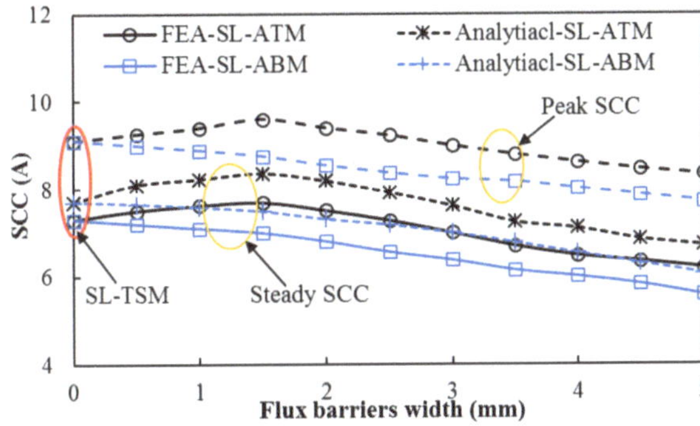

Figure 8. Variation of the peak SCC and the steady SCC with the width of flux barriers at 1000 r/min.

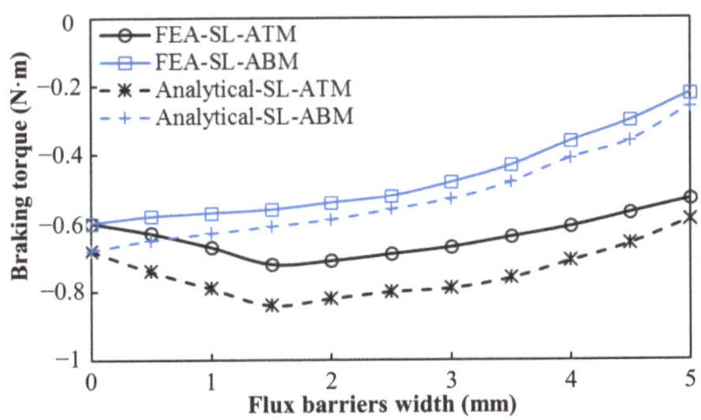

Figure 9. Variation of the braking torque with the width of flux barriers at 1000 r/min.

4.2. Single-Phase Short Circuit Condition

It is known that permanent magnet fault-tolerant machines must achieve magnetic isolation, physical isolation, and thermal isolation between phases, and effectively suppress the SCC. Therefore, the SL winding structure is very necessary. Figure 10 shows the single-phase SCC waveform of the four machines when the phase current is 5A, the speed is 1000 r/min, and the single-phase short-circuit suddenly occurs. Compared with SL winding, DL winding has smaller self-inductance, larger mutual-inductance, and the worst ability to inhibit the SCC. Therefore, the SCC of DL-TSM is the largest. Compared with the SL-TSM, when the width of the flux barriers is 2.5 mm, the self-inductances of SL-ATM and SL-ABM are larger than those without the flux barriers. Although the flux barriers make the PM flux linkage of SL-ATM increase, the SCC is still smaller than SL-TSM. On the contrary, the existence of flux barriers in SL-ABM leads to the decrease of PM flux linkage, so the SCC is inhibited.

Figure 11 shows the variation of the braking torque with speed when single-phase short circuit occurs. It can be seen that the braking torque of all four machines increases first and then decreases with the increase of speed, and there is maximum braking torque at the low speed. The braking torque of DL-TSM is the largest, because the self-inductance of DL winding structure is the smallest. Among SL winding machines, the SL-ATM has the largest braking torque, because the braking torque is not only inversely proportional to the self-inductance, but also proportional to the square of the PM flux linkage.

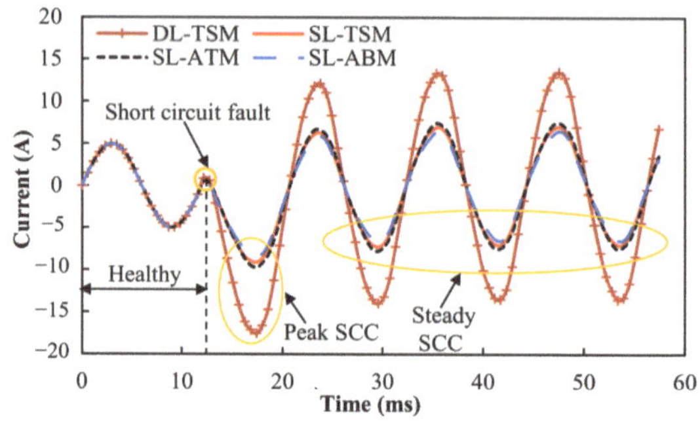

Figure 10. Single-phase SCC waveform of different machines at 1000 r/min.

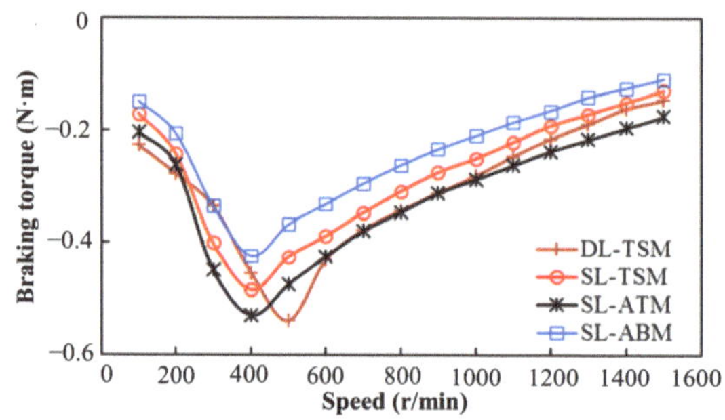

Figure 11. Variation of the braking torque with speed under single-phase short circuit condition.

Figure 12 shows the output torque after the single-phase short circuit fault of different machines. It can be seen that the average output torque of DL-TSM and SL-TSM under healthy conditions is 3.7 N·m, and the torque ripples are 2.2% and 2.8%, respectively. The output torque of the SL-ATM is 4.08 N·m, an increase of 10.27% compared with the traditional machine. However, due to the flux barriers, the 6th-order harmonic component is introduced into the cogging torque, which increases the torque ripple. Therefore, the torque ripple of the SL-ATM is 4.6%. The output torque of the SL-ABM is 3.3 N·m, and the torque ripple is 4.52%. When a single-phase short-circuit fault occurs, not only is the output torque of the machine slightly reduced, but the torque ripple also becomes larger. Taking the SL-ATM as an example, the output torque of the machine is reduced by 26.3%, and the torque ripple is increased to 137.6%. This is because the short circuit phase does not provide output torque, but produces a braking torque, which reduces the total output torque. For the asymmetric short circuit, the fault winding will generate negative sequence current. The magnetic field generated by the negative sequence current is synchronized with the rotor speed, but rotates in the opposite direction, so an even number harmonic of the output torque will be generated after the fault, which increases the torque ripple.

Table 4 is the comparison of electromagnetic performance of four kinds of machines under rated working conditions, in case of health and single-phase short-circuit fault.

Table 4. Electromagnetic performance under single-phase short circuit condition.

Machine Type	Output Torque (N·m)	Torque Ripple (%)	Peak SCC (A)	Steady SCC (A)	Braking Torque (N·m)
DL-TSM	3.7/2.6	2.2/113.3	18.3	14.6	−0.3
SL-TSM	3.7/2.8	2.8/121.6	9.4	7.9	−0.26
SL-ATM	4.1/3.2	4.6/137.6	10.2	8.5	−0.31
SL-ABM	3.3/2.4	4.52/122.2	8.3	7.2	−0.22

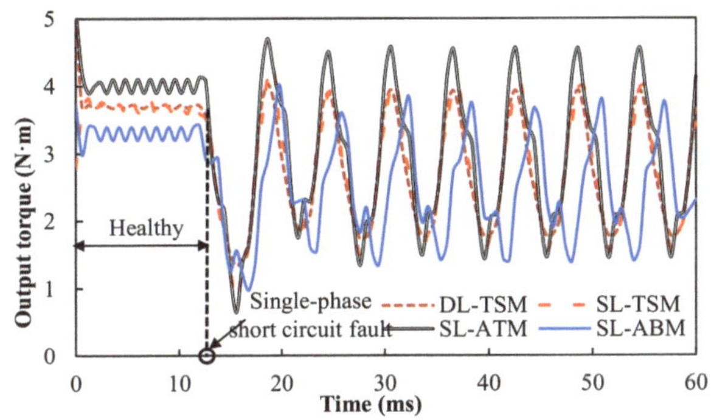

Figure 12. Output torque waveform of the machines when other healthy phases work normally after single-phase short circuit occurs at 1000 r/min.

4.3. Three-Phase Short Circuit Condition

According to the above analysis, the asymmetric short circuit in the winding will produce a large torque ripple. Therefore, for a dual three-phase PM machine, the windings that have a single-phase short-circuit fault can be artificially three-phase symmetrical short-circuited. The three-phase symmetrical short circuit can balance the asymmetry of the magnetic potential generated by the asymmetry of the short-circuit current, thereby reducing the torque ripple generated by the machine short-circuit fault. Then, through another set of healthy three-phase winding overload operations, the machine can reach the rated output torque requirements. Moreover, compared with the single-phase asymmetric short circuit, the active three-phase symmetric short circuit can suppress the SCC. Figure 13 shows the variation of the three-phase SCC of different machines with speed. It can be seen that both the peak SCC and the steady SCC in the three-phase short-circuit are reduced compared with the single-phase short-circuit. Meanwhile, the three-phase SCC of the DL-TSM is still the largest, and the three-phase SCC of the SL-ABM is still the smallest. The variation of the three-phase SCC with the width of the flux barrier is consistent with that of the single-phase SCC, so this paper will not repeat it.

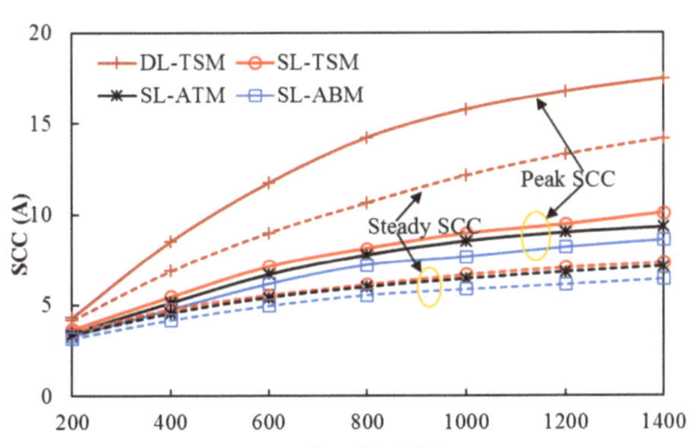

Figure 13. Variation of the three-phase SCC of different machines with speed.

Figure 14 shows the variation of braking torque with speed when three-phase short circuit occurs. Compared with the single-phase short circuit, the braking torque after a three-phase short circuit has a greater increase. However, its variation trend is consistent with a single-phase short-circuit fault, so it will not be described here.

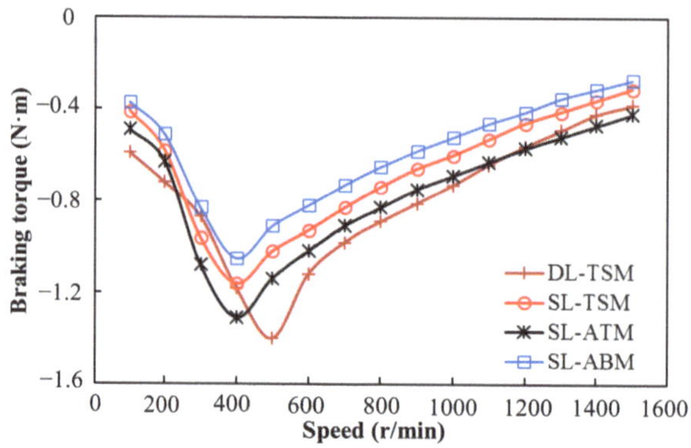

Figure 14. Variation of the braking torque with speed under three-phase short circuit condition.

The output torque of the machines when other healthy phases work normally after three-phase short circuit is shown in Figure 15. Compared with single-phase short circuits, three-phase short circuits can effectively restrain the torque ripple. Among them, the torque ripple of the DL-TSM is 24.6%, the SL-TSM is 36.5%, the SL-ATM is 39.2%, and the SL-ABM is 41.5%. However, the three-phase short circuit fault further reduces the output torque of the machine. At this time, the output torque of SL-ATM is the greatest, at 1.23 N·m. Therefore, although the braking torque of SL-ATM is the largest, the compensation current required for fault-tolerant operation at rated load is the smallest.

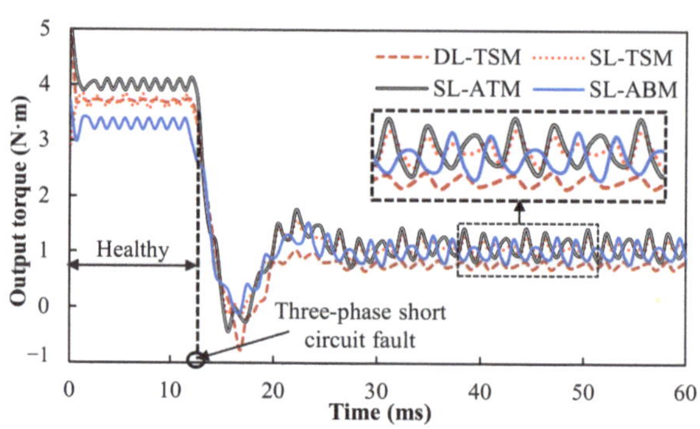

Figure 15. Output torque waveform of the machines when other healthy phases work normally after three-phase short circuit occurs at 1000 r/min.

Table 5 shows a comparison of the electromagnetic performance of four kinds of machines under rated working conditions, both healthy, and with a three-phase short-circuit fault.

Table 5. Electromagnetic performance under three-phase short circuit condition.

Machine Type	Output Torque (N·m)	Torque Ripple (%)	Peak SCC (A)	Steady SCC (A)	Braking Torque (N·m)
DL-TSM	3.7/0.83	2.2/24.6	18.3	14.6	−0.82
SL-TSM	3.7/1.1	2.8/36.5	9.4	7.9	−0.64
SL-ATM	4.1/1.2	4.6/39.2	10.2	8.5	−0.77
SL-ABM	3.3/0.95	4.52/41.5	8.3	7.2	−0.52

5. Experimental Validation

Through the above research, an experimental prototype machine was established, as shown in Figure 16. The specifications and main geometric parameters of the machine are given in Table 1. Figure 16a shows the stator lamination, which is composed of six E-type cores. The stator lamination cannot be fixed by jacket due to the air gap flux barriers between the E-type cores. Therefore, the stator and jacket adopt dovetail groove type overmatch, as shown in Figure 16b. The rotor stack and PMs are shown in Figure 16c. The experimental apparatus consists of dual three-phase inverters with a common DC power supply, and the hardware platform based on dspace-1007 is shown in Figure 16d. The experimental platform of the machine is shown in Figure 16e.

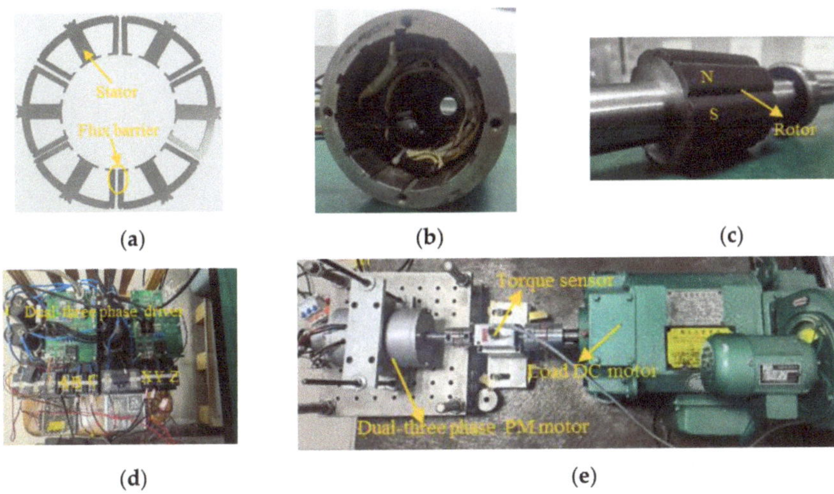

Figure 16. Machine prototype and test rig. (**a**) Stator lamination, (**b**) machine jacket, (**c**) rotor stack with magnets, (**d**) dual three-phase inverter, (**e**) test rig.

Figure 17 shows that the back-EMF with no-load at the speed is 1000 r/min; the measured results are in good agreement with the simulation results. Figure 18 shows the waveform of output torque and A-phase, X-phase current of dual three-phase windings working simultaneously in healthy condition. Figure 19a illustrates the output torque and SCC in the single-phase short-circuit fault. It can be seen that the torque ripple becomes larger and the short-circuit current is 7.5A, which is basically consistent with the simulation. Finally, the output torque waveform and SCC waveform under a three-phase short-circuit fault are shown in Figure 19b. Compared with the single-phase short circuit fault, not only is the torque ripple inhibited, but the SCC also changes from 7.5 A to 6.8 A.

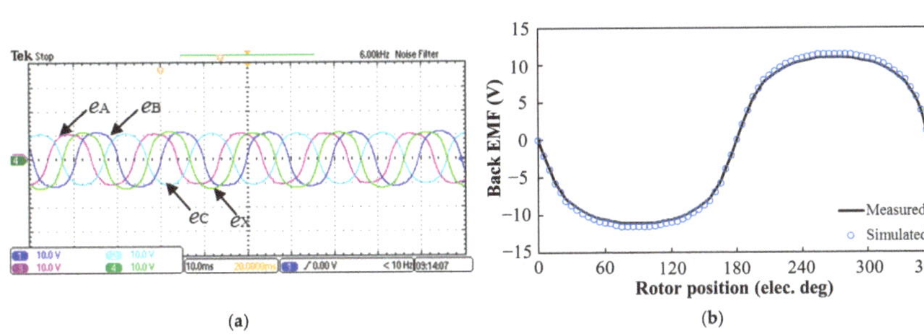

Figure 17. Back-EMF waveform with no-load at the speed is 1000 r/min. (**a**) Measured results, (**b**) simulated results.

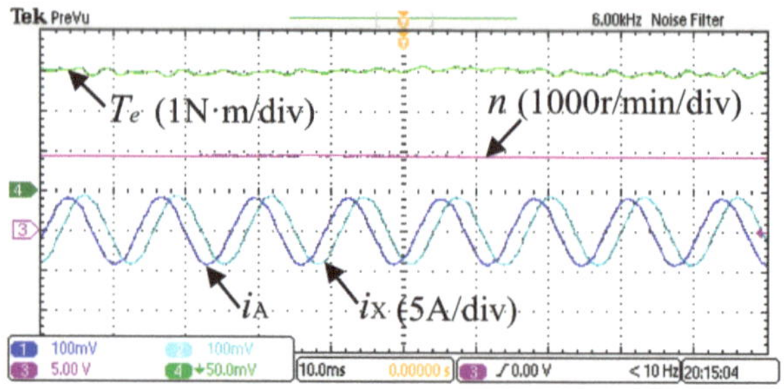

Figure 18. Output torque and phase current waveform under healthy condition.

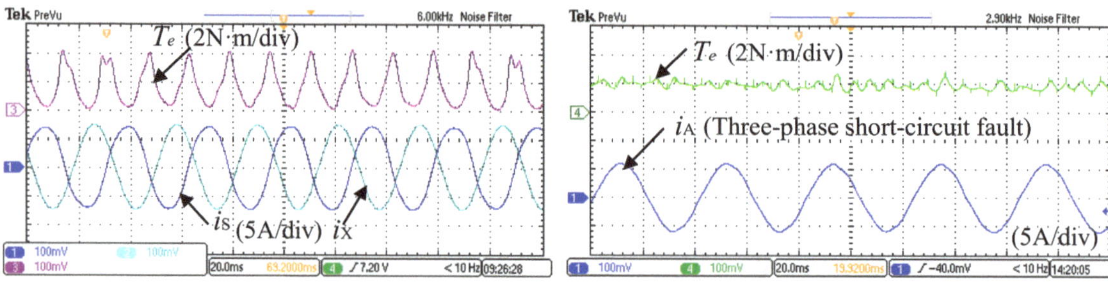

(a) (b)

Figure 19. Output torque and SCC waveform under single-phase and three-phase short circuit fault. (**a**) Single-phase short circuit fault, (**b**) three-phase short circuit fault.

6. Conclusions

This paper systematically analyzes the influence of stator flux barriers on the short-circuit fault tolerance capability of the 12-slot/10-pole dual three-phase PM machine. Compared with conventional machines, the following conclusions can be drawn by theoretical derivation, finite element analysis, and experimental verification.

(1) The appropriate width of the armature tooth flux barriers can increase the PM flux linkage, but the auxiliary tooth flux barriers will reduce the PM flux linkage. Therefore, compared with TSM, the output torque of SL-ATM is increased by 10.3%, while that of SL-ABM is reduced by 10.8%.

(2) Although stator flux barriers can reduce the harmonic leakage inductance of the machine, they can also increase the slot opening leakage inductance and inter-slot leakage inductance. Therefore, the appropriate flux barrier width can improve the self-induction, thus inhibiting SCC and braking torque. Among them, SL-ABM has the smallest SCC and braking torque, followed by SL-ATM.

(3) After entering the fault-tolerant operation mode of three-phase symmetrical short-circuit, in order to keep the machine working at the rated point, the compensation current required by SL-ATM is the minimum, and that of SL-ABM is the maximum.

This paper only studies the influence of the flux barrier width and position on the 12-slot/10-pole dual three-phase PM machine. The influence of the flux barriers on the three-phase or multiphase machines with different pole/slots will be shown in the future research.

Author Contributions: Conceptualization, L.L. and K.W.; methodology, L.L.; software, J.L.; validation, L.L. and L.G.; formal analysis, K.W.; investigation, K.W.; resources, K.W.; data curation, J.L.; writing—original draft preparation, L.L.; writing—review and editing, K.W.; visualization, K.W.;

supervision, K.W.; project administration, K.W.; funding acquisition, K.W. All authors have read and agreed to the published version of the manuscript.

Funding: This research was funded by the National Natural Science Foundation of China under Project 51977109.

Data Availability Statement: Not applicable.

Conflicts of Interest: The authors declare no conflict of interest.

References

1. Li, J.; Wang, K.; Zhang, H. Flux focusing permanent magnet machines with modular consequent-pole rotor. *IEEE Trans. Ind. Electron.* **2020**, *67*, 3374–3385. [CrossRef]
2. Chakraborty, I.; Mavris, D.N.; Emeneth, M.; Schneegans, A. An integrated approach to vehicle and subsystem sizing and analysis for novel subsystem architectures. *Proc. Inst. Mech. Eng. Part G J. Aerosp.* **2016**, *230*, 496–514. [CrossRef]
3. Sarigiannidis, A.G.; Beniakar, M.E.; Kakosimos, P.E.; Kladas, A.G.; Papini, L.; Gerada, C. Fault tolerant design of fractional slot winding permanent magnet aerospace actuator. *IEEE Trans. Ind. Electron.* **2016**, *2*, 380–390. [CrossRef]
4. Scharlau, C.C.; Pereira, L.F.A.; Pereira, L.F.; Haffner, S. Performance of a five-phase induction machine with optimized air gap field under open loop V/f control. *IEEE Trans. Energy Convers.* **2008**, *23*, 1046–1056. [CrossRef]
5. Zhou, H.W.; Liu, G.H.; Zhao, W.X.; Yu, X.D. Dynamic Performance Improvement of Five-Phase Permanent-Magnet Machine with Short-Circuit Fault. *IEEE Trans. Ind. Electron.* **2017**, *52*, 3680–3690.
6. Rahman, A.A.; Galassini, A.; Degano, M.; Gerada, C.; Bozhko, S.; Almurib, H.A.F. Open and Short Circuit Post-Fault Control Strategies for Multi-Three-Phase Interior Permanent Magnet Machines. *IEEE Trans. Energy Convers.* **2022**, *37*, 163–174.
7. Guo, H.; He, X.; Xu, J.Q.; Tian, W. Design of an Aviation Dual-Three-Phase High-Power High-Speed Permanent Magnet Assisted Synchronous Reluctance Starter-Generator with Antishort-Circuit Ability. *IEEE Trans. Power Electron.* **2022**, *37*, 12619–12635. [CrossRef]
8. Zhu, S.; Zhao, W.; Liu, G.; Mao, Y.; Sun, Y. Effect of phase shift angle on radial force and vibration behavior in dual three-phase PMSM. *IEEE Trans. Ind. Electron.* **2021**, *68*, 2988–2998. [CrossRef]
9. Zheng, J.; Zhao, W.; Ji, J.; Zhu, J.; Lee, C.H.T. Sleeve design of permanent-magnet machine for low rotor losses. *Chin. J. Elect. Eng.* **2020**, *6*, 86–96. [CrossRef]
10. Guo, L.Y.; Xu, J.Q.; Wu, S.; Xie, X.M.; Wang, H.M. Analysis and Design of Dual Three-Phase Fractional-Slot Permanent Magnet Motor with Low Space Harmonic. *IEEE Trans. Magn.* **2022**, *58*, 8100112. [CrossRef]
11. Xu, P.; Feng, J.H.; Guo, S.Y.; Feng, S.; Chu, W.; Ren, Y.; Zhu, Z.Q. Analysis of dual three-phase permanent-magnet synchronous machines with different angle displacements. *IEEE Trans. Ind. Electron.* **2018**, *65*, 1941–1954. [CrossRef]
12. Li, F.; Wang, K.; Zhang, J.; Zeng, F. Effect of angle displacements on electromagnetic performance of dual three-phase consequent-pole permanent magnet machine. *IET Electr. Power Appl.* **2020**, *14*, 1177–1185. [CrossRef]
13. Sun, Y.H.; Zhao, W.X.; Ji, J.H.; Zheng, J.Q. Effect of Phase Shift on Inductance and Short-Circuit Current in Dual Three-Phase 48-Slot/22-Pole Permanent-Magnet Machines. *IEEE Trans. Ind. Electron.* **2021**, *65*, 1941–1954. [CrossRef]
14. Barcaro, M.; Bianchi, N.; Magnussen, F. Analysis and tests of a dual three-phase 12-Slot 10-Pole permanent-magnet machine. *IEEE Trans. Ind. Appl.* **2010**, *46*, 2355–2362. [CrossRef]
15. Barcaro, M.; Bianchi, N.; Magnussen, F. Six-Phase supply feasibility using a PM fractional-slot dual winding machine. In Proceedings of the 2010 IEEE Energy Conversion Congress and Exposition, Atlanta, GA, USA, 12–16 September 2011; Volume 47.
16. Wang, W.; Zhang, J.H.; Cheng, M.; Li, S.H. Fault-tolerant control of dual three-phase permanent-magnet synchronous machine drives under open-phase faults. *IEEE Trans. Power. Electron.* **2017**, *32*, 2052–2063. [CrossRef]
17. Dajaku, G.; Xie, W.; Gerling, D. Reduction of low space harmonics for the fractional slot concentrated windings using a novel stator design. *IEEE Trans. Magn.* **2014**, *50*, 1012–1019. [CrossRef]
18. Dajaku, G.; Gerling, D. A novel tooth concentrated winding with low space harmonic content. In Proceedings of the 2013 International Electric Machines & Drives Conference, Chicago, IL, USA, 12–15 May 2013; pp. 755–760.
19. Dajaku, G.; Gerling, D. Eddy current loss minimization in rotor magnets of PM machines using high-efficiency 12-teeth/10-poles winding topology. In Proceedings of the International Conference on Electrical Machines and Systems, Beijing, China, 20–23 August 2011.
20. Li, G.J.; Zhu, Z.Q.; Foster, M.P.; Stone, D.A.; Zhan, H.L. Modular permanent-magnet machines with alternate teeth having tooth tips. *IEEE Trans. Ind. Electron.* **2015**, *62*, 6120–6130. [CrossRef]
21. Li, G.J.; Zhu, Z.Q.; Chu, W.Q.; Foster, M.P.; Stone, D.A. Influence of flux gaps on electromagnetic performance of novel modular PM machines. *IEEE Trans. Magn.* **2014**, *19*, 716–726. [CrossRef]
22. Li, G.J.; Zhu, Z.Q. Analytical modeling of modular and unequal tooth width surface-mounted permanent magnet machines. *IEEE Trans. Magn.* **2015**, *51*, 8107709. [CrossRef]
23. Liu, Z.Q.; Wang, K.; Sun, H.Y.; Zhou, B. DC-Field excitation variable flux reluctance starter generator with modular structure for fault-tolerant capability improvement. *IEEE Trans. Ind. Electron.* **2021**, *68*, 1141–1154. [CrossRef]

24. Wu, F.; Tong, C.; Sui, Y.; Cheng, L.; Zheng, P. Influence of third harmonic back EMF on modeling and remediation of winding short circuit in a multiphase PM machine with FSCWs. *IEEE Trans. Ind. Electron.* **2016**, *63*, 6031–6041. [CrossRef]
25. Liu, L.; Wang, K.; Zhu, S.S.; Liu, C. Split ratio trade-off between output torque and braking torque of surface-mounted permanent-magnet machines. *IET Electr. Power Appl.* **2020**, *14*, 1196–1203. [CrossRef]
26. Du, K.K.; Xu, L.; Zhao, W.X.; Liu, G.H. Analysis and design of a fault-tolerant permanent magnet vernier machine with improved power factor. *IEEE Trans. Ind. Electron.* **2021**, *58*, 1141–1154. [CrossRef]
27. Chen, Q.; Gu, L.C.; Wang, J.B.; Liu, G.; Zhao, W.X. Remedy strategy for five-phase FTPMMs under single-phase short-circuit fault by injecting harmonic currents from third space. *IEEE Trans. Power Electron.* **2022**, *37*, 11152–11163. [CrossRef]
28. Zhao, W.X.; Du, K.K.; Xu, L. Design considerations of fault-tolerant permanent magnet vernier machine. *IEEE Trans. Ind. Electron.* **2020**, *67*, 7290–7300. [CrossRef]

energies

MDPI

Article

Sensorless Control Strategy of Novel Axially Magnetized Vernier Permanent-Magnet Machine

Bowen Xu, Jien Ma *, Qiyi Wu, Lin Qiu, Xing Liu, Chao Luo and Youtong Fang

College of Electrical Engineering, Zhejiang University, Hangzhou 310027, China; xbwen@zju.edu.cn (B.X.); kiwiwu@zju.edu.cn (Q.W.); qiu_lin@zju.edu.cn (L.Q.); xingldl@zju.edu.cn (X.L.); luochao@zju.edu.cn (C.L.); youtong@zju.edu.cn (Y.F.)

* Correspondence: majien@zju.edu.cn; Tel.: +86-137-5716-1686

Abstract: Vernier permanent-magnet machines have been attracted more and more attention because of their high torque density. In this paper, the sensorless control strategy of the novel axially magnetized Vernier permanent-magnet (AMVPM) machine is presented. First, the inductance non-linearity is investigated under different load conditions. Second, the mathematical model is established in cooperation with the finite element method. After that, the back electromotive force based sensorless control strategy is developed according to the state equation of the motor. In the sensorless drive, the model reference adaptive system (MRAS) technique incorporated with the inductance non-linearity is used for the speed estimation. The modified control strategy not only increases the stability but also improves the dynamic response of the system. Finally, the simulation results show that the modified MRAS is of high estimation precision, and the AMVPM machine can be well controlled, and the experimental results validated the theoretical design process.

Keywords: model reference adaptive system; permanent-magnet machine; vector control; Vernier machine

Citation: Xu, B.; Ma, J.; Wu, Q.; Qiu, L.; Liu, X.; Luo, C.; Fang, Y. Sensorless Control Strategy of Novel Axially Magnetized Vernier Permanent-Magnet Machine. *Energies* **2022**, *15*, 5470. https://doi.org/10.3390/en15155470

Academic Editors: Lorand Szabo and Christopher H. T. Lee

Received: 16 June 2022
Accepted: 25 July 2022
Published: 28 July 2022

1. Introduction

Vernier permanent-magnet (VPM) machines have attracted considerable attentions in recent years. It is found that VPM machines have inherent characteristic called magnetic gearing effect, which brings the topologies many benefits, such as high power density, high efficiency, and direct-driving. These advantages make VPM machines a promising candidate in many applications, such as wind power generation, vehicle propulsion, and robotic servo system. The industrial application prospect of VPM machines makes it a research interest in recent years, during which many works have been completed [1–5].

However, the VPM machines with rotor surface-mounted PMs usually have weak mechanical strength. When working in adverse conditions, a sleeve needs to be used on the rotor to reinforce the rotor structure. The adoption of the sleeve complicates the installation process and increases the air-gap width. To solve this problem, researchers focus on the stator-PM structures to enhance the rotor robustness [6–14]. However, among them, the doubly-salient machines [6–8] have quite low torque density and large torque ripple. Additionally, the core lamination shapes of flux-reversal machines [9–11] and switched-flux machines [12–14] are complex, which makes the fabrication and assembly process very difficult.

To this end, a novel axially magnetized VPM (AMVPM) machine structure is proposed [15,16]. The structure installed the PMs directly on the axis to enhance the mechanical strength. The finite element analysis (FEA) and experiment results show that the structure can produce 12.8% larger torque with 22.6% less PM usage compared with regular VPM machine. The expected application for the proposed machine is in a submersible environment, where the position sensor is not recommended. Therefore, it is desirable to use the sensorless control strategy to further improve the performance of the machine.

The sensorless vector control strategies [17–27] can be preliminarily divided into signal injection method [17–19] and back-EMF-based method [20–28]. Among them, the back-EMF-based method is suitable for middle-speed working conditions, and it has been extensively developed in recent years. In order to solve the problem of motor parameters variation, the state observer has been proposed and thoroughly discussed. The state observer mainly includes model reference adaptive system (MRAS) [20–22], artificial neural network [23], extended Kalman filter [24], sliding-mode controller [25], full/reduced order observer [26,27], and extended back-EMF method [28]. Among them, the MRAS method is singled out as having less computing time, easy implementation, and straightforward stability approach.

Motivated by the above observations, in this paper, a modified MRAS online identification method for AMVPM machine is developed as the speed estimation method in the back electromotive force based sensorless control strategy. Meanwhile, the relevant electromagnetic parameters of AMVPM machine is investigated by the FEA method in advance. The inductance non-linearity is investigated under different load conditions. In this way, the accuracy of the reference model can be improved, so the estimation accuracy of the MRAS technique can be enhanced. The main contributions of this paper is: (a) to improve the MRAS speed estimator by incorporating with the non-linear stator inductance calculation table, which is established by FEA method in advance; (b) to demonstrate the feasibility of using MRAS-based sensorless control method to control the AMVPM machine; (c) to apply the improved MRAS-based sensorless control strategy to the AMVPM machine for the first time; and (d) to evaluate the performances of the proposed strategy for AMVPM machine by simulation method in various operating conditions and verify the theoretical analysis results of full paper by experimental validation.

This paper is organized as follows. In Section 2, the structure and operation principle of the proposed machine are demonstrated. In Section 3, the inductance matrix is modeled by the FEA method. In Section 4, the sensorless control system including the MRAS estimation is designed. In Section 5, the simulation results are discussed. Section 6 includes the manufacturing and testing process of the machine prototype to confirm the theoretical analysis. Section 7 concludes this paper.

2. Machine Structure and Operation Principle

The structure of AMVPM machine is shown in Figure 1. As can be seen, the PMs are magnetized along the axis alternately and sandwiched between the rotor and the end caps. In this way, the PMs can be fixed directly on the axis and rotate with it, which can improve the mechanical strength and simplify the machining process. The stator is a conventional open-slot structure.

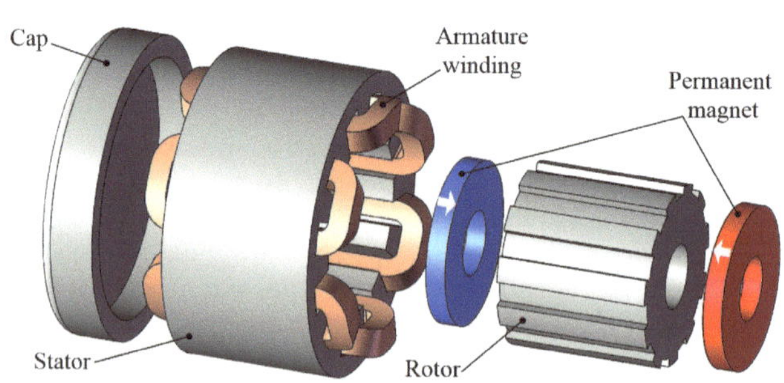

Figure 1. Structure of the axially magnetized VPM machine.

The flux lines distribution is shown in Figure 2. It can be found in the front view (left) that the flux lines come from N-pole of the PMs, pass through the rotor, the air-gap, the stator, the cap, and return to S-pole of the PMs. Therefore, in the top view (right), the flux lines radiate from the rotor and converge at the stator. In this paper, a 12-stator-slot/7-rotor-pole topology is selected as the study objective. The detailed mechanical parameters are given in Table 1.

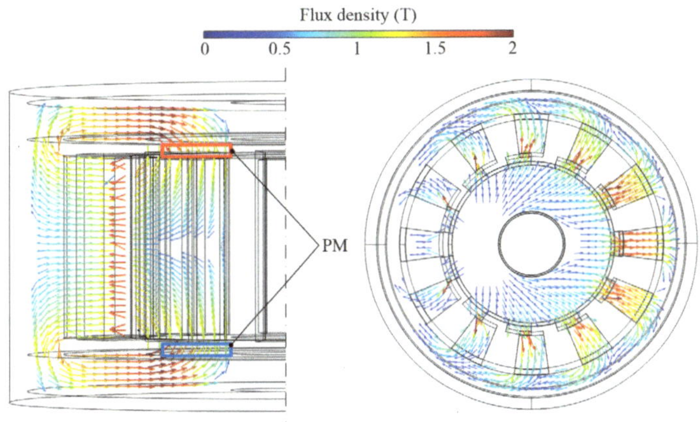

Figure 2. Magnetic field profile distribution of the proposed machine.

Table 1. Design Parameters of the Proposed Machine.

Parameters	Value
Stator outer diameter	140 mm
Split ratio	0.6
Stator slot number	12
Rotor slot number	7
Air-gap length	1 mm
Core length	50 mm
PM thickness	9 mm
PM type	NMX-S38EH (60 °C)
Silicon steel type	35WW400

In order to establish the state equation of the machine, some of the electromagnetic parameters will be investigated in this section. It should be noted that the temperature will influence not only resistance but also characteristic curve of the PM, and the machine loss and temperature variation are both unmeasurable. Therefore, the analysis in this paper is based on the assumption that the temperature is fixed.

According to [29], the permeance density of one-side slotted structure can be expressed as

$$\Lambda(\theta) = \Lambda_0 + \sum_{k=1}^{\infty} \Lambda_k cos(kZ\theta), \quad (1)$$

where

$$\Lambda_0 = \frac{\mu_0}{g}(1 - 1.6\beta c_0), \quad (2)$$

$$\Lambda_k = \frac{2\mu_0\beta}{k\pi g}\frac{0.78}{0.78 - 2c_0^2} sin(1.6k\pi c_0), \quad (3)$$

$$\beta = \frac{1}{2}(1 - \frac{1}{\sqrt{1 + (\frac{o}{2g})^2}}), \quad (4)$$

$$c_0 = \frac{o}{t_d}, \quad (5)$$

where μ_0 represents the permeability of vacuum, Z represents the teeth number, g represents the air-gap length, o represents the slot opening width, and t_d represents the slot pitch. It is noted that, for a well optimized topology, the slot depth can be neglected.

Therefore, the one-side permeance density of stator and rotor can be separately expressed as

$$\Lambda_s(\theta_s) = \Lambda_{s0} + \sum_{i=1}^{\infty} \Lambda_{si} cos(iZ_s\theta_s), \tag{6}$$

$$\Lambda_r(\theta_r) = \Lambda_{r0} + \sum_{j=1}^{\infty} \Lambda_{rj} cos(jZ_r\theta_r). \tag{7}$$

The permeance density of doubly salient structure can be calculated as [30]

$$\Lambda_g(\theta) \approx \Lambda_s(\theta)\Lambda_r(\theta - \theta_0)/\frac{\mu_0}{g}. \tag{8}$$

Substituting (6) and (7) into (8), one can obtain:

$$\begin{aligned}\Lambda_g(\theta) &= \Lambda_{s0}\Lambda_{r0} + \Lambda_{r0}\sum_{i=1}^{\infty}\Lambda_{si}cos(iZ_s\theta) + \Lambda_{s0}\sum_{j=1}^{\infty}\Lambda_{rj}cos(jZ_r(\theta-\omega_m t)) \\ &+ \sum_{i=1}^{\infty}\sum_{j=1}^{\infty}\Lambda_{si}\Lambda_{rj}cos(iZ_s\theta + jZ_r(\theta-\omega_m t)) + \sum_{i=1}^{\infty}\sum_{j=1}^{\infty}\Lambda_{si}\Lambda_{rj}cos(iZ_s\theta - jZ_r(\theta-\omega_m t)).\end{aligned} \tag{9}$$

The air-gap radial flux density can be written as

$$B(\theta) = F \cdot \Lambda(\theta), \tag{10}$$

where F is the magnetic potential difference.

Table 1 shows that in the proposed machine, Z_s is 12 and Z_s is 7. According to the magnetic gearing effect, the pole pair number of armature winding is $(Z_s - Z_r =)5$. Thus, the one-phase stator flux linkage can be expressed as

$$\begin{aligned}\psi &= N_{ph}l_{stk}R_g\left(\int_{-\frac{\pi}{Z_s}}^{\frac{\pi}{Z_s}} F\Lambda(\theta)d\theta - \int_{\pi-\frac{\pi}{Z_s}}^{\pi+\frac{\pi}{Z_s}} F\Lambda(\theta)d\theta\right) \\ &= 2N_{ph}l_{stk}R_g\Big(\sum_{j=1,3,5...}^{\infty}\frac{2\Lambda_{s0}\Lambda_{rj}}{jZ_r}cos(jZ_r\omega_m t)sin(j\frac{Z_r}{Z_s}\pi) \\ &+ \sum_{i=1}^{\infty}\sum_{j=1,3,5...}^{\infty}\frac{\Lambda_{si}\Lambda_{rj}}{iZ_s + jZ_r}cos(jZ_r\omega_m t)sin(i\pi + j\frac{Z_r}{Z_s}\pi) + \sum_{i=1}^{\infty}\sum_{j=1,3,5...}^{\infty}\frac{\Lambda_{si}\Lambda_{rj}}{iZ_s - jZ_r}cos(jZ_r\omega_m t)sin(i\pi - j\frac{Z_r}{Z_s}\pi)\Big).\end{aligned} \tag{11}$$

The one-phase back-EMF can be expressed as

$$\begin{aligned}E = \frac{d\psi}{dt} &= -2N_{ph}l_{stk}R_g\omega_m\Big(\sum_{j=1,3,5...}^{\infty}2\Lambda_{s0}\Lambda_{rj}sin(j\frac{Z_r}{Z_s}\pi) \\ &+ \sum_{i=1}^{\infty}\sum_{j=1,3,5...}^{\infty}\Lambda_{si}\Lambda_{rj}\frac{jZ_r}{iZ_s}\frac{sin(1+\frac{jZ_r}{iZ_s})i\pi}{(1+\frac{jZ_r}{iZ_s})} + \sum_{i=1}^{\infty}\sum_{j=1,3,5...}^{\infty}\Lambda_{si}\Lambda_{rj}\frac{jZ_r}{iZ_s}\frac{sin(1-\frac{jZ_r}{iZ_s})i\pi}{(1-\frac{jZ_r}{iZ_s})}\Big)sin(jZ_r\omega_m t).\end{aligned} \tag{12}$$

Substituting (3) into (12), the amplitude of each order of back-EMF harmonic is shown in Figure 3. The amplitude has been normalized based on the fundamental component value. As can be seen, all the high order harmonic amplitudes are less than 0.08. The analytical results show that the proposed machine has the potential to produce sinusoidal back-EMF. The FEA results also proved the conclusion, as can be seen in Figure 4. The open-circuit back-EMF wave in A-phase coils is very sinusoidal. Additionally, the accuracy of analytical result is proved to be very high. The cogging torque is naturally small because the LCM of stator and rotor poles are very large. As shown in Figure 5, the amplitude of cogging torque is under 0.05 Nm. Therefore, it is concluded that when the phase current is also sinusoidal, the proposed machine has the capacity to produce constant torque, which is the basic function of a vector control system.

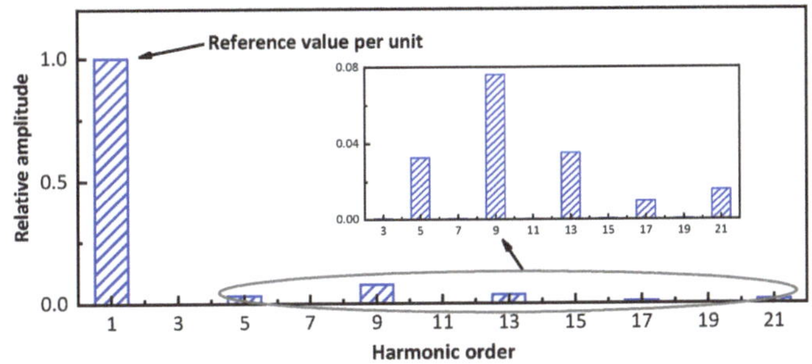

Figure 3. Relationship between normalized amplitude of harmonic component in back-EMF with its order.

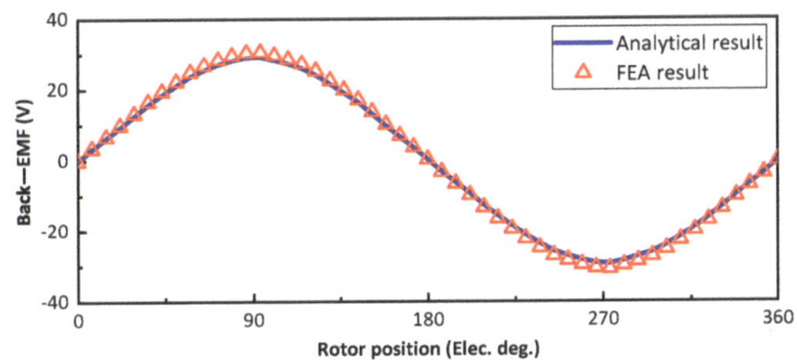

Figure 4. Comparison between FEA and analytical results of induced back-EMF.

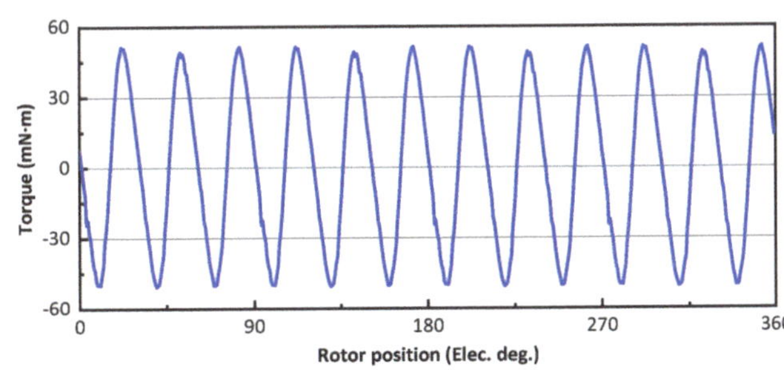

Figure 5. Relationship between cogging torque and rotor position.

3. Inductance Matrix Modeling

The estimation accuracy of the MRAS method is directly affected by the parameter accuracy of the reference model itself. However, in the process of machine design, in order to improve the performance of the motor as much as possible, there will be saturation in the iron core. It causes the stator inductance to be always changing during the operation of the motor. In order to improve the precision of the sensorless speed estimation, the induction matrix under a series of load conditions is calculated by FEA software in advance. In the real-time operation process, the d-axis and q-axis inductance will be calculated by the look-up table.

The calculated L_d and L_q under a set of i_d and i_q are shown in Figure 6. As can be seen, in general, the inductance decreases when the current increases. Considering that i_d is the field current, and i_q is the torque current, the range of i_d is $-10{\sim}10$A and the range of i_q is $0{\sim}20$ A.

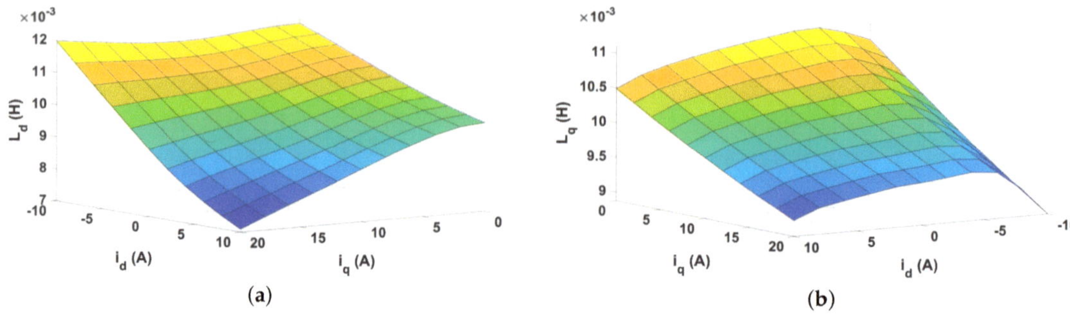

Figure 6. The relationship between inductance and (i_d, i_q). (**a**) L_d; (**b**) L_q.

Additionally, the excitation field generated by the PMs needs to be obtained to complete the state equation. Through FEA method, the three phase open-circuit flux linkage is shown in Figure 7. Since the constant amplitude transformation method is adopted in the control process, the value of ψ_f is 67.38 mWb.

After the stator resistance measured by the RLC testing device, the the state space model of the proposed AMVPM machine can be established.

$$\begin{bmatrix} \psi_d \\ \psi_q \end{bmatrix} = \begin{bmatrix} L_d(i_d, i_q) & 0 \\ 0 & L_q(i_d, i_q) \end{bmatrix} \begin{bmatrix} i_d \\ i_q \end{bmatrix} + \begin{bmatrix} \psi_f \\ 0 \end{bmatrix}, \tag{13}$$

$$\begin{bmatrix} v_d \\ v_q \end{bmatrix} = \begin{bmatrix} R_s + L_d(i_d, i_q)\frac{d}{dt} & -\omega_r L_q(i_d, i_q) \\ \omega_r L_d(i_d, i_q) & R_s + L_q(i_d, i_q)\frac{d}{dt} \end{bmatrix} \begin{bmatrix} i_d \\ i_q \end{bmatrix} + \omega_r \psi_f \begin{bmatrix} 0 \\ 1 \end{bmatrix}, \tag{14}$$

$$T_e = \frac{3Z_r}{2}(\psi_f i_q + (L_d - L_q) i_d i_q). \tag{15}$$

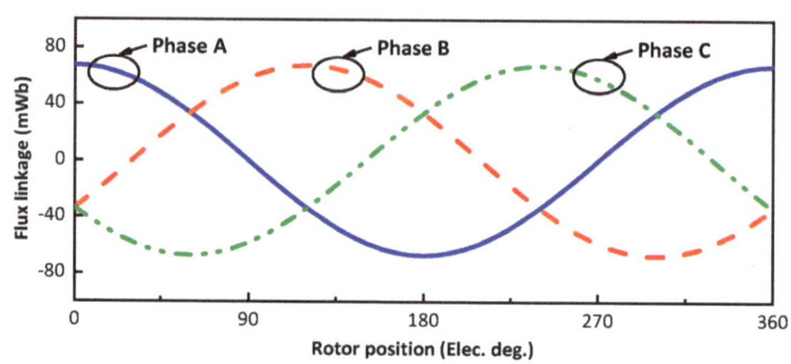

Figure 7. Open-circuit three phase flux linkage.

The other detailed parameters are given in Table 2.

Table 2. Main Electromagnetic Parameters of the Proposed Machine.

Parameters	Value
Stator resistance	0.34 Ω
d-axis inductance	7.1–12 mH
q-axis inductance	9.1–11.1 mH
Excitation flux linkage	0.067 Wb
Rated speed	600 rpm
Open-circuit back-EMF	29.2 V
Rated current	5.2 A
Rated torque	3 Nm
Electrical period	7

4. Structure of Field-Oriented Control System

The whole sensorless control system is shown in Figure 8. As can be seen, the proposed machine is driven by a three-phase inverter, whose switching signals comes from a SVPWM generator. The sampling circuit collects the two-phase current signals. The CLARKE and PARK transformation modules transfer the phase current into synchronous current, during which the rotor speed and position are estimated by the MRAS method. The expected i_q is generated by the speed loop, and the expected i_d is defined by the user. The current loop generates the expected synchronous voltage, which is transferred to the static coordinate as the input of the SVPWM generator. In this paper, the $i_d = 0$ current control strategy is adopted. The switching frequency of IGBTs is 10kHz. Additionally, the inductance L_d and L_q are generated by the inductance estimation module.

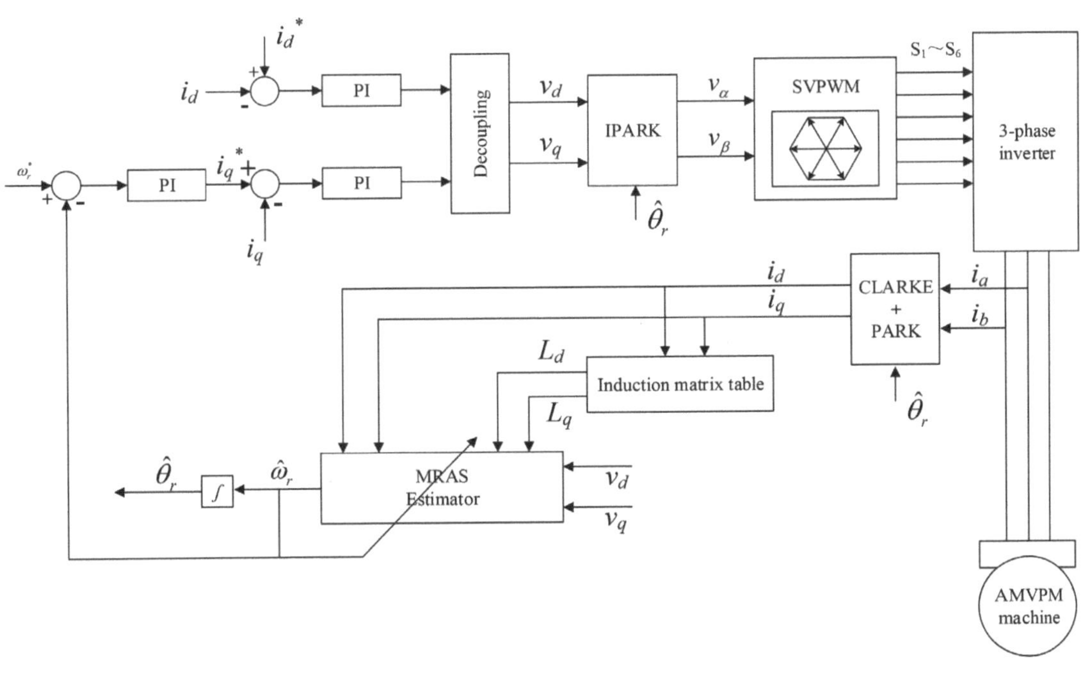

Figure 8. The global sensorless vector control system diagram.

The estimated speed is developed by the MRAS method. The main feature of the MRAS method is to use a reference model to specify the required performance of the system. The method mainly includes a reference model and an adjustable model. Among them, the reference model and the adjustable model are excited by the same input and each produce a response. The difference in their responses reflects the errors in certain parameters in the adjustable model. This response adjusts the parameters of the adjustable model through the adaptive law, so that the response of the adjustable model can approximate the reference model. The structure of MRAS is shown in Figure 9, where the reference model is designed based on (13):

$$\frac{d}{dt}[X] = [A][X] + [B][U] + [C], \tag{16}$$

where

$$\begin{gathered}[X] = \begin{bmatrix} i_d \\ i_q \end{bmatrix}, [A] = \begin{bmatrix} -\frac{R_s}{L_d} & \frac{L_q}{L_d}\omega_m \\ -\frac{L_d}{L_q}\omega_m & -\frac{R_s}{L_q} \end{bmatrix}, \\ [B] = \begin{bmatrix} \frac{1}{L_d} & 0 \\ 0 & \frac{1}{L_q} \end{bmatrix}, [U] = \begin{bmatrix} v_d \\ v_q \end{bmatrix}, [C] = \begin{bmatrix} 0 \\ -\frac{\psi_f}{L_q}\omega_m \end{bmatrix}. \end{gathered} \tag{17}$$

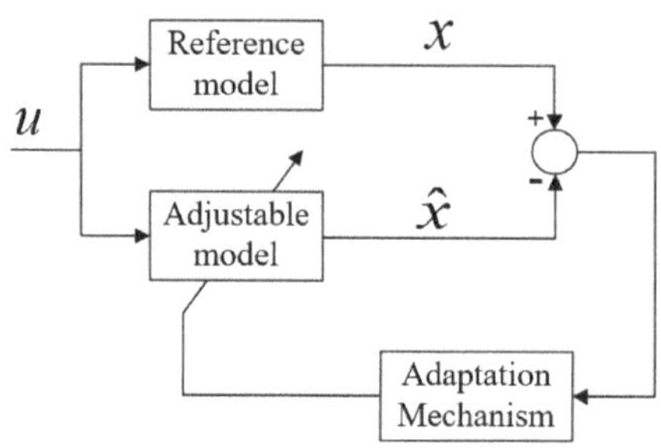

Figure 9. Diagram of MRAS estimation method.

The adjustable model is designed as:

$$\begin{bmatrix} v_d \\ v_q \end{bmatrix} = \begin{bmatrix} R_s + \frac{d}{dt}L_d(i_d, i_q) & -\hat{\omega}_r L_q(i_d, i_q) \\ \hat{\omega}_r L_d(i_d, i_q) & R_s + \frac{d}{dt}L_q(i_d, i_q) \end{bmatrix} \begin{bmatrix} \hat{i}_d \\ \hat{i}_q \end{bmatrix} + \hat{\omega}_r \psi_f \begin{bmatrix} 0 \\ 1 \end{bmatrix}. \tag{18}$$

It is worth mentioning the current changes with high frequency and small amplitude, so the inductance can be taken as constant in (18).

Substracting (18) by (14), the stator current error can be expressed by:

$$\frac{d}{dt}\begin{bmatrix} e_d \\ e_q \end{bmatrix} = \begin{bmatrix} -\frac{R_s}{L_d} & \hat{\omega}_r \frac{L_q}{L_d} \\ -\hat{\omega}_r \frac{L_d}{L_q} & -\frac{R_s}{L_q} \end{bmatrix} \begin{bmatrix} e_d \\ e_q \end{bmatrix} + \begin{bmatrix} \frac{L_q}{L_d}\hat{i}_q \\ -\frac{L_d}{L_q}\hat{i}_d - \frac{\psi_f}{L_q} \end{bmatrix} (\omega_r - \hat{\omega}_r). \tag{19}$$

The state error model of the proposed machine can be written as:

$$\frac{d}{dt}[e] = [A_e][e] + (\hat{\omega}_r - \omega_r)J[\hat{\Phi}], \tag{20}$$

where

$$\begin{gathered}[e] = \begin{bmatrix} e_d \\ e_q \end{bmatrix}, [A_e] = \begin{bmatrix} -\frac{R_s}{L_d} & \hat{\omega}_r \frac{L_q}{L_d} \\ -\hat{\omega}_r \frac{L_d}{L_q} & -\frac{R_s}{L_q} \end{bmatrix}, \\ [J] = \begin{bmatrix} 0 & -1 \\ 1 & 0 \end{bmatrix}, [\hat{\Phi}] = \begin{bmatrix} \frac{L_d}{L_q}\hat{i}_d + \frac{\psi_f}{L_q} \\ \frac{L_q}{L_d}\hat{i}_q \end{bmatrix}. \end{gathered} \tag{21}$$

The equivalent non-linear feedback system is illustrated in Figure 10. According to the Popov hyperstability theory, two conditions needs to be satisfied to prove that the system is asymptotically stable. (a) The linear time-invariant forward path transfer matrix (the gray part) is strictly positive real; (b) The nonlinear feedback element (all the other part) must satisfy the following equation:

$$\eta(0,t_1) = \int_0^{t_1} [e]^T(\hat{\omega}_r - \omega_r)J[\hat{\Phi}]dt \geq -r_0^2. \tag{22}$$

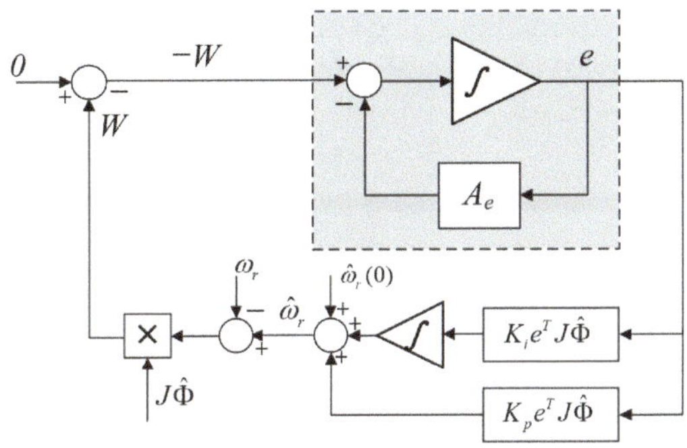

Figure 10. Equivalent non-linear feedback system.

The first condition can be easily proved by the root locus plot of forward path transfer matrix $(s[I] - [A])^{-1}$, which is shown in Figure 11. As is shown, the root locus of the forward path transfer matrix moves within the green area, taking into account the changes of L_d, L_q, and ω_r.

To prove the second condition, $\hat{\omega}_r$ can be set as:

$$\hat{\omega}_r = \int_0^{t_0} K_i[e]^T J[\hat{\Phi}]dt + K_p[e]^T J[\hat{\Phi}] + \hat{\omega}_r(0). \tag{23}$$

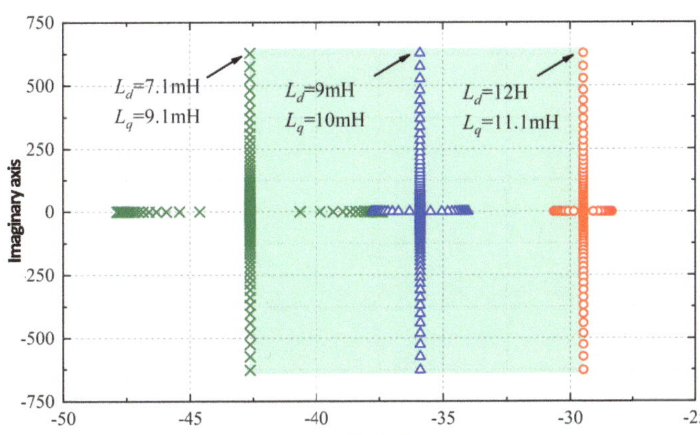

Figure 11. Poles loci for forward path transfer matrix.

Substituting (23) into (22), the inequation becomes:

$$\begin{aligned} \eta(0,t_1) = & \int_0^{t_1} [e]^T\left(\int_0^{t_0} K_i[e]^T J[\hat{\Phi}]d\tau + \hat{\omega}_r(0) - \omega_r\right)J[\hat{\Phi}]dt \\ & + \int_0^{t_1} [e]^T K_p[e]^T J[\hat{\Phi}]dt \geq -r_0^2, \end{aligned} \tag{24}$$

which can be deduced by proving

$$\int_0^{t_1} [e]^T \left(\int_0^{t_0} K_i [e]^T J[\hat{\Phi}] d\tau + \hat{\omega}_r(0) - \omega_r \right) J[\hat{\Phi}] dt \geq -r_1^2 \tag{25}$$

$$\int_0^{t_1} [e]^T K_p [e]^T J[\hat{\Phi}] dt \geq -r_2^2. \tag{26}$$

Equation (25) can be proved by

$$\int_0^{t_1} \frac{df(t)}{dt} kf(t) dt = \frac{k}{2}(f^2(t_1) - f^2(0)) \geq \frac{1}{2} kf^2(0), \tag{27}$$

where

$$\frac{df(t)}{dt} = [e]^T J[\hat{\Phi}], \tag{28}$$

$$kf(t) = \int_0^{t_0} K_i [e]^T J[\hat{\Phi}] d\tau + \hat{\omega}_r(0) - \omega_r. \tag{29}$$

It can be found that (28) is the differential result of (29).

Additionally, (26) can be easily proved if $K_p > 0$. In conclusion, the adaptive law (23) is proved to be asymptotically stable. Meanwhile, it is obtained that $\hat{\omega}_r$ can be estimated as:

$$\hat{\omega}_r = k_p D + \frac{k_i}{s} D, \tag{30}$$

$$D = \frac{L_q}{L_d} i_q e_d - \frac{L_d}{L_q} i_d e_q - \frac{\psi_f}{L_q} e_q. \tag{31}$$

Since $\hat{\omega}_r$ and ω_r are time-varying, it is necessary to investigate the dynamic response of MRAS estimation loop [20]. After linearing the block in Figure 10, the linearized transfer function of $\frac{\Delta e}{\Delta \omega_r}$ can be expressed as:

$$\frac{D}{\Delta \hat{\omega}_r}\Big|_{\Delta \omega_r = 0} = G_p = \frac{\left(\left(\frac{\varphi_d}{L_q}\right)^2 + \left(\frac{\varphi_q}{L_d}\right)^2\right)s + \frac{1}{t_d}\left(\frac{\varphi_d}{L_q}\right)^2 + \frac{1}{t_q}\left(\frac{\varphi_q}{L_d}\right)^2 + \left(1 - \left(\frac{L_q}{L_d}\right)^2\right)\frac{\varphi_d}{L_q} i_q \hat{\omega}_r}{(s + \frac{1}{t_d})(s + \frac{1}{t_q}) + \hat{\omega}_r^2}, \tag{32}$$

where

$$\varphi_d = L_d i_d + \varphi_f, \tag{33}$$

$$\varphi_q = L_q i_q, \tag{34}$$

$$t_d = \frac{L_d}{R_s}, \tag{35}$$

$$t_q = \frac{L_q}{R_s}. \tag{36}$$

Therefore, Figure 10 can be simplified to Figure 12

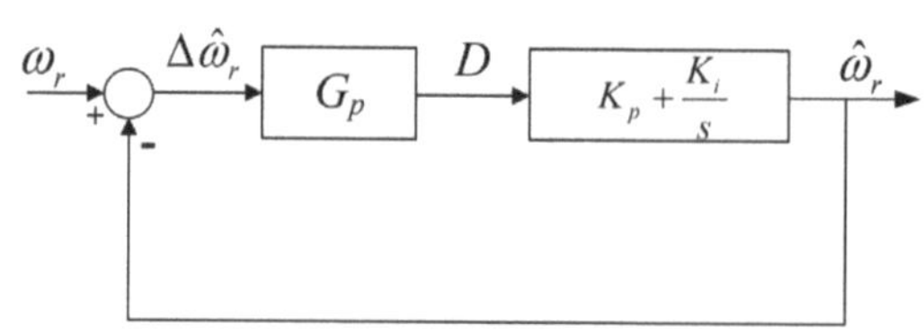

Figure 12. Simplified scheme of speed estimation.

Substitute the parameters in Table 2 into (32), and set the PI controller as K_p = 20, K_i = 10,000, the root locus of the closed-loop poles is shown in Figure 13. As can be seen, when the mechanical speed changes from 0.1 rpm to 1200 rpm, a pair of poles finally stabilize at −524. Another pole is very close to the imaginary axis, and gradually approaches the imaginary axis as the speed increases. The influence of this pole can be effectively reduced by increasing K_i. For example, when ω_r is 1200 rpm, the real value of this pole is −3.79 when K_i is 1000, and the real value of this pole is −30.1 when K_i is 10,000.

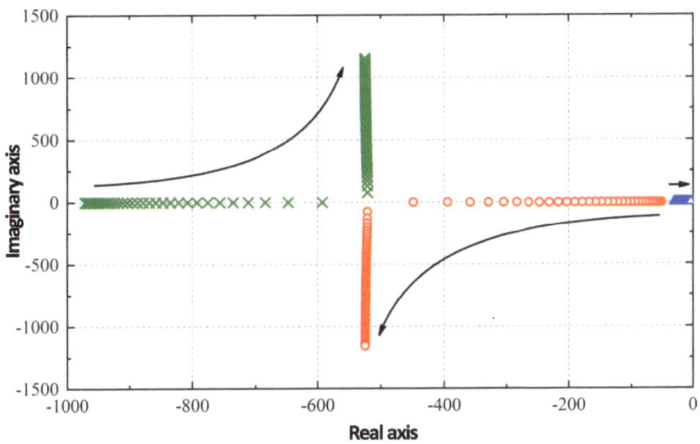

Figure 13. Pole placement of the closed-loop transfer function of speed estimation.

5. Simulation Results

In this section, the simulation calculation is carried out based on the Matlab\Simulink software to verify the feasibility of the designed control strategy. The designed test process is that the initial motor speed is 600 rpm and the load torque is 1.5 Nm (half load). The rotational speed increases with a fixed slope from 1 s and rises to 800 rpm at 1.5 s and remains unchanged. The load torque is switched from 1.5 Nm to 3 Nm (full load) at 2.5 s. During 4 s and 5 s, the speed drops from 800 rpm to 400 rpm.

It can be seen from Figure 14 that the estimated speed is very close to the reference speed. It is worth mentioning that the speed fluctuation looks a little larger than the general simulation results. It is because the proposed machine is a Vernier machine. The stator inductance is inherently larger than normal PMSM, so the stator time constant is quite larger. Therefore, the response speed of proposed machine will be slower. From Figure 14b, it is found that the speed estimation error is around 20rpm. The estimation error at rated speed is ±3.3%, which is acceptable.

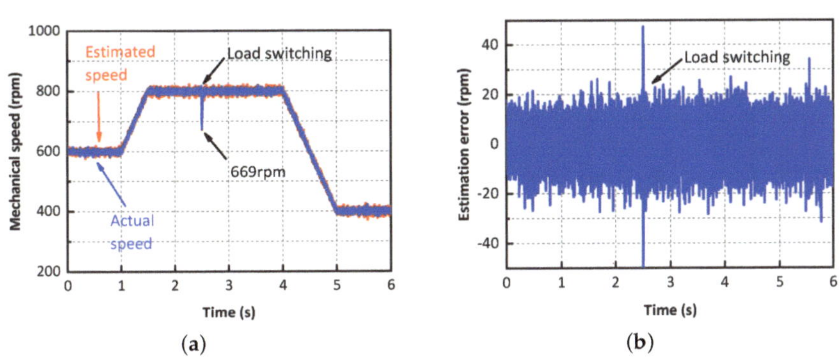

Figure 14. Simulation results of speed using the inductance matrix table. (**a**) Estimated speed and (**b**) estimation error.

Figure 15 shows the estimated and actual electrical rotor position. As can be seen, the estimated position is of high accuracy. The waveform of position estimation error is similar with that of the speed. It is because the estimated position is obtained by integrating the estimated speed, and there is one-cycle delay between the estimated speed and the real speed. The error might be eliminated by algorithmic compensation, which will not be discussed in this paper.

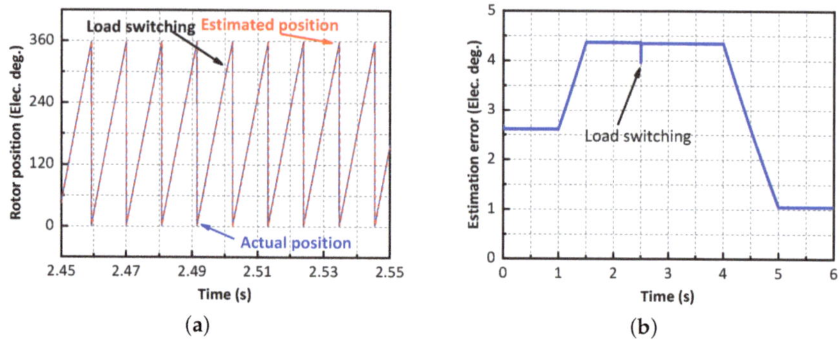

Figure 15. Simulation results of rotor position using the inductance matrix table. (**a**) Rotor position and (**b**) estimation error.

The d- and q-axis current is given in Figure 16. The reference value of i_d is 0, and the reference value of i_q is generated from the speed loop. The simulation results show that the performance of the current loop is very good. Finally, Figure 17 shows the estimated inductance during the control process. Since the current loop has satisfactory performance, the ripple of the current is very small. Therefore, in most of time, the inductance is a constant.

For comparison, both the d- and q-axis inductance are fixed at 10 mH in the following simulation to investigate the control performance without non-linear inductance matrix, during which the inductance in machine model is still current-dependent. The test process is the same, and the simulation results are shown in Figures 18–20. It can be seen that both the speed fluctuation and the estimation error are much larger. The same problem is found in rotor position estimation error and the current fluctuation. It is verified that the addition of inductance estimation can improve the performance of the control strategy.

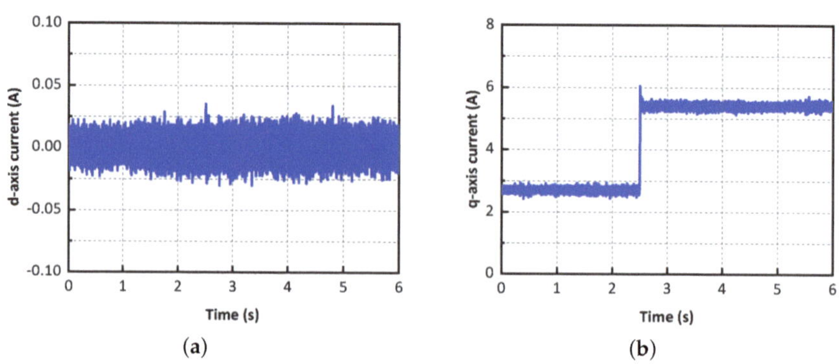

Figure 16. Simulation results of current using the inductance matrix table. (**a**) d-axis current and (**b**) q-axis current.

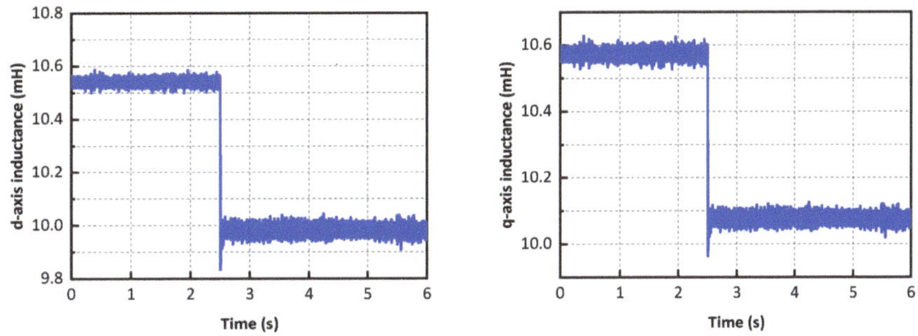

(a) (b)

Figure 17. Estimated inductance according to the inductance matrix table. (**a**) d-axis inductance and (**b**) q-axis inductance.

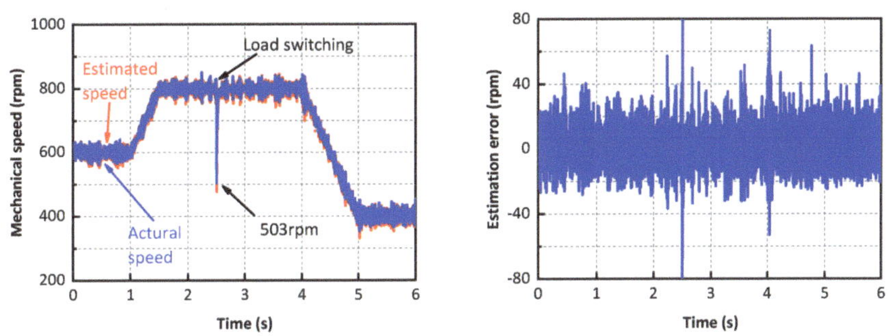

(a) (b)

Figure 18. Simulation results of speed using the constant inductance. (**a**) Estimated speed and (**b**) estimation error.

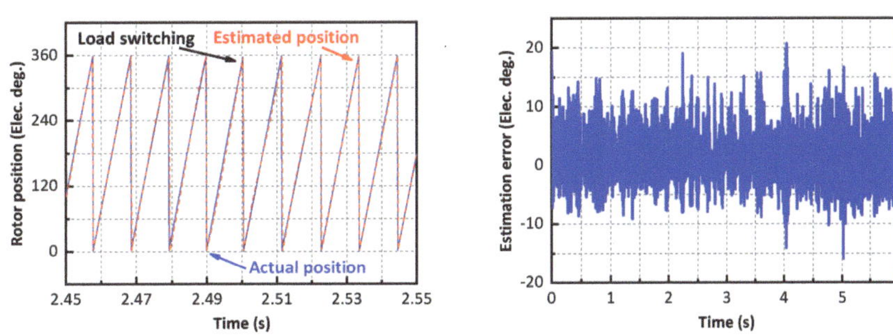

(a) (b)

Figure 19. Simulation results of rotor position using the constant inductance. (**a**) Rotor position and (**b**) estimation error.

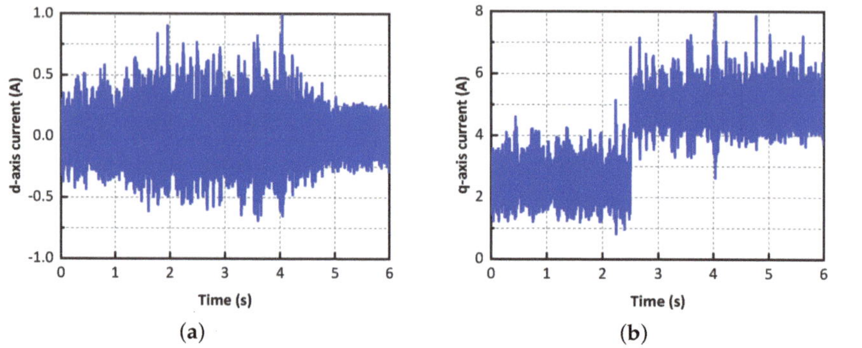

Figure 20. Simulation results without inductance estimation. (**a**) d-axis current and (**b**) q-axis current.

To further investigate the performances of proposed control method on AMVPM machine, an extended EMF-based sensorless control system is also built in the SIMULINK software according to [28]. For fair comparison, the inductance table is also used for inductance estimation.

The speed results are shown in Figure 21, and the rotor position results are shown in Figure 22. From Figure 22b it can be seen that, since the extended EMF-based method focuses on the rotor position (the estimated speed is determined by the position error), its position error is smaller than that of the proposed method. However, as shown in Figure 21b, since the estimated speed is no longer traced, the speed estimation error is larger than that of the proposed method. It is noticed in Figure 21a that the actual speed has a static error of nearly 30 rpm, which is not desired. The phenomenon that the rotor position error is proportional to the speed can also be found in the extended EMF-based method.

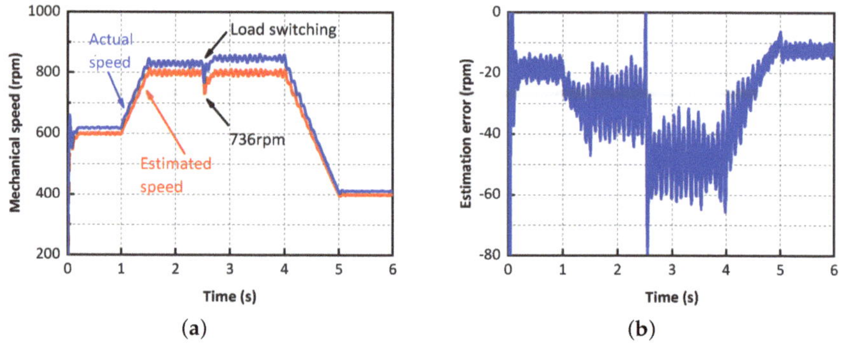

Figure 21. Simulation results of speed by extended EMF-based method. (**a**) Estimated speed and (**b**) estimation error.

Overall, the simulation results show great performance of the MRAS-based sensorless control system for the proposed AMVPM machine. It validates the feasibility of the designed control strategy for real-time working condition.

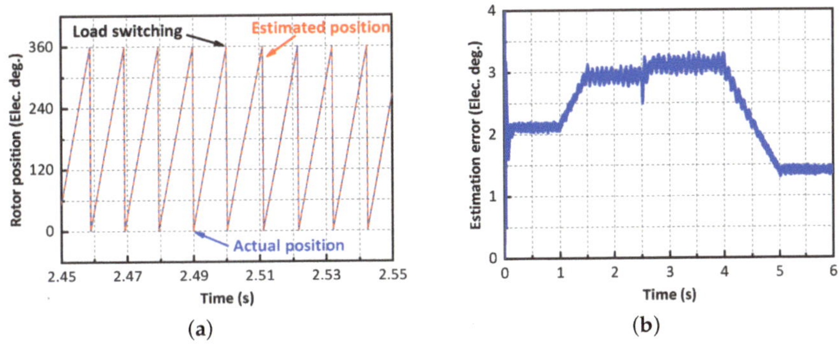

Figure 22. Simulation results of rotor position by extended EMF-based method. (**a**) Rotor position and (**b**) estimation error.

6. Experiment Validation

In order to verify the previous analysis, a test bench is built up as shown in Figure 23. Additionally, the TI HVMTRPFCKIT with TMS320F28335 microcontroller is selected as the inverter, as can be seen in Figure 24. The switching frequency is 10 kHz which is the same as in the simulation. The dead-time is set to 2 μs.

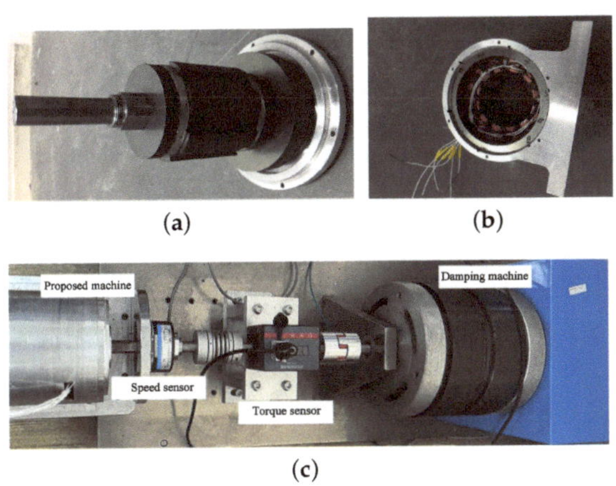

Figure 23. Prototype of the proposed machine. (**a**) Stator; (**b**) rotor; and (**c**) test bench.

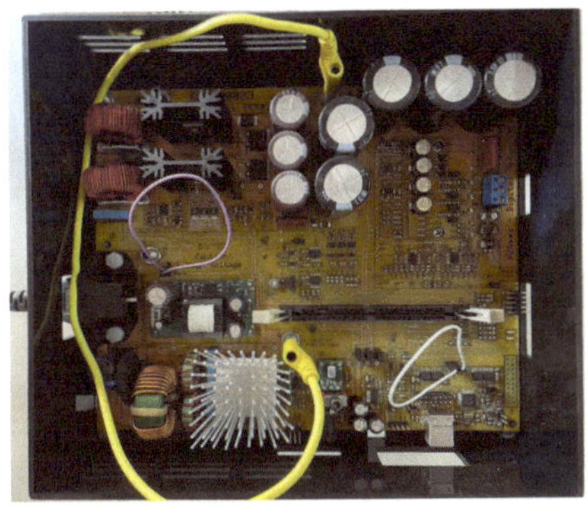

Figure 24. TI HVMTRPFCKIT inverter.

The results of speed estimation performance are shown in Figure 25. It is found that the experiment results are basically consistent with the simulation results. It is noted that in the experiment process there is no speed drop when the load is switched. It is because the magnetic powder dynamometer on the test bench needs to adjust the load manually, the load switching process is not instantaneous. Therefore, the speed loop has enough time to maintain the reference speed. The rotor estimation error in Figure 26 is larger than the simulation results. The main reason is that there are problems, such as motor parameter error, shaft rotation fluctuation, and unsatisfactory controller switch, during the experiment.

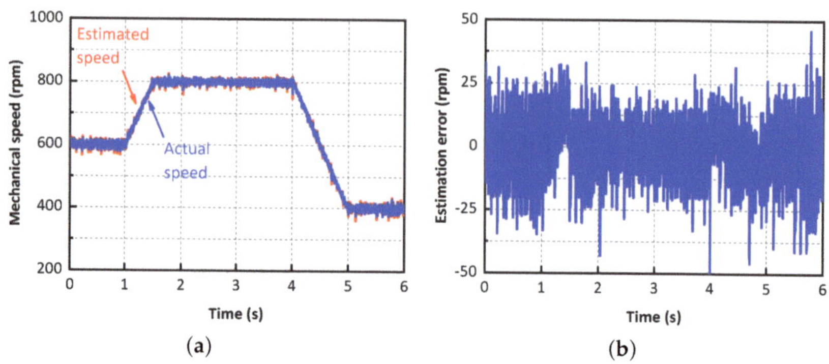

Figure 25. Experimental results.(**a**) Estimated speed and (**b**) estimation error.

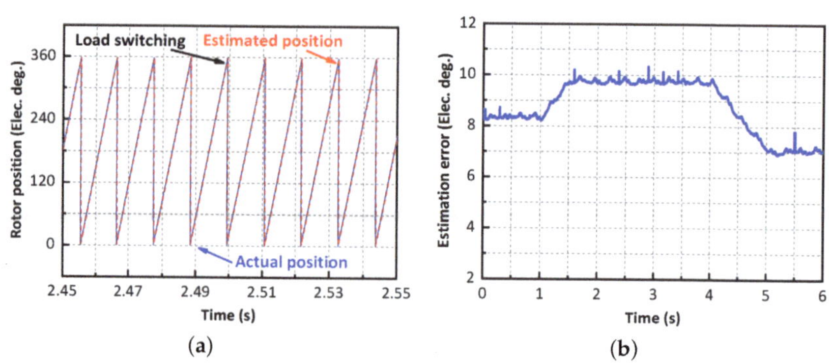

Figure 26. Experimental results.(**a**) Rotor position and (**b**) estimation error.

The current results are shown in Figures 27 and 28. It can be seen from Figure 28 that the d-axis current has a larger glitch after the load is switched. This is mainly due to the coupling between $L_d \frac{d}{dt} i_d$ and $L_q i_q$ in the first line of (14). After the load is switched, i_q becomes larger. Although it can be balanced with larger V_d, the ripple of i_q still increases, which results in larger ripple of i_d. In Figure 28b, the response process of the q-axis current after load switching is also slower than the simulation results. It is also because the magnetic powder dynamometer needs to be adjusted manually. The q-axis current values under half-load and full-load conditions are basically consistent with the simulation results. It can be seen from Figure 27 that when the load torque changes, the adjustment is completed within the 2 electrical periods, reflecting the excellent dynamic performance of the control algorithm.

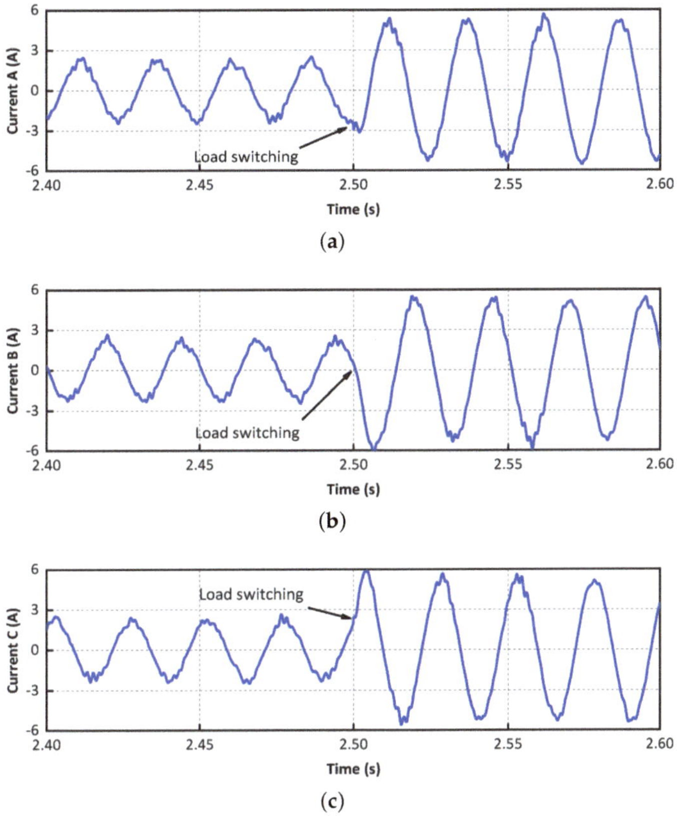

Figure 27. Experimental results.(**a**) Phase A current; (**b**) Phase B current; and (**c**) Phase C current.

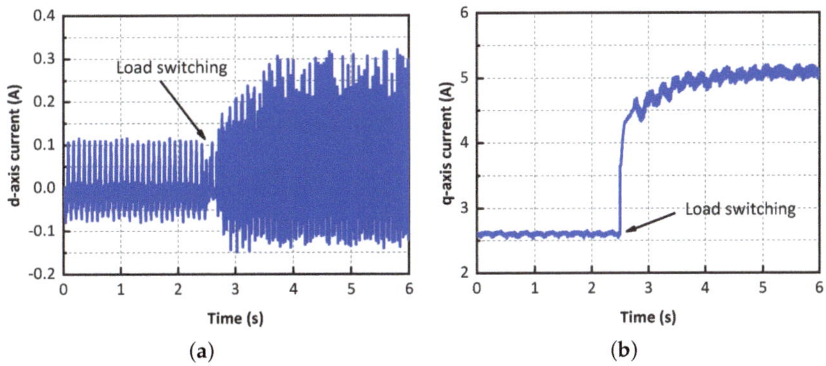

Figure 28. Experimental results.(**a**) d-axis current and (**b**) q-axis current.

Figure 29 shows the estimated value of the stator inductance. Since the q-axis current increases after the load is switched, the saturation degree of the iron core increases. Therefore, the inductance of the d-axis and q-axis decreases.

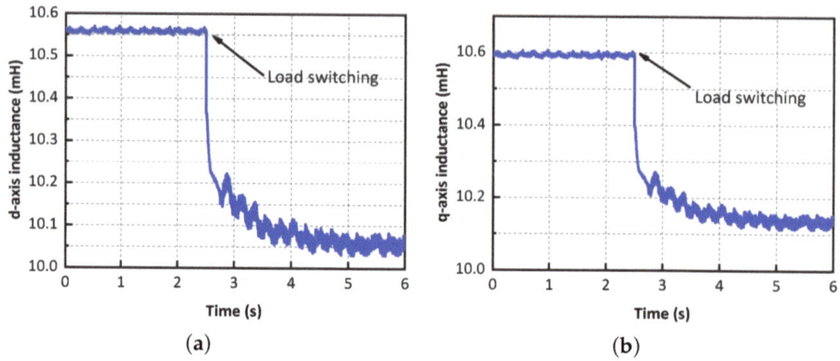

Figure 29. Experimental results.(**a**) d-axis inductance and (**b**) q-axis inductance.

The experiment with constant inductance value is also conducted. It is found that the machine cannot normally rotate. The results once again proved the necessity of combining inductance matrix table with the MRAS estimator.

Overall, the good agreement between the FE model and the manufacturing prototype has been validated.

7. Discussion

In this paper, the inductance non-linearity of the axially magnetized VPM machine under different combinations of d-axis and q-axis currents is investigated. An improved MRAS identification technique for the proposed machine is presented. The electromagnetic characteristic of the proposed machine is calculated. The state equation of the machine is established according to the offline-calculated inductance matrix and other parameters. The simulation results show great performances of the whole system. Finally, the machine prototype, as well as the control system platform, are manufactured, and the simulation conclusions are experimentally validated.

Author Contributions: Conceptualization, Y.F. and B.X.; methodology, J.M.; software, Q.W.; validation, J.M., B.X. and Q.W.; formal analysis, C.L.; investigation, L.Q.; resources, Q.W.; data curation, B.X.; writing—original draft preparation, B.X.; writing—review and editing, X.L.; visualization, X.L.; supervision, B.X.; project administration, J.M.; funding acquisition, Y.F. All authors have read and agreed to the published version of the manuscript.

Funding: This research was funded by National Natural Science Foundation of China, Grant Number 51977193, 51827810.

Institutional Review Board Statement: Not applicable.

Informed Consent Statement: Not applicable.

Data Availability Statement: The data presented in the paper are available upon request from the corresponding author (J.M.).

Conflicts of Interest: The authors declare no conflict of interest.

References

1. Ishizaki, A.; Tanaka, T.; Takasaki, K.; Nishikata S. Theory and optimum design of PM Vernier motor. In Proceedings of the 7th International Conference on Electrical Machines and Drives, Durham, UK, 11–13 September 1995; pp. 208–212.
2. Li, D.; Qu, R.; Lipo, T. A. High-Power-Factor Vernier Permanent-Magnet Machines. *IEEE Trans. Ind. Appl.* **2014**, *50*, 3664–3674. [CrossRef]
3. Jang, D.; Chang, J. Design of a Vernier Machine With PM on Both Sides of Rotor and Stator. *IEEE Trans. Magn.* **2014**, *50*, 877–880. [CrossRef]
4. Li, D.; Qu, R.; Li, J.; Xu, W. Design of consequent pole toroidal winding outer rotor Vernier permanent magnet machines. In Proceedings of the 2014 IEEE Energy Conversion Congress and Exposition, Pittsburgh, PA, USA, 14–18 September 2014; pp. 2342–2349.

5. Kim, B.; Lipo, T.A. Operation and design principles of a PM Vernier motor. *IEEE Trans. Ind. Appl.* **2014**, *50*, 3656–3663. [CrossRef]
6. Liao, Y.; Liang, F.; Lipo, T.A. A novel permanent magnet motor with doubly salient structure. *IEEE Trans. Ind. Appl.* **1995**, *3*, 1069–1078. [CrossRef]
7. Cheng, M.; Chau, K.T.; Chau, C.C.; Zhou, E.; Huang, X. Nonlinear varying-network magnetic circuit analysis for doubly salient permanent-magnet motors. *IEEE Trans. Magn.* **2000**, *36*, 339–348. [CrossRef]
8. Cheng, M.; Chau, K.T.; Chau, C.C. Design and analysis of a new doubly salient permanent magnet motor. *IEEE Trans. Magn.* **2001**, *37*, 3012–3020. [CrossRef]
9. Hua, W.; Zhu, X.; Wu, Z. Influence of coil pitch and stator-slot/rotor-pole combination on back EMF harmonics in flux-reversal permanent magnet machines. *IEEE Trans. Energy Convers.* **2018**, *33*, 1330–1341. [CrossRef]
10. Li, H.; Zhu, Z. Influence of magnet arrangement on performance of flux reversal permanent magnet machine. In Proceedings of the 2017 IEEE International Electric Machines and Drives Conference, Miami, FL, USA, 21–24 May 2017; pp. 1–8.
11. Li, H.; Zhu, Z. Optimal number of magnet pieces of flux reversal permanent magnet machines. *IEEE Trans. Energy Convers.* **2019**, *34*, 889–898. [CrossRef]
12. Rauch, S.E.; Johnson, L.J. Design principles of flux-switching alternators. *AIEE Trans. Power Appl. Syst.* **1955**, *74*, 1261–1268.
13. Chen, J.; Zhu, Z.; Iwasaki, S.; Deodhar, R.P. A novel E-core switched-flux PM brushless ac machine. *IEEE Trans. Ind. Appl.* **2011**, *47*, 1273–1282. [CrossRef]
14. Zhu, Z.; Chen, J.; Pang, Y.; Howe, D.; Iwasaki, S.; Deodhar, R. Analysis of a novel multi-tooth flux-switching PM brushless ac machine for high torque direct-drive applications. *IEEE Trans. Magn.* **2008**, *44*, 4313–4316. [CrossRef]
15. Xu, B.; Wu, L.; Ma, J.; Pierre, D.P.; Huang, X.; Qiu, L.; Fang, Y. A Novel Axially Magnetized Vernier Permanent-Magnet Machine. *IEEE Trans. Magn.* **2021**, *57*, 1–5. [CrossRef]
16. Xu, B.; Ma, J.; Wu, L.; Huang, X.; Fang, Y. Investigation of Novel Axially Magnetized Vernier Permanent-Magnet Machine in Cascade Structure. In Proceedings of the 23rd International Conference on Electrical Machines and Systems, Hamamatsu, Japan, 24–27 November 2020; pp. 1246–1250.
17. Jang, J.; Ha, J.; Ohto, M.; Ide, K.; Sul, S. Analysis of permanent-magnet machine for sensorless control based on high-frequency signal injection. *IEEE Trans. Ind. Appl.* **2004**, *40*, 1595–1604. [CrossRef]
18. Liu, J.; Zhu, Z. Novel Sensorless Control Strategy With Injection of High-Frequency Pulsating Carrier Signal Into Stationary Reference Frame. *IEEE Trans. Ind. Appl.* **2014**, *50*, 2574–2583. [CrossRef]
19. Caruana, C.; Asher, G.; Sumner, M. Performance of HF signal injection techniques for zero-low-frequency vector control of induction Machines under sensorless conditions. *IEEE Trans. Ind. Electron.* **2006**, *53*, 225–238. [CrossRef]
20. Schauder, C. Adaptive speed identification for vector control of induction motors without rotational transducers. *IEEE Trans. Ind. Appl.* **1992**, *28*, 1054–1061. [CrossRef]
21. Rashed, M.; MacConnell, P.; Stronach, A.; Acarnley, P. Sensorless indirect-rotor-field-orientation speed control of a permanent-magnet synchronous motor with stator-resistance estimation. *IEEE Trans. Ind. Electron.* **2007**, *54*, 1664–1675. [CrossRef]
22. Nahid-Mobarakeh, B.; Meibody-Tabar, F.; Sargos, F. Mechanical sensorless control of PMSM with online estimation of stator resistance. *IEEE Trans. Ind. Appl.* **2004**, *40*, 457–471. [CrossRef]
23. Mese, E.; Torrey, D.A. An approach for sensorless position estimation for switched reluctance motors using artifical neural networks. *IEEE Trans. Power Electron.* **2002**, *17*, 66–75. [CrossRef]
24. Shi, Y.; Sun, K.; Huang, L. Online identification of permanent magnet flux based on extended kalman filter for IPMSM drive with position sensorless control. *IEEE Trans. Ind. Electron.* **2012**, *59*, 4169–4178. [CrossRef]
25. Zhao, Y.; Qiao, W.; Wu, L. Improved rotor position and speed estimators for sensorless control of interior permanent-magnet synchronous machines. *IEEE Trans. Emerg. Sel. Top. Power Electron.* **2014**, *2*, 627–639. [CrossRef]
26. Solsona, J.; Valla, M.I.; Muravchik, C. A nonlinear reduced order observer for permanent magnet synchronous motors. *IEEE Trans. Ind. Electron.* **1996**, *2*, 492–497. [CrossRef]
27. Bolognani, S.; Oboe, R.; Zigliotto, M. Sensorless full-digital PMSM drive with EKF estimation ofspeed and rotor position. *IEEE Trans. Ind. Electron.* **1999**, *2*, 184–191. [CrossRef]
28. Morimoto, S.; Kawamoto, K.; Sanada, M.; Takeda, Y. Sensorless control strategy for salient-pole PMSM based on extended EMF in rotating reference frame. *IEEE Trans. Ind. Appl.* **2002**, *38*, 1054–1061. [CrossRef]
29. Heller, B.; Hamata, V. *Harmonic Field Effects in Induction Machines*; Elsevier: Amsterdam, The Netherlands, 1977; pp. 54–67.
30. Gaussens, O.B.; Hoang, E.; Saint-Michel, J.; Manfe, P.; Lecrivain, M.; Gabsi, M. Magnetic field solution in doubly slotted airgap of conventional and alternate field-excited switched-flux topologies. *IEEE Trans. Magn.* **2013**, *49*, 5083–5096. [CrossRef]

energies

MDPI

Article

Design Optimization of a Switched Reluctance Machine with an Improved Segmental Rotor for Electric Vehicle Applications

Yuanfeng Lan [1,2], Mohamed Amine Frikha [1,2], Julien Croonen [1,2], Yassine Benômar [1,2], Mohamed El Baghdadi [1,2] and Omar Hegazy [1,2,*]

[1] MOBI-EPOWERS Research Group, ETEC Department, Vrije Universiteit Brussel (VUB), Pleinlaan 2, 1050 Brussels, Belgium

[2] Flanders Make, Gaston Geenslaan 8, 3001 Heverlee, Belgium

* Correspondence: omar.hegazy@vub.be; Tel.: +32-2-629-2992

Abstract: In this article, a switched reluctance machine (SRM) with six phases and a misaligned segmental rotor is proposed. The segmental rotor has an internal 15-degree misalignment, allowing the SRM structure to be a one-layer 2D structure with a short flux path structure. The proposed SRM produces a relatively low torque ripple by exciting two phases simultaneously. Additionally, an optimization method is applied, allowing for the maximum torque position of one phase to be aligned with the zero-torque position of the adjacent phase. The finite element method (FEM) is used to analyze and design the proposed SRM and to simulate the proposed liquid cooling system. The static torque waveforms are analyzed, and the dynamic torque waveforms are simulated with a drive using SiC MOSFETs. Finally, a prototype is manufactured, and the experiment is performed to validate the design and simulation results.

Keywords: switched reluctance machine; segmental rotor; torque ripple; finite element method; multi-stack; electric vehicles

Citation: Lan, Y.; Frikha, M.A.; Croonen, J.; Benômar, Y.; El Baghdadi, M.; Hegazy, O. Design Optimization of a Switched Reluctance Machine with an Improved Segmental Rotor for Electric Vehicle Applications. *Energies* **2022**, *15*, 5772. https://doi.org/10.3390/en15165772

Academic Editors: Lorand Szabo and Feng Chai

Received: 27 June 2022
Accepted: 6 August 2022
Published: 9 August 2022

1. Introduction

The switched reluctance machine (SRM) has received increasing attention as a potential electric machine for electric vehicle (EV) applications. It has advantages, such as structure simplicity, high robustness, high reliability and low manufacturing costs [1,2]. The main disadvantages of the SRM are the torque ripple and limited torque density. Numerous research studies were conducted to introduce different approaches to reduce the torque ripple or increase the average torque. The SRM with a segmental rotor is one of these approaches. The study presented in [3] demonstrated an SRM with a segmental rotor. This SRM obtained significant performance improvement compared to the conventional one. Furthermore, in [4], an 80 kW SRM with a segmental rotor was developed. The optimization results showed that this SRM is likely to be able to produce an equivalent performance as a class-leading Internal Permanent Magnet (IPM) machine. Moreover, in [5], an SRM with a segmental rotor was proposed. From the simulation results, the average torque increased by 6.5%, and the torque ripple reduced by 2.0% compared to the reference SRM.

The multi-layer structure is another approach to reducing the torque ripple as shown in [6], where a two-layer SRM was proposed. A multi-layer SRM was proposed by [7] to reach a lower torque ripple. A double layer per phase SRM was presented by [8], and a prototype was built and tested. In [9], an SRM with a multilayer structure was proposed. The simulation results show that it is a promising SRM structure. The study [10] analyzed a multi-layer SRM with a magnetic equivalent circuit (MEC) model. A novel shape design method was introduced to decrease the torque ripple of a multi-layer SRM. A combination of a multi-layer and segmental rotor was introduced by [11]. It was a two-step multi-layer rotor from the segmental type. The SRM with a segmental and multi-layer rotor increased the average torque by 10% and reduced the torque ripple by 19% compared to an SRM with a segmental rotor but without a multi-layer rotor.

The asymmetric rotor or stator is an important approach to design an SRM with reduced torque ripple. The article [12] introduced a three-phase 12/10 asymmetric SRM. Finite element method (FEM) simulation was used to analyze the SRM. The study [13] presented a 6/4 asymmetric stator SRM. The air-gap length and the width of stator poles were analyzed. A two-phase SRM with an asymmetric rotor was proposed in [14]. The asymmetric rotor provides the SRM the capability to start from various initial rotor positions. The research [15] proposed a control method for an asymmetric rotor SRM, by focusing on its torque ripple performance. An asymmetric rotor SRM was presented in [16] to reduce the torque ripple. The design method was introduced in the cited study. Compared to a conventional SRM, the presented SRM reduced the torque ripple by 10%. An asymmetric rotor bearingless SRM was introduced in [17] to reduce the torque ripple. The introduced SRM has 20% lower torque ripple compared to a conventional bearingless SRM. In general, the steps on the asymmetric rotor or stator poles increased the total reluctance between the rotor and stator poles when the rotor pole is at the aligned position. The proposed SRM in this paper does not have the step structure and has identical air-gap length when the rotor pole is at the aligned position. It is an advantage to decrease the reluctance and increase the output torque, compared to the asymmetric rotor or stator structure.

The cooling system is an important aspect of SRM design as well. The research [18] introduced a 10 kW liquid cooled SRM and its thermal model. The model was used to optimize the cooling system. A liquid-cooled SRM with a coolant jacket structure was proposed in [19]. A dual-flute helical heat exchanger was used for the cooling jacket. The study [20] proposed a thermal model for an SRM for high-speed applications. An evaporative cooling SRM and a conventional SRM were compared in [21]. The results show that the evaporative cooling method decreases the magnitude of the radial electromagnetic force. An improved thermal lumped parameter model was proposed and validated in [22] for SRMs. An advanced direct coil cooling method for SRMs is modelled and tested in [23]. Compared to a standard water jacket cooling method, the cooling method proposed in [23] indicates a 47.8% power density improvement and 52.8% thermal performance gain.

This article proposes an improved SRM with a misaligned segmental rotor. The rated power of the proposed SRM is 2 kW, which represents a reduced scaled prototype for experimental tests for an 80 kW full scale SRM for EV applications. The rated speed of the proposed SRM is 3000 rpm; the rated torque is 6.5 Nm. It is a combination of a multi-layer SRM and a segmental rotor SRM. Notably, the multi-layer structure is designed for a 2D structure by misaligning the rotor segments with a short flux path configuration. It simplifies the SRM structure to make it easier to analyze and manufacture. With the adopted method where two machine phases are excited simultaneously, the power density of the proposed SRM is increased. With the FEM optimization method, the pole shape of the proposed SRM is designed to achieve lower torque ripple. The structure and method of operation of the proposed SRM are introduced in Section 2. In Section 3, the electromagnetic analysis is conducted with finite element analysis (FEA), and an optimization method is presented. In Section 4, the liquid cooling system is analyzed and tested. In Section 5, the SiC driver and dynamic simulation of the proposed SRM system are analyzed. The prototype of the proposed SRM and the experimental results are presented in Section 6, and conclusions are drawn in Section 7.

2. Structure and Operation Method of the Proposed SRM

This section presents the basic structure and the operation method of the proposed SRM. Figure 1a shows the structure of the proposed SRM, which is a six-phase SRM with a misaligned rotor segment. Two of the rotor segments are misaligned by 15 degrees. With the misaligned rotor segments, the proposed SRM has two phases working simultaneously. It means one out of three phases is working when it is running. The conventional SRM in Figure 1b is the reference SRM, compared to the proposed SRM.

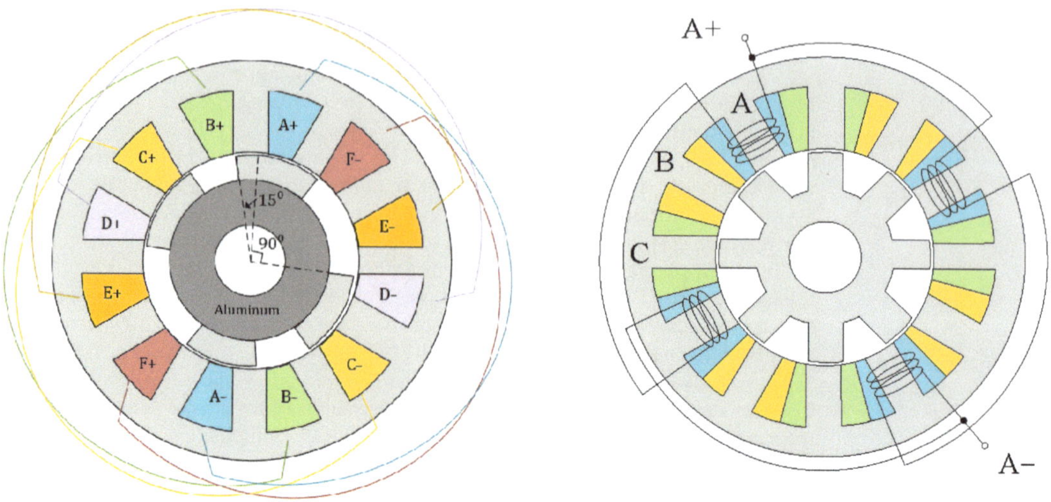

Figure 1. The structure of the SRMs: (**a**) the proposed segmental SRM; (**b**) a conventional SRM.

The conventional SRM is a three-phase SRM. Typically, it has one out of three phases working simultaneously. The operational phase number ratio of the conventional SRM is the same for the proposed SRM. As a result, the conventional three-phase SRM is selected to be the reference SRM.

The parameters of the SRMs are listed in Table 1. The proposed SRM and the conventional SRM have the same volume and air-gap length. They also have the same number of turns and the same number of stator poles.

Table 1. Parameters of the proposed SRM and the conventional SRM.

Parameter	Proposed SRM	Conventional SRM
Stator outer diameter (mm)	139.6	139.6
Lamination thickness (mm)	0.35	0.35
Stator pole height (mm)	23.27	19.3
Stator inner diameter (mm)	71.62	84
Rotor outer diameter (mm)	71.12	83.5
Rotor pole height (mm)	10.72	14.75
Air-gap length (mm)	0.25	0.25
Stack length (mm)	120	120
Number of phases	6	3
Number of phases working simultaneously	2	1
Copper winding length per phase (m)	37.12	32
Number of turns per phase	64	64
Number of stator poles	12	12
Number of rotor poles	4	8
Number of rotor pole pairs	1	4
Maximum average static torque (Nm)	23.9	6.6

From the Table 1, it can be seen that the copper-winding length per phase of the proposed SRM is 16% longer than the conventional SRM. Furthermore, the proposed SRM has six phases, and the conventional SRM has three phases. It means that the proposed SRM contains considerably more copper than the conventional SRM. Due to this reason, the total efficiency of the proposed SRM is lower than the conventional SRM. This is the main disadvantage of the proposed SRM. On the other hand, the torque output performance of the proposed SRM has increased considerably due to the copper increase. With the same maximum current 35 A, the average static torque of the proposed SRM is 23.9 Nm. It is 6.6 Nm for the conventional SRM. Figure 2 presents the windings of the two SRMs.

(a) (b)

Figure 2. Windings of SRMs: (**a**) the proposed segmental SRM; (**b**) the conventional SRM.

Figure 3 demonstrates the inductance of winding for the SRMs from the unaligned to the aligned position. The proposed SRM has higher winding inductance compared to the conventional SRM. The reason is the difference of the winding lengths of the two SRMs.

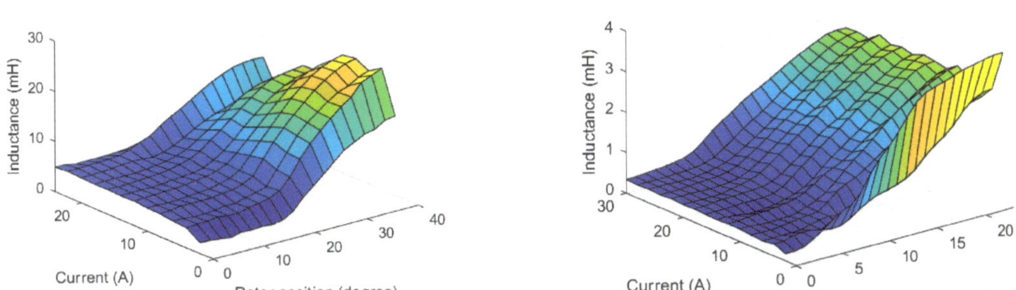

(a) (b)

Figure 3. Measured winding inductance of SRMs: (**a**) the proposed segmental SRM; (**b**) the conventional SRM.

Figure 4b presents the working method of the conventional three phases SRM [24,25]. Figure 4a shows the working method of the proposed SRM and illustrates the method to reduce the torque ripple. Where the T_A, T_B, T_C, T_D, T_E, T_F and T_{total} means the positive torque of phase A, B, C, D, E, F and the total positive torque, respectively. With the 15-degree misalignment and two phases working simultaneously, the torque of each phase can be summed to produce the total output torque and reduce the torque ripple. Furthermore, as there are two phases working together, the power density of the proposed SRM can be doubled compared to the conventional six-phase SRM.

It can be seen from Figure 4b that when the SRM is in commutation from one phase to a different phase, both involved phases produce a low torque. It results in the low sum torque during the commutation process. Figure 4a shows that two phases are working simultaneously for the proposed SRM. When one phase is in commutation, there is always another phase producing high torque. This method of operation reduces the torque ripple.

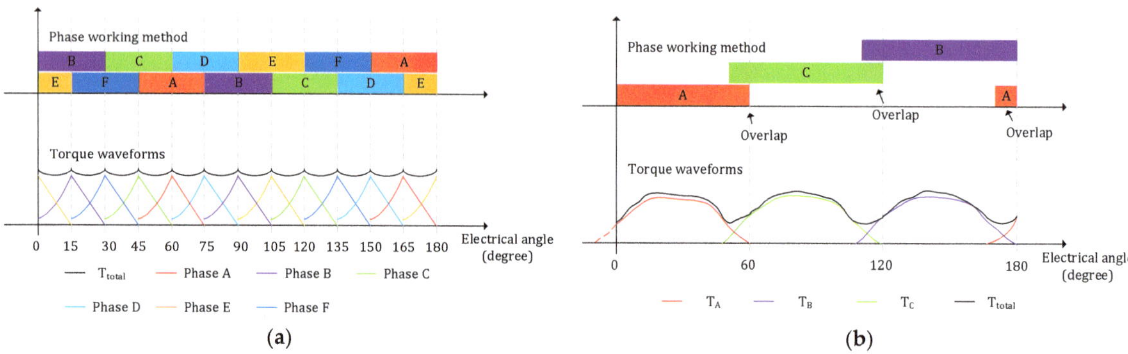

Figure 4. The operation method of the SRMs: (**a**) proposed six-phase SRM; (**b**) conventional three-phase SRM.

The torque curve of each phase is designed by FEM with optimization methods and shifted 30 degrees as the stator poles pitch satisfies the equation:

$$\alpha = 360^\circ/Ns = 360^\circ/12 = 30^\circ \quad (1)$$

where α is the stator pole pitch, and Ns is the number of stator poles.

Figure 5 presents the flux lines of the proposed SRM for phase A. In Figure 5a, phase A is energized, when the previous phase, phase F, is aligned. During operation, the excitation of phase A should start at this position. Figure 5b shows the flux lines when the segment is aligned with phase A. During operation, the demagnetization of phase A should finish at this point.

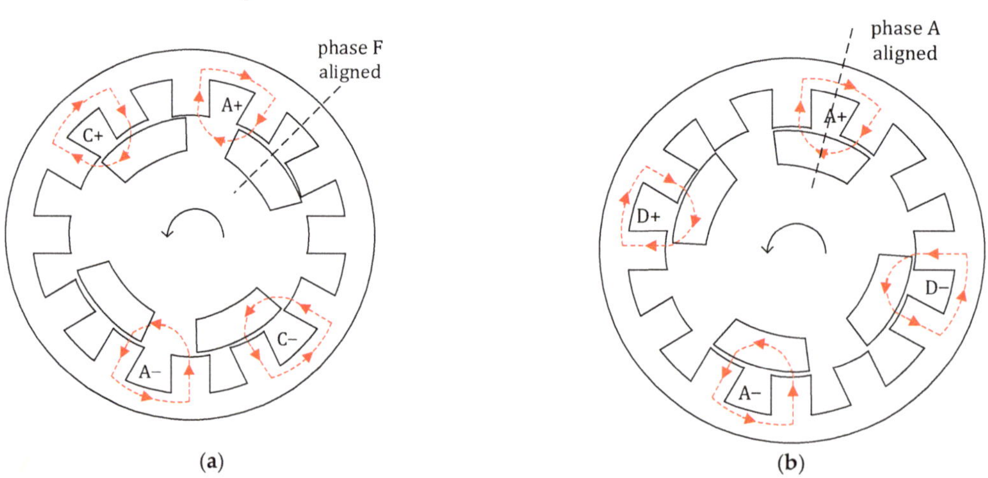

Figure 5. Magnetic fields of the proposed SRM: (**a**) magnetic fields when the previous phase of phase A is aligned; (**b**) magnetic fields when phase A is aligned.

Figure 6 shows the CAD model of the proposed SRM. This CAD model is used for FEM analysis and prototype manufacture.

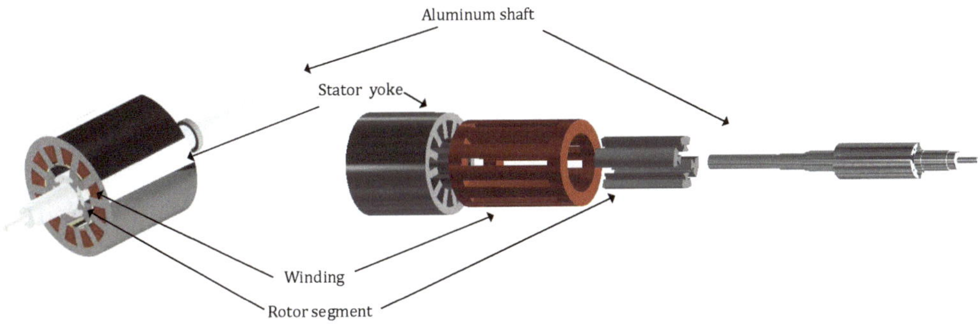

Figure 6. CAD model of the stator and rotor of the proposed SRM.

3. Electromagnetic Analysis and Optimization Method of the SRM

With the 2D FEM technique, the proposed SRM is analyzed and designed. Figure 7 displays the FEM simulation results of the proposed SRM.

The stator and rotor pole shapes are designed with the optimization method, for which the results are shown in Figure 8. It is illustrated in Figure 4 that the distance between the peak torque position and the aligned zero torque position should be 15 degrees which means the maximum torque should appear at a 75-degree point. Figure 8a shows that with the optimization method, the maximum torque point is located at the 75-degree point. This distance is designed to be 15 degrees. Figure 8b illustrates the stator and rotor pole shapes from the optimization.

(a)

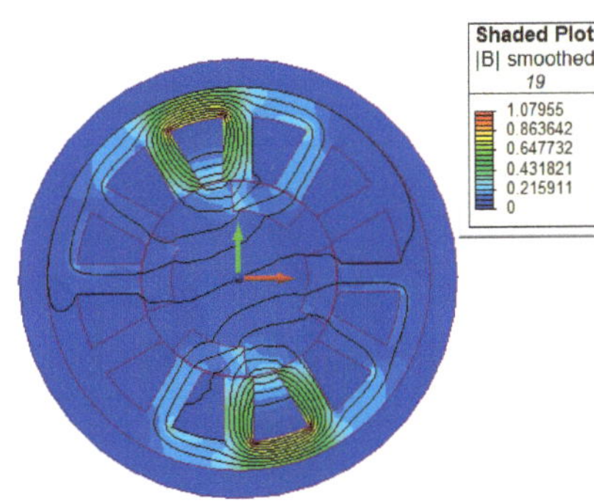

(b)

Figure 7. FEM simulation results of the proposed SRM: (**a**) the mesh distribution; (**b**) the magnitude of the magnetic flux density.

Table 2 lists the details of the optimization method. The objective is to minimize the distance between the maximum torque point and the 75-degree position. When the maximum torque point is fixed to the 75-degree position, the distance between the maximum torque point and the zero-torque point will be 15 degrees. The torque waveforms of the next phase are shifted by 15 degrees. As a result, the maximum torque point of the present phase will be in line with the zero-torque point of the next phase. Therefore, the torque ripple of the proposed SRM is reduced as illustrated in Figure 4. The degree-of-freedom is the width of the stator poles, and the constraint is to keep the slot area constant for the windings. The optimization algorithm used for this problem is the golden section search and parabolic interpolation algorithm [26,27].

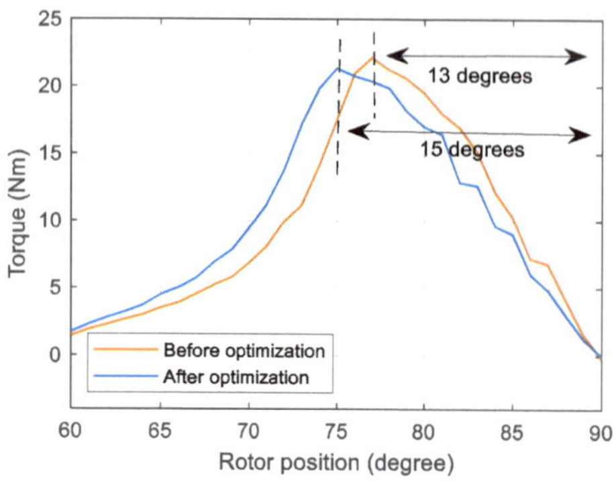

(a)

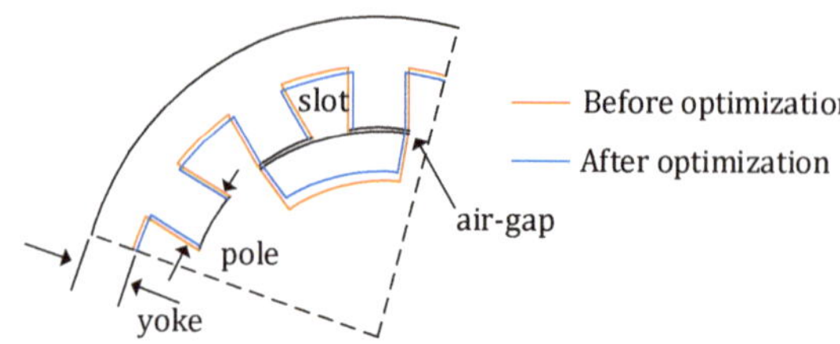

(b)

Figure 8. Optimization of the proposed SRM: (**a**) the optimization results of the torque; (**b**) the optimization result of the pole dimensions.

Table 2. The details of the optimization methods.

Objective	Minimization of the distance between the maximum torque point and the 75-degree position in Figure 8a
Variable	Width of the pole in Figure 8b
Constraint	The slot area is constant in Figure 8b
Optimization algorithm	Golden section search and parabolic interpolation algorithms

The optimization variable is the width of the stator poles in Figure 8b. Both the thickness of the stator yoke and the thickness of the rotor segments are equal to the thickness of the stator poles as shown in Figure 8b. The arc angle of the rotor segments is equal to the arc angle across two stator poles. The thickness of the stator yoke, the thickness of the rotor segments, and the arc angle of the rotor segments can be calculated from the width of stator poles, which is the optimization variable.

The values from the optimization results are listed in Table 3. As the optimization is performed and results obtained, the stator pole width is changed from 5 mm to 5.7 mm, the rotor pole arc is changed from 21.9 degrees to 22.9 degrees, and the rotor inner diameter is changed from 63.5 mm to 60.7 mm.

Table 3. SRM parameters before and after the optimization.

Parameter	Before Optimization	After Optimization
Stator pole width (mm)	5	5.7
Rotor pole arc (degree)	21.9	22.9
Rotor inner diameter (mm)	63.5	60.7

Figure 9 demonstrates the static profiles of the proposed SRM from the FEM simulation. Figure 9a shows that the middle points of the torque profiles are misaligned 15 degrees. Figure 9b presents that the middle points of the flux linkage profiles are misaligned by 15 degrees.

The torque can be expressed as:

$$T_e = \partial W_{em}(\theta, i)/\partial\theta = 1/2 \times i^2 \times dL/d\theta \quad (2)$$

where $W_{em}(\theta, i)$ is the magnetic co-energy, i is the current, θ is the rotor position, and T_e is the torque [28]. All units are according to S.I. units.

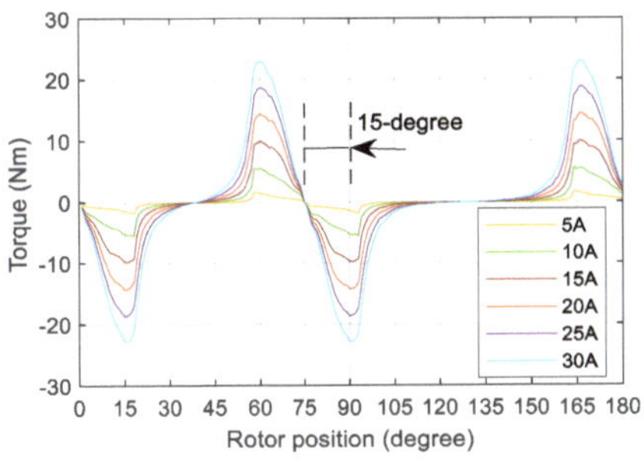

(a)

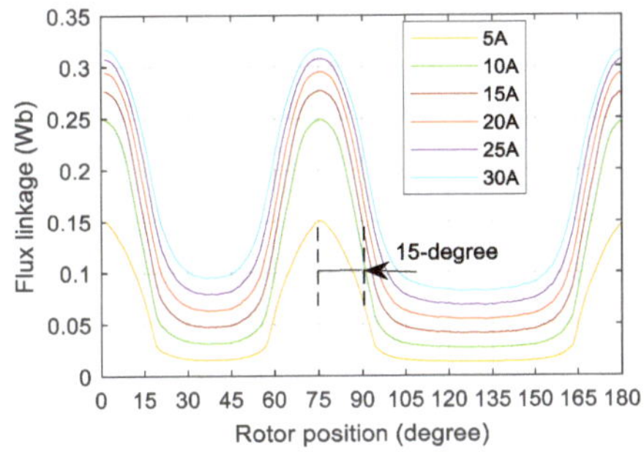

(b)

Figure 9. FEM simulation static torque and flux linkage profiles of the proposed SRM: (**a**) the static torque profiles; (**b**) the static flux linkage profiles.

Figure 10 presents the static performance from the FEM simulation of the proposed SRM. The FEM simulation is performed with the maximum current 35 A. The dashed line in Figure 10 shows the sum of the positive torques of each phase. It explains the method of the proposed SRM to reduce the torque ripple. Once the phases work in the sequence displayed in Figure 4, the output torque ripple will be reduced. Figure 11 displays the FEM simulation results of the conventional SRM with 35 A current. The average value of the static torque for the proposed SRM and the conventional SRM are 23.9 Nm and 6.6 Nm, respectively, as listed in Table 1. The difference is from the inductance difference of the two SRMs.

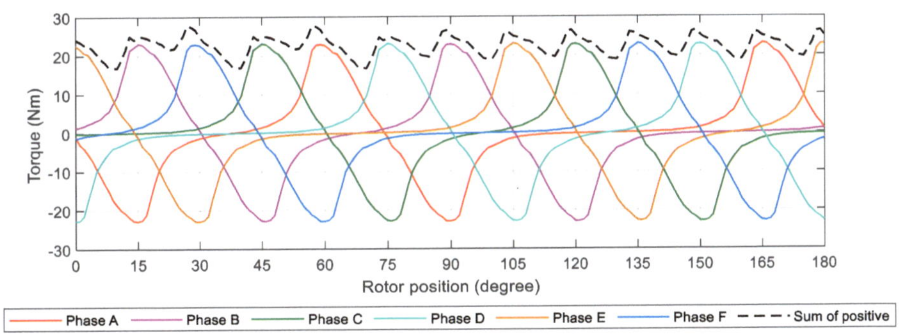

Figure 10. Static characteristics of the proposed SRM from FEM simulation results.

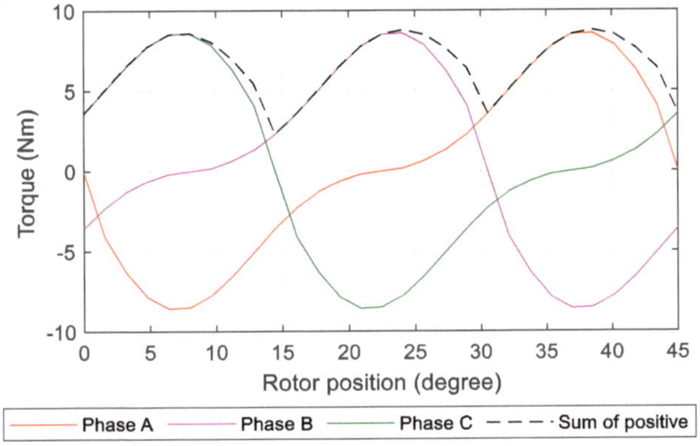

Figure 11. Static characteristics of the conventional SRM from FEM simulation results.

4. Cooling System Design

A water liquid cooling system with an aluminum machine cooling jacket is designed for the proposed SRM. It is used to prevent the SRM from overheating. The FEM method is used to analyze the cooling system. Figure 12 shows the schematic of the liquid cooling system. Figure 13a shows the CAD model of the liquid cooling jacket. Figure 13b shows the mesh distribution of the liquid cooling jacket and the SRM. The cooling jacket is directly connected to the stator yoke. The liquid cooling channel is made directly inside the jacket. Table 4 lists the parameters for the thermal analysis. When the phase current is 20 A, the copper losses are 528 W, and the maximum temperature reaches 29.9 °C. The temperature is in a safe range.

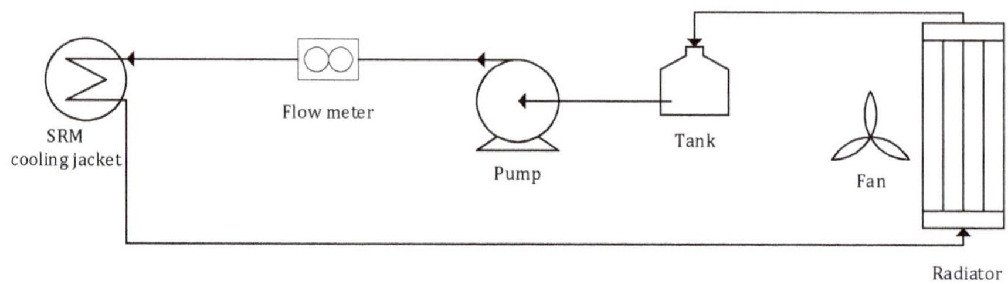

Figure 12. SRM cooling setup schematic.

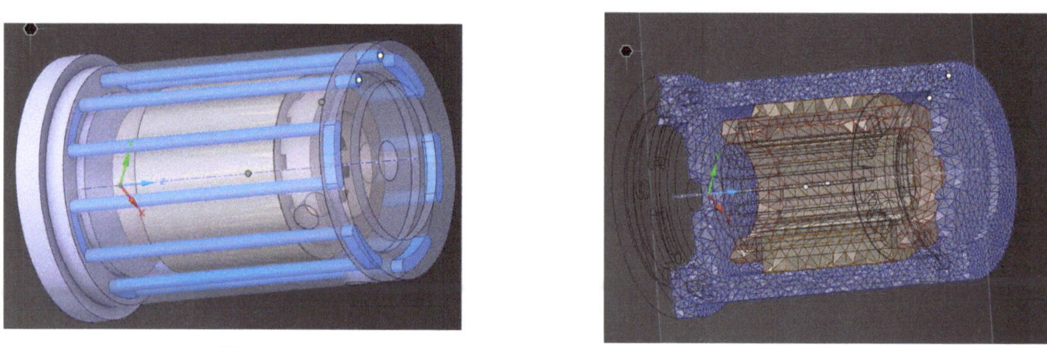

(**a**) (**b**)

Figure 13. Cooling jacket and the FEM analysis: (**a**) the CAD model of the cooling jacket and SRM; (**b**) the mesh distribution for the FEM simulation.

Figure 14 shows the FEM simulation results of the cooling system for the proposed SRM. It illustrates that the highest temperature areas are at the two end-windings zones.

Figure 15a presents the setup of the cooling system. Figure 15b illustrates the measured thermal results. From a hole for the cable terminal on the water jacket, the temperature of the stator can be observed. It is 26.4 °C as shown in Figure 15b.

Table 4. The loss parameters for the thermal analysis.

Phase Current (A)	Copper Losses (W)	Maximum Temperature (°C)
20	528	29.946

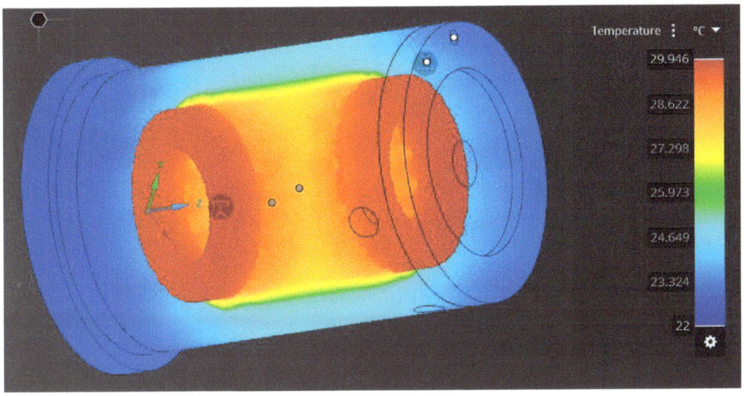

Figure 14. The FEM thermal simulation of the liquid cooling jacket.

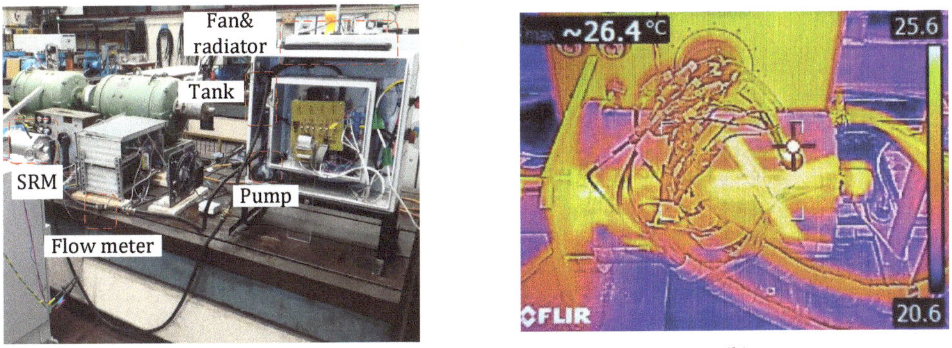

(a) (b)

Figure 15. The test of the cooling system: (**a**) the setup of the cooling system; (**b**) the measured thermal result.

5. Simulation and Analysis

The dynamic simulations are performed to validate the proposed SRM. MATLAB/Simscape™ is used to simulate the SRM systems.

Figure 16 shows the drive and control system for the conventional SRM and the proposed SRM. SiC MOSFETs are used to create the driver as shown in Figure 16a. A dual-loop control system is used in the simulation and experiment. The current controller in Figure 16b is the hysteresis current controller. The speed controller is a PID controller.

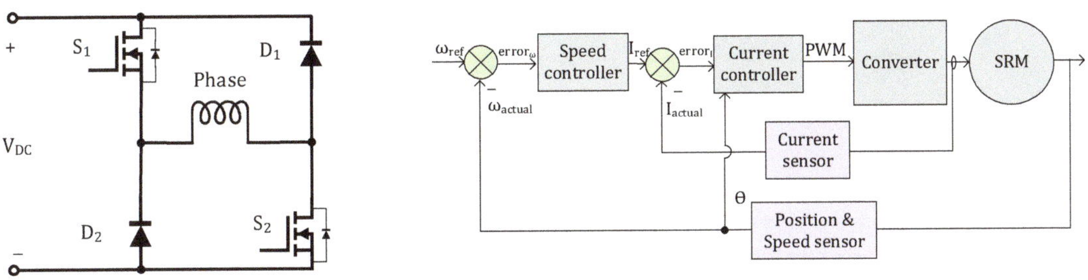

(a) (b)

Figure 16. SiC driver and the control system: (**a**) SiC converter for each phase; (**b**) dual-loop control system diagram.

Figure 17 illustrates the simulated current waveforms and the torque waveform.

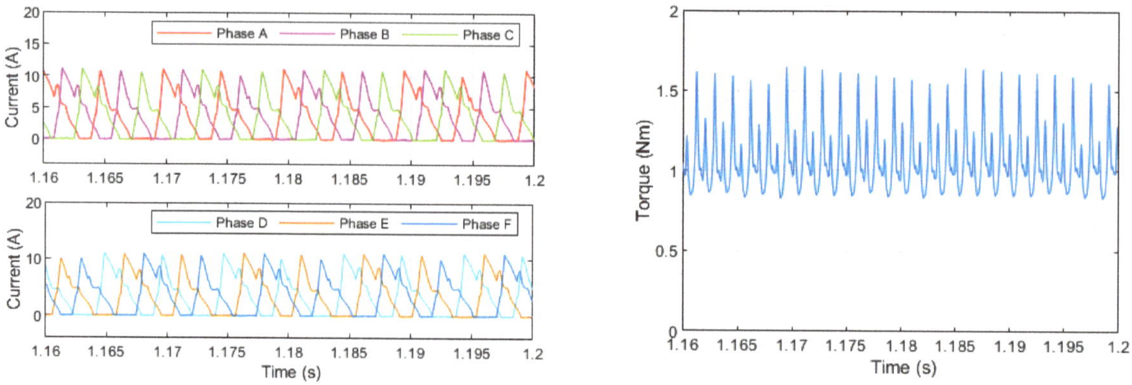

(**a**) (**b**)

Figure 17. MATLAB/Simulink simulation results at 3000 rpm: (**a**) current waveforms. (**top**) Phase A, B and C, (**bottom**) Phase D, E and F; (**b**) torque waveform.

6. Prototype and Experimental Results

To verify the simulation results, a prototype of the proposed SRM is manufactured. A SiC MOSFET driver with a digital signal processor (DSP) controller is used to build the SRM drive system. A DC motor is used as the load machine. The field winding of the DC machine is connected to a DC voltage supply.

Figure 18 presents the prototype and setup of the proposed SRM. The torque sensor is used to measure the static and dynamic torque of the SRM.

Figure 19 illustrates the FEM results and measured static torque and the measured flux linkage for the proposed SRM. The measured results reach a good agreement with the FEM simulation results. The flux calculation is based on the measured torque.

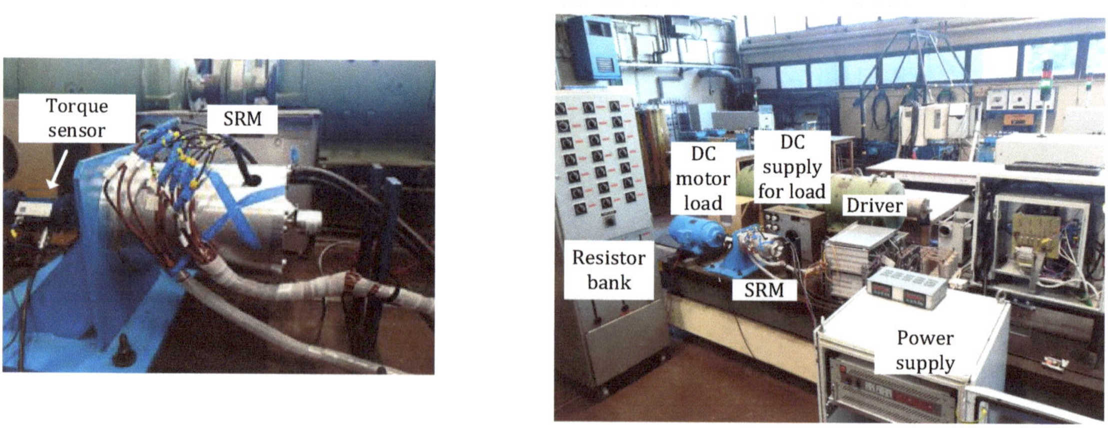

(**a**) (**b**)

Figure 18. Prototype and setup of the proposed SRM: (**a**) the prototype of the proposed SRM; (**b**) the setup of the SRM system for the experiment.

The following equations present the relation between flux and torque [29].

$$W_{em}(\theta_0, i_0) = \int T(\theta, i_0)d\theta + L_0 \times i_0{}^2/2 \quad (3)$$

$$\psi(\theta_0, i_0) = \partial W_{em}(\theta_0, i_0)/\partial i \quad (4)$$

where $T(\theta, i_0)$ is the torque, L_0 is the unaligned inductance of the SRM, which is measured to be 5.3 mH, i_0 is the phase current, $W_{em}(\theta_0, i_0)$ is the magnetic co-energy, and $\psi(\theta_0, i_0)$ is the flux linkage.

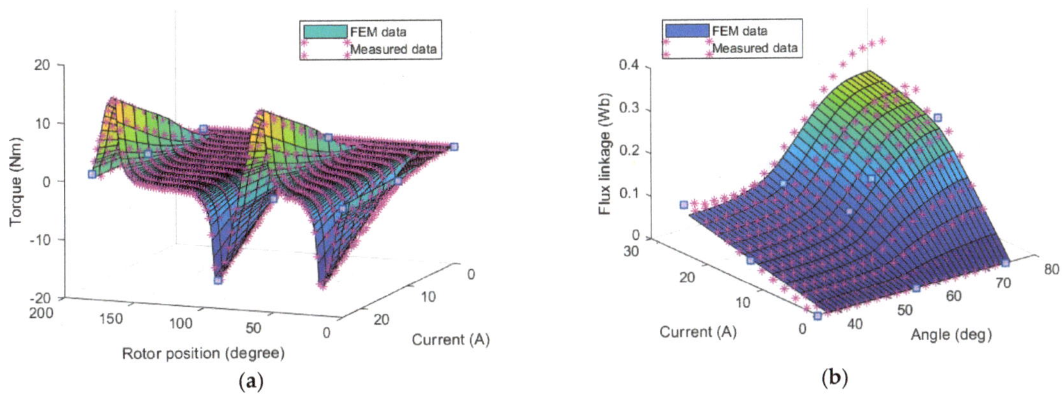

Figure 19. The FEM results and measured static data of the proposed SRM: (**a**) the torque; (**b**) the flux linkage.

Figure 20 presents the torque-speed specification for the 80 kW full-scale SRM and the 2 kW small-scale proposed SRM. For both SRMs, the rated speeds are 3000 rpm, and the maximum speeds are 6000 rpm. For the full-scale SRM, the rated torque is 254 Nm. The rated torque is 6.5 Nm for the small-scale SRM. In Figure 20, the tested points for this paper are marked.

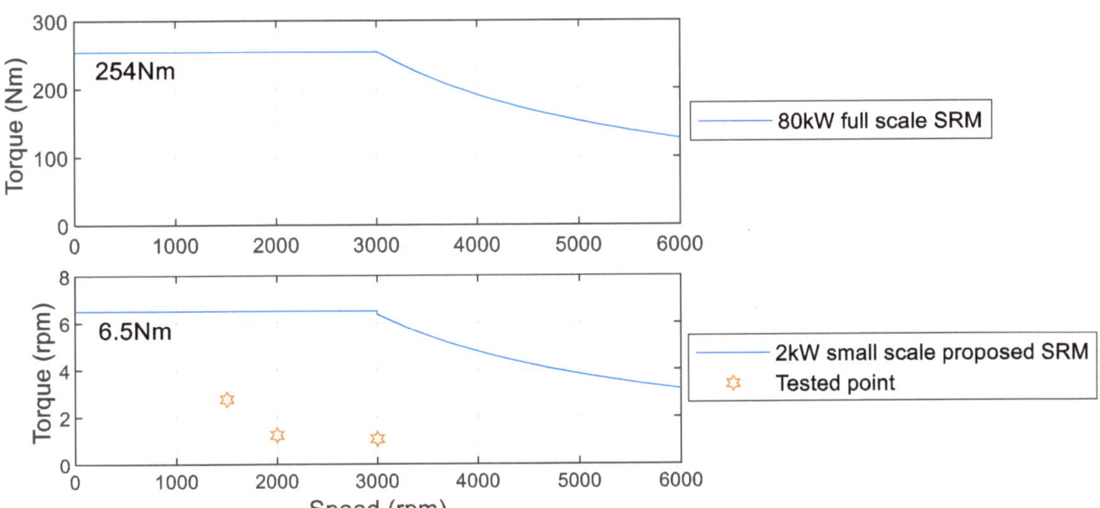

Figure 20. Torque speed specification for the full-scale SRM and the small-scale proposed SRM.

Figure 21a shows the proposed SRM experimental results of the current waveforms for phase A, B and C at a speed of 1500 rpm with load. Figure 21b shows the proposed SRM experimental current waveforms for phases D, E and F at 1500 rpm with load.

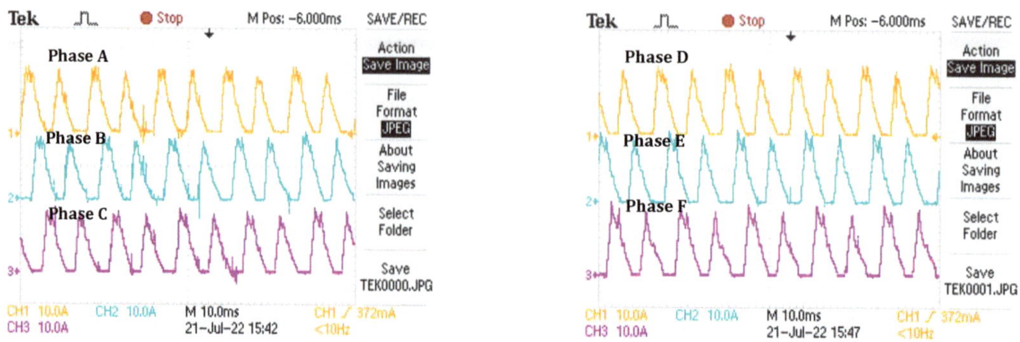

(a) (b)

Figure 21. The proposed SRM experiment results of the phase current waveforms at 1500 rpm with load: (**a**) phase A, B and C current; (**b**) phase D, E and F current.

Figure 22 presents the reference SRM experiment results of the current waveforms at the speed of 1500 rpm with load. The current waveforms are used to reconstruct the accurate torque waveforms.

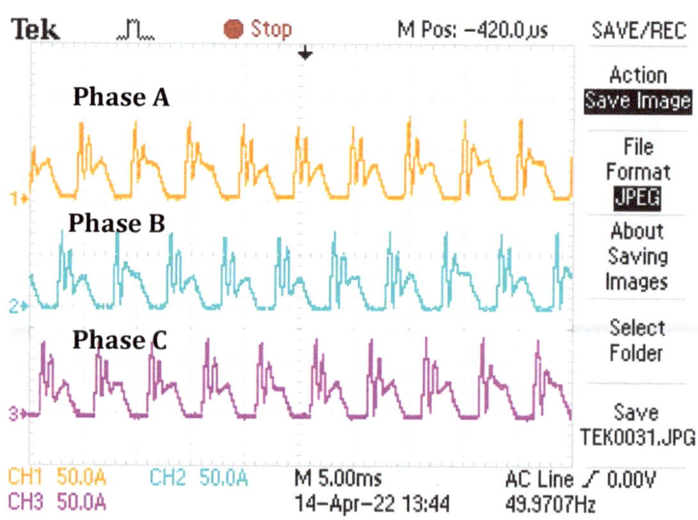

Figure 22. The reference SRM experiment results of the phase current waveforms at 1500 rpm with load.

The (3) expresses the definition of torque ripple:

$$T_r = (T_{max} - T_{min})/T_{avg} \tag{5}$$

where T_r is the torque ripple, T_{max} is the maximum torque, T_{min} is the minimum torque, and T_{avg} is the average torque.

Figure 23a shows the experimental result of the torque waveforms measured from the sensor on the shaft. It illustrates that the proposed SRM reduces the torque ripple by 45% at 1500 rpm compared to the conventional SRM. Figure 23b displays the torque ripple comparison from the sensor on the shaft for different speed values. It shows that the proposed SRM obtains lower torque ripple measured from the sensor on the shaft at different speeds compared to the conventional SRM.

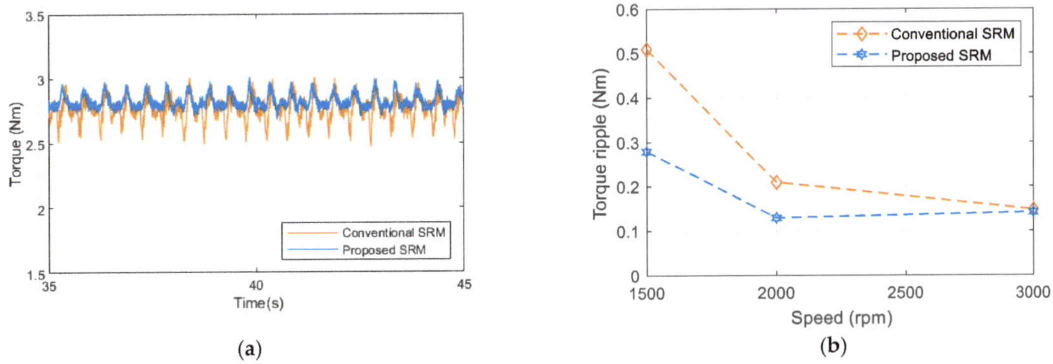

Figure 23. The experimental results from the sensor on the shaft: (**a**) torque waveforms at 1500 rpm; (**b**) torque ripple comparison at different speeds.

However, the torque waveforms measured from the sensor on the shaft are smoothed by the rotor. It should be reconstructed with the experiment current waveforms [30].

A simulink simulation is used to reconstruct the torque waveforms with the current waveforms from Figures 21 and 22. Figure 24 presents the reconstruction results with the 1500 rpm speed and 2.76 Nm torque load. Figure 24a displays the torque waveform and phase A current waveform for the proposed SRM. Figure 24b shows the torque waveform comparison. The torque ripple of the proposed SRM is reduced by 14.86%, compared to the reference SRM. Figure 24c displays the torque and phase A current waveforms for the reference SRM.

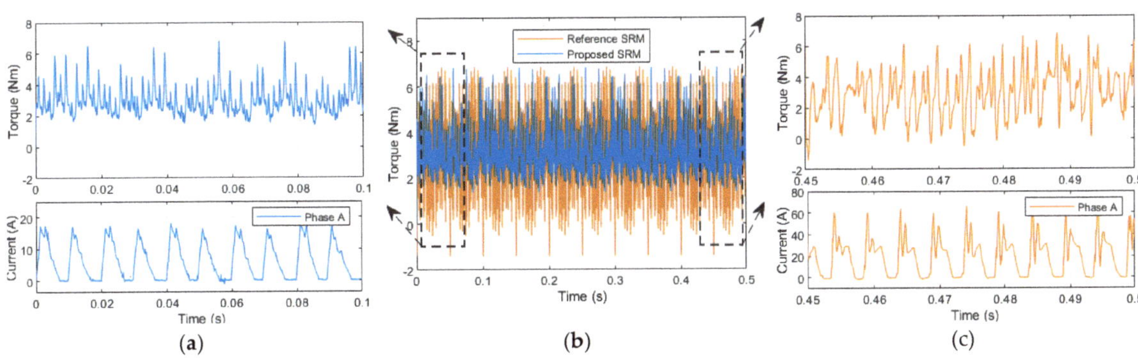

Figure 24. Torque reconstruction based on experiment results with 1500 rpm: (**a**) torque and current for the proposed SRM; (**b**) torque waveforms comparison; (**c**) torque and current for the reference conventional SRM.

Moreover, the mechanical size of the proposed SRM is the same as the conventional SRM as shown in Table 1.The rated power and torque of the proposed SRM are designed to be the same as the conventional SRM, which are 2 kW and 6.5 Nm, respectively.

7. Conclusions and Future Work

A six-phase SRM with a segmental rotor with a 15-degree misalignment is proposed in this paper for EV applications. With this misalignment, the proposed SRM torque ripple is reduced. At 1500 rpm, the torque ripple is reduced by 54.9%, compared to the conventional SRM. The FEM simulation is used to analyze and design the dimension of the proposed SRM, and an optimization method is used to determine the pole shapes of the stator and rotor, resulting in the 15-degree misalignment. A liquid cooling system is designed and analyzed using FEM software, which is subsequently tested with experiments. The dynamic simulation of the SRM system is performed using a drive system with SiC MOSFETs. A dual-loop control system is built with a DSP. The prototype of the proposed SRM was manufactured and built, and experiments were performed to validate the torque ripple improvement of the proposed SRM. The experiment results show that the torque ripple of the proposed SRM is reduced for different speed ranges compared to the conventional SRM.

In the future, an advanced high-quality current controller will be further designed to generate smooth current waveforms. As a result, the torque ripple of the SRM will be further reduced. In addition, the electromagnetic interference (EMI) problem will also be limited. Furthermore, the efficiency and converter cost of the proposed SRMs compared to the interior permanent magnet machine should be analyzed comprehensively in future work. Finally, the copper length of the proposed SRM is long. It may increase copper losses and decrease SRM efficiency. It should be analyzed and updated in the future.

Author Contributions: Conceptualization, Y.L.; methodology, Y.L. and O.H.; software, Y.L.; validation, Y.L. and O.H.; formal analysis, Y.L.; investigation, O.H.; resources, O.H.; data curation, Y.L.; writing—original draft preparation, Y.L.; writing—review and editing, M.A.F., J.C., Y.B. and M.E.B.; visualization, Y.L.; supervision, O.H.; project administration, O.H.; funding acquisition, O.H. All authors have read and agreed to the published version of the manuscript.

Funding: This study was funded by China Scholarship Council (CSC).

Institutional Review Board Statement: Not applicable.

Informed Consent Statement: Not applicable.

Data Availability Statement: Not applicable.

Acknowledgments: Yuanfeng Lan was sponsored by China Scholarship Council. Authors acknowledge Flanders Make for the support to this research group.

Conflicts of Interest: The authors declare no conflict of interest.

References

1. Kang, Z.; Hu, Y.; Sun, D. Multi-mode Drive Control System of Switched Reluctance Motor Based on a Novel N + 2 Power Converter. In Proceedings of the 2021 IEEE 4th Student Conference on Electric Machines and Systems (SCEMS), Huzhou, China, 1–3 December 2021.
2. Lan, Y.; Benomar, Y.; Deepak, K.; Aksoz, A.; Baghdadi, M.E.; Bostanci, E.; Hegazy, O. Switched Reluctance Motors and Drive Systems for Electric Vehicle Powertrains: State of the Art Analysis and Future Trends. *Energies* **2021**, *14*, 2079. [CrossRef]
3. Widmer, J.D.; Mecrow, B.C. Optimized Segmental Rotor Switched Reluctance Machines With a Greater Number of Rotor Segments Than Stator Slots. *IEEE Trans. Ind. Appl.* **2013**, *49*, 1491–1498. [CrossRef]
4. Widmer, J.D.; Martin, R.; Mecrow, B.C. Optimisation of an 80kW Segmental Rotor Switched Reluctance Machine for automotive traction. In Proceedings of the 2013 International Electric Machines & Drives Conference, Chicago, IL, USA, 12–15 May 2013.
5. Lan, Y.; Peng, W.; Aksoz, A.; Benomar, Y.; Van den Bossche, P.; El Baghdadi, M.; Hegazy, O. Design and Modelling of 12/4 Fully-Pitched Segmental Switched Reluctance Motors. In Proceedings of the 2020 Fifteenth International Conference on Ecological Vehicles and Renewable Energies (EVER), Monte-Carlo, Monaco, 10–12 September 2020.
6. Siadatan, A.; Najmi, V.; Asgar, M.; Afjei, E. A new 6/4 two layers switched reluctance motor: Concept, simulation and analysis. In Proceedings of the International Aegean Conference on Electrical Machines and Power Electronics and Electromotion, Joint Conference, Istanbul, Turkey, 8–10 September 2011.
7. Daldaban, F.; Ustkoyuncu, N. Multi-layer switched reluctance motor to reduce torque ripple. *Energy Convers. Manag.* **2008**, *49*, 974–979. [CrossRef]
8. Afjei, E.; Torkaman, H.; Mazloomnezhad, B. A new double layer per phase configuration for switched reluctance motor. In Proceedings of the 2010 IEEE International Conference on Power and Energy, Kuala Lumpur, Malaysia, 29 November–1 December 2010.
9. Siadatan, A.; Najmi, V.; Afjei, E. A novel 4/4 Multilayer Switched Reluctance Motor with 4 magnetically independent layers. In Proceedings of the International Aegean Conference on Electrical Machines and Power Electronics and Electromotion, Joint Conference, Istanbul, Turkey, 8–10 September 2011.
10. Vahedi, P.; Ganji, B.; Afjei, E. Multi-layer switched reluctance motors: Performance prediction and torque ripple reduction. *Int. Trans. Electr. Energy Syst.* **2020**, *30*, e12215. [CrossRef]
11. Higuchi, T.; Ueda, T.; Abe, T. Torque ripple reduction control of a novel segment type SRM with 2-steps slide rotor. In Proceedings of the The 2010 International Power Electronics Conference—ECCE ASIA, Sapporo, Japan, 21–24 June 2010.
12. Khor, M.T.; Sotudeh, R. A 3-phase 12/10 asymmetrical switched reluctance motor. In Proceedings of the 2005 European Conference on Power Electronics and Applications, Dresden, Germany, 11–14 September 2005.

13. Bogusz, P.; Korkosz, M.; Proko, J. Performance analysis of Switched Reluctance Motor with asymmetric stator. In Proceedings of the 2011 IEEE International Symposium on Industrial Electronics, Gdansk, Poland, 27–30 June 2011.
14. Bogusz, P.; Korkosz, M.; Powrózek, A.; Prokop, J. A two-phase switched reluctance motor with asymmetrical rotor for a high-speed drive. In Proceedings of the 2015 International Conference on Electrical Drives and Power Electronics (EDPE), Tatranska Lomnica, Slovakia, 21–23 September 2015.
15. Bogusz, P.; Korkosz, M.; Prokop, J. Control method of high-speed switched reluctance motor with an asymmetric rotor magnetic circuit. *Arch. Electr. Eng.* **2016**, *65*, 685–701. [CrossRef]
16. Torres, J.; Moreno-Torres, P.; Navarro, G.; Blanco, M.; Nájera, J.; Santos-Herran, M.; Lafoz, M. Asymmetrical Rotor Skewing Optimization in Switched Reluctance Machines Using Differential Evolutionary Algorithm. *Energies* **2021**, *14*, 3194. [CrossRef]
17. Xu, Z.; Yu, Q.; Zhang, F. Design and Analysis of Asymmetric Rotor Pole Type Bearingless Switched Reluctance Motor. *CES Trans. Electr. Mach. Syst.* **2022**, *6*, 3–10. [CrossRef]
18. Arbab, N.; Wang, W.; Lin, C.; Hearron, J.; Fahimi, B. Thermal Modeling and Analysis of a Double-Stator Switched Reluctance Motor. *IEEE Trans. Energy Convers.* **2015**, *30*, 1209–1217. [CrossRef]
19. Fairall, E.; Rheberhegen, C.; Rowan, E.; Lo, J.; Bilgin, B.; Emadi, A. Maximizing thermal effectiveness and minimizing parasitic loss in a liquid cooled switched reluctance machine. In Proceedings of the 2016 IEEE Transportation Electrification Conference and Expo (ITEC), Dearborn, MI, USA, 27–29 June 2016.
20. Márquez-Fernández, F.J.; Potgieter, J.H.J.; Fraser, A.G.; McCulloch, M.D. Experimental Validation of a Thermal Model for High-Speed Switched Reluctance Machines for Traction Applications. *IEEE Trans. Ind. Appl.* **2018**, *54*, 3235–3244. [CrossRef]
21. Ru, L. Comparison of Vibration Between an Evaporative Cooling Switched Reluctance Motor and a Conventional Switched Reluctance Motor. In Proceedings of the 2018 21st International Conference on Electrical Machines and Systems (ICEMS), Jeju, Korea, 7–10 October 2018.
22. Nonneman, J.; Schlimpert, S.; T'Jollyn, I.; De Paepe, M. Modelling and Validation of a Switched Reluctance Motor Stator Tooth with Direct Coil Cooling. In Proceedings of the 2020 19th IEEE Intersociety Conference on Thermal and Thermomechanical Phenomena in Electronic Systems (ITherm), Orlando, FL, USA, 21–23 July 2020.
23. Schlimpert, S.; Mrak, B.; Siera, I.; Sprangers, R.; Nonneman, J.; De Paepe, M.; Vanhee, S. Experimental & Modelling Study of Advanced Direct Coil Cooling Methods in a Switched Reluctance Motor. In Proceedings of the 2020 IEEE Vehicle Power and Propulsion Conference (VPPC), Gijon, Spain, 18 November–16 December 2020.
24. Lukman, G.F.; Nguyen, X.S.; Ahn, J.W. Design of a Low Torque Ripple Three-Phase SRM for Automotive Shift-by-Wire Actuator. *Energies* **2020**, *13*, 2329. [CrossRef]
25. Ma, C.; Qu, L.; Tang, Z. Torque ripple reduction for mutually coupled switched reluctance motor by bipolar excitations. In Proceedings of the 2013 International Electric Machines & Drives Conference, Chicago, IL, USA, 12–15 May 2013.
26. Grund, F.; Forsythe, G.E.; Malcolm, M.A.; Moler, C.B. Computer Methods for Mathematical Computations. Englewood Cliffs, New Jersey 07632. Prentice Hall, Inc., 1977. XI, 259 S. *Z. Angew. Math. Und Mech.* **1979**, *59*, 141–142. [CrossRef]
27. Brent, R.P. *Algorithms for Minimization without Derivatives*; Prentice-Hall: Englewood Cliffs, NJ, USA, 1973.
28. Zhou, Y.; Jiang, J.; Hu, P.; Yuan, Y. A Novel Dual-Channel Bearingless Switched Reluctance Motor. *IEEE Access* **2021**, *9*, 122373–122384. [CrossRef]
29. Cailleux, H.; Mouchoux, J.C.; Multon, B.; Hoang, E.; Le Chenadec, J.Y. Comparison of Measurement Methods to Determine the Electromagnetic Characteristics of Switched Reluctance Motors. In Proceedings of the Electric Drive Design and Applications, Lausanne, Switzerland, 19–20 October 1994; pp. 639–644.
30. Lukman, G.F.; Ahn, J.-W. Torque Ripple Reduction of Switched Reluctance Motor with Non-Uniform Air-Gap and a Rotor Hole. *Machines* **2021**, *9*, 348. [CrossRef]

machines

MDPI

Article

Design and Experimental Evaluation of an In-Wheel Flux-Switching Machine for Light Vehicle Application

Gabriel A. Mendonça [1], Diogo P. V. Galo [2], Luís Carlos M. Sales [3], Braz J. Cardoso Filho [4] and Thales A. C. Maia [4,*]

[1] Aperam South America-Praça 1°, de Maio, 9-Centro, Timóteo 30180-018, MG, Brazil
[2] Graduate Program in Electrical Engineering, Universidade Federal de Minas Gerais, Av. Antônio Carlos 6627, Belo Horizonte 31270-901, MG, Brazil
[3] Stellantis Brasil, Av. Contorno, 3455, Bairro Paulo Camilo, Betim 32530-900, MG, Brazil
[4] Electrical Engineering Department, Universidade Federal de Minas Gerais, Av. Antônio Carlos 6627, Belo Horizonte 31270-901, MG, Brazil
* Correspondence: thalesmaia@cpdee.ufmg.br

Abstract: Restrictive regulations regarding emissions and fossil fuel consumption lead to the electric vehicle being an alternative to replace conventional internal combustion engine vehicles. The pure electric powertrain technology and the charging infrastructure are still immature in some markets, where increasing the overall vehicle efficiency by energy harvesting means can be a more viable solution. This paper presents the design and experimental validation of an in-wheel flux-switching machine for regenerative braking in a light passenger vehicle. Later, the energy can be used for fuel handling and reforming, performance enhancement, increasing efficiency, and reducing emissions. Feasibility and technological challenges are also discussed. The Maxwell–Fourier method and a novel steady-state equivalent circuit presented in this paper are used for geometry sensitivity analysis and optimization routine.

Keywords: flux-switching machine; efficiency; regenerative breaking; energy harvesting

Citation: Mendonça, G.A.; Galo, D.P.V.; Sales, L.C.M.; Cardoso Filho, B.J.; Maia, T.A.C. Design and Experimental Evaluation of an In-Wheel Flux-Switching Machine for Light Vehicle Application. *Machines* **2022**, *10*, 671. https://doi.org/10.3390/machines10080671

Academic Editors: Lorand Szabo and Feng Chai

Received: 11 July 2022
Accepted: 1 August 2022
Published: 10 August 2022

1. Introduction

The road transportation sector is responsible for approximately half of world's oil demand, most particulate matter emissions and air pollutants. Vehicle emissions are more critical in urban areas due to health concerns and micro/macro climate changes [1]. Over recent decades, regulatory initiatives, such as the Euro or the US Tier, were introduced and are being constantly updated. The establishment of emissions standards is related to the reduction in transport-related emissions. However, the implementation of more restrictive standards within the next years depends on each market's political, economic and geographic particularities, which demands local technological development [2].

The automotive industry struggles to keep up with upcoming emission standards, which demands increasing the vehicle's efficiency by partially or fully electrifying the powertrain. However, there are several problems related to powertrain electrification yet to be resolved, such as cost, range, technological maturity, safety, charging time, and battery power density. The aforementioned challenges regarding electrification still prevents the widespread adoption of electric vehicles worldwide [3,4].

Automotive energy harvesting systems are usually employed to increase overall vehicle efficiency. One example is thermal recovery, in which an amount of energy is recovered from exhaust gasses and engine coolant. Moreover, keeping the system as simple as possible, reducing weight and minimizing the cost is more important than maximizing power output [5,6]. Heat recovery systems can also be used, combining Rankine cycle systems with electrical generators to create a turbo-generator [6,7]. Energy dissipated on the vehicle suspension is another possibility for energy harvesting. This system recovers waste heat energy from car bumper, while improving vehicle driveability and handling characteristics [8,9]. Electromagnetic-based systems are preferred due to their high efficiency, design simplicity, and controllability [10,11]. Thus, the hybrid electric vehicle (HEV) is a suitable candidate to meet the upcoming emission standards during the electric vehicle technological maturation stage.

HEVs still rely on the use of an internal combustion engines (ICE) as the main source of propulsion, which leverages the well established fueling infrastructure. The use biofuels have advantages for being considered carbon neutral, in addition of benefiting from the existing traditional fueling supply chain [3,12]. As the powertrains obtain higher levels of electrification, ICE efficiency can be increased mainly due to the electric machine contribution [4]. Among different hybrid technologies, it is possible to characterize different levels of hybridization. Vehicles can be classified according to ratio between the electric powertrain power to the total vehicle power [13].

The vehicle with a hybridization level of less than 5%, is known as micro-hybrid. This type of vehicle uses the ICE as the main source of propulsion. It improves performance by using the start-stop functionality, in which the ICE shuts down at idling conditions. Another possible application is the regenerative braking, used to recharge the batteries during vehicle deceleration [14]. Under these conditions the fuel efficiency can reach up to 10%.

The amount of energy stored in vehicle inertia is approximately 6% of the total energy consumed by the vehicle. This energy can be harvested with the aid of regenerative braking instead of being wasted as heat in friction brakes. In micro and mild hybrid electric vehicles, regenerative braking is usually accomplished by using a BSG (Belt-driven Starter Generator) [13]. Considering lower levels of hybridization, the BSG is a cost effective way to reduce CO_2 emissions and lowering the total cost of ownership (TCO) [15]. On the other hand, it demands redesigning most of the electrical system of the vehicle and the front-end accessory drive system. Alternatives based on in-wheel motors are less invasive and demand fewer changes on an existing ICE vehicle platform [16–18]. In-wheel electric motors enable four wheel drive, enhanced vehicle stability and improved acceleration and efficiency [19,20].

This paper presents the design, construction, and validation of an in-wheel flux-switching machine to be used in a regenerative braking system of a light-duty ICE passenger vehicle [21]. The research is based on an existing ICE vehicle platform to reduce costs and increase project feasibility. The main objective is to contribute with a critical analysis regarding the technological challenges faced during the design routine of a flux-switching machine with the aid of a novel equivalent steady-state circuit and the Maxwell–Fourier method during a regenerative breaking operation. The topology is chosen due to its robust structure and the possibility of using its operating characteristics to realize a low cost drive system. In this sense, the electromagnetic modeling, geometry sensibility analysis, design optimization constraints, and construction and validation test of the electrical machine are discussed.

2. State of Art

Regenerative braking is a common feature in hybrid electric vehicles. In micro and mild hybrid vehicles, this solution is obtained by replacing the alternator and starter motor by a single electric machine. This device provides an effective and simple solution using the conventional ICE vehicle platform [22]. The system is usually designed to provide start–stop operation in addition to the regenerative braking functionality. The design of the BSG system is challenging due to its rigorous requirements, such as high starting torque for engine cranking, wide speed range operation, harsh environment, constrained dimensions, and high temperatures [23–25].

Despite the centralized electric drive system being regarded as a natural solution for hybrid and full electric vehicles, there are benefits of exploring alternative configurations. Full electric vehicles can be designed with in-wheel motors to improve transmission efficiency and reduces vehicle weight by eliminating differentials and the drive shaft. Furthermore, it gives more flexibility for vehicle chassis design, allowing more space for components as energy storage system [26]. The distributed drive topology with good bandwidth response provides better vehicle stability, safety, and more effective braking [20,27].

In lower hybridization levels, electric vehicles can be designed with in-wheel motors on the rear axle, with the front axle being propelled by the ICE. This architecture is known as the Through-the-Road (TtR) parallel hybrid. However, unlike a conventional parallel hybrid structure, the TtR configuration can be adapted in any conventional ICE vehicle. The implementation is cheaper and more straightforward since there is no mechanical coupling between the engine and the electrical motor, which makes the complex torque coupling device unnecessary [19,28].

The discussed advantages give in-wheel motors a growing interest within electric vehicles technological development. However, this solution introduces additional challenges that must be addressed. To begin with, the electric machine must be robust enough to support a harsh environment. This includes intense vibrations, repeated shock loads, and impacts from debris. In addition, mechanical friction brakes and on-body ICE powertrain increase the wheel temperature significantly. Furthermore, the increased unsprung mass within the wheels can affect the vehicle driving characteristics (driveability and comfort). Finally, the in-wheel machine must be light and compact enough to be installed inside the wheel hub.

Considering the listed requirements, the choice of the appropriate electric machine topology is of great importance. The design assumptions had led to different motor concepts, but most commercially available solutions focus on permanent magnet (PM) brushless machines. The use of high energy magnet and flexible design for both rotor and stator structures offers a high power and torque density solution with the possibility of tailoring the design according to the application requirements. Examples of current improvements include the study of different PM arrangements to enhance output torque [29–31] or different stator winding arrangements to extend speed range operation [32,33].

On the other hand, rare earth permanent magnets represent a costly component of the electrical machine, responsible for approximately 20% to 30% of its cost [34]. Additionally, the design must consider critical operational aspects, such as demagnetization risk and mechanical harness. These had encouraged design approaches able to reduce or eliminate the use of rare earth magnets [35,36]. The trend towards electrical machines without rare earth magnets are the topologies whose working principle is based on variable reluctance torque capabilities.

Reluctance machines are of simple construction and their rotor is composed of, exclusively, passive elements. This design offers desirable features, such as a robust structure and good thermal performance. The switching reluctance machine (SRM) is a popular topology and has been investigated over the past few years. It offers excellent characteristics for electric vehicle application, such as long constant power range, fault tolerance capability and rugged rotor structure, which allows high-temperature operation [26,35]. This topology has been proposed in the literature as suitable for in-wheel motors. The main disadvantages of SRM are: high torque ripple, noise, vibration, and the need of non-conventional inverter topology with a high current rating for the power semi-conductor modules.

In [37], the authors evaluated the motor performance sensibility to different core designs, focused on reducing the weight of the machine. In [38], a segmented-rotor SRM with concentrated winding is proposed as a high specific torque and efficiency solution. The designed topology presented a torque per active weight of 73% of a brushless DC motor with the same outer diameter. In [39], a double stator axial-flux SRM is designed for a light weight in-wheel application, in which experimental results presented a torque comparable with a base internal PMSM with the same size. In [40], a novel radial field C-core SRM is evaluated, showing good features, such as low excitation current and an easily removable rotor.

Another variable reluctance topology that is emerging is known as flux-switching machine (FSM). It offers great flexibility for both the stator and rotor structures. It can be designed with a hybrid excitation, associating PM with field coils placed in the stator. The hybrid excitation FSM offers good power density and efficiency with flux weakening capability [41]. In [42], different hybrid excitation FSM topologies are evaluated for aircraft applications requiring features, such as reliability, high power and torque densities, and fault tolerance ability. Another type of FSM which adopts only PM excitation has the advantages of being less prone to demagnetization while offering a better PM thermal management. In [43], the performance of a PM-FSM is compared with a surface-mounted PMSM, both designed as an in-wheel motor for a hybrid truck. The evaluated FSM topologies showed higher induced voltage and larger electrical loading capability. In [44], the designed FSM topologies using Dy-free PM or field winding showed great potential for traction application. Eliminating the PM, field-excited FSM offers controllable flux with lower cost. This work focused on robustness, simplicity, and cost effectiveness for the development of a regenerative braking solution. Therefore, other topologies which can be considered as a viable alternative were not considered. For instance, induction machines require complex power electronics and control. Other similar flux-modulation machines, such as the claw pole machine [45] or the synchronous homopolar machine [46], were not suitable due to limited space. In [47], a three-phase field-excited FSM design is demonstrated to be a good candidate for hybrid EV, with the optimized result achieving a power density of 4.8 kW/kg.

FSM also presents a robust structure and, with the adoption of a field winding, can be operated using a simple control circuit, eliminating the need of an active rectifier. The generated voltage is close to a sine waveform and, for a series connected field winding, the field inductance is constant.

Therefore, the machine topology chosen to implement the regenerative braking system in a hybrid electric vehicle is the in-wheel FSM. The FSM proposed for this work is a type of AC machine with salient pole rotor (5 teeth) and stator with field and armature windings distributed over 24 slots, as illustrated in Figure 1. The design variables of the presented topology are given on Table 1.

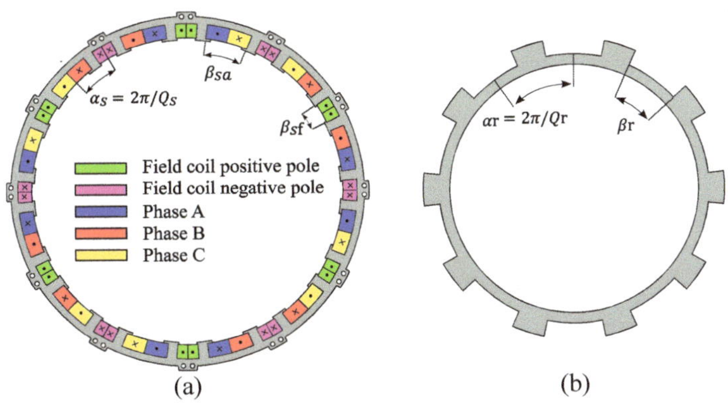

Figure 1. Cross-section of the base geometry used on this paper: (**a**) stator with field and armature windings, and (**b**) salient pole rotor.

The modeling method plays an important role in the analysis and design of electric machines. Numerical methods, such as the finite-element method, are consolidated techniques which are widely used owing to its capacity of modeling complex structures and non-ideal materials properties. On the other hand, these methods are time-consuming and require large computational effort. Analytical and semi-analytical methods provide fast and accurate results on magnetic field calculations. Two types of analytical methods are commonly used for design purposes: magnetic equivalent circuit (MEC) and Maxwell–Fourier method [18]. Regarding reluctance machines, Maxwell–Fourier offers a more flexible solution and is used for geometries which parameters vary in a large scale [48,49]. Thus, this method is used for magnetic field distribution evaluation.

Table 1. FSM design variables.

Parameter	Symbol
Air gap radius	r_{ag}
Armature slot angular aperture	β_{sa}
Armature slot outer radius	r_{sa}
Field slot angular aperture	β_{sf}
Field slot outer radius	r_{sf}
Rotor teeth angular aperture	β_r
Rotor slot inner radius	r_{ri}
Field winding number of coils	N_f
Armature winding number of coils	N_s

3. Methodology

In order to meet the machine requirements of power and performance for the existing light-duty ICE passenger vehicle, authors proposed a methodology based on steady state evaluation and optimization routines. The proposed methodology is presented in Figure 2 and is described as follows. Starting with a draft design, the electrical machine parameters are calculated using a semi-analytical method. In early design stages this approach gives reasonably accurate results with less computational effort when compared with numerical solutions [50]. The generated power is calculated with the aid of a newly developed equivalent circuit. The steady state analysis is obtained considering the effects of variable inductance from the machine double salient structure. The calculated performance gives an indication if the proposed design is consistent, otherwise the dimensions and magnetizing current are modified. The new design goes through a new analysis cycle, which is carried out until the desired characteristics are obtained.

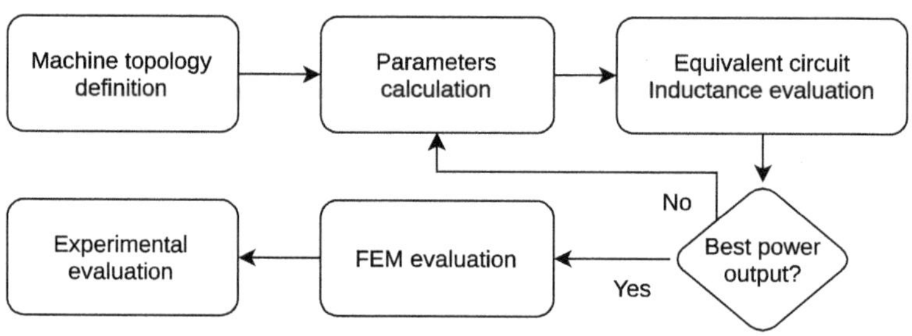

Figure 2. Research work flow methodology.

3.1. Design Specifications and Restrictions

Since the research goal is to promote energy harvesting in an existing light ICE vehicle with the aid of an in-wheel electric machine, the design preserves the original powertrain. Appropriate placement occurs in the rear axle wheels using the available space between the drum brake and inner wheel radius. The geometric constraints are usually restrictive for the number of poles.

In order to define machine rated power constraint two approaches are considered: the amount of power that can be recovered during braking and the average power needed for the auxiliary loads.

Mechanical friction braking systems must coexist with electrical regenerative braking in hybrid and full electric vehicles. During deceleration, 65% of the braking power occurs in the front axle, while only 35% is used in the rear axle [51]. These values are lower for drum-based brake systems. In addition, for a 1000 kg ICE passenger car in the FTP 75 urban drive cycle, the maximum braking power per wheel is around 2 kW and the average power is roughly 750 W. Considering a parallel braking strategy, it is reasonable to account for half of the power, 375 W for the regenerative braking per wheel. The evaluation of the system during the complete drive-cycle is not considered, since it will depends on a specific vehicle platform and its dynamic loads, as well as the storage system capacity and state-of-charge.

The research focuses on the ICE efficiency, so it is reasonable to consider loads that are used to increase ICE efficiency, such as the E-Flex system. The system uses an electric heater for the ethanol with a power demand of 288 W. Therefore, a total of 300 W (150 W per wheel) of electrical regenerative power with air cooling and a maximum of 10 A/mm^2 for current density is considered.

The vehicle's electric system was not modified and the machine is designed to charge the existing 12 V battery. Thus, the electromagnetic project was conducted focusing on this voltage level. The field winding number of turns is chosen considering a rated current below 1 A. This value is compatible with lower wire gauges and manual manufacturing processes. As for the stator winding turns, an initial value is set considering the desired no load induced voltage, but this value is further adjusted with the optimization routine. The available axial length must consider mechanical stator support, miscellaneous, such as bolts, nuts, washers and others, room for electrical connection, and, finally, end-turn coil size. All of those were evaluated with the aid of a 3D CAD software. However, end-turns' length depends on wire thickness, number of coils, and how they are tied together. It is possible to increase or decrease the number of turns keeping the field intensity constant of the field winding. To ensure a minimum end-turn coil length, including a lower curvature radius, many manufacturing trials were performed using a 3D-printed stator structure, as illustrated in Figure 3. The procedure was adequate to define wire gauge for flux and armature coil, tooth-tip geometry, winding fixture procedure and end-turn coil size. The field winging number of turns and wire gauge for both armature and field coils are presented on Table 2.

Figure 3. Manufacturing evaluation using a 3D-printed model, and final result in the built machine.

Although a lower air gap, such as tenths of millimeters, demands lower field current and higher power output, there is a higher chance of direct contact between rotor and stator. In-wheel machines demand mechanical structural evaluation in order to avoid possible catastrophic failure, which is especially troublesome considering the uncertainties of the manufacturing process. Electromagnetic design was evaluated considering different values for air gap, which lead to 0.5 mm as the less troublesome value. The final values are presented in Table 2, which presents some similarities to other published works [41,52].

Table 2. Dimensional limitations for the FSM.

Parameter	Symbol	Value	Unit
Outer radius	r_{out}	155	mm
Inner radius	r_{in}	111	mm
Axial length	l_a	51	mm
Air gap length	l_g	0.5	mm
Armature coil number of turns	N_a	15	-
Field coil number of turns	N_f	840	-
Armature wire gauge	ϕ_a	$2 \times 18 + 3 \times 21$	AWG
Field wire gauge	ϕ_f	27	AWG

3.2. Analytical Modeling

The analysis of the electrical machine performance must consider the double salient structure effect. In the proposed design methodology, a semi-analytical method provides accurate results with lower computational effort as compared to numerical methods. Thus, the magnetic field distribution is calculated using the Maxwell–Fourier method to evaluate the electric machine performance parameters. For early design stages, it offers great flexibility for assessing the electrical machine performance with several geometric parameters variation.

The electromagnetic performance of a machine is usually evaluated in literature considering motoring conditions. Additionally, the reluctance machine is commonly driven by a power converter with current control. In this case, the power converter is regarded as the current source and the magnetic field is calculated with defined current densities. In this case, the FSM behavior is similar to that of a salient pole synchronous machine [53,54].

The present work is based on cost reduction and simplicity. Thus, a passive diode rectifier is adopted to adequately generated three-phase alternating voltages to the vehicle 12 Vdc bus. The voltage level is controlled by manually adjusting the field current using a boost DC–DC converter. Since there is no inverter circuit, the machine performance is reduced due to the lack of a field position control.

The methodology discussed here is proposed for an early design stage, in which it is desirable to evaluate different topologies and geometric parameters. Therefore, the output power is calculated using a simplified model of a generator supplying a three-phase linear load. The challenge faced for analyzing the FSM in such a manner is related to its time-varying inductances that occur due to the double salient structure.

A first approximation is carried out with the field circuit's dynamic being neglected. The voltage equation in machine variables is defined as

$$v_{abc} = -R_s i_{abc} - \frac{d\lambda_{abc}}{dt} + e_{abc}. \tag{1}$$

A per-phase equivalent circuit model for steady state analysis is developed to evaluate the mechanical to electrical power conversion. The Equation (1) is solved for known harmonic components that showed a prominent effect over the generated power.

The generated voltage is evaluated for the FSM using an approximately sinusoidal waveform with a fundamental electrical frequency of ω_1. This allows the armature-field mutual inductance to be simplified considering only its fundamental component, $M_{af,1}$. With a constant field current, i_f, no-load back-EMF is calculated by

$$E_{a,nl} = \omega_1 M_{af,1} I_f \tag{2}$$

The time-varying armature inductances are decomposed into their harmonic components. The amplitude and phase of these components are calculated in relation to the phase a circuit. In a balanced three-phase condition, these relations facilitate the definition of the harmonic effective inductances using a reasoning similar to that of a synchronous machine. The zero order component has a similar effect of the synchronous inductance, and is calculated from the phase a average value, $L_{aa,0}$, that is

$$L_{s,0} = \frac{3}{2} L_{aa,0}. \tag{3}$$

The other inductance harmonic components that affect average output power are the second and fourth order harmonics, which can be simplified into a per-phase effective inductance, calculated by

$$L_{s,2} = \frac{3}{2} L_{aa,2}, \tag{4}$$

$$L_{s,4} = \frac{3}{2} L_{aa,4}. \tag{5}$$

The load is simplified to a linear RL branch. With these approximations, the voltage Equation (1) can be expressed in complex notation as

$$\begin{aligned} E_{a,nl} &- j\omega L_{s,0} I_{a,1} e^{-j\vartheta_{i,1}} \\ &+ j\omega L_{s,2} e^{j2\vartheta_{i,1}} I_{a,1} e^{-j\vartheta_{i,1}} + j\omega L_{s,4} I_{a,5} e^{-j\vartheta_{i,5}} \\ &= (R_{load} + j\omega L_{load}) I_{a,1} e^{-j\vartheta_{i,1}}, \end{aligned} \tag{6}$$

where $I_{a,1}$ and $\vartheta_{i,1}$ are the amplitude and phase of the fundamental component of phase current, $I_{a,1}$ and $\vartheta_{i,1}$, respectively. It can be noticed that despite of the approximated sinusoidal generated voltage, the inductance harmonic component yields a frequency coupling with the fifth order current, $I_{a,5}\underline{/-\vartheta_{i,5}}$. With the considered harmonic components, the voltage equation for this current harmonic is expressed as

$$\begin{aligned} j5\omega L_{s,4} I_{a,1} e^{-j\vartheta_{i,1}} &- j5\omega L_{s,0} I_{a,5} e^{-j\vartheta_{i,5}} \\ &= (R_{load} + j5\omega L_{load}) I_{a,5} e^{-j\vartheta_{i,5}}. \end{aligned} \tag{7}$$

Concerning the evaluation of steady-state power conversion, only the fundamental component is considered. From Equation (6), it can be observed that the harmonic inductance components affect the fundamental circuit behavior. This can be interpreted as a demagnetization effect, which is dependent on the loading conditions. To further clarify this behavior, these demagnetizing terms can be characterized as equivalent RL branches. First, the second order inductance term is evaluated by its admittance frequency response. This transfer function is used to calculate a parallel RL branches with similar behavior, that is

$$\begin{aligned} -\frac{1}{j\omega L_{s,2} e^{j2\vartheta_{i,1}}} &= Y_{eq}(j\omega) \\ &= \frac{1}{R_1 + j\omega L_1} + \frac{1}{R_2 + j\omega L_2}. \end{aligned} \tag{8}$$

The fourth order inductance term is evaluated using Equation (7). This term is related to the fifth harmonic order of the phase current, but the effect is over the fundamental circuit. The resultant per-phase fundamental circuit is presented in Figure 4.

The second order inductance term equivalent RL branch is calculated by

$$R_{eq,2,r1} = \frac{(R_{load} + R_{eq,4})L_{s,2}}{2(L_{s,0} + L_{load} + L_{eq,4} - L_{s,2})}, \tag{9}$$

$$L_{eq,2,r1} = \frac{L_{s,2}}{2}, \tag{10}$$

$$L_{eq,2,r2} = -L_{s,2}. \tag{11}$$

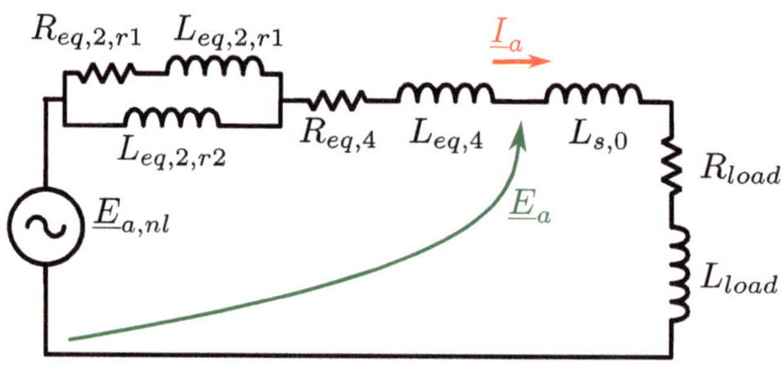

Figure 4. Per-phase FSM equivalent circuit.

The fourth order inductance harmonic effect is evaluated the equivalent RL calculated by

$$R_{eq,4} = \frac{5\omega^2 L_{s,4}^2 R_{load}}{\left(R_{load}^2 + 25\omega^2 (L_{s,0} + L_{load})^2\right)}, \tag{12}$$

$$L_{eq,4} = -\frac{25\omega^2 L_{s,4}^2 (L_{s,0} + L_{load})}{\left(R_{load}^2 + 25\omega^2 (L_{s,0} + L_{load})^2\right)}. \tag{13}$$

This equivalent circuit is used to optimize the generated power. As discussed, the time-varying inductance components have a demagnetizing effect, which reduces the no-load internal generated voltage, $\underline{E}_{a,nl}$. Considering the resultant magnitude and angle variation, the internal induced voltage, $\underline{E}_a$, is calculated. The power converted from mechanical to electrical form is given by

$$P_{conv} = \frac{3}{2}\mathbb{R}(\underline{E}_a \cdot \underline{I}_a^*). \tag{14}$$

The assessment of the proposed steady-state equivalent circuit is validated with a dynamic simulation result for the converted power. The machine voltage equation is solved considering the time-varying inductances using numerical integration method. These inductances are calculated using the Maxwell–Fourier method, but any method can be used to analyze the magnetic field distribution. The field circuit dynamics is discarded and the field current is defined constant. The result is presented in Figure 5. The converted power calculated with the equivalent circuit shows excellent agreement with the average power calculated over an electrical cycle in the dynamic simulation, with an error of less than 0.01%.

Due to the high number of turns for the field winding, each coil has an equivalent resistance of 36.3 Ω. Since the machine was designed targeting micro-hybrid application and, thus, voltage levels below 48 V, the field coils were connected in parallel. Using a series connection would demand voltages up to 200 V. According to the constructive characteristics of the machine, each field coil has a different reluctance path, and each one is mutually coupled and varies unequally over time. That increases the complexity in order to represent the circuit with an equivalent inductance.

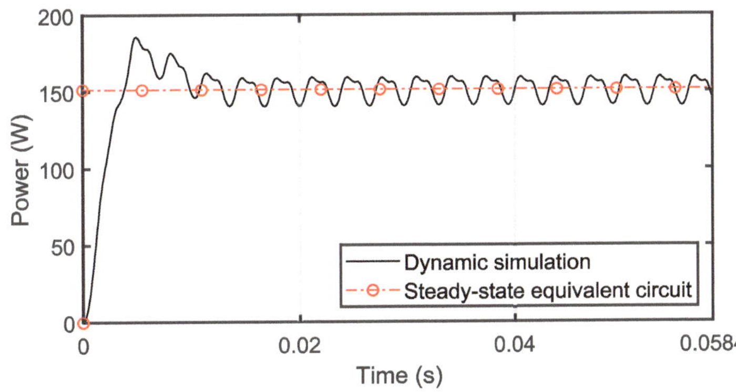

Figure 5. Converted power comparison between proposed steady-state equivalent circuit and dynamic simulation.

Thus, the model to represent this machine requires an individual evaluation of each field coil, leading to an equivalent electrical circuit with 15 variables, 12 for the field circuit and 3 for the armature phases.

The armature winding DC resistance is approximated considering an ideal conductor to evaluate resistive losses. The total armature and field resistance is estimated with Equation (15). It is defined using the winding total number of turn per coil, N_{coil}, the winding fill factor, k_{Cu}, the slot area, A_{slot}, and the electrical resistivity of copper, ρ_{Cu}. The winding fill factor is equal to 0.4, which is in accordance with the adopted value in the literature [55,56].

$$R_{coil} = l_{turn}\frac{\rho_{Cu}}{k_{Cu}(A_{slot}/2)}N_{coil}^2. \tag{15}$$

The average length of a coil turn, l_{turn}, is calculated with stack length, l_s, and an estimate of the end-turn length, l_{end}, that is

$$l_{turn} = 2l_s + 2l_{end}. \tag{16}$$

In this work, the end-turn length is approximated with the average coil circumferential span, l_{circ}, and the axial length extending outward from the machine active length, l_{axis}, being expressed as

$$l_{end} = l_{circ} + 2l_{axis}, \tag{17}$$

where

$$l_{circ} = \frac{1}{2}[2\alpha_s + (2\alpha_s - \beta_{sa})]r_{wind}, \tag{18}$$

and

$$l_{axis} = \frac{\beta_{sa}}{2}r_{wind}. \tag{19}$$

The method adopted for evaluating the winding inductances, the Maxwell–Fourier method, is based on the premise of an iron core with infinity magnetic permeability. The magnetic field distribution is not evaluated in the core, where important phenomena occur. On the other hand, as proposed in [57], stator core teeth and yoke magnetic fluxes can be estimated using an equivalent magnetic circuit.

The tooth flux is obtained by calculating the average air gap flux density over the distance of a tooth pitch, from the mid-point of an adjacent slot opening to the other. With the magnetic fluxes calculated for each tooth, φ_t, the linear equation system can be derived for the yoke fluxes, φ_y. In a 12 slot stator, the equation system is given as

$$\begin{bmatrix} -1 & 1 & 0 & \cdots & 0 \\ 0 & -1 & 1 & & 0 \\ \vdots & & & \ddots & \vdots \\ 0 & 0 & 0 & -1 & 1 \\ 1 & 1 & 1 & 1 & 1 \end{bmatrix} \begin{bmatrix} \varphi_{y1} \\ \varphi_{y2} \\ \vdots \\ \varphi_{y11} \\ \varphi_{y12} \end{bmatrix} = \begin{bmatrix} \varphi_{t1} \\ \varphi_{t2} \\ \vdots \\ \varphi_{t11} \\ 0 \end{bmatrix}. \tag{20}$$

3.3. Topology Considerations

There are some aspects that must be addressed before using optimization routines in the proposed application. Even at early design stages, constructive and material parameters must be accounted for when adopting simplified analytical methodology, such as used in the present work.

The flux-switching machine provides great design flexibility, with different rotor and stator structures. The use of different stator slot and rotor pole number combinations, winding configurations, and rotor structures offers distinct electromagnetic performances. In HEV drive applications, a FSM with 24 stator slots and 10 rotor poles (24s-10p) is proven to be a good candidate [47], and is considered as a base design for the presented work.

It has been discussed that machines presenting fewer stator slots and rotor poles can achieve a higher electromagnetic performance due to the smaller leakage magnetic flux [56]. However, keeping a constant magnetizing flux for a configuration with lower number of poles yields an increased stator and rotor yoke thickness to avoid saturation. Considering the geometric constraints in the present work, the increased stator and rotor yokes shortens the slots' depth, leading to a smaller slot area for placing the conductors. Moreover, adopting a topology with increased number of stator slots and rotor poles allows the use of a deeper slot, but a higher operation frequency increases core and converter switching losses.

Using the base structure of 24s-10p, three different FSM topologies are evaluated for the proposed application. Figure 6 presents the flux density for the configuration: 12s-5p, 24s-10p, and 36s-15p. With a fixed slot opening in relation with the slot pitch angle and a field excitation of 3.0 A/mm^2, the slot depth is adjusted in order to limit the yoke flux density below 1.3 T. These geometries parameters are used to calculated the inductances, and with the equivalent circuit, the converted power are calculated. The results are presented in Table 3.

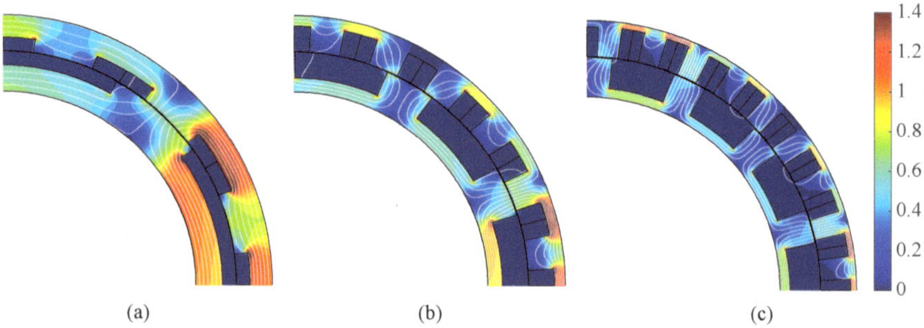

Figure 6. Flux density for three different topologies: (**a**) 12s-5p, (**b**) 24s-10p, and (**c**) 36s-15p.

Table 3. Output power comparison between the studied configurations.

Machine Topology	Output Power	Copper Losses
12s-5p	81 W	62 W
24s-10p	165 W	76 W
36s-15p	148 W	70 W

When evaluating different FSM configurations, some other guide rules are important to address. Using a rotor with an even pole number avoids unbalanced magnetic force, which creates a high mechanical stress, generating noise and vibration. The 10-pole machine has a lower torque ripple when compared to 8-pole and 4-pole configurations [58]. Lastly, the copper losses, as presented in Table 3, are slightly higher than other configurations, but with better output power results.

3.4. Structural and Material Limitations

The realization of initial electromagnetic performance evaluation must consider material properties, mechanical structure stress, and thermal limitations. Despite the chosen methodology for magnetic field analysis, the given geometric flexibility must avoid weak mechanical structures, thermal stresses, and magnetic saturation in order to realize a feasible design. In this context, the present section discusses the constructive constraints and material parameter limitations considered for the design procedure.

It is necessary to determine the slots fill factor, k_{cu}, to define the armature and field winding conductors. In machines that have circular conductors with random winding, the typical fill factor is between 0.3 and 0.5. A fill factor of 0.4 is commonly found in literature in FSM design [55,56] and is used in this work. With Equations (15)–(19), the armature and field winding copper losses are calculated. It is important to have a good estimate for the winding resistivity. The FSM operates at low speeds in direct drive in-wheel applications, in which copper losses are critical and represents a major part of the total machine losses [59,60].

The electric loading capability is determined by the winding insulation material and the ability of the cooling system to dissipate the produced heat. The FSM has the advantage of placing all field sources in the stator, resulting in a better thermal management. With the use of water jacket cooling systems, the use of values of surface current density up to 30 A/mm^2 are observed in the literature [41,47,59]. According to [55], machines with surface current densities higher than 10 A/mm^2 must consider a special cooling system. In the present work, the value of 10 A/mm^2 is adopted as the current density maximum limit.

It is possible to minimize the copper losses using conductors with higher cross section area, which requires slots with greater area. However, to reduce the volume of the magnetic core in order to increase the slot area can lead to core saturation, deteriorating the magnetic performance. For various standard machines (e.g., induction and synchronous machines), there are some permitted flux densities that can be applied in early design stages. Saturation effect is evaluated through magnetic field simulation using the finite element method. At no load condition, the saturation effect over induced voltage of the 24s-10p FSM topology is evaluated for different core magnetic paths, i.e., the teeth and yoke of the stator and rotor core. The ferromagnetic material adopted is the high performance NO30-16HP for electric vehicles application from Aperam South America.

Table 4 shows the magnetization level for different parts of the machine under test, in which the reduction in the electromotive force fundamental was limited to 5% compared to the unsaturated result. This same magnetizing condition applied to an ideal electric steel simulation gives the reference values for this work.

Each row presents the magnetic flux density result at different core parts for each test, where the machine is simulated considering the magnetically non-linear material. Furthermore, the permitted flux densities are obtained from the last column results, which are derived for a machine with an ideal core. This is suitable for simplified methodologies for evaluation of electromagnetic performance at early design stages, such as those proposed in this work.

Considering the approximate teeth and yoke magnetic flux densities calculation, as discussed in the prior section, the stator teeth and yoke magnetization should be limited to 1.8 T and 1.5 T, respectively. According to the simulation results, the saturation on the stator teeth limits the machine design before the rotor teeth does. In addition, considering the flux distribution characteristics, adopting a rotor yoke with a thickness 5% shorter than that of the stator yoke results in a similar flux density. Thus, only the stator flux density is analyzed.

Table 4. Evaluation of the saturation effect on the induced voltage.

Region Tested	Flux Density (T)				
	Rotor Yoke	**Rotor Teeth**	**Stator Yoke**	**Stator Yeeth**	**Linear Model Value**
Rotor yoke	**1.6**	0.77	0.57	1.02	**1.97**
Stator yoke	0.45	0.65	**1.62**	0.88	**1.95**
Stator teeth	0.88	1.16	0.94	**1.61**	**1.73**

Defining the limits, considering both fill factor and core dimensioning to avoid magnetic saturation, allows the design procedure to take into account physical limitations of the machine. Therefore, the semi-analytic optimization of the FSM is carried with criteria, not being limited to the electromagnetic design. The proposed routine for motor design can be optimized using, for instance, formal multi-criteria optimization techniques. These provide solutions to problems with multiple and conflicting objectives.

4. Model Validation

The built prototype was not conceived for road validation, thus no ingress protection was considered. Evaluation tests were conducted on a testbench, and on a roller dynamometer using a real car. Before starting the operational tests the prototype was vacuum impregnated and isolation tests were performed within standard limits [61–63].

Presented results only consider the machine as a generator with electrical loads at the machine's terminals. Voltage and current were measured using oscilloscope probes, and later results were processed to calculate power. Mechanical torque was provided by an induction motor fed by a variable frequency drive responsible to adjust speed of the wheel. Figure 7 presents the test bench and the final setup using the real car.

Figure 7. Machine test bench (**left**) and vehicle evaluation (**right**).

In the worst case scenario, the machine rotates at low speed and flux must increase to produce torque. In that situation, losses are high and energy generation is low, leading to the lowest overall efficiency, as shown in Figure 8. A steady state field current of 7.2 A was drained resulting in a winding temperature of 80 °C.

The simulation results for induced considering parallel field winding connection are compared to the experimental results taken from a test bench. The FSM was driven by an induction motor fed by a variable frequency drive at 500 rpm and with the field circuit fed by an independent current source. Figure 9 shows the results of the no-load induced phase voltage when the total field current is 4.8 A. The parallel connection of the field coils tends to demagnetize the others fields, which leads to a lower induced voltage and output power.

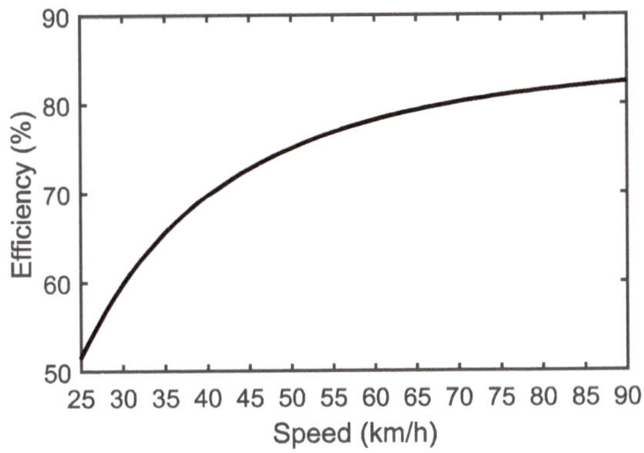

Figure 8. Proposed generator operating within urban driving cycles speed limits.

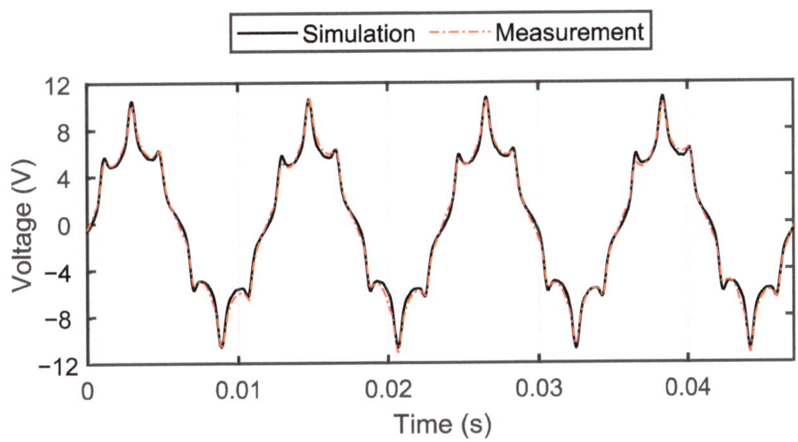

Figure 9. Induced phase voltage for a field current of i_f = 4.8 A parallel connected.

However, a series connection forces the current to be the same in each field coil, avoiding demagnetization, and producing an induced voltage nearly two times higher as compared to the parallel connection, as shown in Figure 10. The fundamental waveform is 82.5% higher, which increases the output power by 3.3 times.

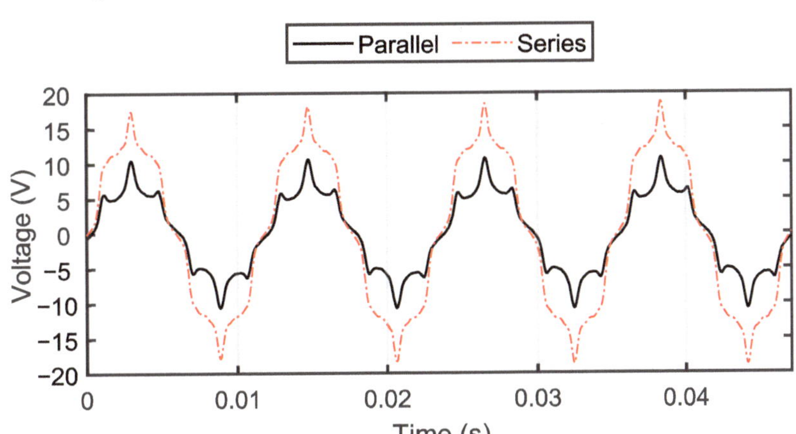

Figure 10. Induced phase voltage considering parallel and series connection of field coils.

Figure 11 shows the results of the induced line voltage for a total field current of 4.8 A with the armature terminals connected to a resistive load of 3.3 Ω connected in delta.

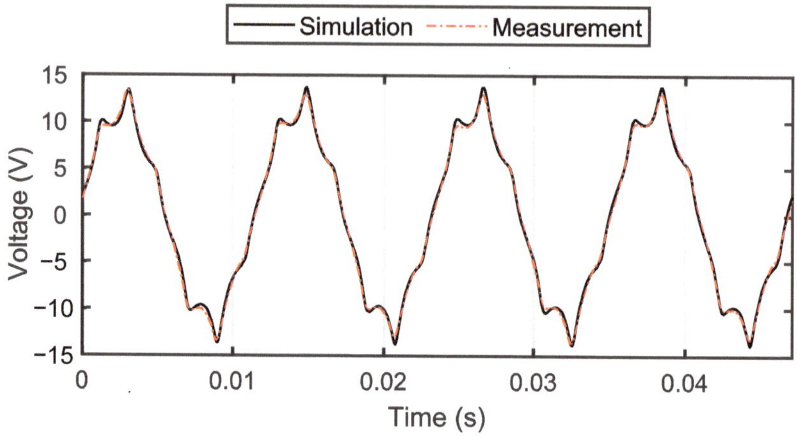

Figure 11. Induced line voltage feeding a 3.3 Ω load.

In order to evaluate a series connected operation, a higher voltage is needed, which limits its capabilities to be boarded. Another alternative is to build a controlled current source for each coil which will be evaluated in future works.

5. Conclusions

In this paper, a flux-switching machine was designed and evaluated for a regenerative breaking operation. The electromagnetic design was calculated using the Maxwell–Fourier analyses. By analysing the presented method, it is possible to obtain the field distribution calculation in a structure with double salience with effectiveness. The calculated field distribution can also be used to evaluate the magnetizing level on the core.

The presented design was constrained by available space inside the wheel, which considered the minimum modifications to the vehicle, preserving the original vehicle drum brake. The machine output power is limited by the internal diameter and the low rotation speed. However, the development and construction process presented show the ground base limitations of the application, which can lead to a better performance for upcoming designs.

The performance was evaluated using the developed equivalent circuit and the machine model was validated using both the finite element method and experimental results. It has been established that keeping the current constant on the field coils prevents field demagnetizing and, consequently, power reduction.

Author Contributions: Conceptualization, G.A.M., L.C.M.S., B.J.C.F. and T.A.C.M.; methodology, G.A.M. and T.A.C.M.; software, G.A.M.; validation, G.A.M. and D.P.V.G.; formal analysis, G.A.M.; investigation, G.A.M. and D.P.V.G.; resources, L.C.M.S., T.A.C.M. and B.J.C.F.; data curation, D.P.V.G.; writing—original draft preparation, G.A.M., D.P.V.G. and T.A.C.M.; writing—review and editing, G.A.M., D.P.V.G. and T.A.C.M.; visualization, G.A.M., D.P.V.G., L.C.M.S., B.J.C.F. and T.A.C.M.; supervision, L.C.M.S., B.J.C.F. and T.A.C.M.; project administration, T.A.C.M. All authors have read and agreed to the published version of the manuscript.

Funding: This research received no external funding.

Institutional Review Board Statement: Not applicable.

Informed Consent Statement: Not applicable.

Data Availability Statement: The data that support the findings of this study are available from the corresponding author upon reasonable request.

Acknowledgments: This work has been supported by the Brazilian agency CAPES (Coordenação de Aperfeiçoamento de Pessoal de Nível Superior). Additionally, we would like to thank Aperam South America for the support in the realization of this work and the Graduate Program in Electrical Engineering—Universidade Federal de Minas Gerais.

Conflicts of Interest: The authors listed above have no conflict of interest to declare. All co-authors have agreed with the contents of the manuscript and there is no financial interest to report. We certify that the submission is an original work and is not under review at any other publication.

References

1. IEA. *Energy and Air Pollution*; Technical Report; International Energy Agency (IEA): Paris, France, 2016.
2. Mulholland, E.; Miller, J.; Braun, C.; Jin, L.; Rodríguez, F. Quantifying the long-term air quality and health benefits from Euro 7/VII standards in Europe. In *The International Council of Clean Transportation. ICCT Briefing June*; ICCT: San Francisco, CA, USA, 2021.
3. Coimbra, M.R.; Barbosa, T.P.; Vasques, C.M. A 3D-Printed Continuously Variable Transmission for an Electric Vehicle Prototype. *Machines* **2022**, *10*, 84. [CrossRef]
4. Nikowitz, M. *Advanced Hybrid and Electric Vehicles*; Springer International Publishing: Berlin/Heidelberg, Germany, 2016. [CrossRef]
5. Ismail, M.M.R.; Wira, J.Y.; Abu, A.; Abidin, M.A.Z. Thermal Energy Harvesting From Automotive Waste Heat. *Adv. Mater. Res.* **2012**, *516–517*, 498–503. [CrossRef]
6. Zhou, F.; Dede, E.; Joshi, S. Application of Rankine Cycle to Passenger Vehicle Waste Heat Recovery—A Review. *SAE Int. J. Mater. Manuf.* **2016**, *9*, 224–235. [CrossRef]
7. Maia, T.A.; Faria, O.A.; Barros, J.E.; Porto, M.P.; Cardoso Filho, B.J. Test and simulation of an electric generator driven by a micro-turbine. *Electr. Power Syst. Res.* **2017**, *147*, 224–232. [CrossRef]
8. Simões, A.N.; Carvalho, D.J.; Morita, E.d.S.; Moretti, H.L.; Vendrameto, H.V.; Fu, L.; Torres, F.; Souza, A.N.d.; Bizzo, W.A.; Mazon, T. A Triboelectric Nanogenerator for Energy Harvesting from Transformers' Vibrations. *Machines* **2022**, *10*, 215. [CrossRef]
9. Carneiro, P.M.; Ferreira, J.A.; Kholkin, A.L.; Soares dos Santos, M.P. Towards Self-Adaptability of Instrumented Electromagnetic Energy Harvesters. *Machines* **2022**, *10*, 414. [CrossRef]
10. Aksu, U.; Halicioglu, R. A review study on energy harvesting systems for vehicles. *Teh. Glas.* **2018**, *12*, 251–259. [CrossRef]
11. Abdelkareem, M.A.A.; Xu, L.; Ali, M.K.A.; Elagouz, A.; Mi, J.; Guo, S.; Liu, Y.; Zuo, L. Vibration energy harvesting in automotive suspension system: A detailed review. *Appl. Energy* **2018**, *229*, 672–699. [CrossRef]
12. Da Silva, J.F.F.G. Challenges for the Electric Mobility Transition: Lessons from Norway and Perspectives for Brazil. Master's Thesis, FGV, Rio de Janeiro, Brazil, 2019.
13. Cardoso, D.S.; Fael, P.O.; Espírito-Santo, A. A review of micro and mild hybrid systems. *Energy Rep.* **2020**, *6*, 385–390. [CrossRef]
14. Chau, K.T.C. *Electric Vehicle Machines and Drives: Design, Analysis and Application*; John Wiley & Sons: Hoboken, NJ, USA, 2015.
15. Picron, V.; Fournigault, D.; Baudesson, P.; Armiroli, P. Cost-Efficient Hybrid Powertrain System with 48 V Network. *ATZ Worldw.* **2012**, *114*, 46–50. [CrossRef]
16. Li, C.; Guo, X.; Fu, J.; Fu, W.; Liu, Y.; Chen, H.; Wang, R.; Li, Z. Design and Analysis of a Novel Double-Stator Double-Rotor Motor Drive System for In-Wheel Direct Drive of Electric Vehicles. *Machines* **2021**, *10*, 27. [CrossRef]
17. Solliec, G.L.; Chasse, A.; Van-Frank, J.; Walser, D. Dual Mode Vehicle with In-Wheel Motor: Regenerative Braking Optimization. *Oil Gas Sci. Technol. Rev. D'IFP Energies Nouv.* **2013**, *68*, 95–108. [CrossRef]
18. Lambert, T.; Biglarbegian, M.; Mahmud, S. A novel approach to the design of axial-flux switched-reluctance motors. *Machines* **2015**, *3*, 27–54. [CrossRef]
19. Sabri, M.F.M.; Danapalasingam, K.A.; Rahmat, M.F. A review on hybrid electric vehicles architecture and energy management strategies. *Renew. Sustain. Energy Rev.* **2016**, *53*, 1433–1442. [CrossRef]
20. Spichartz, P.; Bokker, T.; Sourkounis, C. Comparison of electric vehicles with single drive and four wheel drive system concerning regenerative braking. In Proceedings of the 2017 Twelfth International Conference on Ecological Vehicles and Renewable Energies (EVER), Monte Carlo, Monaco, 11–13 April 2017. [CrossRef]
21. European Commission. Commission Regulation (EU) 2017/1151 of 1 June 2017 supplementing Regulation (EC) No. 715/2007 of the European Parliament and of the Council on type-approval of motor vehicles with respect to emissions from light passenger and commercial vehicles (Euro 5 and Euro 6) and on access to vehicle repair and maintenance information, amending Directive 2007/46/EC of the European Parliament and of the Council, Commission Regulation (EC) No. 692/2008 and Commission Regulation (EU) No. 1230/2012 and repealing Commission Regulation (EC) No. 692/2008. *Off. J. Eur. Union* **2017**, *175*, 1–643.
22. Cui, M.; Zhao, S.; Deguchi, Y.; Chen, C.; Fan, D. Performance of flux switching integrated starter-generator system with dual-mode control circuit. *Int. J. Mechatron. Autom.* **2018**, *6*, 94. [CrossRef]
23. Viorel, I.A.; Szabó, L.; Löwenstein, L. Integrated starter-generators for automotive applications. *Acta Electroteh.* **2004**, *45*, 7.
24. Mirahki, H.; Moallem, M.; Rahimi, S.A. Design Optimization of IPMSM for 42 V Integrated Starter Alternator Using Lumped Parameter Model and Genetic Algorithms. *IEEE Trans. Magn.* **2014**, *50*, 114–119. [CrossRef]

25. Bu, F.; Liu, H.; Huang, W.; Hu, Y.; Degano, M.; Gerada, C.; Rajashekara, K. Induction-Machine-Based Starter/Generator Systems: Techniques, Developments, and Advances. *IEEE Ind. Electron. Mag.* **2020**, *14*, 4–19. [CrossRef]
26. Wang, W.; Chen, X.; Wang, J. Motor/Generator Applications in Electrified Vehicle Chassis—A Survey. *IEEE Trans. Transp. Electrif.* **2019**, *5*, 584–601. [CrossRef]
27. Xu, W.; Zhao, H.; Ren, B.; Chen, H. A regenerative braking control strategy for electric vehicle with four in-wheel motors. In Proceedings of the 2016 35th Chinese Control Conference (CCC), Chengdu, China, 27–29 July 2016. [CrossRef]
28. Rizzo, G.; Naghinajad, S.; Tiano, F.; Marino, M. A Survey on Through-the-Road Hybrid Electric Vehicles. *Electronics* **2020**, *9*, 879. [CrossRef]
29. Zhao, X.; Kou, B.; Huang, C.; Zhang, L. Optimization design and performance analysis of a reverse-salient permanent magnet synchronous motor. *Machines* **2022**, *10*, 204. [CrossRef]
30. Ocana, I.M.; Baker, N.J.; Mecrow, B.C.; Gan, C.; Hilton, C.; Brockway, S. Testing of an in-wheel Halbach array motor. In Proceedings of the 2020 International Conference on Electrical Machines (ICEM), Gothenburg, Sweden, 23–26 August 2020. [CrossRef]
31. Wang, X.; Wu, L.; Zhou, S.; Hu, C.; Zhao, M. Optimization Design of In-wheel Motor Based on Halbach (Ce,Nd)FeB Magnet. In Proceedings of the 2019 22nd International Conference on Electrical Machines and Systems (ICEMS), Harbin, China, 11–14 August 2019. [CrossRef]
32. Gao, P.; Gu, Y.; Wang, X. The Design of a Permanent Magnet In-Wheel Motor with Dual-Stator and Dual-Field-Excitation Used in Electric Vehicles. *Energies* **2018**, *11*, 424. [CrossRef]
33. Akatsu, K.; Fukuda, K. Advanced Control Method of 5-Phase Dual Concentrated Winding PMSM for Inverter Integrated In-Wheel Motor. *World Electr. Veh. J.* **2021**, *12*, 61. [CrossRef]
34. Agamloh, E.; von Jouanne, A.; Yokochi, A. An Overview of Electric Machine Trends in Modern Electric Vehicles. *Machines* **2020**, *8*, 20. [CrossRef]
35. Jahns, T. Getting Rare-Earth Magnets Out of EV Traction Machines: A review of the many approaches being pursued to minimize or eliminate rare-earth magnets from future EV drivetrains. *IEEE Electrif. Mag.* **2017**, *5*, 6–18. [CrossRef]
36. Boldea, I.; Tutelea, L.N.; Parsa, L.; Dorrell, D. Automotive Electric Propulsion Systems With Reduced or No Permanent Magnets: An Overview. *IEEE Trans. Ind. Electron.* **2014**, *61*, 5696–5711. [CrossRef]
37. Fujishiro, S.; Ishikawa, K.; Kikuchi, S.; Nakamura, K.; Ichinokura, O. Design of outer-rotor-type multipolar switched reluctance motor for electric vehicle. *J. Appl. Phys.* **2006**, *99*, 08R324. [CrossRef]
38. Nikam, S.P.; Rallabandi, V.; Fernandes, B.G. A High-Torque-Density Permanent-Magnet Free Motor for in-Wheel Electric Vehicle Application. *IEEE Trans. Ind. Appl.* **2012**, *48*, 2287–2295. [CrossRef]
39. Goto, H. Double Stator Axial-Flux Switched Reluctance Motor for Electric City Commuters. In Proceedings of the 2018 International Power Electronics Conference (IPEC-Niigata 2018—ECCE Asia), Niigata, Japan, 20–24 May 2018. [CrossRef]
40. Patel, N.R.; Shah, V.A.; Lokhande, M.M. A Novel Approach to the Design and Development of 12/15 Radial Field C-Core Switched Reluctance Motor for Implementation in Electric Vehicle Application. *IEEE Trans. Veh. Technol.* **2018**, *67*, 8031–8040. [CrossRef]
41. Ahmad, N.; Khan, F.; Ali, H.; Ishaq, S.; Sulaiman, E. Outer rotor wound field flux switching machine for In-wheel direct drive application. *IET Electr. Power Appl.* **2019**, *13*, 757–765. [CrossRef]
42. Ali, H.; Sulaiman, E.; Kosaka, T. Design and performance analysis of various high torque segmented rotor HE-FSM topologies for aircraft applications. *IET Electr. Power Appl.* **2020**, *14*, 297–304. [CrossRef]
43. Tang, Y.; Motoasca, E.; Paulides, J.J.; Lomonova, E.A. Comparison of flux-switching machines and permanent magnet synchronous machines in an in-wheel traction application. *COMPEL Int. J. Comput. Math. Electr. Electron. Eng.* **2012**, *32*, 153–165. [CrossRef]
44. Raminosoa, T.; El-Refaie, A.M.; Pan, D.; Huh, K.K.; Alexander, J.P.; Grace, K.; Grubic, S.; Galioto, S.; Reddy, P.B.; Shen, X. Reduced Rare-Earth Flux-Switching Machines for Traction Applications. *IEEE Trans. Ind. Appl.* **2015**, *51*, 2959–2971. [CrossRef]
45. Cao, Y.; Zhu, S.; Yu, J.; Liu, C. Optimization Design and Performance Evaluation of a Hybrid Excitation Claw Pole Machine. *Processes* **2022**, *10*, 541. [CrossRef]
46. Dmitrievskii, V.; Prakht, V.; Kazakbaev, V.; Anuchin, A. Comparison of Interior Permanent Magnet and Synchronous Homopolar Motors for a Mining Dump Truck Traction Drive Operated in Wide Constant Power Speed Range. *Mathematics* **2022**, *10*, 1581. [CrossRef]
47. Sulaiman, E.B.; Khan, F.; Kosaka, T. Field-Excited Flux Switching Motor Design, Optimization and Analyses for Future Hybrid Electric Vehicles Using Finite Element Analyses. *Prog. Electromagn. Res. B* **2016**, *71*, 153–166. [CrossRef]
48. Mendonça, G.; Maia, T.; Filho, B.C. Magnetic Field Analytical Solution for Non-homogeneous Permeability in Retaining Sleeve of a High-Speed Permanent-Magnet Machine. *Math. Comput. Appl.* **2018**, *23*, 72. [CrossRef]
49. Mendonça, G.A.; Maia, T.A.C.; de Jesus Cardoso Filho, B. Improved Semi-Analytical Magnetic Field Solution For High-Speed Permanent-Magnet Machines With Permeable Retaining Sleeve Including Diffusion Effect. *Prog. Electromagn. Res. B* **2020**, *88*, 97–118. [CrossRef]
50. Ramakrishnan, K.; Curti, M.; Zarko, D.; Mastinu, G.; Paulides, J.J.H.; Lomonova, E.A. Comparative analysis of various methods for modelling surface permanent magnet machines. *IET Electr. Power Appl.* **2017**, *11*, 540–547. [CrossRef]
51. Ehsani, M.; Gao, Y.; Longo, S.; Ebrahimi, K.M. *Modern Electric, Hybrid Electric, and Fuel Cell Vehicles*; CRC Press: Boca Raton, FL, USA, 2018.

52. Nguyen, H.Q.; Jiang, J.Y.; Yang, S.M. Design of a 12-slot 7-pole wound-field flux switching motor for traction applications. In Proceedings of the 2016 IEEE International Conference on Industrial Technology (ICIT), Taipei, Taiwan, 14–17 March 2016. [CrossRef]
53. Yang, S.M.; Zhang, J.H.; Jiang, J.Y. Modeling torque characteristics and maximum torque control of a three-phase, dc-excited flux-switching machine. *IEEE Trans. Magn.* **2016**, *52*, 1–4. [CrossRef]
54. Balyovski, T.; Ilhan, E.; Tang, Y.; Paulides, J.; Wijnands, C.; Lomonova, E. Control of DC-excited flux switching machines for traction applications. In Proceedings of the 2014 Ninth International Conference on Ecological Vehicles and Renewable Energies (EVER), Monte-Carlo, Monaco, 25–27 March 2014; pp. 1–5.
55. Amara, Y.; Hoang, E.; Gabsi, M.; Lécrivain, M.; Allano, S. Design and comparison of different flux-switch synchronous machines for an aircraft oil breather application. *Eur. Trans. Electr. Power* **2005**, *15*, 497–511. [CrossRef]
56. Zhou, Y.J.; Zhu, Z.Q. Comparison of Wound-Field Switched-Flux Machines. *IEEE Trans. Ind. Appl.* **2014**, *50*, 3314–3324. [CrossRef]
57. Abbaszadeh, K.; Alam, F.R. On-Load Field Component Separation in Surface-Mounted Permanent-Magnet Motors Using an Improved Conformal Mapping Method. *IEEE Trans. Magn.* **2016**, *52*, 1–12. [CrossRef]
58. Sulaiman, E.B.; Kosaka, T.; Matsui, N. Design study and experimental analysis of wound field flux switching motor for HEV applications. In Proceedings of the 2012 XXth International Conference on Electrical Machines, Marseille, France, 2–5 September 2012; pp. 1269–1275.
59. Omar, M.F.; Sulaiman, E.; Jenal, M.; Kumar, R.; Firdaus, R.N. Magnetic Flux Analysis of a New Field-Excitation Flux Switching Motor Using Segmental Rotor. *IEEE Trans. Magn.* **2017**, *53*, 1–4. [CrossRef]
60. Khan, F.; Sulaiman, E.; Ahmad, Z. A novel wound field flux switching machine with salient pole rotor and nonoverlapping windings. *Turk. J. Electr. Eng. Comput. Sci.* **2017**, *25*, 950–964. [CrossRef]
61. *Std 1799-2012*; Recommended Practice for Quality Control Testing of External Discharges on Stator Coils, Bars, and Windings. IEEE: Piscataway, NJ, USA, 2012; pp. 1–48. [CrossRef]
62. *Std 1434-2000*; Guide to the Measurement of Partial Discharges in Rotating Machinery. IEEE: Piscataway, NJ, USA, 2000; pp. 1–64. [CrossRef]
63. *Std 43-2013 (Revision of IEEE Std 43-2000)*; Recommended Practice for Testing Insulation Resistance of Electric Machinery. IEEE: Piscataway, NJ, USA, 2014; pp. 1–37. [CrossRef]

machines

Article

Speed Estimation of PMSM Based on a Super-Twisting Slide Mode Observer

Long Li [1,2], Weidong Zhou [1,*], Xianting Bi [3] and Xiaoping Shi [4]

[1] College of Intelligent Systems Science and Engineering, Harbin Engineering University, Harbin 150009, China
[2] School of Electrical and Electronic Engineering, Harbin University of Science and Technology, Harbin 150080, China
[3] Beijing Institute of Astronautics System Engineering, Beijing 100086, China
[4] School of Astronautics, Harbin Institute of Technology, Harbin 150001, China
* Correspondence: zhouweidong@hrbeu.edu.cn; Tel.: +86-0451-8256-8107

Abstract: To limit the high-frequency chattering originating from the conventional sliding mode observer in a PMSM sensorless control system of vacuum pumps and reduce the steady-state error caused by the phase lag of the coordinate system on the estimated speed, a speed estimation algorithm was proposed based on the super-twisting sliding mode observer. Then, the proposed algorithm was applied to observe an electromotive force. To this end, the Lyapunov function was established to analyze the stability of the current dynamic error equation. In this regard, the estimated phase lag of the rotating coordinate system was considered to study the influence of the speed estimation error. Moreover, a speed estimation algorithm considering the d-axis estimation of back electromotive force is proposed. The performed analyses show that the speed estimation algorithm based on a super-twisting sliding mode observer has promising estimation accuracy. It is concluded that the proposed algorithm can be effectively applied to suppress high-frequency chattering in the system, reduce the influence of phase lag in the estimation coordinate system on the speed estimation error, remove the low-pass filter and phase compensation module, and improve the accuracy of speed estimation.

Keywords: permanent magnet synchronous motor; super-twisting observer; Lyapunov stability; rotating coordinate system; position sensorless control system

Citation: Li, L.; Zhou, W.; Bi, X.; Shi, X. Speed Estimation of PMSM Based on a Super-Twisting Slide Mode Observer. *Machines* **2022**, *10*, 681. https://doi.org/10.3390/machines10080681

Academic Editors: Lorand Szabo, Feng Chai and Alejandro Gómez Yepes

Received: 7 July 2022
Accepted: 8 August 2022
Published: 11 August 2022

1. Introduction

With the advent of permanent magnet synchronous motors (PMSMs), these high-efficiency devices have been widely used in diverse industries such as EV and wind power generation [1]. The unique arrangement and structural characteristics of hub motors have a significant effect on improving the overall performance of vacuum pumps. In this regard, the PMSM has superior characteristics such as a high power density, large torque inertia ratio, and fast dynamic response. Accordingly, this type of motor has become one of the best choices for vacuum pumps. As the direct power source of vacuum pumps, the motor drive performance directly affects the overall performance of the system. Recently, investigating the vacuum pumps' drive technology for industry has attracted many scholars, and it has become a research hot spot [2].

The high-performance drive control strategy for a PMSM relies on accurate rotor position information, which is usually measured by position sensors. However, the use of position sensors in the motor system has many challenges such as high installation accuracy, high costs, and low reliability. Aiming at resolving these problems, the position sensorless control technology was proposed, in which the position and speed of the motor rotor can be obtained without installing expensive mechanical sensors, thereby eliminating the problems caused by position sensors [3,4].

In the PMSM position sensorless control system, the position estimation algorithms mainly include the high-frequency injection method [5–7], sliding mode observer [8–13], model reference adaptive system [14–16], and extended Kalman observer [17]. Studies show that the sliding mode observer is insensitive to the variation of motor parameters and has a strong robustness, low computational cost, and simple implementation, and so it has been widely used in PMSM sensorless control systems.

In [8], the paper proposed an adaptive second order sliding mode observer (SMO) to estimate the position of the surface-mounted PMSM, and VSI non-linearity is considered. In the process, the mismatch between reference and actual voltage was inhibited, especially at a low speed, and the distortion became dominant. L. Ding proposed an improved SMO which adopted an in-current source, converter-fed PMSM. In the paper, the researchers proposed a multi-loop controller having a capacitor voltage control combined with a discrete-time sliding mode observer sensorless strategy and an adaptive filter [9]. It effectively improved the accuracy of the rotor speed and position estimation and the robustness of the system. Ye S. proposed an enhanced SMO method which featured current measurement error compensation, and a hyperbolic tangent switching function-based SMO was introduced with a self-adjusted shape coefficient and a fuzzy logic controller to eliminate the low pass filter and angular compensator of a traditional SMO to reduce the chatter caused by the constant boundary layer thickness [10]. Wang Y. designed a robust adaptive SMO in a rotor reference frame, and an adaptive law was presented in a small signal model, which made the steady-state and dynamic behavior of the sensorless scheme perform excellently [11]. Shen J. Q. proposed a magnetic chain SMO for an interior PMSM in which a soft phase-locked loop flux observer was used to estimate the speed and position of the rotor [12]. The obtained results showed that this method effectively eliminates the low-pass filter and phase compensation module and improves the estimation accuracy of the rotor position. Fan L. proposed a sensorless control method based on an adaptive sliding mode observer in the PMSM drive control of a turboprop aircraft. In this method, the phase-locked loop was used to estimate the rotor angle and eliminate the phase delay caused by the low-pass filter, which improved the performance of conventional sliding mode observers [13].

The performed literature survey indicates that most of the SMOs estimate the speed and position in a static two-phase coordinate system. Therefore, the estimation accuracy can be improved by adding a low-pass filter and a phase compensation module. However, this modification increases the complexity of the control system. Moreover, when there is a phase lag between the estimated and actual coordinate system while the observer is running, estimation errors are inevitable. To solve these problems in the PMSM sensorless control system based on a sliding mode observer, a super-twisting sliding mode observer based on a rotating coordinate system is designed, and a speed estimation algorithm considering the direct axis (d-axis) estimation of back EMF is proposed in the present study. Then, the proposed super-twisting sliding mode observer and speed estimation algorithm are simulated and verified by experiments.

2. Mathematical Model of the PMSM

A three-phase PMSM is considered as the research object. The topology of the PMSM driven with an inverter is shown in Figure 1, where the main circuit contains a three-phase diode rectifier and a 380 V AC source, DC bus, three-phase inverter, PMSM, and hall.

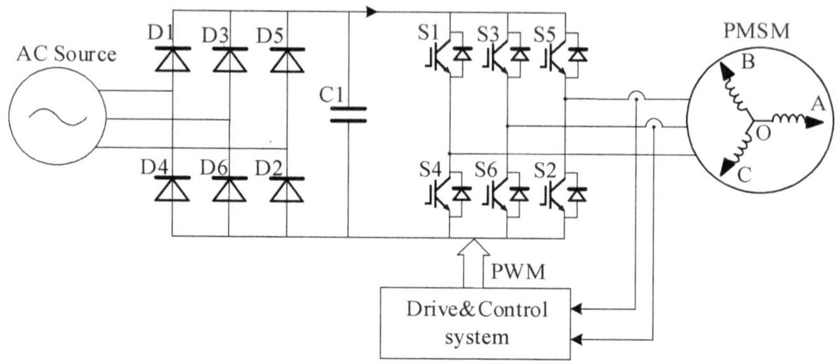

Figure 1. Topology of the PMSM drive system.

To simplify the analysis, it is usually assumed that the distribution of the permanent magnetic field of the rotor in the air gap of the motor has a sine waveform. Neglecting the saturation effect of the stator core, it can be assumed that the magnetic circuit on the stator is linear and the inductance parameters will not change during operation. Moreover, core eddy current and hysteresis losses are ignored, and it is assumed that there is no damping winding on the rotor.

Under the premise of the above assumptions, the following voltage equation can be established for the PMSM in the synchronous rotating frame through the space vector decoupling transformation:

$$\begin{cases} u_d = R_s i_d - \omega_{re} L_q i_q + L_d \frac{di_d}{dt} \\ u_q = R_s i_q + \omega_{re} L_d i_d + \omega_{re} \psi_f + L_q \frac{di_q}{dt} \end{cases} \tag{1}$$

where u_d and id are the voltage and current along the d-axis, u_q and i_q are the voltage and current along the q-axis, and L_d and L_q are the stator inductances along the d-axes and q-axes. In a surface-mounted PMSM, L_d = L_q, R_s is the stator resistance, ω_{re} is the electrical angular speed, and ψ_f is the permanent magnet flux linkage.

After space vector decoupling transformation, the component in the zero-sequence subspace is 0. After performing the Park transform in the d-q frame, the α-β subspace contains the fundamental component and harmonics, which are at the multiples of the $6\ k \pm 1$ th-order. The observed current by the three-phase PMSM is presented in Equations (2) [18] and (3), respectively:

$$\begin{cases} i_d = I_1 \sin(\theta_1) + \sum\limits_{k=1} I_{6k\pm1} \sin((\pm 6k)\omega_e t + \theta_{6k\pm1}) \\ i_q = -I_1 \cos(\theta_1) - \sum\limits_{k=1} I_{6k\pm1} \cos((\pm 6k)\omega_e t + \theta_{6k\pm1}) \end{cases} \tag{2}$$

where θ_1 and I_1 are the initial phases and fundamental amplitude, $\theta_{6k\pm1}$ and the harmonics of $I_{6k\pm1}$ are the initial phases and amplitudes of the harmonics. The comparison of

Equation (2) reveals that the amplitude of harmonics in the d-q axis current observed in the PMSM is smaller, which is conducive to the design of the super-twisting sliding mode observer in the d-q frame.

3. Rotational Speed Estimation Based on the Super-Twisting Sliding Mode Observer

3.1. Design of the Conventional Sliding Mode Observer

The current equation for the PMSM in the α-β frame can be expressed as follows:

$$\frac{d}{dt}\begin{bmatrix} i_\alpha \\ i_\beta \end{bmatrix} = \begin{bmatrix} \frac{1}{L_d} & \frac{1}{L_q} \end{bmatrix}\begin{bmatrix} u_\alpha \\ u_\beta \end{bmatrix} - \begin{bmatrix} \frac{R+\omega\Delta L}{L_d} & \frac{R+\omega\Delta L}{L_q} \end{bmatrix}\begin{bmatrix} i_\alpha \\ i_\beta \end{bmatrix} + \begin{bmatrix} \frac{1}{L_d} & \frac{1}{L_q} \end{bmatrix}\begin{bmatrix} u_\alpha \\ u_\beta \end{bmatrix} - \begin{bmatrix} \frac{1}{L_d} & \frac{1}{L_q} \end{bmatrix}\begin{bmatrix} E_\alpha \\ E_\beta \end{bmatrix} \tag{3}$$

where the $\Delta L = L_d - L_q$ and Equation (3) is for a saliency and non-saliency PMSM, while in this paper, the surface-mounted PMSMs are referred, and so Equation (3) can be simplified to Equation (4):

$$\frac{d}{dt}\begin{bmatrix} i_\alpha \\ i_\beta \end{bmatrix} = \frac{1}{L_d}\begin{bmatrix} -R & 0 \\ 0 & -R \end{bmatrix}\begin{bmatrix} i_\alpha \\ i_\beta \end{bmatrix} + \frac{1}{L_d}\begin{bmatrix} u_\alpha \\ u_\beta \end{bmatrix} - \frac{1}{L_d}\begin{bmatrix} E_\alpha \\ E_\beta \end{bmatrix} \tag{4}$$

The state equation of the conventional sliding mode observer is:

$$\frac{d}{dt}\begin{bmatrix} \hat{i}_\alpha \\ \hat{i}_\beta \end{bmatrix} = \frac{1}{L_d}\begin{bmatrix} -R & 0 \\ 0 & -R \end{bmatrix}\begin{bmatrix} \hat{i}_\alpha \\ \hat{i}_\beta \end{bmatrix} + \frac{1}{L_d}\begin{bmatrix} u_\alpha \\ u_\beta \end{bmatrix} - \frac{1}{L_d}\begin{bmatrix} V_\alpha \\ V_\beta \end{bmatrix} \tag{5}$$

Subtracting Equation (4) from Equation (5), the current error dynamic equation can be obtained in the form below:

$$\frac{d}{dt}\begin{bmatrix} \tilde{i}_\alpha \\ \tilde{i}_\beta \end{bmatrix} = \frac{1}{L_d}\begin{bmatrix} -R & 0 \\ 0 & -R \end{bmatrix}\begin{bmatrix} \tilde{i}_\alpha \\ \tilde{i}_\beta \end{bmatrix} + \frac{1}{L_d}\begin{bmatrix} E_\alpha - V_\alpha \\ E_\beta - V_\beta \end{bmatrix} \tag{6}$$

where

$$\begin{bmatrix} V_\alpha \\ V_\beta \end{bmatrix} = \begin{bmatrix} k\text{sgn}(\hat{i}_\alpha - i_\alpha) \\ k\text{sgn}(\hat{i}_\beta - i_\beta) \end{bmatrix} \tag{7}$$

The estimated expressions for the rotational speed and angle are:

$$\theta = -\arctan(V_\alpha / V_\beta) \tag{8}$$

$$\hat{\omega}_{re} = \frac{\sqrt{V_\alpha^2 + V_\beta^2}}{\psi_f} \tag{9}$$

It is worth noting that the controlled variable in the sliding mode control is a discrete high-frequency switching signal, and so there is chattering in the observed back EMF. The arctangent operation of Equation (8) and the square root operation of Equation (9) will bring the chattering into the speed estimation and angle estimation.

3.2. Design of the Super-Twisting Sliding Mode Observer

To suppress the chattering caused by high-frequency switching signals during the observation process and reduce the load effect on the speed estimation, a super-twisting sliding mode observer in the d-q coordinate system is proposed. The super-twisting algorithm [19] is a second-order sliding mode control algorithm which can be mathematically expressed as follows:

$$\begin{cases} \frac{d}{dt}X_1 = X_2 + k|e_1|^{1/2}\text{sgn}(e_1) \\ \frac{d}{dt}X_2 = a\text{sgn}(e_1) \end{cases} \tag{10}$$

where k and a are the gain coefficients of the super-twisting sliding mode observer.

The current state equation of the PMSM in the d-q frame is:

$$\begin{bmatrix} \dot{i}_d \\ \dot{i}_q \end{bmatrix} = \begin{bmatrix} -\frac{R_s}{L_d} & \frac{L_q}{L_d}\omega_{re} \\ -\frac{L_d}{L_q}\omega_{re} & -\frac{R_s}{L_q} \end{bmatrix}\begin{bmatrix} i_d \\ i_q \end{bmatrix} + \begin{bmatrix} \frac{1}{L_d} & 0 \\ 0 & \frac{1}{L_q} \end{bmatrix}\begin{bmatrix} u_d - E_d \\ u_q - E_q \end{bmatrix} \tag{11}$$

Where $E_d = 0$ and $E_q = \omega_{re}\psi_f$ are the back EMFs in the d-q frame.

The state equation of the super-twisting sliding mode observer in the d-q frame can be expressed as the following:

$$\begin{bmatrix} \dot{\hat{i}}_d \\ \dot{\hat{i}}_q \end{bmatrix} = \begin{bmatrix} -\frac{R_s}{L_d} & \frac{L_q}{L_d}\hat{\omega}_{re} \\ -\frac{L_d}{L_q}\hat{\omega}_{re} & -\frac{R_s}{L_q} \end{bmatrix}\begin{bmatrix} \hat{i}_d \\ \hat{i}_q \end{bmatrix} + \begin{bmatrix} \frac{1}{L_d} & 0 \\ 0 & \frac{1}{L_q} \end{bmatrix}\begin{bmatrix} u_d - V_d \\ u_q - V_q \end{bmatrix} \tag{12}$$

The estimated back EMF in the super-twisting sliding mode observer is:

$$\begin{bmatrix} V_d \\ V_q \end{bmatrix} = \begin{bmatrix} k|s_d|^{1/2}\text{sat}(s_d) + a\int \text{sat}(s_d) \\ k|s_q|^{1/2}\text{sat}(s_q) + a\int \text{sat}(s_q) \end{bmatrix} \tag{13}$$

where $\hat{i}_d$ is the estimated value of the d-axis current, $\hat{i}_q$ is the estimated value of the q-axis current, $\hat{\omega}_{re}$ is the estimated value of the rotational speed, $s_d = \hat{i}_d - i_d$, $s_q = \hat{i}_q - i_q$, and sat is a saturation function with a limiting amplitude of 1. Compared with Equation (8), when using Equation (13) to observe the back EMF, the integral term in the super-twisting sliding mode control algorithm suppresses the chattering of the saturation function and the square root term of the absolute value of the error improves the convergence rate.

Compared with the conventional SMO as shown in Equations (4)–(9), which are based on the α-β frame, the observer of the super-twisting sliding mode observer based on the d-q frame, as shown in Equations (10)–(13), depends on the decoupling voltage relationship of the PMSM. According to super-twisting sliding mode observer shown in Equation (13), it can be seen that the observer is not dependent on the PMSM parameters such as L_d and L_q, and while it depends on the estimated value and real value of the current, the observer guarantees the robustness of the system. This characteristic of the super-twisting sliding mode observer can be widely used in a PMSM system such as a surface-mounted PMSM and an interior permanent magnet (IPM) motor.

When the estimated coordinate system coincides with the actual coordinate system, $V_d = 0$ and $\hat{\omega}_{re} = V_q/\psi_f$. However, when the estimated rotational speed at the current moment is applied to the state equation at the next moment, the observation delay will be generated. After the current signal passes through the operational amplifier and the analog-to-digital converter, it will result in a sampling delay. The observation delay and sampling delay will cause the angular deviation $\Delta\theta$ between the estimated and actual coordinate systems, thereby affecting the estimated value of the back EMF. Meanwhile, the basic relationship between the estimated back EMF and $\Delta\theta$ can be expressed as:

$$\begin{bmatrix} \Delta V_d \\ \Delta V_q \end{bmatrix} = \begin{bmatrix} -\hat{\omega}_{re}\psi_f \sin\Delta\theta \\ \hat{\omega}_{re}\psi_f \cos\Delta\theta \end{bmatrix} \tag{14}$$

When the rotational speed is estimated using the expression $\hat{\omega}_{re} = V_q/\psi_f$, the term $\cos\Delta\theta$ will increase the estimated value. After controlling the speed's closed-loop, the actual speed of the motor reduces. To eliminate the estimation error, the influence of the estimated d-axis back EMF on the speed is considered using calculations. Based on Equations (12) and (13), the current dynamic error equation can be expressed as follows:

$$\dot{\boldsymbol{S}} = \boldsymbol{AS} + \boldsymbol{B}\overline{\omega}_{re} - \boldsymbol{P} \tag{15}$$

where $\boldsymbol{A} = \begin{bmatrix} -\frac{R_s}{L_d} & \hat{\omega}_{re} \\ -\hat{\omega}_{re} & -\frac{R_s}{L_d} \end{bmatrix}$, $\boldsymbol{B} = \begin{bmatrix} i_q & -i_d \end{bmatrix}^{\mathrm{T}}$, $\boldsymbol{P} = \frac{1}{L_q}\begin{bmatrix} V_d & V_q - \omega_{re}\psi_f \end{bmatrix}^{\mathrm{T}}$, $\overline{\omega}_{re} = \hat{\omega}_{re} - \omega_{re}$, and $\boldsymbol{S} = \begin{bmatrix} \hat{i}_d - i_d & \hat{i}_q - i_q \end{bmatrix}^{\mathrm{T}}$.

3.3. Stability Analysis of the Observer

In this section, the stability of Equation (16) is analyzed by the Lyapunov stability theory. To this end, the Lyapunov function is constructed as follows:

$$V = \frac{1}{2}\boldsymbol{S}^{\mathrm{T}}\boldsymbol{S} \tag{16}$$

Taking the derivation of Equation (16) yields the following expression:

$$\dot{V} = \boldsymbol{S}^{\mathrm{T}}\dot{\boldsymbol{S}} \tag{17}$$

Substituting Equation (16) into Equation (17), we can obtain:

$$\begin{aligned}\dot{V} &= \boldsymbol{S}^{\mathrm{T}}(\boldsymbol{A}\boldsymbol{S} + \boldsymbol{B}\overline{\omega}_{re} - \boldsymbol{P}) \\ &= (-\tfrac{R_s}{L_d}s_d^2 - \tfrac{R_s}{L_q}s_q^2) + (\hat{\omega}_{re} - \omega_{re})(i_q s_d - s_q i_d) - \\ &\tfrac{1}{L_q}(V_d s_d + s_q V_q - s_q \psi_f \omega_{re})\end{aligned} \tag{18}$$

Considering the influence of the d-axis back EMF on the speed estimation, Equation (18) can be divided into three parts for analysis. To ensure the global asymptotic stability of the current dynamic error equation, Equation (19) should have a negative value. This can be mathematically expressed as follows:

$$(-\frac{R_s}{L_d}s_d^2 - \frac{R_s}{L_q}s_q^2) < 0 \tag{19}$$

$$(\hat{\omega}_{re} - \omega_{re})(i_q s_d - s_q \frac{\psi_f}{L_q} - s_q i_d) = 0 \tag{20}$$

$$\frac{1}{L_q}(V_d s_d + s_q V_q - s_q \psi_f \hat{\omega}_{re}) = 0 \tag{21}$$

If the current dynamic error equation is globally asymptotically stable, the estimated value of the rotational speed is equal to the actual value, and so $\hat{\omega}_{re} - \omega_{re} = 0$, and Equation (21) holds. Meanwhile, to ensure the stability of the entire current dynamic error equation, the estimated speed can be derived as follows:

$$\hat{\omega}_{re} = \frac{(V_d s_d + s_q V_q)}{s_q \psi_f} = \frac{V_q}{\psi_f} + \frac{s_d V_d}{s_q \psi_f} \tag{22}$$

Substituting Equation (13) into Equation (22), we can obtain:

$$\hat{\omega}_{re} = \frac{1}{\psi_f}(\frac{s_d}{s_q}(k|s_d|^{1/2} + \frac{a}{s})\mathrm{sat}(s_d) + (k|s_q|^{1/2} + \frac{a}{s})\mathrm{sat}(s_q)) \tag{23}$$

Equation (23) indicates that the rotational speed $\hat{\omega}_{re}$ can be estimated using the super-twisting sliding mode algorithm. The integral term in the super-twisting sliding mode control algorithm filters the harmonics in the current errors s_d and s_q and suppresses the residual error in the observer. The square root of the absolute error value is added to the proportional gain to improve the convergence speed of the speed estimation. Meanwhile, V_d suppresses the speed estimation error caused by the phase lag of the estimated coordinate system.

The estimated angle can be obtained by integrating the estimated rotational speed as follows:

$$\hat{\theta} = \int \hat{\omega}_{re} dt \tag{24}$$

4. Simulation and Experimental Analysis

A super-twisting sliding mode observer is established in the d-q coordinate system. The motor speed value can be estimated using Equation (24). The PMSM is controlled through two current loops and a one-speed loop. A SVPWM modulation algorithm is adopted for the closed-loop control of a PMSM sensorless control system. Considering the PWM delay in the closed loop control system, if the motor is running in a high-speed area or a high electrical frequency area, the delay caused by the switching frequency should be considered, while if the PMSM is operating at a low speed or middle speed area, the delay caused by the PWM can be neglected and only the error of the estimated angle in Equation (14) is taken into account. It should be indicated that these methods are used to verify the effectiveness of the proposed algorithm. The parameters of the PMSM are R_s = 50 mΩ, $L_d = L_q$ = 1.03 mH, ψ_f = 0.171 Wb, and the number of pole pairs p = 4.

4.1. Simulation Analysis

To simulate the PMSM sensorless control system, the model was established in Matlab/Simulink and he switch frequency f_{sw} is 10 kHz. The two working conditions of the loading and speed mutation are analyzed. In the former analysis, the rated speed is set to 1000 rpm and a torque of 50 N·m is applied at t = 0.1 s. In the latter analysis, the initial speed is set to 1000 rpm, and then the speed is increased to 1500 rpm at t = 0.2 s and finally returned to 1000 rpm at t = 0.3 s. Figure 2 shows the estimated rotational speed curves for the two observers and two working conditions.

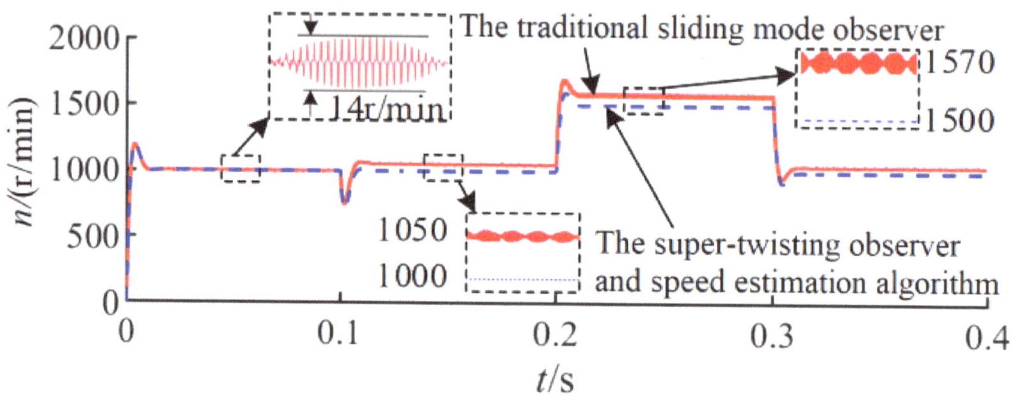

Figure 2. Simulation curves of speed for the two observers.

Figure 2 reveals that in the steady-state conditions, the sensorless control system equipped with the conventional sliding mode observer has a high-frequency chattering with a peak value of 14 rpm, and there are residual errors of 50 rpm and 70 rpm in the loading and speed mutation conditions, respectively. Compared with the conventional sliding mode observer, the estimated speed in the sensorless control system using the super-twisting observer and the proposed speed estimation algorithm can quickly and accurately follow the actual speed. In this case, there is no high-frequency chattering or estimation error in the estimated speed, which demonstrates the estimation accuracy of the proposed algorithm.

Figure 3 shows the d-q axis back EMF waveform. It can be seen that during the running process of the system, the estimated d-axis back EMF has an error that varies depending on the working conditions. Moreover, Figure 4 illustrates the comparison curves between the estimated speed and the actual speed, with and without considering V_d in the calculations, and the lower error when the speed or load is stepped when V_d is considered. As shown in Figure 5 in the estimation error curve, the effect of V_d should be taken into account, and it can be seen that there is a 7 rpm speed error between the real speed and the reference value when it not considered, which is greater in a high-speed area. The speed error is nearly 0 in a steady state when the effect of V_d is taken into account.

The speed curves in Figure 4 and the speed estimation error curve in Figure 5 reveal that when only considering V_q for speed estimation, the actual speed can be followed. However, there is an error of 0.8% at 1000 rpm. When the operating speed increases to 1500 rpm, the rotation speed error decreases to 0.6%. It is inferred that the correction of V_d can reduce the speed estimation error and improve the accuracy of the speed estimation.

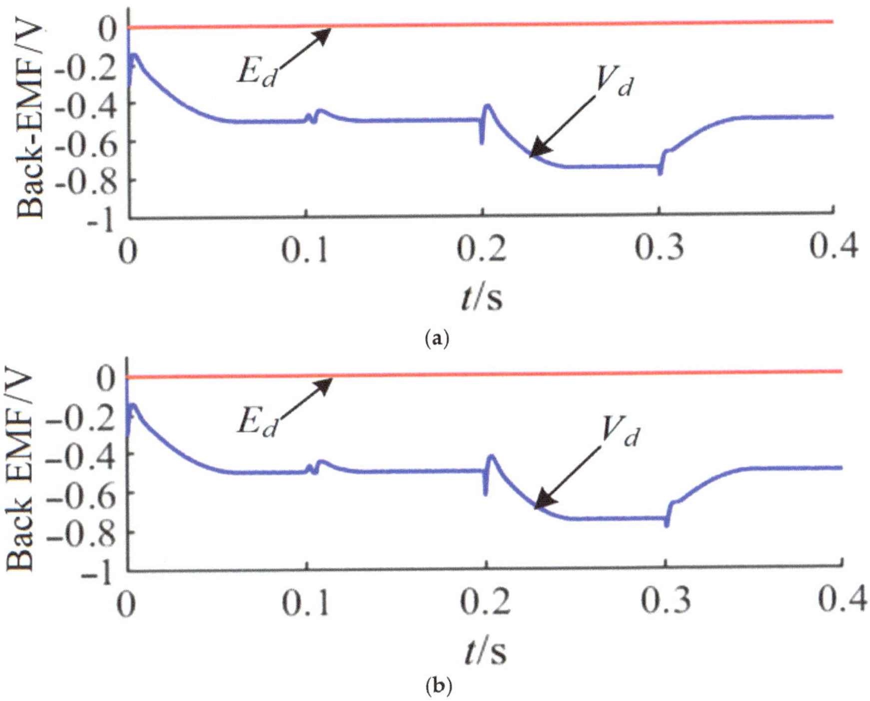

Figure 3. The d-q axis back EMF curve. (**a**) The d-axis back EMF curve. (**b**) The q-axis back EMF curve.

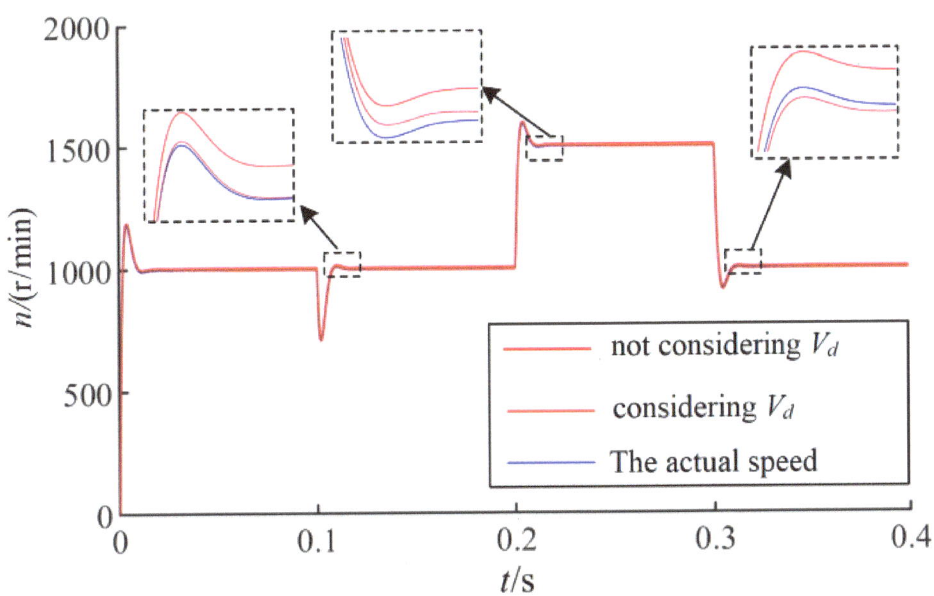

Figure 4. Speed simulation curve.

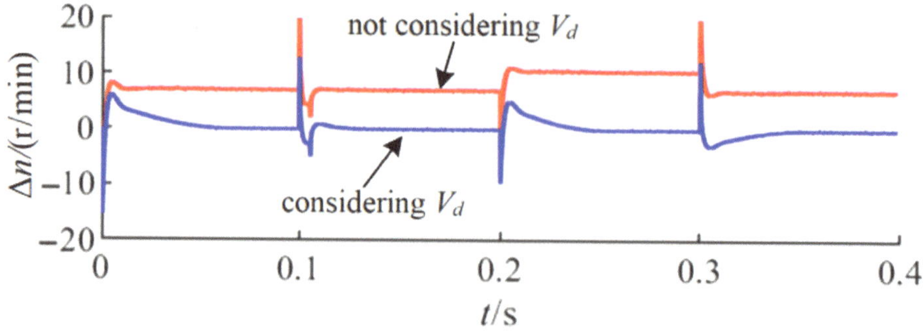

Figure 5. Simulation curve of the speed error.

4.2. Experimental Analysis

In order to further verify the effectiveness of the proposed algorithm, an experimental platform is built. The experimental platform, shown in Figure 6, mainly consists of a controller (Haawking DSP HX320 F28034, China), an IGBT switch device (Infineon FF450 R17 ME4, Germany), a current Hall sensor (LEM CASR15-NP, Swiss), and a dynamometer system which is composed of a tension measurement and control (Zhangli 610 L, China) and an eddy current dynamometer (Guangzhong MTS-II, China). High-speed serial communication is used to change the speed command, read the back EMF and speed information of the motor, and record the read data.

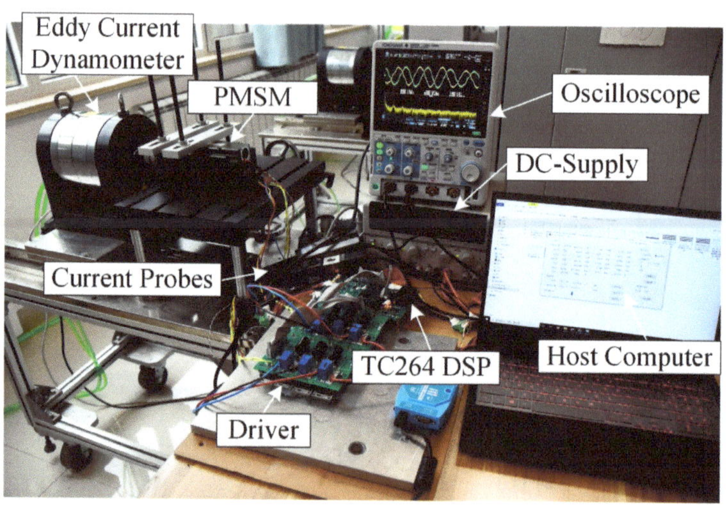

Figure 6. Integrated three-phase PMSM experimental platform.

Figure 7 shows the speed curve under the condition of the speed mutation. It is observed that the estimated speed is consistent with the actual speed in real-time when the actual speed changes suddenly. However, the variation trends of V_d and V_q in Figure 8 reveal that there is a rapid convergent oscillation in the convergence process of the speed estimation error. Finally, it becomes stable, and the estimation error is $\pm$ 1 rpm.

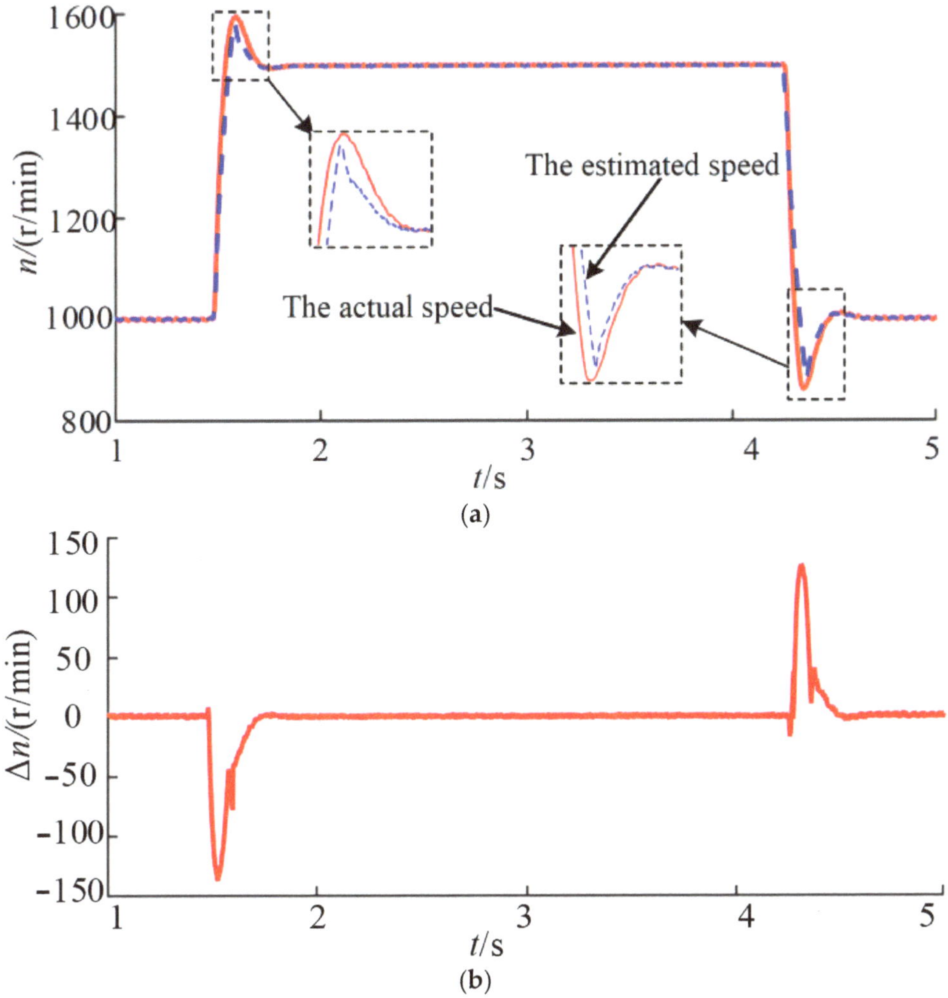

Figure 7. Distributions of the speed mutation. (**a**) Experimental curve of the speed. (**b**) Experimental curve of the rotational speed error.

Figure 8 shows the back EMF curve under the condition of the speed mutation. It is observed that as the motor speed increases, the estimated back EMF V_q is consistent with the actual speed and its stable amplitude rises from 73.0 V to 106.9 V. Moreover, the V_d peaks increase and the amplitude decreases from −1 V to −2 V after stabilization. Similarly, there is an error between the estimated and actual back EMF as the operating speed decreases. Due to the coupling in the d-q axis voltage equation, s_d and s_q in the super-twisting observation have different trends, and so the trends of V_d and V_q are different.

Figure 9 shows the speed curve and back EMF curve under loading conditions. It is observed that after loading, the estimated speed is consistent with the actual speed, but the speed estimation error has a peak of 28 rpm and quickly converges to a stable value. Furthermore, it is found that V_q and V_d have amplitude changes of about 0.5 V.

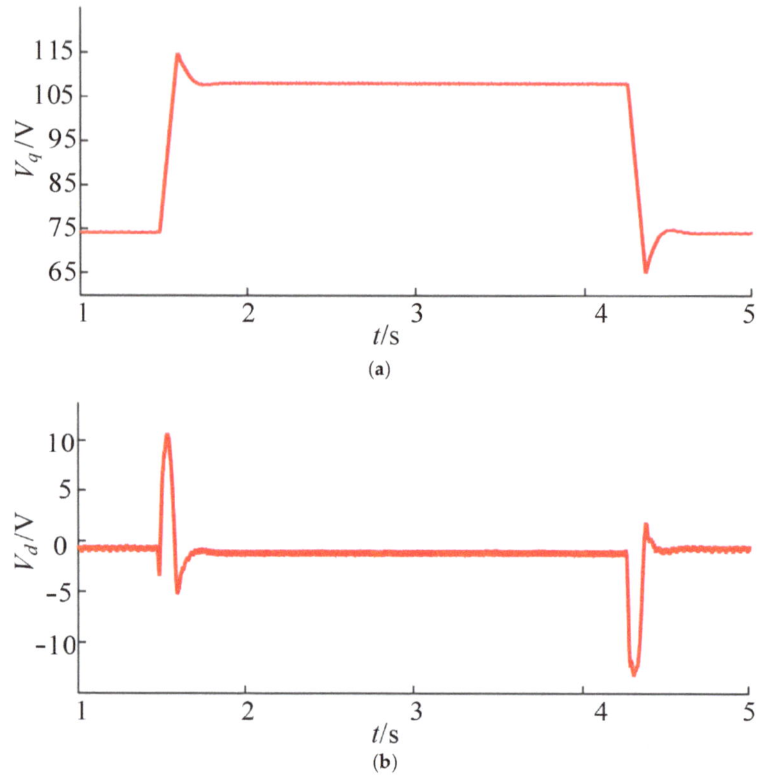

Figure 8. Estimated back EMF curve at the speed mutation. (**a**) V_q curve. (**b**) V_d curve.

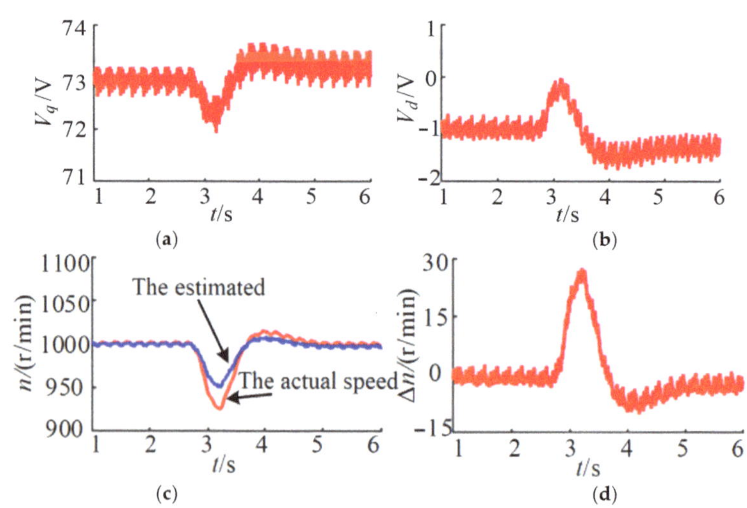

Figure 9. Experimental curves under the loading conditions. (**a**) V_q curve. (**b**) V_d curve. (**c**) Speed curve. (**d**) Speed estimation error curve.

The detected amplitude changes of the rotational speed and estimated back EMF may be attributed to an increase in s_q in the super-twisting sliding mode observer when i_q increases. The gain coefficient of the super-twisting sliding mode includes the square root of the absolute value of s_q, resulting in the amplitude change of V_q. Meanwhile, V_d changes due to the current coupling in the d-q axis voltage equation. According to Equation (22), the estimated speed will change when V_d and V_q change.

Figure 10 shows the estimated angle curve under the loading and rotational speed mutation conditions. The rotor's initial position angle is obtained by the pre-position, and the estimated angle is the integral value of the estimated rotational speed. Figure 10 reveals that the angle estimated using the proposed observer is consistent with the actual angle, but there is a certain phase lag which may originate from the sampling delay and observation delay. This lag is mutually verified with the error of V_d due to the angle lag in Figures 8 and 9.

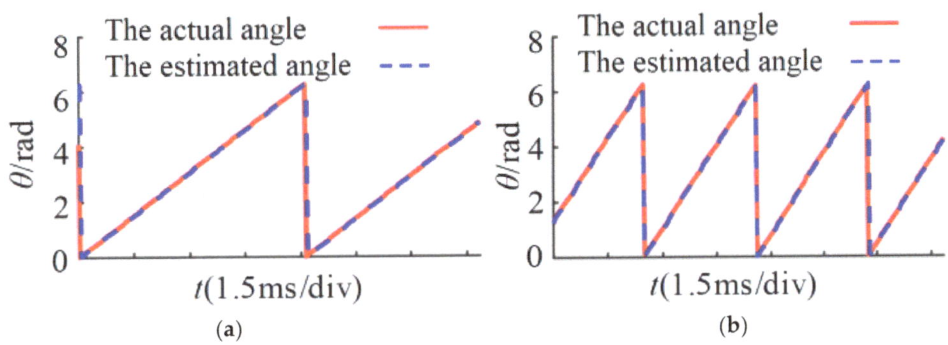

Figure 10. The estimated angle curve under the loading and rotational speed mutation conditions. (**a**) Load. (**b**) Speed mutation.

5. Conclusions

In the present study, a super-twisting sliding mode observer based on a rotating coordinate system is designed, and a new speed estimation algorithm is proposed. Then, the proposed algorithm is applied to the PMSM sensorless control system of vacuum pumps. The feasibility and effectiveness of the algorithm are verified through simulations and experiments. The obtained results demonstrate that the proposed algorithm has the following advantages over the conventional schemes:

(1) The d-q axis current is used as the state variable to design the super-twisting sliding mode observer and remove the low-pass filter and phase compensation module. This algorithm can be applied to suppress high-frequency chattering in the conventional sliding mode observer and improve the accuracy of the speed estimation without increasing the system's complexity.

(2) Considering the influence of the phase lag of the estimated coordinate system on the speed estimation error, the introduction of the d-axis estimated back EMF in the speed estimation algorithm reduces the speed estimation error under the loading and speed mutation.

It is concluded that compared with the conventional sliding mode observer, the proposed algorithm can prevent the energy loss caused by system chattering, improve the speed estimation accuracy, and improve the dynamic performance of the PMSM sensorless control system of vacuum pumps.

Author Contributions: Conceptualization, W.Z.; methodology, L.L.; validation, X.B.; validation; formal analysis, X.S. All authors have read and agreed to the published version of the manuscript.

Funding: This research received no external funding.

Conflicts of Interest: The authors declare no conflict of interest.

References

1. Wang, C.; Gou, L.; Dong, S.; Zhou, M.; You, X. Sensorless Control of IPMSM Based on Super-Twisting Sliding Mode Observer With CVGI Considering Flying Start. *IEEE Trans. Transp. Electrif.* **2022**, *8*, 2106–2117. [CrossRef]
2. Liao, Z.L.; Jie, J.Y.; Yuan, D.; Zhao, Q.J. Research progress on position sensorless technology for hub motor of military vehicle. *J. Ordnance Equip. Eng.* **2021**, *42*, 1–7.
3. Ilioudis, V.C. Sensorless Control of Permanent Magnet Synchronous Machine with Magnetic Saliency Tracking Based on Voltage Signal Injection. *Machines* **2020**, *8*, 14. [CrossRef]
4. Liu, J.L.; Xiao, F.; Shen, Y.; Mai, Z.; Li, C. Position- sensorless control technology of permanent-magnet synchronous motor—A review. *Trans. China Electrotech. Soc.* **2017**, *32*, 76–88.
5. Li, H.; Zhang, X.; Yang, S.; Liu, S. Unified Graphical Model of High-Frequency Signal Injection Methods for PMSM Sensorless Control. *IEEE Trans. Ind. Electron.* **2020**, *67*, 4411–4421. [CrossRef]
6. Almarhoon, A.H.; Zhu, Z.Q.; Xu, P.L. Improved pulsating signal injection using zero-sequence carrier voltage for sensorless control of dual three-phase PMSM. *IEEE Trans. Energy Convers.* **2017**, *32*, 436–446. [CrossRef]
7. Medjmadj, S.; Diallo, D.; Mostefai, M.; Delpha, C.; Arias, A. PMSM Drive Position Estimation: Contribution to the High-Frequency Injection Voltage Selection Issue. *IEEE Trans. Energy Convers.* **2015**, *30*, 349–358. [CrossRef]
8. Liang, D.; Li, J.; Qu, R.; Kong, W. Adaptive Second-Order Sliding-Mode Observer for PMSM Sensorless Control Considering VSI Nonlinearity. *IEEE Trans. Power Electron.* **2017**, *33*, 8994–9004. [CrossRef]
9. Ding, L.; Li, Y.W.; Zargari, N.R. Discrete-Time SMO Sensorless Control of Current Source Converter-Fed PMSM Drives with Low Switching Frequency. *IEEE Trans. Ind. Electron.* **2020**, *68*, 2120–2129. [CrossRef]
10. Ye, S.; Yao, X. An Enhanced SMO-Based Permanent Magnet Synchronous Machine Sensorless Drive Scheme with Current Measurement Error Compensation. *IEEE J. Emerg. Sel. Top. Power Electron.* **2020**, *9*, 4407–4419. [CrossRef]
11. Wang, Y.; Xu, Y.; Zou, J. Sliding-Mode Sensorless Control of PMSM With Inverter Nonlinearity Compensation. *IEEE Trans. Power Electron.* **2019**, *34*, 10206–10220. [CrossRef]
12. Lee, H.; Lee, J. Design of Iterative Sliding Mode Observer for Sensorless PMSM Control. *IEEE Trans. Control. Syst. Technol.* **2013**, *21*, 1394–1399. [CrossRef]
13. Fan, L.; Yang, T.; Rashed, M.; Bozhko, S. Comparative study of back EMF based sensorless control methods for dual three-phase PMSM. In Proceedings of the 2018 IEEE International Conference on Electrical Systems for Aircraft, Railway, Ship Propulsion and Road Vehicles & International Transportation Electrification Conference, Nottingham, UK, 7–9 November 2018; pp. 1–6.
14. Li, W.; Wen, X.; Zhang, J. Comparison of MRAS and SMO method for sensorless PMSM drives. In Proceedings of the 2019 22nd International Conference on Electrical Machines and Systems, Harbin, China, 11–14 August 2019; pp. 1–4.
15. Zhang, Z.; Liu, Y.; Liang, W.; Peng, J. MRAS Sensorless Control Strategy of PMSM based on SiC-MOSFET Three-Phase Inverter. In Proceedings of the 2020 IEEE 9th International Power Electronics and Motion Control Conference, Nanjing, China, 29 November–2 December 2020; pp. 1488–1492.
16. Hassan, M.; Mahgoub, O.; Shafei, A.E. ANFIS based MRAS speed estimator for sensorless control of PMSM. In Proceedings of the 2013 Brazilian Power Electronics Conference, Gramado, Brazilian, 27–31 October 2013; pp. 828–835.
17. Taheri, A.; Ren, H.P.; Song, C.H. Sensorless direct torque control of the six-phase induction motor by fast reduced order extended Kalman filter. *Complexity* **2020**, *2020*, 8985417. [CrossRef]
18. Yang, S.Y.; Liu, S.Y.; Li, H.Y.; Liu, S.H.; Zhang, X. Research on harmonic suppression strategy of position sensorless control for permanent magnet synchronous motor. *Proc. CSEE* **2019**, *39*, 6075–6084+6188.
19. Levant, A. Sliding order and sliding accuracy in sliding mode control. *Int. J. Control* **1993**, *58*, 1247–1263. [CrossRef]

machines

Article

A Novel Rotor Position Observer for Sensorless Control of Permanent Magnet Synchronous Motor Based on Adaptive Generalized Second-Order Integrator

Jie Gao [1], Yanyang Wang [1], Meng Xu [1,*] and Yanqing Mi [2]

1 College of Electronic Information and Automation, Civil Aviation University of China, Tianjin 300300, China
2 Tianjin Internal Combustion Engine Research Institute, Tianjin 300192, China
* Correspondence: 2020022212@cauc.edu.cn

Abstract: In order to realize sensorless control of permanent magnet synchronous motor (PMSM) with high performance in low speed region, a novel rotor position observer scheme based on finite control set model predictive control (FCS-MPC) is presented in this paper. Firstly, the FCS-MPC is used to predict the current and drive the PMSM by selecting the optimal control quantity that minimizes the cost function. Next, an adaptive second-order generalized integrator (ASOGI) with adaptive center frequency adjustment was designed to replace the band-pass filter (BPF) in the rotor position observer. The ASOGI can calculate the high frequency value that can be used for position estimation by the controller switching frequency. The current ripple inherent in the FCS-MPC is considered as the response current obtained by the high frequency injection (HFI) method. The current ripple after ASOGI filtering is input to the phase-locked loop (PLL) for phase locking to obtain the estimated rotor position. In addition, adaptive linear (Adaline) neural networks are used to identify sensitive motor parameters online to avoid mismatch of model parameters, which causes degradation of control performance. Simulation experiments and hardware experiments show that this scheme is excellent in both static and dynamic conditions.

Keywords: permanent magnet synchronous motor (PMSM); sensorless control; adaptive second-order generalized integrator; finite control set model predictive control (FCS-MPC); parameter correction; phase-locked loop

Citation: Gao, J.; Wang, Y.; Xu, M.; Mi, Y. A Novel Rotor Position Observer for Sensorless Control of Permanent Magnet Synchronous Motor Based on Adaptive Generalized Second-Order Integrator. *Machines* **2022**, *10*, 751. https://doi.org/10.3390/machines10090751

Academic Editors: Lorand Szabo and Feng Chai

Received: 2 August 2022
Accepted: 27 August 2022
Published: 31 August 2022

1. Introduction

Permanent magnet synchronous motor (PMSM) has the advantages of high power density, quick dynamic response, and great efficiency [1,2]. With the advancement in microprocessors, predictive control is widely applied to the high-performance control of PMSM [3]. Finite control set model predictive control (FCS-MPC) uses the discrete quality of the inverter switching states to control the motor by selecting the switching state that minimizes the value of the cost function from a finite number of alternative switching states [4,5], eliminating the need for pulse width modulation (PWM) modulation and generating the drive signal directly. Although the model predictive control has many advantages, in practical applications, the computational delay of the digital implementation brings degradation to the system performance. In [6], the one-step delay compensation method in [7] was used to calculate the current compensation term using Heun's method for the action delay problem, which improved the prediction accuracy of current values. The reference voltage is obtained by Deadbeat principle and the number of calculations for the non-zero vector is reduced from 18 to 4 when selecting the optimal voltage vector and its duty cycle, which greatly reduces the computational burden. In [8], the prediction model is updated with the current in the α-β subspace at moment $k + 1$ as the compensation current to obtain the prediction model at moment $k + 2$. Therefore, it is necessary to take measures to compensate the MPC controller. Similar to field-oriented control, rotor position is also very important in model predictive control [9]. However, traditional mechanical position sensors have problems such as high installation cost and low reliability [10]. Therefore, sensorless control has gradually become a new control method for PMSM [11]. Due to the small back electromotive force (EMF) of PMSM in the low speed region, the rotor position is usually estimated using the high-frequency injection (HFI) method [12,13]. Ref. [14] adopts a hybrid control strategy to achieve smooth switching from low speed to high speed. The high-frequency signal injection method is used to effectively extract the back EMF signal at a low speed, and then the finite-position-set-phase-locked loop (FPS-PLL) with high accuracy and robustness is used at high speed. However, the HFI will inevitably generate torque ripple and increased loss, and the dynamic performance will be insufficient when the load torque and speed suddenly change [15,16]. The choice of discrete switching states by the FCS-MPC creates an inherent high-frequency current ripple [17], and this inherent characteristic can replace the additional signal in the sensorless control which can be regarded as the response current generated on the q-axis when the d-axis is excited, so that the motor produces a saturated saliency [18–20]. In [21], the traditional vector control is replaced by the more advanced model predictive control. The combination of MPC and FPS-PLL makes the controller structure simple and has better dynamic stability performance. The FPS-PLL can directly search for the rotor position angle without adjusting the parameters of the PI controller. Using filters to process the inherent current ripple is an essential step in the rotor position estimation. In [22], a PLL with a modified control function is combined with FCS-MPC, which avoids the wrong estimation during speed reversal conditions as well as singularities. Unlike the fixed frequency injection signal of HFI, the switching frequency of FCS-MPC is not constant. There are a large number of changing high-frequency harmonics and noise, which makes it difficult to extract the signal at a specific frequency accurately in the current ripple with changing frequency by directly using the band-pass filter (BPF). The phase offset of the BPF will cause deviations in the estimation result of the rotor position [23]. The second-order generalized integrator (SOGI) has a simple structure without time delay which, combined with PLL, can eliminate direct current (DC) offset and harmonics well to realize accurate estimation of rotor position [24]. The SOGI can improve the dynamic performance of sensorless control at low speeds [25,26].

In addition, MPC performs predictive control based on an accurate mathematical model, and changes in the external environment and motor operating conditions may cause changes in motor parameters, resulting in model mismatch [27–29]. It is necessary to identify and modify the sensitive parameters such as the inductance and flux linkage of the motor online to correct the preset parameters of the model [30].

An adaptive second-order generalized integrator (ASOGI)-PLL structure is proposed to realize sensorless control of PMSM in this paper. The main contributions are shown as below:

1. The rotor position is estimated using the inherent current ripple of the FCS-MPC instead of the externally injected high-frequency signal.
2. The center frequency of the ASOGI can be adaptively varied to adaptively filter the inherent current ripple with large frequency fluctuations, and it is used instead of the BPF.
3. The parameters of the controller are corrected online using Adaptive linear (Adaline) neural network.

The experimental results show that the proposed method can accurately estimate the rotor position at low and even ultra-low speeds of PMSM. The PMSM has good load carrying and anti-disturbance capability.

This paper is organized as follows: Section 2 analyzes the mathematical model and the inherent current ripple of the PMSM at finite control set model predictive control. Section 3 presents the ASOGI-PLL structure for rotor position estimation and uses Adaline to correct the controller parameters. The simulation and experimental results are presented in Section 4. Finally, Section 5 gives the conclusion of this paper.

2. MPC Mathematical Model of Permanent Magnet Synchronous Motor

In this section, the mathematical model of the PMSM under finite control set model predictive control is first established. In order to extract the rotor position using the intrinsic current ripple, the inherent current ripple is further analyzed.

2.1. FCS-MPC Mathematical Model

The coordinate system of permanent magnet synchronous motor is shown in Figure 1. *ABC* is the three-phase coordinate system, α-β is the stationary coordinate system, d-q is the rotational coordinate system, θ is the angle between the rotor, and the $A(\alpha)$ axis, the $\hat{d}$ axis is defined as the estimated d-axis and the $\hat{q}$ axis is the estimated q-axis, $\hat{\theta}$ is the angle between the $\hat{d}$ axis and the $A(\alpha)$ axis, the estimated angular error is $\Delta\theta = \theta - \hat{\theta}$, N and S are the polarities.

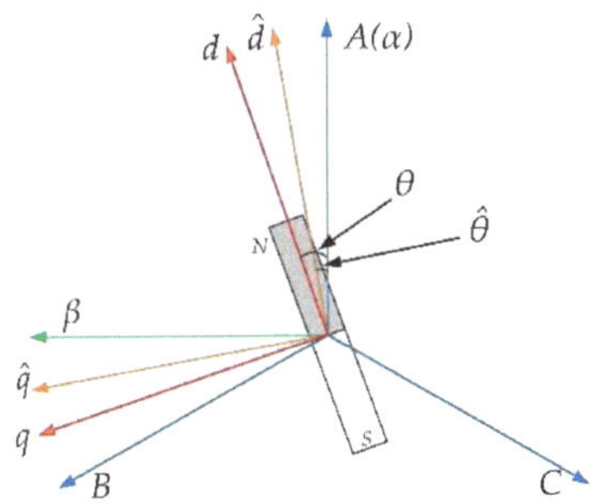

Figure 1. Coordinate system of PMSM.

The voltage equation of PMSM in the d-q rotating coordinate system can be expressed as

$$\begin{cases} u_d = L_d \frac{di_d}{dt} - \omega L_q i_q + R i_d \\ u_q = \omega L_d i_d + L_q \frac{di_q}{dt} + \omega \varphi_f + R i_q \end{cases} \tag{1}$$

where u_d and u_q are the d and q axes voltages, i_d and i_q are the d and q axes currents in the synchro-nous rotation coordinate system, respectively, φ_f is the rotor permanent magnet flux linkage, ω is the rotor angular velocity, L_d and L_q are the motor straight-axis inductance and cross-axis inductance, respectively, and R is the stator resistance.

The schematic diagram of the model predictive control of the PMSM is shown in Figure 2. The i_d^* and i_q^* are reference currents on the d and q axes, respectively.

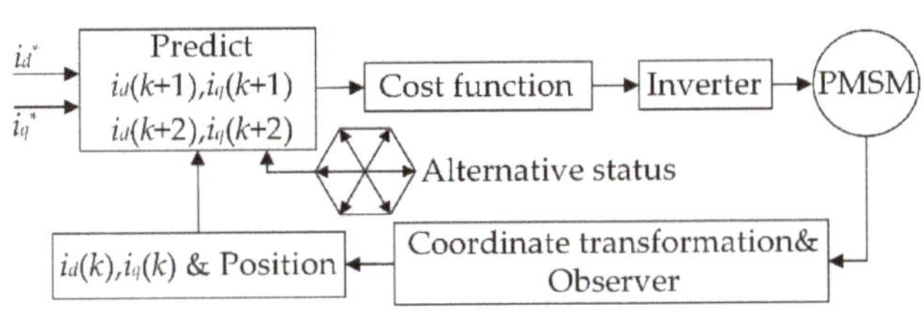

Figure 2. PMSM predictive control schematic.

In the finite set model predictive control, the controlled object is a discretized model, and the d and q axes currents are chosen as state variables, and the predicted currents are obtained by discretizing (1) using Euler method at the k moment.

$$\begin{cases} i_d\left(k+1\right) = i_d^M(k) + \frac{T_s}{L_d}\left[u_d(k) - R i_d^M(k) + \omega(k) L_q i_q^M(k)\right] \\ i_q\left(k+1\right) = i_q^M(k) + \frac{T_s}{L_q}\left[u_q(k) - R i_q^M(k) + \omega(k) L_d i_d^M(k) - \omega(k)\varphi_f\right] \end{cases} \tag{2}$$

where the superscript M denotes the measured value, T_s is the sampling period. The output of the inverter circuit S_{abc} represents the eight switching states of the inverter under discrete control, consisting of six non-zero voltage vectors and two zero voltage vectors, expressed as

$$V \in \{V_0, V_1, V_2, V_3, V_4, V_5, V_6, V_7\} \tag{3}$$

where V denotes the alternative switching state, V_1 to V_6 are non-zero voltage vectors, and V_0 and V_7 denote zero voltage vectors.

Due to the computational delay in the digital implementation, a two-step approach is generally used to compensate for the one-step controller delay. After using Equation (2) to obtain the predicted current at moment k + 1, we then used Equation (3) to predict the current at moment k + 2 for each of the eight S_{abc} states.

$$\begin{cases} i_d\left(k+2\right) = i_d^M(k+1) + \frac{T_s}{L_d}[u_d(k+1) - Ri_d^M(k+1) + \omega(k+1)L_q i_q^M(k+1)] \\ i_q\left(k+2\right) = i_q^M(k+1) + \frac{T_s}{L_q}[u_q(k+1) - Ri_q^M(k+1) + \omega(k+1)L_d i_d^M(k+1) - \omega(k+1)\varphi_f] \end{cases} \tag{4}$$

The controller traverses the discrete switching states. Since the sampling interval is extremely short, the current loop is given a reference current i_d^*, i_q^* constant for adjacent moments. Construct the cost function based on the value at moment k and the predicted value at moment k + 2

$$J = [i_d^* - i_d\left(k+2\right)]^2 + \lambda[i_q^* - i_q(k+2)]^2 \tag{5}$$

where λ is the weighting factor, which is generally taken as 1.

The voltage vector that minimizes the cost function is selected as the optimal voltage vector, and the corresponding $u_d(k+1)$, $u_q(k+1)$ and the optimal switch combination are obtained for rolling the next control cycle roll. Since the set of alternative switching states is fixed and discrete, it is finite control set as discrete control.

2.2. Inherent Current Ripple of FCS-MPC

The discrete switching control traverses each switching state at each control cycle in Figure 3, corresponding to seven different spatial voltage vectors. The processor needs to calculate the cost function which corresponds to the predicted current generated by each voltage vector, which inevitably creates the high-frequency current ripple inherent in FCS-MPC control.

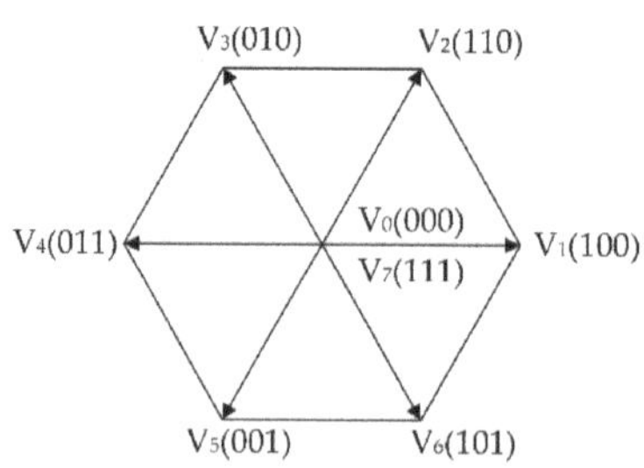

Figure 3. Spatial voltage vector.

In the motor synchronous rotation coordinate system, the current relationship between d-q axis and $\hat{d} - \hat{q}$ axis is

$$\begin{cases} \hat{i}_d = i_d \cos\Delta\theta - i_q \sin\Delta\theta \\ \hat{i}_q = i_d \sin\Delta\theta + i_q \cos\Delta\theta \end{cases} \tag{6}$$

Similarly, the voltage relationship between *d*-*q* axis and $\hat{d} - \hat{q}$ axis is

$$\begin{cases} \hat{u}_d = u_d \cos \Delta\theta - u_q \sin \Delta\theta \\ \hat{u}_q = u_d \sin \Delta\theta + u_q \cos \Delta\theta \end{cases} \tag{7}$$

The term $u_d(k) - Ri_d(k) + \omega(k)L_q i_q(k)$ in (2) is written as U_D and the term $u_q(k) - Ri_q(k) + \omega(k)L_d i_d(k) - \omega(k)\varphi_f$ is written as U_Q, therefore the predicted current on the *d*-*q* axis at moment $k + 1$ is

$$\begin{cases} i_d(k+1) = i_d(k) + U_D T_s / L_d \\ i_q(k+1) = i_q(k) + U_Q T_s / L_q \end{cases} \tag{8}$$

The PMSM produces saturated convex polarity under high frequency excitation, so that the inherent current ripple can be used as a high frequency characteristic instead of external high frequency voltage injection to produce high frequency response currents, and thus the position observer can be designed to extract the rotor position contained in the current ripple. In order to make the current ripple high frequency signal more significant and stable, the frequency distribution range of the current ripple is usually reduced using the adjacent vector voltage selection method. As shown in Table 1, the optimal switching state for the next moment that minimizes the cost function is selected from two adjacent valid zero vectors using the switching combination of the previous moment.

Table 1. Switch Status Table.

Previous States	Alternate States		
000 (V_0)	100	010	001
100 (V_1)	000	110	101
110 (V_2)	010	100	111
010 (V_3)	110	000	011
011 (V_4)	111	001	010
001 (V_5)	000	011	101
101 (V_6)	001	111	100
111 (V_0)	011	101	110

We started the PMSM from static to 150 r/min (10% of the rated speed) with 50% of the rated load. The *d*-*q* axis currents at moment *k* were obtained by Park's transformation after collecting three-phase currents, and the FFT analysis of motor phase currents is shown in Figure 4. The current contains high-frequency current components that are much higher than the fundamental frequency. It can be seen that this method makes the frequency of the high-frequency part of the current ripple mostly concentrated in 500 Hz~2000 Hz. However, due to the discrete switching characteristics, the frequencies of the currents are not fixed, and the harmonic dispersion is more chaotic, which causes some difficulties in the selection and extraction of the filter cutoff frequency and affects the estimation of rotor position.

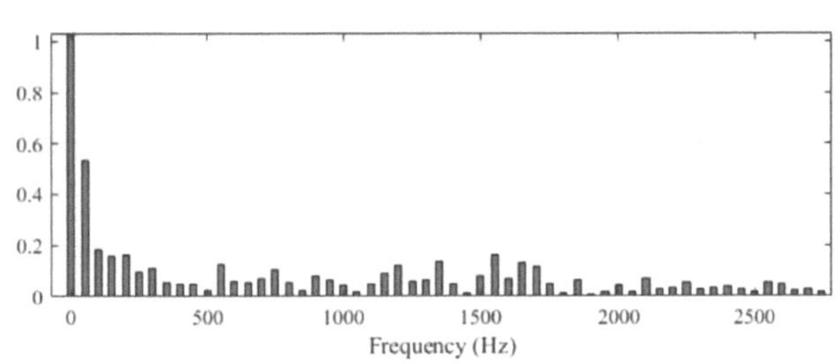

Figure 4. A-phase current FFT analysis.

3. ASOGI-PLL Rotor Position Observer

In this section, a novel rotor observation structure is proposed. To solve the problem that the BPF cannot adapt to the frequent changes of the inherent current ripple frequency, the ASOGI with adaptive center frequency selection is designed. The ASOGI-PLL can accurately estimate the rotor position. Furthermore, Adaline linear neural network is used to correct the MPC parameters to avoid the system performance degradation.

3.1. Conventional PLL Rotor Position Observer

In the $\hat{d}-\hat{q}$ axis, the predicted current expression from (2) is

$$\begin{cases} \hat{i}_d\left(k+1\right) = \hat{i}_d^M(k) + \frac{T_s}{L_d}[\hat{u}_d(k) - R\hat{i}_d^M(k) + \hat{\omega}(k)L_q\hat{i}_q^M(k)] \\ \hat{i}_q\left(k+1\right) = \hat{i}_q^M(k) + \frac{T_s}{L_q}[\hat{u}_q(k) - R\hat{i}_q^M(k) + \hat{\omega}(k)L_d\hat{i}_d^M(k) - \hat{\omega}(k)\varphi_f] \end{cases} \tag{9}$$

From (6), the measured value of the current in the $\hat{d}-\hat{q}$ axis can be expressed as

$$\begin{cases} \hat{i}_d^M(k+1) = i_d(k+1)\cos\Delta\theta - i_q(k+1)\sin\Delta\theta \\ \hat{i}_q^M(k+1) = i_d(k+1)\sin\Delta\theta + i_q(k+1)\cos\Delta\theta \end{cases} \tag{10}$$

In the present control cycle, the current error between the estimated value and the measured value is obtained from (9) and (10) as

$$\Delta i_d(k+1) = \hat{i}_d^M(k+1) - \hat{i}_d(k+1) \tag{11}$$

$$\Delta i_q(k+1) = \hat{i}_q^M(k+1) - \hat{i}_q(k+1) \tag{12}$$

Due to the extremely short sampling time, the k and $k+1$ moments can be considered as steady states, and the current error on the q-axis is obtained by substituting (6) to (10) into (12) and simplifying

$$\Delta i_q(k+1) = T_s(\frac{U_D}{L_d} - \frac{U_D}{L_q})\sin\Delta\theta \tag{13}$$

The position error information is included in (13), and the motor parameters can be regarded as constants within a control cycle, U_D and U_Q can be used as control constants without the interference of nonlinearity, and the interference of the cosine term is excluded by establishing the q-axis current error correlation term.

Considering the rotor position information contained in the q-axis error current and the effect of the d-axis current, the rotor position angle error term is established and derived according to (8) and (13).

$$\varepsilon(k) = [\hat{i}_q^M(k) - \hat{i}_q(k)]\hat{U}_D = T_s\left[\frac{1}{L_d} - \frac{1}{L_q}\right]U_D^2\sin\Delta\theta\cos\Delta\theta - T_s\left[\frac{1}{L_d} - \frac{1}{L_q}\right]U_DU_Q\sin^2\Delta\theta \tag{14}$$

Due to the orthogonality of d-axis and q-axis, U_D and U_Q are not correlated with each other. In the steady state, U_D and U_Q are not constant at zero value in the discrete switch control steady state, but their average values $\overline{U}_D = 0$, $\overline{U}_Q = 0$. Therefore, the second term in (14) is approximately equal to zero, giving a function on $\Delta\theta$

$$\varepsilon(\Delta\theta) = \frac{T_s}{2}\left[\frac{1}{L_d} - \frac{1}{L_q}\right]U_D^2\sin(2\Delta\theta) = K_e\sin(2\Delta\theta) \tag{15}$$

where K_e is a constant.

When the estimated rotor position is true, the $\hat{d}$ axis coincides with the $\hat{q}$ axis, $\Delta\theta = 0$, and (15) is also zero. Use the PLL in Figure 5 to lock the phase contained in (15).

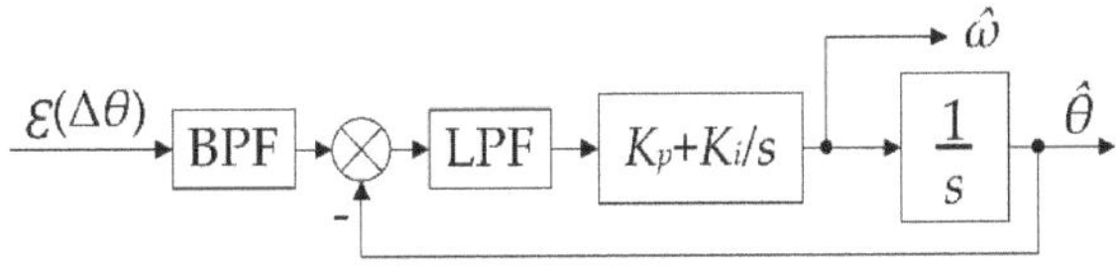

Figure 5. Phase locked loop structure diagram.

In the HFI method, the high frequency signal is usually processed by BPF first, and the center frequency of the BPF is kept the same as the frequency of the injected signal. However, in FCS-MPC, the frequency of the high-frequency signal is not stable and is accompanied by high-frequency harmonics. The conventional second-order BPF has the defects of phase shift and fixed center frequency, so it is necessary to design a filter that can adapt to the high-frequency signal with varying current ripple and minimize the phase shift, and then perform phase locking through the PLL structure to obtain the rotor position.

3.2. ASOGI-PLL Rotor Position Observer

The SOGI structure is shown in Figure 6. The transfer function can be expressed as

$$D(s) = \frac{v'(s)}{v(s)} = \frac{l\omega_c s}{s^2 + l\omega_c s + \omega_c{}^2} \tag{16}$$

where l is the damping factor, ω_c is the center frequency, v is the input signal, v' and qv' are output signal.

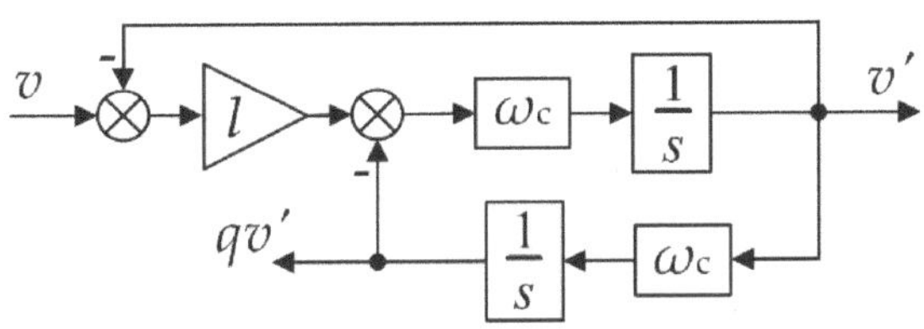

Figure 6. SOGI structure diagram.

Figure 7 reveals the Bode diagrams of the SOGI transfer function $D(s)$ with l = 0.05, 0.1, 0.3, 0.5, respectively. The central frequency of SOGI in Figure 7 is 50 Hz (100π rad/s). The bandwidth of $D(s)$ is different for different l values. As l increases from 0.05 to 0.5, the bandwidth widens, and the change of phase frequency characteristic curve also becomes slow. It can be seen that there is no delay in the phase frequency characteristic curve of $D(s)$. Obviously, the filtering effect is better when the bandwidth is narrower, but from a practical point of view, the passband need not be too narrow.

When the frequency of the input signal v coincides with the center frequency of the SOGI, the output signals v' and qv' are two quadrature signals of the same amplitude, v' overdetermined qv' in phase $\pi/2$. $D(s)$ can be considered as a second-order BPF. Let the quality factor $Q_D = 1/l$, (16) can be modified as

$$D(s) = \frac{\omega_c/Q_D s}{s^2 + \omega_c/Q_D s + \omega_c{}^2} \tag{17}$$

As can be seen from (17), the quality factor is related to the damping factor, and the smaller the l, the better filtering performance will be, but the longer dynamic response time of the system will be, generally taking the value about $\sqrt{2}$.

For the unstable current ripple frequency, a second-order generalized integrator with adaptive center frequency adjustment is designed, and the structure is shown in Figure 8.

The switching selection frequency of the FCS-MPC controller is sampled with a sampling period of

$$t_{sa} = \hat{\omega}eT_s \tag{18}$$

where t_{sa} is the sampling interval time, $\hat{\omega}e$ is the estimated electro-mechanical angular velocity.

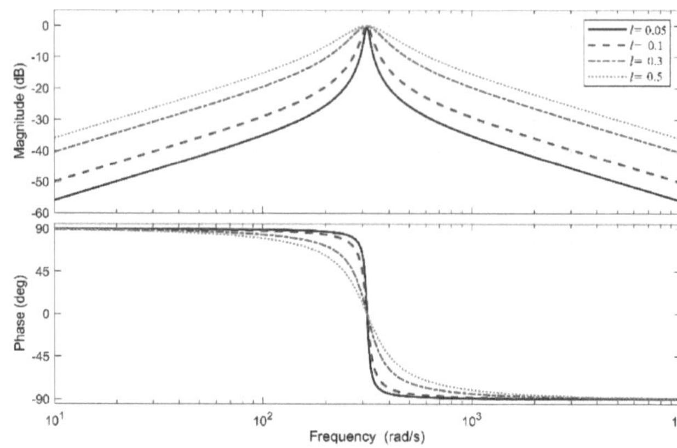

Figure 7. Bode diagram of SOGI.

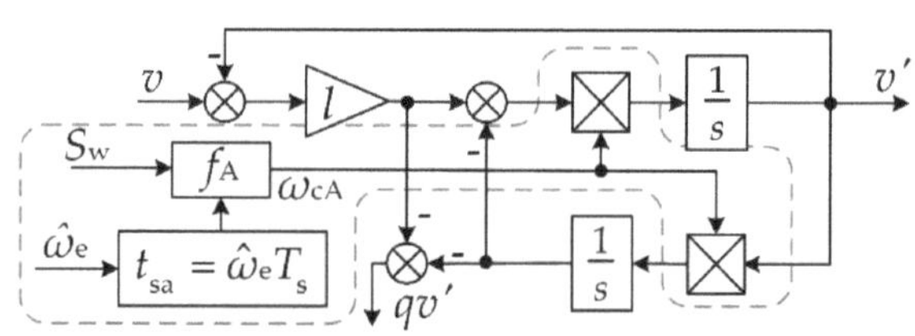

Figure 8. ASOGI structure diagram.

Due to the short sampling period, the average value of the current cycle switching times can be regarded as the current switching times, and the average switching frequency of the current cycle is

$$f_A = \frac{S_w}{t_{sa}} \tag{19}$$

where S_w is the number of MPC switch selections in one cycle.

From this, the center frequency of the ASOGI-PLL can be calculated

$$\omega_{cA} = \frac{\pi S_w}{180 t_{sa}} \tag{20}$$

As can be seen from (20), the center frequency is related to the number of switch selections and sampling interval, and the average frequency within t_{sa} of MPC controller operation can approximately represent the high-frequency of the current cycle. As shown in Figure 8, compared to the SOGI with a fixed center frequency, it is possible to dynamically adjust the value of the center frequency to keep it consistent with the high frequency of the current ripple. Compared with the conventional BPF, there is no phase shift, and it is able to extract the changing high frequency current ripple in FCS-MPC.

The $D_A(s)$ transfer function of ASOGI is

$$D_A(s) = \frac{l\omega_{cA}s}{s^2 + l\omega_{cA}s + \omega_{cA}{}^2} \tag{21}$$

When the calculated value of ω_{cA} is changed, the center frequency of ASOGI is also adjusted. The Bode diagrams of $D_A(s)$ are shown in Figure 9. For different frequency input signals, ASOGI can adjust the center frequency adaptively. When ω_{cA} is small, the amplitude-frequency characteristic curve and the phase-frequency characteristic curve of $D_A(s)$ will be shifted to the right and conversely to the left. These actions reflect that the filter is changing the filter interval with the external signal while there is no delay in phase. When l is changed, the −3 dB bandwidth is also changed. For the universal case, 0.707 is taken, the signal frequency is generally higher in this scheme, and the value of l should be reduced appropriately.

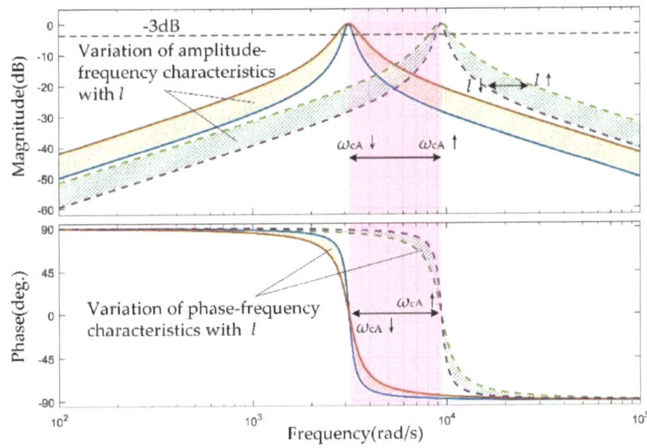

Figure 9. Bode of $D_A(s)$ when the center frequency changes.

Once the current passes through the ASOGI, the PLL can be used for phase locking. The structure of the composed ASOGI-PLL rotor position observer is shown in Figure 10.

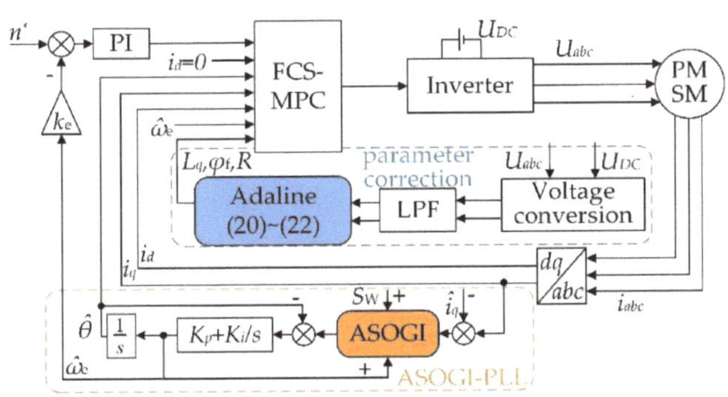

Figure 10. Overall block diagram.

3.3. MPC Parameter Correction

The operating point of the motor will change due to temperature and operating condition changes, etc. Deviations between the prediction model of MPC and the actual parameters of the motor will cause parameter mismatch, resulting in inaccurate current prediction and poor control performance. As (1) and (2) show, the q-axis inductance, stator resistance R, and flux linkage φ_f are critical to the performance of the MPC. In this section, Adaline is used for online identification of motor parameters and real-time correction of MPC prediction model parameters. Adaline neural network has a simple structure. It is a linear adjustable neural network with only the operations of addition, subtraction, and multiplication. The Adaline input–output relationship is expressed as

$$O(W_i, X_i) = \sum_{i=0}^{n} W_i X_i \tag{22}$$

where W_i is the weight of the neural network, the weights are the identification quantities L_q, R, and φ_f, X_i is the input signal, $d(k)$ is the network target input, and $O(k)$ is the output of the network excitation function.

Adjusting network weights using least mean square (LMS)

$$\begin{cases} W_i(k+1) = W_i(k) + 2\eta X_i \delta(k) \\ \delta(k) = d(k) - O(k) \end{cases} \tag{23}$$

where η is the weight adjustment factor, $\varepsilon(k)$ is the error between the neural network output and the target.

When using the Adaline to identify the parameters of a motor, η should satisfy

$$0 < 2\eta |X(k)|^2 < 1 \tag{24}$$

The motor parameter identification model is constructed, and the cross-axis inductance and flux linkage are identified first. The expression of the cross-axis inductance L_q is calculated as

$$\begin{cases} X_{L_q}(k) = -\omega_e(k) i_q(k) \\ O_{L_q}(k) = L_q(k) X_{L_q}(k) \\ d_{L_q}(k) = U_{rd}(k) - R i_d(k) \\ L_q(k+1) = L_q(k) + 2\eta L_q X_{L_q}(k) \delta_{L_q}(k) \end{cases} \tag{25}$$

The calculated expression for flux linkage φ_f is

$$\begin{cases} X_{\varphi_f}(k) = \omega_e(k) \\ O_{\varphi_f}(k) = \varphi_f(k) X_{\varphi_f}(k) \\ d_{\varphi_f}(k) = U_{rq}(k) - \omega_e(k) L_d i_d(k) - R i_q(k) \\ \varphi_f(k+1) = \varphi_f(k) + 2\eta_{\varphi_f} X_{\varphi_f}(k) \delta_{\varphi_f}(k) \end{cases} \tag{26}$$

where U_{rd} and U_{rq} are the valid voltage values.

The temperature of the motor changes relatively slowly within a short period of time when the motor is started, and the resistance value does not change significantly, the calculated expression for the resistance R is

$$\begin{cases} X_R(k) = i_q(k) \\ O_R(k) = R(k) X_R(k) \\ d_R(k) = u_q(k) - \varphi_f \omega_e(k) \\ R(k+1) = R(k) + 2\eta X_R(k) \delta_R(k) \end{cases} \tag{27}$$

The overall control block diagram is shown in Figure 10.

4. Simulation and Hardware Experiments

In this section, simulation and physical platform experiments are designed and the experimental results are analyzed.

The performance of the proposed scheme was verified using a PMSM with parameters as shown in Table 2. Experiments were conducted for low speed operation, sudden variable speed, and load disturbance of the PMSM, respectively, and the experimental results were also analyzed.

Table 2. Parameters of PMSM.

Parameter	Value
Bus voltage	220 V
Pole pairs	4
Stator resistance	1.2 Ω
d-axis inductance	8.5 mH
q-axis inductance	8.5 mH
Moment of inertia	0.008 kg·m^2
Permanent magnet flux	0.117 Wb
Rated speed	1500 r/min

4.1. Variable Speed Experiment at Low Speed

To verify the effectiveness of ASOGI-PLL for rotor position estimation at low speeds, the speed is started at a given speed of 5 r/min, the speed is increased to 10 r/min suddenly at $t = 20$ s, and then the speed is increased to 15 r/min suddenly after 15 s of stable operation. The speed waveform is shown in Figure 11, the speed estimation error of PLL is within 2 r/min, and the speed estimation error of ASOGI-PLL is within 1.2 r/min. The "PLL" in the legend of the experimental results refers to the BPF-PLL structure shown in Figure 5.

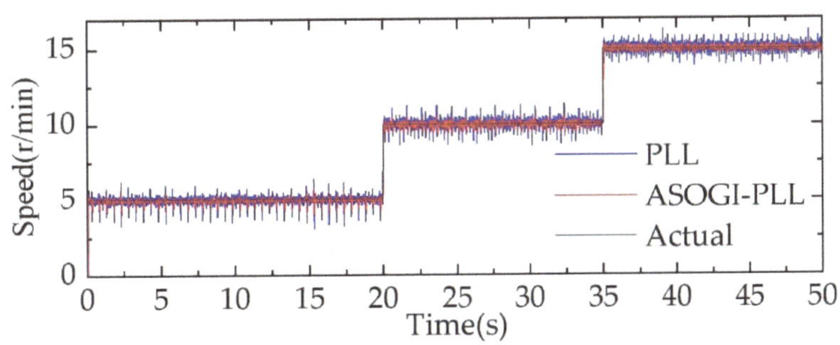

Figure 11. Speed waveform at speed up.

Figure 12a illustrates the rotor angle waveform during the speed change, and Figure 12b shows the rotor position estimation error. The maximum error of rotor estimation by PLL during the speed change is within 3.5° (0.06 rad), the maximum error of rotor position estimated by ASOGI-PLL is within 0.93° (0.016 rad), and the error is reduced by 73.33% relative to PLL. Figure 12c,d show the local enlargement for the speed change at $t = 20$ s and $t = 30$ s, and it can be seen that ASOGI-PLL is also able to achieve good tracking of the rotor position at the instant of speed change.

Figure 13 shows the variation of the ASOGI center frequency with the FCS-MPC switching frequency for different speeds. It is worth noting that the center frequency f_{cA}, obtained from the calculations in the experimental results, needs to be converted to the radian system when used for ASOGI. At different speeds, the f_{cA} varies with the switching frequency of the controller.

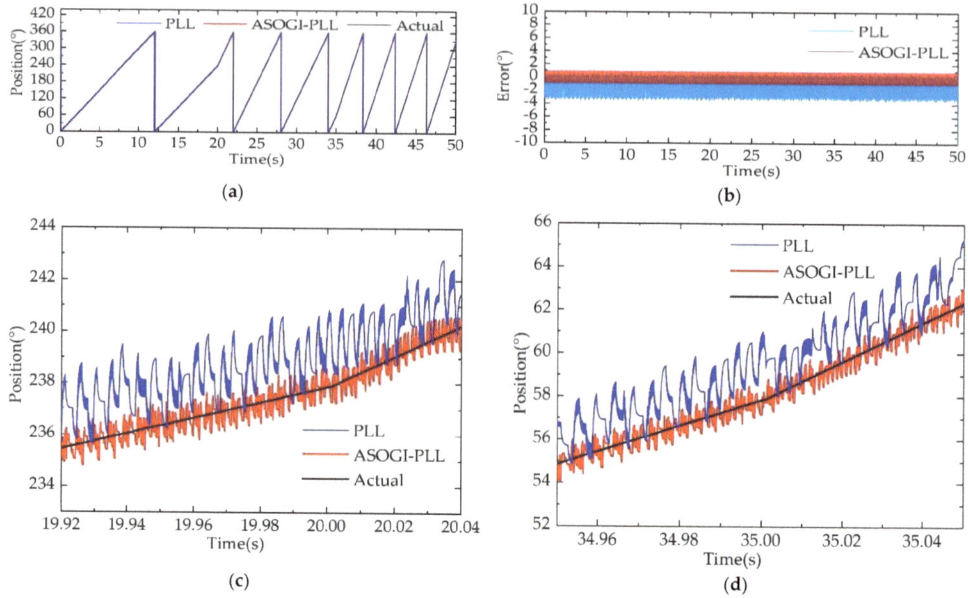

Figure 12. Speed up waveform: (**a**) Rotor position waveform; (**b**) Rotor position estimation error; (**c**) Rotor position local enlarged waveform at t = 20 s; (**d**) Rotor position local enlarged waveform at t = 35 s.

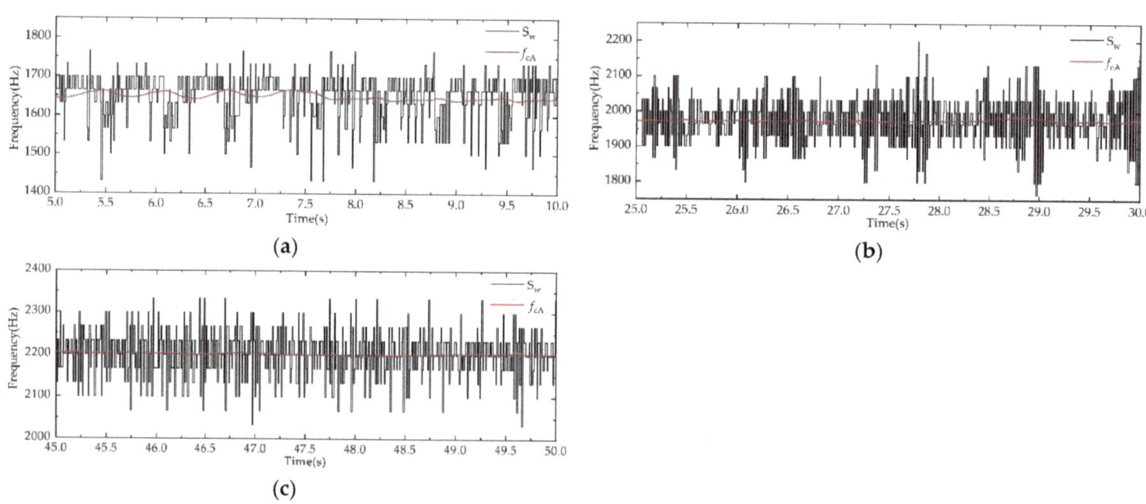

Figure 13. Switching frequency of FCS-MPC and calculated value of ASOGI center frequency: (**a**) At the speed of 5 r/min; (**b**) At the speed of 10 r/min; (**c**) At the speed of 15 r/min.

Generally, within 5% of the rated speed (75 r/min) is the ultra-low speed [31], and within 20% of the rated speed (300 r/min) is called the low speed range [32]. To test the performance of the proposed strategy in the whole low speed range, we varied the speed from 0 r/min to 300 r/min and load from 0 N·m to 14 N·m. The corresponding rotor position error changes, as shown in Figure 14a. Figure 14b shows the calculated values of the center frequency f_{cA} for different speeds and loads. f_{cA} varies with the selected frequency of the controller when the speed is stable. These show that the method has excellent performance at low speeds.

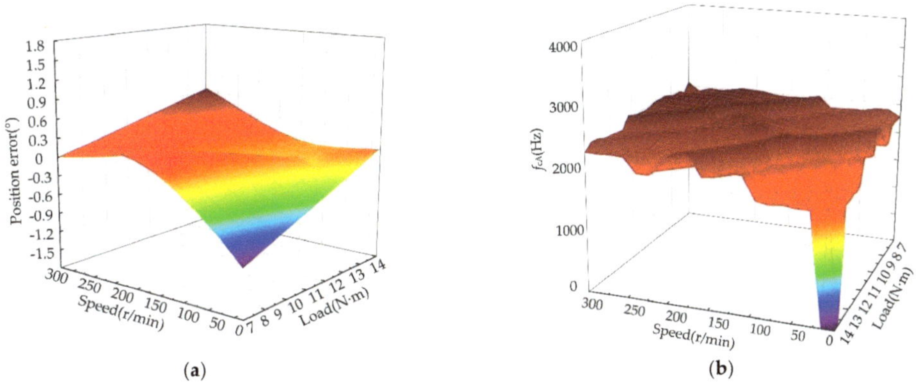

Figure 14. Results within 300 r/min: (**a**) Rotor position error at different speeds and loads; (**b**) Waveform of center frequency f_{cA} variation.

4.2. Motor Reversal Experiment

The motor is started at a given speed of 10 r/min under rated load, and the given speed is abruptly changed to −10 r/min at t = 15 s. The speed waveform is shown in Figure 15a, and the maximum speed estimation error is within 1.5 r/min for PLL and within 0.7 r/min for ASOGI-PLL. It converges to the stable value about 0.08 s after the speed reversal.

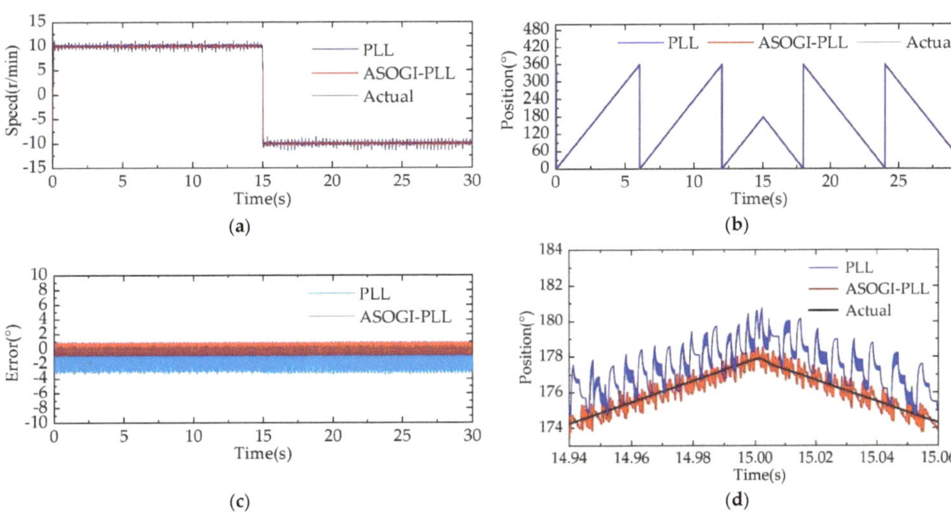

Figure 15. Rotor position waveform during motor reversal: (**a**) Speed waveform in reverse rotation; (**b**) Rotor position waveform; (**c**) Rotor position estimation error waveform; (**d**) Rotor position local enlarged waveform at t = 15 s.

The maximum error of the rotor position estimated by PLL is within 4° (0.058 rad), while the maximum error of the rotor position estimated by ASOGI-PLL is within 1.15° (0.07 rad), and the error is reduced by 71.13% compared to PLL, as shown in Figure 15b,c. Figure 15d shows the local magnified waveform of the rotor angle at the moment of reversal. The rotor angle is reversed when the motor speed is reversed, and the ASOGI-PLL is more stable than the PLL rotor position estimation.

4.3. Load Mutation Experiments

To verify the usability of the low speed control strategy under sudden load changes, the speed was started and stabilized to 15 r/min, and the load was suddenly increased from 7 N·m to 14 N·m at t = 5 s and suddenly decreased from 14 N·m to 7 N·m at t = 10 s. The speed waveform is shown in Figure 16. There is a small change in speed when the load changes, but it quickly returns to the given speed.

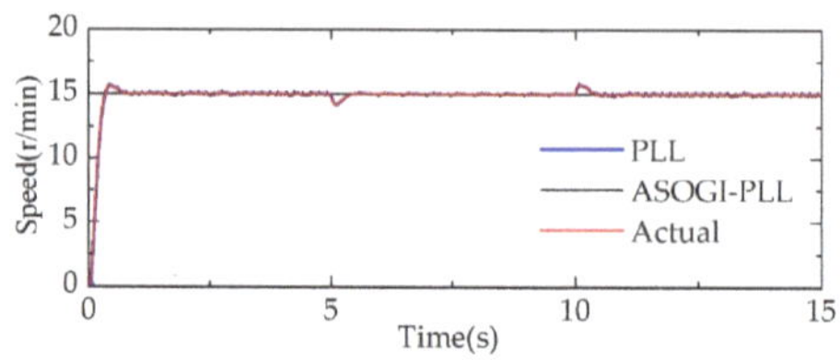

Figure 16. Speed waveform during sudden load change.

Figure 17a shows the rotor position waveform during sudden load change, and Figure 17b shows the rotor position error waveform. The maximum error of rotor position estimated by PLL is within 3.3° (0.058 rad), and the maximum error of rotor position estimated by ASOGI-PLL is within 0.93° (0.016 rad). Figure 17c,d shows the local enlarged view of the sudden load addition and sudden load reduction, and it can be seen that the error of rotor position estimation by ASOGI-PLL is reduced by 1.41° compared with PLL during sudden load addition and 1.33° compared with PLL during sudden load reduction. The ASOGI-PLL is able to track the rotor position accurately during sudden load changes and is more stable.

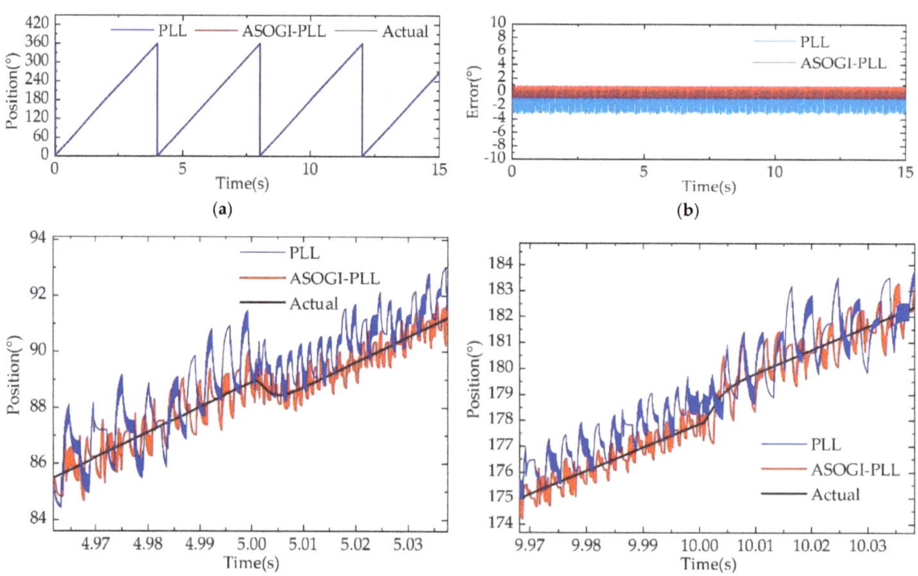

Figure 17. Sudden load change: (**a**) Rotor position waveform; (**b**) Rotor position estimation error waveform; (**c**) Rotor position local enlarged waveform at t = 5 s; (**d**) Rotor position local enlarged waveform at t = 10 s.

Figure 18 shows the variation of the ASOGI center frequency with the FCS-MPC switching frequency for different loads. At different loads, the calculated value of the center frequency varies with the selected frequency of the controller.

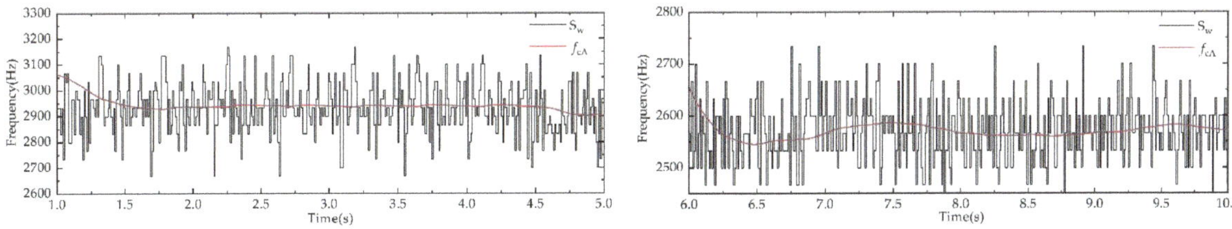

(a) **(b)**

Figure 18. Switching frequency of FCS-MPC and calculated value of ASOGI center frequency: (**a**) At the load of 7 N·m; (**b**) At the load of 14 N·m.

4.4. Online Parameter Identification Experiment

The PMSM parameters for different operating conditions were identified online using the Adaline neural network. Figure 19a shows the parameter identification results for the PMSM speed at t = 5 s with a sudden increase from 5% to 10% of the rated speed and then a sudden decrease to 5% of the rated speed at t = 10 s. The load during the speed sudden change is 50% of the rated load. The inductance and flux linkage decrease when the rotational speed increases. They increase when the rotational speed decreases. The resistance is closely related to the temperature, so the resistance value is increasing. Figure 19b shows the parameter identification results when the PMSM load increases suddenly from 50% to 100% of rated load at t = 5 s and then decreases suddenly to 50% of rated load at t = 10 s. The speed during the load sudden change experiment is 100 r/min. Both inductance and flux linkage decrease when the load increases, and they increase when the load decreases. There is an increasing trend of resistance. The Adaline neural network is accurate and fast for parameter identification under different operating conditions.

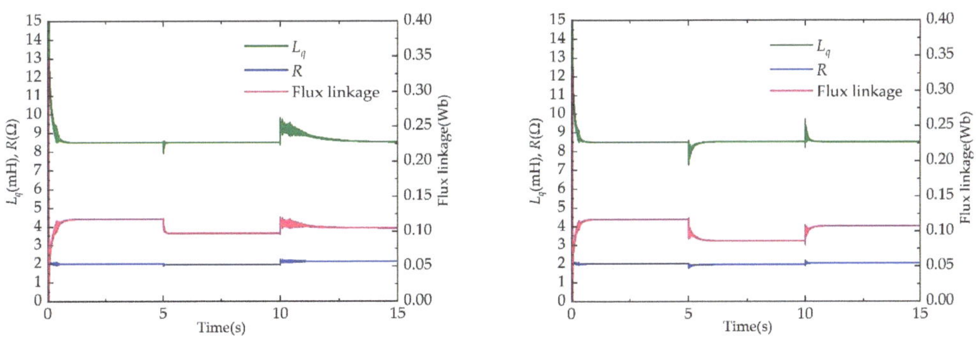

(a) **(b)**

Figure 19. Parameter identification results: (**a**) Results of PMSM parameters identification during speed change; (**b**) Results of PMSM parameters identification during load change.

4.5. Hardware Experiments

To verify the viability and effectiveness of the proposed method, an experimental platform of a 3 kW three-phase PMSM was conducted, as shown in Figure 20. The parameters of the PMSM are shown in Table 2. The control system includes an upper computer, DC power supply, ARM STM32F103, inverter, oscilloscope, PMSM, driver board, voltage sensors, and current sensor. The sampling time used in the experimental test is 1×10^{-5} s. The sampling rate of the current sensor is 10 kHz.

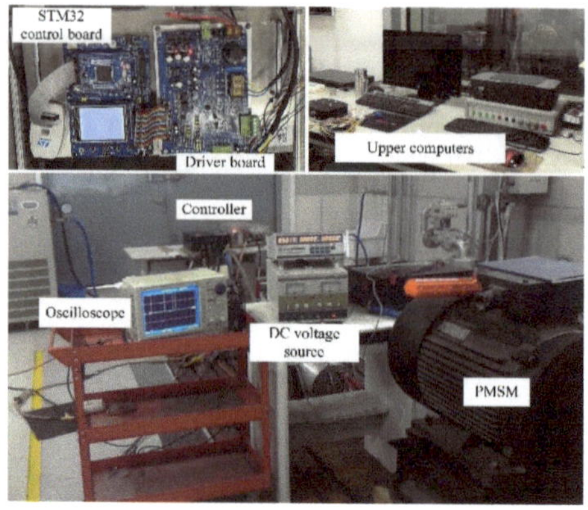

Figure 20. The experimental platform.

Figure 21a shows the rotor position and phase current waveforms when the PMSM is running at 10 r/min. The system operation is stable, and the rotor position is estimated accurately. Figure 21b shows the waveforms of rotor position and phase current when the load is suddenly increased. When the load increases, the current becomes larger, and the motor quickly resumes running at the given speed after a brief decrease in speed. Figure 21c shows the waveforms of rotor position and phase current when the PMSM is running at higher speed with increased load and then rapidly decreasing the load, and the system has good anti-disturbance capability and static stability. Figure 21d shows the change of the controller selection switching frequency when the speed increases. The selective switching frequency increases with the speed and load.

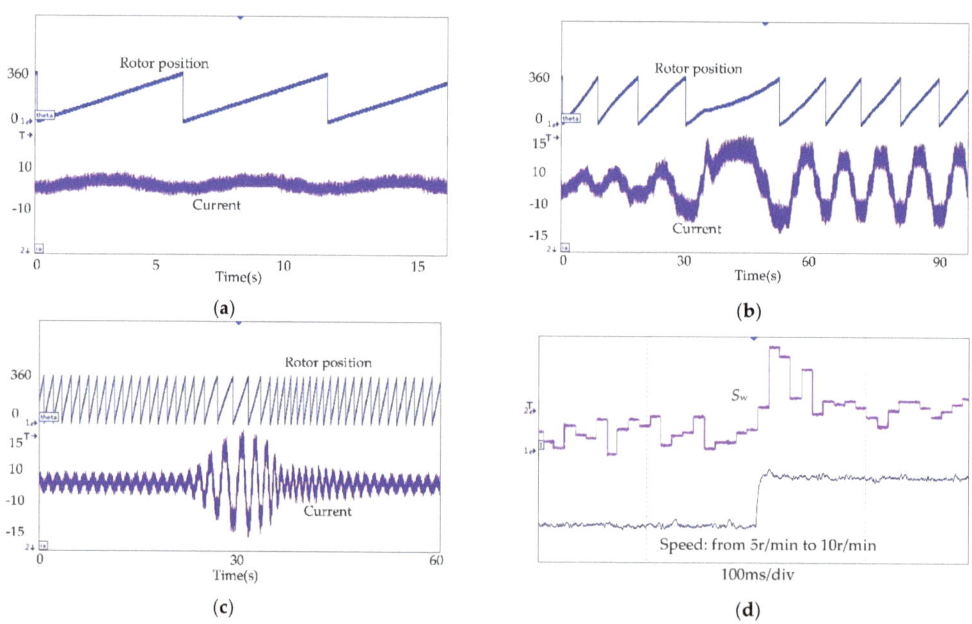

Figure 21. Hardware experimental results: (**a**) Rotor position and stator current at low speed; (**b**) Rotor position and stator current during sudden loading; (**c**) Rotor position and stator current during load disturbance; (**d**) Variation of the controller selection switching frequency when speed changes from 5 r/min to 10 r/min.

5. Conclusions

In this paper, an observer composed of ASOGI and PLL is presented to estimate the rotor position of PMSM in FCS-MPC. The inherent current ripple is generated internally by the controller, and using it instead of HFI can avoid affecting the system's dynamic performance and reduce the system complexity. The frequency of inherent current ripple is not stable. In the proposed scheme, ASOGI is used instead of BPF, and it adjusts the center frequency adaptively according to the input signal. The irrelevant frequency signal in the high-frequency current will be filtered out by ASOGI. The ASOGI-PLL structure will estimate the rotor position accurately. PMSM parameters may change with operating conditions, and sensitive parameters can cause parameter mismatch in the controller. Adaline neural network corrects controller parameters through online parameter identification, the prediction variables will be more accurate. The experimental results demonstrate the excellent performance of the proposed scheme in the low speed or even ultra-low speed region of the PMSM.

Although this scheme has excellent position-free control performance at low PMSM speeds, the fundamental frequency has a negative impact on rotor position estimation as the speed increases, so this scheme may be limited at high speed. The strategy of combining multiple methods for different speeds may be considered more suitable for full speed high performance control of PMSM.

Author Contributions: Conceptualization, J.G. and Y.W.; methodology, Y.W.; software, Y.W. and Y.M.; validation, J.G. and M.X.; formal analysis, Y.W.; investigation, Y.W. and Y.M.; resources, M.X. and Y.M.; data curation, J.G. and Y.W.; writing—original draft preparation, Y.W.; writing—review and editing, J.G., M.X. and Y.W.; visualization, M.X.; supervision, M.X.; project administration, J.G.; funding acquisition, J.G. All authors have read and agreed to the published version of the manuscript.

Funding: This research was funded by the National Natural Science Foundation of China, grant number 51707195 and 62173331.

Institutional Review Board Statement: Not applicable.

Informed Consent Statement: Not applicable.

Data Availability Statement: The data presented in this study are available on request from the corresponding author.

Conflicts of Interest: The authors declare no conflict of interest.

References

1. Wang, C.-S.; Guo, C.-W.C.; Tsay, D.-M.; Perng, J.-W. PMSM Speed Control Based on Particle Swarm Optimization and Deep Deterministic Policy Gradient under Load Disturbance. *Machines* **2021**, *9*, 343. [CrossRef]
2. Zhao, L.; Chen, Z.; Wang, H.; Li, L.; Mao, X.; Li, Z.; Zhang, J.; Wu, D. An Improved Deadbeat Current Controller of PMSM Based on Bilinear Discretization. *Machines* **2022**, *10*, 79. [CrossRef]
3. Ma, Z.; Saeidi, S.; Kennel, R. FPGA Implementation of Model Predictive Control with Constant Switching Frequency for PMSM Drives. *IEEE Trans. Ind. Informs.* **2014**, *10*, 2055–2063. [CrossRef]
4. Stellato, B.; Geyer, T.; Goulart, P.J. High-Speed Finite Control Set Model Predictive Control for Power Electronics. *IEEE Trans. Power Electron.* **2017**, *32*, 4007–4020. [CrossRef]
5. Farhan, A.; Abdelrahem, M.; Hackl, C.M.; Kennel, R.; Shaltout, A.; Saleh, A. Advanced Strategy of Speed Predictive Control for Nonlinear Synchronous Reluctance Motors. *Machines* **2020**, *8*, 44. [CrossRef]
6. Sun, X.; Li, T.; Yao, M.; Lei, G.; Guo, Y.; Zhu, J. Improved Finite-Control-Set Model Predictive Control with Virtual Vectors for PMSHM Drives. *IEEE Trans. Energy Convers.* **2021**, *37*, 1885–1894. [CrossRef]
7. Sun, X.; Wu, M.; Lei, G.; Guo, Y.; Zhu, J. An Improved Model Predictive Current Control for PMSM Drives Based on Current Track Circle. *IEEE Trans. Ind. Electron.* **2020**, *68*, 3782–3793. [CrossRef]
8. Sun, X.; Li, T.; Tian, X.; Zhu, J. Fault-Tolerant Operation of a Six-Phase Permanent Magnet Synchronous Hub Motor Based on Model Predictive Current Control with Virtual Voltage Vectors. *IEEE Trans. Energy Convers.* **2021**, *37*, 337–346. [CrossRef]
9. Chen, Z.; Qiu, J.; Jin, M. Prediction-Error-Driven Position Estimation Method for Finite-Control-Set Model Predictive Control of Interior Permanent-Magnet Synchronous Motors. *IEEE Trans. Emerg. Sel. Topics Power Electron.* **2019**, *7*, 282–295. [CrossRef]
10. Anayi, F.J.; Al Ibraheemi, M.M.A. Estimation of Rotor Position for Permanent Magnet Synchronous Motor at Standstill Using Sensorless Voltage Control Scheme. *IEEE ASME Trans. Mechatron.* **2020**, *25*, 1612–1621. [CrossRef]

11. Morimoto, S.; Kawamoto, K.; Sanada, M.; Takeda, Y. Sensorless Control Strategy for Salient-Pole PMSM Based on Extended EMF in Rotating Reference Frame. *IEEE Trans. Ind. Appl.* **2002**, *38*, 1054–1061. [CrossRef]
12. Li, H.; Zhang, X.; Yang, S.; Liu, S. Unified Graphical Model of High-Frequency Signal Injection Methods for PMSM Sensorless Control. *IEEE Trans. Ind. Electron.* **2020**, *67*, 4411–4421. [CrossRef]
13. Ilioudis, V.C. Sensorless Control of Permanent Magnet Synchronous Machine with Magnetic Saliency Tracking Based on Voltage Signal Injection. *Machines* **2020**, *8*, 14. [CrossRef]
14. Sun, X.; Cao, J.; Lei, G.; Guo, Y.; Zhu, J. Speed Sensorless Control for Permanent Magnet Synchronous Motors Based on Finite Position Set. *IEEE Trans. Ind. Electron.* **2019**, *67*, 6089–6100. [CrossRef]
15. Liang, D.; Li, J.; Qu, R.; Kong, W. Adaptive Second-Order Sliding-Mode Observer for PMSM Sensorless Control Considering VSI Nonlinearity. *IEEE Trans. Power Electron.* **2018**, *33*, 8994–9004. [CrossRef]
16. Chen, S.; Luo, Y.; Pi, Y. PMSM sensorless control with separate control strategies and smooth switch from low speed to high speed. *ISA Trans.* **2015**, *58*, 650–658. [CrossRef]
17. Wang, X.; Fang, X.; Lin, S.; Lin, F.; Yang, Z. Predictive Common-Mode Voltage Suppression Method Based on Current Ripple for Permanent Magnet Synchronous Motors. *IEEE Trans. Emerg. Sel. Topics Power Electron.* **2019**, *7*, 946–955. [CrossRef]
18. Rovere, L.; Formentini, A.; Gaeta, A.; Zanchetta, P.; Marchesoni, M. Sensorless Finite-Control Set Model Predictive Control for IPMSM Drives. *IEEE Trans. Ind. Electron.* **2016**, *63*, 5921–5931. [CrossRef]
19. Nalakath, S.; Preindl, M.; Babak, N.M.; Emadi, A. Low speed position estimation scheme for model predictive control with finite control set. In Proceedings of the IECON 2016—42nd Annual Conference of the IEEE Industrial Electronics Society, Florence, Italy, 23–26 October 2016; pp. 2839–2844. [CrossRef]
20. Nalakath, S.; Sun, Y.; Preindl, M.; Emadi, A. Optimization-Based Position Sensorless Finite Control Set Model Predictive Control for IPMSMs. *IEEE Trans. Power Electron.* **2018**, *33*, 8672–8682. [CrossRef]
21. Sun, X.; Li, T.; Zhu, Z.; Lei, G.; Guo, Y.; Zhu, J. Speed Sensorless Model Predictive Current Control Based on Finite Position Set for PMSHM Drives. *IEEE Trans. Transp. Electr.* **2021**, *7*, 2743–2752. [CrossRef]
22. Kumar, S.; Sreejith, R.; Singh, B. Sensorless PMSM EV Drive using Modified Enhanced PLL Based Sliding Mode Observer. In Proceedings of the International Conference on Sustainable Energy and Future Electric Transportation (SEFET), Hyderabad, India, 21–23 January 2021; pp. 1–6. [CrossRef]
23. Peng, W.; Qiao, M.; Jiang, C.; Lu, X.; Zhu, X. Vibration analysis and dynamic performance improvement of high-frequency injection method. *J. Power. Electron.* **2021**, *21*, 364–375. [CrossRef]
24. Tian, B.; Molinas, M.; An, Q.; Zhou, B.; Wei, J. Freewheeling Current-Based Sensorless Field-Oriented Control of Five-Phase Permanent Magnet Synchronous Motors Under Insulated Gate Bipolar Transistor Failures of a Single Phase. *IEEE Trans. Ind. Electron.* **2022**, *69*, 213–224. [CrossRef]
25. Sreejith, R.; Singh, B. Sensorless Predictive Current Control of PMSM EV Drive Using DSOGI-FLL Based Sliding Mode Observer. *IEEE Trans. Ind. Electron.* **2021**, *68*, 5537–5547. [CrossRef]
26. Xu, W.; Jiang, Y.; Mu, C.; Blaabjerg, F. Improved Nonlinear Flux Observer-Based Second-Order SOIFO for PMSM Sensorless Control. *IEEE Trans. Power Electron.* **2019**, *34*, 565–579. [CrossRef]
27. Ahn, H.; Park, H.; Kim, C.; Lee, H. A Review of State-of-the-art Techniques for PMSM Parameter Identification. *J. Electr. Eng. Technol.* **2020**, *15*, 1177–1187. [CrossRef]
28. Wang, Z.; Chai, J.; Xiang, X.; Sun, X.; Lu, H. A Novel Online Parameter Identification Algorithm Designed for Deadbeat Current Control of the Permanent-Magnet Synchronous Motor. *IEEE Trans. Ind. Appl.* **2022**, *58*, 2029–2041. [CrossRef]
29. Liu, X.; Pan, Y.; Wang, L.; Xu, J.; Zhu, Y.; Li, Z. Model Predictive Control of Permanent Magnet Synchronous Motor Based on Parameter Identification and Dead Time Compensation. *Prog. Electromagn. Res. C* **2022**, *120*, 253–263. [CrossRef]
30. Yu, H.; Wang, J.; Xin, Z. Model Predictive Control for PMSM Based on Discrete Space Vector Modulation with RLS Parameter Identification. *Energies* **2022**, *15*, 4041. [CrossRef]
31. Wang, S.; Yang, K.; Chen, K. An Improved Position-Sensorless Control Method at Low Speed for PMSM Based on High-Frequency Signal Injection into a Rotating Reference Frame. *IEEE Access* **2019**, *7*, 86510–86521. [CrossRef]
32. Gu, S.-M.; He, F.-Y.; Zhang, H. Study on Extend Kalman Filter at Low Speed in Sensor less PMSM Drives. In Proceedings of the 2009 International Conference on Electronic Computer Technology, Macau, China, 20–22 February 2009; pp. 311–316. [CrossRef]

machines MDPI

Article

Design, Analysis and Experimental Research of Dual-Tendon-Driven Underactuated Gripper

Yunzhi Zhang [1], Dingkun Xia [2], Qinghua Lu [1,*], Qinghua Zhang [1], Huiling Wei [1] and Weilin Chen [1]

[1] School of Mechatronic Engineering and Automation, Foshan University, Foshan 528225, China
[2] Dalian Hi-Think Computer Technology, Corp., Dalian 116000, China
* Correspondence: qhlu@fosu.edu.cn

Abstract: To improve the adaptive clamping performance of traditional single-tendon-driven underactuated grippers for grasping multiple categories of objects, a novel dual-tendon-driven underactuated gripper is proposed in this paper. First, two independent tendons with different winding paths are designed in the gripper to realize the changeable resultant moment of the end knuckle rotating joint and the movement sequences of gripper knuckles driven by different tendons are analysed too. Then, some kinematic analysis and dynamical simulations are carried out to verify the validation of the knuckle structure and dual-tendon winding path design. At last, a prototype of the novel gripper is manufactured and some grasping experiments are carried out on multiple categories of objects, with different sizes and shapes. The experimental results show that all the objects can be clamped tightly. Compared with the traditional single-tendon-driven gripper, the novel one can achieve a more flexible grasping operation and a larger end clamping force, which are more suitable for the adaptive grasping requirements of robotic automatic sorting.

Keywords: tendon-driven gripper; underactuated gripper; adaptive grasp; robotic automatic sorting; industrial automation

Citation: Zhang, Y.; Xia, D.; Lu, Q.; Zhang, Q.; Wei, H.; Chen, W. Design, Analysis and Experimental Research of Dual-Tendon-Driven Underactuated Gripper. *Machines* **2022**, *10*, 761. https://doi.org/10.3390/machines10090761

Academic Editors: Lorand Szabo and Feng Chai

Received: 3 August 2022
Accepted: 30 August 2022
Published: 2 September 2022

1. Introduction

In the past few decades, to meet the diverse needs of different customers, the traditional single production mode has been transformed into the intelligent manufacturing mode, which puts forward high requirements for the flexible production capacity of a production line [1]. For flexible manufacturing, it is necessary to improve the universal clamping capability for multi-category targets of industrial robot grippers [2]. With simple structure, convenient control and strong adaptive grasping ability, tendon-driven underactuated grippers have been studied for a long time and many research results have been obtained [3–5]. In addition, more application scenarios for tendon-driven grippers have been developed, such as industrial field defect detection [6] and agricultural fruit picking [7].

According to the literature, the underactuated grippers have been researched since the 1970s. In 1977, Hirose proposed and designed an underactuated gripper for a multi-finger tendon driver. By studying the relationship between the number of gripper knuckles and the gripping torque, the relationship between the end clamping force and the length of the driving tendon was verified at the theoretical level [8]. In 2008, Gosselin et al. [9] designed a gripper using a tandem rotatable pulley as an intermediate transmission unit. By adjusting the position of the rotatable pulley in stages, the input torque can be evenly distributed to the end of each finger, thereby minimizing the variation in the clamping force. In 2011, Odhner [10,11] designed a two-finger gripper with flexible joints and multiple drive sources, in which each knuckle is controlled by a separate drive source. Further, an algorithm is used to control the cooperative action of each driving source to realize the multi-finger cooperative grasping action, which is difficult to achieve using general grippers. Aukes et al. [12,13] increased the grasping stiffness in a gripper by adding electronic brakes at the joints of each finger, preventing the gripper from loosening or dislodging the target. In 2013, Ciocarlie et al. [14,15] added a restraining tendon and a stretching tendon to each finger and used a gradient optimization algorithm to optimize the parameters as tendon stiffness, tendon path and gripper size to enable the gripper to perform a variety of clamping actions in a larger clamping range. However, due to the single transmission path at the end clamping force, this kind of gripper cannot achieve a wide range of end clamping force adjustments. In addition, the shape memory alloy material can also be used to change the clamping stiffness of the gripper. In 2015, Firouzeh et al. [16,17] used shape memory alloy material to make the rotary joint of the gripper and adjusted the rotating joint stiffness of the gripper by changing the temperature in the shape memory alloy to achieve variable stiffness gripping. To improve the clamping accuracy of a tendon-driven gripper, Palli et al. [18] proposed a friction compensation mechanism, which can be used to eliminate the influence of position errors at the joints of the gripper on the end positioning accuracy. In addition, a variety of optimization research on the structural parameters has also been carried out to improve the clamping performance of tendon-driven grippers. In 2018, a genetic algorithm [19,20] was used to optimize the finger joint length, tendon winding path, guide wheel diameter and also the shape of the gripper fingertip, to meet the universal grasping requirements of more types of targets and improve the grasping stability. In 2020, Ma Tao et al. [21] designed and optimized a three-finger tendon-driven underactuated gripper, which can achieve both fingertip clamping and envelope clamping actions. In the same year, Bao Jialei et al. [22] designed an underactuated gripper based on metamorphic theory, which improved the gripping stability by making the end knuckle of the gripper completely passively enveloped. The metamorphic theory was also used by Li et al. [23] to design a three-knuckle gripper, which has two working modes. When grasping some tiny objects, the gripper works in linear parallel pinching mode and can be regarded as a two-joint finger with better stability; when grasping some larger objects, the gripper works in an adaptive grasping mode and can be regarded as a three-joint finger with larger clamping range. On the other hand, in order to expand the clamping range of a tendon-driven gripper, a length-adjustable linkage mechanism [24] is integrated between two rotatory joints of the gripper knuckle, which is controlled by an antagonistic tendon pair.

In summary, optimizing the structural parameters and the control algorithm can improve the clamping performance of existing single-tendon-driven underactuated grippers, but the simple and rigid clamping postures cannot meet the universal grasping requirement of multi-category targets in the automatic robotic sorting scenarios. Although installing additional drive motors at the gripper joints or integrating deformable mechanism into the gripper knuckles can increase the degrees of freedom in the gripper and, thus, realize complex grasping operation, this will increase the control complexity and production cost of the gripper, which is not conducive to the popular application of the gripper.

Therefore, a novel dual-tendon-driven underactuated gripper is proposed in this paper, which can achieve more types of clamping postures and larger range of end clamping force. Different from the traditional single-tendon-driven gripper, the novel gripper is driven by two independent tendons, which can be switched by forward or reverse rotation of the driving motor. Due to the different winding paths, for each driving tendon, the resultant moment of the end knuckle rotating joint is different, which will result in a different rotating direction for the end knuckle and, thereby, a different movement sequence of gripper knuckles.

The main accomplishments of this paper are listed as follows:

1. We designed a novel dual-tendon-driven underactuated gripper, which is driven by two independent tendons with different winding paths and can achieve four kinds of clamping postures.

2. We calculated and compared the resultant moments of end knuckle rotary joints driven by different tendons and analysed the conditions of knuckles moving when driven by different tendons.
3. We simulated the end clamping force and movement of gripper knuckles with different clamping postures and verified the validation of the structure and tendon winding path design.
4. We manufactured a prototype of the novel gripper and conducted some grasping experiments. The experimental result shows that four kinds of objects can be clamped tightly by the gripper proposed in this paper.

2. Gripper Structure Design

2.1. Calculate the Degrees of Freedom

Since the structure of the two-finger gripper is completely symmetrical from left to right, it is convenient to analyse and calculate the degrees of freedom for only the right-half part of the gripper.

As shown in Figure 1, the right-half part of the gripper consists of a frame, two mechanical knuckles and two plane rotary joints, which means that each finger of the gripper consists of one frame, two linkages and two rotating pairs, so its degrees of freedom can be calculated by the following equation:

$$Dof = 3 \times n - 2 \times P_L = 2 \tag{1}$$

where n represents the number of gripper linkages (frame is also considered as one linkage) and P_L represents the number of rotating pairs.

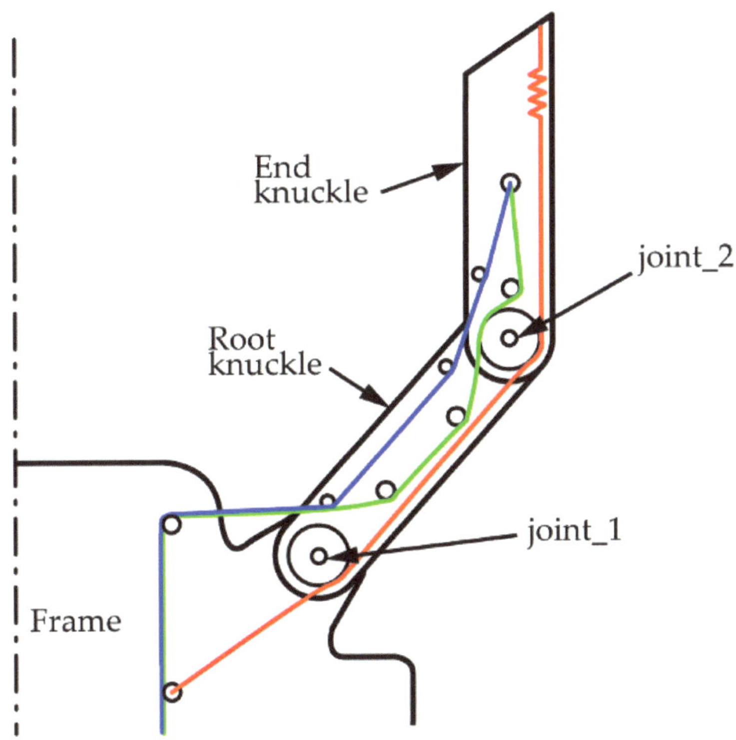

Figure 1. The structure of the right-half part of dual-tendon-driven gripper. The red line is the recovery tendon, the blue and green lines indicate the two driving tendons.

Since the gripper is driven by only one motor, the gripper designed in this paper is a typical underactuated mechanism.

2.2. Calculation of the Moment Acted on the Joint

Each knuckle is driven by the resultant moments of the rotary joint, which is generated by the driving tendon. Therefore, to study the clamping postures of the tendon-driven gripper, it is necessary to investigate the mechanism of moment generation at the knuckle rotary joints.

Figure 2 shows a schematic diagram of the two contact types between the tight driving tendon and the rotary joint of the knuckles, where the red line represents the driving tendon, the black circle with radius R represents the guide wheel mounted at the rotary joint, P_1 and P_2 are the driving tendon winding points and D_{arm} is the moment arm of the driving tendon tension.

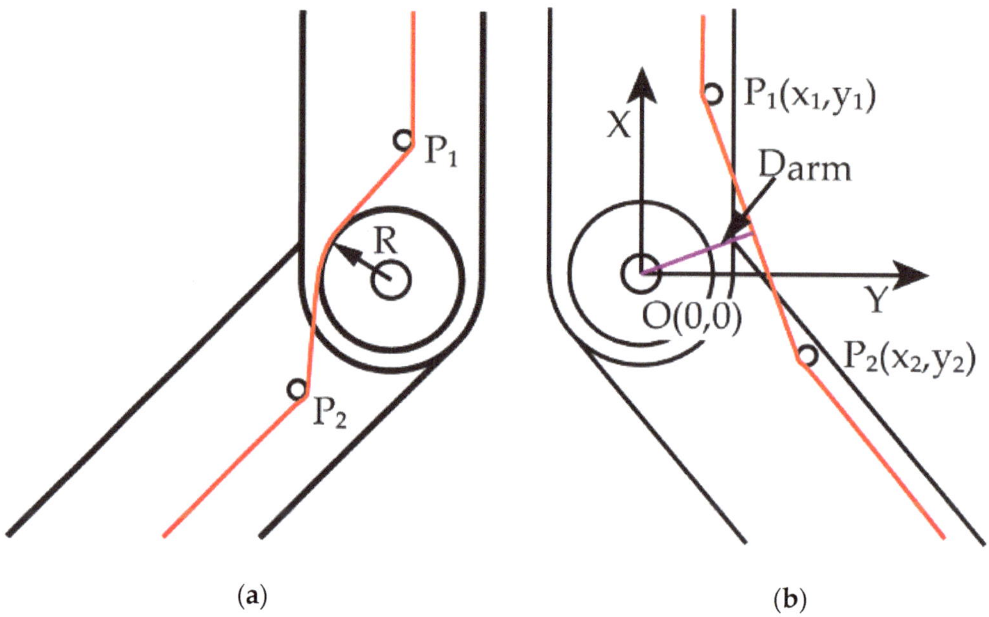

Figure 2. Two contact types between the tight driving tendon and guide wheel at rotary joint. The red line represents the driving tendon, the black circle with radius R represents the guide wheel mounted at the rotary joint. D_{arm} is the moment arm of driving tendon tension at rotatory joint, which is also explained in Appendix A. (**a**) The driving tendon is in close contact with the guide wheel; (**b**) the driving tendon is out of contact with the guide wheel.

It can be seen from Figure 2 that the moment arm of a tendon tension varies with the contact situation of the driving tendon and guide wheel. When the driving tendon is in contact with the guide wheel, the moment arm of a tendon tension is R. When the driving tendon is out of contact with the guide wheel, the moment arm of the driving tendon can be calculated by the following steps:

The formula for the line between P_1 and P_2 is

$$\frac{x - x_1}{x_2 - x_1} = \frac{y - y_1}{y_2 - y_1} \tag{2}$$

It can be transformed as

$$(y_2 - y_1)x - (x_2 - x_1)y + x_2y_1 - x_1y_2 = 0 \tag{3}$$

According to the distance formula from point to line, the moment arm of the driving tendon tension can be calculated as follows:

$$D_{arm} = \frac{x_2 y_1 - x_1 y_2}{\sqrt{(y_2 - y_1)^2 + (x_2 - x_1)^2}} \tag{4}$$

Above all, the value of the moment arm can be summarized as follows:

$$D_{arm} = \begin{cases} R \\ \frac{x_2 y_1 - x_1 y_2}{\sqrt{(y_2 - y_1)^2 + (x_2 - x_1)^2}} \end{cases} \tag{5}$$

Therefore, the moment generated by the driving tendon tension on the joint is:

$$T = F \cdot D_{arm} \tag{6}$$

where F is the value of the driving tendon tension.

2.3. Force Analysis of End Knuckle at joint_2

As shown in Figure 3, the green line represents the driving tendon and the red line represents the recovery tendon. F_d, D_d, F_r, D_r represent the tension and moment arm of the driving and recovery tendons at joint_i, respectively. F_e is the reaction force of root knuckle acting on the end knuckle at joint_2.

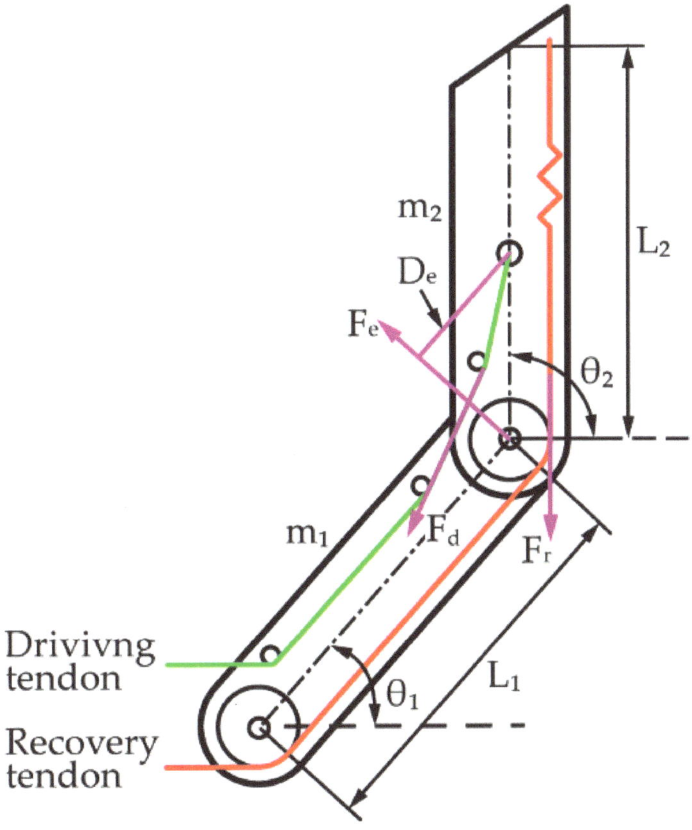

Figure 3. The force analysis of end knuckle at joint_2. The meanings of all the variables are listed in Appendix A.

When the gripper works normally, the driving and recovery tendons exert two opposite-direction tensions at the knuckle rotary joint. Therefore, the knuckle can rotate around joint_i (i = 1,2) in the corresponding direction according to the resultant moment, which can be calculated as:

$$T_i = F_d \cdot D_d - F_r \cdot D_r \ (i = 1, 2) \tag{7}$$

When the root knuckle and the end knuckle rotate together around joint_1, the moment of inertia of the whole finger can be calculated as:

$$J_t = \frac{1}{3}m_1 L_1^2 + \frac{1}{12}m_2 L_2^2 + m_2\left(L_1\cos\theta_1 + \frac{1}{2}L_2\cos\theta_2\right)^2 + m_2\left(L_1 sin\theta_1 + \frac{1}{2}L_2 sin\theta_2\right)^2 \tag{8}$$

where m_1 and m_2 represent the mass of the root and end knuckles, respectively, L_1 and L_2 represent the length of the root and end knuckle, respectively, and θ_1 and θ_2 represent the angle between each knuckle and the horizontal line, respectively.

When the root knuckle is rotating, the angular acceleration α_r can be calculated as

$$\alpha_r = T_1 / J_t \tag{9}$$

where T_1 is the combined torque of the root knuckle and the end knuckle at joint_1.

Without considering the end knuckle, if the root knuckle is rotating with an angular acceleration α_r, its own rotational inertia J_r and the required driving torque T_{req} can be calculated as below:

$$J_r = \frac{1}{3}m_1 L_1^2 \tag{10}$$

$$T_{req} = J_r \alpha_r \tag{11}$$

Therefore, the moment of the end knuckle acting on the root knuckle is:

$$T_{er} = T_1 - T_{req} \tag{12}$$

Since the forces of the root knuckle and the end knuckle at joint_2 are reaction forces, the force acting on the end knuckle by the root knuckle can be calculated as:

$$F_e = T_{er} / L_1 \tag{13}$$

The direction of F_e is perpendicular to the centreline of the root knuckle.

2.4. Movement Sequence Analysis

As shown in Figure 3, the end knuckle is enacted by three forces, such as F_d, F_r and F_e, which are enacted by the driving tendon, recovery tendon and root knuckle, respectively. Therefore, the rotating direction of the end knuckle is determined by comparing the magnitude of the torque T_d, T_r and T_e at joint_2, which are generated by F_d, F_r and F_e, respectively; the direction of T_d is anticlockwise and the directions of T_r and T_e are clockwise. The movement sequence of the end knuckle enacted by different resultant moment at joint_2 is analysed as follows:

1. If $T_d > T_r + T_e$, the resultant moment acting on the end knuckle at joint_2 is counter clockwise. Under this condition, the movement sequence of two knuckles is: the end knuckle first rotates counterclockwise around joint_2, while the root knuckle remains stationary. When the end knuckle moves to the limit position, the relative position of the root knuckle and the end knuckle no longer changes. Then, the root knuckle and the end knuckle rotate counterclockwise around joint_1 synchronously to achieve the grasping of the object.
2. If $T_d < T_r + T_e$, the resultant moment acting on the end knuckle at joint_2 is clockwise. Under this condition, the movement sequence of two knuckles is: the root knuckle

rotates counterclockwise around joint_1, while the end knuckle rotating clockwise around joint_2 simultaneously, so that the relative posture of the end knuckle to the frame remains constant. When the root knuckle is in contact with the object, the root knuckle stops rotating and the end knuckle rotates counterclockwise around joint_2 until the object is clamped.

Based on the above analysis, under the premise of constant tendon tension, if two driving tendons are designed to realize two difference moment arms at joint_2, the magnitude and direction of the resultant moment at joint_2 can both be adjusted and, thereby, the movement sequence of each knuckle can also be changed.

2.5. Tendon Winding Path Design

To achieve two different driving torques at joint_2, two different tendon winding paths are designed in this paper.

As shown in Figure 4, three tendons are arranged in the gripper, of which the blue line in driving tendon_1, the green line is driving tendon_2, and the red line is recovery tendon. Different from the fact that driving tendon_2 is tightly attached to the guide wheel, driving tendon_1 is farther away from the centre of joint_2, which means that under the premise of constant tendon tension, the moment of driving tendon_1 acting on joint_2 is larger than that of driving tendon_2. The other ends of tendon_1 and tendon_2 are wound in opposite directions on the driving motor shaft, so when motor turns in one direction, one tendon tightens and the other loosens and, thereby, the driving tendons can be switched by the forward and reverse rotation of the drive motor.

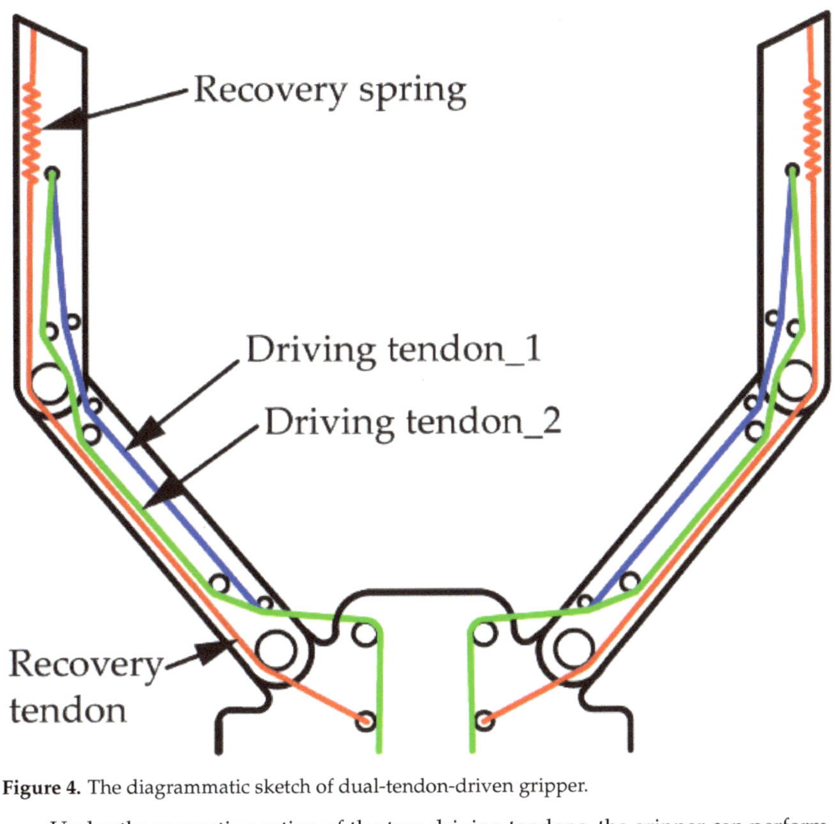

Figure 4. The diagrammatic sketch of dual-tendon-driven gripper.

Under the respective action of the two driving tendons, the gripper can perform four clamping postures, such as parallel clamping, fingertip clamping, envelope clamping and general clamping. As shown in Figure 5, the fingertip clamping and general clamping postures are achieved by driving tendon_1 and the parallel clamping and envelope clamping postures are achieved by driving tendon_2.

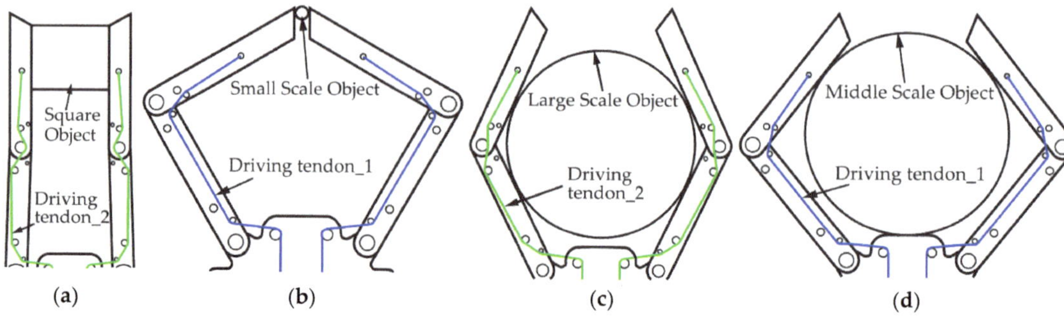

Figure 5. The clamping postures of gripper when grasping different objects. (**a**) parallel clamping; (**b**) fingertip clamping; (**c**) envelope clamping; (**d**) general clamping.

3. Dynamic Simulation of Gripper

3.1. Movement Sequence Simulation

To verify the rationality of the designed driving tendon winding paths, some dynamic simulations of the knuckles' movement sequence driven by different tendons have been carried out in this paper. As an example, the dynamic simulation of parallel clamping driven by tendon_2 is shown in Figure 6. The simulation results of four clamping postures are shown in Figure 7, where α represents the angle between the centrelines of the end knuckle and root knuckle and β represents the obtuse angle between the root knuckle and the horizontal line. According to Figure 3, α and β can be calculated as follows:

$$\alpha = \pi - \theta_2 + \theta_1 \tag{14}$$

$$\beta = \pi - \theta_1 \tag{15}$$

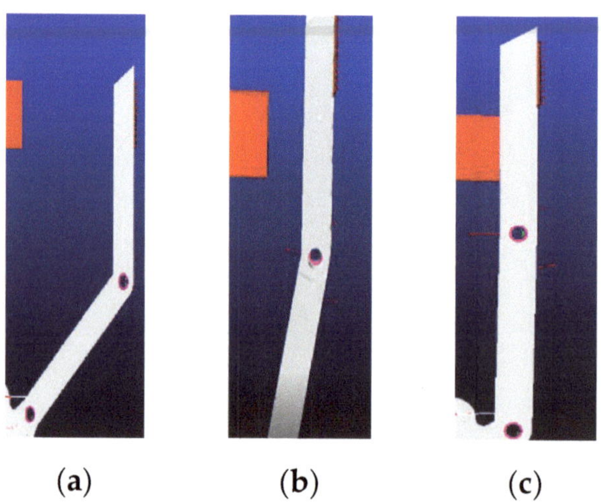

Figure 6. The dynamic simulation of the parallel clamping. (**a**) The gripper is in initial position; (**b**) the root knuckle rotates counterclockwise, while the end knuckle rotates clockwise simultaneously, thereby, the end knuckle maintains a vertical posture until it touches the target; (**c**) when the end knuckle contacts with the target, the root knuckle keeps rotating counterclockwise until the gripper clamps the target tightly.

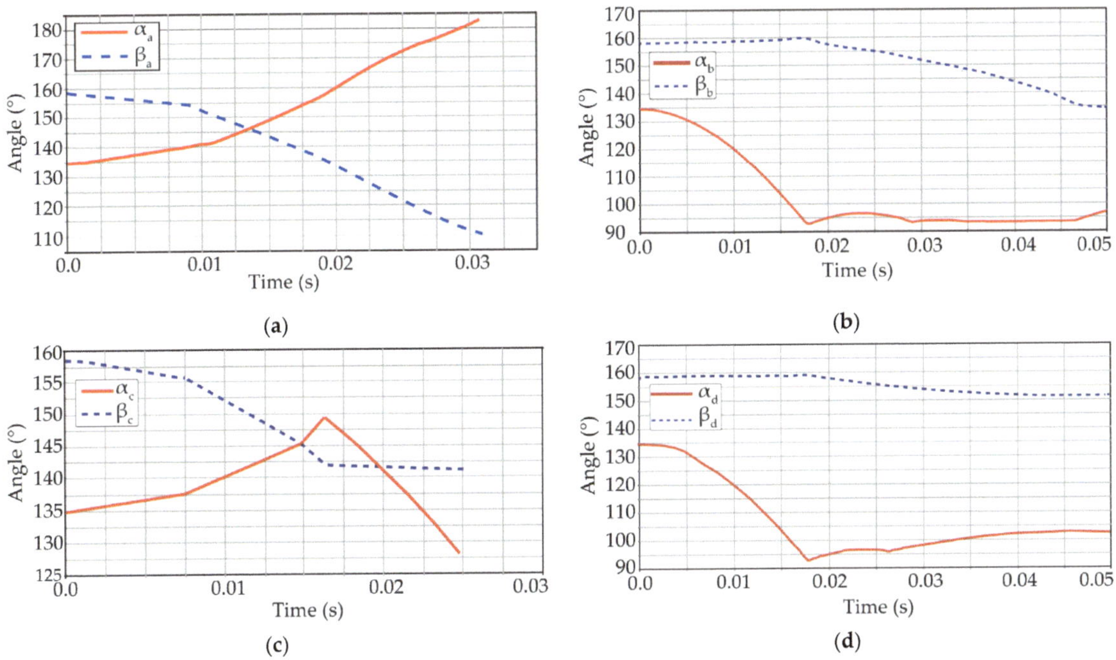

Figure 7. Motion curves of gripper knuckles under different clamping attitudes. (**a**) Parallel Clamping; (**b**) Fingertip Clamping; (**c**) Envelope Clamping; (**d**) General Clamping.

Figure 7a shows the simulated angle curves when the gripper is driven by tendon_2 for parallel clamping. When driving tendon_2 is tightened, the end knuckle starts to rotate clockwise around joint_2, while the root knuckle rotates counterclockwise around joint_1 simultaneously, so α_a keeps increasing and β_a keeps decreasing slowly. When the end knuckle contacts the target, the root knuckle keeps rotating counterclockwise around joint_1, while the end knuckle remains stationary. Therefore, α_a keeps increasing while β_{a1} keeps decreasing until the two knuckles reach their limit position, respectively, thereby, achieving tight clamping.

Figure 7b shows simulated angle curves of fingertip clamping driven by tendon_1. When tendon_1 is tightened, the end knuckle rotates counterclockwise around joint_2, while the root knuckle remains stationary, so α_b keeps decreasing and β_b remains constant. When the end knuckle reaches the limit position, the relative position between the root and end knuckles will be fixed, which means that α_b tends to be stable. As the tendon continues to be tightened, the root knuckle and the end knuckle rotate counterclockwise around joint_1 simultaneously, so β_b keeps decreasing until the target is tightly clamped.

Figure 7c shows the simulated angle curves of envelope clamping driven by tendon_2. When tendon_2 is tightened, the end knuckle starts to rotate clockwise around joint_2, while the root knuckle rotates counterclockwise around joint_1 simultaneously, so α_c keeps increasing while β_c keeps decreasing. Since the target is big, the root knuckle touches the target first, then it will stop, so β_c remain constant. As the tendon continues to be tightened, the end knuckle starts to rotate counterclockwise, which means that the angle α_c will turn to decrease, until the target is clamped tightly.

Figure 7d shows the simulated angle curves of general clamping driven by tendon_1. When tendon_1 is tightened, the end knuckle rotates counterclockwise around joint_2, while the root knuckle remains stationary, so α_d keeps decreasing while β_d keeps increasing. When the end knuckle touches the target, the clamping range formed by two knuckles will be extended due to the larger size of the target. It is specifically expressed as the counterclockwise rotation of the root knuckle and the clockwise rotation of the end knuckle. Therefore, α_d keeps increasing while β_d keeps decreasing until a stable clamping is achieved.

3.2. Simulation of End Clamping Force

Under the premise of constant tendon tension, the resultant moment at joint_2 can be adjusted by switching the driving tendon and, thereby, the end clamping force of the gripper can be changed too. To compare and analyse the end clamping force driven by different tendons, envelope clamping and general clamping are chosen and simulated to grasp four different sizes of objects. The simulated results are shown in Figure 8.

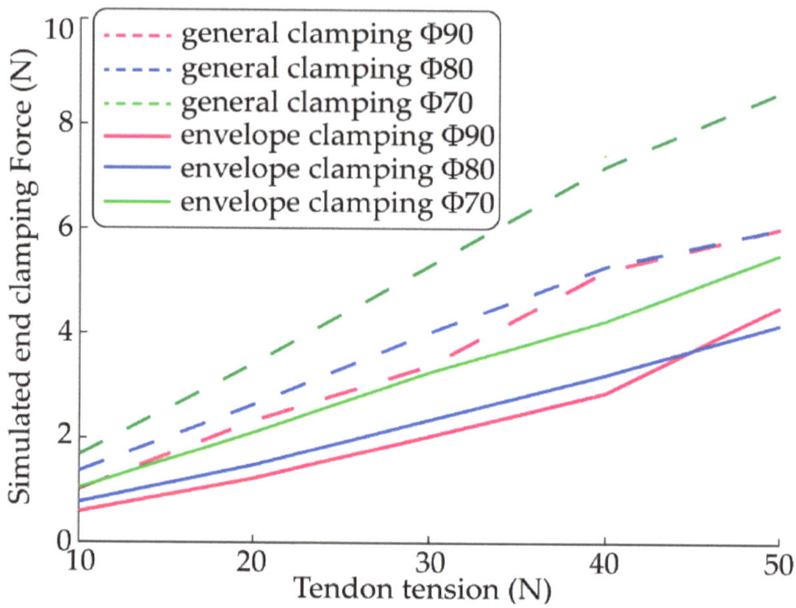

Figure 8. The simulated end clamping force of two clamping postures when grasping four different sizes of objects.

It can be seen from the simulation results that, for grasping a same-sized object, the end clamping forces of general clamping are generally larger than those of envelope clamping. Therefore, it is theoretically possible to adjust the end clamping force by changing the driving tendon.

4. Prototype Experimental Testing

4.1. Measurement of End Clamping Force

To verify the influence of the dual-tendon winding path design on the end clamping force, an experimental platform of end clamping force measurement was built and some relevant clamping experiments were carried out.

As shown in Figure 9, the experimental platform consists of a protype of the gripper, two end pressure sensors, a digital pull force gauge, a tendon tension adjuster and several objects with different sizes. Information on the sensors and equipment used in the experimental platform is shown in Table 1.

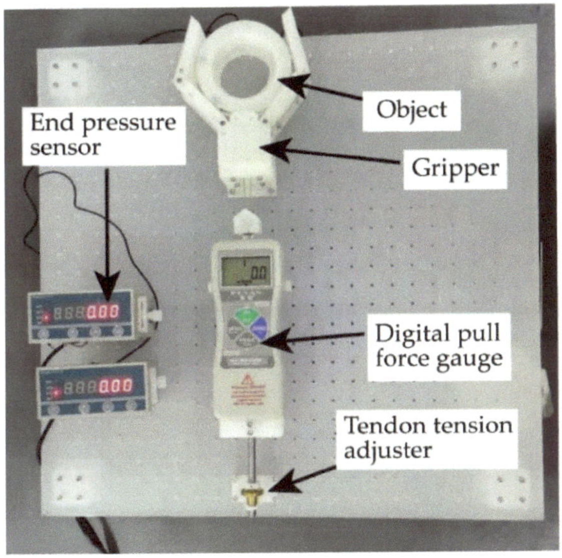

Figure 9. The experimental platform of end clamping force measurement.

Table 1. Sensor and other equipment information on the experiment platform.

Equipment	Type
End pressure sensor	(DYHW-108, DAYSEN-SOR, China)
Digital pull force gauge	(DS2-500N, PUYAN, China)
Driving motor	(42BYG34-401A, China)

Similar to the simulation, when the prototype gripper grasps an object in a specific clamping posture, for each specific tendon tension, which is adjusted by the adjuster, the relevant end clamping force is repeatedly measured ten times by the end pressure sensors. The mean values and standard deviations of FoE (End Clamping Force of Envelope Clamping) and FoG (End Clamping Force of General Clamping) are listed in Table 2 and the curves are shown in Figure 10.

Table 2. The end clamping forces measured by the sensors.

Object Size	Tendon Tension (N)	FoE (N)	FoG (N)	Increase
Φ70	10	0.76 ± 0.05	0.85 ± 0.07	11.8%
	20	1.38 ± 0.21	1.57 ± 0.13	13.8%
	30	2.51 ± 0.17	3.26 ± 0.46	29.9%
	40	3.13 ± 0.25	5.68 ± 0.74	81.5%
	50	3.77 ± 0.41	7.44 ± 0.98	97.3%
Φ80	10	0.60 ± 0.04	1.21 ± 0.13	101.7%
	20	1.38 ± 0.16	2.13 ± 0.18	54.3%
	30	2.51 ± 0.32	3.38 ± 0.27	34.7%
	40	3.14 ± 0.25	5.78 ± 0.81	84.1%
	50	3.76 ± 0.41	7.09 ± 0.89	88.6%
Φ90	10	0.61 ± 0.09	0.86 ± 0.09	41.0%
	20	1.47 ± 0.12	2.37 ± 0.21	61.2%
	30	2.59 ± 0.19	3.59 ± 0.47	38.6%
	40	3.25 ± 0.49	4.80 ± 0.43	47.7%
	50	3.58 ± 0.27	6.11 ± 0.86	70.7%

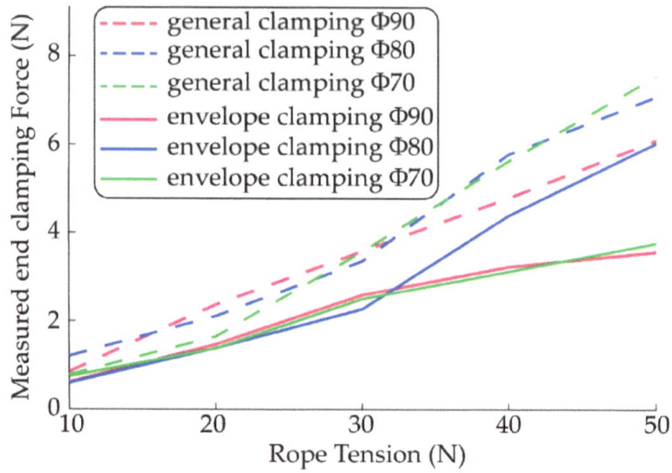

Figure 10. The measured end clamping force for grasping objects of different sizes in general clamping and envelope clamping postures.

It can be seen from the experimental results that, for grasping a same-sized object, the end clamping force of general clamping is larger than the envelope clamping, increasing by at least 11% and at most 101%, which is consistent with the simulation results. Therefore, it is feasible to adjust the end clamping force of the gripper by switching the tendon winding path.

4.2. Test of Grasping Objects

To test the performance of grasping objects of different sizes, a prototype dual-tendon-driven gripper (as shown in Figure 11) was fabricated by 3D printing and four kinds of objects with different shapes and sizes were selected as the grasping targets. By switching the driving tendon, all the targets are grasped tightly for at least 10 s by four different clamping postures, as shown in Figure 12. The dimension and mass of each target are shown in Table 3.

Figure 11. The prototype dual-tendon-driven gripper.

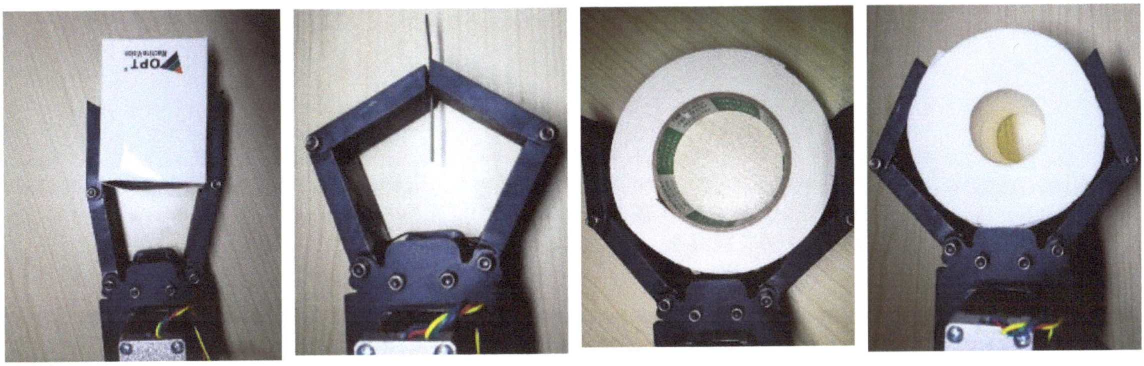

(a) (b) (c) (d)

Figure 12. Four kinds of objects are grasped by the prototype gripper. (**a**) parallel clamping; (**b**) fingertip clamping; (**c**) envelope clamping; (**d**) general clamping.

Table 3. The results of clamping experiments.

Serial Number	Target	Size (mm)	Mass (g)
Figure 12a	Square box	64.42	9.13
Figure 12b	Spanners	1.47	4.31
Figure 12c	Tapes	129.90	12.49
Figure 12d	Paper rolls	57.89	155.53

5. Conclusions

In order to meet the adaptive clamping requirements of multi-category objects on the occasion of robotic autonomous sorting, a novel dual-tendon-driven underactuated gripper is proposed in this paper. Different from the traditional single-tendon-driven gripper, the novel gripper is driven by two tendons. Since each tendon winding path is different, the resultant moment at the rotating joint of the end knuckle enacted by each driving tendon is different, which will change the movement sequence of the two knuckles. Compared with the traditional tendon-driven gripper, the novel gripper enables more flexible grasping operations and is more suitable for the sorting of multi-category objects. In addition to that, the end clamping force of the novel gripper can also be adjusted in a larger range. When grasping some objects with similar dimensions but large weight differences, the end clamping force can be enlarged by switching the driving tendon. According to the experimental results in this paper, by switching the driving tendon, the end clamping force of the gripper prototype can be increased by at least 11% and at most 101%. The specific increase magnitude may be different due to the difference in friction and structural assembly process in another experiment platform.

Although the novel gripper proposed in this paper performs well in the laboratory environment, it still has some shortcomings and limitations. The existing problems will continue be researched in subsequent studies:

1. Although the resultant moment at the rotating joint of the end knuckle can be adjusted by switching the driving tendon, this kind of method is a little discrete; a more stepless adjusting method should be studied in the future.
2. The end clamping force of a tendon-driven gripper can also be enlarged by optimising the geometric parameters of the gripper, such as the length and width of the knuckle, which is another area of research and has already been studied for some time.
3. The end clamping force of a tendon-driven gripper is hardly influenced by the magnitude of a tendon tension; slight fluctuations in tendon tension may cause a sudden release by the gripper. Therefore, a kind of anti-loosen mechanism should be added to the gripper.

4. The grasping operation of the gripper is not smooth enough, so it is necessary to improve the rotating joint structure of each knuckle to improve the clamping efficiency.

Author Contributions: Conceptualization, Y.Z. and Q.L.; Data curation, D.X.; Formal analysis, H.W.; Funding acquisition, Q.L.; Investigation, Q.Z.; Methodology, Y.Z. and D.X.; Project administration, Y.Z.; Resources, D.X.; Software, D.X.; Supervision, Q.L.; Validation, D.X. and W.C.; Visualization, Y.Z. and D.X.; Writing—original draft, Y.Z. and D.X.; Writing—review and editing, Y.Z. and Q.Z. All authors have read and agreed to the published version of the manuscript.

Funding: This research was funded by the Guangdong Province Key Field R&D Program Project: Grant No. 2021B0101410002, 2020B0404030001; Foshan City Key Field Science and Technology Research Project: Grant No. 2020001006282, 2020001006509, 2020001006297; Natural Science Foundation of Guangdong Province: Grant No. 2020B1515120070, 2021B1515120017; National Natural Science Foundation of China: Grant No. 62106048; Project of colleges and universities in Guangdong Province: Grant No. 2021KTSCX118.

Institutional Review Board Statement: Not applicable.

Informed Consent Statement: Not applicable.

Data Availability Statement: Not applicable.

Conflicts of Interest: The authors declare no conflict of interest.

Appendix A

Table A1. Index of variables and abbreviated words.

Name	Meaning	Position
R	The radius of guide wheel at the rotary joint of knuckle	Figure 2a
D_{arm}	The moment arm of driving tendon tension at rotatory joint	Figure 2b
$m_1(m_2)$	The mass of root knuckle (end knuckle)	Figure 3
$L_1(L_2)$	The length of root knuckle (end knuckle)	Figure 3
$\theta_1(\theta_2)$	The angle between the root (end) knuckle and the horizontal line	Figure 3
$F_d(F_r)$	The tension of driving tendon (recovery tendon)	Figure 3
F_e	The reaction force of root knuckle acting on the end knuckle	Figure 3
$D_d(D_r, D_e)$	The moment arm of F_d (F_r, F_e)	Figure 3
$T_d(T_r, T_e)$	The torque at joint_2 generated by F_d (F_r, F_e)	Section 2.4
α	The angle between the centrelines of end knuckle and root knuckle	Section 3.1
β	the obtuse angle between the root knuckle and the horizontal line	Section 3.1
FoE	End Clamping Force of Envelope Clamping	
FoG	End Clamping Force of General Clamping	

References

1. Agnieszka, R.; Arne, B.; Marcel, B.; Erik, S.M. The Smart Factory: Exploring Adaptive and Flexible Manufacturing Solution. *Procedia. Eng.* **2014**, *69*, 1184–1190.
2. Laliberte, T.; Birglen, L.; Gosselin, C. Underactuation in robotic grasping hands. *Jpn. J. Mach. Intell. Robot. Control.* **2002**, *4*, 1–11.
3. Nazma, E.; Suhaib, M. Tendon driven robotic hands: A review. *Int. J. Mech. Eng. Robot. Res.* **2012**, *3*, 350–357.
4. Licheng, W.; Yanxuan, K.; Xiali, L. Review and Research Issues on Underactuated Finger Mechanism. In Proceedings of the 2015 Chinese Intelligent Automation Conference, Fuzhou, China, 8–10 May 2015.
5. Vincent, B.; Clement, G. Mechanisms for Robotic Grasping and Manipulation. *Annu. Rev. Control. Robot. Autom. Syst.* **2021**, *4*, 573–593.
6. Qinghua, L.; Junmeng, L.; Lufeng, L.; Yunzhi, Z.; Wenbo, Z. A supervised approach for automated surface defect detection in ceramic tile quality control. *Adv. Eng. Inf.* **2022**, *53*, 101692.
7. Lufeng, L.; Wei, Y.; Zhengtong, N.; Jinhai, W.; Huiling, W.; Weilin, C.; Qinghua, L. In-field pose estimation of grape clusters with combined point cloud segmentation and geometric analysis. *Comput. Electron. Agric.* **2022**, *200*, 107197.
8. Hirose, S.; Umetani, Y. The development of soft gripper for the versatile robot hand. *Mech. Mach. Theory* **1978**, *13*, 351–359. [CrossRef]
9. Gosselin, C.; Pelletier, F.; Laliberte, T. An anthropomorphic underactuated robotic hand with 15 dofs and a single actuator. In Proceedings of the IEEE International Conference on Robotics and Automation, Pasadena, CA, USA, 19–23 May 2008.
10. Odhner, L.U.; Jentoft, L.P.; Claffee, M.R.; Corson, N.; Tenzer, Y.; Ma, R.R.; Buehler, M.; Kohout, R.; Howe, R.D.; Dollar, A.M. A compliant, underactuated hand for robust manipulation. *Int. J. Robot. Res.* **2014**, *33*, 736–752. [CrossRef]

11. Odhner, L.U.; Dollar, A.M. Dexterous manipulation with underactuated elastic hands. In Proceedings of the IEEE International Conference on Robotics and Automation, Shanghai, China, 9–13 May 2011.
12. Aukes, D.M.; Kim, S.; Garcia, P.; Edsinger, A.; Cutkosky, M.R. Selectively compliant underactuated hand for mobile manipulation. In Proceedings of the IEEE International Conference on Robotics and Automation, St. Paul, MN, USA, 14–18 May 2012.
13. Aukes, D.M.; Heyneman, B.; Ulmen, J.; Stuart, H.; Cutkosky, M.R.; Kim, S.; Garcia, P.; Edsinger, A. Design and testing of a selectively compliant underactuated hand. *Int. J. Robot. Res.* **2014**, *33*, 721–735. [CrossRef]
14. Ciocarlie, M.; Hicks, F.M.; Stanford, S. Kinetic and dimensional optimization for a tendon-driven gripper. In Proceedings of the 2013 IEEE International Conference on Robotics and Automation, Karlsruhe, Germany, 6–10 May 2013.
15. Ciocarlie, M.; Hicks, F.M.; Holmberg, R.; Hawke, J.; Schlicht, M.; Gee, J.; Stanford, S.; Bahadur, R. The Velo gripper: A versatile single-actuator design for envelopeing, parallel and fingertip grasps. *Int. J. Robot. Res.* **2014**, *33*, 753–767. [CrossRef]
16. Firouzeh, A.; Paik, J. Grasp mode and compliance control of an under-actuated origami gripper using adjustable stiffness joints. *IEEE-Asme Trans. Mechatron.* **2017**, *22*, 2165–2173. [CrossRef]
17. Firouzeh, A.; Salehian, S.S.M.; Billard, A.; Paik, J. An under actuated robotic arm with adjustable stiffness shape memory polymer joints. In Proceedings of the IEEE International Conference on Robotics and Automation, Seattle, WA, USA, 26–30 May 2015.
18. Palli, G.; Melchiorri, C. Friction compensation techniques for tendon-driven robotic hands. *Mechatronics* **2014**, *24*, 108–117. [CrossRef]
19. Dong, H.; Asadi, E.; Qiu, C.; Dai, J.; Chen, I.-M. Grasp analysis and optimal design of robotic fingertip for two tendon-driven fingers. *Mech. Mach. Theory* **2018**, *130*, 447–462. [CrossRef]
20. Dong, H.; Asadi, E.; Qiu, C.; Dai, J.; Chen, I.-M. Geometric design optimization of an under-actuated tendon-driven robotic gripper. *Robot. Comput. Manuf.* **2018**, *50*, 80–89. [CrossRef]
21. Ma, T.; Yang, D.; Zhao, H.; Ai, N. Grasp analysis and aptimal design of a new underactuated manipulator. *Chin. J. Robot.* **2020**, *42*, 354–364.
22. Bao, J.; Han, K.; Zheng, C.; Cao, Y.; Sun, K. Design of underactuated manipulator based on metamorphic theory. *Chin. J. Mech. Drive* **2020**, *44*, 90–93.
23. Li, X.; Guo, J.; Sun, W.; Shang, D.; Wen, B. Design and Contact Force Analysis of Under-actuated Manipulator with Hybrid Working Mode. *Chin. J. Mech.-Eng.* **2021**, *57*, 8–18.
24. Schee, J.Y.; Minsik, C.; Bomin, J.; Yong, L.P. Elongatable Gripper Fingers With Integrated Stretchable Tactile Sensors for Underactuated Grasping and Dexterous Manipulation. *IEEE Trans. Robot.* **2022**, *38*, 2179–2193.

applied sciences

MDPI

Article

A Novel Variable-Proportion Desaturation PI Control for Speed Regulation in Sensorless PMSM Drive System

Zihan Wei, Mi Zhao *, Ximu Liu and Min Lu

College of Mechanical and Electrical Engineering, Shihezi University, Shihezi 832003, China; weizihan615@163.com (Z.W.); lzy3301357367@163.com (X.L.); lm_shz@163.com (M.L.)

* Correspondence: zhaomi@shzu.edu.cn

Featured Application: Elimination of integral saturation and optimization of dynamic performance of sensorless drive system of permanent magnet synchronous motor based on SMO-PLL observation.

Abstract: The problem of integral saturation greatly restricts the engineering application of PID controller in AC speed regulator systems, which also affects the control performance of sensorless drive system with the phase-locked loop (PLL) structure. To address this, this paper proposes a novel variable-proportion desaturation PI (VPDPI) control method for permanent magnet synchronous motor (PMSM) sensorless drive system. Firstly, an overall scheme of sensorless control system with sliding mode observer (SMO)-PLL observer is presented, which includes the equivalent mathematical model and its parameter identification model of PMSM. Secondly, the control principle of the VPDPI algorithm is innovatively designed. Note that the novel regulator scheme consists of two interconnected components in terms of the concepts of threshold segmentation, i.e., the variable-proportion link and desaturation PI part. Meanwhile, the stability analysis of the regulator is further discussed by the root locus and the Bode diagram analysis. Finally, the numerical simulations of speed regulation are carried out under the various operation conditions, and the effectiveness of the proposed method is further verified by experimental platform. Several groups of comparative experiments reveal that the proposed method has a more higher control performance for the sensorless drive system, especially in the ability of overshoot suppression.

Keywords: permanent magnet synchronous motor (PMSM); sensorless; variable-proportion desaturation PI (VPDPI); integral saturation; observation performance

Citation: Wei, Z.; Zhao, M.; Liu, X.; Lu, M. A Novel Variable-Proportion Desaturation PI Control for Speed Regulation in Sensorless PMSM Drive System. *Appl. Sci.* **2022**, *12*, 9234. https://doi.org/10.3390/app12189234

Academic Editors: Lorand Szabo and Feng Chai

Received: 3 August 2022
Accepted: 7 September 2022
Published: 15 September 2022

1. Introduction

Permanent magnet synchronous motor (PMSM) has been widely used in electric vehicle, ship electric propulsion system and other driving fields because of its advantages of simple structure, high power density and high operating efficiency [1,2]. With the development of electronic technology, PMSM drive structures are also constantly updated to adapt to the needs of different operating conditions [3,4]. In the detection of traditional drive system, the installation position of mechanical sensors such as photoelectric encoder and rotary transformer is not accurate. In addition, the aging of sensors restricts the drive system to achieve satisfactory control performance, which reduces the system reliability and economics of production [5,6]. In order to solve the influence of traditional mechanical sensors on the drive system, the sensorless idea is developed and applied in PMSM drive system. PMSM sensorless drive system can construct the rotor position and speed information required by detecting the voltage and current signals of the motor port [7,8].

At present, the sensorless control technology of permanent magnet synchronous motor can be roughly divided into two types based on fundamental wave observation and salient pole effect [9,10]. The observation method based on salient pole effect has better observation performance for low speed operation condition [11,12]. However, it is a pity that this method has some limitations for hidden pole machine [13,14]. Therefore, the method based on fundamental wave observation is still used to obtain PMSM rotor information in more control occasions. The fundamental wave observation method is divided into four types: arc-tangent method, phase-locked loop (PLL) method, model reference adaptive method and intelligent algorithm according to the different methods of obtaining rotor information [15,16]. Among them, the model reference adaptive method has the characteristics of simple structure. However, the accuracy of the observation is closely related to the design rules of the adaptive law, which is difficult to be applied in practical systems [17,18]. In addition, it is difficult to realize in the actual system. Meanwhile, intelligent algorithm observation rotor information accuracy is closely related to the number of training samples, the control structure is very complex [19,20]. On the contrary, the arc-tangent method can use the sliding mode variable structure method to obtain the inverse electromotive force information of the motor rotor, and directly construct the arc-tangent equation to obtain the required position information [21,22]. However, this method has a low signal noise ratio (SNR) and is greatly affected by back EMF harmonics [23,24].

Fortunately, the closed-loop feedback of rotor back EMF information based on phase-locked loop method is constructed by introducing PI controller [25,26]. In the presence of harmonics, closed-loop self-adjustment process can be adopted to alleviate the disadvantages of arc-tangent method [27,28]. It can be seen that the phase-locked loop method is an observation method with relatively good comprehensive performance. Meanwhile, PI regulator introduced by phase-locked loop method also aggravates the problem of integral saturation and integral drift of the whole system. Therefore, in order to achieve better performance of the sensorless system, it is necessary to perform integral saturation suppression on the driving system. In terms of integral saturation suppression, most engineering uses anti-saturation methods, such as variable speed integration and integral separation, to suppress the integral saturation state of the drive system [29–31]. In [32], the influence of saturation phenomenon has been suppressed to a certain extent, nonetheless it does not eliminate the accumulated integral error. Hence, the anti-saturation method is easy to fall into integral saturation again. In order to completely eliminate the integral saturation of the system, the desaturated PI control structure is developed based on the idea of anti-saturation [33–35]. The method can completely remove the integral from the saturation state through the reverse accumulation of errors. However, it should be noted that if the reverse error accumulation is too large, it is easy to weaken the role of the integral link, so that the drive system has steady-state error and increases the response time of the system.

In order to make PMSM sensorless system have better observation and drive performance, the original contributions of this paper are shown as follows:

(1) Compared with the traditional model parameter equivalence, this paper uses MATLAB/System Identification toolbox to realize the parameter identification of high-order system, and the identification accuracy can reach 91.22%. Meanwhile, root locus and Bode diagram are used to complete the stability analysis of the closed-loop system and the selection of VPDPI regulator parameters.
(2) In terms of the ideal response curve as the design objective, this paper analyzes the causes of system saturation phenomenon. Moreover, the influence of saturation phenomenon on dynamic and steady performance of sensorless system is considered in this paper, especially the ability to suppress overshoot, and the VPDPI regulator is innovatively proposed. Meanwhile, the control rules of the proposed regulator are further given.
(3) Simulations and experiments under various working conditions are presented in detail, and the effectiveness and reasonability of the proposed method are both amply verified.

In Section 2, the overall structure and mathematical model of PMSM sensorless drive system are presented step by step. In Section 3, the requirements of ideal response curve are reviewed, and the design rules and threshold parameter selection rules of VPDPI regulator are given in detail. Section 4 provides the comprehensive test results of the proposed method on MATLAB/Simulink experimental platform. Meanwhile, the validity of the proposed method on hardware platform is further verified in Section 5. Finally, the brief conclusions are drawn in Section 6.

2. The Overall Structure of PMSM Sensorless Drive System

2.1. Equivalent of PMSM Mathematical Model

It is well known that PMSM sensorless system is composed mainly of the motor ontology, the inverse model, the current controller and the speed controller. Accurate equivalence of each component model is the basis of stability control system.

Generally speaking, the voltage equation of the PMSM in the $d-q$ rotary transformer can be expressed as (1):

$$\begin{cases} u_d = Ri_d + L_d\frac{di_d}{dt} - \omega_e L_q i_q \\ u_q = Ri_q + L_q\frac{di_q}{dt} + \omega_e(L_d i_d + \psi_f) \end{cases} \tag{1}$$

Among them, u_d and u_q are the $d-q$ axis component of the stator voltage. i_d and i_q are the $d-q$ axis component of the stator current; R is the stator resistance. ω_e is electric angular velocity. L_d and L_q are the inductance component of $d-q$ axis. ψ_f stands for permanent magnet flux.

When the $i_d = 0$ control strategy is used, the voltage equation can be reduced as shown in (2):

$$\begin{cases} u_d = -\omega_e L_q i_q \\ u_q = Ri_q + L_q\frac{di_q}{dt} + \omega_e\psi_f \end{cases} \tag{2}$$

When there is no external load disturbance, the torque equation and the motion equation of the motor will be written as [36]:

$$\begin{cases} J\frac{d\omega_m}{dt} = T_e + T_L - B\omega_m \\ T_e = \frac{3}{2}P_n i_q \psi_f \end{cases} \tag{3}$$

In (3), ω_m is mechanical angular velocity. B is the damping coefficient. J is usually thought of as the moment of inertia. T_e and T_L are electromagnetic torque and load torque, respectively. P_n is polar logarithm of PMSM.

After the laplace-transform, the transfer function of the PMSM electrical part G_{PMSMe} and mechanical part G_{PMSMm} can be equivalent to:

$$\begin{cases} G_{PMSMm}(s) = \frac{1}{L_q s + R} \\ G_{PMSMm}(s) = \frac{1}{Js + B} \end{cases} \tag{4}$$

It is obviously that PMSM are a second order system composed of two inertial processes.

Meanwhile, the phase voltage of the inverter module output is not in response to the input vector pulse synchronization, and the inverter does not do the voltage amplitude scale transformation. Hence, the inverter gain $K_{in} = 1$, and it is the initial equivalent of the driving part:

$$G_{SVPWM}(s) = \frac{K_{in}}{T_s s + 1} \tag{5}$$

where T_s is defined the switching frequency.

Unfortunately, the inverter is generally used to switch dead areas and delay in action. Therefore, the small inertia link is compensated for (5). The transfer function for an inertia element can be written:

$$G_{ine}(s) = \frac{1}{t_d s + 1} \tag{6}$$

In (6), t_d is the inertia time constant.

If the current controller is equivalent to a differential link and an integral link, the transfer function block diagram of PMSM sensorless system can be expressed in Figure 1.

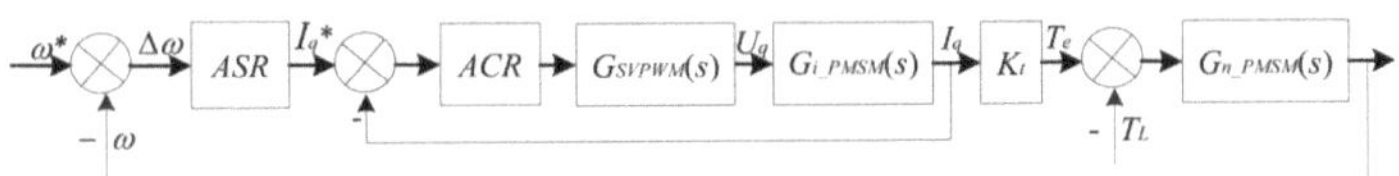

Figure 1. The transfer function block diagram of PMSM sensorless system.

It is worth noting that the speed regulator control object $G_{obj}(s)$ is made up of five poles and a zero. Among them, ASC and ACR are the automatic speed regulator and the automatic current regulator, respectively.

2.2. Parameter Identification of the Drive System

In order to better fit the actual mathematical model of the sensorless system, the sensorless system is equivalent to a higher-order system in this paper. However, due to the appropriate condition neglect and approximation in the equivalent process, there is a certain gap between the transfer function obtained directly according to the parameters and the real value. Therefore, MATLAB system identification toolbox can be utilized to identify the known parameters of the zero-pole system. Figure 2 is used as the equivalent control circuit for identifying the unknown parameters.

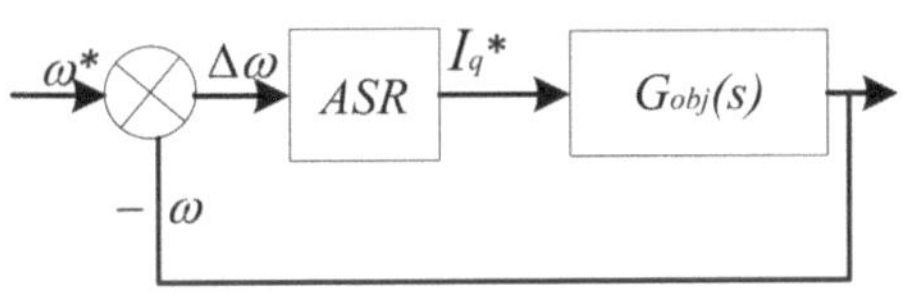

Figure 2. The equivalent control circuit.

Where $G_{obj}(s) = \frac{k(s-z)}{\prod(s-p_i)}(i = 1\ldots5)$.

General steps to realize unknown parameter Identification of equivalent control object using MATLAB/System Identification toolbox:

Step 1: obtaining identification input signal I_{qref} and output signal ω_m, as shown in Figure 3;

Step 2: calling toolbox and import data;

Step 3: configuring the zeros and poles of the identification system (In terms of Section 2.1, the number of poles = 5, and the number of zeros = 1);

Step 4: running and analyzing the feasibility of identification results.

On the basis of the same input data, the Zero-Pole-Gain (ZPK) model of the post-identification controlled system is shown as (7), and the response is shown in Figure 4.

$$\begin{cases} z = -10.4774 \\ p = [-13405, -16.77, -92.5 \pm 368.8i, -0.0015] \\ k = 8.0228 \times 10^{11} \end{cases} \quad (7)$$

From the response curve estimated by using "sysinit" as template, it can be seen that the similarity between the response and the actual data after identification parameters is 91.22%.

Generally, it can be utilized as the equivalent transfer function of the control object in the parameter debugging process of the controller.

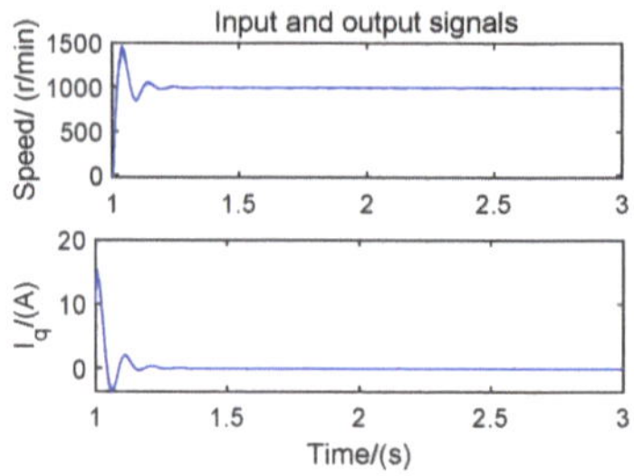

Figure 3. The output and input data of the system to be identified.

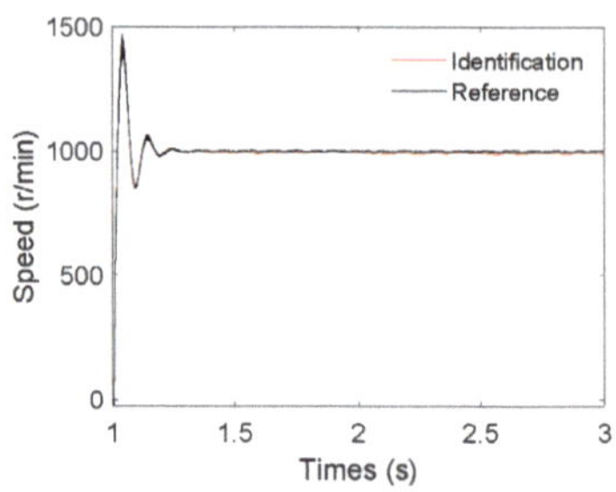

Figure 4. Identification result response curve.

2.3. The Observer Design of the Rotor Speed and Position Angle

Since the sensorless observation of arctangent method is greatly affected by the back electromotive force, the observation accuracy of position and speed is not ideal. Therefore, this paper adopts phase-locked loop to ensure the accuracy of estimation of rotor speed and rotor position Angle [37].

The construction of reliable sliding surface is the basis of effective sensorless realization. Based on PMSM voltage equation in (1), the observed current equation is constructed as follows:

$$\begin{cases} \frac{d\tilde{i_d}}{dt} = \frac{1}{L_d}(-R\tilde{i_d} + u_d + L_q\omega_e\tilde{i_q} - V_d) \\ \frac{d\tilde{i_q}}{dt} = \frac{1}{L_q}(-R\tilde{i_q} + u_q + L_d\omega_e\tilde{i_d} - V_q) \end{cases} \tag{8}$$

where V_d and V_q are defined the observed electromotive force .

If the observed electromotive force is constructed by (9), it can be used as the input of the PLL [37].

$$\begin{cases} V_d = ksgn(\tilde{i_d} - i_d) \\ V_q = ksgn(\tilde{i_q} - i_q) \end{cases} \tag{9}$$

Hence, the SMO-PLL structure is shown in Figure 5.

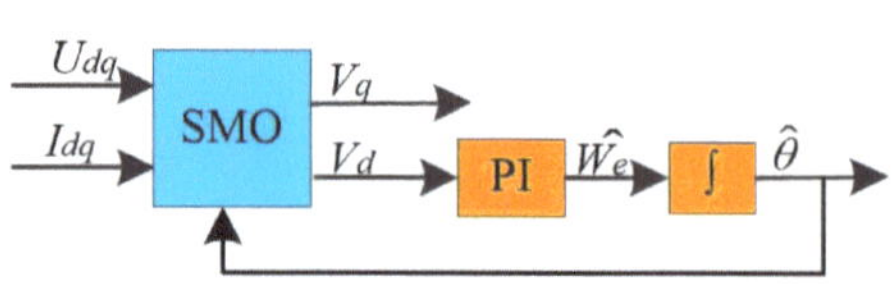

Figure 5. The structural diagram of SMO-PLL technology.

It can not be ignored that the phase-locked loop uses PI controller to achieve accurate position Angle tracking, but also aggravates the system integral saturation phenomenon, which will enhance the speed control performance of system overshoot and stability.

3. Proposed VPDPI Speed Regulator

3.1. The Design of the VPDPI Regulator

There is no doubt that the proportion link can quickly amplify the speed error signal, and it can quickly adjust the speed to the reference value. Unfortunately, the single proportion link will make the system have a certain steady-state error, which can not meet the performance requirements of the most control systems. Fortunately, the integration link is extremely effective in eliminating static errors. However, the control system has a new integral saturation problem since the accumulation of error signals, which leads to the increase of overshoot and even out of control. In order to ensure the control performance of PI controller, it is highly necessary to eliminate the integral saturation phenomenon in the integral link. At present, the anti-saturation ideas is widely used to suppress integral saturation phenomenon, which weakens the negative impact of saturation problem but has not been fundamentally solved. Different from the traditional controller design ideas, a new regulator designed in this paper is based on thoroughly eliminating integral saturation phenomenon and the actual performance requirements of fast response. The ideal response curve with fast response and no overshoot is shown in Figure 6. Therefore, a novel VPDPI regulator is further designed.

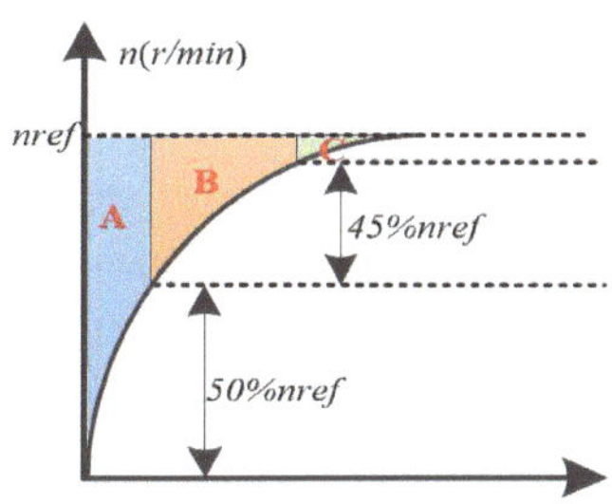

Figure 6. The curve of the ideal response.

In details, the dynamic response process of the controlled system can be simply divided into three stages according to the ideal response of the system (i.e., Stage A, B and C). The Stage A is called as the fast rising zone, in which the speed error signal is large and the integration process rapidly accumulates the speed error. Therefore, this stage will be the main process leading to the phenomenon of integral saturation. The Stage B is defined as the steady-state regulation zone, in which the system response transitions from dynamic regulation to steady-state process, and the integral regulator in the steady-state regulation zone still accumulates a certain speed error, which can play the role of the integral link in this stage to eliminate the system deviation. The Stage C is generally considered to be the stability zone, in which the system response is within the error band and runs stably around a given speed.

In order to realize the actual output tracking in terms of the ideal output curve, the primary tasks in Stage A have to ensure that the integral will not fall into saturation state. Hence, the logical threshold method is used to judge the speed error in real time, and the feedback compensation coefficient γ ($\gamma < 0$) is introduced in Stage A. In Stage B, the rotation speed should be adjusted smoothly to the steady-state interval, and the traditional PI control idea can be adopted without additional compensation process. Stage C is in the steady-state interval, and the response speed can be appropriately improved at this stage. On the one hand, the effect of integral link is not obvious due to too much desaturation, and on the other hand, the response speed of the system can be significantly improved. The structure diagram of VPDPI control method proposed in this paper is shown in Figure 7.

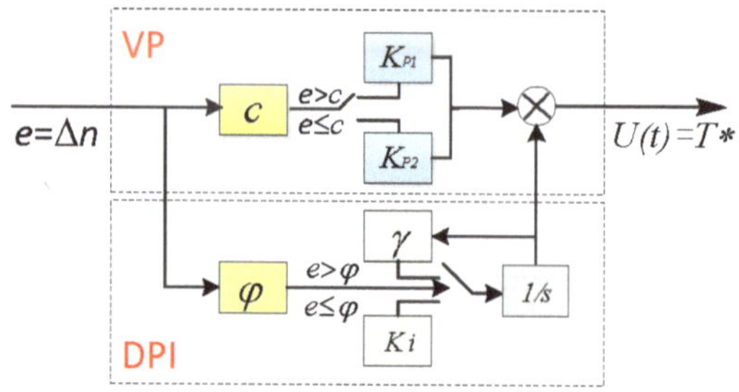

Figure 7. The structure diagram of VPDPI regulator control algorithm.

Where K_{p1} and K_{p2} are proportional coefficients, respectively. K_i is the integral coefficient. In terms of the ideal response curve, c is the threshold used to determine whether to change the proportionality coefficient. Meanwhile, the threshold φ is used to judge the saturation condition.

In practical application, the control parameter configuration of the proposed VPDPI regulator can usually be divided into the following steps.

(1) completing the mathematical model equivalence of PMSM drive system;
(2) analyzing the influence of different control parameters (such as K_{p1}, K_{p2} and K_i) on the stability;
(3) selecting the control parameters according to different drive performance requirements and observation performance;
(4) verifying the validity of the selected parameters under different conditions.

Remark 1. *The specific steps for selecting threshold c, threshold φ and the feedback compensation coefficient γ are as follows.*

(1) According to the ideal curve, the threshold c_{ref} should be selected around $0.95n_{ref}$. Meanwhile, the threshold φ_{ref} should be selected near $0.5n_{ref}$.
(2) The effects of different threshold values and the feedback compensation coefficient on response performance and observation performance of the system should be analyzed.
(3) Fine-tune the selected threshold near the reference value c_{ref} and φ_{ref}, respectively.
(4) Adjust the feedback compensation coefficient γ ($\gamma < 0$) to ensure that the cumulative error is eliminated.
(5) Verify the effect of the selected thresholds under different conditions.

It is obviously that the VPDPI regulator can be divided into VP regulator and DPI regulator. When the speed error e is higher than the threshold φ, the compensation coefficient γ ($\gamma < 0$). The rest is governed by the normal integral coefficients K_i. Similarly, when the speed error e is higher than the threshold c, VP link moves. Meanwhile, K_{p2} acts as the proportional coefficient of PI controller ($K_{p2} > K_{p1}$).

Therefore, the VPDPI control law can be expressed as shown in (10):

$$u(t) = [\rho(K_{p2} - K_{p1}) + K_{p1}]e + \gamma K_i \int edt \tag{10}$$

In (10), ρ is defined the selection coefficient, and if $e > c$ then $\rho = 0$, otherwise $\rho = 1$.

3.2. The Stability Analysis of Regulator

Based on the transfer function block diagram in Figure 1 and the system identification results G_{obj} in (7), it can be seen that the open-loop transfer function $G_{open}(s)$ of the sensorless system can be equivalent to below:

$$G_{open}(s) = \frac{K_p(s + K_i)}{s} G_{obj}(s) \tag{11}$$

According to (11), the driving system designed in this paper is equivalent to a sixth-order system, which has six open-loop poles and two open-loop zeros. Therefore, it is a better choice to use root locus analysis system stability. When the PI parameters are standardized, the open-loop transfer function of the system can be further written as follow:

$$G_{open}(s) = \frac{K_c(8.023 \times 10^{10}s^2 + 9.208 \times 10^{11}s + 8.406 \times 10^{11})}{s^6 + 1.4 \times 10^4 s^5 + 2.9 \times 10^6 s^4 + 2 \times 10^9 s^3 + 3.3 \times 10^{10} s^2 + 4.8 \times 10^7 s} \quad (12)$$

In (12), K_c is the amplification factor.

The root locus of the closed system is obtained by using the root trajectory drawing rule, as shown in Figure 8. Among them, Figure 8a is the overall root trajectory, and Figure 8b,c are magnified zero-pole images.

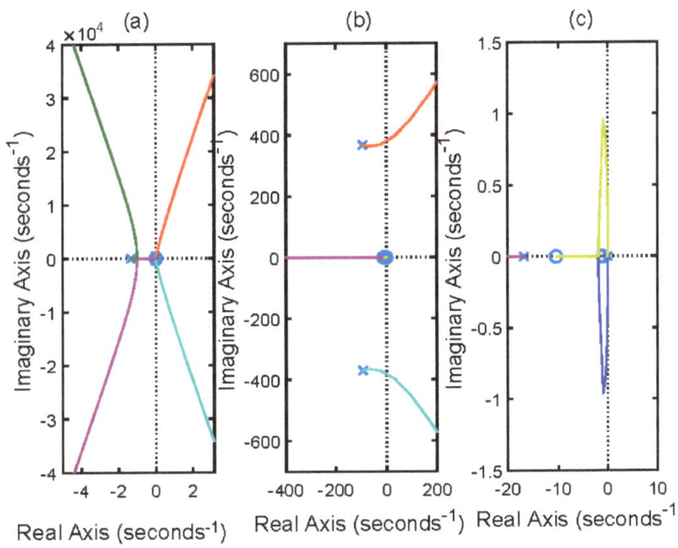

Figure 8. (**a**) The overall trend of the root locus. (**b**,**c**) show enlarged images of different proportions of poles, respectivelyof the sensorless system.

It is well known that, to ensure the stability of the system, all root trajectories should be restricted to the left half plane of the complex frequency domain. However, it can be seen in Figure 8b that when K_c is greater than 0.45, the root locus crosses the imaginary axis and enters the right half-plane of the complex frequency domain. This means that the system is stable when the limit K_c is less than the critical value.

When the integral coefficient $K_i = 1$, the Bode diagram under different proportionality coefficients is shown in Figure 9, and the corresponding amplitude-frequency margin and phase frequency margin information are shown in Table 1.

Table 1. The characteristics with different proportionality coefficients.

Proportionality Coefficient K_p	Magnitude Margin G_m	Phase Margin P_m	Stability (Yes/No)
0	20.3 (at 65.8 rad/s)	11.1 (at 18.8 rad/s)	Yes
0.1	12.9 (at 377 rad/s)	81 (at 41.1 rad/s)	Yes
0.2	6.89 (at 378 rad/s)	83.8 (at 85.5 rad/s)	Yes
0.3	3.39 (at 378 rad/s)	78.8 (at 140 rad/s)	Yes
0.4	0.897 (at 379 rad/s)	18.4 (at 349 rad/s)	Yes
0.5	−1.04 (at 379 rad/s)	−11.9 (at 398 rad/s)	No

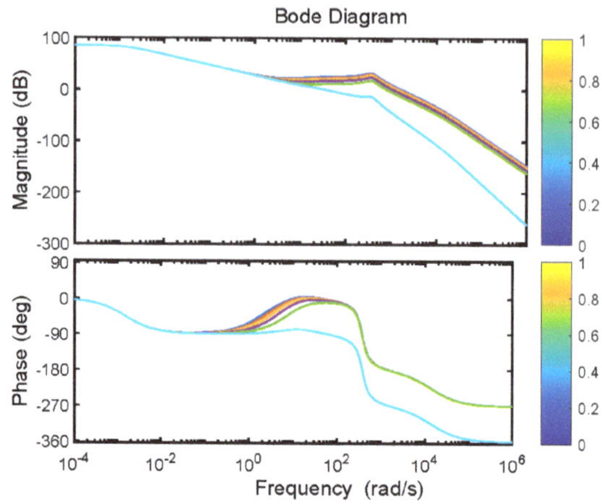

Figure 9. The bode diagram under the different k_p.

As above aforementioned results, K_p has a positive phase margin in the range of $0 \sim 0.4$, and the control system is more stable. To ensure the stability of the sensorless system, the proportionality coefficient K_{p1} and K_{p2} should be less than 0.45 when the integration coefficient K_i is set to 1. Therefore, When the median of this range $K_p = 0.2$ is selected, the different K_i is selected to discuss the optimal value of K_i in the selected data. The Bode diagram under different integral coefficients is shown in Figure 10, and the corresponding amplitude-frequency margin and phase frequency margin information are shown in Table 2.

In terms of the above results, when the system is purely proportional control, it is an unstable system. However, with the increase of the integral coefficient from 0 to 1, the stability margin of the system decreases slightly, but it is always stable. Therefore, the influence of proportion coefficient and integral coefficient on position sensorless system is comprehensively considered in this paper, and the final control parameters are shown in Table 3.

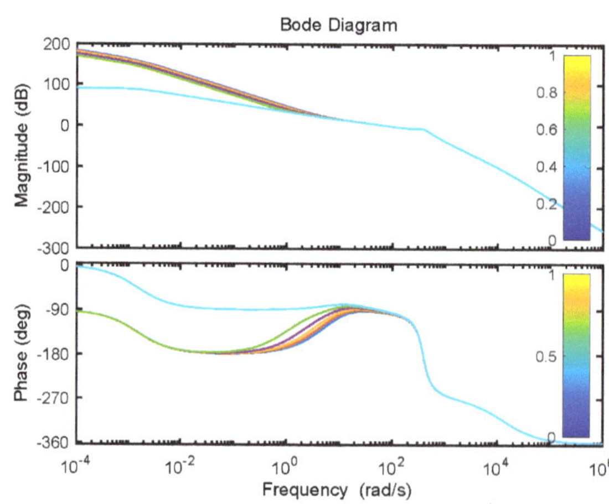

Figure 10. The bode diagram under the different K_i.

Table 2. The characteristics with different integral coefficients.

Proportionality Coefficient K_i	Magnitude Margin G_m	Phase Margin P_m	Stability (Yes/No)
0	6.94(at 379 rad/s)	87.1(at 85.7 rad/s)	No
0.2	6.93(at 379 rad/s)	86.5(at 85.7 rad/s)	Yes
0.4	6.92(at 379 rad/s)	85.8(at 85.7 rad/s)	Yes
0.6	6.91(at 378 rad/s)	85.1(at 85.7 rad/s)	Yes
0.8	6.9(at 378 rad/s)	84.5(at 85.8 rad/s)	Yes
1	6.89(at 378 rad/s)	83.8(at 85.5 rad/s)	Yes

Table 3. The setting of regulators parameters.

Regulator Type	Parameter	Value
PI	Proportionality coefficient K_p	0.2
	Integral coefficient K_i	1
VPDPI	Basic proportionality coefficient K_{p1}	0.2
	Changing proportionality coefficient K_{p2}	0.4
	Integral coefficient K_{i1}	1
	Proportional switching threshold c	50
	Compensate switching threshold φ	500
	Compensation coefficient γ	−14

4. Simulation Verification and Result Analysis

To demonstrate the correctness of the proposed method in the sensorless control system of PMSM, the offline simulation and the real-time experiment are used to verify the performance of the proposed method under various operating conditions, respectively.

4.1. Simulation Experiment Platform and Key Parameters Setting

Based on the overall structure of PMSM sensorless drive system in Section 2, the structure diagram is built as shown in Figure 11.

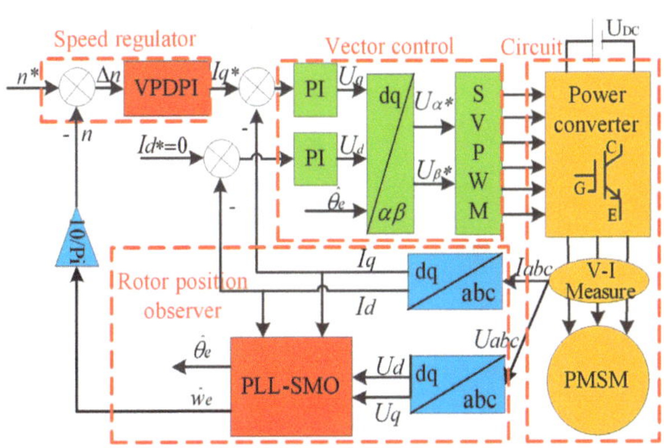

Figure 11. The structure diagram of sensorless PMSM drive system.

Moreover, the simulation model of sensorless drive system is built on the MATLAB/Simulink simulation platform. Meanwhile, the VPDPI regulator and the conventional PI regulator are tested under the same circuit parameters configuration, respectively. Table 3 present the parameters configuration of controller. The parameters configuration of PMSM are listed in Table 4.

Table 4. The setting of PMSM parameters.

Parameter	Value
Stator phase resistance R_s (Ω)	0.011
d-axis and q-axis phase inductances L_d, L_q (mH)	1.6, 1
Moment of inertia J (kg· m^2)	0.0008
Rotor pole pairs p_n	3
Reference speed n_{ref} (rpm = r/min)	1000
Flux linkage ψ_f (Wb)	0.077

4.2. Performance Testing under Multiple Operating Conditions

It is worth noting that the sensorless drive system desires to be operated at rated conditions. However, the control system may run in an unsatisfactory mode when the influence of disturbance or fault factors are considered in the actual site. Therefore, on the basis of analyzing the performance of the motor at rated speed, it is necessary to test and analyze the performance of PMSM at low speed (80% of rated speed) and high speed (120% of rated speed). Meanwhile, the detection accuracy and stability of feedback signals have gradually become the focus to evaluate the performance of sensorless system. Therefore, the dynamic performance, steady-state performance and detection accuracy between the VPDPI regulator and conventional PI regulator are comprehensively compared and analyzed under multiple working conditions based on the SMO-PLL sensorless control structure.

4.2.1. Performance Verification at Rated Speed

As shown in Table 4, all the comparative experiments are conducted on the same control parameters, and the performance can be fairly compared with different control algorithms. Meanwhile, all the simulation results are obtained under the same running operations, which refers to the load that changes from no-load to 2 N·m at 0.4 s.

When the speed is set at rated speed (1000 r/min), the speed response and the system estimated speed error using the proposed VPDPI regulator are shown in Figure 12a,b, respectively. Figure 13a,b show the speed response and the estimated speed error using the traditional PI regulator, respectively. Figures 14a and 15a record PMSM rotor position Angle information under VPDPI regulator and PI regulator, respectively. The Angle tracking error of the sensorless system under the above two regulators are shown in Figures 14b and 15b, respectively. Moreover, the output torque and three-phase current state of PMSM are shown in Figures 16 and 17, respectively.

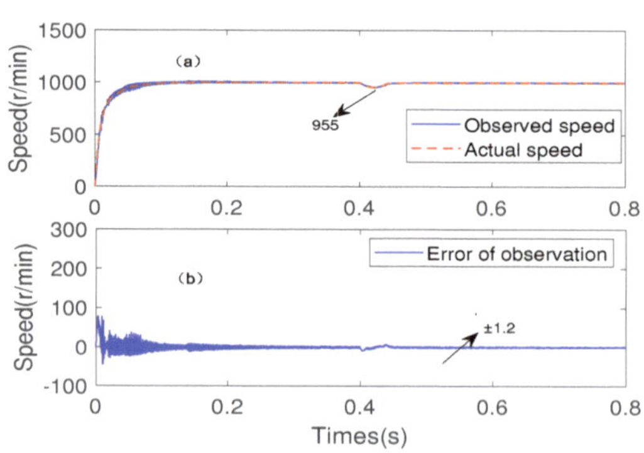

Figure 12. The speed response of VPDPI regulator under 1000 r/min. (**a**) Observed speed and actual speed. (**b**) Estimation error curve.

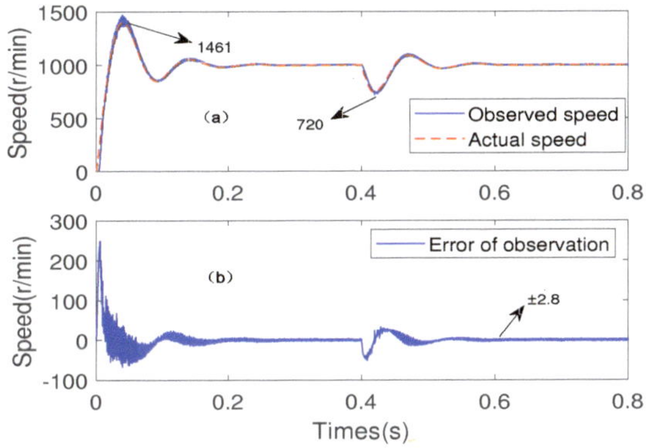

Figure 13. The speed response of PI regulator under 1000 r/min. (**a**) Observed speed and actual speed. (**b**) Estimation error curve.

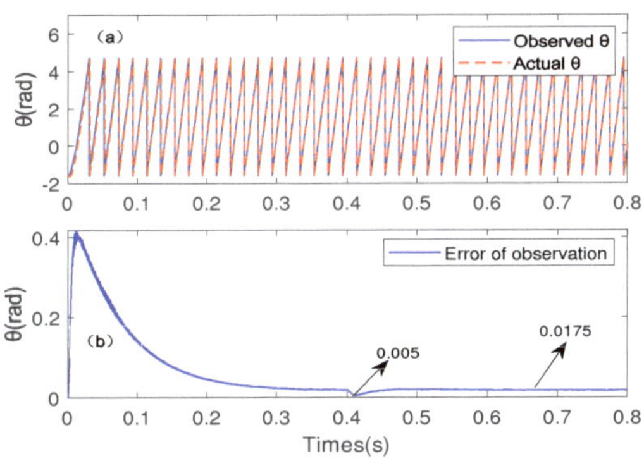

Figure 14. The estimation of rotor position Angle using VPDPI. (**a**) Observed and actual rotor position Angle. (**b**) Rotor position Angle observation error.

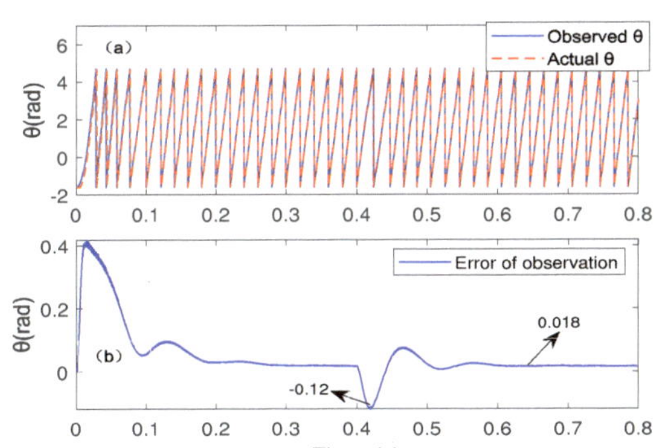

Figure 15. The estimation of rotor position Angle using PI. (**a**) Observed and actual rotor position Angle. (**b**) Rotor position Angle observation error.

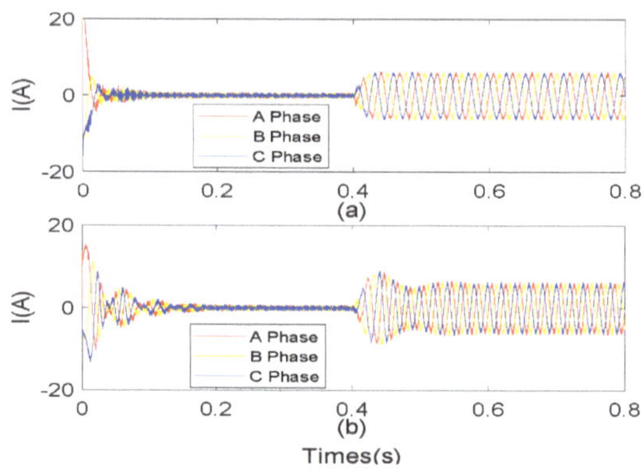

Figure 16. Three phase current response process under two control methods. (**a**) Current response under the proposed VPDPI control. (**b**) Current response under the PI control method

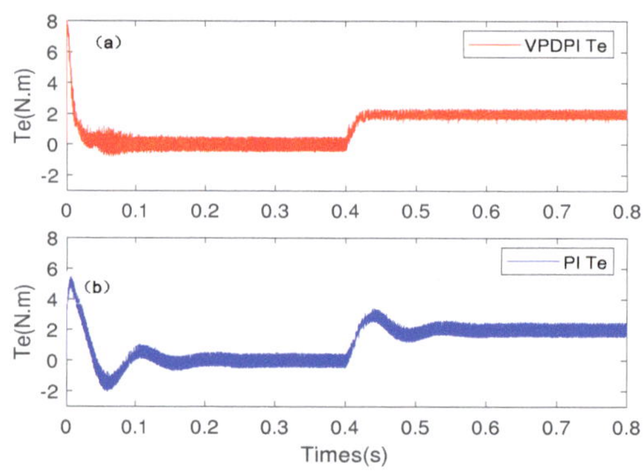

Figure 17. Torque response process under two control methods. (**a**) Torque response under the proposed VPDPI control. (**b**) Torque response under the PI control method

From Figures 12 and 13, It is worth noting that the VPDPI regulator has a only 0.05 s adjusting time, whereas the adjustment time of the traditional PI regulator is about 0.113 s. Meanwhile, the sensorless system is no overshoot or steady state error under the control of the VPDPI regulator. However, the overshoot can reach 461 r/min using the traditional PI regulator. On the other hand, the traditional PI regulator experiences a 280 r/min dynamic landing when the load changed, and a 45 r/min dynamic landing is happening in the VPDPI regulator. The adjustment process of sensorless system can also be further seen from the current and torque response results in Figures 16 and 17, respectively.

In the observation accuracy, the observation speed controlled by the VPDPI is about ±1.2 r/min. On the contrary, the observation speed of the traditional PI regulator has a more higher fluctuation, which even can reach ±2.8 r/min. Meanwhile, the observation error curves in Figures 14 and 15 show that the Angle tracking error on the basic of VPDPI regulator is more stable, which ensures the observation accuracy of PMSM sensorless system. In summary, the performance advantages of VPDPI regulator can be effectively verified at rated speed.

There is no doubt that dynamic performance and steady-state performance are important performance indicators of the regulating ability of the reaction regulator. However, in order to meet the complexity and uncontrollability of the actual field conditions, robustness has gradually become an important reference in measuring the performance of the regulator. Therefore, Appendix A tests the robustness of the proposed regulator at rated speed when parameters R_s or J do not match. Since R_s is configured as 0.011 in Table 4,

VPDPI regulation performance is tested when R_s is 0.0011, 0.021 and 0.031, respectively. On the other hand, since J is configured as 0.0008 in Table 4, VPDPI regulation is tested when J is 0.0006, 0.001 and 0.0012, respectively.

4.2.2. Performance Verification at Various Speed

The running state of the rated speed is the pursuit of the most driver systems. Due to the disturbance, it will need the sensorless system to run smoothly around the rated speed. Therefore, testing near the rated speed will be more comprehensive to verify the performance of the proposed regulator.

(1) Performance verification at 80% rated speed:

When the speed is set at 800 r/min, the estimated speed and speed error curves under two regulator are shown in Figures 18 and 19, respectively. Figures 20a,b record the response of position Angle and the response error under VPDPI regulator, respectively. The response of position Angle and the response error under VPDPI regulator are presented in Figures 21a,b, respectively. Moreover, the output torque and three-phase current state of PMSM are shown in Figures 22 and 23, respectively.

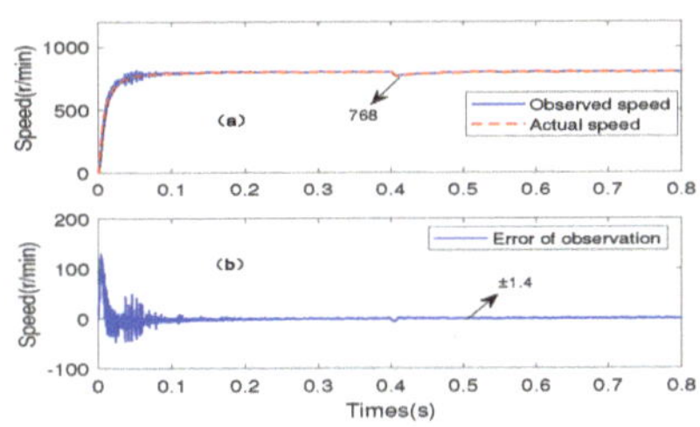

Figure 18. The speed response of VPDPI regulator under 800 r/min. (**a**) Observed speed and actual speed. (**b**) Estimation error curve.

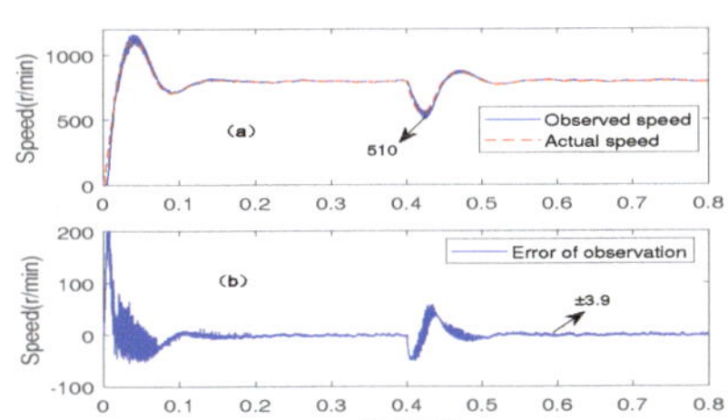

Figure 19. The speed response of PI regulator under 800 r/min. (**a**) Observed speed and actual speed. (**b**) Estimation error curve.

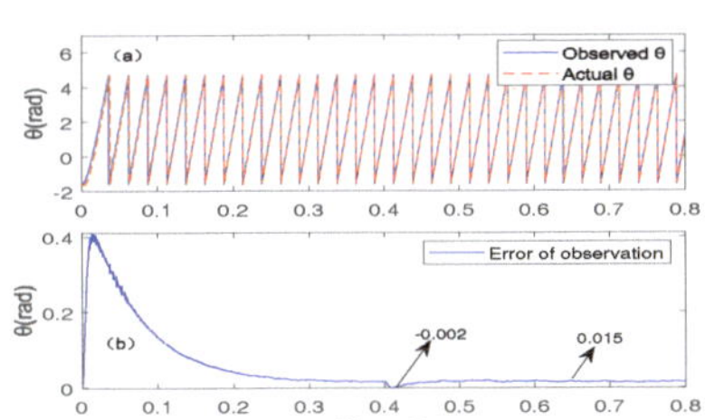

Figure 20. The estimation of rotor position Angle using VPDPI. (**a**) Observed and actual rotor position Angle. (**b**) Rotor position Angle observation error.

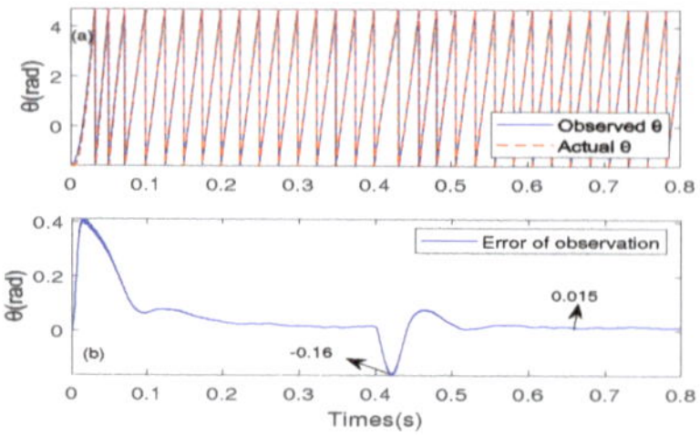

Figure 21. The estimation of rotor position Angle using PI. (**a**) Observed and actual rotor position Angle. (**b**) Rotor position Angle observation error.

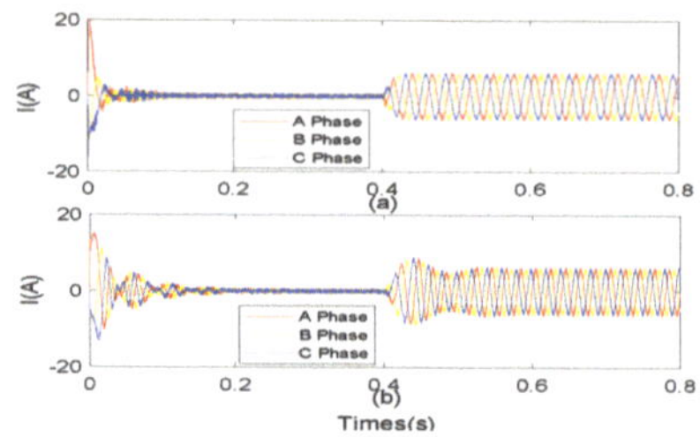

Figure 22. Phase current response process under two control methods. (**a**) Current response under the proposed VPDPI control. (**b**) Current response under the PI control method.

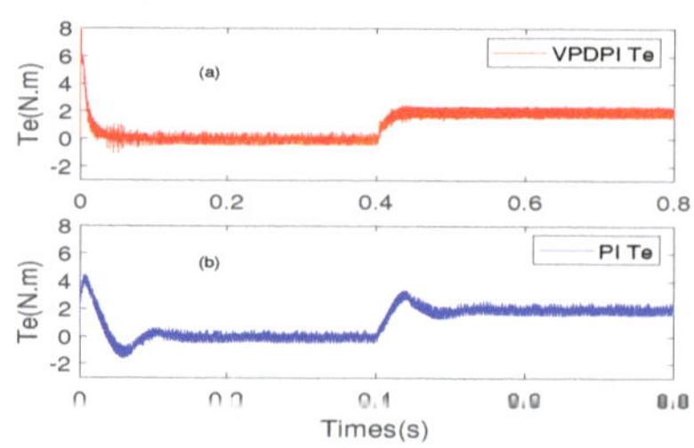

Figure 23. Torque response process under two control methods. (**a**) Torque response under the proposed VPDPI control. (**b**) Torque response under the PI control method.

The operation results in Figures 18 and 19 can reveal the control performance of the proposed control method from two perspectives. From the perspective of sensorless observation: the estimation errors generally exists in the starting stage. After a period of adjustment, the speed estimation errors tend to be stable. When the liability torque is added to the original system, the speed estimation error of VPDPI regulator changes little, and it soon achieves stable again. However, the control system using PI regulator once again experience an oscillation process, and it finally reaches a steady state process. From Figures 18 and 19, the proposed method controls the error range within ±1.4 r/min, and it improves the error observation accuracy by 64.1% than the traditional PI regulator (±3.9 r/min). Therefore, the dynamic performance and steady-state performance of the traditional PI regulator have improved in the observation process, especially the suppression of overshoot.

From the perspective of speed response curve performance: the traditional PI control system has a large overshoot since the integral saturation influence. Through a adjusting process, the traditional PI control system is gradually stabilize within the error band at 0.13 s (Δ = 5%). On the contrary, the proposed VPDPI method takes into account the influence of integral saturation. The VPDPI control system is more faster and more stable entering the error band, which greatly weakens the influence of overshoot on system balance. After the load is added to the sensorless system, the dynamic landing under the traditional PI regulator reaches 290 r/min, which go through a long period of disturbance rejection process. Unexpectedly, the control system under the VPDPI regulator only has a dynamic landing of 32 r/min, and the response quickly recovered to a stable state. Therefore, it is worth noting that the proposed method has more advantages in response curve performance analysis.

In addition to the emphasis of the observation effect, the accuracy of the observation is also an important indicator of the control of the sensorless system. As shown in Figures 20 and 21, the system estimation under the control of steady state time is similar. Nonetheless the position Angle error of the VPDPI regulator is more smoother. Meanwhile, the dynamic landing of VPDPI system is more smaller when the load is added.

In Figures 22 and 23, it can be seen that the VPDPI regulator converges to the steady-state value with a more shorter adjustment time than the traditional PI regulator.

In conclusion, when the speed is set at 800 r/min, the sensorless system using VPDPI regulator has a better dynamic performance, steady-state performance and observation effect than the traditional PI system.

(2) Performance verification at 120% rated speed:

When the speed is set at 1200 r/min, the speed and speed error curves under the control of VPDPI regulator and PI regulator are shown in Figures 24 and 25, respectively. Figures 26 and 27 record PMSM rotor position Angle information under the control of the two regulators and the error of tracking the rotor position Angle of the actual system, respectively. Moreover, the output torque and three-phase current of PMSM are shown in Figures 28 and 29, respectively.

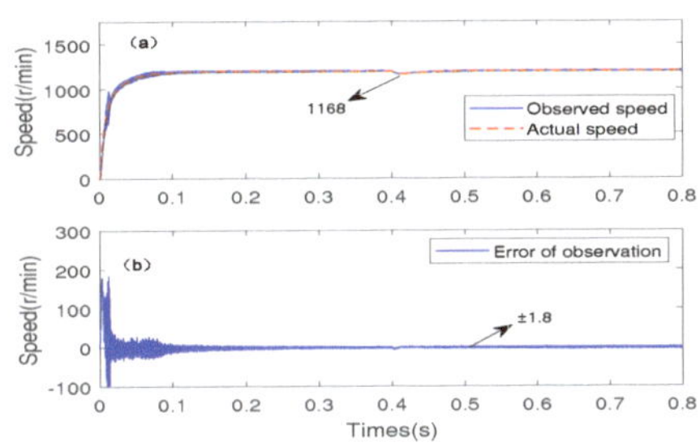

Figure 24. The speed response of VPDPI regulator under 1200 r/min. (**a**) Observed speed and actual speed. (**b**) Estimation error curve.

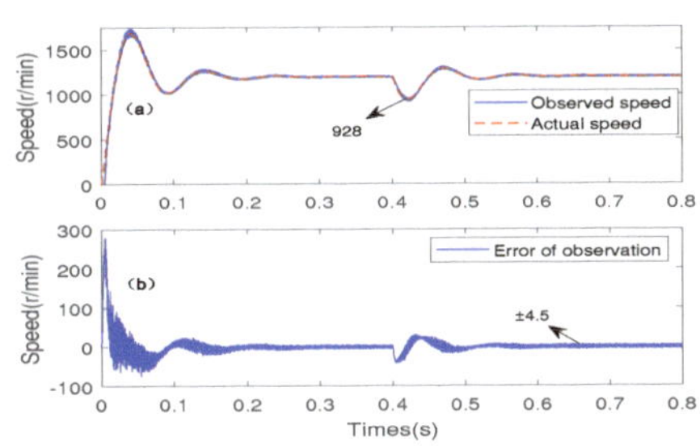

Figure 25. The speed response of PI regulator under 1200 r/min. (**a**) Observed speed and actual speed. (**b**) Estimation error curve.

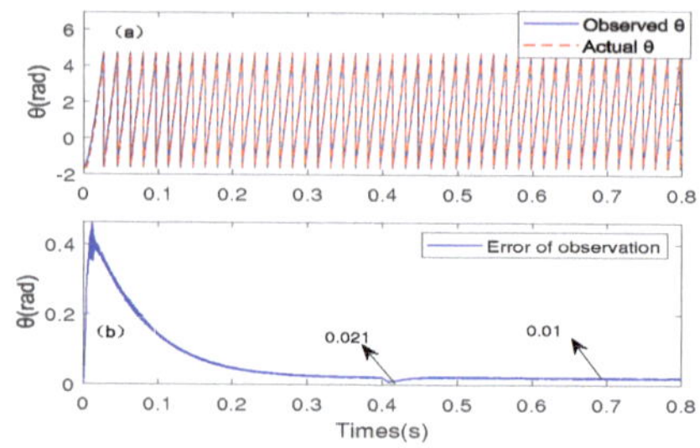

Figure 26. The estimation of rotor position Angle using VPDPI. (**a**) Observed and actual rotor position Angle. (**b**) Rotor position Angle observation error.

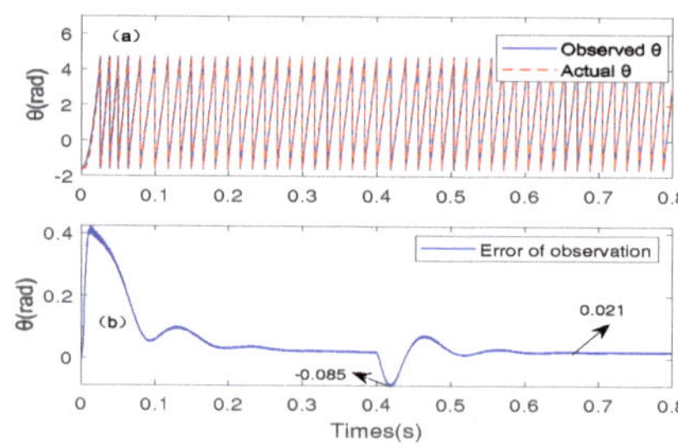

Figure 27. The estimation of rotor position Angle using PI. (**a**) Observed and actual rotor position Angle. (**b**) Rotor position Angle observation error.

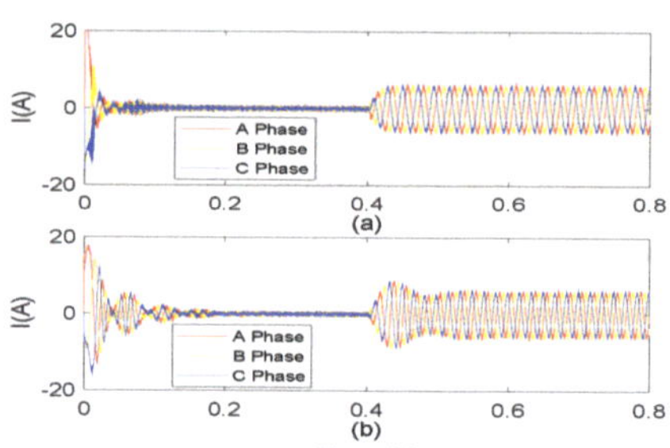

Figure 28. Three phase current response process under two control methods. (**a**) Current response under the proposed VPDPI control. (**b**) Current response under the PI control.

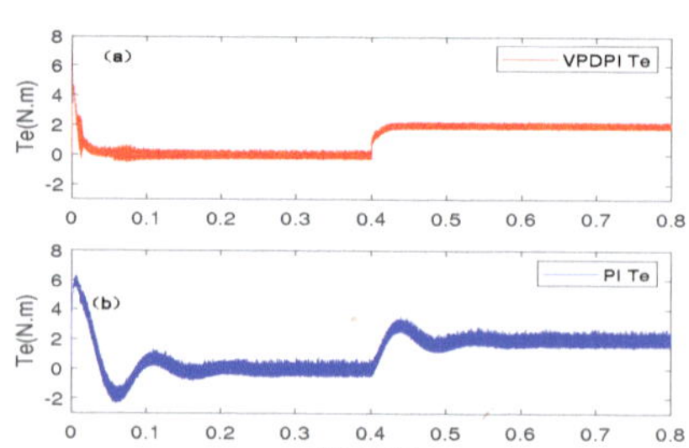

Figure 29. Torque response process under two control methods. (**a**) Torque response under the proposed VPDPI control. (**b**) Torque response under the PI control.

From the above experimental results, the control performance of the two regulators is similar to the above two working conditions.

The maximum overshoot of the traditional PI regulator is 548 r/min, and the adjustment time is about 0.115 s. When the load mutation occurred, a dynamic landing of 272 r/min occurred. In terms of observation accuracy, the traditional PI observation error pulsation reaches 4.5 r/min. On the contrary, the proposed VPDPI regulator does not have overshoot, and the adjustment time is only 0.053 s. The dynamic landing under load mutation is only 32 r/min. Meanwhile, the steady-state observation error fluctuation is only 1.8 r/min. The position Angle observation error is more gentle, and the observation result is more stable. Therefore, the capability of VPDPI regulator to solve integral saturation in PMSM sensorless control system is proved again.

5. Experimental Verification and Results

After the effectiveness of the proposed control method is tested through simulation experiments, the VPDPI control method is further loaded into the hardware-in-the-loop experimental platform to better verify the reliability of the proposed algorithm. The structure of the experimental platform is shown in Figure 30. In addition, the experimental platform is clearly presented in Figure 31. Hence, the performance differences of various control algorithms can be accurately compared. Without losing generality, the simulation control parameters are completely consistent with the experimental control parameters.

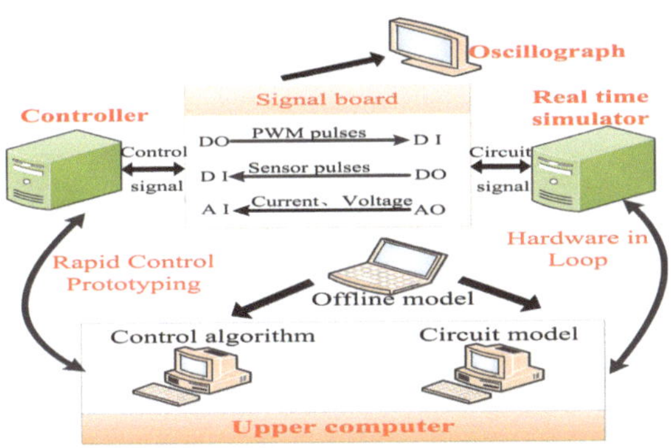

Figure 30. The structure of the experimental platform.

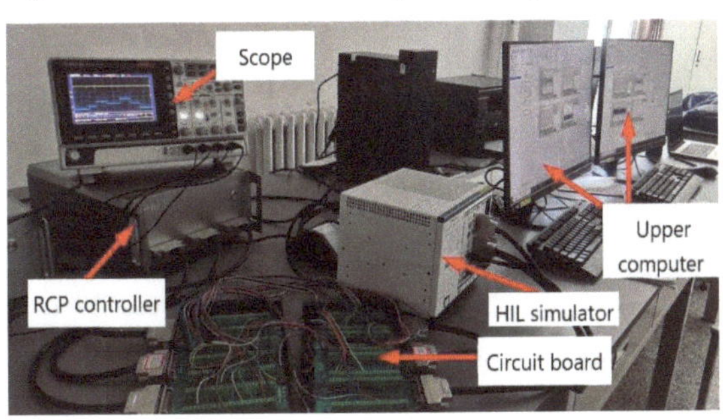

Figure 31. Experimental test platform for the proposed algorithm

5.1. Experimental Verification at Rated Speed

When the speed is set to 1000 r/min, the observed speed response of the two regulators under no-load and $2N\dot{m}$ load are presented in Figures 32 and 33.

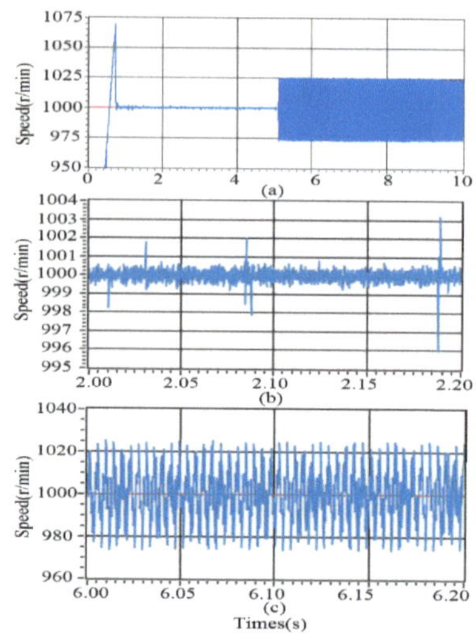

Figure 32. The observed speed using PI at rated speed. (**a**) The overall response diagram of observed speed at 1000 r/min. (**b**) Observed speed under no load. (**c**) Observed speed under 2 N·m load.

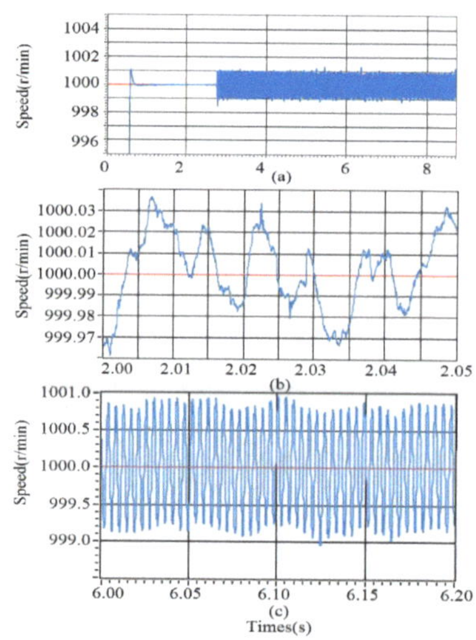

Figure 33. The observed speed using VPDPI at rated speed. (**a**) The overall response diagram of observed speed at 1000 r/min. (**b**) Observed speed under no load. (**c**) Observed speed under 2 N·m load.

From Figures 32 and 33, it is worth noting that the adjustment time of the proposed VPDPI regulator is clearly shorter, while almost no overshoot. However, the observed speed overshoot under the traditional PI regulator is about 75 r/min. Meanwhile, the speed pulsation of the proposed regulator is more smaller under both no-load and load conditions, especially with load condition. In addition, the observation accuracy is more higher than the traditional PI regulator when the system runs stably. Hence, the proposed regulator has outstanding dynamic performance and precision of speed estimation, which proves the validity of the method again.

5.2. Experimental Verification at Various Speed

(1) Experimental verification at 80% rated speed

When the speed is set to 800r/min, Figures 34 and 35 show the observed speed response of the two regulators under no-load and 2 N·m load, respectively.

From Figures 34 and 35, the observed speed under the traditional PI regulator has a large overshoot in the dynamic response process, and the result also exists a large speed ripple when the response is stable. In contrast, the maximum overshoot of the VPDPI regulator is about 0.3% of the maximum overshoot of the traditional PI. Compared with the traditional PI regulator, the speed pulsation of the proposed method is reduced about 90% under no-load condition. Moreover, the traditional PI speed pulsation becomes more serious under changing load condition, and the pulsation is up to ±25 r/min. Fortunately, the proposed method can control the pulsation within ±5 r/min, which reflects a more higher stability.

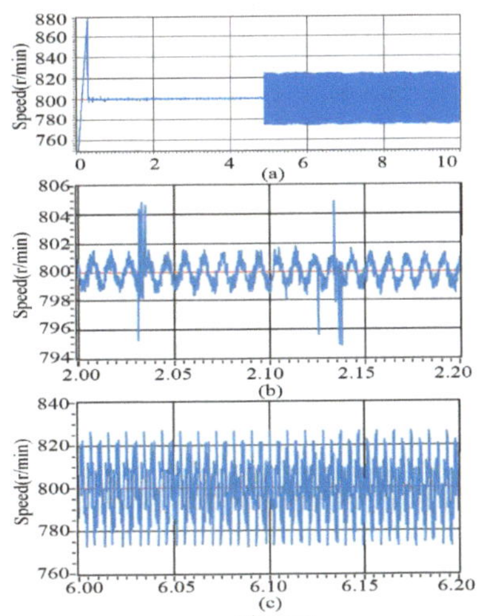

Figure 34. The observed speed using PI at 800 r/min. (**a**) The overall response diagram of observed speed at 800 r/min. (**b**) Observed speed under no load. (**c**) Observed speed under 2 N·m load.

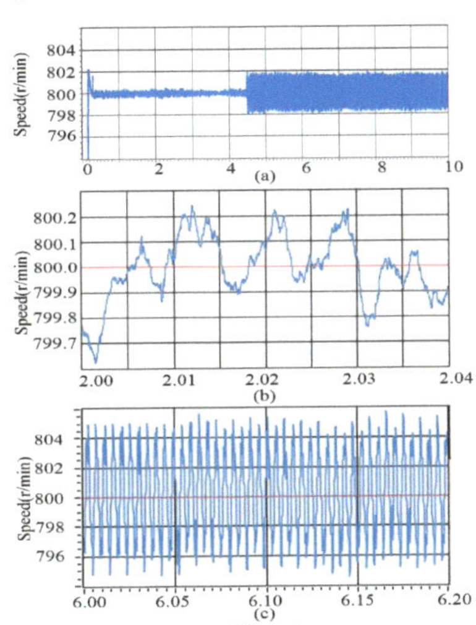

Figure 35. The observed speed using VPDPI at 800 r/min. (**a**) Observed speed at 800 r/min. (**b**) Observed speed under no load. (**c**) Observed speed under 2 N·m load.

(2) Experimental verification at 120% rated speed:

When the speed is set to 1200 r/min, the observed speed response of the two regulators under no-load and 2 N·m load are presented in Figures 36 and 37.

From Figures 36 and 37, it can be seen clearly that the overshoot of the traditional PI regulator is about 64 r/min. However, the overshoot of the VPDPI regulator is only 1.6% of the traditional PI regulator, which indicates a significant overshoot suppression effect. Meanwhile, the observed speed pulsations using two regulators are obviously reduced than others operation conditions, especially the VPDPI regulator. Nevertheless, it can still be detected that the pulsation of the traditional PI regulator is as high as 25 r/min with 2 N·m load condition. It is worth noting that the pulsation of VPDPI regulator is less than 0.15 r/min. In brief, the experimental results under this condition are similar to the performance reflected before.

The experimental results in this section verify the effectiveness of the proposed method to suppress integral saturation, which further improve the control performance and observation accuracy of the traditional sensorless system.

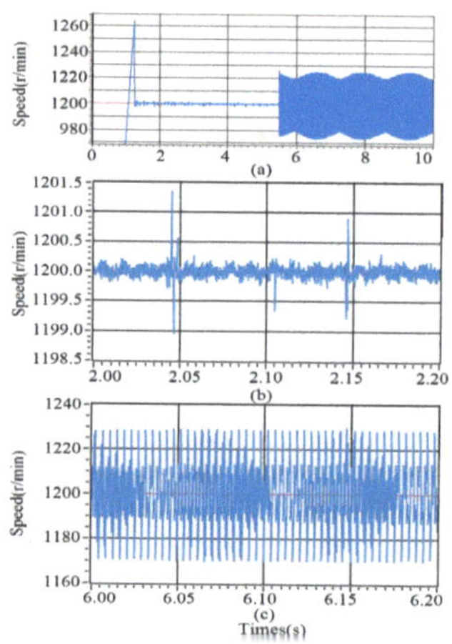

Figure 36. The observed speed using PI at 1200 r/min. (**a**) The overall response diagram of observed speed at 1200 r/min. (**b**) Observed speed under no load. (**c**) Observed speed under 2 N·m load.

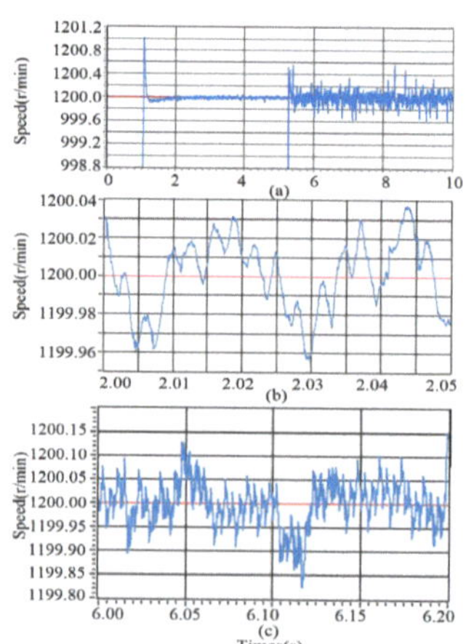

Figure 37. The observed speed using VPDPI at 1200 r/min. (**a**) The overall response diagram of observed speed at 1200 r/min. (**b**) Observed speed under no load. (**c**) Observed speed under 2 N·m load.

6. Conclusions

In this article, a VPDPI control method is creatively proposed to eliminate the integral saturation phenomenon of PMSM sensorless system, and its original contributions can be extracted as follows:

(1) The essence of integral saturation in traditional PI controller is unexpectedly revealed based on the classical control theory, and the corresponding control scheme is further developed from the perspective of dynamic response characteristic. Consequently, the novel VPDPI contains two interconnected regulators with independent effects: (a) a DPI regulator stably eliminates integral saturation by utilizing feedback compensation coefficients γ, and (b) a VP regulator ensures the system response speed by switching the proportional coefficient between k_{p1} and k_{p2}.

(2) According to the distribution of root trajectory curve, the stability of the system is analyzed. Meanwhile, the parameters range of speed regulator is given by using frequency domain analysis method based on the identified model. Thereinto, the system has a higher stability margin when the proportional coefficient is selected in the range of 0~0.4. When the integral coefficients are set as 0~1, it can be found that the system is always in a stable state, but the amplitude-frequency margin and phase-frequency margin are slightly weakened.

In addition, the simulation and hardware experiments with different speed regulators are tested under the same control parameters. On the one hand, the simulation results can demonstrate the effectiveness of the new regulator under random operation conditions. Compared with the traditional PI regulator, the adjusting time of the proposed regulator is about 0.05 s under rated speed with no−load torque, which is less than 44.2% of the tradition PI regulator. More importantly, the overshoot is totally eliminated, which is about 461 r/min in traditional PI regulator. When the system exists a low load disturbance, the dynamic landing is reduced about 16% than that of traditional PI regulator. Moreover, the observed speed error fluctuation is also decreased about 42.9%. Similarly, the control performance remains consistent under variable speed. On the other hand, the experimental results further verify the dynamic performance, accuracy and stability of the novel regulator under multiple working conditions, especially in the suppression abilities of overshoot and the estimated speed ripple. In summary, the effectiveness and reasonability of the proposed method are both amply verified. In the future, some intelligent algorithms will be combined with the proposed method to achieve automatic adjustment of threshold parameters.

Author Contributions: Z.W. wrote this article, designed the control method, implemented the hardware platform, drew the figures, and performed the simulation as well as the experiment. M.Z. supervised the study, coordinated the investigations, and checked the manuscript's logical structure. X.L. validated the simulation and hardware experimental data and reviewed the writing and editing. M.L. provided practical suggestions and evaluated the feasibility of the application. All authors have read and agreed to the published version of the manuscript.

Funding: This work was Supported in part by the Natural Science Foundation of China under Grant No. 61563045, and in part by the International Cooperation Projects of Shihezi University under Grant GJHZ202003.

Institutional Review Board Statement: Not applicable.

Informed Consent Statement: Not applicable.

Data Availability Statement: Not applicable.

Acknowledgments: We sincerely thank Shanghai Yuankuan Energy Technology Co., Ltd. for providing technical guidance for the use of the hardware experimental platform.

Conflicts of Interest: The authors declare no conflict of interest.

Abbreviations

The following abbreviations are used in this paper:

PMSM	Permanent Magnet Synchronous Motor
PI	Proportional Integral
VPDPI	Variable-proportion Desaturation Proportional Integral
ACR	Automatic Current Regulator
ASR	Automatic Speed Regulator
SNR	Signal Noise Ratio
ZPK	Zero Pole and Gain
PLL	Phase-locked Loop
SMO	Sliding Mode Observer
RCP	Rapid Control Prototype

Nomenclature

The following nomenclatures are used in this manuscript:

u_d/u_q	stator voltage under d/q-axis
V_d/V_q	observed back-EMF under d/q-axis
i_d/i_q	stator current under d/q-axis
$\tilde{i_d}/\tilde{i_q}$	observed stator current under d/q-axis
L_d/L_q	stator inductance under d/q-axis
J	moment of inertia
R	stator resistance
B	damping coefficient
P_n	polar logarithm
T_e	electromagnetic torque
T_L	load torque
γ	feedback compensation coefficient
c	The judgment threshold of proportional coefficient selection
φ	The judgment threshold of compensation coefficient selection
ρ	The selection coefficient

Appendix A. Robustness Test Considering the Mismatch of R_s or J

Figures A1–A3 show the test results when the parameters R_s are not matched at rated speed. Figures A4–A6 show the test results when the coefficient J is not matched.

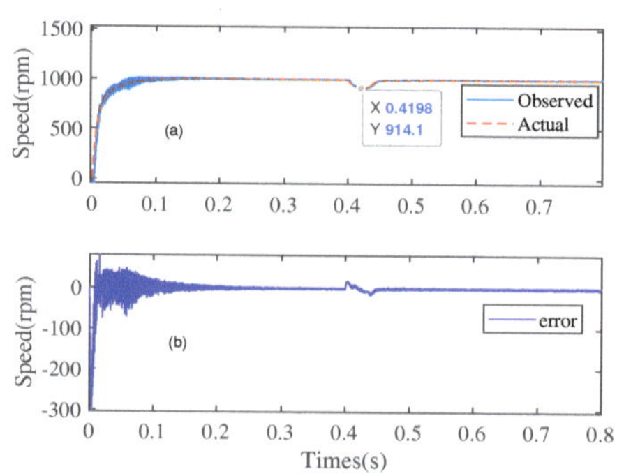

Figure A1. The speed response of VPDPI regulator when $R_s = 0.0011$ and $J = 0.0008$. (**a**) Observed speed and actual speed. (**b**) Estimation error curve

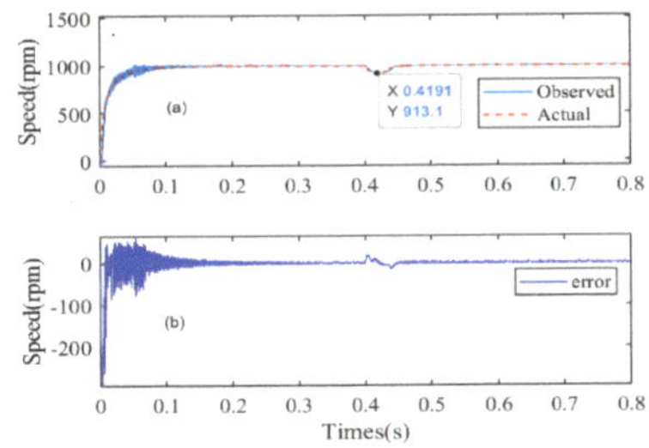

Figure A2. The speed response of VPDPI regulator when $R_s = 0.021$ and $J = 0.0008$. (**a**) Observed speed and actual speed. (**b**) Estimation error curve

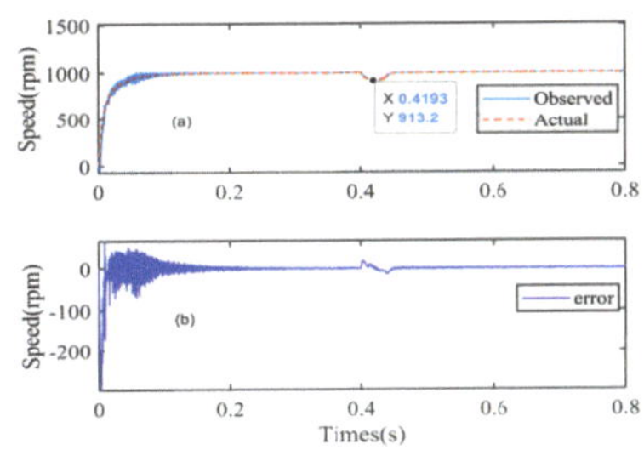

Figure A3. The speed response of VPDPI regulator when $R_s = 0.031$ and $J = 0.0008$. (**a**) Observed speed and actual speed. (**b**) Estimation error curve

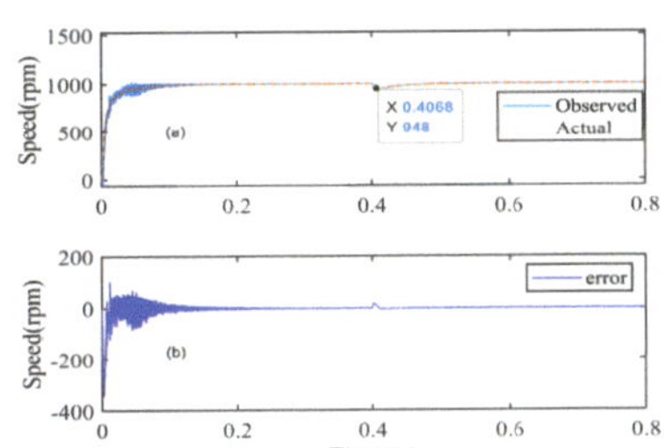

Figure A4. The speed response of VPDPI regulator when $J = 0.0006$ and $R_s = 0.011$. (**a**) Observed speed and actual speed. (**b**) Estimation error curve

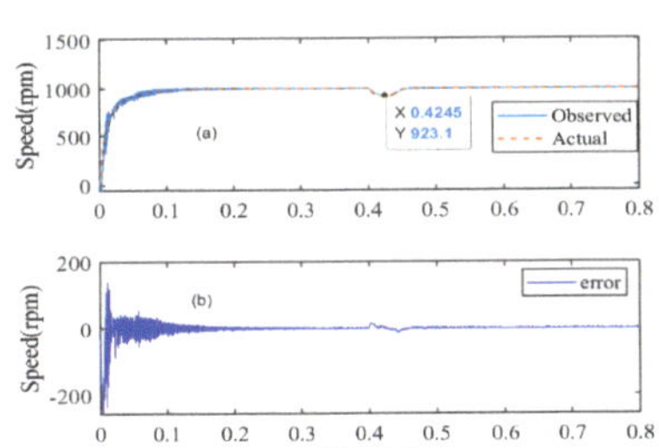

Figure A5. The speed response of VPDPI regulator when $J = 0.001$ $R_s = 0.011$. (**a**) Observed speed and actual speed. (**b**) Estimation error curve

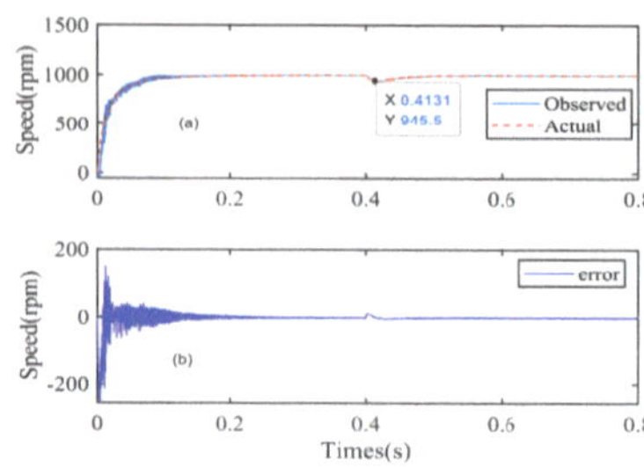

Figure A6. The speed response of VPDPI regulator when J = 0.0012 R_s = 0.011. (**a**) Observed speed and actual speed. (**b**) Estimation error curve

According to the test results of the appeal, it can be seen that when the parameters fluctuate, the adjustment performance of the proposed VPDPI regulator changes little, which ensures that the system without position sensor can still run stably when the parameters change, and further verifies the robustness of the proposed regulator.

References

1. Zhao, Y.; Liu, X.; Zhang, Q. Predictive Speed-Control Algorithm Based on a Novel Extended-State Observer for PMSM Drives. *Appl. Sci.* **2019**, *9*, 2575. [CrossRef]
2. Chen, H.; Gao, C.; Si, J.; Nie, Y.; Hu, Y. A Novel Method for Diagnosing Demagnetization Fault in PMSM Using Toroidal-Yoke-Type Search Coil. *IEEE Trans. Instrum. Meas.* **2022**, *71*, 1–12. [CrossRef]
3. ThatDong, T.; MinFu, H. A Deadbeat Current and Flux Vector Control for IPMSM Drive with High Dynamic Performance. *Appl. Sci.* **2022**, *12*, 3789. [CrossRef]
4. Rocha, L.R.; de Araujo, M.B.; Silva, E.C.; Scher, L.H.O.; Vieira, R.P. Speed and Position Observer for Non-sinusoidal PMSM Based on a PSD+PLL Structure. *J. Control Autom. Electr. Syst.* **2022**, *33*, 304–314. [CrossRef]
5. Hosseyni, A.; Trabelsi, R.; Mimouni, M.F.; Iqbal, A.; Alammari, R. Sensorless sliding mode observer for a five-phase permanent magnet synchronous motor drive. *ISA Trans.* **2015**, *58*, 462–473. [CrossRef]
6. Ye, S. Fuzzy sliding mode observer with dual SOGI-FLL in sensorless control of PMSM drives. *ISA Trans.* **2019**, *85*, 161–176. [CrossRef]
7. Yang, Z.; Ding, Q.; Sun, X.; Lu, C.; Zhu, H. Speed sensorless control of a bearingless induction motor based on sliding mode observer and phase-locked loop. *ISA Trans.* **2021**, *123*, 346–356. 10.1016/j.isatra.2021.05.041. [CrossRef]
8. Wang, G.; Wang, H. A new speed adaptive estimation method based on an improved flux sliding-mode observer for the sensorless control of PMSM drives. *ISA Trans.* 2021, *in press*. [CrossRef]
9. Li, A.; Jiang, D.; Sun, X.; Liu, Z. Online Drive Topology Conversion Technology for PMSM Speed Range Extension. *IEEE Trans. Power Electron.* **2022**, *37*, 7113–7121. [CrossRef]
10. Liu, S.; Liu, C.; Huang, Y.; Xiao, Y. Direct Modulation Pattern Control for Dual Three-Phase PMSM Drive System. *IEEE Trans. Ind. Electron.* **2022**, *69*, 110–120. [CrossRef]
11. Wang, Y.; Xu, Y.; Zou, J. Online Multiparameter Identification Method for Sensorless Control of SPMSM. *IEEE Trans. Power Electron.* **2020**, *35*, 10601–10613. [CrossRef]
12. Wang, H.; Lu, K.; Wang, D. Simple and Effective Online Position Error Compensation Method for Sensorless SPMSM Drives. *IEEE Trans. Ind. Appl.* **2020**, *56*, 1475–1484. [CrossRef]
13. Lee, J.; Kwon, Y.; Sul, S. Identification of IPMSM Flux-Linkage Map for High-Accuracy Simulation of IPMSM Drives. *IEEE Trans. Power Electron.* **2021**, *36*, 14257–14266. [CrossRef]
14. Choi, J.; Nam, K.; Bobtsov, A.A.; Ortega, R. Sensorless control of IPMSM based on regression model. *IEEE Trans. Power Electron.* **2019**, *34*, 9191–9201. [CrossRef]
15. Wang, Q.; Wang, S.; Chen, C. Review of sensorless control techniques for PMSM drives. *IEEJ Trans. Electr. Electron. Eng.* **2019**, *14*, 1543–1552. [CrossRef]
16. Ganapathy, R.; Santha, K. Review of sliding mode observers for sensorless control of permanent magnet synchronous motor drives. *Int. J. Power Electron. Drive Syst.* **2018**, *9*, 46–54.
17. Abd, Samat, A.A.; Ishak, D.; Omar, A.M.; Iqbal, S.; Razak, M.A. A New Speed Sensorless Field Oriented Controller for PMSM Based on MRAS and PSO. *J. Electr. Syst.* **2016**, *12*, 566–573.
18. Khlaief, A.; Boussak, M.; Chari, A. A MRAS-based stator resistance and speed estimation for sensorless vector controlled IPMSM drive. *Electr. Power Syst. Res.* **2014**, *108*, 1–15. [CrossRef]
19. Tan, L.; Cong, T.; Cong, D. Neural Network Observers and Sensorless Robust Optimal Control for Partially Unknown PMSM With Disturbances and Saturating Voltages. *IEEE Trans. Power Electron.* **2021**, *36*, 12045–12056. [CrossRef]

20. Moujahed, M.; Azza, H.B.; Jemli, M.; Boussak, M. Extended Kalman Filter for Sensorless Fault Tolerant Vector Control of PMSM with stator resistance estimation. *Int. J. Electr. Eng. Informatics* **2017**, *9*, 207–221. [CrossRef]
21. Wang, L.; Zhang, Y.; Hao, S.; Song, B.; Hao, M.; Tang, Z. Study on high precision magnetic encoder based on PMSM sensorless control. *Sens. Rev.* **2016**, *36*, 267–276. [CrossRef]
22. Lin, S.; Zhang, W. An adaptive sliding-mode observer with a tangent function-based PLL structure for position sensorless PMSM drives. *Int. J. Electr. Power Energy Syst.* **2017**, *88*, 63–74. [CrossRef]
23. Chen, Z.; Zhao, Y.; Liu, X.; Peng, K.; Pu, H. Embedded Position Detecting Method for Permanent Magnet Linear Motor Systems. *IEEE Trans. Instrum. Meas.* **2021**, *70*, 1–10. [CrossRef]
24. Chen, G.R.; Yang, S.C.; Hsu, Y.L.; Li, K. Position and Speed Estimation of Permanent Magnet Machine Sensorless Drive at High Speed Using an Improved Phase-Locked Loop. *Energies* **2017**, *10*, 1571. [CrossRef]
25. Chen, S.; Zhang, X.; Wu, X.; Tan, G.; Chen, X. Sensorless Control for IPMSM Based on Adaptive Super-Twisting Sliding-Mode Observer and Improved Phase-Locked Loop. *Energies* **2019**, *12*, 1225. [CrossRef]
26. Yin, Z.; Zhang, Y.; Cao, X.; Yuan, D.; Liu, J. Estimated Position Error Suppression Using Novel PLL for IPMSM Sensorless Drives Based on Full-Order SMO. *IEEE Trans. Power Electron.* **2022**, *37*, 4463–4474. [CrossRef]
27. Ye, S. A novel fuzzy flux sliding-mode observer for the sensorless speed and position tracking of PMSMs. *Optik* **2018**, *171*, 319–325. [CrossRef]
28. Wei, H.; Zhang, Y.; Chu, J. New Rotor Position Estimation Strategy for Permanent Magnet Synchronous Motor Based on Rotor Flux Model. *Trans. China Electrotech. Soc.* **2018**, *33*, 2963–2971.
29. Hoo, C.L.; Haris, S.M.; Chung, E.C.; Nik Mohamed, N.A. New Integral Antiwindup Scheme for PI Motor Speed Control. *Asian J. Control* **2015**, *17*, 2115–2132. [CrossRef]
30. Li, H.; Song, B.; Chen, T.; Xie, Y.; Zhou, X. Adaptive fuzzy PI controller for permanent magnet synchronous motor drive based on predictive functional control. *J. Frankl. Inst.* **2021**, *358*, 7333–7364. [CrossRef]
31. Seep, P.; Anjali, J.; Prakash, D. Anti-windup fractional order PI^{λ}-PD^{μ} controller design for unstable process: A magnetic levitation study case under actuator saturation. *Arab. J. Sci. Eng.* **2017**, *42*, 5015–5029.
32. Chiah, T.; Hoo, C. Hardware Simulation of Semi-Decoupled Tuning Gain Anti-windup PI Controllers for Motor Speed Application. *AIP Conf. Proc.* **2019**, *2137*, 1–8.
33. Cong, T.; Cheon, J. Development of Anti-windup PI Control and Bumpless Control Transfer Methodology for Feedwater Control System. *Ann. Nucl. Energy* **2019**, *131*, 233–241.
34. Eldigair, Y.; Garelli, F.; Kunusch, C.; Ocampo-Martinez, C. Adaptive PI control with robust variable structure anti-windup strategy for systems with rate-limited actuators. *Trans. China Electrotech. Soc.* **2020**, *96*, 104282–104294.
35. Seuret, A.; Prieur, C.; Tarbouriech, S.; Zaccarian, L. LQ-based event-triggered controller co-design for saturated linear systems. *Automatica* **2016**, *74*, 47–54. [CrossRef]
36. Ugur, K.; Mustafa, G.; Mikail, K.; Ozan, G. Modulated Model Predictive Control of Permanent Magnet Synchronous Motors with Improved Steady-State Performance. In Proceedings of the 2021 3rd Global Power, Energy and Communication Conference (GPECOM), Antalya, Turkey, 5–8 October 2021; pp. 67–72.
37. Mu, F.; Xu, B.; Shi, G.; Ji, W.; Ding, Q. Sensorless Vector Control of PMSM Based on Improved Sliding Mode Observer. *Lect. Notes Electr. Eng.* **2016**, *405*, 323–334.

machines

Article

Intramedullary Skeletal Distraction Robot: Novel Design and Optimization of Implantable Lengthening Nail

ShiKeat Lee [1,2,3], Zhenguo Nie [1,2,3,*], Handing Xu [1,2,3], Kai Hu [4], Hanwei Lin [1,2,3], Fugui Xie [1,2,3] and Xin-jun Liu [1,2,3]

1 Department of Mechanical Engineering, Tsinghua University, Beijing 100084, China
2 State Key Laboratory of Tribology in Advanced Equipment, Beijing 100084, China
3 Beijing Key Laboratory of Precision/Ultra-Precision Manufacturing Equipments and Control, Tsinghua University, Beijing 100084, China
4 Jiangsu Key Laboratory of Special Robot Technology, Hohai University, Changzhou 213022, China
* Correspondence: zhenguonie@tsinghua.edu.cn

Abstract: Leg Length Discrepancy (LLD) is a critical problem which not only impacts the quality of life but also causes other physical ailments such as limping and severe back pain. Most people had a non-negligible difference in the length of their two legs, and 0.1% of the population with differences of 20 mm or more are diagnosed as a difference in limb length. The Ilizarov external fixator is commonly used in treating LLD as conventional therapeutic equipment, but often causes serious complications that cannot be prevented. Therefore, intramedullary leg-lengthening treatment has become popular in distraction osteogenesis to eliminate extracorporeal surgery. This paper presents a study on the design and optimization of a novel electromagnetic-driven Intramedullary Skeletal Distraction Robot (ISDR) with robust mechanical stiffness and surplus electromagnetic driving force. Compared with PRECICE, the split structure of ISDR eases the optimal design and manufacturing difficulties to strengthen mechanical stiffness, and the electromagnetic configuration allows improving the distraction force by adjusting the Permanent Magnet Brushless Direct Current (PMBLDC) motor parameters. ISDR, which is implanted in the medullary cavity, has the Von-Mises stress of 952.15 MPa, and the first mode of natural frequency is 28.823 Hz indicating that it can withstand the load during the walking gait phases. On the other hand, the ISDR distraction force encounters resistance from muscle fibers, and an average driving torque of 9 Nmm ensures its distraction. Based on the results, ISDR is proven secure and reliable during and after leg-lengthening treatment, which can significantly reduce lifestyle disruption and medical complications.

Keywords: Leg Length Discrepancy; bone distraction; ISDR; design and optimization

Citation: Lee, S.; Nie, Z.; Xu, H.; Hu, K.; Lin, H.; Xie, F.; Liu, X.-J. Intramedullary Skeletal Distraction Robot: Novel Design and Optimization of Implantable Lengthening Nail. *Machines* **2022**, *10*, 843. https://doi.org/10.3390/machines10100843

Academic Editors: Lorand Szabo and Feng Chai

Received: 23 August 2022
Accepted: 15 September 2022
Published: 23 September 2022

1. Introduction

Leg Length Discrepancy (LLD), also known as anisomelia, is a condition of noticeably unequal length of the paired lower extremity limbs that is commonly caused by congenital, developmental, or posttraumatic conditions, such as trauma from birth, the malformed socket of the hip joint, and arthritis of the joints [1]. Abnormal growth, hip dysplasia, scoliosis, limping, back pain, and osteoarthritis have been deemed the complications of LLD [2]. Guichet et al. found that 0.1% of the population went for surgery when the difference in leg length exceeded 2 cm [3]. Generally, the methods frequently used in the clinical treatment of LLD include non-surgical treatment (wearing a shoe lift) and surgical treatment (shortening longer limbs or lengthening shorter limbs) [4]. However, limb shortening is disallowed in some instances since it leads to muscle weakening.

An orthopedic surgeon, Gavriil A. Ilizarov, proposed the theory of tensions, which has become the fundamental principle in limb lengthening and reshaping treatment. According to this theory, the Ilizarov external fixator was widely promoted in the 1980s and gradually became the primary treatment apparatus in this field, which showed revolutionary results [5]. However, the Ilizarov external fixator has accentuated the problem of inconvenience and discomfort that may cause certain complications such as pin site infection, axial deviation, joint stiffness, soft tissue incarceration, and delayed union of the docking site [6–9]. As a result, an implantable lengthening nail (ILN) [10] was invented to solve these issues.

The implantable lengthening nail is a kind of intramedullary distraction device that has been widely used in limb lengthening and replaced the mainstream therapy apparatus due to its effectiveness in treating LLD without complications. The first implantable lengthening nail, the Bliskunov intramedullary nail, was driven by a mechanical ratchet system and controlled via the rod bolted to the pelvis [5]. Baumgart et al. proposed a motorized implantable lengthening nail with a subcutaneous antenna, where the electric currents for powering the motor are generated via an induction method [11,12]. However, the wire between the antenna and the motor might fail to function due to daily routine or tissue fluid corrosion.

The Orthofix company designed an implantable lengthening nail, the Intramedullary Skeletal Kinetic Distractor (ISKD), which has a magnet encased within the tip of the threaded rod showing a new driving mechanism concept. However, the Food and Drug Administration (FDA) banned ISKD from the market because of its unstable and uncontrollable distraction rate [13–17]. A. Soubierian proposed an internal rotating magnet implantable lengthening nail (Phenix) driven by the circular motion of a hand-held magnet around the limb. Even though Phenix demonstrates outstanding clinical results, it is less known [18].

The NuVasive company designed an internal magnet implantable lengthening nail and two external magnets housed in a computer-controlled power assembly. PRECICE (P1), a welded-split structure implantable lengthening nail, was the first implantable lengthening nail marketed in the United States. However, P1 was unbearable for weight-bearing, resulting in a rift appearing at the welding part after implantation [16]. Therefore, an upgrade version of P1, PRECICE2 (P2), utilized a seamlessly connected structure, giving higher bending strength than P1 [10,17]. P2 achieved great accomplishment in LLD clinical treatment but still had some shortcomings, e.g., full weight-bearing before the consolidation phase is prohibited and temporarily decelerating or stopping distraction [19,20]. NuVasive's latest and largest weight-bearing implantable lengthening nail, PRECICE STRYDE, allowed the patient to proceed with their daily routine quicker. Yet, its material was unendurable to tissue fluid, and its non-hermetically seal design corroded the internal and external of the implantable lengthening nail [21–24]. Despite its remarkable success, several problems still exist in PRECICE, such as insufficient rigidity, unreliable distraction rate, and material corrosion.

In this study, we propose an implantable lengthening nail based on an electromagnetic configuration, Intramedullary Skeletal Distractor Robot (ISDR). ISDR possesses a high reduction ratio (exceeding 150), large distraction force (exceeding 1000 N), and sturdy structure stiffness characteristics for treating LLD. The essay is organized in the following way. Section 2 describes the electromagnetic configuration synthesis in terms of the human lower extremity and presents the conceptual design of the ISDR. Next, Section 3 investigates the mechanical stiffness and distraction forces that commonly trouble the clinician during limb lengthening. Section 4 shows the simulation and experiment results of the designed ISDR, while Section 5 optimizes the electromagnetic properties. Finally, Section 6 concludes.

2. Configuration Synthesis of ISDR

In this section, the configuration synthesis of ISDR is investigated by setting its design guidelines based on lower extremity characteristics for conceptual design.

2.1. Design Guidelines for ISDR Based on Human Lower Extremity Characteristics

According to medical reports, the diameter of the femur in Asian men is 11.08 ± 1.9 mm, the length of the femur is 418 ± 20 mm, and the circumference of the thigh is 53.9 ± 6.3 cm [25,26].

Ji et al. showed that the frequency of men walking has a range of 1.77 ± 0.36 Hz [27]. The values of the lower extremity are chosen nearly to their extreme value for general purpose (listed in Table 1).

Table 1. Lower extremity of the subject.

Category	Represent	Value
Human body	Weight	70 Kg
	Isthmus diameter	10 mm
Characteristics	Femur length	400 mm
	Thigh circumference	60 cm
Motion	Walking frequency	2 Hz

Generally, the human walking gait cycle is divided into six phases: heel strike (HS), foot-flat (FF), mid stance (MS), heel-off (HO), toe-off I (TO-I), and toe-off II (TO-II) [28] (shown in Figure 1). The maximal force occurs during the MS phase, around 40% to 60% of the stance phase [29]. Based on this, we estimate the forces acting on the femur in three directions (frontal force, lateral force, and axial force) among six phases, as listed in Table 2.

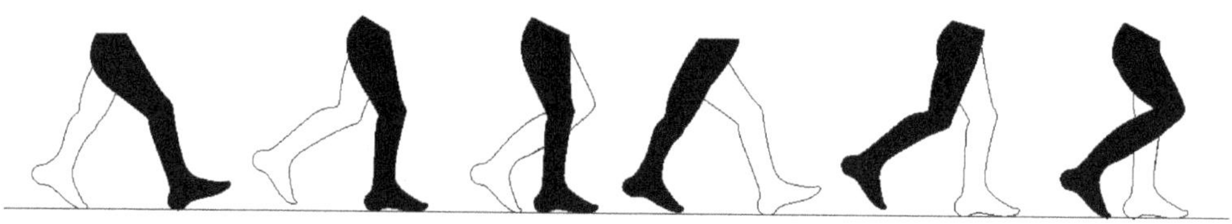

Figure 1. Six phases of the human walking gait cycle.

Table 2. Average frontal, lateral, and axial forces acting on the femur within six gait phases.

Gait Phases	Frontal Force (N)	Lateral Force (N)	Axial Force (N)
HS	83.79	−64.26	349.66
FF	83.79	−139.23	598.38
MS	83.79	−64.26	789.87
HO	83.79	−64.26	557.45
TO-I	251.37	0	0
TO-II	83.79	0	359.03

On the basis of these prerequisites, the priority of ISDR design guidelines is the body characteristics. After setting design guidelines, the driving method becomes our main concern. Although magnetic configurations achieve massive success in treating LLD, the limitation of external magnets is that their properties are unadjustable in various cases. Therefore, in this study, electromagnetic configurations are used in designing ISDR.

2.2. Conceptual Design of ISDR

The configuration of the intramedullary lengthening nail system includes two parts: an implantable lengthening nail implanted inside the medullary canal and an internal magnet driver placed outside the limb. The internal magnet driver generates rotating magnetic fields that spin the permanent magnet of the implantable lengthening nail to endow a force for bone distraction.

Clinicians make an incision near the hip joint and femoral shaft before implanting the implantable lengthening nail into the limb. Then, they use a cannulated drill to enlarge the medullary canal to facilitate implantation. After completing the enlargement procedure, they apply an osteotomy by using the osteotome, followed by the insertion of the implantable lengthening nail into the medullary canal and fixed tightly by bone screws at both the upper and lower ends. Once the clinicians complete implantation surgery, they can place the internal magnet driver near the limb to drive the implantable lengthening nail magnet for treating LLD.

There are some differences between ISDR and PRECICE (Figure 2a–c). PRECICE places its permanent magnet near the proximal femoral shaft, while ISDR places its permanent magnet near the distal femoral shaft. This design signifies that the air gap between the permanent magnet and the driver is minor, revealing the driving torque for the bone distraction is much more stable and forceful. Moreover, the transmission component of ISDR has a longer lifespan compared to PRECICE because the transmission of the axial load from the distraction nail towards the thrust bearing and protective shell occurs without passing through the gearbox. Last but not least, the distraction force of ISDR is adjustable and controllable by tuning the electromagnetic driver parameters akin to the Permanent Magnet Brushless Direct Current (PMBLDC) motor model.

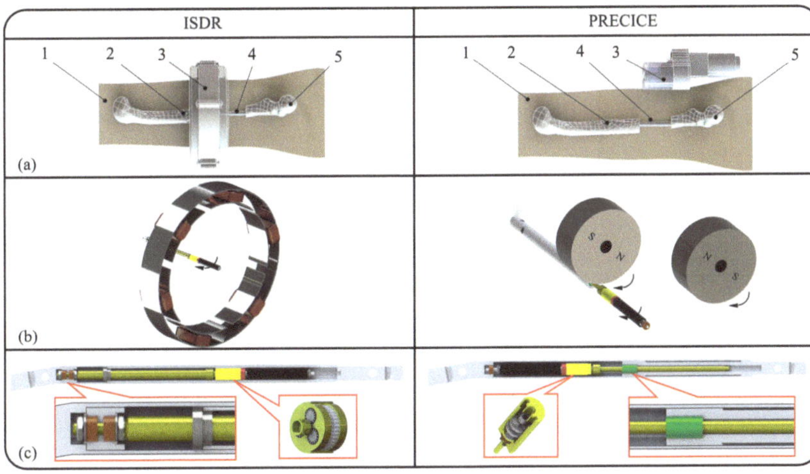

Figure 2. The comparison figure between ISDR and PRECICE. (**a**) The schematics of intramedullary lengthening nail system between ISDR and PRECICE. 1—thigh, 2—bone screws, 3—internal magnet driver, 4—implantable lengthening nail, 5—femur; (**b**) the schematics of driving method between ISDR and PRECICE; (**c**) the schematics of intramedullary lengthening nail between ISDR and PRECICE.

2.2.1. Implantable Lengthening Nail

Figure 3 illustrates that the proposed implantable lengthening nail distraction mechanism is well-protected by a protective shell. The mechanical stiffness of implantable lengthening nail is guaranteed by a titanium alloy with good bio-compatibility (non-toxic and high resistance to oxidization), lightweight, and rigid characteristics, which is commonly used in surgical transplants [30–32].

The implantable lengthening nail adopts a distraction structure design, where the distraction mechanism is placed within the distraction nail. The distraction mechanism consists of an internal permanent magnet, a gearbox that can be either a planetary gearbox or Rotate Vector (RV) gearbox, a lead screw, and its nut. The end of the lead screw is connected to the shafting system, which is fit or screwed tightly into the protective shell. This design provides a rotational degree of freedom for the lead screw, and its nut moves freely along with the distraction nail inside the protective shell since they are transition fit. The screw holes are designed in conventional patterns at both ends of the protective shell and distraction nail for fixation on the extremity medullary canal.

Once the electromagnetic driver is energized, the rotating magnetic field starts to drive the internal permanent magnet to rotate. The driving torque is amplified by the gearbox and transmitted to the lead screw. The thrust bearing is placed on the shafting tube so that the axial load acts directly on the protective shell without passing through the distraction mechanism. The distal bone begins to distract with the assistance of the distraction nail, meaning bone lengthening occurs as long as sufficient torque is provided.

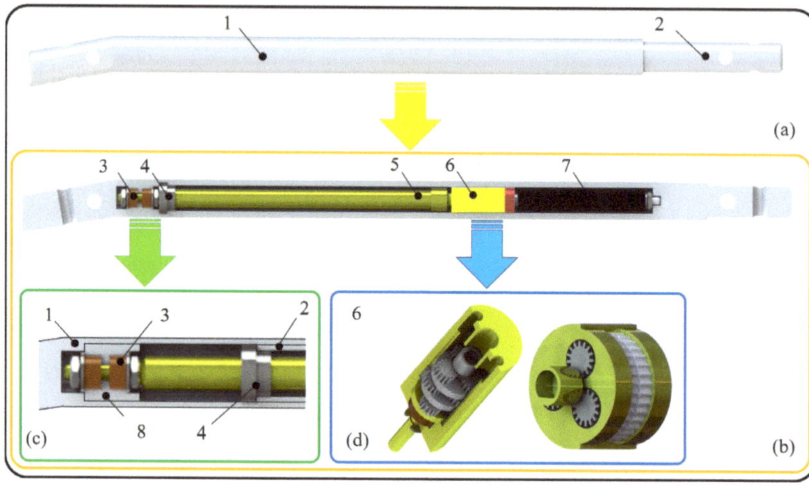

Figure 3. Prototype of ISDR implantable lengthening nail. 1—protective shell, 2—distraction nail, 3—thrust bearing, 4—lead screw nut, 5—lead screw, 6—gearbox, 7—internal permanent magnet, 8—shafting tube. (**a**) The schematics of implantable lengthening nail; (**b**) cross-section of implantable lengthening nail; (**c**) axial load acts on the shafting system in the implantable lengthening nail; (**d**) cross-section of planetary gearbox and RV gearbox.

2.2.2. Electromagnetic Driver

The electromagnetic driver acts as a PMBLDC motor stator group, providing the external rotating magnetic field required to drive the internal permanent magnet, as shown in Figure 4.

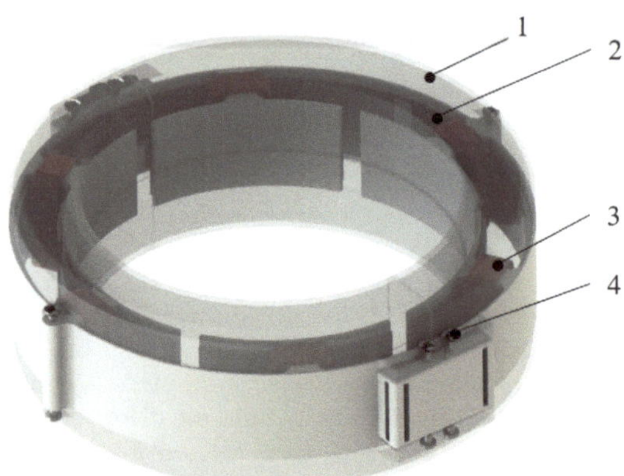

Figure 4. Prototype of electromagnetic driver. 1—nylon protective layer, 2—stator core, 3—copper wire, 4—self-tapping screws.

The electromagnetic driver mainly consists of a stator core, self-tapping screws, nylon protective layers, and copper wire coils. The electromagnetic driver is designed as a tuck-in design for the convenience of wearing. The stator core and middle nylon protective layer are attached firmly by screwing them together with the self-tapping screws. The function of the nylon protective layer is to prevent the sharp edges of silicon steel sheets from accidentally cutting the patient or current leakage. Copper wire coils wind around the stator core slot with a distributed half-coiled pattern. The upper and lower nylon protective layers are screwed after winding, forming a closed ring structure.

3. Mechanics and Dynamics Evaluation Index of ISDR

In this section, the mechanics and dynamics of ISDR are evaluated via mechanical stiffness and torques, and mechanistic indexes are quantitatively simulated by finite element analysis (FEA).

3.1. Mechanical Stiffness of the Lengthening Nail

Mechanical stiffness is determined by examining that the product stress value is lower than its material yield tensile strength. Therefore, the mechanical stiffness of implantable lengthening nails is verified via FEA, which is frequently used in computer-aided engineering (CAE) [33,34] to ensure its capability for withstanding forces under walking gait phases.

FEA is an approximating function (Interpolating functions/Shape functions) that expresses the unknown field variables by dividing the solution regions into small elements [35]. The Von-Mises stress obtained via FEA must not exceed the material yield tensile strength, which can be written as [36]:

$$\sigma \leqslant [\sigma], \tag{1}$$

where σ is the Von-Mises stress, and $[\sigma]$ is the yield tensile strength.

On the other hand, the dynamic behavior is determined via modal analysis in the patterns of damping factors, natural frequencies, and mode shapes [37]. Resonant damage must be avoided as it shortens the lifespan of implantable lengthening nails and thus endangering the patient. Hence, the walking frequency of the patient must keep away from the natural frequency range of implantable lengthening nails. With the aid of FEA, the natural frequency of implantable lengthening nails is also determined.

3.2. Torques of ISDR

3.2.1. Lifting Torque for Bone Distraction by Implantable Lengthening Nail

The distraction force of ISDR is a crucial indicator when the clinician is treating LLD. Based on the research work of Zhang et al. [38], the distraction rate in the range of 0.91 ± 0.41 mm/day is more effective for the treatment. On the other hand, the distraction period (days) relates logarithmically to the distraction force of body mass (N/kg) [39]. In other words, if the body mass is heavier or the distraction period is longer, surgery failure is more likely to happen due to insufficient distraction force.

Sufficient lifting torque is a must to achieve the distraction force required. The lifting torque T_L for a lead screw can be written as follows [40]:

$$T_L = \frac{Fd_m}{2}\left(\frac{l + \pi\mu d_m}{\pi d_m - \mu l}\right), \tag{2}$$

where F is the distraction force, d_m is the mean diameter, μ is the coefficient of friction for the thread, and l is the screw lead.

Due to the presence of the planetary gearbox, even with some efficiency loss in the planetary gearbox, the driving torque generated from the electromagnetic driver is much smaller than the required lifting torque. Therefore, the driving torque T_D is obtained in terms of Equation (3) [41–43]:

$$T_D = T_L \cdot \left(\frac{1}{e_1 i_1} \cdot \frac{1}{e_2 i_2} \cdots \frac{1}{e_n i_n}\right), \tag{3}$$

where e is efficiency, i is reduction ratio, and n is the number of stages. The driving torque T_D is rewritten as Equation (4) since the reduction ratio and efficiency of the planetary gearbox in each stage are equal. The relationship between lifting torque, driving torque and distraction force are illustrated in Figure 5 for better understanding.

$$T_D = T_L \cdot \frac{1}{(ei)^n}. \tag{4}$$

Although reduction ratio and the number of stages are effective ways of improving the distraction force, the total weight of the implantable lengthening nail became heavier. Hence, raising the driving torque of the electromagnetic driver is an alternative method.

Figure 5. The schematic of the relationship between T_L, T_D, and F.

3.2.2. Driving Torque for Actuation by Electromagnetic Driver

The electromagnetic driver and internal permanent magnet combination are equivalent to PMBLDC motors since their rotors rotate synchronously when the stator provides a rotating magnetic field. Therefore, the equations for the PMBLDC motor are fully applicable to the electromagnetic driver [44]. The torque of the PMBLDC motor is described as:

$$T_D = 2PB_g I_s n_s L R_{si}, \tag{5}$$

where P is the number of poles, B_g is the air gap flux density in the middle 120° of a pole, I_s is the DC source current, n_s is the number of turns per slot, L is the active length of the motor, and R_{si} is the stator inner radius. Based on Equation (5), source current and number of turns are the most productive way of driving the torque increment. Yet, these variables are insufficient to determine the geometric parameters of the electromagnetic driver, so their relations with torque are given by:

$$A_{slot} = \frac{1}{2}(w_{st} + w_{sb})d_s = \frac{n_s}{K_{fill}} A_{Cu}, \tag{6}$$

$$A_{Cu} = \frac{\pi d_{Cu}^2}{4}, \tag{7}$$

where A_{slot} is the slot area, w_{st} is the top slot width, w_{sb} is the bottom slot width, d_s is the slot depth, K_{fill} is the slot fill factor, A_{Cu} is the coil cross-sectional area, and d_{Cu} is the copper wire diameter. In the meantime, the remaining geometric parameters in the stator core are obtainable from the function of three geometric parameters in the slot area.

$$\tau_{slot} = \frac{2\pi R_{si}}{N_{slot}}, \tag{8}$$

$$\tau_c = 2\pi\left(R_{si} + \frac{1}{2}d_s\right)\frac{1}{P}, \tag{9}$$

$$w_{sb} = \tau_{slot} - w_t, \tag{10}$$

$$d_s = R_{so} - R_{si} - 1.5w_{bi}, \tag{11}$$

$$w_{st} = \frac{2\pi(R_{si} + d_s)}{N_{slot}} - w_t, \tag{12}$$

where τ_{slot} is the slot pitch, N_{slot} is the number of slots, τ_c is the coil pitch, w_t is the tooth width, R_{so} is the stator outer radius, and w_{bi} is the back iron length. These geometric parameters are presented in Figure 6.

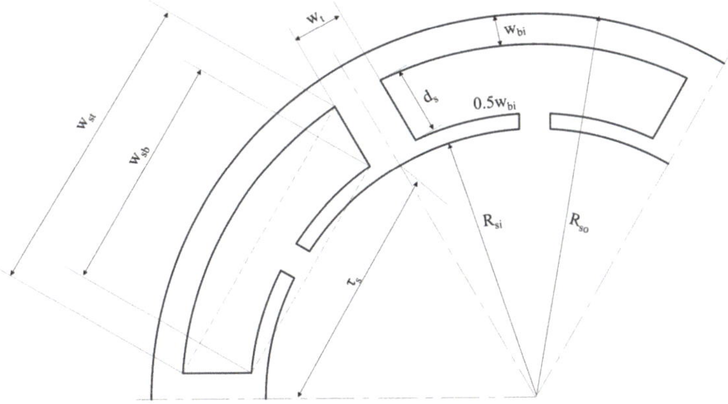

Figure 6. Geometric parameters of the electromagnetic driver stator core.

4. Simulation and Experimental Validation

In this section, both mechanical stiffness and torques are analyzed and simulated while the driving torque is tested with the torque measurement system.

4.1. Analysis and Simulation of Mechanical Stiffness

Referring to the characteristics given in Table 1, the implantable lengthening nail is designed according to the parameters listed in Table 3.

Table 3. Implantable lengthening nail and its components' design specifications.

Product/Component	Represent	Value/Parameter
Implantable lengthening nail	Material	Ti-6Al-4V
	Diameter	10 mm
	Total length (before distraction)	245 mm
	Maximum allowable distraction length	80 mm
Internal permanent magnet	Material	NdFeB
	Inner diameter	2.5 mm
	Outer diameter	7.5 mm
	Coating material	Raw Epoxy (BE)
	Length	40 mm
Planetary gearbox	Reduction ratio per stage	5.33
Lead screw	Screw specification	M6

The implantable lengthening nail starts to distract along with the distal bone after clinicians perform an osteotomy. In our case, the maximal allowable distraction length showed an 80 mm gap between the proximal and distal bone at the end of the distraction period and before the consolidation phase (shown in Figure 7). Three principal forces acting on the femur throughout six gait phases (shown in Table 2) are applied at the lateral and medial condyle, while fixed support is applied at the femoral head. FEA is used to examine the implantable lengthening nail maximal stress and minimal natural frequency.

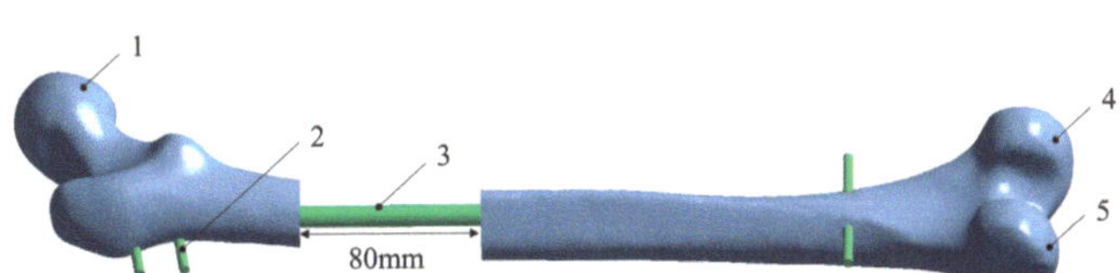

Figure 7. Boundary condition applied in FEA. 1—femoral head, 2—bone screw, 3—implantable lengthening nail, 4—medial condyle, 5—lateral condyle.

According to the results of Table 4 and Figure 8, we noticed that the highest value of Von-Mises Stress occurs around the 80 mm gap, especially during the fifth phase of the walking gait cycle (around 952.15 MPa). The material has a slightly higher yield tensile strength (1100 MPa) than the maximal stress, so it is encouraged to use a crutch before the consolidation phase. Based on the boundary condition given for the static structure, the basis natural frequency is 28.823 Hz, a much larger value than the walking frequency (2 Hz), as shown in Table 5.

Table 4. The maximum and average stress on implantable lengthening nail during walking gait cycle.

Gait Phases	Stress Max. (MPa)	Stress Avg. (MPa)
HS	773.99	68.363
FF	936.24	93.788
MS	838.36	76.365
HO	780.27	72.147
TO-I	952.15	145.6
TO-II	538.59	53.641

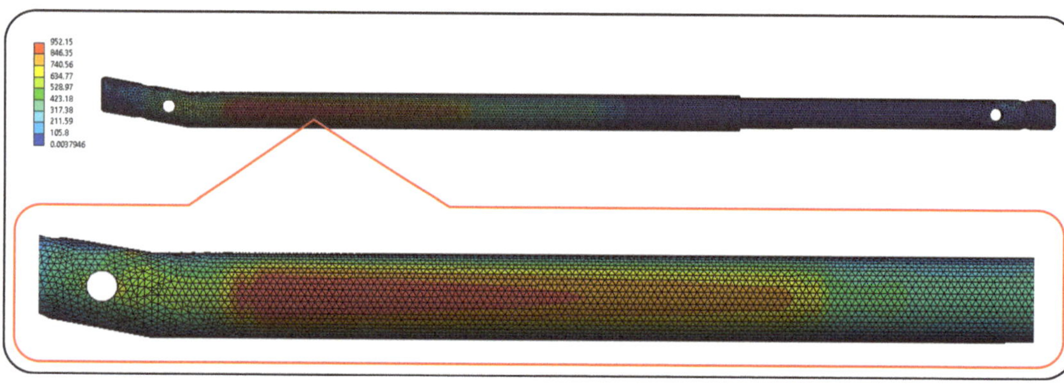

Figure 8. Result of static structural analysis based on walking gait cycle.

Table 5. The first 6 modes of implantable lengthening nail's natural frequencies.

Mode	Frequency (Hz)
1	28.823
2	29.467
3	345.79
4	365.75
5	402.24
6	1111.3

4.2. *Analysis and Simulation of Torques*

4.2.1. Lifting Torque of Implantable Lengthening Nail

The distraction rate commonly used by clinicians is 1 mm/day, and the distraction force of body mass is around 9.5 N/kg at the end of the distraction period (80 days) [39]. Therefore, we refer to the bodyweight in Table 1 and obtain the minimum distraction force required is 665 N. However, a safety factor should be considered since the distraction force of body mass might vary among the patients. So, the safety factor is set to 1.5, which gives the final distraction force of approximately 1000 N.

The geometric sizes of the M6 lead screw $d_m = 5.1$ mm, $l = 0.75$ mm, and $\mu = 0.1$ are substituted into Equation (2), giving us a value of lifting torque of 376.1 Nmm. Once the lifting torque is confirmed, we can compute the driving torque required for bone distraction via Equation (4). Generally, the efficiency of planetary gear per stage is $e = 0.97$, giving us a driving torque of approximately 2.721 Nmm, which indicates the minimum value to guarantee the distraction.

4.2.2. Driving Torque of Electromagnetic Driver

According to the design guidelines set in Section 2, we design a three-phase six salient poles stator core to serve as the electromagnetic driver and two magnet poles to be used as the internal permanent magnet of the implantable lengthening nail. The design specifications of the electromagnetic driver components are listed in Table 6.

Table 6. Electromagnetic driver's components design specifications.

Product/Component	Represent	Value/Parameter
Stator core	Material	M27_26G
	Inner radius (R_{si})	95 mm
	Outer radius (R_{so})	127.5 mm
	Length (L)	40 mm
	Top slot width (w_{st})	100 mm
	Bottom slot width (w_{sb})	75 mm
	Slot depth (d_s)	13 mm
	Number of slots (N_{slot})	6
	Tooth width (w_t)	29.72 mm
	Back iron width (w_{bi})	10 mm
Copper wire coils	Wire diameter (d_{Cu})	1.725 mm
	Number of turns in slot (n_s)	75
	Number of strands	1
	Slot fill factor (K_{fill})	0.16
	Winding method	Distributed half-coiled

A microcontroller and, alternatively, analog or digital circuits can implement the commutation of PMBLDC motor current source even if it is a DC. The PMBLDC motor commutation order is one winding energized positive, one winding energized negative, and one winding non-energized. According to the wire gauge standard, an allowable current for a 1.725 mm diameter wire is 9.2 A. Based on safety and power consumption considerations, we restrict the electric current of the electromagnetic driver to around 7 A, which turned out to be sufficient to generate the driving torque.

Some parameters of the PMBLDC motor model shown in Equation (5), such as the active length of the motor and the inner radius of the stator core, are highly dependent on the lower extremity characteristic. Therefore, the DC source current and number of turns have become the key factors to adjust the driving torque. After trial and error in adjusting these parameters, the simulation showed that the average driving torque of the magnetic drive is 8.2 Nmm (shown in Figure 9) which is extremely large compared to the minimum lifting torque of 2.721 Nmm.

4.3. Experiment Validation of Driving Torque

A simple driving torque experiment is conducted, and the experimental setup is shown in Figure 10. A microcontroller unit is adopted to control the commutation order of the stator core while the power supply powers them. The permanent magnet is erected in the middle of the stator core. As soon as the external rotating magnetic field is generated, a torque measurement system is utilized to measure the driving torque, and the computer records the following data.

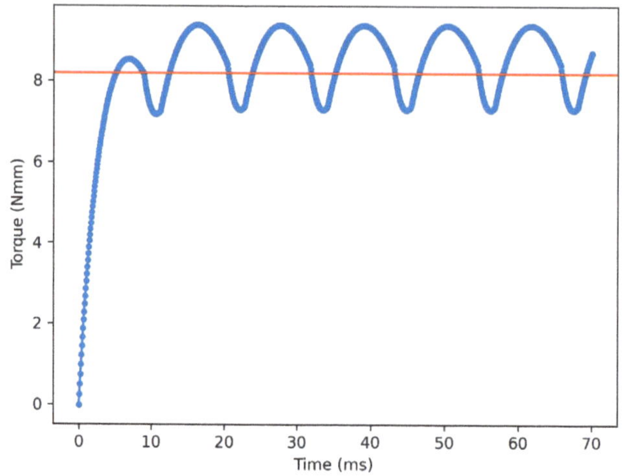

Figure 9. Torque generated by PMBLDC motor model under 6.7 A DC current source.

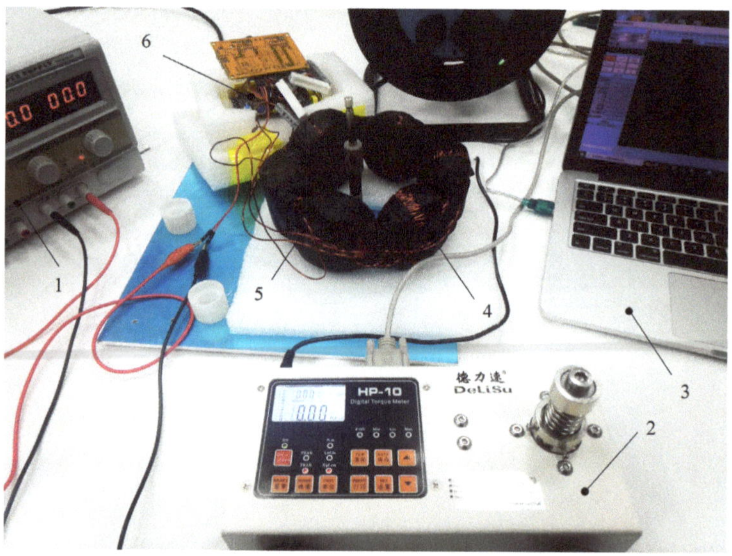

Figure 10. Experiment setup to test permanent magnet driving torque. 1—power supply, 2—torque measurement system, 3—computer, 4—stator core, 5—permanent magnet, 6—microcontroller unit.

The maximum and minimum values of experimental driving torque data are around 9.8 Nmm, and 4.9 Nmm, respectively, and its average value is about 6 Nmm, as shown in Figure 11. The reasons that cause differences in both outcomes may be the frictional force in the permanent magnet shaft, the copper winding is winded manually, the permanent magnet is not in the center position, and the torque measuring system is handheld. The driving torque can be measured more accurately if the problem stated is completely solved.

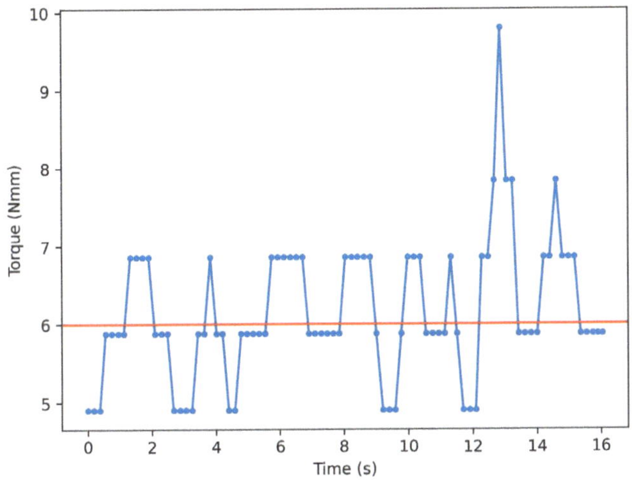

Figure 11. The experimental results of permanent magnet driving torque.

5. Optimization of the Electromagnetic Driver Design

Although the driving torque is sufficient for bone distraction, there are sacrifices in the electromagnetic driver's overall weight and power loss. Therefore, the genetic algorithm (GA) procedure is introduced to achieve the optimal value of driving torque.

5.1. Problem Formulation

The goal is to find the optimum geometric parameters by minimizing the objective function when the requirements are satisfied. The main emphasis of electromagnetic drivers is driving torque capability, followed by the apprehension of wearable design overall weight and the PMBLDC motor model power loss. In addition to these criteria, other objectives such as magnetic flux leakage and cogging torque minimization are also suitable for optimization.

Generally, the objectives are reorganized into a mathematical expression as a function of geometric parameters. As mentioned above, these objective variables are driving torque (T_D), overall weight (W_{total}), and power loss (P_{total}). The driving torque of the electromagnetic driver is reconstructed as a relation with geometric parameters by referring to Equations (5)–(7):

$$T_D = \frac{K_m I_s (w_{st} + w_{sb}) d_s}{A_{Cu}}, \tag{13}$$

where $K_m = PB_g L R_{si} K_{fill}$. These geometric parameters are further replaced by back iron length (w_{bi}), number of slots (N_{slot}), and tooth width (w_t) as is shown in Equations (8)–(12). Hence, these basic parameters and the copper wire diameter (d_{Cu}) are selected as the variables of driving torque, namely the genes in a chromosome for the GA procedure.

The overall weight of the electromagnetic driver is obtained by summing the mass of the stator core and the copper winding together, after multiplying steel density and copper density with their volume, respectively. Since the dimensions of the stator core and copper winding rely on the electromagnetic driver geometry, the following expression is written:

$$W = W_{Fe} + W_{Cu} = \rho_{Fe} V_{Fe} + \rho_{Cu} V_{Cu}, \tag{14}$$

$$V_{Fe} = \pi (R_{so}^2 - R_{si}^2) L - N_{slot} A_{slot} L, \tag{15}$$

$$V_{Cu} = n_s N_{slot} A_{Cu} (L + \tau_c), \tag{16}$$

where W_{Fe} is the mass of silicon steel sheet, W_{Cu} is the mass of copper winding, ρ_{Fe} is the density of silicon steel sheet, and ρ_{Cu} is the density of copper. After that, the substitution of variables is conducted to achieve the identical variable as the driving torque.

On the other hand, power loss of the PMBLDC motor model is divided into three categories: electrical, magnetic, and mechanical [45]. Mechanical loss is negligible in the power loss since efficiency loss in the planetary gearbox and safety factors were previously considered in the lifting torque. Thus, electrical loss and magnetic loss are discussed here. Electrical loss originates from the resistance of copper winding as well as magnetic loss comes from hysteresis and eddy current losses, which vary nonlinearly with frequency and magnetic flux density [44]. The power loss is obtained through the summation of electrical loss and magnetic loss, as shown in the equation below.

$$P = P_{elec} + P_{mag}, \tag{17}$$

$$P_{elec} = \frac{3I_s^2 P n_s \zeta_{Cu}(L + \tau_c)}{A_{Cu}}, \tag{18}$$

$$P_{mag} = L(f, B_{max})W_{Fe}, \tag{19}$$

and

$$I_s = \frac{v}{r} = \frac{vA_{Cu}N_{slot}}{12\zeta_{Cu}n_s(L + \tau_c)}, \tag{20}$$

where P_{elec} is the electrical loss, P_{mag} is the magnetic loss, ζ_{Cu} is the resistivity of copper, $L(f, B_{max})$ is the core loss of silicon steel, v is the source voltage, and r is the resistance of each phase. The geometric parameters are changed to typical variables in driving torque.

5.2. Optimization Procedure

5.2.1. Objective Functions

The optimization variables that need to be optimally discovered are selected and represented as a vector of x. As the previous section mentioned, the basic parameters are chosen as the optimization variables, which can be written as follows:

$$x = [w_{bi} \quad N_{slot} \quad w_t \quad d_{Cu}]^T. \tag{21}$$

The lower and upper bound values of each variable are set, which are denoted as x_{min} and x_{max}, respectively. The pattern of an objective function may vary according to the application and requirement of the PMBLDC motor. In this paper, the objective function comprises the overall weight, power loss, and inverse of the driving torque that are meant to be minimized. Both weighting factors and normalization are considered in the objective function to adjust the significance of each objective and scale their values to an equivalent order of magnitude.

$$\min f(x) = \lambda_1 \frac{T_{d0}}{T_d(x)} + \lambda_2 \frac{W(x)}{W_0} + \lambda_3 \frac{P(x)}{P_0}, \tag{22}$$

where λ_1, λ_2 and λ_3 are the weighting factors, T_{d0}, W_0 and P_0 are the randomization values of driving torque, overall weight and power loss, respectively.

Apart from determining the objective function, constraints are added based on the PMBLDC motor limitations of heat, electrical and mechanical energy. Since the copper wire diameter is an optimization variable, the source current is set as an inequality constraint that satisfies the allowable current for a particular wire gauge. Hence, the following objective function is rewritten as [45]:

$$\min f(x) = \lambda_1 \frac{T_{d0}}{T_d(x)} + \lambda_2 \frac{W(x)}{W_0} + \lambda_3 \frac{P(x)}{P_0} + \frac{1}{\epsilon}\left[f_u(1 - \frac{I_{allow}}{I_s})\right], \tag{23}$$

where ϵ is a small constant, I_{allow} is the allowable current for a specific wire gauge, and

$$f_u(x) = \frac{1}{1 + e^{-\sigma x}}, \tag{24}$$

where σ is a large constant. The penalty function is introduced in Equation (24) to override the multi-objective function result whenever I_s exceeds I_{allow} giving larger values that deviate from the local or global optimum.

5.2.2. Genetic Algorithms

GA is an evolutionary algorithm that aims to find the optimal solution based on inspiration from a natural selection such as selection, crossover, and mutation, which are discovered in nature [46]. An individual represents the solution domain, and a fitness function evaluates the solution domain.

First, the values of the objective parameters are randomly selected within their limits to form the initial populations for minimizing the objective functions by examining the fitness value. If the fitness value satisfies the stopping criterion, the procedure ends immediately. On the contrary, new populations are formed by selection, crossover, and mutation in each generation and adopted in the upcoming iterations until the fitness value is optimally found or reaches the maximum number of iterations. The fittest parameters are utilized in the electromagnetic driver to promote its performance. The flowchart of a simple GA procedure is shown in Figure 12.

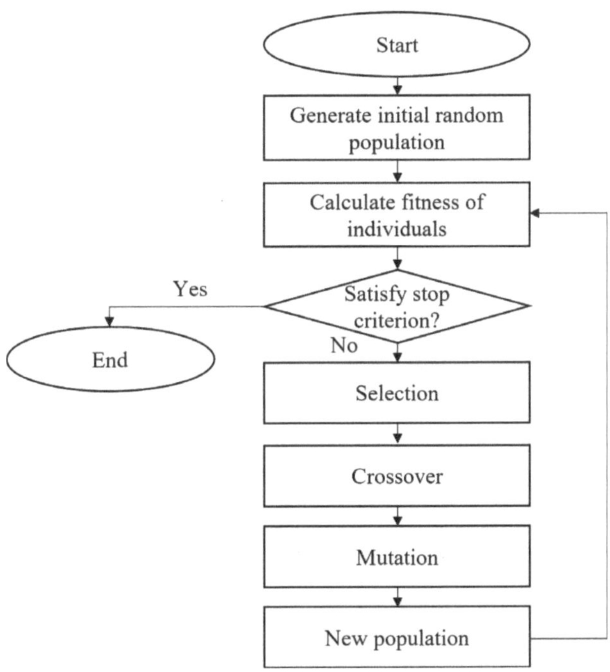

Figure 12. GA procedure flowchart.

5.3. Optimization Result

Before solving the optimization problem, the constant parameters and optimization technical parameters needed for initialization of the objective function are listed in Table 7. The lower and upper bounds of the optimization variables are confirmed, which contributed to the optimum value, see Table 8. Other characteristics of the optimized stator core are listed in Table 9. The geometric parameters and variables from the GA procedure are utilized in the simulation and gave the results shown in Figure 13 and Table 10.

Table 7. List of constant parameters and their values.

Parameter	Value	Parameter	Value
R_{si}	95 mm	K_{fill}	0.16
R_{so}	127.5 mm	B_g	2 mT
R_{ri}	2.5 mm	ρ_{Fe}	7650 kgm^{-3}
R_{ro}	7.5 mm	ρ_{Cu}	8960 kgm^{-3}
L	40 mm	ζ_{Cu}	1.72×10^{-8} Ωm
P	2	$L(f, B_max)$	3.76 Wkg^{-1}
λ_1	1.0	V	24 V
λ_2	0.1	ϵ	0.2
λ_3	0.1	σ	1000

Table 8. Optimum, lower, and upper bounds of the optimization variables.

Variables	Min.	Max.	Optimum
w_{bi}	10 mm	19.5 mm	10 mm
N_{slot}	6	30	6
w_t	10 mm	19.5 mm	10.5 mm
D_c	0.75 mm	2 mm	1.45 mm

Table 9. Characteristics of the optimized stator core using GA.

Characteristics	Values	Characteristics	Values
w_{st}	107.5 mm	A_{slot}	1719.375 mm^2
w_{sb}	89 mm	A_{Cu}	1.651 mm^2
d_s	17.5 mm	τ_{slot}	99.5 mm
n_s	175	τ_c	325.94 mm

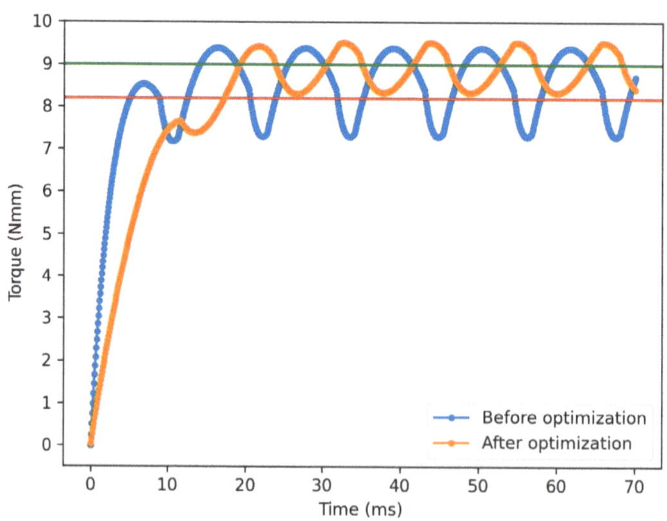

Figure 13. The driving torque of electromagnetic driver before and after optimization.

Table 10. Objective functions before and after optimization.

Obj. Functions (Bef.)	Values	Obj. Functions (Aft.)	Values
T_d	8.2 Nmm	T_d	9 Nmm
W	8.24 kg	W	9.455 kg
P	71.624 W	P	66.045 W

As attested by the results, the performance of the electromagnetic driver shows significant improvement in driving torque and power loss by adjusting the stator core slot area, the number of turns, and the copper wire diameter. Nevertheless, the total weight increases but is still in the acceptable range. Moreover, we notice that the start-up takes more time than the previous one, but it is not the primary concern. Overall, the electromagnetic driver becomes more efficient than before optimization.

6. Conclusions

In this paper, we proposed an inverted distraction structure design implantable lengthening nail where the internal permanent magnet is placed nearer to the distal femoral shaft, which enables it to receive the driving torque without effort. Furthermore, the magnetic configuration is substituted with an electromagnetic one for ease of controlling the driving torque required for bone distraction. To this end, we designed an implantable lengthening nail and its electromagnetic driver based on the human lower extremity. Then, the mechanical stiffness of the implantable lengthening nail is verified by simulating the human walking gait cycle granting us a value of nearly 950 MPa. Next, the lifting torque required for the implantable lengthening nail to conduct bone distraction is calculated, followed by analyzing and experimenting with the driving torque of the electromagnetic driver. Lastly, the parameters of the electromagnetic driver are adjusted via the genetic algorithm to improve its driving torque to 9 Nmm, which is increased by 8.89% compared with the trial and error method.

In conclusion, the proposed ISDR meets the requirements that clinicians are eager for, but a crutch is recommended to avoid unpredictable accidents. In future works, we will manufacture the prototype of ISDR and verify its capability. In addition, an animal experiment will be conducted to prove its effectiveness in treating LLD.

Author Contributions: Conceptualization, Z.N. and X.-j.L.; methodology, S.L. and Z.N.; formal analysis and validation, S.L. and K.H.; investigation, S.L. and Z.N.; resources, Z.N., F.X. and X.-j.L.; data curation, S.L., Z.N. and H.X.; experiment and validation, H.X. and H.L.; writing—original draft preparation, S.L.; writing—review and editing, Z.N. and H.X.; visualization, S.L.; supervision, Z.N., F.X. and X.-j.L.; project administration, Z.N. and X.-j.L.; funding acquisition, Z.N., F.X. and X.-j.L. All authors have read and agreed to the published version of the manuscript.

Funding: This work is supported by Zhongguancun (ZGC) open laboratory proof-of-concept (POC) projects under Grant 20210421084.

Data Availability Statement: Not applicable.

Conflicts of Interest: The authors declare no conflict of interest.

References

1. Shi, B.; Shu, H. Research progress of fully implantable intramedullary lengthening nail. *Chin. J. Orthop.* **2019**, *12*, 58–64.
2. Vogt, B.; Gosheger, G.; Wirth, T.; Horn, J.; Rödl, R. Leg length discrepancy—Treatment indications and strategies. *Dtsch. Ärzteblatt Int.* **2020**, *117*, 405–411. [CrossRef] [PubMed]
3. Guichet, J.M.; Spivak, J.M.; Trouilloud, P.; Grammont, P.M. Lower limb-length discrepancy. An epidemiologic study. *Clin. Orthop. Relat. Res.* **1991**, *272*, 235–241. [CrossRef]
4. Steen, H.; Terjesen, T.; Bjerkreim, I. Anisomelia. Clinical consequences and treatment. *Tidsskr. Nor. Laegeforen. Tidsskr. Prakt. Med. Raekke* **1997**, *117*, 1595–1600.
5. Birch, J.G. A brief history of limb lengthening. *J. Pediatr. Orthop.* **2017**, *37*, S1–S8. [CrossRef]
6. Fragomen, A.T.; Miller, A.O.; Brause, B.D.; Goldman, V.; Rozbruch, S.R. Prophylactic postoperative antibiotics may not reduce pin site infections after external fixation. *HSS J.* **2017**, *13*, 165–170. [CrossRef]

7. Bhave, A.; Shabtai, L.; Woelber, E.; Apelyan, A.; Paley, D.; Herzenberg, J.E. Muscle strength and knee range of motion after femoral lengthening: 2-to 5-year follow-up. *Acta Orthop.* **2017**, *88*, 179–184. [CrossRef]
8. Kazmers, N.H.; Fragomen, A.T.; Rozbruch, S.R. Prevention of pin site infection in external fixation: A review of the literature. *Strateg. Trauma Limb Reconstr.* **2016**, *11*, 75–85. [CrossRef]
9. Liu, Y.; Yushan, M.; Liu, Z.; Liu, J.; Ma, C.; Yusufu, A. Complications of bone transport technique using the Ilizarov method in the lower extremity: A retrospective analysis of 282 consecutive cases over 10 years. *BMC Musculoskelet. Disord.* **2020**, *21*, 354. [CrossRef]
10. Paley, D. PRECICE intramedullary limb lengthening system. *Expert Rev. Med. Devices* **2015**, *12*, 231–249. [CrossRef]
11. Black, S.R.; Kwon, M.S.; Cherkashin, A.M.; Samchukov, M.L.; Birch, J.G.; Jo, C.H. Lengthening in congenital femoral deficiency: A comparison of circular external fixation and a motorized intramedullary nail. *J. Bone Jt. Surgery. Am. Vol.* **2015**, *97*, 1432–1440. [CrossRef]
12. Accadbled, F.; Pailhé, R.; Cavaignac, E.; de Gauzy, J.S. Bone lengthening using the Fitbone® motorized intramedullary nail: The first experience in France. *Orthop. Traumatol. Surg. Res.* **2016**, *102*, 217–222. [CrossRef]
13. Cole, J.D.; Justin, D.; Kasparis, T.; DeVlught, D.; Knobloch, C. The intramedullary skeletal kinetic distractor (ISKD): First clinical results of a new intramedullary nail for lengthening of the femur and tibia. *Injury* **2001**, *32*, 129–139. [CrossRef]
14. Lee, D.H.; Ryu, K.J.; Song, H.R.; Han, S.H. Complications of the Intramedullary Skeletal Kinetic Distractor (ISKD) in distraction osteogenesis. *Clin. Orthop. Relat. Res.* **2014**, *472*, 3852–3859. [CrossRef]
15. Mahboubian, S.; Seah, M.; Fragomen, A.T.; Rozbruch, S.R. Femoral lengthening with lengthening over a nail has fewer complications than intramedullary skeletal kinetic distraction. *Clin. Orthop. Relat. Res.* **2012**, *470*, 1221–1231. [CrossRef]
16. Schiedel, F.M.; Vogt, B.; Tretow, H.L.; Schuhknecht, B.; Gosheger, G.; Horter, M.J.; Rödl, R. How precise is the PRECICE compared to the ISKD in intramedullary limb lengthening? Reliability and safety in 26 procedures. *Acta Orthop.* **2014**, *85*, 293–298. [CrossRef]
17. Paley, D.; Harris, M.; Debiparshad, K.; Prince, D. Limb lengthening by implantable limb lengthening devices. *Tech. Orthop.* **2014**, *29*, 72–85. [CrossRef]
18. Thaller, P.H.; Fürmetz, J.; Wolf, F.; Eilers, T.; Mutschler, W. Limb lengthening with fully implantable magnetically actuated mechanical nails (PHENIX®)—Preliminary results. *Injury* **2014**, *45*, S60–S65. [CrossRef]
19. Nasto, L.A.; Coppa, V.; Riganti, S.; Ruzzini, L.; Manfrini, M.; Campanacci, L.; Palmacci, O.; Boero, S. Clinical results and complication rates of lower limb lengthening in paediatric patients using the PRECICE 2 intramedullary magnetic nail: A multicentre study. *J. Pediatr. Orthop. B* **2020**, *29*, 611–617. [CrossRef]
20. Iliadis, A.D.; Palloni, V.; Wright, J.; Goodier, D.; Calder, P. Pediatric lower limb lengthening using the PRECICE nail: Our experience with 50 cases. *J. Pediatr. Orthop.* **2021**, *41*, e44–e49. [CrossRef]
21. Galal, S.; Shin, J.; Principe, P.; Khabyeh-Hasbani, N.; Mehta, R.; Hamilton, A.; Rozbruch, S.R.; Fragomen, A.T. STRYDE versus PRECICE magnetic internal lengthening nail for femur lengthening. *Arch. Orthop. Trauma Surg.* **2021**. [CrossRef]
22. Jellesen, M.S.; Lomholt, T.N.; Hansen, R.Q.; Mathiesen, T.; Gundlach, C.; Kold, S.; Nygaard, T.; Mikuzis, M.; Olesen, U.K.; Rölfing, J.D. The STRYDE limb lengthening nail is susceptible to mechanically assisted crevice corrosion: An analysis of 23 retrieved implants. *Acta Orthop.* **2021**, *92*, 621–627. [CrossRef]
23. Hothi, H.; Bergiers, S.; Henckel, J.; Iliadis, A.D.; Goodier, W.D.; Wright, J.; Skinner, J.; Calder, P.; Hart, A.J. Analysis of retrieved STRYDE nails. *Bone Jt. Open* **2021**, *2*, 599–610. [CrossRef]
24. Frommer, A.; Roedl, R.; Gosheger, G.; Hasselmann, J.; Fuest, C.; Toporowski, G.; Laufer, A.; Tretow, H.; Schulze, M.; Vogt, B. Focal osteolysis and corrosion at the junction of Precice Stryde intramedullary lengthening device: Preliminary clinical, radiological, and metallurgic analysis of 57 lengthened segments. *Bone Jt. Res.* **2021**, *10*, 425–436. [CrossRef]
25. Xu, W.; Wang, M.; Jiang, C.M.; Zhang, Y.M. Anthropometric equation for estimation of appendicular skeletal muscle mass in Chinese adults. *Asia Pac. J. Clin. Nutr.* **2011**, *20*, 551–556.
26. Thiesen, D.M.; Ntalos, D.; Korthaus, A.; Petersik, A.; Frosch, K.H.; Hartel, M.J. A comparison between Asians and Caucasians in the dimensions of the femoral isthmus based on a 3D-CT analysis of 1189 adult femurs. *Eur. J. Trauma Emerg. Surg.* **2021**, *48*, 2379–2386. [CrossRef]
27. Ji, T.; Pachi, A. Frequency and velocity of people walking. *Struct. Eng.* **2005**, *84*, 36–40.
28. Song, M.; Guo, S.; Oliveira, A.S.; Wang, X.; Qu, H. Design method and verification of a hybrid prosthetic mechanism with energy-damper clutchable device for transfemoral amputees. *Front. Mech. Eng.* **2021**, *16*, 747–764. [CrossRef]
29. Duda, G.N.; Schneider, E.; Chao, E.Y. Internal forces and moments in the femur during walking. *J. Biomech.* **1997**, *30*, 933–941. [CrossRef]
30. Hansen, D.C. Metal corrosion in the human body: The ultimate bio-corrosion scenario. *Electrochem. Soc. Interface* **2008**, *17*, 31. [CrossRef]
31. Manam, N.S.; Harun, W.S.W.; Shri, D.N.A.; Ghani, S.A.C.; Kurniawan, T.; Ismail, M.H.; Ibrahim, M.H.I. Study of corrosion in biocompatible metals for implants: A review. *J. Alloy. Compd.* **2017**, *701*, 698–715. [CrossRef]
32. Hermawan, H.; Ramdan, D.; Djuansjah, J.R. Metals for biomedical applications. *Biomed. Eng.-Theory Appl.* **2011**, *1*, 411–430.
33. Nie, Z.; Wang, G.; Wang, L.; Rong, Y. A coupled thermomechanical modeling method for predicting grinding residual stress based on randomly distributed abrasive grains. *J. Manuf. Sci. Eng.* **2019**, *141*, 081005. [CrossRef]
34. Nie, Z.; Jung, S.; Kara, L.B.; Whitefoot, K.S. Optimization of part consolidation for minimum production costs and time using additive manufacturing. *J. Mech. Des.* **2020**, *142*, 072001. [CrossRef]

35. Bhavikatti, S.S. *Finite Element Analysis*; New Age International: New Delhi, India, 2005; pp. 1–4.
36. Hou, X.; Liu, Z.; Wang, B.; Lv, W.; Liang, X.; Hua, Y. Stress-strain curves and modified material constitutive model for Ti-6Al-4V over the wide ranges of strain rate and temperature. *Materials* **2018**, *11*, 938. [CrossRef] [PubMed]
37. Fu, Z.F.; He, J. *Modal Analysis*; Elsevier: Oxford, UK, 2001; pp. 1–11.
38. Zhang, J.; Zhang, Y.; Wang, C.; Qin, S. Research progress of intramedullary lengthening nail technology. *Chin. J. Reparative Reconstr. Surg.* **2021**, *35*, 642–647. [CrossRef]
39. Lauterburg, M.T.; Exner, G.U.; Jacob, H.A. Forces involved in lower limb lengthening: An in vivo biomechanical study. *J. Orthop. Res.* **2006**, *24*, 1815–1822. [CrossRef]
40. Budynas, R.G.; Nisbett, J.K. *Shigley's Mechanical Engineering Design*; Tata McGraw-Hill: New York, NY, USA, 2011; Volume 9, pp. 400–408.
41. Uicker, J.J.; Pennock, G.R.; Shigley, J.E.; Mccarthy, J.M. *Theory of Machines and Mechanisms*; Oxford University Press: New York, USA, 2003; pp. 188–201.
42. Pawar, P.V.; Kulkarni, P.R. Design of two stage planetary gear train for high reduction ratio. *Int. J. Res. Eng. Technol.* **2015**, *4*, 150–157.
43. Levai, Z. Structure and analysis of planetary gear trains. *J. Mech.* **1968**, *3*, 131–148. [CrossRef]
44. Carunaiselvane, C.; Jeevananthan, S. Generalized procedure for BLDC motor design and substantiation in MagNet 7.1.1 software. In Proceedings of the International Conference on Computing, Electronics and Electrical Technologies (ICCEET 2012), Kumaracoil, India, 21–22 March 2012; pp. 18–25.
45. Rahideh, A.; Korakianitis, T.; Ruiz, P.; Keeble, T.; Rothman, M.T. Optimal brushless DC motor design using genetic algorithms. *J. Magn. Magn. Mater.* **2010**, *322*, 3680–3687. [CrossRef]
46. Mitchell, M. *An Introduction to Genetic Algorithms*; MIT Press: Cambridge, MA, USA, 1998; pp. 1–16.

machines

MDPI

Article

A Hybrid-Excitation Synchronous Motor with a Change in Polarity

Chiara Contò and Nicola Bianchi *

Department of Industrial Engineering, University of Padova, 35122 Padova, Italy; chiara.conto@phd.unipd.it
* Correspondence: nicola.bianchi@unipd.it

Abstract: The hybrid-excitation permanent magnet (HEPM) motor is a synchronous motor characterized by a rotor that includes both permanent magnets and excitation windings. Thus, an active change in the magnetic flux by means of the excitation current is possible. In particular, the hybrid-excited rotor studied in this paper can be used to change the number of poles of the rotor itself. At the same time, the stator winding is also designed to perform a polarity change. The change in polarity allows users to obtain different torque versus speed characteristics with the same motor. In particular, a configuration with a lower pole number exhibits low torque at high speeds, while a configuration with a higher pole number produces high torque at low speeds. In this way, a single HEPM motor behaves like two different machines, extending the usual operating speed range of synchronous motors. In this paper, an HEPM rotor configuration is designed, and its performance is analyzed through finite element electromagnetic and thermal simulations.

Keywords: permanent magnet synchronous motor (PMSM); synchronous machine; change in polarity; motor design

Citation: Contò, C.; Bianchi, N. A Hybrid-Excitation Synchronous Motor with a Change in Polarity. *Machines* **2022**, *10*, 869. https://doi.org/10.3390/machines10100869

Academic Editors: Lorand Szabo and Feng Chai

Received: 15 July 2022
Accepted: 21 September 2022
Published: 27 September 2022

1. Introduction

In recent years, due to significant environmental concerns, electric vehicles have gained more and more popularity. Electric machines are generally required to have high torque density, an extended flux-weakening region, high efficiency in the entire range of operation, high reliability, and high fault-tolerance capability.

Permanent magnet (PM) synchronous motors are widely used in automotive applications as they allow for high torque density and highly efficient operation [1]. In the last few years, the trend has been to adopt interior permanent magnet (IPM) motors [2] in applications requiring a wide operating speed range. However, the increasing cost and shortage of rare-earth materials (such as NdFeB and SmCo) will alter this trend. As a solution, different types of rare-earth-free machines have been analyzed, such as the ferrite PM machine [3] and synchronous reluctance machines [4]; sometimes, such designs are assisted by low-cost magnets [5].

An innovative solution consists in designing a synchronous motor whose rotor includes both PMs and excitation windings [6,7]; PMs produce a constant flux component, while excitation coils create a variable flux. This machine is called a hybrid-excited permanent magnet (HEPM) synchronous motor, and it can be considered a combination of classical PM and wound rotor machines. There are two main subcategories of HEPM machines: parallel and series HEPM machines [8]. The former is characterized by PM rotor poles and excitation coil rotor poles; thus, the flux of the PMs and the flux of the excitation winding follow different paths. The latter is characterized by PMs and excitation coils in each pole such that the paths of the two fluxes are the same.

In general, excitation coils can either increase or decrease the PM flux, adding a further degree of freedom for motor operation control. An advantage of the HEPM motor is the possibility of enhancing the flux weakening (FW) capacity thanks to an adequate excitation current supply. As shown in [8–10], the possibility of modulating the rotor flux allows users to overcome the speed limitations of standard PM machines and maintain high torque and power at a wide range of speeds.

In the literature, the memory motor represents another example of a hybrid-excited machine with enhanced FW operation. The name memory is related to the nature of the aluminum–nickel–cobalt (AlNiCo) PMs in the motor, which can be online magnetized or demagnetized to various magnetization levels through a temporary DC current pulse [11]. Typical solutions include AlNiCo PMs in the stator core and regulate the air-gap flux to improve torque versus speed capabilities [12–15]. In fact, by applying a temporary current pulse to a small magnetizing winding, the magnetization of AlNiCo PMs is online tuned. Consequently, the machine can provide flexible air-gap flux control, thus achieving high efficiency at different speeds and loads.

Such PMs can be employed in combination with traditional NdFeB PMs. However, the AlNiCo PM exhibits a low coercivity; thus, the machine power rating is limited to avoid accidental demagnetization.

In this paper, an HEPM machine configuration is employed in designing a synchronous motor that is able to perform a change in polarity [16]. A change in polarity is a technology already in use for induction motors [17–19], in particular, in squirrel cage induction motors, since the rotor currents automatically adapt to the stator pole number. In contrast, when considering a synchronous motor, the number of poles for both the stator and the rotor have to be changed at the same time.

One example of a two-speed synchronous machine is the line-start motor presented in [20]. Its configuration includes a low pole number that increases its starting torque and a high pole number that enhances its synchronization capability. The start-up is carried out by means of two-pole induction torque. Since the PM flux is neutralized, the starting torque is increased, and the braking torque is cancelled. When the two-pole motor reaches four-pole speed, the number of poles is changed to four. Thus, normal operation is conducted as a four-pole synchronous PM motor, without rotor losses and with a higher power factor.

An improvement upon the solution described above is presented in [21]. The rotor is able to adapt to the number of poles of the stator thanks to a dual-polarity PM arrangement: a four-pole configuration is overlaid onto an eight-pole configuration.

Another example of a dual-polarity machine in the literature is described in [22]. The pole-changing electronic motor presented is characterized by a rotor with PMs with low coercive force (variable magnetization magnets) arranged radially and embedded at the iron core. The motor is able to switch from an eight-pole to a four-pole configuration thanks to the magnetizing field. The armature winding polarity is modified using power-switching devices. When the d-axis pulse current flows in the four-pole armature windings, the magnets in the rotor undergo variable magnetization due to the magnetic field generated by the pulse current. Thus, the rotor number of the poles switches from an eight-pole to a four-pole configuration.

Generally, synchronous machines operate at variable speeds using flux-weakening controls. The flux-weakening current, which decreases voltage at high speeds, causes significant copper and core losses. Because core loss is significant at high speeds, the efficiency of the motor is lower. If the number of poles in the motor is reduced, the lower frequency in the motor decreases its core loss. Therefore, pole-changing enables high-efficiency motors.

This paper presents the design of a dual-polarity HEPM motor characterized by both PMs and excitation coils in the rotor. A stator change in polarity is achieved through a particular supplying technique: each winding phase is split into two parts, one supplied by a first inverter and the other by a second inverter. When the current supplied by the two inverters is identical, a configuration with a higher pole number is achieved. In contrast, when the current supplied by the second inverter is reversed, a configuration with a lower pole number is obtained. This solution is tolerated in high-power applications and is better discussed in Section 2.

At the same time, the rotor polarity also has to change. This is carried out by means of hybrid excitation. In particular, a parallel HEPM rotor configuration [8] is considered herein. Excitation coils are supplied with a direct current, and the change in rotor polarity is achieved by reversing that current. This is explored more deeply in Section 3.

Here, a rotor typology is designed; its initial model is shown in Figure 1. Since the HEPM motor is characterized by a rotor with interior PMs, it is identified herein as an HE-IPM motor. Section 4 presents the finite element analysis of the HE-IPM designed in this work, highlighting the most relevant simulation results. The finite element analysis is performed to clarify the basic pole-changing motor operation and to verify the feasibility of the machine designed. Section 5 includes a parametric analysis of the rotor geometry, and Section 6 explains the performance of the dual-polarity HE-IPM (torque and power versus speed), comparing it to a traditional IPM motor. Finally, Section 7 presents a simple two-dimensional thermal analysis, and conclusions are drawn in the last section.

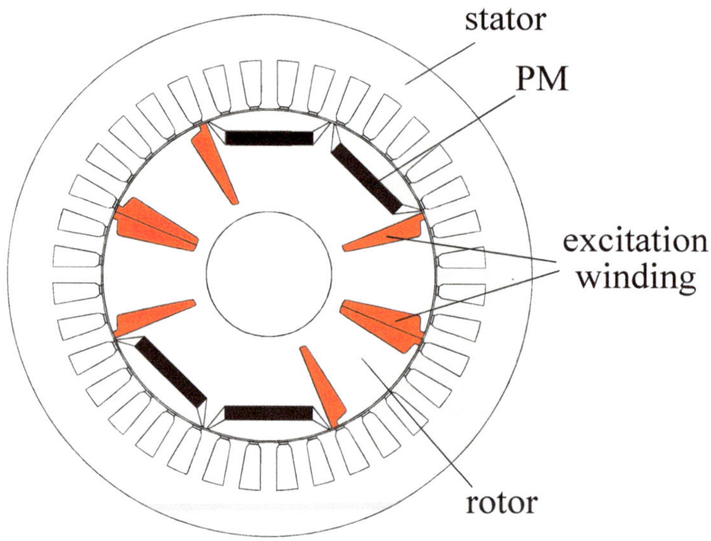

Figure 1. Sketch of the HEPM motor with interior PMs.

2. Double-Polarity Stator Winding

There are many winding configurations that allow for dual-polarity operation, as reported in [23]. Commonly, a change in polarity is realized by a switch that modifies the winding connections. In contrast, in this paper, the change in polarity is achieved by means of two inverters. As represented in Figure 2, in phase A, each phase of the stator winding is split into two parts. Half of the stator winding phase is supplied by inverter 1, and the other half by inverter 2.

The operating principle of this particular technique is described by means of the example in Figure 3, where four-coil winding is shown. This simple winding is supplied by two inverters, as described previously. In the top part of the figure, the two inverters supply the same current, giving rise to the magneto motive force (MMF) waveform reported below, which corresponds to a four-pole configuration. In the second part of the figure, the second inverter supplies an opposite current. The resulting MMF is modified and corresponds to a two-pole configuration.

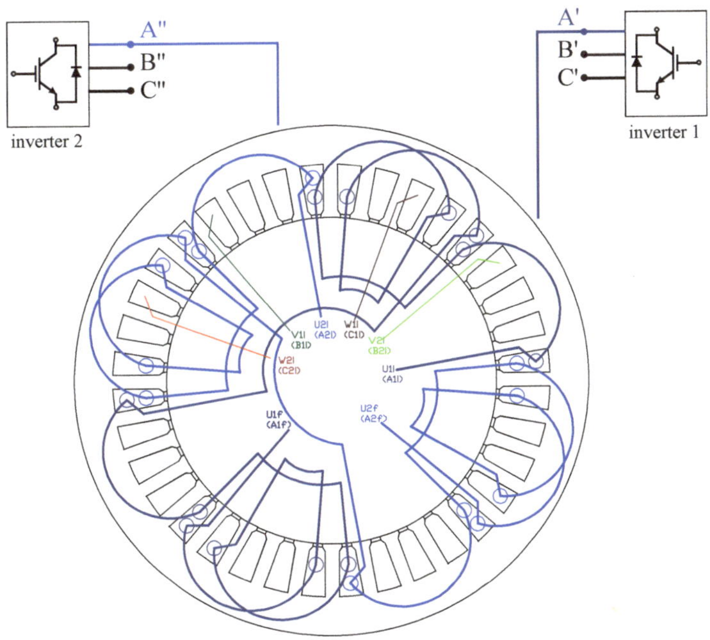

Figure 2. Supply condition schematic for phase A.

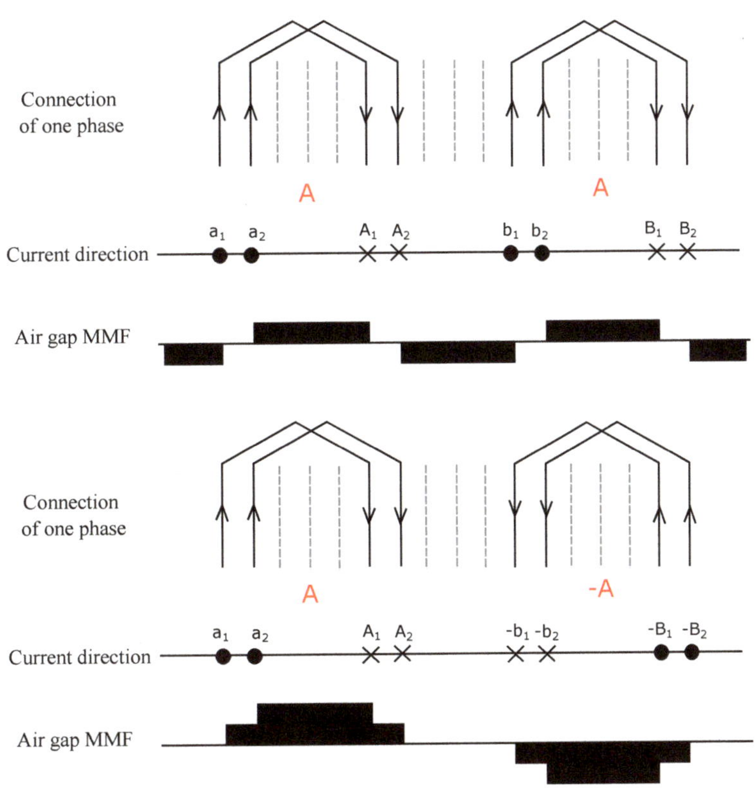

Figure 3. Sketch of four-pole and two-pole configurations.

The actual winding designed herein is a 36-slot double-layer winding. Its polarity can change from eight to four poles. Considering the eight-pole configuration, the slot angle is:

$$\alpha_s^e = p \cdot \frac{360^\circ}{Q_s} = 40^\circ$$

The number of slots per pole per phase is:

$$q_s = \frac{Q_s}{2p \cdot m} = 1.5$$

This means that it is a fractionary winding. The winding throw is $y_q = \frac{Q_s}{2p} = 4.5$. The value of $y_q = 4$, resulting in a chording angle of $\beta_r^e = 0.5 \cdot \alpha_s^e = 20^\circ$. This particular angle allows a reduction in the seventh harmonic MMF. The resulting winding coefficient k_w is calculated from the product of the pitch factor k_p and the distribution factor k_d:

$$k_p = cos(\frac{\beta_r^e}{2}) = 0.985$$
$$k_d = \frac{sin(q \cdot \frac{\alpha_s^e}{2})}{q \cdot sin(\frac{\alpha_s^e}{2})} = 0.975$$
$$k_w = k_p \cdot k_d = 0.96$$

When the current of the second inverter is reversed, the winding shifts to the four-pole configuration. The slot angle becomes:

$$\alpha_s^e = p \cdot \frac{360^\circ}{Q_s} = 20^\circ$$

And the number of slots per pole per phase is as follows:

$$q_s = \frac{Q_s}{2p \cdot m} = 3$$

The winding throw would be $y_q = 9$, but the connections cannot be modified. Consequently, the chording angle is $\beta_r^e = 5 \cdot \alpha_s^e = 100^\circ$. The resulting winding coefficient becomes much lower with this configuration:

$$k_p = cos(\frac{\beta_r^e}{2}) = 0.643$$
$$k_d = \frac{sin(q \cdot \frac{\alpha_s^e}{2})}{q \cdot sin(\frac{\alpha_s^e}{2})} = 0.96$$
$$k_w = k_p \cdot k_d = 0.62$$

Consequently, since the winding is designed as an eight-pole configuration, the winding factor is higher when a high number of poles is achieved.

The winding connections are described hereafter.

Referring to phase A' (connected to the first inverter) and phase A'' (connected to the second inverter), the coil side distribution within slots is represented by a slot matrix. Such a matrix contains as many elements as the number of slots. Each element represents the location of the phase coil side within each slot: the value is 1 if the coil side fills the whole slot (i.e., two layers), 0.5 if it occupies only half a slot (i.e., only one layer), and 0 if there is no coil side for that phase in the slot. The sign represents the reference current direction. The phase A' and phase A'' slot matrices are as follows:

$$k_{A'} = [0.5\,0\,0\,0\ -1\ -0.5\,0\,0\,0.5\,0.5\,0\,0\,0\,0\,0\,0\,0\,0 \ldots$$
$$0.5\,0\,0\,0\ -1\ -0.5\,0\,0\,0.5\,0.5\,0\,0\,0\,0\,0\,0\,0\,0]$$
$$k_{A''} = [0.5\,0\,0\,0\,0\,0\,0\,0\,0\,0\,0.5\,0\,0\,0\ -1\ -0.5\,0\,0\,0.5 \ldots$$
$$0.5\,0\,0\,0\,0\,0\,0\,0\,0\,0\,0.5\,0\,0\,0\ -1\ -0.5\,0\,0\,0.5]$$

For phases B', B'', C', and C'', the slot matrices must display 120° electrical degrees; this corresponds to three slots (three elements of the vector). However, it is necessary to shift the 120° of phase B by 360° ($3 + 9 = 12$ slots) with respect to phase A. In fact, only by adding this shift can a correct arrangement of the windings be achieved when the polarity is changed to four. Figure 4 shows the eight-pole winding connections chosen herein.

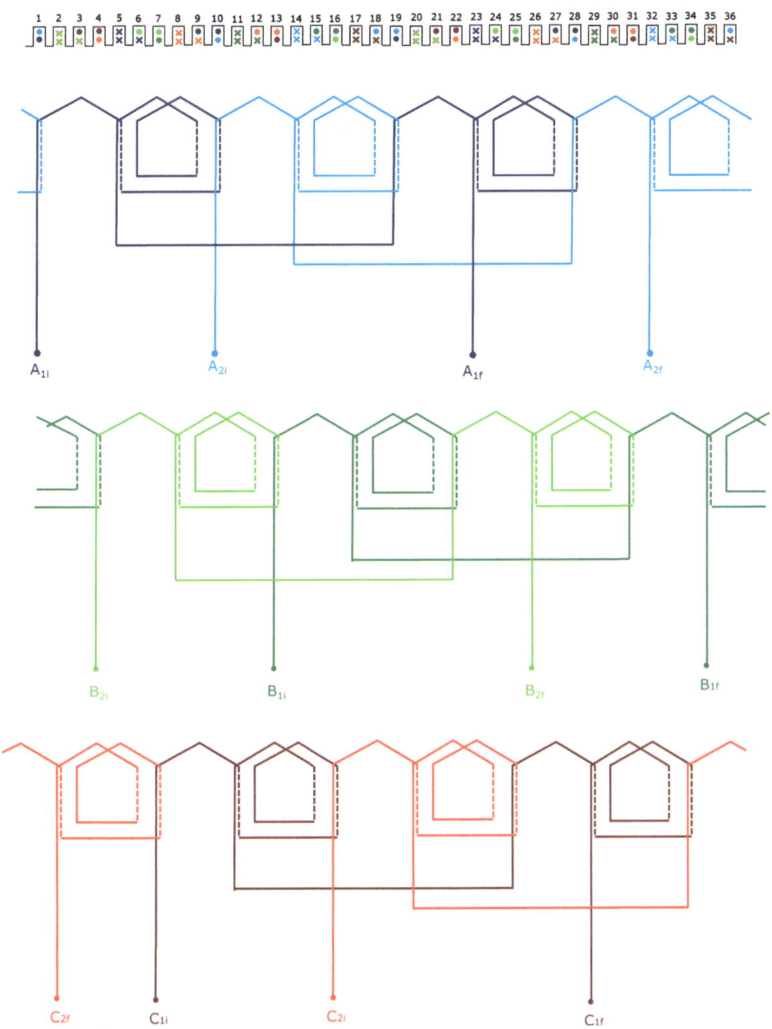

Figure 4. Eight-pole winding connections.

When the current is reversed to change the pole number, both phase *B* and phase *C* are still connected to the same inverter (1 or 2), i.e., phases *B*′ and *C*′ are still connected to the first inverter, and phases *B*″ and *C*″ are still connected to the second inverter. This is due to the design choice of adding 360 electrical degrees in order to shift phase *B* with respect to phase *A*. Figure 5 shows the four-pole configuration. It is worth noting that phases *B*′ and *C*′ and phases *B*″ and *C*″ switch with each other; however, they are still connected to the same inverter as in the eight-pole configuration.

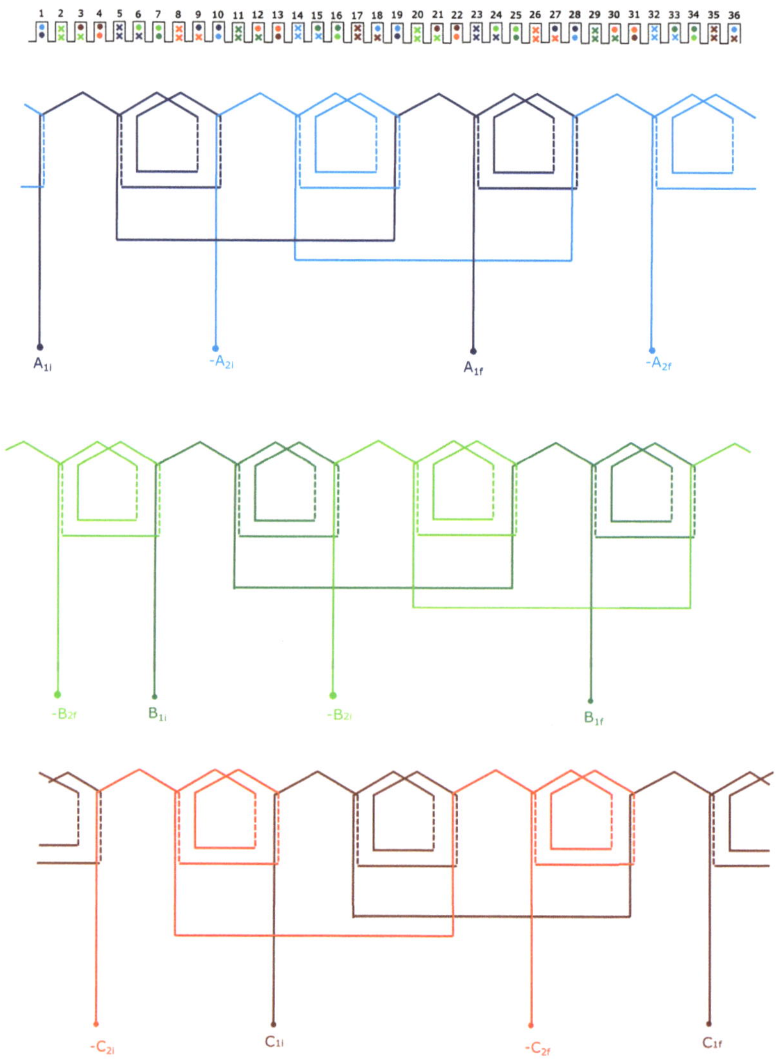

Figure 5. Four-pole winding connections.

The total slot matrix for the winding, considering both inverter phase connections, is:

$$k = \begin{bmatrix} k_{A'} + k_{A''} \\ k_{B'} + k_{B''} \\ k_{C'} + k_{C''} \end{bmatrix} = \begin{bmatrix} k_A \\ k_B \\ k_C \end{bmatrix}$$

The MMF is analyzed considering the current supplied in Figure 6. The eight-pole current has half the four-pole current period. To obtain the eight-pole configuration, the first and second inverters have to supply the motor with two equal direct-sequence three-phase currents. The four-pole configuration is achieved when the second inverter current is reversed. In addition, the two inverters have to supply the motor with inverse-sequence three-phase currents. The four-pole configuration requires the inverse sequence of the three-phase current supply since, as observed in Figures 4 and 5, phases B' and C' (and B'' and C'') switch with each other. Thus, the supply condition for dual-polarity operation can be summarized as:

$$\text{Eight-pole configuration: } \begin{cases} i_{A'} \ i_{B'} \ i_{C'} \\ i_{A''} \ i_{B''} \ i_{C''} \end{cases} \quad \text{Four-pole configuration: } \begin{cases} i_{A'} \ i_{C'} \ i_{B'} \\ -i_{A''} \ -i_{C''} \ -i_{B''} \end{cases}$$

Figure 7 shows the MMF waveforms of the double-polarity winding, together with their harmonic content (from Fourier series expansion). The MMF is drawn for a rotor position that varies from 0° to 360° mechanical degrees, while the current phasor electrical angle is 0°. It is worth noting that the eight-pole waveform is not symmetrical with respect to the x-axis, as expected for fractional slot winding. Concerning the Fourier analysis, both configurations exhibit good harmonic content thanks to the chorded double-layer winding design; this is the case even though the four-pole configuration presents a higher number of harmonics. This influences the torque ripple of this configuration.

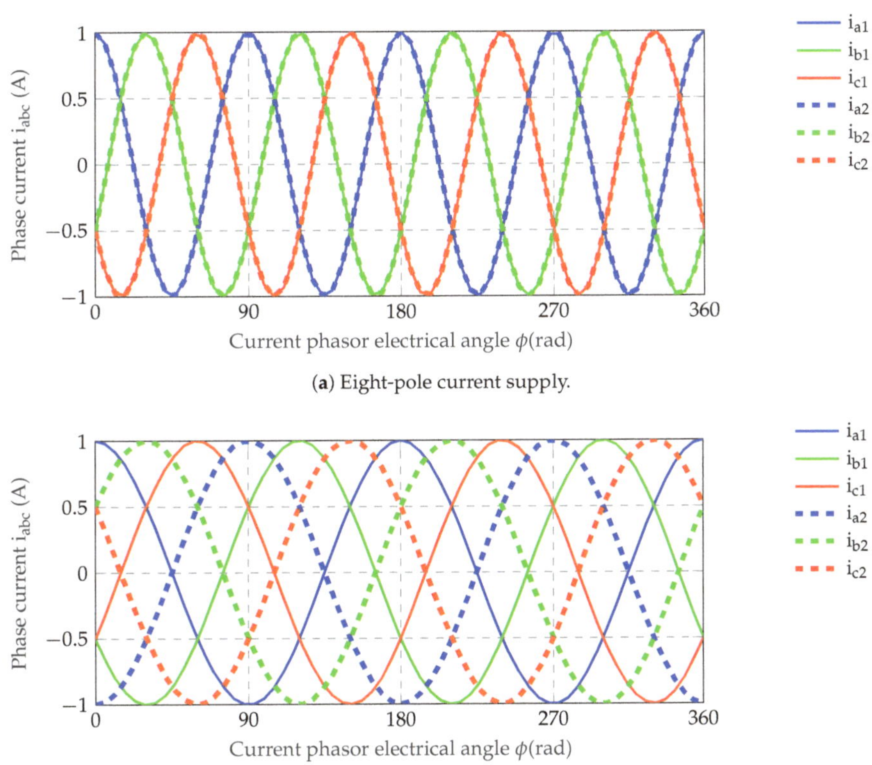

(**a**) Eight-pole current supply.

(**b**) Four-pole current supply.

Figure 6. Current supplied by two inverters with eight-pole and four-pole configurations.

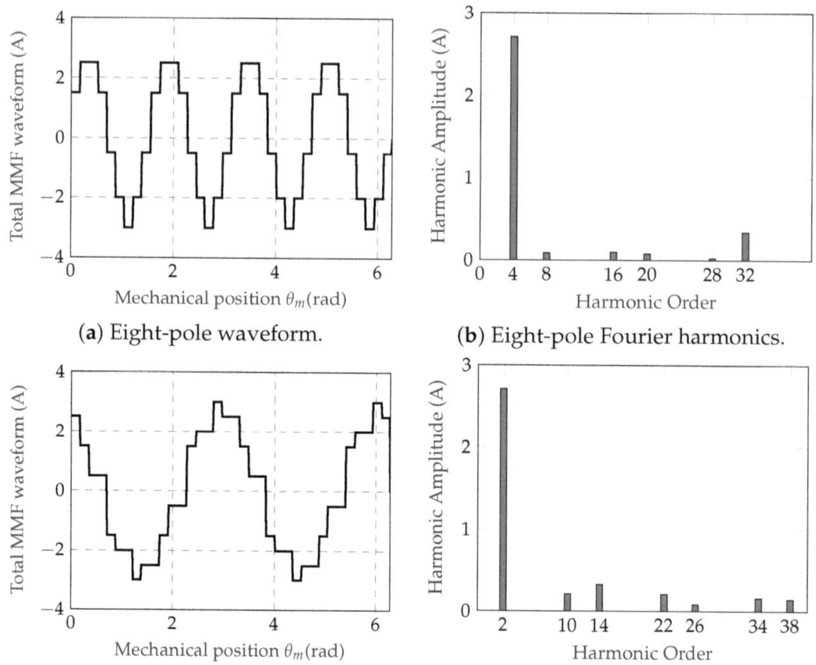

(**a**) Eight-pole waveform. (**b**) Eight-pole Fourier harmonics. (**c**) Four-pole waveform. (**d**) Four-pole Fourier harmonics.

Figure 7. MMF waveform due to the stator currents and harmonic content of both configurations.

Table 1 summarizes stator geometric data and stator winding data.

Table 1. Stator data.

Parameter	Symbol	Value	Unit
External diameter	D_e	150	mm
Internal diameter	D_s	96	mm
Slot number	Q_s	36	-
Tooth width	w_t	13.14	mm
Slot cross-section area	S_{slot}	97.7	mm^2
Stack length	L_{stk}	40	mm
Iron pack length	L_{fe}	38.8	mm
Air gap width	g	0.5	mm
Poles	$2p$	8/4	-
Number of conductors per slot	$n_c = n_{cs}$	100	-
Conductor diameter	d_c	0.56	mm
Fill factor	k_{fill}	0.35	-
Nominal current	I_N	1.11	A
Current density	J_s	5	A/mm^2

3. Double-Polarity Rotor

3.1. Operating Principle

As the stator's number of poles is changed, the rotor needs to adapt to the selected number of poles. This is carried out by means of parallel hybrid excitation thanks to the possibility of regulating the rotor coil current. Since PMs generate a constant magnetization flux, the excitation coils are supplied with a direct current. The rotor change in polarity is achieved by reversing that current.

Figure 8 shows the change in polarity principle. A rotor with only excitation coils is compared to the hybrid configuration considered in this study. The excitation-only rotor is able to switch from an eight-pole to a four-pole configuration by reversing the current of two consecutive coils. It is necessary to reverse the current of half the coils, as observed in the figure. Consequently, the supply condition of the other four coils does not change.

Since some coils are supplied by the same constant direct current in both an eight-pole and a four-pole configuration, they can be substituted with PMs, i.e., the constant magnetizing flux is the same. In this way, rotor losses decrease and, at the same time, the amount of rare earth material used is limited.

It is worth noting that, when a four-pole configuration is achieved, both PMs and excitation coils contribute to generating the rotor flux.

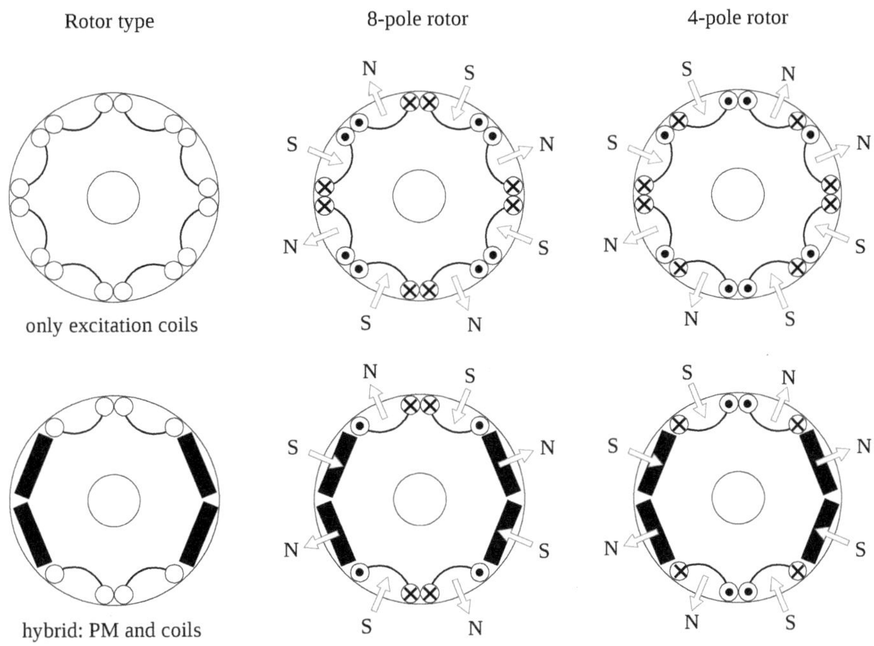

Figure 8. Sketch of the rotor change in polarity from eight-pole to four-pole.

3.2. Design Considerations

As mentioned before, a rotor change in polarity is obtained with the HEPM motor configuration in Figure 1. Since this is a theoretical study, the initial design employed simple geometry based on the magnetic equivalent circuit (MEC) of the motor. The magnetic equivalent circuit method is useful for simple designs with fast and intuitive parameters, but it cannot derive the distribution of the magnetic field. Only the finite element method can provide an accurate magnetic field distribution; however, it requires a higher computational time. In this study, the MEC method is employed to design the initial rotor structure and, in particular, the PM dimensions.

The PM pole no-load MEC is represented in Figure 9. Considering the desired no-load air-gap flux and the magnetic material properties of *NdFeBUH*35, the PM dimensions can be computed from the MEC equation. The parameters of the magnetic circuit are computed as follows:

$$\mathcal{R}_g = \frac{g}{\mu_0 \cdot S_g} = \frac{g}{\mu_0 \cdot \frac{\pi D_s}{2p} \cdot L_{stk}} \tag{1}$$

$$\mathcal{R}_m = \frac{t_m}{\mu_{rec} \cdot \mu_0 \cdot S_m} = \frac{t_m}{\mu_{rec} \cdot \mu_0 \cdot l_m \cdot L_{stk}} \tag{2}$$

$$\Phi_{rem} = B_{rem} \cdot S_m = B_{rem} \cdot l_m \cdot L_{stk} \tag{3}$$

$$\Phi_{sat} = 2 \cdot B_{sat} \cdot t_{rib} \cdot L_{fe} \tag{4}$$

$$\Phi_g = B_g \cdot S_g = B_g \cdot \frac{\pi D_s}{2p} \cdot L_{stk} \tag{5}$$

The air gap is represented by the reluctance $\mathcal{R}_g$. The PM is represented by the parallel between the remaining reluctance associated with the material $\mathcal{R}_m$, the remaining flux Φ_{rem}, and the saturation flux Φ_{sat} associated with the PM ribs. The value of the air-gap flux Φ_g is set considering a flux density at the air gap equal to $B_g = 0.8$ T.

g is the air gap, S_g is the air gap surface, t_m is the PM thickness, S_m is its flux surface, B_{rem} is the remaining flux of the PM employed, B_{sat} is the saturation flux, t_{rib} is the rib thickness, and L_{fe} is the iron pack length.

Since all geometric parameters and material properties are known, the only unknown variables are the PM dimensions (radial length l_m and thickness t_m). Considering a fixed ratio $l_m/t_m = 6$, the MEC equation is solved as follows:

$$(\mathcal{R}_g + \mathcal{R}_m) \cdot \Phi_g - \mathcal{R}_m \cdot (\Phi_{rem} - \Phi_{sat}) = 0$$

$$\Rightarrow \Phi_{rem} = \frac{\mathcal{R}_g}{\mathcal{R}_m} \cdot \Phi_g + \Phi_g + \Phi_{sat} \tag{6}$$

Then, l_m is computed from the resulting Φ_{rem}, and the result is equal to 26 mm. Values of $l_m = 25$ mm and $t_m = 4$ mm are chosen. Thus, by the solution of the magnetic equivalent circuit of the PM pole, the PM geometry is defined.

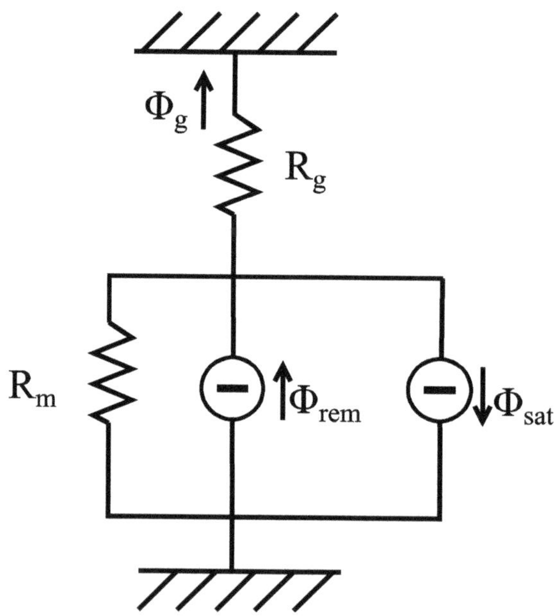

Figure 9. Magnetic equivalent circuit of the PM pole.

Considering the excitation coils, their design is strictly constrained by the limited space in the rotor. In fact, rotor slot shape is defined with an iterative method in order to limit iron saturation, both along the rotor tooth and close to the rotor shaft. In addition, the rotor slot area has to be large enough to limit the current density of the rotor. The resulting current density is equal to 10 A/mm^2, which requires a water cooling system. In fact, heat removal is more challenging for the rotor than for the stator. Thermal aspects are considered in Section 7.

3.3. Excitation Current

The hybrid-excited motor designed herein is characterized by a rotor magnetization flux generated by both the PMs and the excitation coils. Since PMs produce a constant magnetization flux, the excitation current supplied has to be constant, and its magnitude has to be modulated in order to generate the same amount of flux as the PMs.

When a positive excitation current is supplied, the generated flux induces an alternate flux density distribution in the air gap, resulting in an eight-pole configuration. Then, when the excitation current is negative, the flux produced is reversed, so that it assumes the same direction of the flux produced by the PM of the adjacent pole. The resulting flux density distribution in the air gap corresponds to a four-pole configuration. Figure 10 shows how both configurations are achieved for the HE-IPM motor under study.

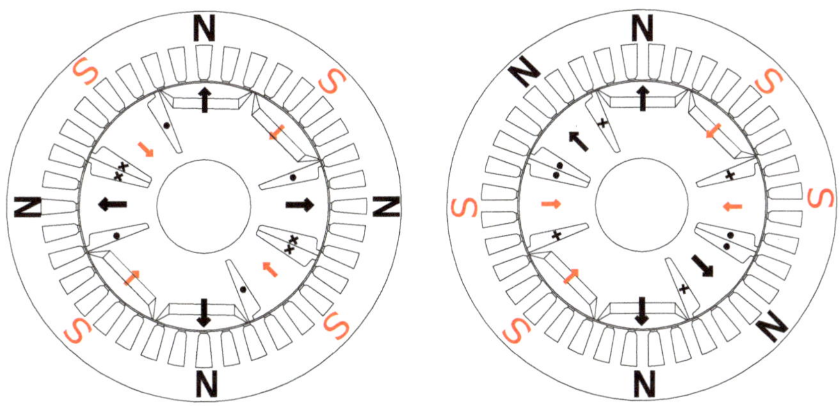

(**a**) Eight poles. (**b**) Four poles.

Figure 10. Change in polarity with reversing excitation currents.

Both PMs and excitation coils are sources of the no-load rotor flux:

$$\Phi_g = \Phi_{PM} + \Phi_{exc}$$

Thus, the excitation current is non-zero during the no-load operation. Therefore, the excitation current is chosen according to the no-load analysis of the motor. In particular, the air-gap flux density due to the PMs has to be equal to the air-gap flux density due to the excitation coils. An initial value is computed considering the magnetic equivalent circuit of both the PMs and the excitation coils, as seen in Figure 11. The two MECs are characterized by the same air-gap flux Φ_g. The parameters of the magnetic circuit are computed as already described in Equations (1)–(4). To achieve a constant air-gap flux Φ_g, ampere-turns $N_e I_e$ yield a result of 500 A-turns. Then, this value is adjusted to attain a proper air-gap flux density waveform. In addition, the value is chosen to reduce the harmonic content of the four-pole no-load flux linkage. This is performed by adjusting the excitation current and evaluating the flux linkage. Thus, the excitation ampere-turns are as follows:

$$N_e I_e = 450 \text{ A} - \text{turns}$$

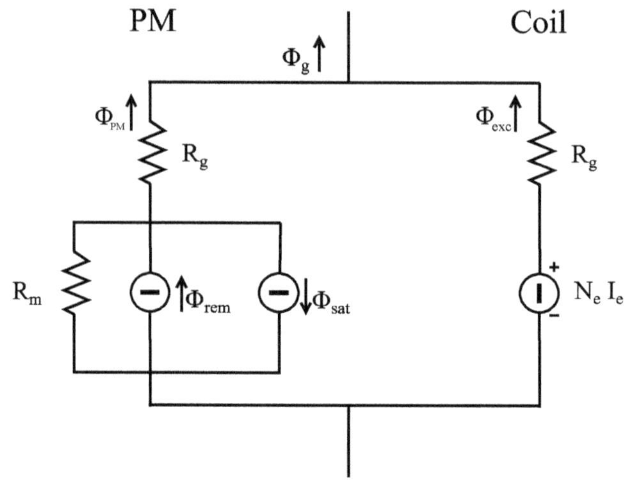

Figure 11. Magnetic equivalent circuit of hybrid excitation.

Table 2 summarizes the rotor geometric data and rotor winding data. Both the stator and rotor laminations are of *M530 50A* type, and the PMs are *NdFeB UH35*.

Figure 12 shows the flux lines and the flux density map for the HE-IPM motor in the no-load condition. The flux line distributions are different for the four-pole configuration and the eight-pole configuration, as expected due to the difference in pole numbers. The back-iron flux density is much higher when the four-pole configuration is employed. Thus, to reduce saturation and iron power losses, the back-iron thickness has to be designed according to the low polarity flux.

Figure 13 shows the no-load flux density waveform along the air gap achieved with four-pole and eight-pole excitation.

Table 2. Rotor data.

Parameter	Symbol	Value	Unit
Rotor diameter	D_r	95	mm
PM thickness	t_m	4	mm
PM length	l_m	25	mm
Distance PM air gap	d_{gm}	10	mm
Excitation length	h_{exc}	25	mm
Excitation upper width	$w_{exc,u}$	6	mm
Excitation lower width	$w_{exc,l}$	2	mm
Number of conductors per slot	N_e	220	-
Nominal current	I_e	2.05	A
Conductor diameter	$d_{c,r}$	0.5	mm
Fill factor	$k_{fill,r}$	0.44	-
Current density	J_e	10	A/mm^2

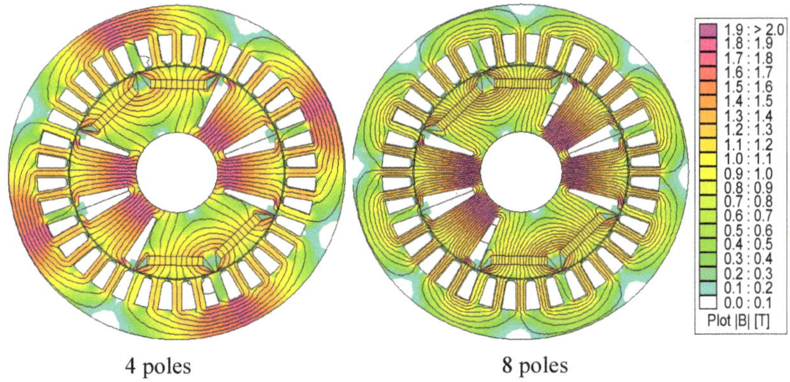

Figure 12. No-load flux density map and flux lines.

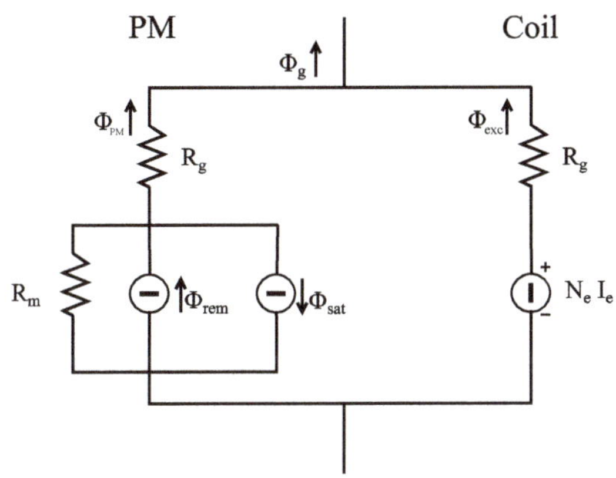

Figure 13. No-load flux density waveform along the air gap.

4. Finite Element Analysis

Two-dimensional magneto static finite element simulations are carried out to verify the dual-polarity operation of the motor designed as described in Section 3. The finite element analysis software employed is *FEMM 4.2*. *FEMM* is a suite of programs for solving low-frequency electromagnetic problems in two-dimensional domains. The pre- and post-processing programs employed were written by the University of Padova research team.

4.1. No-Load Analysis

The first step is the no-load finite element analysis of the HE-IPM motor. After the alignment of the IPM rotor, no-load flux linkages are computed as:

$$\lambda_j = n_{cs} \cdot L_{fe} \cdot \sum_{q=1}^{Q} k_{j,q} \cdot \frac{1}{S_{slot}} \int_{S_{slot}} A_z dS \qquad \text{j = A, B, C} \tag{7}$$

where $k_{j,q}$ is the slot matrix value of phase j for the slot q, and A_z is the magnetic scalar potential. Since the two-inverter slot matrices are defined separately, the fluxes linked by the first and second inverter are computed separately as well:

$$\lambda_{j'} = n_{cs} \cdot L_{fe} \cdot \sum_{q=1}^{Q} k_{j',\,q} \cdot \frac{1}{S_{slot}} \int_{S_{slot}} A_z \ dS \tag{8}$$

$$\lambda_{j''} = n_{cs} \cdot L_{fe} \cdot \sum_{q=1}^{Q} k_{j'',q} \cdot \frac{1}{S_{slot}} \int_{S_{slot}} A_z \; dS \tag{9}$$

Figure 14 shows the no-load flux linkages. The flux linkages of the first and second windings almost overlap during eight-pole operation. On the contrary, they are opposite in the four-pole configuration. As expected, the eight-pole configuration period is exactly half that of the four-pole configuration period. In addition, the four-pole configuration is characterized by more distorted waveforms. The greater distortion of the four-pole flux linkage waveform compared to the eight-pole one is related to the nature of the four-pole configuration itself. In fact, the four-pole configuration is obtained from the eight-pole geometry, with a reversed excitation current. The current inversion affects the four-pole air-gap flux density waveform and, consequently, the flux linkage waveform. When the number of poles changes from eight to four, the fundamental harmonic winding factor $k_{w,1}$ is reduced, but the higher harmonic winding factor $k_{w,v}$ increases.

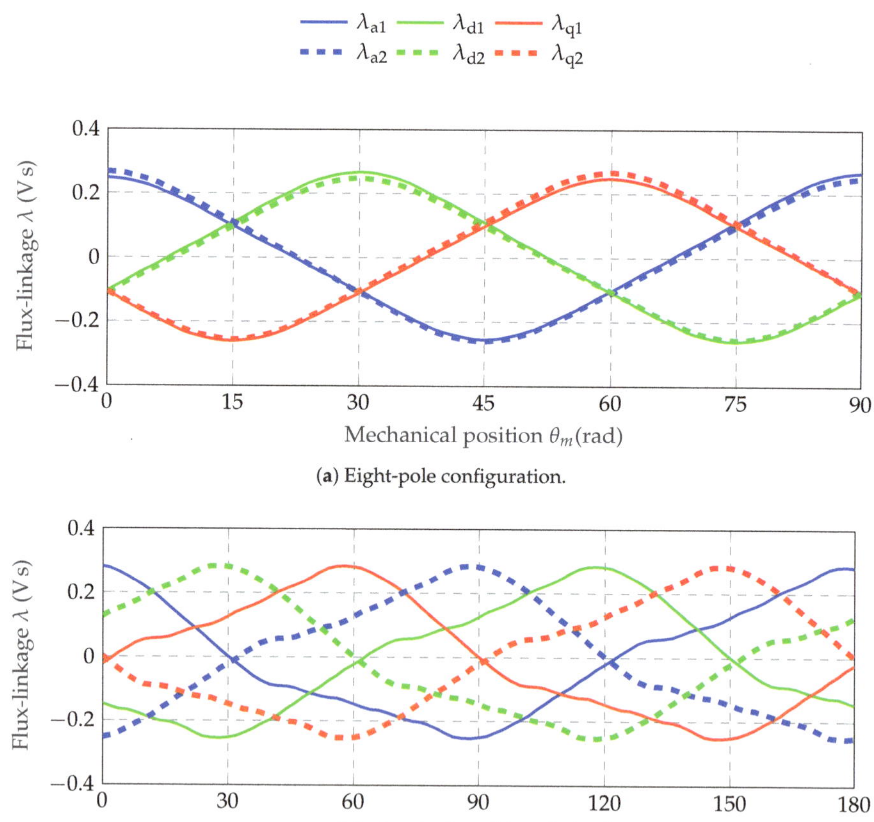

(**a**) Eight-pole configuration.

(**b**) Four-pole configuration.

Figure 14. No-load flux linkage waveforms.

4.2. On-Load Analysis

The on-load finite element analysis is carried out considering the specific rotor alignment for each of the two polarities. Indeed, the d-axis and the q-axis have different relative positions depending on the number of poles of the machine. Both motors operate with the maximum torque per ampere (MTPA), which also differs between the eight-pole and four-pole configurations. The four pole torque is lower than the eight-pole torque. As such, the current phasor electrical angle for MTPA operation is slightly different due to the different saliency factor (i.e., a different ratio between the q-axis and d-axis inductances).

Considering the MTPA supply condition, the torque is computed for different rotor mechanical positions to obtain the average torque in a period, as well as the oscillations. Figure 15 shows the torque for one period. The torque computed by means of Maxwell's stress tensor is indicated as τ_{Mxw}, while its average value is indicated by $<\tau>$. The eight-pole torque waveform period is 15 mechanical degrees, while the four-pole torque waveform period is exactly twice the period of the configuration with a higher pole number (i.e., 30 mechanical degrees), as expected. The four-pole configuration is characterized by half the average torque of the eight-pole configuration, along with a higher torque ripple, as expected from the higher harmonic content highlighted above. Indeed, the four-pole flux density and flux linkage waveforms are not perfectly sinusoidal, which is the case for those of the eight-pole.

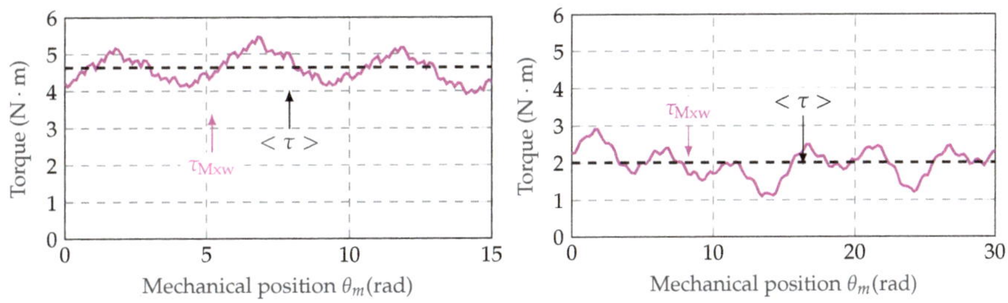

(**a**) Eight-pole configuration. (**b**) Four-pole configuration.

Figure 15. Torque ripple in MTPA conditions.

The resulting mean torque is equal to 4.6 N · m for the eight-pole configuration and 1.9 N · m for the four-pole configuration. The torque ripple is 34% for the eight-pole machine and 90% for the four-pole machine.

5. Parametric Analysis

As examined previously, the resulting torque ripple is too high, in particular when the four-pole configuration is employed. To achieve a smooth torque waveform, several design solutions can be taken into consideration:

- Skewing of the stator pack;
- Skewing of the rotor pack;
- Rotor pole shaping.

Concerning the first solution, the skewing of the stator pack is barely used due to winding manufacturing difficulties. In contrast, the use of rotor skewing is more common, despite the fact that it is more complex, with a hybrid-excitation rotor. Indeed, the presence of excitation coils requires continuous skewing, which is very expensive for PMs. Herein, it was decided to employ step-skewing of the PMs [24], inserting them into a continuously skewed hole, as depicted in Figure 16. Since the skewing angle is equal to 10°, a step-skewing equal to 5 is chosen. Thus, a single PM is divided into five parts, which represent the five steps of the discrete skewing. In this way, the PM width $w_{PM,skew}$ has an additional constraint, since it has to be smaller than the PM housing width w_{PM}.

To compute the step-skewing 2D simulation, five different simulations are carried out, one for each section of the skewed geometry. In other words, each of the five simulations is carried out with a different rotor shift (-4°, -2°, 0°, 2°, and 4°). More precisely, the same motor model magneto static simulation is carried out five times with the additional constant rotor angle $\theta_{m,skew}$. Thus, the rotor position of each jth−simulation varies with the law:

$$\theta_m^j = \theta_{m,skew}^j + \theta_m \quad \text{with } j = 1, 2, 3, 4, 5$$

where $\theta^{j}_{m,skew}$ is equal to the additional skewing angle, which corresponds to the mean position of the section considered with respect to the central PM that serves as the reference (with $\theta^{0}_{m,skew} = 0°$). Figure 17 shows such a simulation setup.

The skewing simulation gives the result for a step-skewed machine, even if, for rotor excitation, the skewing is continuous.

Thus, considering the HE-IPM with the skewing solution described, the resulting torque ripple is reduced, as shown in Figure 18. The results of the skewing solution are a mean torque of 4.4 N · m for the eight-pole configuration and 2 N · m for the four-pole configuration. The configuration with the higher number of poles decreases from a 34 % torque ripple to 3.5 %, while the configuration with the lower number of poles is reduced from 90 % to 22 %.

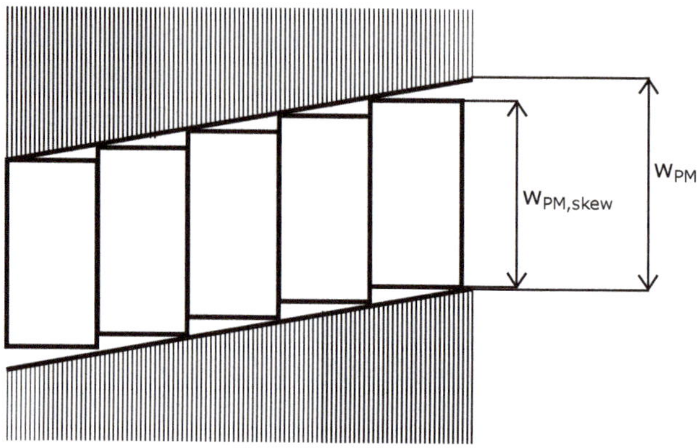

Figure 16. Permanent magnet step-skewing.

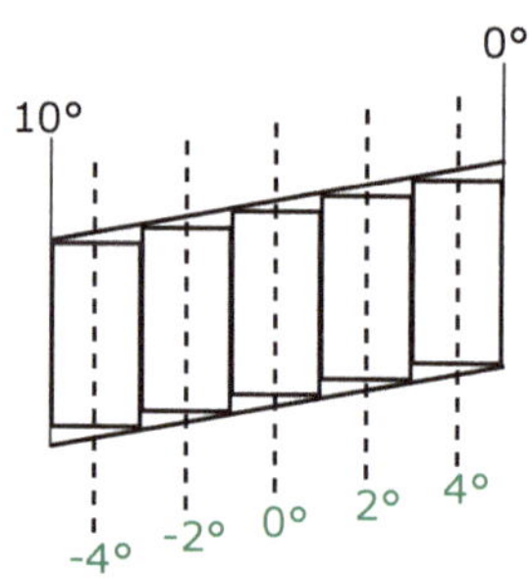

Figure 17. Step-skewing simulation conditions.

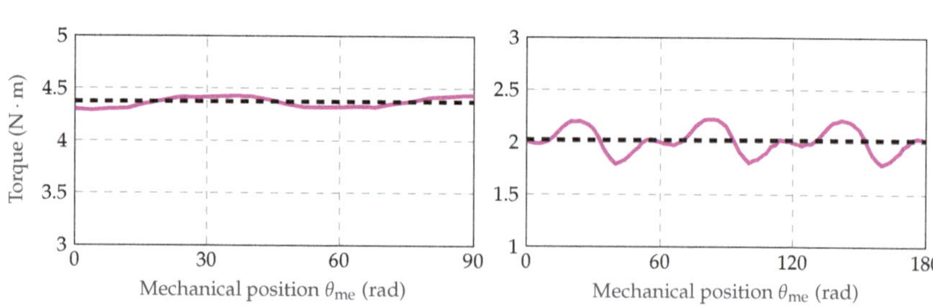

(**a**) Eight-pole configuration. (**b**) Four-pole configuration.

Figure 18. HE-IPM torque ripple with rotor skewing solution.

The third solution consists in modifying the HE-IPM motor to properly shape the rotor. Figure 19 shows two parameters that can be modified to properly shape the rotor: the maximum distance to the stator $d_{s,max}$ and the arc angle α_s. As an example, four different rotor configurations, with the shaping parameters described in Table 3, are analyzed. It is worth noting that the torque waveform becomes smoother as the rotor shaping increases. Nevertheless, as the shaping increases, the air gap volume increases as well; thus, the torque mean value is slightly decreased. Nevertheless, the eight-pole configuration performance results are always better than those of the four-pole configuration due to the fact that both flux linkage and torque waveform worsen when the excitation current is reversed. The best configurations for eight-pole and four-pole torque ripple (n. 4) are presented in Figure 20.

Table 3. Torque ripple with the shaped rotor solution.

HE-IPM	Parameter	8 Poles	4 Poles
non-shaped $d_{s,max} = 47.5°$ $\alpha_s = 0.5$ mm	Average torque $<\tau>$ Percentage ripple	4.6 N · m 34%	1.9 N · m 90%
shaping n.1 $d_{s,max} = 47°$ $\alpha_s = 1$ mm	Average torque $<\tau>$ Percentage ripple	4.21 N · m 14.9%	1.9 N m 45.2%
shaping n.2 $d_{s,max} = 30°$ $\alpha_s = 1$ mm	Average torque $<\tau>$ Percentage ripple	4.65 N m 9.2%	2 N m 36%
shaping n.3 $d_{s,max} = 35°$ $\alpha_s = 1.5$ mm	Average torque $<\tau>$ Percentage ripple	4.7 N m 5.9%	2 N m 30.8%
shaping n.4 $d_{s,max} = 30°$ $\alpha_s = 1.5$ mm	Average torque $<\tau>$ Percentage ripple	4.35 N m 5.9%	1.88 N m 23%

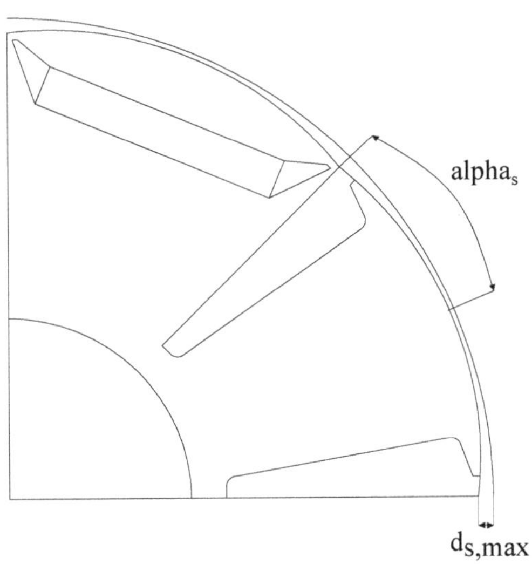

Figure 19. HE-IPM shaping parameters.

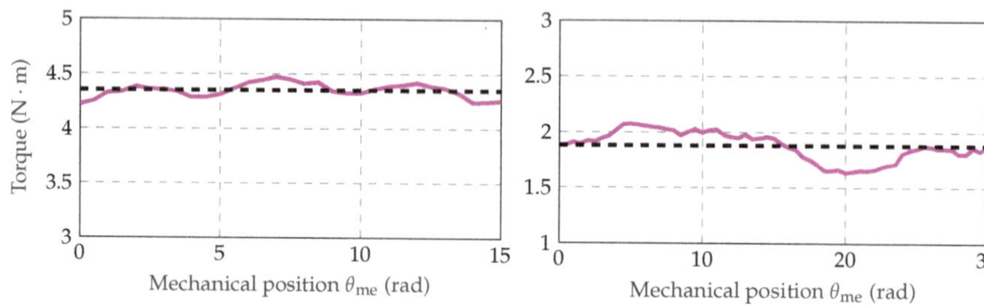

(**a**) Torque ripple, shaping n. 3. (**b**) Torque ripple, shaping n. 3.

Figure 20. HE-IPM torque ripple with the rotor-shaping solution.

6. Mechanical Characteristics

6.1. Torque and Power versus Speed of the HE-IPM Motor

Another finite element analysis is conducted to examine a possible electric drive for this machine. Torque and power versus speed characteristics can be obtained through magneto static simulations under different loading conditions. The eight-pole and four-pole HEPMs are considered separately, and they are fed with a constant current amplitude and a variable current phasor angle. The rotor excitation current is maintained at a constant, and it is reversed only when the four-pole configuration is analyzed.

The speed reached for each operating point is computed from the d-axis and q-axis flux linkage values, considering a constant DC bus voltage V_{DC} with PWM control and a certain drop in voltage (ΔV):

$$\omega_B = \frac{1/\sqrt{3} \cdot V_{DC} - \Delta V}{\sqrt{\lambda_D^2 + \lambda_Q^2}} \tag{10}$$

where λ_D and λ_Q are the flux linkages computed from the ABC values to the d- and q-axis values, respectively.

The MTPA current is supplied until the base speed is reached, resulting in a constant torque region. Then, the current angle is increased, reducing the torque value until it reaches zero, when the maximum speed is achieved. The power increases linearly until the base speed is achieved; then, it reaches its maximum at a higher speed and becomes zero at the maximum speed.

Both the torque and power mechanical characteristics of the HE-IPM motor under analysis are reported in Figure 21. When the eight-pole configuration is employed, the motor exhibits high torque at low speeds, while the four-pole configuration is characterized by a lower level of torque but a wider speed range. Thus, the eight-pole configuration is suitable for start-up, as it has a high starting torque. However, its speed range is limited. It is possible to switch to the four-pole configuration in order to extend the range of speeds. The main performance parameters are summarized in Table 4.

Table 4. Mechanical characteristics: data comparison.

Parameter	8 Poles	4 Poles
Base speed ω_B	1010 r/min	2810 r/min
Maximum speed ω_M	1330 r/min	3090 r/min
MTPA torque τ	4.6 N m	1.9 N m

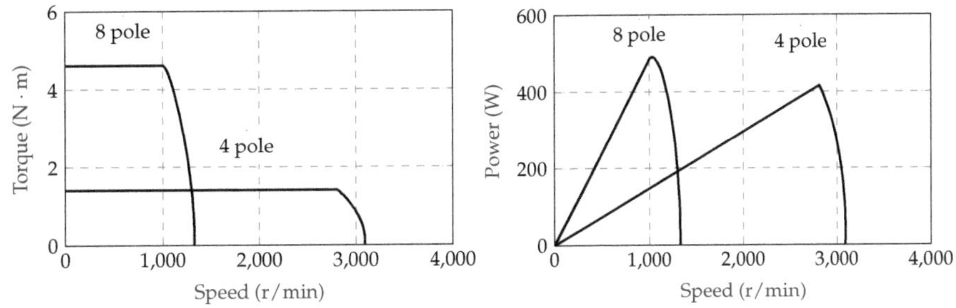

(**a**) Torque vs. speed characteristic. (**b**) Torque vs. speed characteristic.

Figure 21. Eight-pole and four-pole HE-IPM torque and power versus speed characteristics.

6.2. Comparison between HE-IPM and a Traditional IPM Motor

In order to underline the advantage of this machine, the dual-polarity HEPM motor is compared to a traditional IPM motor. Two different magnetizing flux IPM motors are considered and compared to the non-shaped and non-skewed HE-IPM motor. The first IPM's geometry is designed in the same way as the HE-IPM analyzed but without excitation coils. All poles are made of *NdFeB* PMs of the same type and dimensions as the HEPM PMs, as represented in Figure 22a.

The other IPM is designed with a lower PM volume, as represented in Figure 22b. A lower volume of magnetic material leads to a lower magnetizing flux and, thus, a lower torque with respect to the requirements of the eight-pole motor. However, the reduction in the magnetizing flux allows having an extended FW region, i.e., a higher maximum speed. Consequently, the resulting motor has an intermediate performance between high torque and high speed.

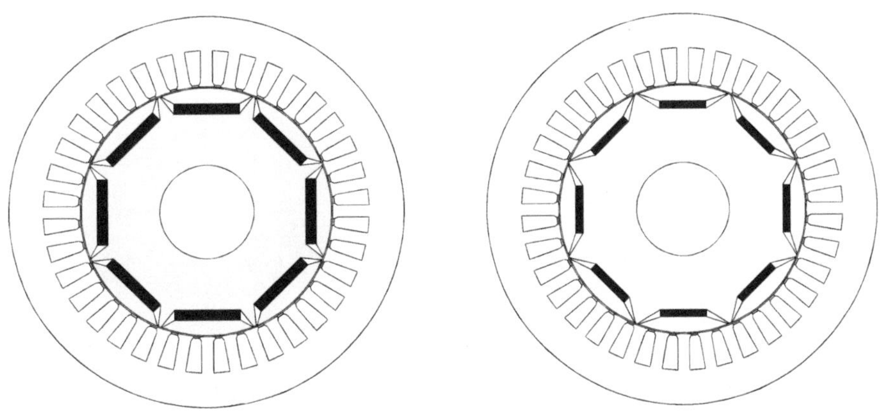

(**a**) IPM 1. (**b**) IPM 2.

Figure 22. Traditional IPM motors.

The comparison between the HE-IPM and the two traditional IPM motors is shown in Figure 23. The machine in Figure 22a exhibits the same performance as the eight-pole HE-IPM since it is designed by substituting rotor excitation coils with identical PMs. The machine in Figure 22b is characterized by an extended flux-weakening operation; thus, it has a higher maximum speed, but its torque is lower. This is related to the lower PM volume of the machine. In fact, the reduction in the magnetization flux leads to a lower torque. At the same time, the lower flux linkage allows reaching higher speeds when supplying the same DC bus voltage.

Table 5 shows the performance comparison between the HE-IPM motor and the first IPM motor when the rated speed is reached. Motor losses are computed as described in Appendixes A and B, then the efficiency is obtained as in Appendix C. The HE-IPM motor has a lower efficiency at the rated speed due to the additional joule losses of the rotor coils. Such rotor joule losses are high in this small-sized motor but have a very small influence on the total losses in high-power applications. In fact, large-sized motors that use rotor excitation current are characterized by high efficiency, despite the additional rotor losses.

As shown in Figure 24, the IPM motor efficiency drops for speeds higher than 1250 r/min, while the HE-IPM continues to operate with higher efficiency. In addition, for speeds higher than 1320 r/min, the IPM motor does not provide power, while the HE-IPM continues to operate.

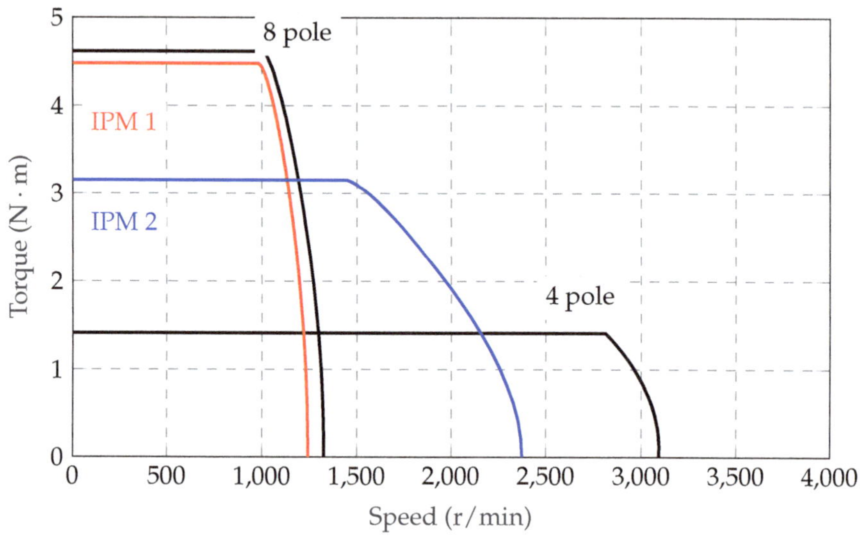

(**a**) Torque–speed comparison.

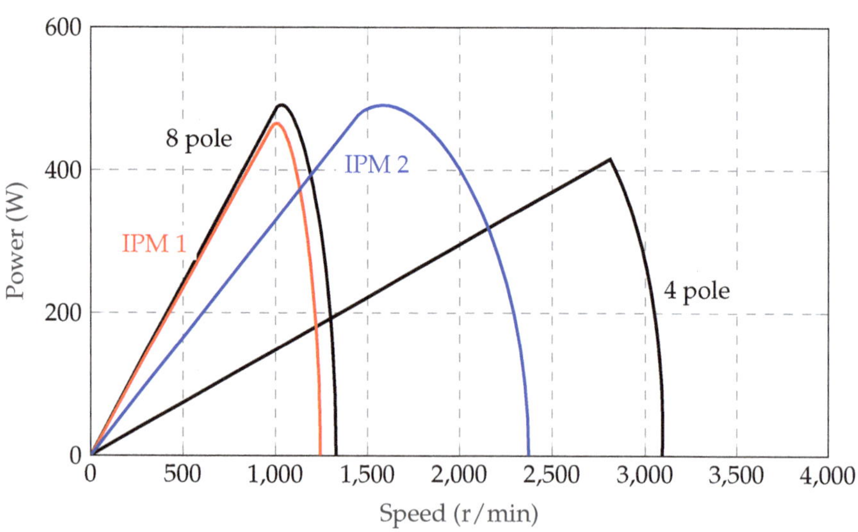

(**b**) Power–speed comparison.

Figure 23. IPM and HE-IPM torque and power comparisons.

Table 5. Rated speed performance comparison between HE-IPM and IPM.

Parameter	8-Pole HE-IPM	4-Pole HE-IPM	IPM Case 1	IPM Case 2
Rated speed ω_B	1010 r/min	2810 r/min	1000 r/min	2810 r/min
MTPA torque τ	4.4 N m	1.9 N m	4.3 N m	0 N m
Efficiency η	76 %	67 %	89 %	0
Joule losses P_J	141.5 W	260 W	520 W	0 W
Iron losses P_{fe}	8 W	24 W	3.5 W	0 W
Output power P_{out}	470 W	560 W	460 W	0 W
Power factor PF	0.97	0.99	0.93	0

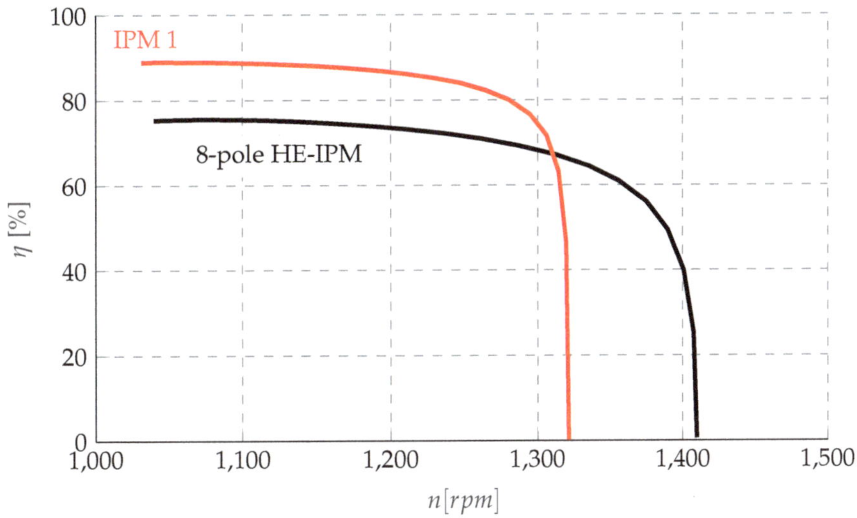

Figure 24. Motor efficiency during flux weakening: a comparison between HE-IPM and IPM.

In addition, the HE-IPM machine displays improved torque versus speed capabilities thanks to the possibility of changing the number of poles. In fact, by regulating the excitation current and without changing the rotor geometry, the speed range can be extended up to more than double the maximum speed of the eight-pole configuration. In contrast, the speed range of traditional IPM motors can be regulated only by changing PM dimensions, i.e., by modifying the motor geometry. The dual-polarity HEPM machine allows it to perform like two machines with only one hybrid-excited configuration.

In addition, it is possible to vary the HE-IPM field during flux-weakening operation to enhance performance. An example of this control strategy is presented in Figure 25. The rotor current is reduced, lowering the rotor flux. Thus, the torque slightly decreases while the speed range is gradually extended. Consequently, the motor efficiency increases, since rotor joule losses decrease as the excitation current is reduced.

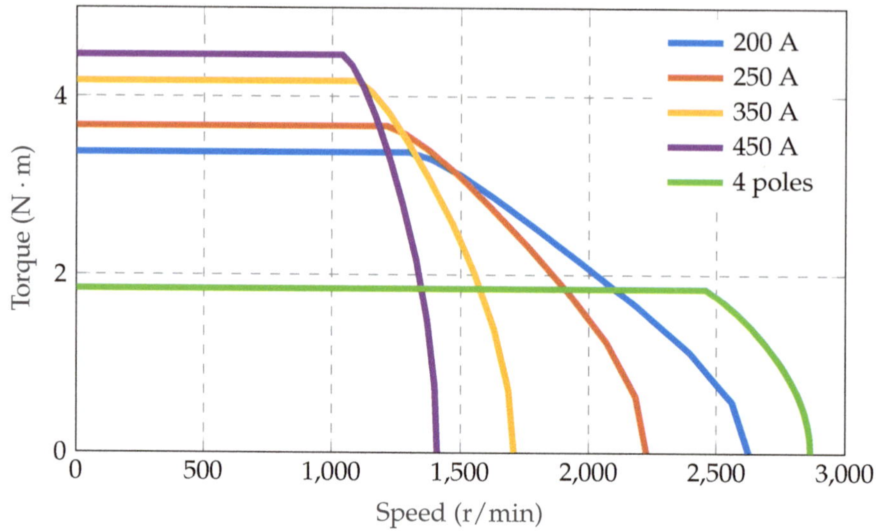

Figure 25. Flux-weakening operation with different excitation currents.

7. Thermal Considerations

In the literature, several thermal studies on wound rotor machines have been conducted [25–28]. Salient pole synchronous machines provide high efficiency, low noise, and low vibration [29–31]. In addition, they should have excellent thermal stability, especially because windings are placed in the rotor as well. To consider the thermal stability of the machine under study, a thermal analysis is carried out in parallel to the electromagnetic study. In this work, a simple two-dimensional finite element thermal study is presented. For the thermal finite element analysis, the motor model is the same model used for the magneto static simulations.

Two different cooling systems are considered, and their performance is compared. The first is a water cooling system, with water flowing in tubes along the motor case. The second is based on a classical finned case, cooled by forced air, but with an additional cooling system in the motor shaft that consists in oil flowing along the shaft itself [32,33].

Simulations are carried out in order to determine the heat exchange parameters such that the temperature rise of the rotor winding remains lower than:

$$\Delta T_{winding,r} = 125\ ^{\circ}\mathrm{C}$$

Calculations are performed considering class H of insulation, with a maximum environmental temperature of 40 °C.

To solve the thermal finite element problem, boundary conditions have to be imposed to specify the convection heat transfer coefficient h and impose an environmental temperature T_0 equal to 0 K. Thus, the simulation result represents the temperature increase for each motor region. Since half the machine is simulated, periodic boundary conditions are imposed along the lines that divide the motor in half.

In addition to the thermal study of the HE-IPM motor, other thermal simulations are presented in this section to compare the thermal performance of the IPM and HEPM machines.

7.1. Water Cooling System

The thermal finite element simulation model is represented in Figure 26. The external holed layer represents the motor aluminium case through which the cooling water flows. Between the motor and the case, there is a 0.02 mm still air layer. To avoid a too-small mesh size, this layer is modelled as a 2 mm layer with a thermal conductivity 100 times higher than that of still air (0.02 W/(m · K)), i.e., 2 W/(m · K). All other regions, depending on the material and the possible fluid motion, have different thermal conductivities that are summarized in Table 6.

Figure 26. Motor model for the thermal finite element simulation of the HE-IPM motor with a water cooling system.

Table 6. Parameters of the thermal finite element simulation.

Region	Thermal Conductivity λ W/(m · K)	Heat Generation P/Vol W/m^3
Shaft steel	22	0
Stator iron stack	30	25,160
Rotor iron stack	30	0
Permanent magnet	10	0
Still air (ribs)	0.02	0
Still air (stator case)	2	0
Rotor copper winding	1	328,6482
Winding insulation	1	0
Air gap air	0.1	0
Stator copper winding	1	601,619
Aluminium case	200	0

It is worth noting that the air gap is characterized by a thermal conductivity five times higher than that of still air due to rotor motion. In fact, the air between the rotor external surface and the stator internal surface is not still, but the fluid layer that is in contact with the rotor is moving at the same speed as the rotor. Based on the study of the thermal flow in a salient pole motor, for a similar size motor at a rated speed of 1000 r/min, a feasible convection heat transfer coefficient h is 300 W/(m^2 · K). Thus, considering the surface S through which the heat flow occurs, balancing the heat exchanged through conduction and the heat exchanged through convection along the air gap:

$$\frac{g}{\lambda \cdot S} = \frac{2}{h \cdot S}$$

the thermal conductivity of the air gap λ is computed:

$$\lambda = \frac{h \cdot g}{2} \approx 0.1$$

In addition to the thermal problem settings, the heat generated by each region is also specified in Table 6. The heat generated by the motor is computed by a single magneto static simulation at the rated speed ω_B. Both the rotor and stator windings produce heat due to joule losses; thus, their contributions to the temperature increase are considered separately as $P_{J,s}$ and $P_{J,r}$. In addition, stator iron losses P_{fe} are taken into account for the thermal analysis. In the thermal problem, each region that generates heat flux is defined by specific power losses, i.e., the ratio between power losses and the volume of the region. All data are summarized in Table 7.

Table 7. Power losses and specific losses of the eight-pole HE-IPM at the rated speed $\omega_B = 1010\ \mathrm{r/min}$.

Parameter	Value
Stator joule losses $P_{J,s}$	51.84 W
Rotor joule losses $P_{J,r}$	89.61 W
Total joule losses P_J	141.5 W
Iron losses P_{fe}	8 W
Stator winding copper volume Vol_s	2.39×10^{-6} m^3
Rotor winding copper volume Vol_r	3.41×10^{-6} m^3
Stator iron stack volume $Vol_{fe,s}$	3.18×10^{-4} m^3
Specific stator joule losses $P_{J,s}/(Vol_s \cdot Q_s)$	601,619 W/m^3
Specific rotor joule losses $P_{J,r}/(Vol_r \cdot Q_r)$	3,286,482 W/m^3
Specific iron losses P_{fe}/Vol_{fe}	25,160 W/m^3

Finally, boundary conditions are imposed to solve the finite element problem. Periodic boundary conditions are imposed along the lines that divide the motor in half. On the external case surface, an air-free convection heat exchange is imposed, with coefficient:

$$h = 5\ \mathrm{W/(m^2 \cdot K)}$$

Then, the thermal simulation is carried out to verify the fluid condition required to reach acceptable temperatures. Figure 27 shows the simulation result. The temperature increase is very low along the aluminium case and also in the stator, thanks to the water cooling system. Due to the high current density of the rotor excitation coils, the temperature increase in the rotor is higher than in the stator.

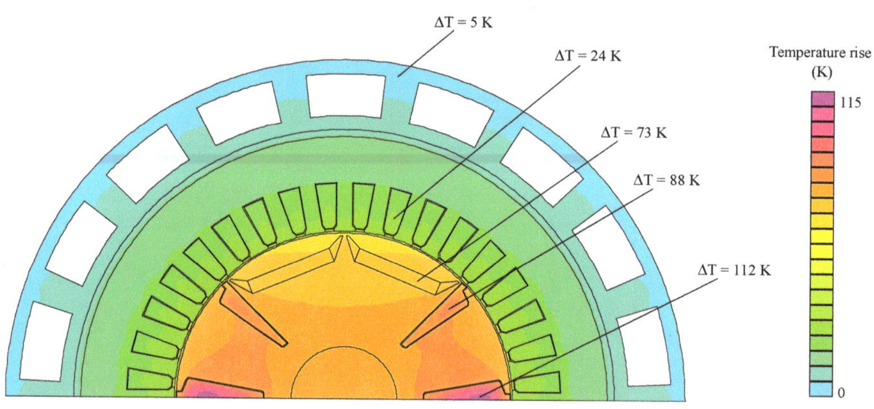

Figure 27. Temperature increase map with a water cooling system.

The rotor slots that face each other are characterized by the highest temperature increases, while the rotor slots close to the PMs reach lower temperatures. The highest temperature increase is equal to 112 K, which is acceptable considering the class *H* of insulation. In addition, it is worth noting that the temperature of the PMs is acceptable as well. In fact, the magnetic material adopted, PM *NdFeB UH*35, can sustain a maximum temperature of 180 °C due to demagnetization problems.

The resulting convective coefficient obtained from the simulations is equal to:

$$h = 500\ \mathrm{W/(m^2 \cdot K)}$$

This corresponds to water flowing in tubes in a cooling system selected for such a high rotor current density motor. This coefficient is imposed as a boundary condition along all the lines that represent pipes filled with water. The other boundary condition imposed is the temperature, equal to $T_0 = 0$ K, as explained in the introduction to this section.

7.2. Air and In-Shaft Oil Cooling System

The alternative cooling system model considered is represented in Figure 28. The external aluminium case is completely finned in order to favour heat transfer. The 0.02 mm layer is modelled as in the water cooling model described previously. The thermal conductivities of all regions are the same as in the water cooling model, as summarized in Table 6. In addition, the air gap thermal conductivity is the same since both simulations refer to the rated speed operation.

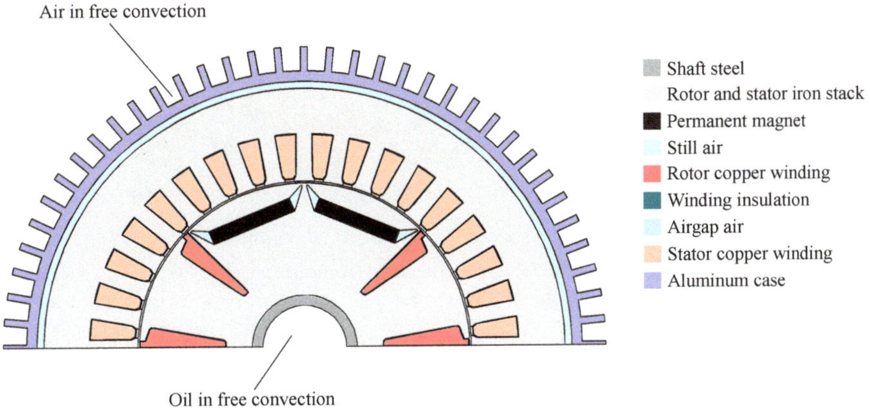

Figure 28. Motor model for the thermal finite element simulation of the HE-IPM motor with air cooling and oil flux along the shaft.

Along the finned surface of the case, the boundary condition of forced air convection is imposed:

$$h = 20\ \mathrm{W/(m^2 \cdot K)}$$

Then, the thermal simulation is carried out to verify the fluid condition required to reach acceptable temperatures. Figure 29 shows the simulation result. Similar to the water cooling system, the highest temperature increase in the rotor windings is equal to 114 K, which is acceptable considering the class H of insulation. The rotor slots that face each other are characterized by the highest temperature increases, while the rotor slots close to the PMs reach lower temperatures. In addition, the temperature of the PMs is acceptable and is the same as in the water cooling model.

The main difference between this water-and-oil cooling system and the previously analyzed water cooling system is the gradient of the temperature increase. In fact, the external air does not have the same cooling effectiveness as water. In contrast, the rotor heat is effectively dissipated by the oil flowing inside the motor shaft. As a result, the temperature increase of the rotor is comparable between the two models, while the temperature increase reached on the external case for the water-and-oil model is equal to 58 K; in the stator slots, it is 68 K (versus the water cooling results of 5 K and 24 K, respectively).

The resulting convective coefficient obtained from the simulations is equal to:

$$h = 350\ \mathrm{W/(m^2 \cdot K)}$$

This corresponds to oil flowing with free convection inside the motor shaft. Such a coefficient is imposed as a boundary condition along the internal shaft line.

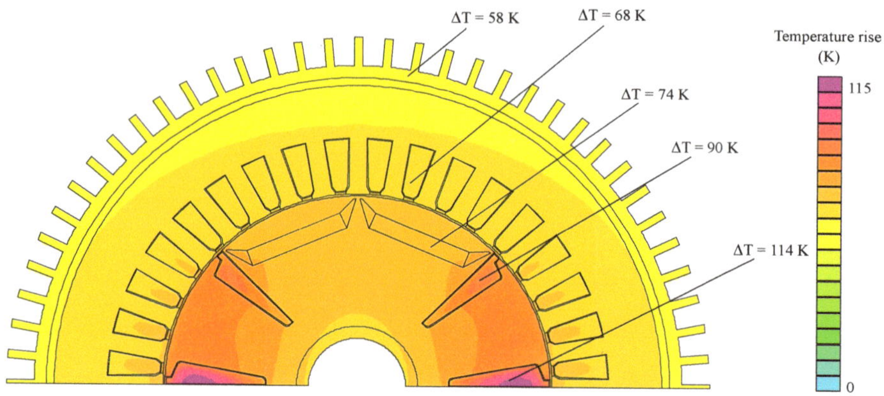

Figure 29. Temperature increase map with air cooling and oil flux along the shaft.

7.3. Comparison with the IPM Motor: Air Cooling System

The traditional IPM motor studied in Section 6.2 is considered here for a comparison of thermal behaviour. A forced-air cooling system is adopted, as represented in the model in Figure 30. With respect to the rated point operation considered, the thermal conductivity values for all regions of the motor and the thermal losses are reported in Table 6. Along the external case surface, which is finned and cooled by forced air, a convective heat transfer boundary condition is imposed, with the following parameters:

$$h = 15\ \mathrm{W/(m^2 \cdot K)}$$

Figure 31 shows the thermal simulation result. The temperature increase varies from 70 K to 80 K. The stator winding temperature increase is equal to 80 K, similar to that of the PMs and the whole rotor core.

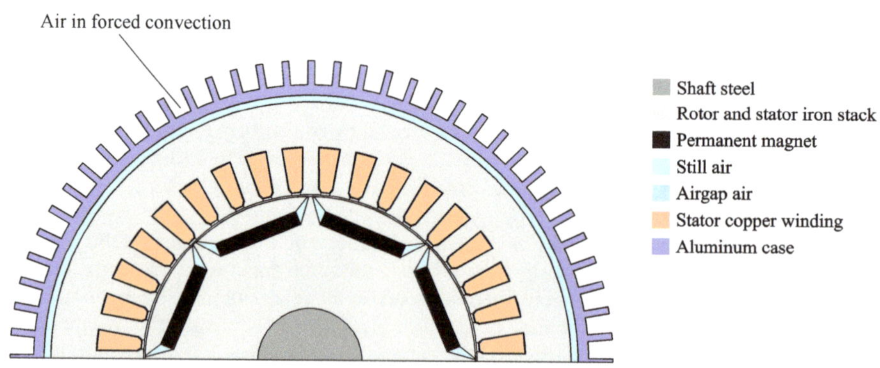

Figure 30. Motor model for the thermal finite element simulation of the IPM motor with a forced-air cooling system.

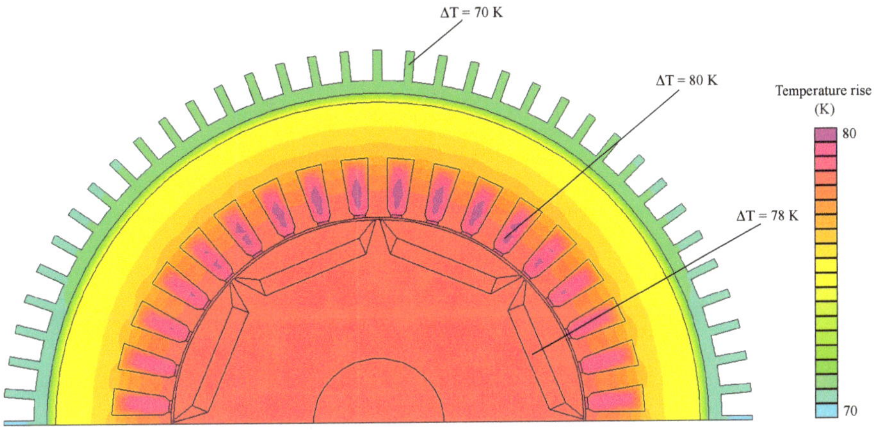

Figure 31. Temperature increase map for the HEPM motor with a forced-air cooling system.

Figure 32 shows the thermal simulation result for the HEPM motor with the same air cooling system as the IPM motor. The temperature increase is too high, leading to the burning of windings insulation and the demagnetization of PMs. It is worth noting that the rotor temperatures are higher than those of the stator. This is due to the low air gap thermal conductivity, which limits heat transfer from the rotor to the stator. Hence, the air cooling system is effective for the stator but does not allow for good rotor heat dissipation.

This thermal simulation proves that the hybrid-excited motor cannot be cooled using the same strategy as an IPM motor. In fact, the rotor windings generate additional heat in the rotor core; this heat is hard to dissipate due to the air gap.

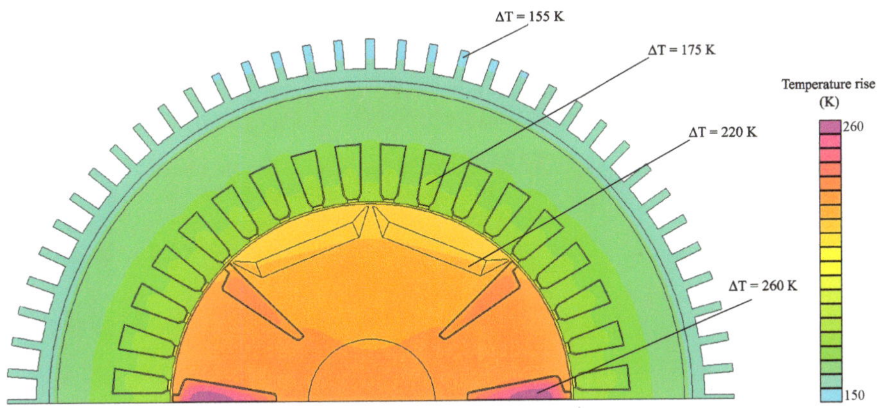

Figure 32. Temperature increase map for the HEPM motor with a forced-air cooling system.

Considering the presence of additional rotor windings, the HEPM motor cooling system has to be improved. Thus, an additional rotor cooling system is adopted that forces oil flow inside the shaft. Figure 33 shows the HEPM motor temperature increases with this improved cooling system. To reach a maximum temperature increase of about 120 K in the rotor windings, the convective heat transfer coefficient imposed for the inner surface of the shaft is:

$$h = 350\ \mathrm{W/(m^2 \cdot K)}$$

It is worth noting that the PM temperature increase is sufficiently low to avoid demagnetization (80 K); in addition, the rotor windings are limited to an acceptable temperature increase (120 K).

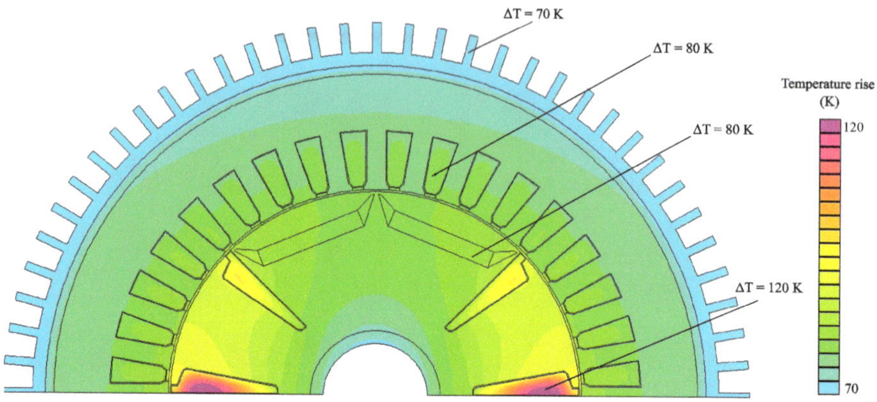

Figure 33. Temperature increase map for the HEPM motor with a forced-air cooling system and oil flux along the shaft.

Table 8 compares the thermal performance of the IPM and HEPM motors when considering simulations carried out with the same losses, i.e., the same heat generated (by the stator winding). The IPM and HEPM motors perform similarly, with different efficiencies, as described in Section 6.2. Temperature increases are comparable only when the improved cooling system is adopted for the HEPM motor.

Table 8. Comparison of the thermal performance of the IPM and HEPM motors with the same stator losses.

Parameter	IPM Motor Stator Water Jacket	HEPM Motor Stator Water Jacket	HEPM Motor Stator Water Jacket + Shaft Oil Cooling
Stator winding ΔT	80 °C	175 °C	80 °C
PMs ΔT	78 °C	220 °C	80 °C
Rotor winding ΔT	/	260 °C	120 °C
Rated speed ω_B	1000 r/min	1010 r/min	1010 r/min
MTPA torque τ	4.3 N m	4.4 N m	4.4 N m
Efficiency η	89 %	76 %	76 %
Joule losses P_J	520 W	141.5 W	141.5 W
Iron losses P_{fe}	3.5 W	8 W	8 W
Output power P_{out}	460 W	470 W	470 W
Power factor PF	0.93	0.97	0.97

7.4. Comparison with the IPM Motor: Water Cooling System

In order to improve the motor performance, a water cooling system is adopted, as represented in the model in Figure 34. The thermal conductivity values for all regions of the motor are reported in Table 6; the estimated value of the heat generated by the stator winding is selected on the basis of the temperature increase of the stator copper winding. The cooling system is modelled by imposing a convective heat transfer coefficient along the pipes of the water jacket equal to:

$$h = 500\ \text{W}/(\text{m}^2 \cdot \text{K})$$

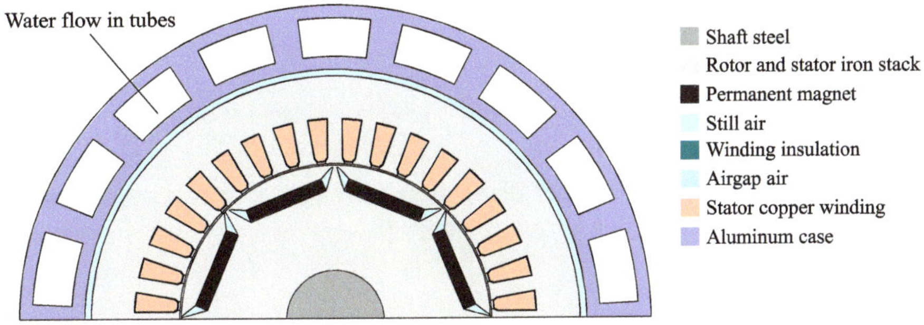

Figure 34. Motor model for the thermal finite element simulation of the IPM motor with a water cooling system.

Figure 35 shows the thermal simulation result. The stator copper winding temperature increase is equal to 107 K, which is the target temperature rise considering a class *H* insulation and with a maximum water temperature of 40 °C. It is worth noting that the temperature increase for the external case is low (23 K) and that the increase for the PMs is sufficiently low to avoid demagnetization (93 K).

The thermal simulation proves again that the hybrid-excited motor cannot be cooled with the same strategy as an IPM motor, even though water cooling is more effective than the forced-air system adopted in the simulation described in Section 7.3. The water cooling system is more effective in the stator; the temperatures reached are similar to those observed for the IPM motor, as shown in the thermal simulation result in Figure 36. In contrast, the rotor temperatures are higher than in the IPM motor because the rotor is thermally isolated with respect to the stator. In addition, PM temperature increases can cause demagnetization (more than 140 K), and the rotor winding maximum temperature increase reaches 190 K.

Considering the presence of additional rotor windings, the HEPM motor cooling system has to be improved. Thus, another thermal simulation is carried out that also includes a rotor cooling system. Figure 37 shows the HEPM motor temperature increase when the rotor is cooled by oil flowing inside the shaft. To reach a maximum temperature increase of about 120 K in the rotor windings, the convective heat transfer coefficient is equal to $h = 350\ \mathrm{W}/(\mathrm{m}^2 \cdot \mathrm{K})$.

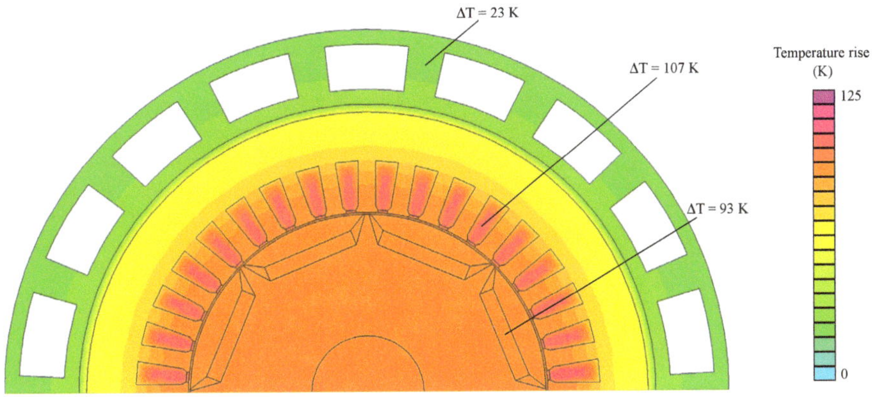

Figure 35. Temperature increase map for the IPM motor with a water cooling system.

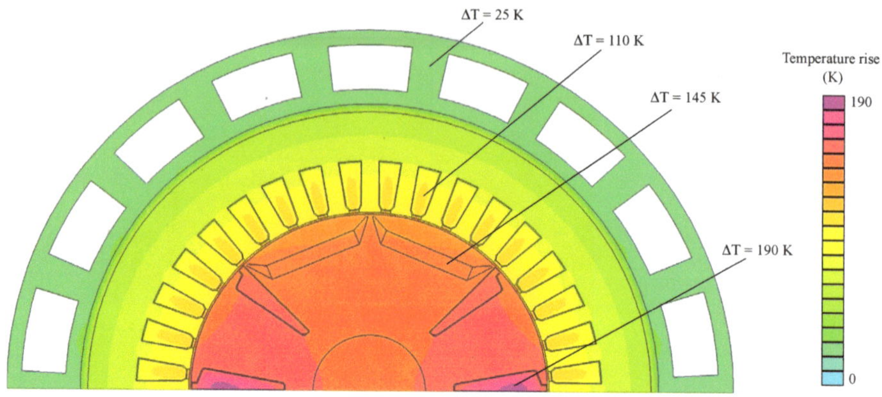

Figure 36. Temperature increase map for the HEPM motor with a water cooling system.

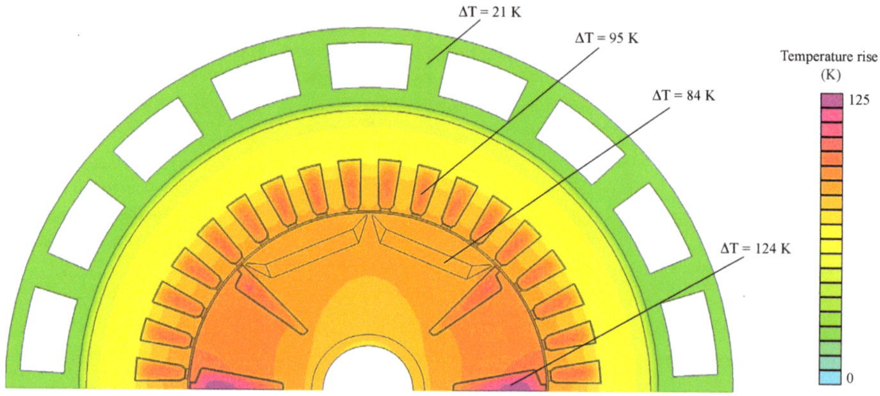

Figure 37. Temperature increase map for the HEPM motor with a water cooling system with oil flux along the shaft.

8. Conclusions

The hybrid-excitation motor is a new type of synchronous machine that includes both permanent magnets and excitation coils in the rotor.

This special type of machine requires a special manufacturing procedure. First, an additional winding has to be wrapped in the rotor. Then, a slip ring and brushes are needed to supply the excitation current. Concerning the stator winding, the two-inverter power supply system is suitable for high-power applications and allows for a redundant power source. Both inverters are designed to supply half the total required power.

Despite all the additional components, such a motor is characterized by a requirement for less magnetic material than traditional IPM motors. Designs that allow for a reduction in the need for rare-earth materials are currently a trend due to the high cost and poor availability of such materials.

The HEPM motor with a change in polarity exhibits several advantages. Firstly, the reduced need for magnetic material is useful. In addition, other advantages are its high torque at low speeds, which is achieved by means of the eight-pole configuration, and its extension of the operating speed range, which is achieved by means of the four-pole configuration. It is also possible to modulate the flux to improve the power factor of the HE-IPM machine.

The main drawback underlined in this study is the high torque ripple. However, it is shown herein that proper strategies exist to greatly reduce it.

In addition to the electromagnetic and mechanical performance of the dual-polarity HEPM motor, its thermal behaviour is also analyzed in this work, and a possible cooling system is suggested. Comparison with a conventional IPM motor shows the need for improved heat transfer in the rotor due to the additional losses associated with the excitation coils.

Author Contributions: C.C.: Investigation, Simulation, Reporting, Writing review and editing. N.B.: Conceptualization, Supervision, Writing review, Funding acquisition, Project administration. All authors have read and agreed to the published version of the manuscript.

Funding: This research received no external funding.

Data Availability Statement: Not applicable.

Conflicts of Interest: The authors declare that they have no known competing financial interests or personal relationships that could have appeared to influence the work reported in this paper.

Appendix A. Joule Losses

The HEPM machine is characterized by both stator and rotor joule losses due to the presence of excitation windings. Considering the geometric data and motor parameters summarized in Tables 1 and 2, stator joule losses are computed as:

$$P_{J,s} = J_s^2 \cdot V_{Cu,s} \cdot \frac{1}{\sigma_{Cu,120^\circ}} \tag{A1}$$

where $V_{Cu,s}$ is the volume of stator copper material, and $\sigma_{Cu,120^\circ}$ is the copper conductivity at 120° temperature, equal to 37.5 MS/m.

The current density J_s is calculated as:

$$J_{stat} = \frac{I_{max}}{\sqrt{2}} \cdot \frac{1}{S_{slot} \cdot k_{fill}} \tag{A2}$$

The maximum current for each FW point is $I_{max} = n_{cs} \cdot \sqrt{I_d^2 + I_q^2}$.

The volume of copper material in the stator is evaluated considering an approximated end winding length $L_{ew} \approx 2.5 \cdot D_s/p$:

$$V_{Cu,s} = k_{fill} \cdot S_{slot} \cdot Q_s \cdot (L_{stk} + L_{ew}) \tag{A3}$$

Similarly, rotor joule losses are computed by (A1), considering the excitation windings' current density J_e instead of J_s and rotor copper volume $V_{Cu,e}$ instead of $V_{Cu,s}$. The current density depends on the excitation current supplied at the specific FW operating point I_{exc}. The volume of rotor copper material is estimated as:

$$V_{Cu,exc} = k_{fill} \cdot S_{exc} \cdot Q_{exc} \cdot (L_{stk} + L_{ew,exc}) \tag{A4}$$

Appendix B. Iron Losses

Iron losses are computed using the Steinmetz formulation, considering both hysteresis and eddy current effects. The iron losses in the back-iron are computed using an increase coefficient $k_{magg,bi} = 1.5$, as follows:

$$P_{fe,bi} = k_{magg,bi} \cdot p_{s,bi} \cdot G_{bi} \tag{A5}$$

where the back-iron length is computed as $h_{bi} = (D_e - D_s)/2 - h_s$, considering the slot height h_s.

Specific back-iron losses $p_{s,bi}$ are computed considering the maximum flux density value in the back-iron core and the specific FW frequency compared to the reference flux density $B_{fe}^* = 1.5$ T and the frequency $f^* = 50$ Hz, respectively. For such reference values, hysteresis accounts for 70 % of iron losses (hysteresis coefficient $k_h = 0.7$), while eddy currents account for 30 % of iron losses (eddy current coefficient $k_{ec} = 0.3$). Thus, specific back-iron losses are calculated as follows:

$$p_{s,bi} = p_{s,fe} \cdot \left(\frac{\hat{B}_{bi}}{B^*_{fe}}\right)^2 \cdot \left[k_h \cdot \left(\frac{f}{f^*}\right) + k_{ec} \cdot \left(\frac{f}{f^*}\right)^2\right] \tag{A6}$$

The back-iron weight is:

$$G_{bi} = \gamma_{fe} \cdot \pi(D_e - 2 \cdot h_{bi})h_{bi} \cdot L_{stk} \tag{A7}$$

Similarly, iron losses in the teeth are computed considering an increase coefficient $k_{magg,t} = 2$, as follows:

$$P_{fe,bi} = k_{magg,t} \cdot p_{s,t} \cdot G_t \tag{A8}$$

Specific teeth iron losses are computed by (A5) considering the maximum flux density value in the tooth $\hat{B}_t$ instead of $\hat{B}_{bi}$. Teeth weight is:

$$G_t = \gamma_{fe} \cdot Q_s \cdot h_s \cdot \omega_t \cdot L_{stk} \tag{A9}$$

Total iron losses account for both back-iron and teeth losses:

$$P_{fe} = P_{fe,bi} + P_{fe,t} \tag{A10}$$

Appendix C. Efficiency

Total losses of the motor are computed as follows:

$$P_{loss} = P_{fe} + P_{J,s} + P_{J,r} \tag{A11}$$

Considering the output power P_{out}, which depends on the torque–speed performance of the motor, the efficiency is:

$$\eta = \frac{P_{out}}{P_{loss} + P_{out}} \tag{A12}$$

References

1. Melfi, M.J.; Evon, S.; McElveen, R. Induction versus permanent magnet motors. *IEEE Ind. Appl. Mag.* **2009**, *15*, 28–35. [CrossRef]
2. Papini, F.; Osama, M. Electromagnetic Design of an Interior Permanent Magnet Motor for Vehicle Traction. In Proceedings of the 2018 XIII International Conference on Electrical Machines (ICEM), Alexandroupoli, Greece, 3–6 September 2018; pp. 205–211. [CrossRef]
3. Vagati, A.; Boazzo, B.; Guglielmi, P.; Pellegrino, G. Design of Ferrite-Assisted Synchronous Reluctance Machines Robust Toward Demagnetization. *IEEE Trans. Ind. Appl.* **2014**, *50*, 1768–1779. [CrossRef]
4. Vagati, A.; Pastorelli, M.; Francheschini, G.; Petrache, S. Design of low-torque-ripple synchronous reluctance motors. *IEEE Trans. Ind. Appl.* **1998**, *34*, 758–765. [CrossRef]
5. Zhao, W.; Chen, D.; Lipo, T.A.; Kwon, B.I. Performance Improvement of Ferrite-Assisted Synchronous Reluctance Machines Using Asymmetrical Rotor Configurations. *IEEE Trans. Magn.* **2015**, *51*, 8108504. [CrossRef]
6. Wardach, M.; Palka, R.; Paplicki, P.; Prajzendanc, P.; Zarebski, T. Modern hybrid excited electric machines. *Energies* **2020**, *13*, 5910. [CrossRef]
7. Zhu, Z.Q.; Cai, S. Overview of Hybrid Excited Machines for Electric Vehicles. In Proceedings of the 2019 Fourteenth International Conference on Ecological Vehicles and Renewable Energies (EVER), Monte-Carlo, Monaco, 8–10 May 2019; pp. 1–14. [CrossRef]
8. Michieletto, D.; Cinti, L.; Bianchi, N. Hybrid Excitation PM Synchronous Motors: Part II —Finite Element Analysis. *IEEE Trans. Energy Convers.* **2021**, *37*, 495–504. [CrossRef]
9. Michieletto, D.; Cinti, L.; Bianchi, N. Hybrid Excitation PM Synchronous Motors: Part I Per Unit Analysis. *IEEE Trans. Energy Convers.* **2021**, *37*, 487–494. [CrossRef]
10. Cinti, L.; Michieletto, D.; Bianchi, N.; Bertoluzzo, M. A Comparison between Hybrid Excitation and Interior Permanent Magnet Motors. In Proceedings of the 2021 IEEE Workshop on Electrical Machines Design, Control and Diagnosis (WEMDCD), Modena, Italy, 8–9 April 2021; pp. 10–15. [CrossRef]
11. Liu, H.; Lin, H.; Fang, S.; Zhu, Z.Q. Permanent Magnet Demagnetization Physics of a Variable Flux Memory Motor. *IEEE Trans. Magn.* **2009**, *45*, 4736–4739. [CrossRef]

12. Yu, C.; Chau, K.T. Design, Analysis, and Control of DC-Excited Memory Motors. *IEEE Trans. Energy Convers.* **2010**, *26*, 479–489. [CrossRef]
13. Li, W.; Chau, K.T.; Gong, Y.; Jiang, J.Z.; Li, F. A New Flux-Mnemonic Dual-Magnet Brushless Machine. *IEEE Trans. Magn.* **2011**, *47*, 4223–4226. [CrossRef]
14. Yang, H.; Lin, H.Y.; Fang, S.H.; Huang, Y.K. Novel flux-regulatable dual-magnet vernier memory motors for electric vehicle propulsion. In Proceedings of the 2013 IEEE International Conference on Applied Superconductivity and Electromagnetic Devices, Beijing, China, 25–27 October 2013; pp. 359–360. [CrossRef]
15. Yang, H.; Lin, H.; Zhu, Z.Q.; Fang, S.; Huang, Y. Novel Flux-Regulatable Dual-Magnet Vernier Memory Machines for Electric Vehicle Propulsion. *IEEE Trans. Appl. Supercond.* **2014**, *24*, 0601205. [CrossRef]
16. Ostovic, V. Pole-changing permanent magnet machines. In Proceedings of the Conference Record of the 2001 IEEE Industry Applications Conference, 36th IAS Annual Meeting (Cat. No.01CH37248), Chicago, IL, USA, 30 September–4 October 2001; Volume 4, pp. 2570–2576. [CrossRef]
17. Latif, T.; Husain, I. Design and analysis of an induction motor for an enhanced constant power region with electronic pole changing. In Proceedings of the 2021 IEEE International Electric Machines Drives Conference (IEMDC), Virtual, 16–18 May 2021; pp. 1–6. [CrossRef]
18. Umanand, L. Pole Changing Wide Speed Range Induction Motor Drive for Electric Vehicles. In Proceedings of the 2018 IEEE International Conference on Power Electronics, Drives and Energy Systems (PEDES), Chennai, India, 18–21 December 2018; pp. 1–5. [CrossRef]
19. Cistelecan, M.V.; Melcescu, L.M.; Cosan, H.B.; Popescu, M. Induction motors with changeable pole windings in the ratio 1:4. In Proceedings of the International Aegean Conference on Electrical Machines and Power Electronics and Electromotion, Joint Conference, Istanbul, Turkey, 8–10 September 2011; pp. 781–786. [CrossRef]
20. Damaki Aliabad, A.; Mirsalim, M.; Farrokhzad Ershad, N. Line-Start Permanent-Magnet Motors: Significant Improvements in Starting Torque, Synchronization, and Steady-State Performance. *IEEE Trans. Magn.* **2010**, *46*, 4066–4072. [CrossRef]
21. Ghoroghchian, F.; Damaki Aliabad, A.; Amiri, E. Two-Speed Line Start Permanent Magnet Synchronous Motor With Dual Magnetic Polarity. *IEEE Trans. Ind. Appl.* **2018**, *54*, 4268–4277. [CrossRef]
22. Niizuma, T.; Sakai, K. Electronic motors capable of pole-changing and variable machine constants. In Proceedings of the 2012 15th International Conference on Electrical Machines and Systems (ICEMS), Sapporo, Japan, 21–24 October 2012; pp. 1–4.
23. Crisci, G. *Design, Schemes and Computation of Windings for Electrical Machines*, 15 ed.; S.T.E.M.: online, 1978; pp. 460, 472, 493–499. 516, 544–550. 553–557. 559. (In Italian)
24. Ocak, O.; Aydin, M. An Innovative Semi-FEA Based, Variable Magnet-Step-Skew to Minimize Cogging Torque and Torque Pulsations in Permanent Magnet Synchronous Motors. *IEEE Access* **2020**, *8*, 210775–210783. [CrossRef]
25. Ghahfarokhi, P.S.; Kallaste, A.; Podgornov, A.; Belahcen, A.; Vaimann, T.; Kudrjavtsev, O. Thermal Analysis of Salient Pole Synchronous Machines by Multiple Model Planes Approach. In Proceedings of the 2020 International Conference on Electrical Machines (ICEM), Virtual , 23–26 August 2020; Volume 1, pp. 1511–1517. [CrossRef]
26. Wang, Y.; Nuzzo, S.; Gerada, C.; Zhao, W.; Zhang, H.; Galea, M. 3D Lumped Parameter Thermal Network for Wound-Field Synchronous Generators. In Proceedings of the 2021 IEEE Workshop on Electrical Machines Design, Control and Diagnosis (WEMDCD), Modena, Italy, 8–9 April 2021; pp. 5–9. [CrossRef]
27. Vahedi, P.; Ganji, B.; Afjei, E. Lumped thermal model for the multi-layer switched reluctance motor. *IET Electr. Power Appl.* **2020**, *14*, 1873–1885. [CrossRef]
28. Mellor, P.; Roberts, D.; Turner, D. Lumped parameter thermal model for electrical machines of TEFC design. In *IEE Proceedings B (Electric Power Applications)*; IET: London, UK, 1991; Voume. 138, pp. 205–218.
29. Dorrell, D.G. Are wound-rotor synchronous motors suitable for use in high efficiency torque-dense automotive drives? In Proceedings of the IECON 2012—38th Annual Conference on IEEE Industrial Electronics Society, Montreal, QC, Canada, 25–28 October 2012; pp. 4880–4885. [CrossRef]
30. Bash, M.L.; Pekarek, S. Analysis and Validation of a Population-Based Design of a Wound-Rotor Synchronous Machine. *IEEE Trans. Energy Convers.* **2012**, *27*, 603–614. [CrossRef]
31. Liu, H.; Xu, L.; Shangguan, M.; Fu, W.N. Finite Element Analysis of 1 MW High Speed Wound-Rotor Synchronous Machine. *IEEE Trans. Magn.* **2012**, *48*, 4650–4653. [CrossRef]
32. Wang, H.; Liu, X.; Kang, M.; Guo, L.; Li, X. Oil Injection Cooling Design for the IPMSM Applied in Electric Vehicles. *IEEE Trans. Transp. Electrif.* **2022**, *8*, 3427–3440. [CrossRef]
33. Gai, Y.; Chong, Y.C.; Adam, H.; Goss, J.; Popescu, M. Power Losses and Thermal Analysis of a Hollow-Shaft Rotor Cooling System. In Proceedings of the 2019 22nd International Conference on Electrical Machines and Systems (ICEMS), Harbin, China, 11–14 August 2019; pp. 1–6. [CrossRef]

energies

Article

Multi-Objective Optimization Design of a New Permanent Magnet Synchronous Motor Based on the Taguchi Method

Chunyu Qu, Ziqi Guo *, Yongzhuang Hu, Xiuping Wang and Fangxu Han

School of Electric Power, Shenyang Institute of Engineering, Shenyang 110136, China
* Correspondence: 13028279718@163.com

Abstract: To improve the performance and stability of the permanent magnet synchronous motor (PMSM), a new type of built-in permanent magnet synchronous motor (IPMSMB) is proposed. Firstly, the performance indexes of IPMSM, IPMSMA, and IPMSMB are compared by finite element analysis. The results show that the effective harmonic of the air-gap magnetic density of the motor increases when the rotor outer diameter is piecewise eccentric. At the same time, torque ripple and cogging torque decrease. Then the permanent magnet structure of the motor is changed on the basis of IPMSMA to form IPMSMB, which improves the output torque of the motor. Secondly, the Taguchi method is used to optimize the structural parameters of IPMSMB. After optimization, the output torque of IPMSMB is increased by 4.6%. The cogging torque and torque ripple are decreased by 45.5% and 25.7%, respectively. The consumption of permanent magnets is reduced by 7.74%. Finally, the rationality of the motor design is verified by the prototype experiment.

Keywords: permanent magnet synchronous motor; air gap flux density; cogging torque; torque ripple; structure optimization; prototype testing

Citation: Qu, C.; Guo, Z.; Hu, Y.; Wang, X.; Han, F. Multi-Objective Optimization Design of a New Permanent Magnet Synchronous Motor Based on the Taguchi Method. *Energies* **2022**, *15*, 7347. https://doi.org/10.3390/en15197347

Academic Editors: Lorand Szabo and Feng Chai

Received: 10 September 2022
Accepted: 29 September 2022
Published: 6 October 2022

1. Introduction

With the rapid development of the new energy vehicle industry, the permanent magnet synchronous motor for vehicles stands out among the electric vehicle drive motors due to its advantages of high power, high torque density, simple structure, and high reliability. At present, since pure electric passenger vehicles use permanent magnet synchronous motors as the main technical route, how to further improve their performance has become an important issue in the industry. In general, permanent magnet synchronous motors for electric vehicles need to have large starting torque, high power density, and small cogging torque and torque ripple [1–3]. It brings certain challenges to the structural design and optimization of permanent magnet synchronous motors for electric vehicles.

Reference [4] shows that the harmonic content in the air gap magnetic density can be reduced by optimizing the shape of the permanent magnet and the shape of the rotor surface core respectively, so as to effectively weaken the cogging torque, suppress torque ripple and improve the overall performance of the motor. Reference [5] proposes an asymmetric V-type built-in permanent magnet synchronous motor and carries out multi-objective optimization. The motor structure is shown in Figure 1. The results show that the improved asymmetric rotor structure can reduce torque ripple and improve torque density. In reference [6], the eccentric air gap is obtained by changing the shape of the pole shoe. The simulation results show that the air gap magnetic field waveform of the motor is improved. Reference [7] presents a segmented permanent magnet structure, which improves the electromagnetic torque of the motor. It also achieves a wide constant power speed range and shows good performance at high speed. Reference [8] proposes a method of adding auxiliary slots on the rotor surface to optimize the air-gap magnetic field of PMSM. The iron loss and torque ripple caused by harmonics are reduced. The performance of the motor is greatly improved. In reference [9], an IPMSM motor with a flared-shape rotor structure is proposed and compared with the basic IPMSM motor. Simulation results show that the proposed flared-shape IPMSM motor has higher efficiency, lower torque ripple, and higher reliability. In reference [10], each permanent magnet of the surface-mounted PMSM is divided into eight parts, which effectively reduces the cogging torque of this motor. In reference [11], holes are cut on the rotor surface of the IPMSM motor. Then the size, number, and position of holes are optimized. The results show that the cogging torque of the motor is reduced and the average torque is increased.

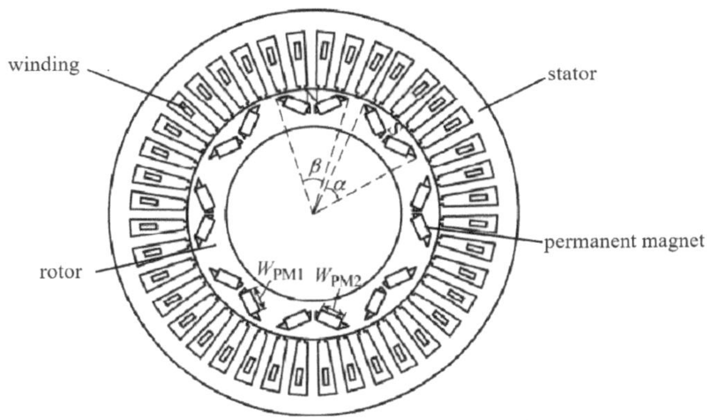

Figure 1. Schematic diagram of IPMSM.

In recent years, the Taguchi method proposed by Dr. Genichi Taguchi from Japan has been widely used in the field of motor optimization design. Although the Taguchi method is a local optimization method, its objective function is relatively simple to establish, the calculation period is short, and the optimal combination parameters for multi-objective optimization can be quickly and effectively searched. The optimization efficiency is high [12]. Reference [13] uses the Taguchi method to optimize the design of the magnetic pole parameters by taking the ratio of the electromagnetic torque pulsation of the motor to the average torque value and the levitation force pulsation and the average value of the levitation force as evaluation criteria.

To solve the problems of torque ripple and large cogging torque currently existing in the permanent magnet synchronous motor of electric vehicles and improve the running stability of electric cars. We propose a permanent magnet synchronous motor (IPMSMB) with a novel rotor structure and a novel permanent magnet topology. Compared with the basic IPMSM, IPMSMB increases the output torque of the motor and reduces the cogging torque and torque fluctuation of the motor under rated conditions. Finally, the rationality and feasibility of the motor structure design are verified by the experimental test of the prototype.

2. The Structure of the Motor

In this paper, an IPMSM with 3 phases, 8 poles, and 48 slots is selected as the research object, as shown in Figure 2. The basic design parameters of the IPMSMB designed in this paper are shown in Table 1.

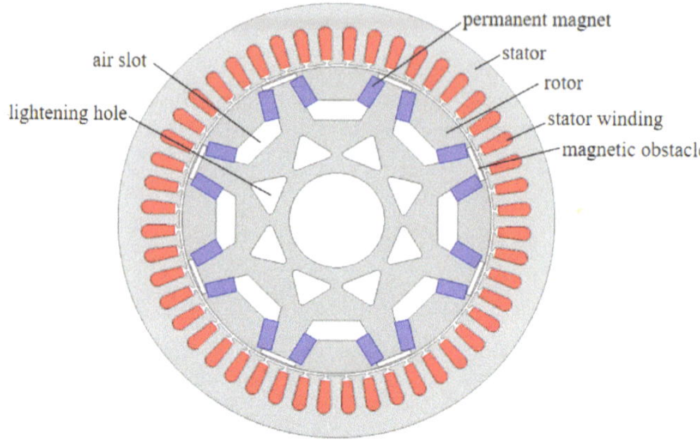

Figure 2. Schematic diagram of IPMSM.

Table 1. The basic parameters of IPMSM.

Parameter	Symbol	Unit	Value
Rated power	P	Kw	20
Rated voltage	U	V	380
Rated current	I	A	20
Frequency	f	Hz	100
Rotating speed	n	r/min	1500
Number of pole pairs	P	-	4
Outer diameter of the stator	R_g	mm	185
Inner diameter of the stator	R_s	mm	132.6
Rotor outside diameter	R_o	mm	130.6
Inner diameter of the rotor	R_i	mm	40
Axial length of the iron core	L_a	mm	160
Length of single permanent magnet	h	mm	12.4
Thickness of single permanent magnet	d_1	mm	6

First, we eccentric the rotor outer diameter of the IPMSM in Figure 2 with an eccentricity of 9 mm, thus obtaining IPMSMA. Figure 3 shows the schematic diagram of the IPMSM and IPMSMA rotor structure. Figure 4 shows IPMSMA. Table 2 shows the basic parameters of IPMSMA.

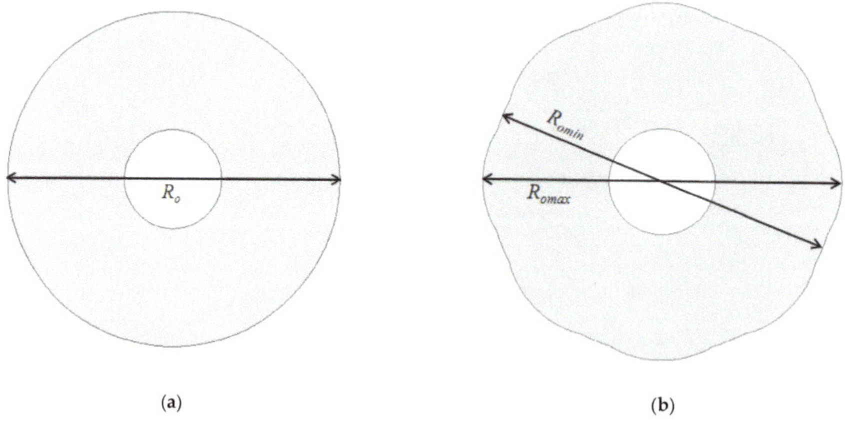

Figure 3. (**a**) Rotor structure diagram of IPMSM; (**b**) rotor structure diagram of IPMSMA.

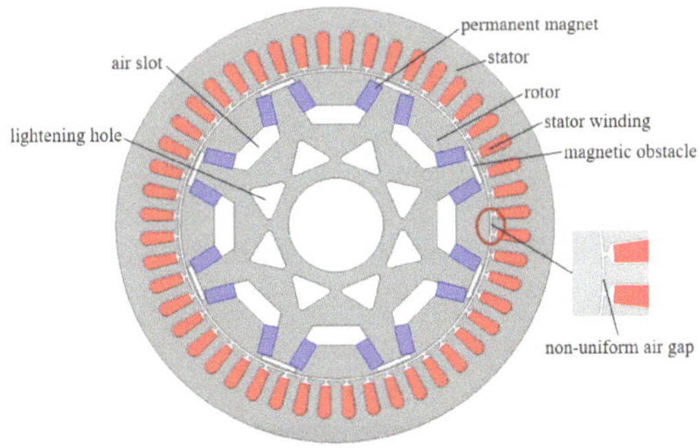

Figure 4. Schematic diagram of IPMSMA.

Table 2. The basic parameters of IPMSMA.

Parameter	Symbol	Unit	Value
Rated power	P	Kw	20
Rated voltage	U	V	380
Rated current	I	A	20
Frequency	f	Hz	100
Rotating speed	n	r/min	1500
Number of pole pairs	p	-	4
Outer diameter of the stator	R_g	mm	185
Inner diameter of the stator	R_s	mm	132.6
Maximum value of rotor outer diameter	R_{omax}	mm	131.5
Minimum value of rotor outer diameter	R_{omin}	mm	130.6
Inner diameter of the rotor	R_i	mm	40
Axial length of the iron core	L_a	mm	160
Length of single permanent magnet	h	mm	12.4
Thickness of single permanent magnet	d_1	mm	6

Then, we disassemble each integral permanent magnet of IPMSMA into two small permanent magnets of equal size. The width of each adjacent small permanent magnet becomes half of the original whole permanent magnet, which results in IPMSMB. The dotted line in Figure 5 is a single permanent magnet in IPMSMA. IPMSMB is formed by dividing a single permanent magnet equally at a certain angle on its wide side [14,15]. To provide a fair and reasonable comparison, all motor key parameters other than the air gap (inner and outer diameter, axial length, and stator slot width) are set equal. Figure 6 shows the structure of IPMSMB. Table 3 shows the basic parameters of IPMSMB.

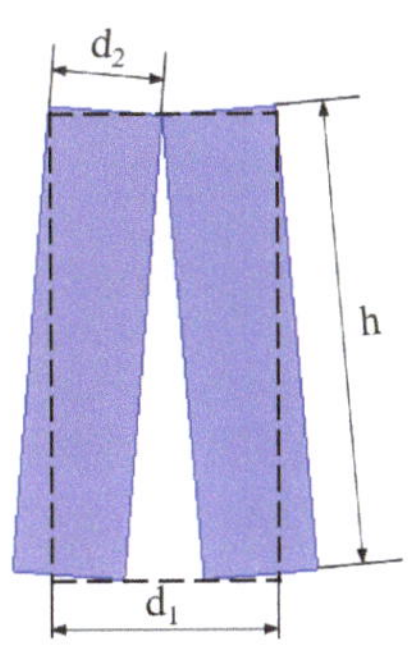

Figure 5. Structure diagram of a single permanent magnet for IPMSMB.

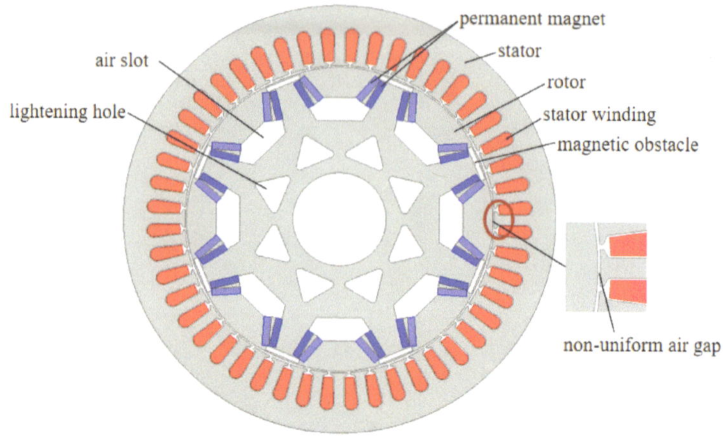

Figure 6. Schematic diagram of IPMSMB.

Table 3. The basic parameters of IPMSMB.

Parameter	Symbol	Unit	Value
Rated power	P	Kw	20
Rated voltage	U	V	380
Rated current	I	A	20
Frequency	f	Hz	100
Rotating speed	n	r/min	1500
Number of pole pairs	p	-	4
Outer diameter of the stator	R_g	mm	185
Inner diameter of the stator	R_s	mm	132.6
Maximum value of rotor outer diameter	R_{omax}	mm	131.5
Minimum value of rotor outer diameter	R_{omin}	mm	130.6
Inner diameter of the rotor	R_i	mm	40
Axial length of the iron core	L_a	mm	160
Length of single permanent magnet	h	mm	12.4
Thickness of single permanent magnet	d_2	mm	3

3. Electromagnetic Characteristic Analysis and Structural Parameter Optimization

Firstly, the structure of the IPMSM is optimized in this paper, and the outer diameter of the rotor of the IPMSM is designed by segmental eccentricity to form an uneven air gap. The structure of the non-uniform air gap can reduce the harmonic content of the air gap magnetic density, increase the effective harmonic content, and improve the performance of the motor. When designing the rotor structure with an uneven air gap, as shown in Figure 7, the rotor is divided into $2p$ parts. The LL' part of the outer circle of the rotor of each region is eccentric to the same degree while the rest of the elements remain unchanged. Thus, a motor rotor with a segmented abnormal structure is formed. The black dotted line in Figure 7 is the outer surface of the rotor before eccentricity. r is the outer diameter of the rotor before eccentricity and r' is the radius of the eccentric circle. The position of the extreme circle center of the rotor under the non-uniform air gap structure is different from the axial center of the rotating shaft. a is the eccentric distance between the center of the eccentric circle and the rotor axis. In this paper, the eccentric distance a = 9 mm, the maximum air gap length $\delta_{\max}$ = 1 mm, and the minimum air gap length $\delta_{\min}$ = 0.54 mm.

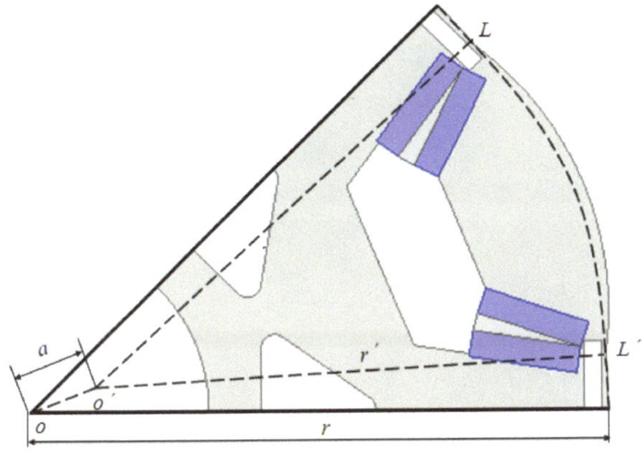

Figure 7. Rotor structure with an uneven air gap.

Secondly, this paper optimizes the permanent magnets of the rotor and divides each permanent magnet of the IPMSMA into two equal parts. The two small permanent magnets are arranged at an angle α. The optimized magnetic pole structure is shown in Figure 8. In this paper, the length of a single permanent magnet is h = 12.4 mm, the width of the permanent magnet is d = 3 mm, the angle between the permanent magnets is α = 9.8°, and the distance between the two small permanent magnets outside each pole is the V-shaped angle β = 63°.

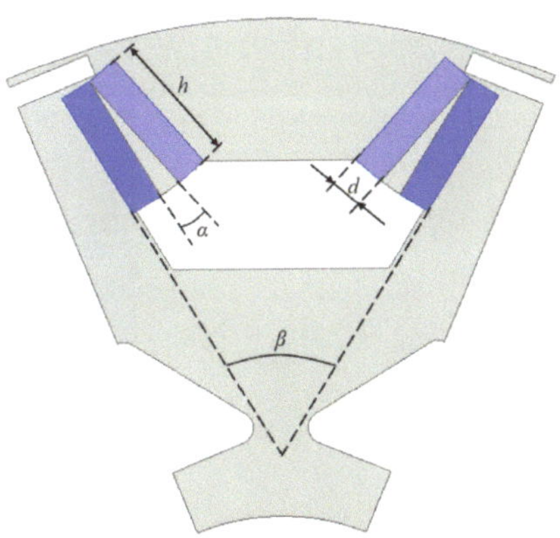

Figure 8. Optimized permanent magnet structure.

3.1. Magnetic Flux Density Analysis

The magnetic field density of the motor at no-load is shown in Figure 9, where Figure 9a represents PMSM, Figure 9b is IPMSMA, and Figure 9c is IPMSMB. By comparison, it is found that the amplitudes of the magnetic flux density at the stator teeth and the magnetic poles of IPMSMB are the largest, and the IPMSMB has the largest average magnetic flux density.

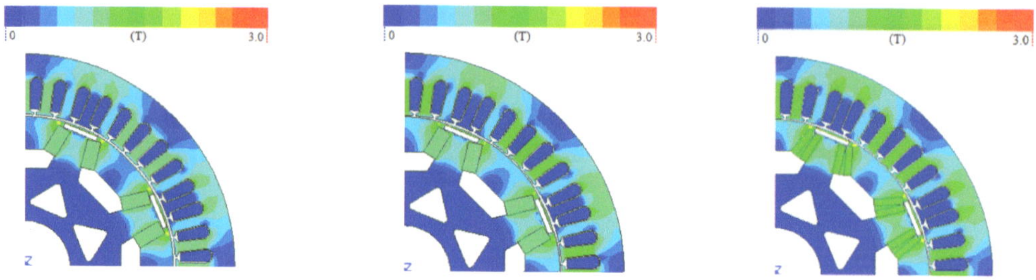

(a) (b) (c)

Figure 9. Comparison of magnetic flux density of three types of motors. (**a**) IPMSM; (**b**) IPMSMA; (**c**) IPMSMB. T is the unit of magnetic induction, that is, tesla.

3.2. No-Load Air Gap Flux Density Harmonic Analysis

Figure 10 shows the harmonic analysis diagram of the no-load air-gap flux density of the motor before and after optimization. It can be seen from the figure that the amplitude of the 4th harmonic component is the highest. This is because the number of pole pairs of the permanent magnets of the motor designed in this chapter is four. The uneven air gap is formed between the stator and the rotor by decentering the rotor section by section, which improves the air gap flux density of the motor at no-load. The 4th harmonic component of the motor is increased to a certain extent, while the higher harmonic component is decreased. The wave components have decreased. After the optimization of the permanent magnet structure, the fourth harmonic of IPMSMB is further enhanced, but the higher harmonic components are not significantly improved. In this paper, the IPMSM adopts a uniform air gap. IPMSMA and IPMSMB employ non-uniform air gaps. We need to be careful that the greater the eccentricity is not the better.

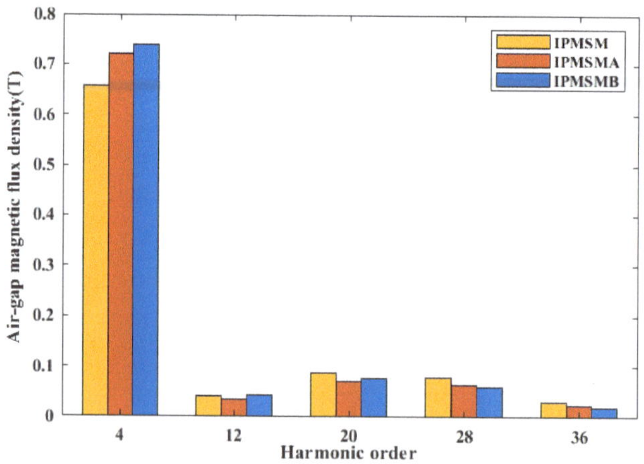

Figure 10. Fourier Decomposition of No-Carrier Gas Magnetic Density for Three Structure Motors.

3.3. No-Load Air Gap Flux Density Harmonic Analysis

The no-load back EMFs of IPMSM, IPMSMA, and IPMSMB are shown in Figure 11.

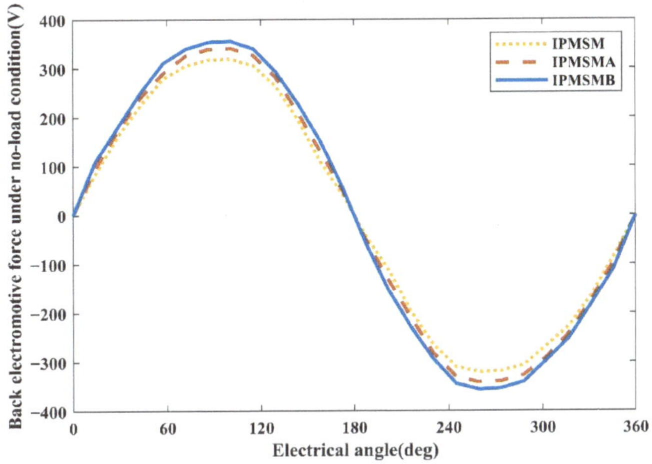

Figure 11. No-load back EMF of three types of motors.

Under the rated conditions, as shown in Figure 12, when the air gap is uniform, the amplitude of the cogging torque is 1210 mNm. When there is an uneven air gap for the motor, the amplitude of the cogging torque is 480 mNm. The amplitude of the cogging torque is reduced by 60.3%, the cogging torque is weakened to a certain extent. This is because the non-uniform air gap changes the air gap magnetic density of the motor, which affects the size of the cogging torque [16–19]. At the same time, after optimizing its magnetic pole structure based on IPMSMA, the amplitude of cogging torque of IPMSMB has been further reduced to 330 mNm.

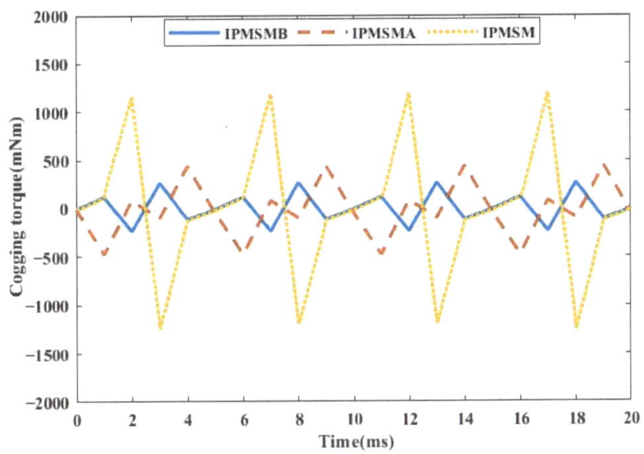

Figure 12. Cogging torque of three types of motors.

3.4. Analysis of Load Torque

Under the rated conditions, the output torque of the motor is shown in Figure 13. By comparing IPMSM and IPMSMA, it is shown that the amplitude of the output torque of the motor has been improved after the non-uniform air gap is adopted. Meanwhile, the torque ripple is changed from 8.72–6.5%. This is due to the uneven air gap between the stator and rotor. Therefore, the waveform of the air-gap magnetic field is sinusoidal, and the effective harmonic content of the air-gap magnetic density is improved [20–22]. Compared with IPMSMA, the average output torque of IPMSMB is increased by 4.36 Nm, while the torque ripple is reduced to 6.19%. This is because this arrangement of permanent magnets improves the magnetic field distribution on the stator and rotor. At the same time, the output torque of the motor is improved.

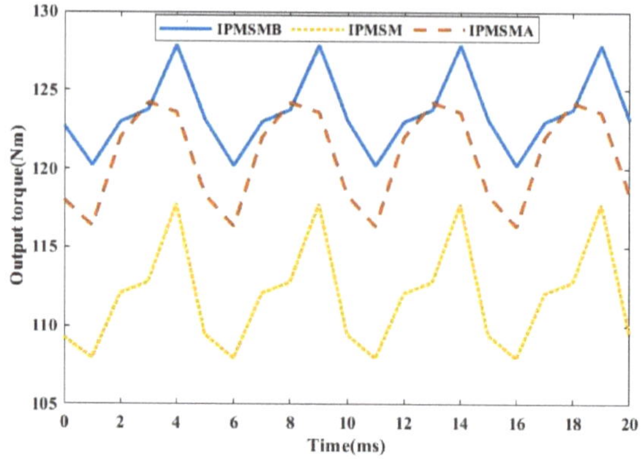

Figure 13. Output torque of three types of motors.

Table 4 shows the performance comparison of the three kinds of motors under rated conditions.

Table 4. Performance comparison of three kinds of motors under rated conditions.

Parameter	IPMSM	IPMSMA	IPMSMB
Amplitude of no-load back potential/V	319.97	341.38	355.80
Output torque/Nm	112.90	120.32	124.98
Cogging torque/Nm	1.21	0.48	0.33
Torque ripple/%	8.72	6.50	6.19

4. Multi-Objective Optimization of Rotor Structure Based on the Taguchi Method

In this paper, we use the Taguchi method for multi-objective optimization of the motor. The flowchart is shown in Figure 14.

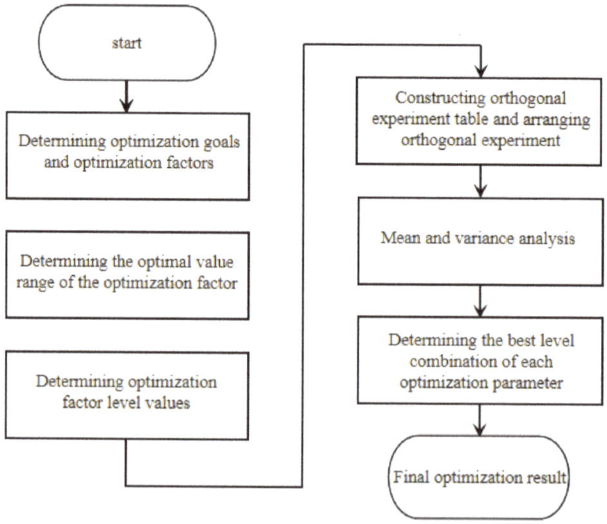

Figure 14. Flow chart of optimization design of Taguchi method.

4.1. Determining Optimization Goals and Optimization Factors

Since the PMSM of electric vehicles needs to meet its performance requirements of high torque density and power density, stable operation, and low noise, the average output torque T_{av}, cogging torque T_c, and torque ripple K are selected as the optimization objectives in this paper. Larger is better for the first indicator and smaller is better for the last two indicators. Due to the eccentricity of the rotor outer diameter, the structure size of the permanent magnet and the position of the permanent magnet in the rotor have a great influence on the motor. Based on the above factors, we choose five parameter variables as optimization factors: the eccentricity a, the length d of the permanent magnet, the thickness h of the permanent magnet, the Angle α of the permanent magnet, and the Angle β of the V-shape.

4.2. Determining the Optimal Value Range and Level Value of the Optimization Factor

In this paper, we use Ansys Maxwell finite element analysis software to adjust the optimization factor parameters many times and a large number of simulation calculations. As a result, the optimal value range of the optimization factor is obtained, as shown in Table 5. We select four influence levels equidistant within the optimal value range of each factor. Then we arrange the four-level values according to the ordering principle from small to large and named them as level values 1, 2, 3, and 4, respectively. The values of each level are shown in Table 6.

Table 5. Selection range of optimization parameters.

Optimization Factor	Preferred Range
Eccentricity/a	8–11
Length of permanent magnet/d	12–13.2
Thickness of permanent magnet/h	2.6–3.8
Angle of permanent magnet/α	8.8–11.8
V-shaped angle/β	61–67

Table 6. Number of levels for the optimization factor.

Optimization Factor	Level 1	Level 2	Level 3	Level 4
Eccentricity/a	8	9	10	11.80
Length of permanent magnet/d	12	12.4	12.8	13.22
Thickness of permanent magnet/h	2.6	3	3.4	3.8
Angle of permanent magnet/α	8.8	9.8	10.8	11.8
V-shaped angle/β	61	63	65	67

4.3. Determining the Optimal Value Range and Level Value of the Optimization Factor

According to the five optimization factors selected in Table 6 and the four level values determined by each optimization factor, we establish the orthogonal table by using the construction principle of the orthogonal table and the experimental design principle of the Taguchi method. In this way, we only need $4^2 = 16$ experiments to realize the multi-objective optimization of the motor, which greatly reduces the simulation times and time. Then we use Ansys simulation software to establish 16 motor models and their finite element analysis calculation. The solution results are shown in Table 7.

Table 7. Orthogonal table and finite element simulation results.

Test Number	Experiment Matrix					T_{av}/Nm	T_c/Nm	K/%
	a	d	h	α	β			
1	1	1	1	1	1	118.35	0.14	11.10
2	1	2	2	2	2	123.52	0.20	6.56
3	1	3	3	3	3	126.83	0.93	9.82
4	1	4	4	4	4	132.17	2.13	8.06
5	2	1	2	3	4	123.83	0.67	6.20
6	2	2	1	4	3	121.92	0.54	5.30
7	2	3	4	1	2	126.82	1.02	8.85
8	2	4	3	2	1	128.94	0.68	5.12
9	3	1	3	4	2	124.25	0.58	4.43
10	3	2	4	3	1	127.78	0.28	2.58
11	3	3	1	2	4	125.11	0.54	6.17
12	3	4	2	1	3	129.92	0.67	4.64
13	4	1	4	2	3	124.80	0.84	10.45
14	4	2	3	1	4	128.70	0.62	4.05
15	4	3	2	4	1	128.75	0.14	5.59
16	4	4	1	3	2	126.86	0.31	8.62

4.4. Mean Analysis

To obtain the influence of all optimization factors on the optimization target and the proportion of the parameter changes of each factor, we carry out the proportion of the influence on the optimization target according to the above orthogonal experiment table and finite element analysis results. Firstly, the average value is analyzed. Then the variance is analyzed. Finally, the optimization result is obtained.

According to Formula (1), we calculate the average value of the finite element simulation results of all optimization objectives. The results are shown in Table 8.

$$m = \frac{1}{n}\sum_{i=1}^{n} S_i \tag{1}$$

Table 8. Average results for each optimization objective.

-	T_{av}/Nm	T_c/Nm	K/%
m	126.16	0.67	6.72

In the formula, m is the average result of each optimization objective in Table 7; n is the number of trials, n = 16 in this paper; S_i is the i-th optimization objective value. (i is one of 1–16).

Secondly, the average value of the optimization factor at different levels is analyzed. According to Formula (2), when all the optimization factors matter at different levels, the average calculation results of the average output torque T_{av}, cogging torque T_c, and torque ripple K are shown in Table 9.

$$m_{ij}(f) = \frac{1}{4}[f(1) + f(2) + f(3) + f(4)] \tag{2}$$

Table 9. The average value of motor performance indexes of various optimization factors at different levels.

Optimization Factor	Number of Levels	T_{av}/Nm	T_c/Nm	K/%
a	1	125.22	0.85	8.89
	2	125.38	0.73	6.37
	3	126.77	0.52	4.46
	4	127.28	0.48	7.18
d	1	122.80	0.56	8.05
	2	125.48	0.41	4.62
	3	126.88	0.66	7.61
	4	129.47	0.95	6.61
h	1	122.81	0.38	7.80
	2	126.51	0.54	5.75
	3	127.18	0.70	5.86
	4	127.89	1.07	7.49
α	1	125.95	0.61	7.16
	2	125.59	0.57	7.08
	3	126.33	0.55	6.81
	4	126.77	0.85	5.85
β	1	125.96	0.43	6.10
	2	125.36	0.53	7.12
	3	125.87	0.75	7.55
	4	127.45	0.99	6.12

In Formula (2), i is one of the optimization factors a, d, h, α, and β. j is the number of levels. f is one of the optimization objectives T_{av}, T_c, and K.

4.5. Variance Analysis

According to Table 8 of the average value of each optimization target and Table 9 of the average value of each optimization factor under different levels of motor performance indicators, the variance of each optimization factor under each motor performance indicator and its impact proportion on each motor performance can be calculated according to Formula (3). Table 10 shows the influence of each optimization factor on different motor performance indicators.

$$D = \frac{1}{4}\sum_{j=1}^{4}\left[m_{ij}(f) - m(f)\right]^2 \tag{3}$$

Table 10. Performance comparison of three motors.

Optimization Factor	T_{av}/Nm		T_c/Nm		K/%	
	D	Proportion/%	D	Proportion/%	D	Proportion/%
a	0.78	7	0.02	10	2.54	44
d	5.81	52	0.04	20	1.75	29
h	3.84	34	0.07	35	0.86	15
α	0.19	2	0.02	10	0.27	5
β	0.61	5	0.05	25	0.40	7
In total	11.23	100	0.2	100	24.77	100

In Formula (3), D is the variance, j is the number of levels, i is one of the optimization factors a, d, h, α, and β, and f is one of the optimization targets T_{av}, T_c, and K.

According to the data in Table 10, eccentricity a has the greatest impact on torque ripple K, so we chose a = 10 mm; The permanent magnet length d has a relatively important influence on T_{av} and T_c, and its influence on T_{av} is the most obvious, so we chose d = 13.2 mm; the thickness of the permanent magnet h has a great influence on T_c. According to the calculation results in the table, we should choose the scheme with the smallest thickness, so we chose h = 2.6 mm; the influence of the angle α of the permanent magnet on the parameters is not particularly large, mainly affecting T_c. We chose $\alpha = 9.8°$ or $\alpha = 10.8°$ to obtain a relatively optimal value. From the perspective of improving the output torque and reducing the cogging torque, we finally chose $\alpha = 10.8°$; the included angle β has the greatest impact on T_c, so $\beta = 61°$ is selected [23–25]. The final best combination of motor main parameters is shown in Table 11. The comparison results of performance indexes before and after optimization are shown in Table 12.

Table 11. Optimization factor parameters before and after optimization.

Parameter	a/mm	d/mm	h/mm	α/°	β/°
Before optimization	9	12.4	3	9.8	63
After optimization	10	13.2	2.6	10.8	61

Table 12. Comparison of various performance indexes before and after optimization.

Parameter	T_{av}/Nm	T_c/Nm	K/%
Before optimization	124.98	0.33	6.19
After optimization	130.67	0.18	4.84

4.6. Comparative Analysis of Motor Performance Simulation

Under the rated conditions, the above-optimized motor model is simulated by finite element analysis, including the simulation analysis of torque characteristics and efficiency. The comparison of performance indexes before and after optimization is shown in Table 12. According to Table 12, the output torque of the motor before the optimization is 124.98 Nm, the cogging torque is 0.33 Nm, and the torque ripple is 6.19%. The optimized performance indexes are: output torque is 130.67 Nm, cogging torque is 0.18 Nm, and torque ripple is 4.84%. After optimization, the average output torque of the motor increases by 4.6%, the cogging torque decreases by 45.5%, and the torque ripple decreases by 25.7%. After optimization, the output torque of the motor increases, and the cogging torque and torque ripple decrease to a certain extent, which improves the operational performance and stability of the motor. The output torque comparison diagram and cogging torque comparison diagram of the motor before and after optimization are shown in Figures 15 and 16, respectively.

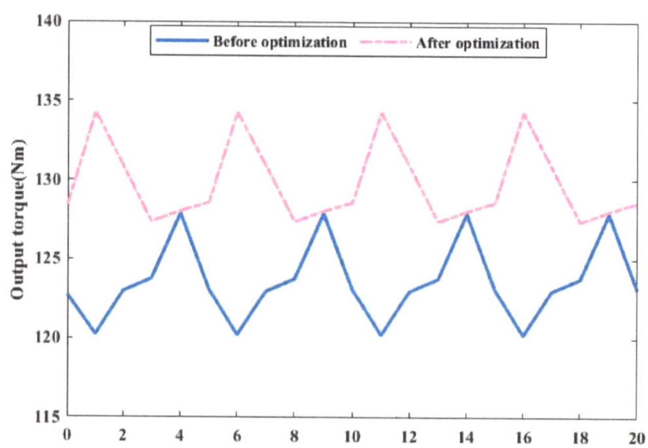

Figure 15. Comparison of motor output torque before and after optimization.

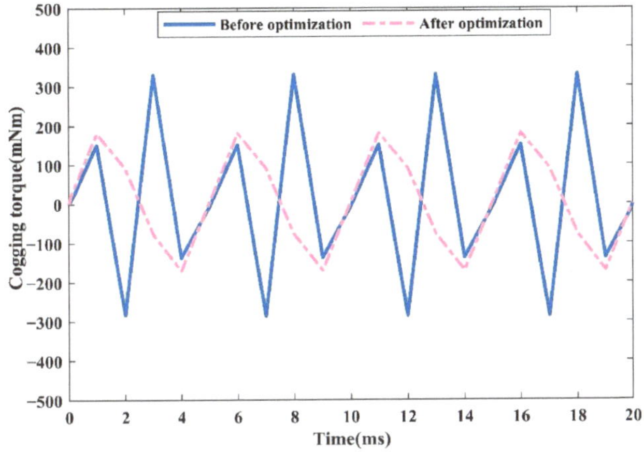

Figure 16. Comparison of cogging torque of motor before and after optimization.

4.7. On Load Performance Analysis of Optimized Motor

Figure 17 shows the relationship between the output torque and the current of the motor. According to the figure, the output torque of the motor gradually increases with the increase in the load current of the motor.

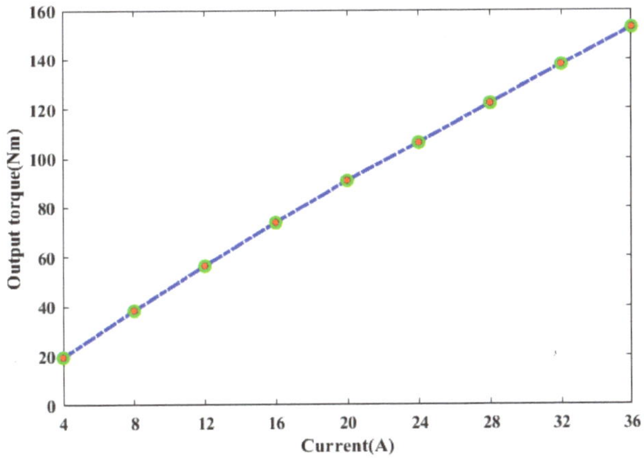

Figure 17. Influence of current on torque.

The relationship between the electromagnetic torque of the motor and the current angle is shown in Figure 18. Given the current of the motor of 30 A, the set current angle varies from $-90°$ to $90°$, and the current angle of the motor gradually increases. With the increase in the current angle of the motor, the electromagnetic torque of the motor gradually increases first and then decreases. It can be seen from the curve that when the current angle of the motor is $0°$, the electromagnetic torque of the motor reaches the maximum value.

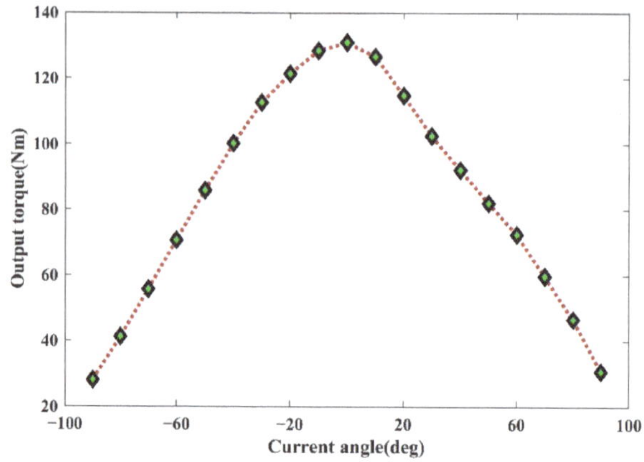

Figure 18. Influence of current angle on torque.

The comparison of the economic indicators of the motor before and after optimization is shown in Table 13. After optimization, the consumption of permanent magnets of the motor is reduced by 7.74%, the utilization rate of permanent magnets is increased by 13.3%, and the comprehensive economy is improved to a certain extent.

Table 13. Comparison of economic indexes of the motor before and after optimization.

Parameter	Volume of Permanent Magnet/mm^3	Utilization of Permanent Magnets/Nm/mm^3	Cost of Permanent Magnets/RMB (yuan)
Before optimization	190,464.00	6.562×10^{-4}	539.29
After optimization	175,718.40	7.436×10^{-4}	497.54

5. Experimental Platform and Experimental Research

5.1. Prototype and Experimental Platform

To verify the rationality of the above motor design and finite element simulation analysis, we use the prototype to carry out experimental tests. The main parameters of the prototype are shown in Table 14.

Table 14. Main parameters of the prototype.

Parameter	Value
Rated power/Kw	20
Rated voltage/V	380
Rated current/A	20
Rated frequency/f	100
Rated power factor	0.96
Wiring mode	Y
Number of motor poles	8

The prototype experimental platform is shown in Figure 19, mainly composed of a permanent magnet motor system, speed and torque instrument, power analyzer, and frequency converter.

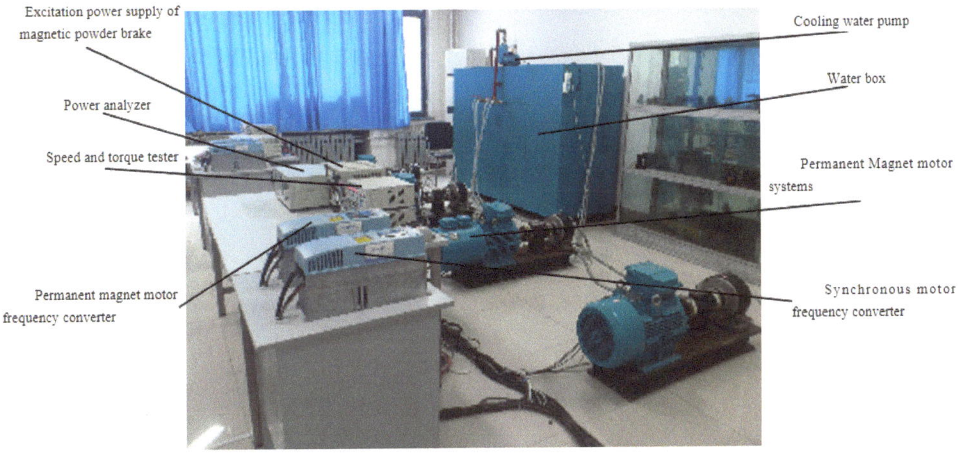

Figure 19. Prototype experiment platform.

5.2. Prototype Experiment Platform

In order to test the rationality and feasibility of the motor designed by us, we use the load experiment to test the load capacity of the prototype. First, we run the motor no-load at rated conditions. Then, we continuously increase the load of the motor until the rated load. Second, we record the power, output torque, and other electrical parameters of the motor with different sizes of load. Finally, we calculate the efficiency of the motor under different conditions to judge the load capacity of the motor.

We measure the output power and efficiency after changing the load. The results are shown in Table 15. According to Figure 20, to show the difference between the calculated value and the actual value of efficiency more intuitively, we import the experimental results into the line chart.

Table 15. Load power and efficiency.

-	Measured Value		Calculated Value	
Adjusting Torque/Nm	Output Power/Kw	Efficiency/%	Output Power/Kw	Efficiency/%
0	0.305	26.5	0.305	22.4
27	2.782	76.9	2.781	74.2
54	5.451	86.1	5.508	84.0
81	8.244	88.4	8.243	87.2
108	11.008	89.0	10.954	88.7
135	13.728	90.9	13.638	90.2
162	16.324	91.8	16.324	91.2
189	18.917	92.9	18.916	91.3

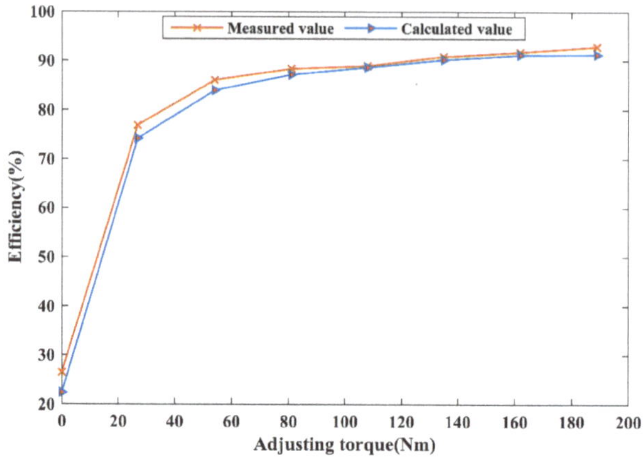

Figure 20. Variation of efficiency with adjusting torque.

The simulation results in Figure 20 show that the actual efficiency of the prototype test is slightly lower, which may be caused by the increase in the core loss of the motor caused by the temperature rise during the operation of the motor.

6. Conclusions

To improve the performance and reliability of a permanent magnet synchronous motor for a vehicle, we have designed a new type of permanent magnet synchronous motor for a vehicle, IPMSMB, and carried out multi-objective optimization by the Taguchi method. Finally, we have used the prototype experiment platform to verify the rationality of the design. The research conclusions are as follows:

(1) Through finite element analysis, we have compared the air-gap magnetic field, back potential, and torque characteristics of three structural motors. The results show that the waveform of air-gap magnetic density can be effectively improved and the coat-torque of the motor can be reduced by adopting a piecewise eccentric design on the external surface of the motor rotor under the rated conditions. The optimization of permanent magnet structure can improve the electromagnetic torque of the motor to a certain extent.
(2) The Taguchi method has been used to optimize the key parameters affecting the performance of the motor, and the simulation verification has been carried out. The results show that the permanent magnet consumption of the optimized IPMSMB has been reduced by 7.74% under the rated conditions. Meanwhile, the output torque has been increased by 4.6%, the cogging torque has been reduced by 45.5%, and the torque ripple has been reduced by 25.7%.
(3) The experimental measurement results of the prototype are very close to the simulation calculation values, which verifies the rationality of the motor design.
(4) In future work, we will strive to reduce the loss of the motor and improve the efficiency of the motor by using low-loss magnetic materials or trying to make the iron core into a laminar structure.

Author Contributions: Investigation, Z.G. and F.H.; Methodology, C.Q. and Z.G.; Project administration, C.Q. and X.W.; Software, Y.H.; Writing—original draft, C.Q. and Z.G.; Writing—review & editing, Z.G. and X.W. All authors have read and agreed to the published version of the manuscript.

Funding: This work is supported in part by the Liaoning Natural Science Foundation under Grant 2020-MS-240 and 2020-MS-242, Shenyang Young and Middle-aged Scientific and Technological Innovation Talents under Grant RC200192, Liaoning Provincial Department of Education Scientific Research Project under Grant LJKZ1085.

Conflicts of Interest: The authors declare no conflict of interest.

References

1. Xu, Y.; Ai, M.; Xu, Z.; Liu, W.; Wang, Y. Research On Interior Permanent Magnet Synchronous Motor Based on Performance Matching of Electric Bus. *IEEE Trans. Appl. Supercond.* **2021**, *31*, 1–4. [CrossRef]
2. Zhang, Y.; Cao, W.; McLoone, S.; Morrow, J. Design and Flux-Weakening Control of an Interior Permanent Magnet Synchronous Motor for Electric Vehicles. *IEEE Trans. Appl. Supercond.* **2016**, *26*, 1–6. [CrossRef]
3. Sun, X.; Shi, Z.; Lei, G.; Guo, Y.; Zhu, J. Analysis and Design Optimization of a Permanent Magnet Synchronous Motor for a Campus Patrol Electric Vehicle. *IEEE Trans. Veh. Technol.* **2019**, *68*, 10535–10544. [CrossRef]
4. Wang, K.; Sun, H.; Zhang, L.; Liu, C.; Zhu, Z. Summary of rotor pole optimization technology for permanent magnet synchronous motors. *Proc. Chin. Soc. Electr. Eng.* **2017**, *37*, 7304–7317+7445.
5. Liu, G.; Wang, Y.; Chen, Q. Multi-objective optimization design of asymmetric V-shaped built-in permanent magnet synchronous motor. *J. Electrotech. Technol.* **2018**, *33*, 385–393.
6. Xu, Y.; Li, Q.; Wang, T. Waveform optimization of no-load air gap flux density for permanent magnet synchronous motor. *J. Southwest Jiaotong Univ.* **2009**, *44*, 513–516+540.
7. Zhao, X.; Kou, B.; Huang, C.; Zhang, L. Optimization Design and Performance Analysis of a Reverse-Salient Permanent Magnet Synchronous Motor. *Machines* **2022**, *10*, 204. [CrossRef]
8. Wang, S.; Tian, X.; Sun, C.Z.; Ma, X.H. Optimization Design of Auxiliary Grooves on Rotor of Permanent Magnet Synchronous Motor for Electric Vehicles. *Appl. Mech. Mater.* **2014**, *467*, 536–541. [CrossRef]
9. Yoon, K.-Y.; Baek, S.-W. Performance Improvement of Concentrated-Flux Type IPM PMSM Motor with Flared-Shape Magnet Arrangement. *Appl. Sci.* **2020**, *10*, 6061. [CrossRef]
10. Ghasemi, A. Cogging Torque Reduction and Optimization in Surface-mounted Permanent Magnet Motor Using Magnet Segmentation Method. *Electr. Power Compon. Syst.* **2014**, *42*, 1239–1248. [CrossRef]
11. Ramezani, A.; Ghiasi, M.; Dehghani, M.; Niknam, T.; Siano, P.; Alhelou, H.H. Reduction of Ripple Toothed Torque in the Internal Permanent Magnet Electric Motor by Creating Optimal Combination of Holes in the Rotor Surface Considering Harmonic Effects. *IEEE Access* **2020**, *8*, 215107–215124. [CrossRef]
12. Cheng, P.; Yang, X.; Lan, H.; Hong, Y.; Dai, Q. Design and efficiency optimization of synchronous generator based on finite element method and Taguchi method. *J. Electr. Mach. Control* **2019**, *23*, 94–96.
13. Zhu, R.; Cheng, Y. Rotor optimization design of bearingless Permanent Magnet Synchronous Motor Based on combined magnetic poles. *J. Mot. Control* **2020**, *24*, 123–130.
14. Lee, S.-H.; Yang, I.-J.; Song, S.-W.; Kim, W.-H.; Jang, I.S. Study on IPMSM rotor shape for reduction of harmonics and vibration. In Proceedings of the IEEE Energy Conversion Congress and Exposition (ECCE 2021), Virtual Conference, 10–14 October 2021; pp. 4449–4453.
15. Chai, F.; Liang, P.; Pei, Y.; Cheng, S. Analytical Method for Iron Losses Reduction in Interior Permanent Magnet Synchronous Motor. *IEEE Trans. Magn.* **2015**, *51*, 1–4. [CrossRef]
16. Liu, F.; Wang, X. Design of a 35kW Permanent Magnet Synchronous Motor for Electric Vehicle Equipped with Nonuniform Airgap Rotor. In Proceedings of the 2020 IEEE 3rd Student Conference on Electrical Machines and Systems (SCEMS 2020), Jinan, China, 4–6 December 2020; pp. 179–183.
17. Zhao, S.; Chen, J.; Zhang, C.; Li, Z.; Meng, Y. Analytical calculation of magnetic field of surface mounted permanent magnet synchronous motor with uneven air gap. *J. Electrotech.* **2022**, *37*, 3502–3513.
18. Ma, S.; Zhang, X.; Geng, H.; Zhang, Y.; Meng, X. Non uniform air gap modeling and performance analysis of a new permanent magnet synchronous motor for electric vehicles. *J. Xi'an Jiaotong Univ.* **2019**, *53*, 70–76.
19. Ji, P.; Wang, X.; Wang, D.; Yang, Y. Research on cogging torque of surface permanent magnet motor with static eccentricity of rotor. *Proc. Chin. Soc. Electr. Eng.* **2004**, *24*, 192–195.
20. Guo, H.; Qian, H. Robust design of permanent magnet synchronous motor with low torque ripple. *Proc. Chin. Soc. Electr. Eng.* **2012**, *32*, 88–95+15.
21. Dong, P.; Huang, X.; Zhou, C. Effect of non-uniform air gap on torque, electromagnetic force, mechanical stress and vibration. In Proceedings of the 2019 22nd International Conference on Electrical Machines and Systems (ICEMS), Harbin, China, 11–14 August 2019; pp. 1–4.
22. Gao, F.; Li, X.; Qi, X.; Tao, C.; Gao, P. Analysis of torque ripple of built-in permanent magnet synchronous motor with asymmetric v-pole offset. *J. Electr. Mach. Control* **2021**, *25*, 112–120.
23. Lan, Z.; Yang, X.; Wang, F.; Zheng, C. Application of Taguchi method in optimization design of embedded sine wave permanent magnet synchronous motor. *J. Electrotech.* **2011**, *26*, 37–42.
24. Hwang, C.-C.; Chang, C.-M.; Liu, C.-T. A Fuzzy-Based Taguchi Method for Multi-objective Design of PM Motors. *IEEE Trans. Magn.* **2013**, *49*, 2153–2156. [CrossRef]
25. Sun, X.; Shi, Z.; Zhu, J. Multi-objective Design Optimization of an IPMSM for EVs Based on Fuzzy Method and Sequential Taguchi Method. *IEEE Trans. Ind. Electron.* **2020**, *68*, 10592–10600. [CrossRef]

energies

MDPI

Article

Speed Control of Magnetic Drive-Trains with Pole-Slipping Amelioration

Xiaowen Liao [1,2,*], Chris Bingham [3,*] and Tim Smith [3]

1 Guangdong University of Petrochemical Technology (GDUPT), Maoming 525000, China
2 The Joint Research Lab between GDUPT and University of Lincoln, Lincoln LN6 7TS, UK
3 School of Engineering, University of Lincoln, Lincoln LN6 7TS, UK
* Correspondence: liaoxiaowen@gdupt.edu.cn (X.L.); cbingham@lincoln.ac.uk (C.B.)

Abstract: The paper introduces new techniques to reduce the potential for pole-slipping induced by control systems and presents a low-cost pole-slipping detection and recovery scheme for magnetic drive-trains (MDTs). For the first time, the paper shows that a combination of electromagnetic and load-torque excitations which individually are not greater than the maximum coupling torque can initiate pole-slipping. For applications where acceleration feedback is unavailable, the motor-side inertia is virtually increased with a tracking differentiator to provide feedback of acceleration. Subsequently, controller design and parameter optimization are discussed. Experimental measurements on a custom test facility verify the presented principles that low-bandwidth controller designs with low inertia ratios can accommodate a wider range of on-load startup torque and load-torque disturbances without pole-slipping. To address overload issues, a pole-slipping detection method based on the kurtosis of electromagnetic torque and a recovery strategy based on converting the state of pole-slipping into that of on-load startup are presented. Experimental results demonstrate that detecting slip anomalies without load-side information, and recovery from pole-slipping without auxiliary mechanical devices are both feasible.

Keywords: magnetic drive-trains; dynamic analysis; speed control; pole-slipping detection; recovery

Citation: Liao, X.; Bingham, C.; Smith, T. Speed Control of Magnetic Drive-Trains with Pole-Slipping Amelioration . *Energies* **2022**, *15*, 8148. https://doi.org/10.3390/en15218148

Academic Editors: Lorand Szabo and Feng Chai

Received: 22 September 2022
Accepted: 28 October 2022
Published: 1 November 2022

1. Introduction

Due to the physical isolation of motor- and load-side shafts when employing magnetic couplings or magnetic gears, magnetic drive-trains (MDTs) can overcome the inherent problems of traditional couplings/gearboxes, such as overload failures, jamming, mechanic fatigue, and regular maintenance requirements [1]. Since the first modern magnetic gear solution was presented in 2001 [2], much published work has discussed various topologies, computational models, and design aspects to improve torque densities and gearing ratios for magnetic-torque transmission devices [3–5]. However, studies around the high-performance control of MDTs, with pole-slipping amelioration, remain limited.

MDTs, which present with variable nonlinear stiffness, can be regarded as a variant of two-mass systems. Montague [1] linearized the model of MDTs at the no-load condition to derive a constant torsional stiffness; MDTs are then modelled as classic two-mass systems. Recently reported work [6] shows that linearsing the MDT model at 75% of the maximum coupling torque is more appropriate. Using the linearized model, Refs. [1,6] adopt basic pseudo derivative feedback (PDF) and 2-degree-of-freedom PI as the control strategy, respectively. By contrast, Refs. [7–9] employ observers to achieve a full-state feedback control of MDTs. Among these studies, only [8] introduces a proportional gain to reduce the probability of controller-induced slip; however, as in [1,7,9], specifications such as the maximum starting load-torque or load-torque disturbance that designed controllers can accommodate, have not been specified in [8].

Pole-slipping detection and recovery are crucial for the control design of MDTs. Montague [1,10] observed that pole-slipping imposes a modulation onto motor-side speed and that pole-slipping can then be observed through the speed oscillations of motor-side components. However, [1,10] do not give a detection criterion due to the amplitude of oscillations being affected by inertia ratios and design coupling torque. Ref. [11] calculates the relative mechanical displacement angle to identify pole-slipping, but it requires sensors to be installed on both motor- and load-side shafts. If load-side information is obtained through estimations, high-resolution sensors (to obtain motor-side information) and high-performance computational units are required to be used.

Ref. [1] proposes a slip recovery strategy based on reconfiguring PDF controllers. The pole-slipping recovery approach presented in [11] requires measuring or observing the load-side position of MDTs. The pole-slipping recovery experimental results given by [1,11] show that the speed of the load-side is almost zero during the overload pole-slipping. This means the MDTs often require extra mechanic devices to isolate the shafts during overload conditions and then re-couple them when recovery begins.

If speed oscillations can be suppressed, the possibility for pole-slipping can be reduced. Methodologies [12–14] adopted to inherently reduce speed oscillations of two-inertia systems indicate that effective oscillation suppression can be achieved by adjusting the inertia ratio with additional acceleration or torsional torque feedback loops. In contrast, speed oscillation attenuation with inertia ratio regulation has yet to be studied in the control methods proposed for MDTs.

Moreover, previously reported work [1,6,11,15] has only considered pole-slipping in response to overload or aggressive control effort induced by references or load-torque disturbances. However, load-startup failures caused by the mismatch of load- and motor-side inertias have yet to be discussed. The primary aim of this paper is therefore to provide a mathematical analysis for pole-slipping and then use the developed principles to improve the operational robustness of MDTs. The benefits presented in this paper include.

(1) Through a formal mathematical derivation, the paper firstly highlights that a MDT can be led into a pole-slipping regime with electromagnetic and load-torque excitations that are not individually greater than the maximum coupling torque. Moreover, the analysis also shows that the robustness of MDTs to actuation induced pole-slipping can be improved by decreasing the inertia ratio and reducing the bandwidth of designed controllers.
(2) The kurtosis of electromagnetic torque is introduced to provide a pole-slipping detection criterion that doesn't rely on the information of load-side position/speed. Moreover, a novel pole-slipping recovery is presented by transforming the pole-slipping regime into that of an on-load startup problem. The pole-slipping remedial strategy ensures MDTs recover from failures without the need for additional mechanical intervention/devices, as have been required in previous studies, to facilitate recovery.

2. Dynamics of MDTs

2.1. Experimental Rig

A magnetically-coupled drive-train test facility has been setup as shown in Figure 1 to provide a practical focus to the investigation. The motor-side position sensor is a 12 bit incremental encoder. The magnetic coupling has 5 pole pairs (p), and the maximum (pull-out) coupling torque (T_G) is 1.6 N·m. The test rig has the load- and motor-side friction coefficients (B_L and B_M) of 0.003 N·m/(rad/s) and the load- and motor-side inertias (J_L and J_M) are 0.001 kg·m^2.

For the driving unit, the motor-side speed ω_M and position θ_M can be measured, but the load-side speed ω_L and position θ_L are assumed unavailable for control purposes. This test facility emulates the most common application scenarios for MDTs, where ω_L and θ_L are ideally not measured. Control effort is electromagnetic torque (T_{em}), and hence the prime mover is operating in a torque control mode. The driven unit is used to emulate load (T_L) variations.

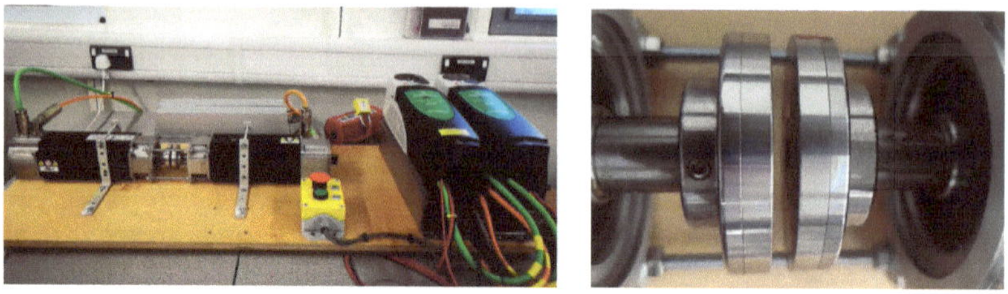

(a) (b)

Figure 1. Experimental test rig (**a**) motor- and load-side actuation (**b**) magnetic coupling.

2.2. Dynamic Model

Defining T_C as the torsional torque developed between the two magnetically-coupled shafts, T_C is then given by

$$T_C = T_G \sin(p\theta_D) \tag{1}$$

where $\theta_D = \theta_M - \theta_L$ is the relative mechanical displacement angle.

If damping torque is not taken into consideration, MDTs are described by the following set of equations [1].

$$\begin{aligned} \dot{\theta}_M &= \omega_M \\ \dot{\omega}_M &= \frac{1}{J_M}(T_{em} - T_C - B_M\omega_M) \\ \dot{\theta}_L &= \omega_L \\ \dot{\omega}_L &= \frac{1}{J_L}(T_C - T_L - B_L\omega_L). \end{aligned} \tag{2}$$

$\theta_M, \omega_M, \theta_L, \omega_L$ are considered as state variables, T_{em} and T_L as inputs, and θ_D as the output. Linearizing (2) at an operating point, and letting K_{lin} represent the resulting torsional stiffness ($K_{\text{lin}} = T_G p \cos(p\theta_D)$), the transfer functions from T_{em} and T_L to θ_D are then, respectively, given by

$$\frac{\theta_D(s)}{T_{em}(s)} = \frac{J_L s + B_L}{D(s)}, \tag{3}$$

$$\frac{\theta_D(s)}{T_L(s)} = \frac{J_M s + B_M}{D(s)} \tag{4}$$

where $D(s) = J_M J_L s^3 + (J_M B_L + B_M J_L)s^2 + (B_L B_M + (J_L + J_M)K_{\text{lin}})s + (B_M + B_L)K_{\text{lin}}$.

The anti-resonance frequency (ω_a) and natural resonance frequency (ω_n) are then given by

$$\omega_a = \sqrt{\frac{K_{\text{lin}}}{J_L}}, \qquad \omega_n = \sqrt{\frac{K_{\text{lin}}}{J_L}(1+R)} \tag{5}$$

where $R = J_L/J_M$. Table 1 enumerates ω_a and ω_n, which ranges from no-load to the 99% of the pull-out torque.

Table 1. ω_a and ω_n of the MDT rig subject to various percentages of pull-out torque conditions.

Percentage (%)	**0**	**50**	**75**	**90**	**99**
ω_n (rad/s)	126.0	119.0	103.0	71.1	48.4
ω_a (rad/s)	89.4	83.2	72.7	59.1	33.6

If $B_L B_M$ and $(B_M + B_L)K_{\text{lin}}$ are negligible, then (3) and (4) simplify to

$$\frac{\theta_D(s)}{T_{em}(s)} = \frac{1/J_M}{s^2 + 2\zeta\omega_n s + \omega_n^2}, \tag{6}$$

$$\frac{\theta_D(s)}{T_L(s)} = \frac{1/J_L}{s^2 + 2\zeta\omega_n s + \omega_n^2} \tag{7}$$

where $\zeta = \frac{J_M B_L + B_M J_L}{2J_M J_L \omega_n}$.

2.3. On-Load Startup And Criteria For MDTs To Maintain Normal Running

On-load startup is the most demanding operating condition that can result in actuation induced pole-slipping, where the relative angle of both shafts is simultaneously developed by the transients of electromagnetic torque demands and load-torque.

If T_{em} and T_L are considered as step inputs ($T_{em}(s) = T_{em}/s$, and $T_L(s) = T_L/s$), then θ_D resulting from T_{em} and T_L can be, respectively, rewritten based on (6) and (7), and are given by

$$\theta_D(s)|_{T_{em}} = \frac{T_{em}/J_M}{s(s^2 + 2\zeta\omega_n s + \omega_n^2)}, \tag{8}$$

$$\theta_D(s)|_{T_L} = \frac{T_L/J_L}{s(s^2 + 2\zeta\omega_n s + \omega_n^2)}. \tag{9}$$

Inverse Laplace transforming of (8) and (9) yields

$$\theta_D(t)|_{T_{em}} = \frac{T_{em}}{\omega_n^2 J_M}\left(1 - (1 + \frac{\zeta\omega_n}{\omega_d})e^{-\zeta\omega_n t}\sin(\omega_d t)\right), \tag{10}$$

$$\theta_D(t)|_{T_L} = \frac{T_L}{\omega_n^2 J_L}\left(1 - (1 + \frac{\zeta\omega_n}{\omega_d})e^{-\zeta\omega_n t}\sin(\omega_d t)\right) \tag{11}$$

where $\omega_d = \sqrt{1-\zeta^2}\omega_n$. Since ζ is small, $\omega_n \approx \omega_d$. Referring to (10) and (11), the maximum increment of θ_D results from when T_{em} and T_L occur simultaneously and approximates to

$$\theta_{Dm} = \frac{T_{em}R + T_L}{\omega_n^2 J_L}\left(1 + e^{-\frac{3\pi}{2}\zeta}\right). \tag{12}$$

It can be seen from (12) that increasing J_M improves the on-load startup capability provided that J_L is fixed ($R = J_L/J_M$). Moreover, if T_{em} output exhibits first-order dynamics compared to T_L which is usually assumed to be a step input, θ_{Dm} would take a much lower value. This also means that low-bandwidth controllers can potentially accommodate a higher on-load startup torque and load-torque disturbances than high-bandwidth designs.

A MDT will enter a pole-slip operating regime if (13) or (14) holds.

$$|\theta_D| > \frac{\pi}{2p}, \tag{13}$$

$$T_{em} > T_G + T_{fm}, \text{ or } T_L > T_G + T_{fl} \tag{14}$$

where T_{fm} and T_{fl} represent motor- and load-side friction torque. Now, assuming θ_{D0} is the relative angle of both shafts while MDTs are running under normal conditions, then a criterion for a MDT to avoid pole-slipping is given by

$$\theta_{D0} + \theta_{Dm} < \frac{\pi}{2p}. \tag{15}$$

The experimental study of [6] shows that selecting 75% of the pull-out torque as the operating point can accommodate a wider range of load-torque disturbance. Here, we will show how to use Equations (12), (13), and (15) to estimate the maximum on-load startup torque and the permitted load-torque disturbance that a MDT can accommodate through two example scenarios.

Example 1: supposing the MDT rig is operating at 800 r/min under no-load conditions ($\theta_{D0} = \arcsin(0.18)/p$), 75% of the pull-out torque is selected as the operating point ($\omega_n = 103.0\,\text{rad/s}$), and the increment of T_{em} can immediately follows T_L to find the maximum load-torque disturbance that the MTD rig accommodate without pole-slipping.

According to (12) and (15), we have

$$T_L < \frac{\frac{\pi}{2p} - \theta_{D0}}{4}\omega_n^2 J_L. \tag{16}$$

Substituting the values of θ_{D0}, ω_n, and J_L into (16) yields $T_L = 0.43T_G$. However, due to the delay induced by current/speed control loops, the permitted load-torque disturbance can conceivably be greater than $0.43T_G$.

Example 2: Assuming 75% of the pull-out torque is selected as the operating point, and T_{em} is equal to T_G, find the maximum on-load starting torque (T_{Lm}) that the facility can accommodate without pole-slipping.

According to (12) and (13), T_{Lm} for the rig discussed in the paper is 0.66 T_G. If friction torque is taking into consideration, then, $T_{Lm} = 0.48\,T_G$.

3. Controller Design And Pole-Slipping Amelioration

3.1. Virtually Increasing J_M Using Acceleration Feedback

From the analysis above, we know that increasing J_M (effectively decreasing R, see (12)) decreases the possibility of pole-slipping related to on-load startup. Figure 2 shows a model of the MDT with motor-side acceleration feedback (K_a is the feedback coefficient). Clearly, with acceleration feedback, the equivalent motor-side inertia $\hat{J}_M$ is given by

$$\hat{J}_M = J_M + K_a. \tag{17}$$

Accordingly, the inertia ratio is adjusted to

$$R = \frac{J_L}{\hat{J}_M}. \tag{18}$$

Letting T'_{em} represent the actual torque developed by the drive, then

$$T'_{em} = T_{em} - K_a\dot{\omega}_M. \tag{19}$$

Equation (19) indicates the aggressive control effort T_{em} can be modified by K_a, and this mechanism can be used to prevent θ_D exceeding its constraint under on-load startup and inertia mismatch.

The drive for a practical facility usually integrates period or frequency measurement methods to calculate speed signals with feedback from encoders. However, most drives do not provide acceleration information. Calculating acceleration from speed measurements at a particular sampling frequency will inevitably amplify noise components.

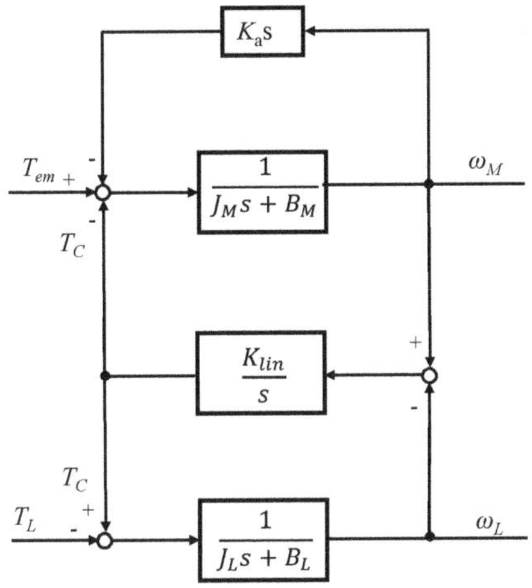

Figure 2. Block diagram of MDTs with acceleration feedback.

A tracking differentiator (TD) [16] can effectively suppress the amplification of derivative noise while quickly tracking the derivative of input signals without relying on model information. Here, the TD is adopted to obtain an estimation of acceleration.

Supposing v_1 tracks the motor-side speed ω_M, v_2 is the derivative of v_1, then the TD is described by

$$\begin{aligned} \dot{v}_1 &= v_2 \\ \dot{v}_2 &= \text{fst}(v_1, v_2, \omega_M, r, h) \end{aligned} \tag{20}$$

where r dictates tracking speed, and h is a parameter that determines the filtering effect ($h \geq T_s$, T_s is the sampling period), and $\text{fst}(v_1, v_2, \omega_M, r, h)$ is given by

$$\text{fst} = \begin{cases} -ra, & |a| \leq d \\ -r\text{sign}(a), & |a| > d \end{cases} \tag{21}$$

where $d = rh$, and a is given by

$$a = \begin{cases} v_2 + y/h, & |y| \leq d_0 \\ v_2 + \frac{1}{2}(a_0 - d)\text{sign}(y), & |y| > d_0 \end{cases} \tag{22}$$

with $d_0 = dh$, $y = v_1 - \omega_M + hv_2$, $a_0 = \sqrt{d^2 + 8r|y|}$.

3.2. Controller Design and Parameter Optimization

The paper now describes the design of a generalized controller (fractional-order controller) represented by

$$F_C(s) = K_p + \frac{K_i}{s^\lambda} + K_d s^\mu \tag{23}$$

where $\{K_p, K_i, K_d\}$ represent the proportional, integral, and derivative gains and $\{\lambda, \mu\}$ are the order of integrals and derivative items. Of note is that traditional integer PID controllers can be regarded as fractional-order PID controllers (FOPID) with $\lambda = \mu = 1$. Hence, the following design process is also applicable for the design of integer-order controllers. Figure 3 shows the principle of optimizing parameters for speed controller using integral of time and absolute error as a performance metric. Specifically, the objective function $f(x,t)$ is considered:

$$f(x,t) = \int_0^\infty t|e(t)|dt \tag{24}$$

where $x = (K_p, K_i, K_d, \lambda, \mu)$ and $e(t)$ is the error of the control loop at time t.

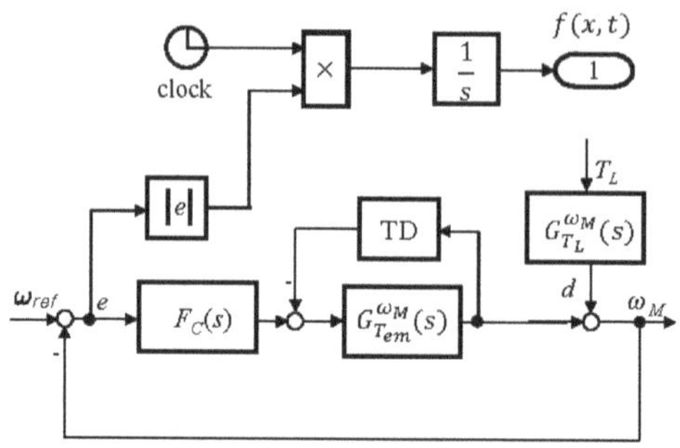

Figure 3. Block diagram to optimize controller parameters.

The purpose of parameter optimization is to find an x subject to constraints to ensure $f(x,t)$ has a minimum value, described by

$$\min_{x} f(x,t) \quad \text{such that} \begin{cases} c(x) < 0 \\ l_b \le x \le u_b \end{cases} \tag{25}$$

where $c(x)$ represents general nonlinear inequality constraints; and l_b and u_b are the lower- and upper-boundaries for x, respectively. Normally, $c(x)$ can be constructed with a combined specifications of gain crossover frequency, phase margin, gain margin, sensitivity function, and complementary sensitivity function.

From the design perspective of two-mass systems, the gain cross-over frequency (ω_{gc}) should be set much smaller than ω_a to obtain responses similar to that of a low-pass filter, i.e. ideally with no overshoot. It can be seen from Table 1 that $\omega_a = 59.1\,\text{rad/s}$ while the MDT rig is running at 90% of the pull-out load-torque. Moreover, the analysis of Section 2.3 shows that a low-bandwidth design can accommodate a wider range of load-torque disturbances. Here, ω_{gc} is set to 5 rad/s.

The sensitivity function $S(j\omega)$ represents the transfer function from the disturbance (d) to ω_M and the speed reference (r) to e $\left(S(j\omega) = 1/\left(1 + F_C(s)G^{\omega_M}_{Tem}(s)\right), \text{see Figure 3}\right)$. Equations (10) and (11) show that θ_D under excitations takes the form of a step plus a damped oscillation which is regarded as disturbances in this design. To obtain a small control error, $|S(j\omega)|$ is also required to be set small at low frequencies since references usually take the form of relatively low-frequency inputs. Hence, the sensitivity specification for controller design can therefore be given by

$$\omega_s = 5\,\text{rad/s}, \; |S(j\omega)| \le -20\text{dB}, \; \forall\omega \le \omega_s. \tag{26}$$

Note: here, ω_s is assumed to be a value approximately equal to ω_{gc}.

The complementary sensitivity function $T(j\omega)$ relates speed measurement noise to motor-side speed ω_M. Typically, the cut-off frequency of shaft speed filters is usually designed to be 3~5ω_n ($\omega_n = 126\,\text{rad/s}$ while the MDT rig is running at no-load conditions, see Table 1) to meet dynamic processing requirements.The specifications for the complementary sensitivity function are then given by

$$\omega_t = 5\omega_n, \quad |T(j\omega)| \le -20\,\text{dB}, \; \forall\omega \ge \omega_t. \tag{27}$$

Previously reported work [17] shows that satisfactory performance can be obtained by setting the gain margin $GM > 6\,\text{dB}$ and the phase margin $30^0 \le PM \le 60^0$. Here, PM and GM are chosen to be

$$\text{PM} = 45^0, \quad \text{GM} = 10\,\text{dB}. \tag{28}$$

The parameters for the TD are $r = 500$ and $h = 0.004$ (the sampling frequency for running speed signals is set to 250 Hz). In this case, K_a is set to be 0.1 (see Figure 2). The optimization is carried out using the open-source software FOMCON [18], and the optimized parameters for the design controller are $K_p = 3.97, K_i = 3.17, K_d = 3.8 \times 10^{-5}, \lambda = 1$, and $\mu = 0.12$. The optimized controller can be further approximated to a classical PI structure as $(3.97s + 3.17)/s$.

3.3. A Criterion For Pole-Slipping Detection

Here, an overload pole-slipping detection methodology is considered. MDTs will enter a pole-slipping regime when the relative mechanical displacement angle between the magnetically-coupled shafts exceeds the designed range. During pole-slipping, ω_M will rise for a short period and then return to the reference speed since the speed controller remains in operation. The control effort T_{em} will therefore converge to be roughly equal to the friction torque in steady state ($T_{em} \approx T_{fm}$).

The abrupt change of T_{em} can be measured by kurtosis, which is a dimensionless indicator widely used for evaluating the characteristics of vibrations. The kurtosis of T_{em} is derived from [19]

$$K_{ur} = \frac{\frac{1}{n}\sum_{i=1}^{n}(x_i - \bar{x})^4}{\left(\frac{1}{n}\sum_{i=1}^{n}(x_i - \bar{x})^2\right)^2} \tag{29}$$

where x_i represents the control effort T_{em} at time i, and $\bar{x}$ is the mean value of x_i.

A pole-slipping detection criteria can then be described as when

$$T_{em} \approx T_{fm}, K_{ur} > 5. \tag{30}$$

3.4. A Strategy for Pole-Slipping Recovery

It can be seen from the paragraph below Equation (2) that the equivalent torsional stiffness (K_{lin}) varies with the displacement angle (θ_D) between the magnetically-coupled shafts. Once the load-side shaft synchronizes with the motor-side, the displacement angle is constant, resulting in a stable coupling torque ($T_c = K_{lin}\theta_D$). Hence, the variance of ω_M during synchronization is much smaller than the commensurate value when the motor- and load-side speeds are different. The proposed pole-slipping recovery strategy considered here is based on the assumption that T_{em} can be greater than T_G, and T_L can restore from the state $T_L > T_G$ to $T_L < T_G$. The proposed recovery strategy is summarised as follows:

- At the onset of pole-slipping, reverse the running direction of the motor-side shaft and control the speed of it up to the maximum specification speed, then;
- Control the electromagnetic torque to be zero ($T_{em} = 0$); $|\omega_M|$ will consequently then decrease;
- As $|\omega_M|$ decreases, the motor-side shaft will re-synchronize with the load-side shaft (the variance of ω_M is now used as a synchronization detection indicator). The PI plus TD configuration can then be reconfigured to control the MDT to the reference speed.

4. Simulation And Experimental Results

To verify the theoretical analysis in Section 2.3, this Section will initially consider three cases of on-load startup (first case: $T_L = 0.4T_G$, second case: $T_L = 0.5T_G$, third case: $T_L = 0.6T_G$) through simulation. For the discussed cases, the electromagnetic torque is assumed to be saturated immediately ($T_{em} = T_G$). Note that in this paper, T_{em} and T_L are normalized at T_G for both simulation and experimental trials.

Figure 4a shows the simulation results for $R = 1$. It can be seen that slip between two shafts of the MDT occurs while T_L is $0.5T_G$ or above. Although increasing J_M results in a slower transient response, Figure 4b shows that the on-load starting torque can be up to $0.5T_G$ by modifying J_M to $2J_L$. However, as shown in Figure 4c, by setting J_M to $0.5J_L$, the two sides of the MDT are out of synchronization in all studied cases. It is clear that decreasing the inertia ratio through increasing J_M improves on-load startup capability. In practice, low-cost drives often cannot provide sufficient bandwidth for their current control loops. Hence, the output electromagnetic torque and load-toque are assumed to exhibit first-order dynamics. Under these circumstances, from Equation (12) and subsequent discussion, the permitted on-load startup torque can be a higher value compared with the simulation studies.

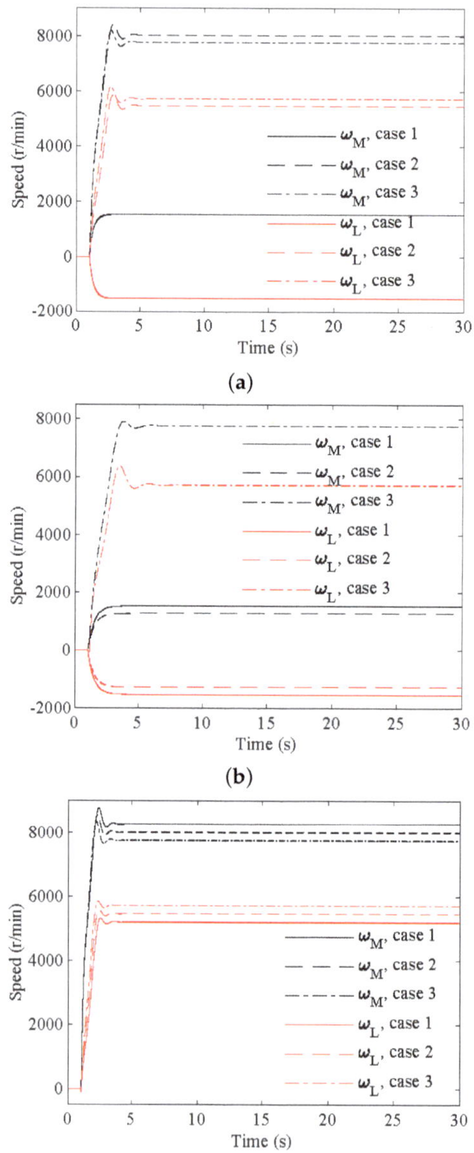

Figure 4. Simulation results of on-load startup: (**a**) R = 1 (J_M = 0.001 kg·m^2); (**b**) R = 0.5 (J_M = 0.002 kg·m^2); (**c**) R = 2 (J_M = 0.0005 kg·m^2).

Figure 5 shows tshows the measured load rejection dynamics for the optimized controller. As shown in Figure 5a, the speed of load-side coupling follows the motor-side without speed oscillations even when the MDT rig is subject to 2 significant load disturbances. Figure 5b shows that by adjusting the inertia ratio with $K_a = 0.1$ offers first-order filter dynamics for the reactive control effort T_{em} compared to the step load-torque disturbance occurring at 13 s. This characteristic greatly improves the range of load-torque disturbance and on-load startup torque that the MDT rig can accommodate while referring to (10), (11), and (12). It can also be seen from Figure 5b that the peak control effort resulting from the speed reference is about $0.7T_G$. Assuming $K_a = 0$ (remove the acceleration feedback) and T_{em} can immediately follow T_L without overshoot, the maximum on-load startup torque would be around $0.48T_G$ if friction torque is included (as shown in Figure 5b, $T_{fm} = T_{fl} \approx 0.18T_G$ at around 800 r/min).

The on-load startup torque is now set to $0.63T_G$. Figure 6 shows that virtually increasing J_M through acceleration feedback overcomes the pole-slipping problem associated with on-load startup and that a starting load-torque that is greater than $0.63T_G$ can lead the MDT to pole-slipping (T_{em} slightly exceeds T_G at around 1.8 s; however, the effective coupling torque is still a slightly lower than T_G due to the existence of motor-side friction). Of note is that if load-side friction torque (T_{fl}) is considered, the maximum on-load starting torque is $0.81T_G$ in this case.

It can also be seen from the experimental trials that the inertia ratio is set to 0.09 ($K_a = 0.1$) to prevent pole-slipping resulting from on-load startup and inertia mismatch. By contrast, previously reported work [12,13,20] shows that effective oscillation suppression is achieved by setting the resonance ratio (ω_n/ω_a) of two-mass systems to be $\sqrt{2} \sim \sqrt{5}$; thus, R ranges between 1 and 4 (see (5)). Although MDTsare traditionally regarded as two-mass systems, it can be seen from the simulations and experimental trials that the recommended inertia ratio for MDTs should generally be set much lower than that of classical two-inertia systems to prevent MDTs from entering pole-slipping initiated by the combined effects of T_{em} and T_L.

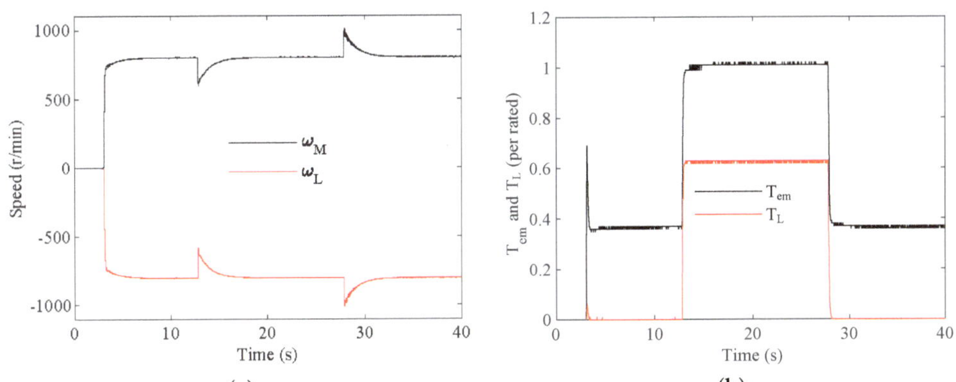

Figure 5. Experimental responses using the optimized controller with $K_a = 0.1$ (speed reference ω_{ref} is imposed at 1 s; T_L is set to 0%, 63%, and 0% of the pull-out torque at 1–13 s, 13–28 s, and 28–40 s, respectively): (**a**) shaft speed of on both sides of the magnetic coupling; (**b**) T_{em} and T_L.

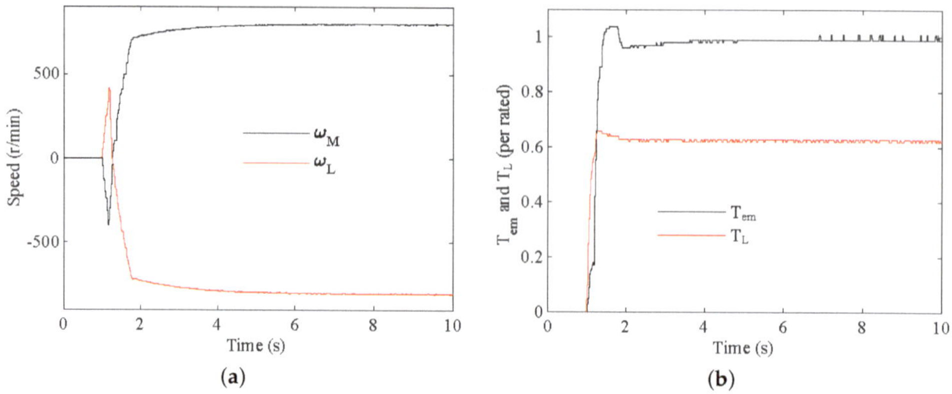

Figure 6. Experimental measurement while starting with 63% of the pull-out torque (ω_{ref} and T_L are imposed at 1 s, simultaneously): (**a**) speed of both side shafts; (**b**) T_{em} and T_L.

The efficacy of pole-slipping detection and recovery strategies are now investigated. Figure 7b shows that the kurtosis of T_{em} is lower than three, when the MDT is operating in normal conditions. However, the kurtosis rises dramatically to 8.71 during the period of 10.42 s to 12.64 s (pole-slipping occurs at 12.1 s). It can also be seen from Figure 7a that $T_{em} \approx T_{fm}$ at around 12.5 s. Hence, the proposed method can effectively detect the transition to pole-slipping.

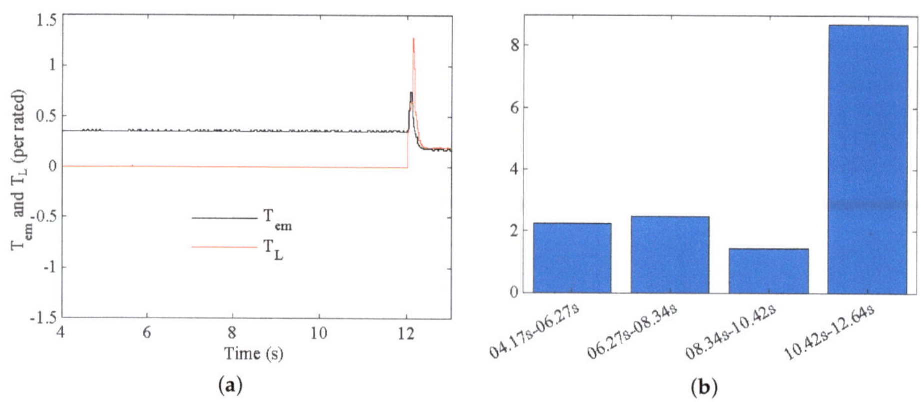

Figure 7. Pole-slipping detection using kurtosis: (**a**) the output of T_{em} with respect to T_L; (**b**) kurtosis of T_{em} (pole-slipping occurs at 12.1 s, and kurtosis is calculated individually with 200 samples of T_{em} output).

The experimental results shown in Figure 8 provide an example of the recovery strategy. As shown in the figure, pole-slipping is induced at 12 s. At the beginning of pole-slipping, the load-side shaft reverses its direction due to T_L being temporarily higher than the maximum coupling torque and T_{em}. Following that, the load-side shaft loses synchronization with the primary-side, running at around 1600 r/min during pole-slipping. After detecting pole-slipping with the kurtosis methodology discussed above, the motor-side drive reverses its running direction from 14 s and controls ω_M to 2000 r/min, a setting higher than ω_L. It can be seen from Figure 8 that the motor-side speed varies around 2000 r/min from at 16 s and that the variance of ω_M increases considerably when both side speeds are different. From 25 s to 30 s, T_{em} is set to 0, $|\omega_M|$ then decreases. Following that, the motor-side shaft synchronizes with the load-side shaft from roughly 27 s. Now, the coupling torque is relative constant; therefore, speed oscillations are reduced. This is a situation that can be regarded as on-load startup. The reconfigured scheme can then be used to recover the MDT rig from pole-slipping. It can be seen from the figure that the MDT recovers after 30 s. According to Table 2, the variance of ω_M during the period of 19–21 s ($\omega_L = 1600\,\text{r/min}$, $\omega_M = 2000\,\text{r/min}$) is far higher than the variance calculated from 9 s–11 s or 27 s–29 s ($\omega_L = \omega_M$). Hence, the variance of ω_M is a suitable indicator for synchronisation.

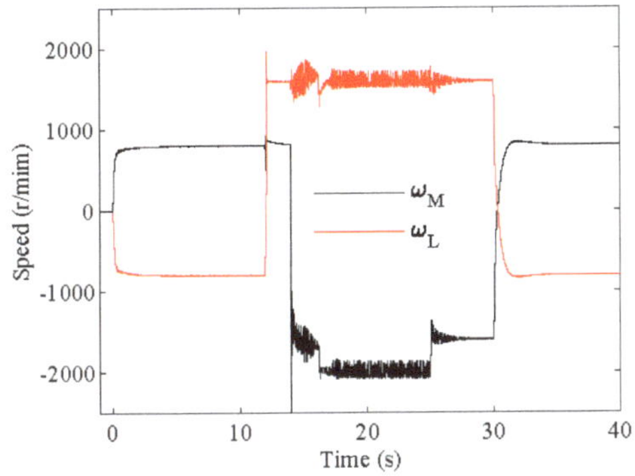

Figure 8. Experimentally measured speed of both coupled shafts components before, during, and after the overload condition.

Table 2. Variance of ω_M.

Time (s)	9–11	11–13	19–21	25–27	27–29
σ	2.01×10^{-6}	0.88	0.02	0.71	4.70×10^{-5}

5. Conclusions

In this paper, pole-slipping anomalies induced by electromagnetic inputs and load-torque disturbances are investigated. Experimental trials have been used to verify that a low-bandwidth controller design incorporating a tracking differentiator to virtually increase motor-side inertia significantly improves the on-load startup performance and load-torque disturbance handling range for the MDT where acceleration feedback is unavailable. MDTs are traditionally classified as two-inertia systems; however, the experimental trials show that an appropriate inertia ratio for MDTs to maintain normal operation is much lower than that of traditional dual-inertia control systems. Experimental measurements have confirmed that the proposed amelioration methods can effectively detect pole-slipping without using load-side position/speed measurements and can overcome the requirement for including additional mechanic devices to facilitate recovery.

Author Contributions: Conceptualization, X.L., C.B., and T.S.; formal analysis, X.L.; project administration, C.B.; software, X.L.; supervision, C.B.; experiment investigation, X.L., C.B., and T.S.; writing—original draft, X.L.; writing—review and editing, C.B.. All authors have read and agreed to the published version of the manuscript.

Funding: This research received no external funding.

Institutional Review Board Statement: Not applicable.

Data Availability Statement: Not applicable.

Acknowledgments: The authors would like to thank Argyrios Zolotas for inspiring to employ fractional-order control as the control scheme.

Conflicts of Interest: The authors declare no conflict of interest.

Abbreviations

ω_a	Anti-resonance frequency (rad/s)
ω_d	The frequency of damped sine signal (rad/s)
ω_L	Load-side angular velocity (rad/s)
ω_M	Motor-side angular velocity (rad/s)
ω_n	Natural resonance frequency (rad/s)
θ_L	Load-side angle (rad)
θ_M	Motor-side angle (rad)
θ_D	$\theta_M - \theta_L$ (rad)
ζ	Damping coefficient (kg/s), and $\zeta = \frac{J_M B_L + B_M J_L}{2 J_M J_L \omega_n}$
λ	Integer order
μ	Derivative order
B_L	Load-side friction coefficient N·m/(rad/s)
B_M	Motor-side friction coefficient N·m/(rad/s)
J_L	Load-side inertia (kg·m^2)
J_M	Motor-side inertia (kg·m^2)
$\hat{J}_M$	The equivalent value of motor-side inertia (kg·m^2)
K_a	Feedback coefficient
K_d	Derivative gain
K_i	Integral gain
K_p	Proportional gain
K_{lin}	Linearized torsional stiffness (N·m/rad)
p	Pole-pairs
R	Inertia ratio (J_L/J_M)
T_C	Coupling torque (N·m)
T_{em}	Electromagnetic torque (N·m)
T_{fl}	Load-side friction torque (N·m)
T_{fm}	Motor-side friction torque (N·m)
T_L	Load torque (N·m)
T_G	The maximum coupling torque (N·m)

References

1. Montague, R.; Bingham, C.; Atallah, K. Servo control of magnetic gears. *IEEE/Asme Trans. Mechatron.* **2011**, *17*, 269–278. [CrossRef]
2. Atallah, K.; Howe, D. A novel high-performance magnetic gear. *IEEE Trans. Magn.* **2001**, *37*, 2844–2846. [CrossRef]
3. Wang, Y.; Filippini, M.; Bianchi, N.; Alotto, P. A Review on Magnetic Gears: Topologies, Computational Models and Design Aspects. *IEEE Trans. Ind. Appl.* **2019**, *55*, 4557–4566. [CrossRef]
4. Rasmussen, P.O.; Andersen, T.O.; Jorgensen, F.T.; Nielsen, O. Development of a high-performance magnetic gear. *IEEE Trans. Ind. Appl.* **2005**, *41*, 764–770. [CrossRef]
5. Jorgensen, F.T.; Andersen, T.O.; Rasmussen, P.O. The Cycloid Permanent Magnetic Gear. *IEEE Trans. Ind. Appl.* **2008**, *44*, 1659–1665. [CrossRef]
6. Liao, X.; Bingham, C.; Zolotas, A.; Zhang, Q.; Smith, T. Servo Control of Drive-Trains Incorporating Magnetic Couplings. *IEEE Trans. Ind. Appl.* **2022**, *58*, 3674–3684. [CrossRef]
7. Montague, R.; Bingham, C.; Atallah, K. Dual-observer-based position-servo control of a magnetic gear. *IET Electr. Power Appl.* **2011**, *5*, 708–714. [CrossRef]
8. Bouheraoua, M.; Wang, J.; Atallah, K. Rotor position estimation of a pseudo direct-drive pm machine using extended kalman filter. *IEEE Trans. Ind. Appl.* **2016**, *53*, 1088–1095. [CrossRef]
9. Bouheraoua, M.; Wang, J.; Atallah, K. Design and implementation of an observer-based state feedback controller for a pseudo direct drive. *IET Electr. Power Appl.* **2013**, *7*, 643–653. [CrossRef]
10. Montague, R.; Bingham, C.; Atallah, K. Magnetic gear dynamics for servo control. In Proceedings of the MELECON 2010–2010 15th IEEE Mediterranean Electrotechnical Conference, Valletta, Malta, 26–28 April 2010; pp. 1192–1197.
11. Bouheraoua, M.; Wang, J.; Atallah, K. Slip recovery and prevention in pseudo direct drive permanent-magnet machines. *IEEE Trans. Ind. Appl.* **2014**, *51*, 2291–2299. [CrossRef]
12. O'Sullivan, T.M.; Bingham, C.M.; Schofield, N. High-performance control of dual-inertia servo-drive systems using low-cost integrated SAW torque transducers. *IEEE Trans. Ind. Electron.* **2006**, *53*, 1226–1237. [CrossRef]
13. Saarakkala, S.E.; Hinkkanen, M. State-space speed control of two-mass mechanical systems: Analytical tuning and experimental evaluation. *IEEE Trans. Ind. Appl.* **2014**, *50*, 3428–3437. [CrossRef]
14. Szabat, K.; Orlowska-Kowalska, T. Vibration Suppression in a Two-Mass Drive System Using PI Speed Controller and Additional Feedbacks—Comparative Study. *IEEE Trans. Ind. Electron.* **2007**, *54*, 1193–1206. [CrossRef]

15. Montague, R.G.; Bingham, C.; Atallah, K. Magnetic gear pole-slip prevention using explicit model predictive control. *IEEE/ASME Trans. Mechatronics* **2012**, *18*, 1535–1543. [CrossRef]
16. Han, J. From PID to active disturbance rejection control. *IEEE Trans. Ind. Electron.* **2009**, *56*, 900–906. [CrossRef]
17. Ogata, K. *Modern Control Engineering*; Prentice Hall: Upper Saddle River, NJ, USA, 2010; Volume 5.
18. Tepljakov, A. FOMCON: Fractional-order modeling and control toolbox. In *Fractional-Order Modeling and Control of Dynamic Systems*; Springer: Berlin/Heidelberg, Germany, 2017; pp. 107–129.
19. Immovilli, F.; Cocconcelli, M.; Bellini, A.; Rubini, R. Detection of generalized-roughness bearing fault by spectral-kurtosis energy of vibration or current signals. *IEEE Trans. Ind. Electron.* **2009**, *56*, 4710–4717. [CrossRef]
20. Hori, Y.; Sawada, H.; Chun, Y. Slow resonance ratio control for vibration suppression and disturbance rejection in torsional system. *IEEE Trans. Ind. Electron.* **1999**, *46*, 162–168. [CrossRef]

energies

Technical Note

Toward the Application of Pulse Width Modulated (PWM) Inverter Drive-Based Electric Propulsion to Ice Capable Ships

Eric Christopher Renz [1] and James Turso [2,3,*]

1 Ford Motor Company, Dearborn, MI 48126, USA
2 Newport News Shipbuilding, Newport News, VA 23607, USA
3 Department of Marine Engineering, United States Merchant Marine Academy, Kings Point, NY 11024, USA
* Correspondence: tursoja@email.gwu.edu

Abstract: The current world geopolitical state has driven significant interest in expanding influence in the polar regions. In order to develop new trade routes and defense of these areas, specialized ice-capable vessels are needed. Traditional propulsion technology utilized for these vessels has featured diesel and turbo electric drives with cycloconverter based technology used for speed regulation. Cycloconverter technology has the major disadvantage of characteristic output voltage waveforms that are very non-sinusoidal which induce harmonic rich currents with their resulting negative effects on total power factor and efficiency. There is also the potential to induce harmonic currents that may increase heat production in power cables, stressing insulation and reducing overall component life. Furthermore, sensitive loads such as control system power supplies may also be affected. The output waveform of a PWM drive has harmonic content that is both less severe, and more easily analyzed with current modeling tools. There are significant technical challenges that need to be assessed when considering PWM drive deployment in ice-capable vessels. This is due to the severe load torques, and torque variation imposed on propeller blades during ice-milling operations. To the authors' knowledge, an analysis of the behavior of vector-controlled PWM drives to this type of service has not been published in the open literature. This paper presents an initial modeling and simulation study that investigates whether such a PWM drive and associated controller may be effective for ice milling and other operational evolutions for these types of ships.

Keywords: ice breaker; PWM inverter electric drive; DQ control of propulsion motors

Citation: Renz, E.C.; Turso, J. Toward the Application of Pulse Width Modulated (PWM) Inverter Drive-Based Electric Propulsion to Ice Capable Ships. *Energies* **2022**, *15*, 8217. https://doi.org/10.3390/en15218217

Academic Editors: Lorand Szabo and Feng Chai

Received: 30 August 2022
Accepted: 31 October 2022
Published: 3 November 2022

1. Introduction

There are two general types of electric propulsion drive arrangements: dedicated and integrated. Dedicated systems are as the name implies, dedicated to propulsion needs only. Integrated systems serve all of the vessel loads, both propulsion and non-propulsion related. The key technical challenge for integrated systems comes from the need to control the speed of such a large electric load as a propulsion motor, while simultaneously providing constant and stable AC bus voltage, frequency, and low harmonic content. A typical AC/AC cycloconverter (CCV) system (Figure 1) includes multiple AC generators feeding a medium voltage bus. In the cycloconverter, the incoming voltage from each phase feeds a switched silicon-controlled rectifier (SCR) circuit. In so doing the frequency of the output voltage is varied. The output voltage wave form produced by cycloconverters is useful for propulsion, but is no longer a pure sinusoidal waveform. It becomes rich in harmonic content throughout the frequency spectrum such that tuned filters (harmonic trap filters) are ineffective in restoring power quality. The main method used to mitigate this phenomenon is the use of a 12 pulse cycloconverter circuit. This approach is accepted and has been adopted for ice-capable vessels.

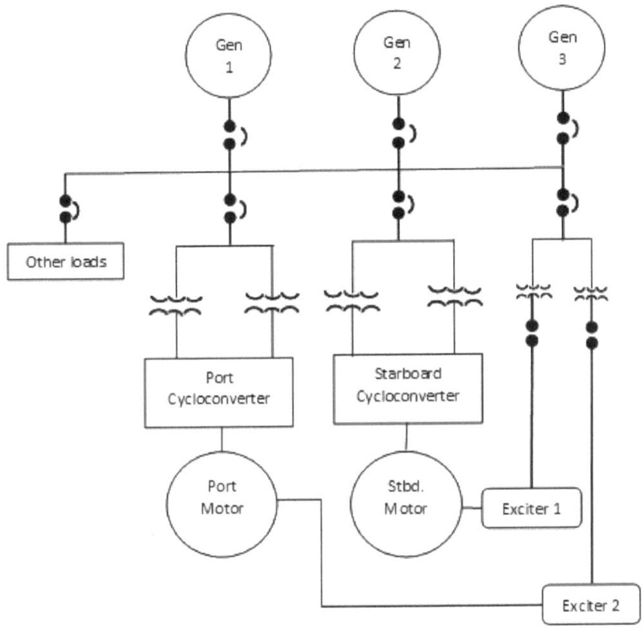

Figure 1. Typical Cycloconverter (CCV) System, [1] (p. 317).

The commercial availability of medium voltage Insulated Gate Bipolar Transistors (IGBT) and integrated gate-commutated thyristors (IGCT) has enabled the use of larger PWM motor drives needed for propulsion applications using induction motors and synchronous motors, including permanent magnet synchronous motors. The PWM drive is an AC/DC/AC technology. A typical 2-level drive topology is presented in Figure 2.

This type system needs more components than the AC/AC cycloconverter system described above. There are generators, a rectifier, a DC bridge that includes a filter/capacitor, and an inverter. The inverter is actively controlled to commutate the DC bridge voltage to a controlled waveform that consists of a series of rectangular pulses.

When applied to a sufficiently inducive load these pulses produce a current that is nearly sinusoidal, and consequently has a low harmonic content. The most effective way to control the pulsed waveform for a synchronous motor application is to commutate them to maximize torque for a given current. This results in greater overall efficiency of the PWM drive, and produces fuel savings which may justify the additional components and associated costs needed to make it work. While economic viability is of great importance for any new application of technology, the focus of this study considers only technical viability, with the hope that proven technical viability will be a natural precursor to consideration by industry including a conceptual design and economic viability study.

Large mechanical and electrical systems are expensive and complex. Performance testing can generate massive forces, with the potential to destroy the system itself if poorly designed. The objective of the system model is to emulate the performance of a real system, without having to build a prototype system. MATLAB/Simulink™ is a tool that has been developed to model such complex systems.

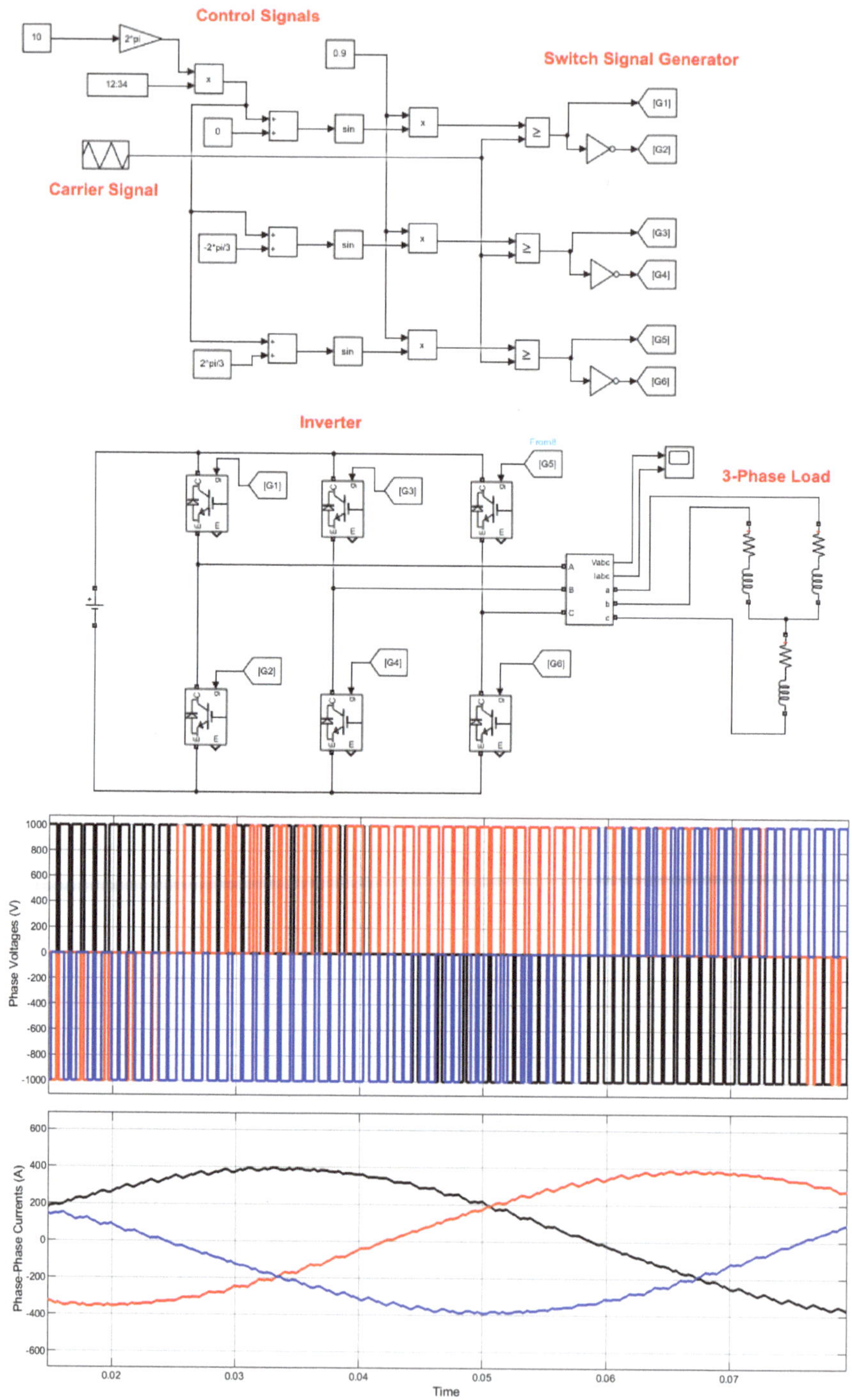

Figure 2. Typical PWM Inverter System and Output Waveforms.

2. Problem Description

The complex system chosen for this study is a functionally complete diesel electric marine power plant including generator speed control, voltage regulation, rectifier, DC link/filter, inverter, PWM drive controller, permanent magnet synchronous machine type propulsion motor, shaft, and propeller as shown in the following functional diagram (Figure 3). Dynamic modeling of the shaft dynamics, hull dynamics and propeller dynamics are based on the ice-capable vessel USCGC Healy.

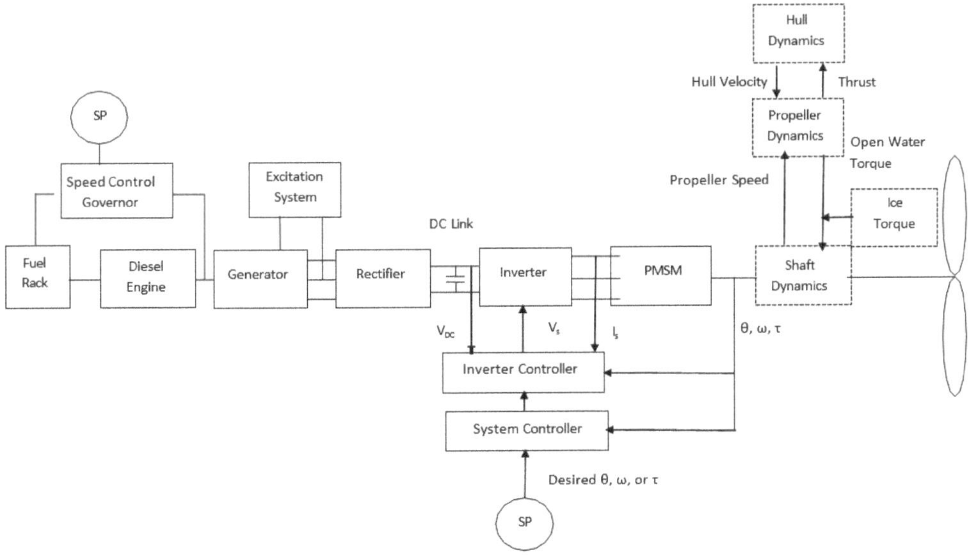

Figure 3. Functional Diagram of the Full System Model.

The USCGC Healy was chosen as the subject of study because it has an AC/AC cycloconverter drive which has been the subject of significant study in Reference [2] and has significant reference data available for comparison with a PMSM powered by a PWM drive system. Although the primary objective is to assess the ability of this type of system to sustain ice-milling torque (short term disruption), the author realizes that the model may be used to assess other, long-term events of potential significant interest. These include full system performance testing, such as an emergency astern test (e.g., "crash-back"). A permanent magnet synchronous motor model was chosen for the PWM model due to its being the technology of choice for modern marine propulsion systems and its simplicity and reliability due to it not requiring a separate excitation system. This has the added benefit of making the associated closed-loop space vector control system relatively straight-forward to implement, compared to comparably sized induction motor drive controllers. The entire power generation and propulsion drive system is modeled in Simulink^{TM}. The propulsion motor is modeled as a salient pole Permanent Magnet Synchronous Machine with five (5) pole pairs. The motor currently installed on USCGC Healy is a dual winding motor. The single winding motor model is effectively a proxy for one motor winding which serves one half of the external motor torque.

2.1. The Magnetic Flux Space Vector [2]

Consider a fixed stator with three coils installed 120 degrees apart. The three coils draw current which produce a magnetic field which depends on the magnitude and direction of the current. Given the three coils are fed from a balanced three-phase sinusoidal voltage source, a current space vector, i_s, for a balanced 3-phase load with current amplitude I is defined as:

$$i_s = I\left[\cos(wt)e^0 + \cos\left(wt - \frac{2\pi}{3}\right)e^{-\frac{2\pi}{3}} + \cos\left(wt + \frac{2\pi}{3}\right)e^{+\frac{2\pi}{3}}\right]$$

The stator current space vector describes the equivalent current that would produce the same total magnetic field magnitude and direction as the three phase currents produce at a specified point in time [3]. By controlling the frequency and magnitude of individual currents to each phase, implicitly, the magnitude and direction of the resulting current space vector and internal magnetic field space vector, are also controlled. Spec vector control uses PWM motor control technology to dynamically position the flux vector within the PMSM, and can be used for speed, position, or direct torque control of the machine.

2.2. *The D-Q Reference Frame [2]*

The origin of the rotating current space vector is the motor rotational centerline, and is coincident with the origin of the direct-quadrature (D-Q) reference frame. The D-Q reference frame is fixed to the rotor, and rotates along with it. By adopting this reference frame, 3 phase quantities appear to be DC in steady-state, and are easier to control compared to their 3-phase counterparts. Indeed, use of a control system that manipulates motor winding voltage demand in the DQ reference frame enables a PMSM to be controlled as if it were a separately-excited DC motor. Figure 4 shows the DQ axes superimposed upon a (simplified) rotor permanent magnet. The permanent magnet flux vector is colinear with the D-axis.

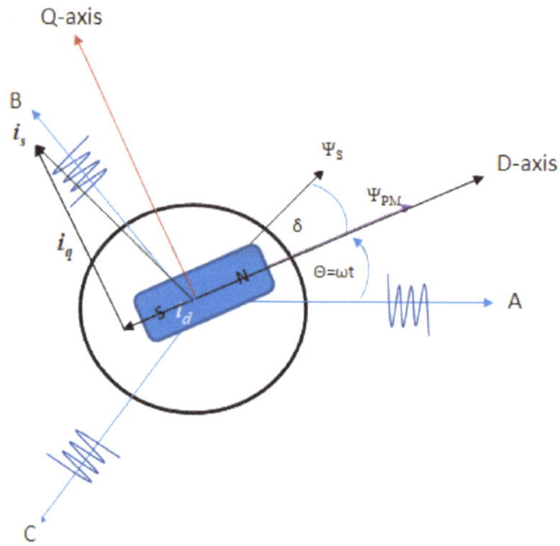

Figure 4. Rotor (i.e., DQ) Reference frame with current and flux vectors shown.

The transformation from the stationary ABC reference frame to the rotating (synchronous) DQ reference frame is provided in Equation (1).

$$x_{dq0} = \sqrt{\frac{2}{3}}\begin{bmatrix} cos(\theta) & cos(\theta - \frac{2\pi}{3}) & cos(\theta + \frac{2\pi}{3}) \\ -sin(\theta) & -sin(\theta - \frac{2\pi}{3}) & -sin(\theta + \frac{2\pi}{3}) \\ \sqrt{\frac{1}{2}} & \sqrt{\frac{1}{2}} & \sqrt{\frac{1}{2}} \end{bmatrix}\begin{bmatrix} a \\ b \\ c \end{bmatrix} \quad (1)$$

The transformation from the rotating (synchronous) DQ reference frame to the stationary ABC reference frame is provided in Equation (2).

$$x_{abc} = \sqrt{\frac{2}{3}} \begin{bmatrix} cos(\theta) & -sin(\theta) & \sqrt{\frac{1}{2}} \\ cos(\theta - \frac{2\pi}{3}) & -sin(\theta - \frac{2\pi}{3}) & \sqrt{\frac{1}{2}} \\ cos(\theta + \frac{2\pi}{3}) & -sin(\theta + \frac{2\pi}{3}) & \sqrt{\frac{1}{2}} \end{bmatrix} \begin{bmatrix} d \\ q \\ 0 \end{bmatrix} \quad (2)$$

The stator current space vector $i_{s,}$ can be thought of as the vector sum of the individual phase currents. The resulting magnetic field is shown as the stator field space vector, lagging the stator current space vector by 90 degrees. The stator current space vector can lead or lag the Q-axis. This would affect the power angle, which describes the angle difference between the rotor and stator flux vectors. The power angle δ dictates the mode of the synchronous machine (motor, δ lease the PM flux vector). The stator current space vector i_s can be decomposed into components i_d and i_q. If i_d is zero, then $i_s = i_q$. Given that the direct axis current is responsible for the rotor magnetic field, a permanent magnet machine requires no i_d.

As mentioned previously, the DQ transformation makes the 3-phase motor "appear" to be a DC motor from the perspective of the control system. With a PM motor, field current is unnecessary ($i_d = 0$) thus, the motor current is controlled by controlling i_q to some setpoint. For motor speed control, this setpoint would be provided by an outer speed control loop.

2.3. Voltage and Flux Linkage Equations [4]

The voltage equations of a PMSM given in the d-q rotor reference frame are provided below [2].

$$v_{sd} = R_s i_{sd} + \frac{d\psi_{sd}}{dt} - \omega\psi_{sq} \quad (3)$$

$$v_{sq} = R_s i_{sq} + \frac{d\psi_{sq}}{dt} + \omega\psi_{sd} \quad (4)$$

$$\psi_{sd} = L_{sd} i_{sd} + \psi_{PM} \quad (5)$$

$$\psi_{sq} = L_{sq} i_{sq} \quad (6)$$

For modeling of the Healy's PMSM, these equations are implemented in a Simulink™ Simscape™ Permanent Magnet Machine block.

2.4. PMSM Model Parameters

PMSM model parameters include the flux of the permanent magnet, reference currents, torque constant, and motor winding impedances. The motor winding impedance values were based on those given in Reference [3]:

$$L_d = 0.0298\ H$$

$$L_q = 0.0159\ H$$

$$R_S = 0.0133\ \Omega$$

According to [3], the USCGC Healy has two 15,000 HP dual winding propulsion motors. The Simulink™ platform supports a single winding PMSM only, so it was instead modeled as a single 7500 HP (5593 kW) motor with an associated scalar multiplier (2) used to adjust performance for the full 15,000 HP per propulsion shaft. The motor torque is proportional to the rated power at rated shaft speed as follows:

$$T = \frac{P}{160RPM * (2\pi/60)} = 3338008\ N - m \quad (7)$$

where: T = Torque, Newton-meters. P = Power, (5,593,000 Watts)

The Delta-connected motor winding voltage at full power is 2320 $Vrms_{,L\text{-}N}$. The corresponding line current at rated power factor (0.7)

$$Power = 3 * V_{LN} I_{line} * \ Power\ Factor \tag{8}$$

$$I_{line} = \frac{P}{3 * V_{LN} * PF} = 1194\ A \tag{9}$$

where: I_{line} = Line current, Amperes. P = Power, (5,593,000 Watts). PF = power factor (0.7)

The motor torque constant may be calculated now as the ratio of rated torque to peak current is known.

$$Torque\ Constant = \frac{T}{3 * I_{LN}} = 93.2\ N - m/Amp \tag{10}$$

where: T = Torque, (333,800 Newton-meters). I_{LN} = Line current, (1194 Amperes).

Motor torque for a PMSM can be characterized according to the following equation [4]:

$$T_e = \frac{3}{2} p \left[\psi_{PM} i_{sq} - \left(L_{mq} - L_{md} \right) i_{sq} i_{sd} \right] \tag{11}$$

For a surface mounted PMSM, L_{mq} and L_{md} are nearly equal. PMSM motors generally do not have damper coils in them, therefore i_d = 0. As i_d is controlled to zero as well, the above torque formula is simplified to:

$$T_e = \frac{3}{2} p \left[\psi_{PM} i_{sq} \right] \tag{12}$$

This relationship can be used to calculate the current reference values for the PMSM model. Since the current reference has been calculated already, as well as the torque, the required flux of the permanent magnet can be calculated as follows:

$$\psi_{PM} = \frac{2 * T_e}{3 * p * \sqrt{2}\, i_{sq}} = 26.3\ \text{Webber} \tag{13}$$

where: ψ_{PM} = magnetic flux, Webber. T_e = Torque, (333,800 Newton-meters). p = number of poles (5). i_{sq} = q component current, (1194 Amperes)

The number of pole pairs and the magnetic flux linkage of the permanent magnet are constant values, thus, with i_d = 0, the electromagnetic torque is directly proportional to i_{sq}, in effect having a 3-phase permanent magnet synchronous motor appear as a separately-excited DC motor (to the control system), with the amount of torque directly proportional to the magnitude of the stator current space vector, i_{sq} applied to it (as presented in Equation (12)).

The control method for the PMSM needs to control i_q for torque production and $i_d = 0$ (or as close to it as possible during transient), regardless of the potentially massive and sudden external torque loads which may be encountered during ice-capable service. The control method requires rotor position (as measured by an encoder) to be relayed constantly to the controller, where field commutation is adjusted continuously. For ice capable vessels, external torque loads encountered during ice milling are a concern because these external forces are relatively large, intermittent, and repetitive. To the authors' knowledge, an analysis of the behavior of vector-controlled PWM drives for this type of service has not been published in the open literature. These forces act to disturb the rotor position, knowledge of which is critical to successful application of the vector control method—this analysis seeks to investigate the effects of these disturbance torques on the closed-loop behavior of the system.

2.5. Motor Control System Overview

The system is based on speed control—the desired propulsion shaft speed is input as a throttle command to a cascaded current control loop, where an outer PI controlled speed loop provides a Q-axis current demand signal. The output of the inner current loop is inverter output voltage demand to the inverter, where the inverter output voltage across the motor stator windings modulates the stator current space vector as required to increase or reduce motor torque as necessary to match the desired shaft speed. There are several other sub-systems that support this action and are necessary to make it work properly. The stator current space vector is the vector sum of the individual phase currents at any given time. The magnitude and the phase of these field currents are controlled by controlling the magnitude and phase of the voltage supplied to the stator windings. i_d is not a measured quantity that is available directly from measurements, therefore the 3-phase currents must be measured and transformed to the dq reference frame. Both the D-axis and Q-axis currents are controlled using Proportional-plus-Integral (PI) controllers with the D-axis current demand being equal to zero. It was found during the analysis that acceptable demand following and disturbance rejection were achieved using PI controller with proportional and integral gains (respectively) $K_p = 6000$ and $K_i = 1000$. These values were determined via iterative testing and successive tuning, given that inner current loop gains are typically orders of magnitudes higher than speed control gains due to the system electrical dynamics responding mush faster than the mechanical characteristics (e.g., mechanical inertia), making the current response is virtually instantaneous compared to the system's mechanical response. Figure 5 presents the flux vector controller implemented in Simulink^{TM}. The voltage demands calculated for the d and q axes are in-turn inverse transformed to the abc reference frame, normalized to the value of the DC link voltage, and converted to logic signals used to actuate the IGBT switches in the inverter. Figure 2 illustrates the process of generating switch signals from a demand signal. The control voltages would be generated from the inverse-transformed d and q axis demand signals from the control system shown in Figure 5.

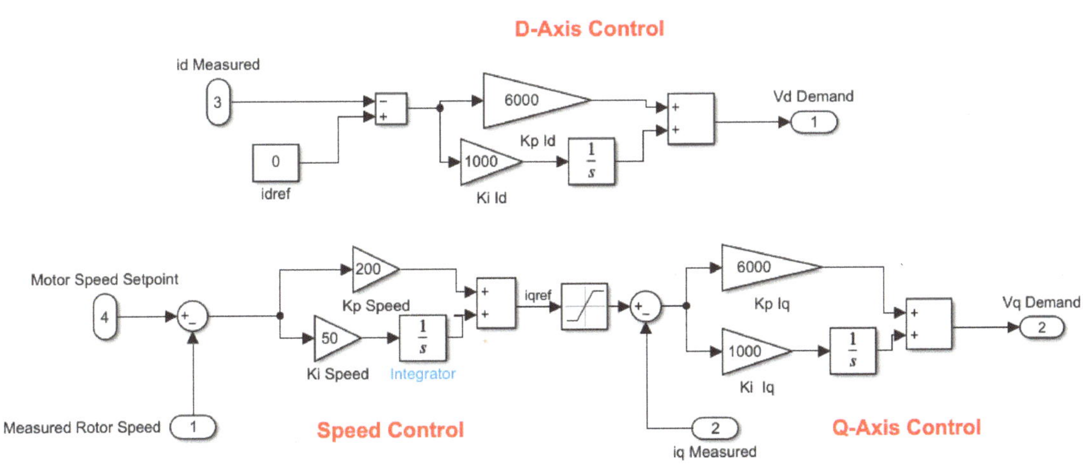

Figure 5. Cascade control of PMSM current.

2.6. USCGC Healy Simulink Model

The main objective of the study is to verify the stability of the controlled system during a variety of operational scenarios encountered by ice-capable ships. This model would be considered an end-to-end model in that the Diesel generator through propellor has been modeled to varying degrees of complexity consistent with the goals and objectives of this study. The model developed supports analysis of the impact of propulsion motor loads on a theoretical shared electric distribution system and identify and quantify potential impacts to power quality by the PWM drive, if any. Figures 6 and 7 present the end-to-end Simulink^{TM} model used for this analysis.

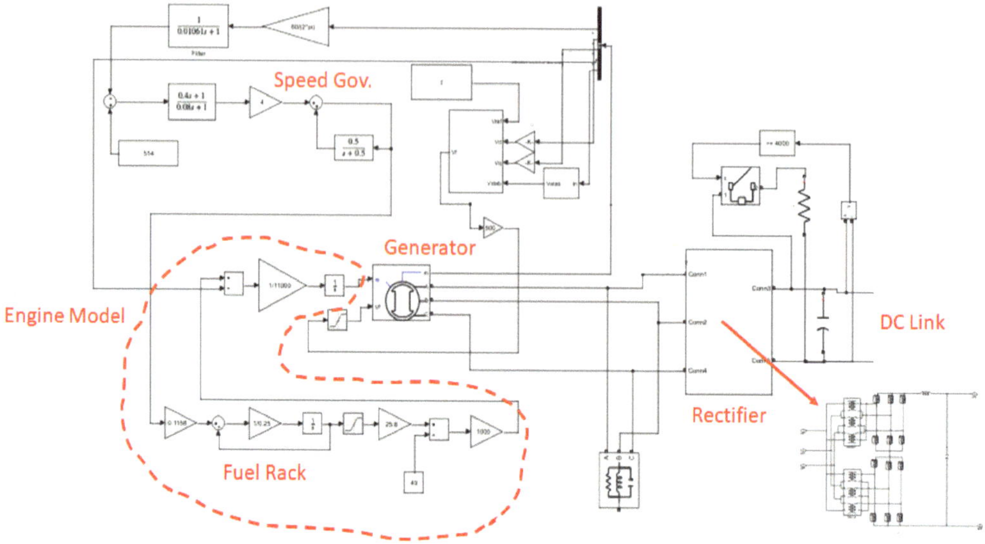

Figure 6. Diesel Engine/Generator/Rectifier Simulink^{TM} Model.

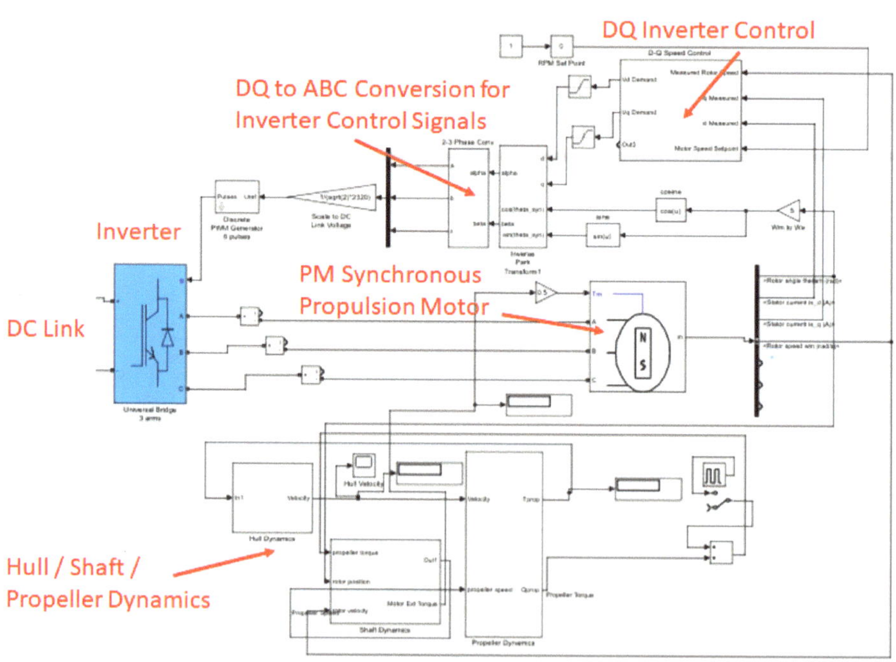

Figure 7. DQ Controller/Inverter/Propulsion Simulink^{TM} Model.

2.7. Engine/Generator

The generator model is modeled as a diesel driven synchronous machine with the properties inspired from the USCGC Healy—four 7.2 MW, 6.6 kV, 0.7 power factor, three-phase, 514 RPM engines as documented in Reference [2].

The engine is modeled as a first-order system, where engine acceleration is function of the net total torque of the engine/generator system divided by the inertia of the rotating assembly.

$$\dot{\omega}_{engine} = (Q_{engine} - Q_{generator}) / J_{total} \tag{14}$$

The engine torque ultimately is proportional to the fuel rack position and any first order lags associated with the turbocharger dynamics. Generator electrical torque due to load are calculated using a wound-field synchronous generator Simulink model, using standard impedance consistent with a machine of its size. The angular acceleration value from this subsystem is integrated to obtain the rotor speed of the Diesel engine/synchronous generator assembly.

Reference [2] included a functional diagram of the engine speed control governor system including proportional and integral gains, a speed signal filter, and actuator both with respective first order time constants. As originally modeled a dual proportional gain was included. This approach was abandoned and a single proportional gain was used to promote better stability in the new end-to-end model.

A separate (non-engine driven) excitation system was modeled after the IEEE Type 1 Synchronous Machine Voltage Regulator Model. Although shaft-driven DG generator type excitation systems are still in use, modern practice has favored separate, solid state excitation systems. This approach has the added benefit of decreasing complexity of the model while mimicking likely system performance.

2.8. Rectifier Circuit & DC Link/Filter

A 12-pulse type rectifier circuit was selected (instead of 6-pulse or 24 pulse) as typical AC/DC conversion technology used for PWM motor drive applications. The multi-pulse rectifier provides for lower harmonic content in the AC system upstream of the rectifiers. Parallel delta-wye and delta-delta transformers were used. A line inductor of 0.002 H was found to provide adequate performance with low voltage ripple. The DC link is fitted with a 0.005 F capacitor. All of these values were iteratively adjusted to get acceptable system dynamic and harmonic performance.

2.9. Inverter/Motor Drive

The inverter used for this study is a 2-level, 6 device (IGBT) inverter, fed from the DC link. The DC link provides a relatively smooth DC voltage at the approximately the line-to-line peak voltage of the generator output. Using sine-triangle pulse-width modulation (SPWM—shown in Figure 2), the output from the dq current controller to converted to the abc reference frame, as a control signal. The control signal is normalized to the DC voltage and compared to a triangle wave generated at a designated switching frequency. As shown in Figure 2, the normalized control signal and triangle wave are logically compared, with the output of the comparison being the IGBT actuation signal (conduct or not). The motor windings are powered by the pulse-width-modulated inverter output voltage.

2.10. Hull Characteristics/Propellor Shaft Dynamics

A traditional marine AC power generation and PWM drive system seldom experiences transient load events imposed on an ice-capable ships. Accordingly, additional models must be included that describe the dynamics of the propeller, shaft, and hull subject to external forces that impose a time-varying load torque on the propulsion motor. The equations describing hull characteristics and propellor shaft dynamics used for this study were derived from Reference [2].

2.11. Hull Dynamics

The hull dynamics model is based on linear (ahead and astern) motion only where the acceleration of the ship mass is determined by the net thrust divided by the vessel mass in accordance with Newton's second law of motion:

$$F_{net} = M \times \dot{V} \quad (15)$$

$$F_{net} = F_{eff} - R_T \quad (16)$$

$$F_{eff} = N_{prop} \times F_{prop} \times (1 - t) \quad (17)$$

where: F_{net} = net force (fore and aft) on the vessel, Newtons. F_{eff} = effective total propeller thrust, Newtons. F_{prop} = propeller thrust of each propeller, Newtons. N = number of propellers. t = thrust reduction fraction (dimensionless). R_T = total hull resistance, Newtons. M = effective vessel mass, kg. $\dot{V}$= vessel acceleration.

The thrust reduction fraction, t, typically ranges from 0.1 to 0.2 for twin propellor shaft ships, with the larger value typical for ships with bossings [5]. A value of 0.2 was included the model. The design displacement was assumed to be 16,000 LT [2]. The hull resistance, R_T, is different depending on whether the vessel is in open water or ice.

Figure 8 shows the hull dynamics model. The resistance model may use either ice breaking resistance or open water resistance but not both concurrently.

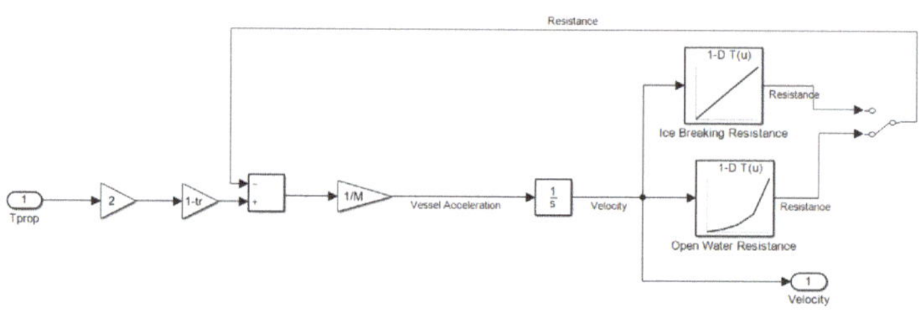

Figure 8. Hull dynamics model implemented in Simulink^{TM}.

2.12. Hydrodynamic Resistance

Both open water hull resistance and ice-braking resistance must be considered. Reference [3] provided typical curves for each of these cases (Figure 9). These curves were approximated in the Simulink model using lookup tables.

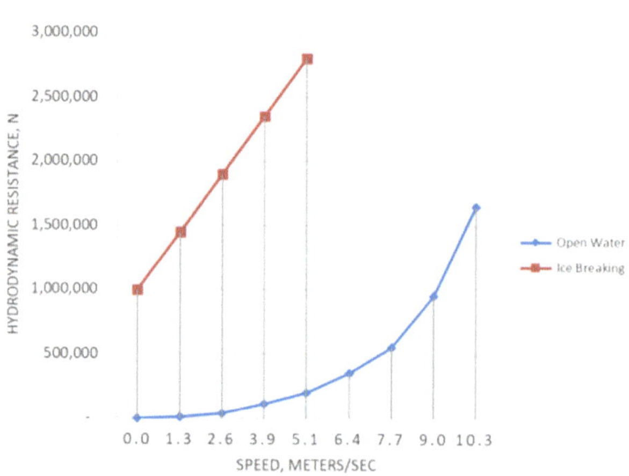

Figure 9. Hull resistance curves.

2.13. Shaft & Propeller Dynamics

The propulsion shaft (Figure 10) is a complex mechanical system that has many forces acting on it together. Reviewed in order from the motor outboard end to the stern tube bearing, the motor windings provide electromechanical torque that varies with the magnitude of the current space vector.

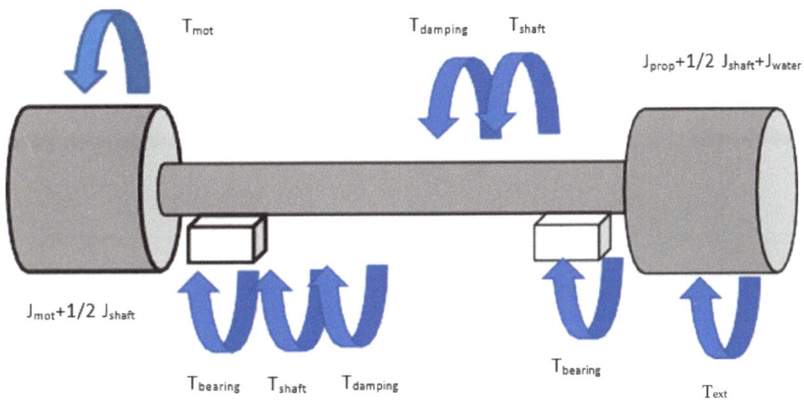

Figure 10. Shaft dynamics diagram.

Opposing this motor torque is the inertia of the rotor and shaft, motor bearing friction, shaft elastic torque, and internal shaft damping torque.

From the propeller end there is the shaft torque plus the damping force less the stern tube bearing friction, inertia of the shaft, including the propeller and entrained water, and finally the external torque from the hydrodynamic resistance. If the vessel is in ice, then the propeller is also subject to external torque loads from impacting the ice.

The propulsion shaft system model is best represented visually:

The sum of the torques on the motor side of the line shaft may be represented mathematically as follows:

$$T_{mot} - \left(J_{mot} + \frac{1}{2} J_{shaft}\right)\ddot{\theta}_m - D_{brg}\dot{\theta}_m - \left(\dot{\theta}_m - \dot{\theta}_p\right)C_{shaft} - (\theta_m - \theta_p)K_{shaft} \tag{18}$$

$$(\theta_m - \theta_p)K_{shaft} - \left(J_{prop} + J_{water} + \frac{1}{2} J_{shaft}\right)\ddot{\theta}_p - D_{brg}\dot{\theta}_p + \left(\dot{\theta}_m - \dot{\theta}_p\right)C_{shaft} - T_{ext} \tag{19}$$

where: C_{shaft} = Internal damping coefficient, 71.4 kN-m-s/rad, per Reference [3]. K_{shaft} = Shaft torque coefficient, 112 MN-m/rad, per Reference [3]. D_{brg} = Bearing coefficient of friction, 500 kN-m-s/rad, assumed for both motor and stern tube bearings. T_{mot} = Motor electromagnetic torque, N-m, variable. T_{ext} = Propeller external torque, N-m, variable. J_{mot} = Motor moment of inertia, 18,100 kg-m^2, per Reference [3]. J_{shaft} = Shaft moment of inertia, 18,100 kg-m^2, per Reference [2]. J_{prop} = Propeller moment of inertia, 41,200 kg-m^2, per Reference [2]. J_{water} = Entrained water moment of inertia, 5100 kg-m^2, per Reference [3]. $\ddot{\theta}_m$ Motor end angular acceleration, rad/s^2. $\dot{\theta}_m$ = Motor end angular velocity, rad/s. θ_m = Motor end rotor angular position, rad. $\ddot{\theta}_p$ = Propeller angular acceleration, rad/s^2. $\dot{\theta}_p$ = Propeller angular velocity, rad/s. θ_p = Propeller angular position, rad.

The PMSM Simulink^{TM} block uses load torque calculated from the model above and calculates rotor speed and angular position. The shaft model sums the external and internal system torques into a total motor external torque that is experienced by the propulsion motor. Since only one 3-phase set of windings of the propulsion motor is modeled, only one half of the shaft torque is used as the total external torque input to the PMSM model.

2.14. Hydrodynamic Propeller Torque and Thrust

The hydrodynamic propeller torque is the torque developed due to the reaction of the propeller blades relative to their relative motion with respect to the surrounding water. This torque is a function of the water density, propeller rational speed, propeller diameter, relative rotational efficiency, and vessel velocity. The propeller thrust is a function of these same quantities less the relative rotative efficiency.

These quantities are related according to the following per Reference [3]:

$$Propeller\ Torque = T_{prop} = C_Q \rho D^3 \left(V_A^2 + n^2 D^2\right) / \eta_R \tag{20}$$

$$Propeller\ Thrust\ Force = F_{prop} = C_T \rho D^2 \left(V_A^2 + n^2 D^2\right) \tag{21}$$

$$Apparent\ Velocity\ of\ Ship = V_A = V(1 - w_t) \tag{22}$$

$$Advance\ Coefficent = U = nD / \left(V_A^2 + n^2 D^2\right)^{1/2} \tag{23}$$

where: T_{prop} = *Propeller torque*, N-m. F_{prop} = *Propeller thrust force*, N. ρ = Density of seawater, kg/m^3. n = Propeller speed, revolutions/s. D = Propeller diameter, m. η_R = relative rotative efficiency, percent. V_A = Apparent hull velocity, m/s. V = Hull velocity, m/s. w_T = Effective Tailor wake fraction based on thrust identity. Percent. U = Advance coefficient, calculated. C_Q = Torque coefficient, dimensionless. C_T = Thrust coefficient, dimensionless.

Reference [3] uses a technique that indexes C_T (Figure 11) and C_Q (Figure 12) directly as a function of U. This method is very efficient for modelling and was incorporated into the model in the form of lookup tables that are switched on depending propulsion mode (forward or reverse) and hull velocity. By using this same technique, it is possible to compare performance cases for the PWM drive with the AC/AC drive model used in Reference [2].

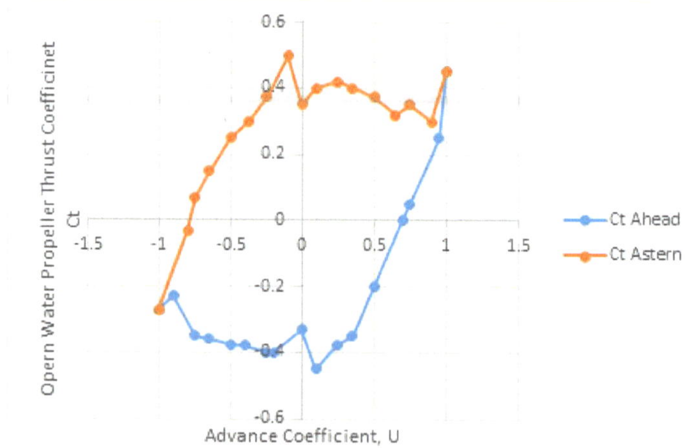

Figure 11. Shaft thrust coefficient vs. advance coefficient.

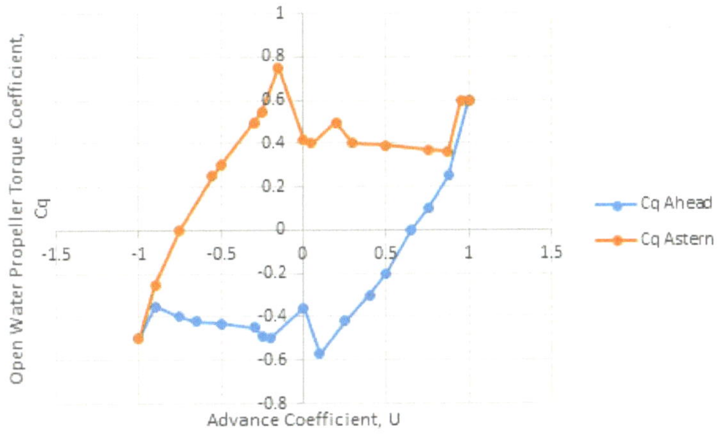

Figure 12. Shaft torque coefficient vs. advance coefficient.

2.15. Ice Milling Torque

The torque model employed in Reference [3] is a series of steep triangular spikes with a peak magnitude of 1540 kN-m based on MARITECH Report No. 82-116 (1982).

These spikes are modeled as four equidistant pulses per revolution to coincide with the four propeller blades. The corresponding Simulink model is shown in Figure 13.

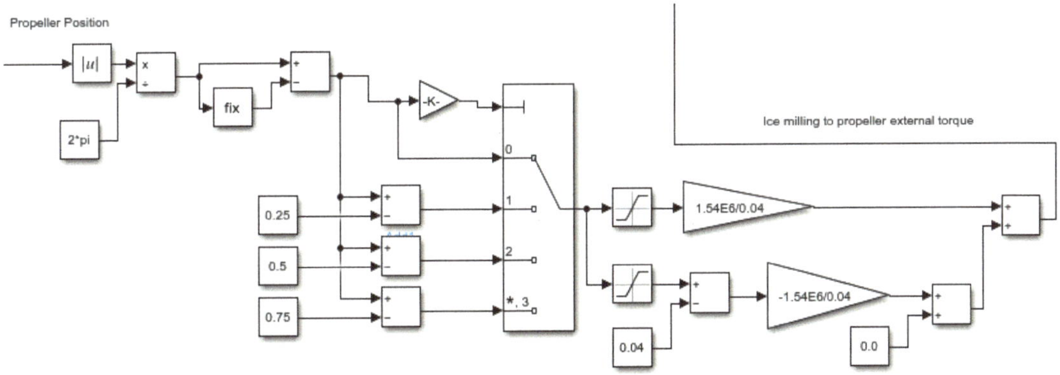

Figure 13. Ice milling external torque model.

3. Results

The aforementioned physical model described includes the major components of the electric generation, power conversion and propulsion systems, including the major diesel generator speed control governor, diesel engine, synchronous generator, excitation system, rectifier circuit, DC-link, filter, inverter, controller, permanent magnet synchronous motor, shaft and propeller. The external dynamics such as hull resistance and ice milling torque have also been accounted for. All of these components/subsystems were modeled at a level of complexity consistent models found in the literature and the goals of this initial analysis. Future studies may warrant higher-fidelity and complexity models depending on the study objectives.

3.1. Ahead Operation

The simplest and perhaps most frequent maneuvering operation is the ahead command (Figure 14). During this evolution the shaft is initially motionless. A two second delay period is included to allow the generator excitation system to stabilize and build voltage in the DC-link. At t = 2 s the desired shaft speed set point is ramped up from 0 RPM to 143 RPM. In this scenario the hull, propeller and shaft dynamics are all set for open water operation, with the propeller torque coefficient varying with shaft and vessel speed. At t = 25 s the speed demand reaches its final value, though steady state shaft speed takes an additional 10 s to achieve due to system inertia. The motor phase current I_a is sinusoidal with frequency and magnitude varying as commanded by the controller to match the desired shaft speed amid the variable external torque.

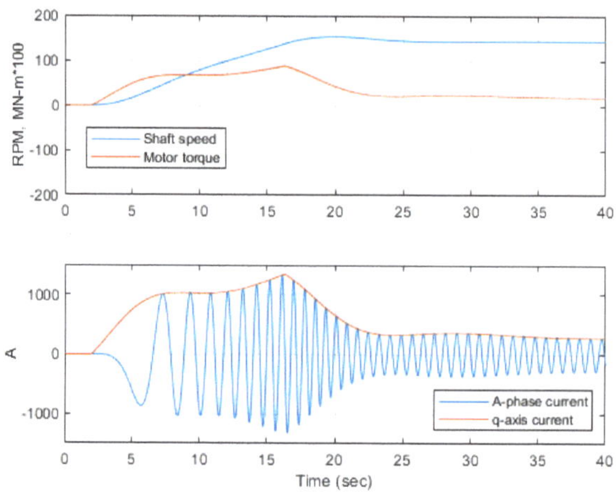

Figure 14. Ahead operation as calculated with the Healy end-to-end model.

The system appears to be functioning as expected for this type of operation, with higher currents drawn by the motor during system acceleration, where inertial torques are significant, with the current decreasing during steady-state operation, where motor torque is needed to overcome bearing and propellor hydraulic drag. Total harmonic distortion of the current was calculated to be less than 1%, demonstrating that harmonically, SPWM-type vector control is superior than use of a Cycloconverter-type motor drive.

3.2. Harmonic Distortion

A major advantage of utilizing a PWM drive over the Cycloconverter drive of the Healy is the potential reduction of harmonic currents in the propulsion motor power cables and windings. Harmonic have the effects of reducing torque and efficiency. The harmonic effect of the rectifier on the generator outputs appears to be relatively minor, as is the effect of the load on the PMSM, propagating back to the generator as an electrical load through the motor drive.

3.3. Ice Breaking

When the vessel encounters ice the open water resistance curves are no longer valid. For this test the same speed command timing is used as for the ahead test but the desired speed is increased to 143 RPM and the time range is expanded to 65 s in order to give the simulation adequate time for the ship's propulsion system to develop sufficient thrust to counter the increased hull resistance imposed by the ice.

This test was of relatively short duration, but sufficient to demonstrate the effect of ice breaking at steady state. As can be seen from Figure 15, the PMSM motor currents reach their maximum value before the speed set point is reached. This analysis demonstrates stable operation during ice breaking.

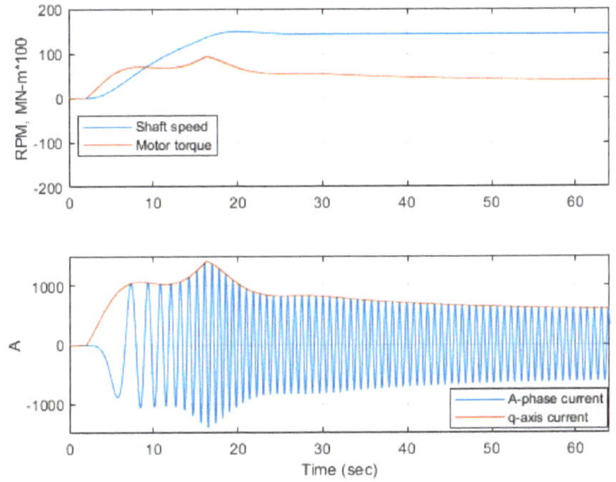

Figure 15. Motor current performance during ice breaking scenario.

3.4. Ice Milling

Ice milling is considered to be one of the more challenging operational scenarios for ice capable ships with PWM drives. The primary concern of this technology is whether the system can maintain stable motor drive and system control, and reject the substantial disturbance of ice milling loads.

The ice milling scenario assumes the vessel starts with a slow astern command of −38 RPM (−4 rad/s), followed by ice—impact and resultant milling. This is a likely scenario for a ship maneuvering in ice to change direction. Hydrostatic hull and propeller resistance, derived from Reference [2], are also included in this analysis.

The speed response, shown in blue in Figure 16, demonstrates the initial speed response from the moment the slow astern bell is given. The desired shaft speed is achieved in 3 s before the ice-milling torque (shown in red), is applied. Both speed and torque are shown as negative, indicating reverse operation.

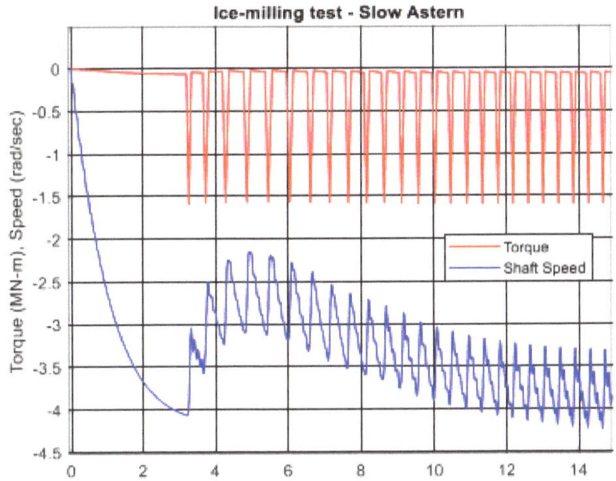

Figure 16. Ice Milling Shaft Speed and Torque—Slow Astern Operation.

It is clear that the initial torque has the effect of dramatic speed reduction. The system has been disturbed, clearly and repeatedly, causing additional speed reduction up to t = 5 s. However, despite the consistent, repetitive external torque loads, the propulsion motor recovers and corrects to achieve the desired speed set point. Steady state response is achieved despite continuous transient disruption from the external ice milling torque pulses.

As previously discussed, dq PMSM proportional-plus-integral control maintains $i_d = 0$ (or very close to it during transients). i_q is controlled to the demand value calculated by the motor speed controller (also proportional-plus-integral control). The i_d and i_q currents are shown in Figure 17 was observed during the same test run, shown below:

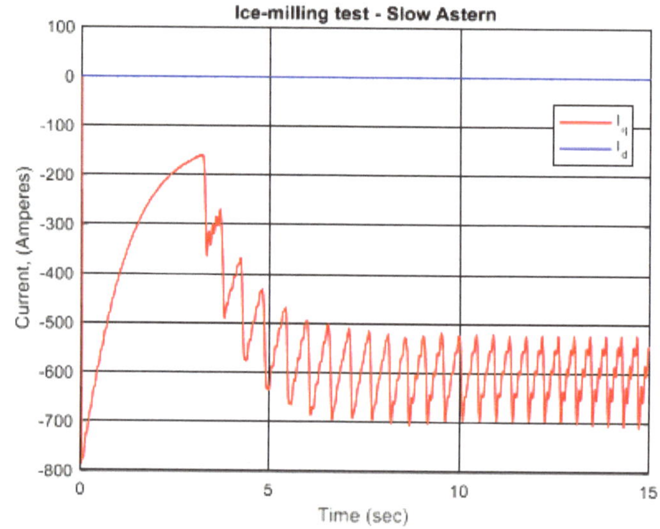

Figure 17. Ice milling motor dq currents.

As shown in Figure 17, i_d is effectively controlled to zero. i_q is effectively controlled to zero, and i_q performs as expected and is responsible for motor torque production. There is an initial current spike on the speed initial speed command that slowly fades as the desired shaft speed is reached. This is followed by more spikes as speed drops off due to external torque disturbance. Eventually, a semblance of steady state is achieved, but with repeating transient disruption from the continued, repetitive ice milling torque.

3.5. Crash Back (Emergency Crash Astern)

Although not necessary for ice breaking or ice milling, ship builder sea trials usually typically include an emergency astern test or "crash back" test at sea. To simulate this test the model was initialized with an initial hull velocity of 8 m/s in the ahead direction with throttle command at 143 RPM (15 rad/s). At t = 10 s a −143 RPM (full reverse) throttle command is given. The results of this test run are shown in Figure 18.

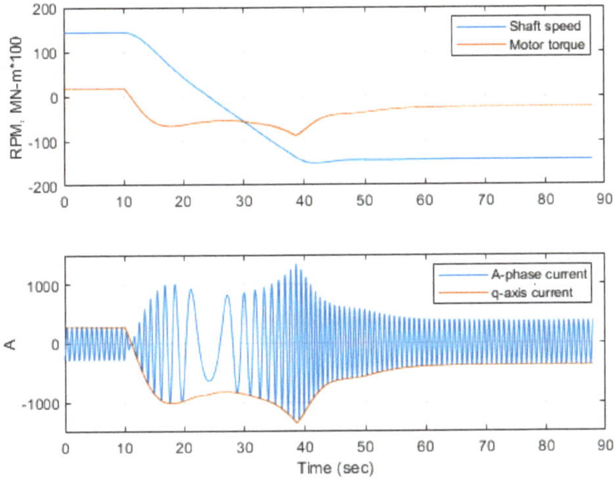

Figure 18. Motor performance during crash back test.

The USCGC Healy specification calls for a full power reversal within 25 s. Maximum Torque was achieved within this range, however the shaft speed continued to accelerate until more than 35 s after the reverse command was given in response to the hull dynamics. At t = 15 s after the full-astern command is given, a reversal is observed, as expected, in the A-phase and i_q currents (i_q should be the amplitude of the phase current) resulting is a change in the direction of shaft rotation. The shaft speed ultimately achieved the demanded −143 RPM, the torque performance is acceptable given that this maneuver would rarely, if ever, be encountered by this type of vessel.

4. Discussion

The results of this preliminary study demonstrate that a PWM-based inverter and dq current controller supplying power to a PMSM is a technically viable option for ice-capable vessel propulsion systems. The performance test simulations achieve stable operation throughout the load range, and successfully reject the most severe forms of external torque disruption that are expected to be encountered in typical ice service.

Moreover, the motor currents are shown to be sinusoidal during the different test conditions simulated unlike those of a cycloconverter motor drive. The reduced harmonic distortion allows the PMSM to be utilized most efficiently as demonstrated in the performance test results for the maneuvers analyzed. The notional PMSM chosen for the study is a 3-phase, surface-mounted machine, similar to those found in other ship propulsion applications, yet significantly different than the cycloconverter-based drive deployed on the Healy. In addition to the harmonic benefits of sine-triangle pulsed-width modulation over cycloconverter drives, application of PMSM provide for shipboard installations that are considerably more power-dense, smaller volume, and lighter weight than a wound-field, cycloconverter-fed synchronous motors.

The characterization of any operational savings available from the superior performance and the potential costs associated with the alternate equipment was outside of the scope of this analysis. As with any new design, these factors should be considered during the initial development phases.

Author Contributions: Conceptualization, E.C.R. and J.T.; methodology, E.C.R. and J.T.; software, E.C.R. and J.T.; validation, E.C.R. and J.T.; formal analysis, E.C.R. and J.T.; investigation, E.C.R.; resources, E.C.R. and J.T.; data curation, E.C.R.; writing—original draft preparation, E.C.R.; writing—review and editing, E.C.R. and J.T.; visualization, E.C.R. and J.T.; supervision, J.T.; project administration, J.T. All authors have read and agreed to the published version of the manuscript.

Funding: This research received no external funding.

Data Availability Statement: Data provided upon request.

Conflicts of Interest: The authors declare no conflict of interest.

References

1. John, A. Beverley, Chapter VIII Electric Propulsion Drives. In *Marine Engineering*; Roy, L.H., Ed.; Society of Naval Architects and Marine Engineers: Jersey City, NJ, USA, 1992.
2. Ned, M. *Electric Machines and Drives, A First Course*; John Wiley & Sons, Inc.: Hoboken, NJ, USA, 2012.
3. Lecourt, E.J., Jr. Using Simulation to Determine the Maneuvering Performance of the WAGB-20. *Nav. Eng. J.* **1998**, *110*, 171–188. [CrossRef]
4. Krause, W.; Sudhoff, P. *Analysis of Electric Machinery and Drive Systems*, 3rd ed.; Wiley: Hoboken, NJ, USA, 2013.
5. Chester, L. Long, Chapter X Propellers, Shafting, and Shafting System Vibration Analysis. In *Marine Engineering*; Roy, L.H., Ed.; Society of Naval Architects and Marine Engineers: Jersey City, NJ, USA, 1992.

energies

MDPI

Article

Analysis of the Simulation of the Operation of a Wheel Hub Motor Mounted in a Hybrid Drive of a Delivery Vehicle

Piotr Dukalski [1,*], Jan Mikoś [1] and Roman Krok [2]

1 Łukasiewicz Research Network-KOMEL Institute of Electric Drives and Machines, 40-203 Katowice, Poland
2 Department of Mechatronics, Faculty of Electrical Engineering, Silesian University of Technology, 44-100 Gliwice, Poland
* Correspondence: piotr.dukalski@komel.lukasiewicz.gov.pl

Abstract: The article presents the analysis of operational parameters and thermal analysis of the wheel hub motor during operation in the car drive. The authors proposed an analysis of the operation of the wheel hub motor mounted in a hybrid car, during Artemis driving cycles and while driving on the road with different slopes. The simulations were carried out in the Ansys Motor-CAD program. The calculations are based on coupled models of the electromagnetic circuit and thermal models of the motor. The conducted research is a proposal of an approach to the design of electric vehicle propulsion motors, which allows us to consider problems related to predicting at the motor design stage what are its possibilities and what risks during operation in a real drive. The analysis also includes the impact of the applied motor control strategy and the variation of the supply voltage. These are aspects that are extremely important in wheel hub motors, as they are weight-optimized motors with a limited volume and a relatively high power and torque density.

Keywords: wheel hub motor; electric drive; permanent magnet synchronous motor

Citation: Dukalski, P.; Mikoś, J.; Krok, R. Analysis of the Simulation of the Operation of a Wheel Hub Motor Mounted in a Hybrid Drive of a Delivery Vehicle. *Energies* **2022**, *15*, 8323. https://doi.org/10.3390/en15218323

Academic Editors: Lorand Szabo and Feng Chai

Received: 5 October 2022
Accepted: 31 October 2022
Published: 7 November 2022

1. Introduction

Nowadays, the dynamic development of electromobility focuses on ever greater requirements for vehicle drives, resulting in the search for new designs and topology of drive systems and continuous increase in performance.

One of them is the wheel hub motor, the use of which allows for the elimination of elements mediating the transmission of the torque. Motors of this type can be used in various applications, and research on various designs [1–6] and design methods [7–10] is the subject of many studies and scientific descriptions. The use of compact wheel hub motors in the automotive industry, due to the working conditions and location, is a challenge for the design of the motors and the technology of manufacture, related to mechanical strength, sealing, cooling system concept, and control [11–15]. One of the well-known problems with wheel motors is the large unsprung mass. The motors mounted in the car in this way constitute an additional unsprung mass, which has a negative impact on the operation of the shock-absorbing system components and on the vehicle's steerability [16–21]. The public research and conclusions suggest that depending on the vehicle and the ratio of unsprung and sprung masses, the impact of using this type of solution when driving the vehicle may be limited to the impact imperceptible for the average driver. Both the research carried out by the authors of the project and other publications indicate that the impact of a suitably reduced unsprung mass of motors can be compensated for by some changes in the structure of the shock-absorbing system. This solution offers a number of possibilities related to driving configurations for the front, rear, or both drive axles. An interesting application also seems to be the use of this type of motor in a hybrid drive, because the motors do not take up space in the vehicle and can offer various, switchable modes of drive operation. Drives of this type can be used in many different applications in many industries, such as passenger cars, buses, mining transporters, unmanned vehicles for uniformed services, and space rovers. Motors of this type are of interest not only to scientific but also to business entities [22–26].

In the Łukasiewicz Research Network-Institute of Electric Drives and Machines KOMEL is implementing a project aimed at designing, manufacturing, and testing, in laboratory and road conditions, a direct drive with wheel hub motors of a hybrid car. The drive under development involves the use of two electric motors mounted in the rear drive axle of the car. The developed electric drive is to co-create a hybrid drive with the original internal combustion engine. Due to practical and economic aspects, a delivery vehicle weighing up to 3.5 t Fiat Ducatto was selected as the target vehicle. According to the assumptions, the vehicle should be able to move in the urban and suburban areas at speeds of up to 90 km/h using only the electric drive system, while outside the city, the system of two electric motors is only to support the drive system with a combustion engine.

The article presents an important aspect of motor and drive design, concerning the simulation of motor operation in various car cycles, taking into account the control strategy and the selection of the supply voltage. It is important to properly select the vehicle driving cycles that reflect the actual dynamics of the car in various driving scenarios, not used to analyze the cycle for testing the emissions of internal combustion cars. Traction motors for installation in wheels of vehicles must have a relatively high torque at lower rotational speeds and a sufficiently high torque necessary to overcome resistance to motion at maximum speed and to maintain appropriate driving dynamics in the entire speed range. In the case of wheel hubs motors, an extremely important aspect is the operating temperature of sensitive elements such as windings or permanent magnets. These are compact motors, due to the very limited space in which they must fit, therefore their efficiency in terms of the required power, especially torque, must be increased in relation to standard motors. Due to their shape and the requirements for operating parameters, they are multi-pole motors, which are associated with higher frequency and higher losses in the magnetic core, permanent magnets, and AC losses in the winding. Available publications lack simulations of the wheel hub motors' operation, which was designed for real dimensions, resulting from the available space and real mechanical requirements imposed by mechanical designers. The works carried out by the research team have been conducted for several years. Some of the results were presented in publications [27,28]. The concept of the motor under consideration is based on the analyzes and results of the work carried out earlier, including laboratory tests of prototype motors, calibration of computational models, and experience gained in designing in terms of mechanical aspects. There is a lack of traction calculations related to thermal calculations, which are key aspects of designing such demanding and specific motors. Figure 1 shows the Fiat Ducato with fittings of the 3D-printed model of the motor to the rim and the drive axle.

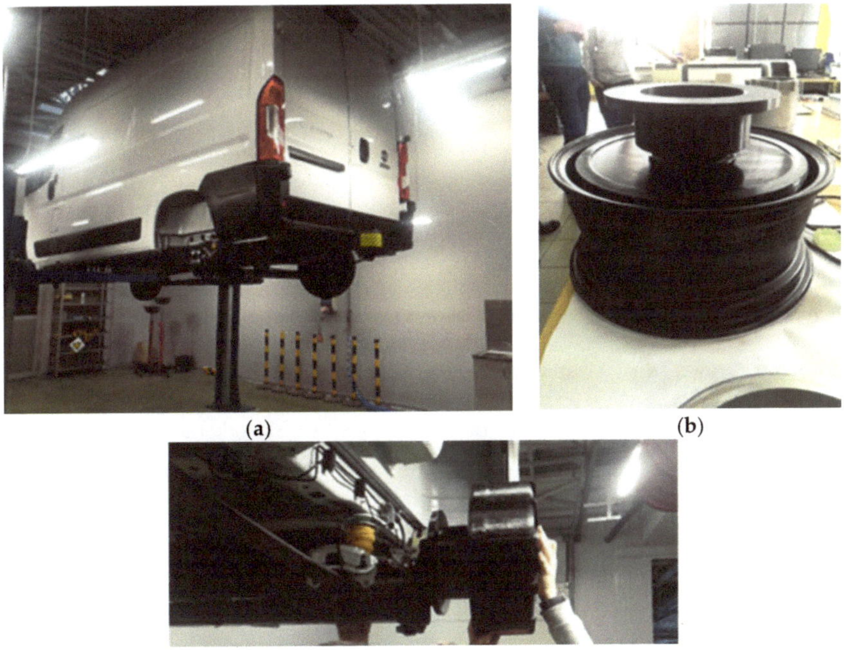

Figure 1. Fitting 3D printed model of motor: (**a**) Fiat Ducato car; (**b**) in the rim; (**c**) at the drive axle.

The authors presented an aspect of wheel hub motor design including a thermal analysis of key motor components under conditions that reflect the predicted actual operating conditions. The selection of the appropriate driving cycles allows the assessment of the motor's performance in driving conditions and the assessment of the influence of the selected supply voltage and control strategy. Such complete analyzes are lacking in scientific publications. The presented research results are also important due to the fact that the considered drive is hybrid. The analysis allows us to assess the extent to which the drive will be able to work only with electric motors, and in which operating ranges it will be necessary to cooperate with the internal combustion engine. Carrying out such tests at the stage of developing motors, drive components and power supply system is necessary, in particular, due to the fact that the weight and volume of the machine are limited. In this case, the most advantageous is the use of permanent magnet motors, the price of which, due to the cost of permanent magnets, may be higher than DC, induction, or reluctance motors, but they are characterized by a higher torque density to the total mass.

The use of permanent magnets requires such an analysis, because magnets are also an element whose permissible operating temperature may limit the operating range of the motor. The operating temperature of the magnets depends not only on the temperature of the winding and magnetic core, but also on the power losses that are generated in them, and these in turn depend not only on the design of the electromagnetic circuit but also on the operating point of the motor (current intensity, operating frequency of the electromagnetic circuit) and control strategy. Due to the application, the traction motor is characterized by a very wide range of changes in operating parameters, which exposes the motor structure to thermal tests in various ways. The authors performed an analysis that may be part of the methodology for designing motors of this type.

2. Calculation Model of the Motor

The motor under development is a synchronous motor with permanent magnets. The motor is an outer rotor with a surface permanent magnet (SPM) structure. The motor is liquid-cooled. The design assumes that the motor will fill the entire rim space. In this way, the full available space will be used to build and test the motor. In the future, it is assumed that if the driving parameters are achieved, work will be carried out to reduce the length of the machine. The motor winding is a double-layer winding with machine-wound concentrated coils. In this way, the overhang of the end windings (in relation to the distributed winding) can be significantly reduced, which allows for the reduction of the power losses generated in the winding and the reduction of the mass of the motor's electromagnetic circuit. The stator slots are open. The winding is potted in a special epoxy resin, which improves heat dissipation from the stator winding. Figure 2 shows the coil, magnetic core and wound stator, built in the Łukasiewicz Research Network-KOMEL Institute. The magnets are axially segmented, each consisting of five segments along the length of the magnetic core in order to reduce the power losses generated in the magnets [29–34]. The power losses generated in permanent magnets decrease the efficiency of the motor and increase their operating temperature, which causes deterioration of their parameters, and in extreme cases may even lead to their demagnetization, especially in the second motor control zone (flux weakening) [35–38] are influenced by an external magnetic field.

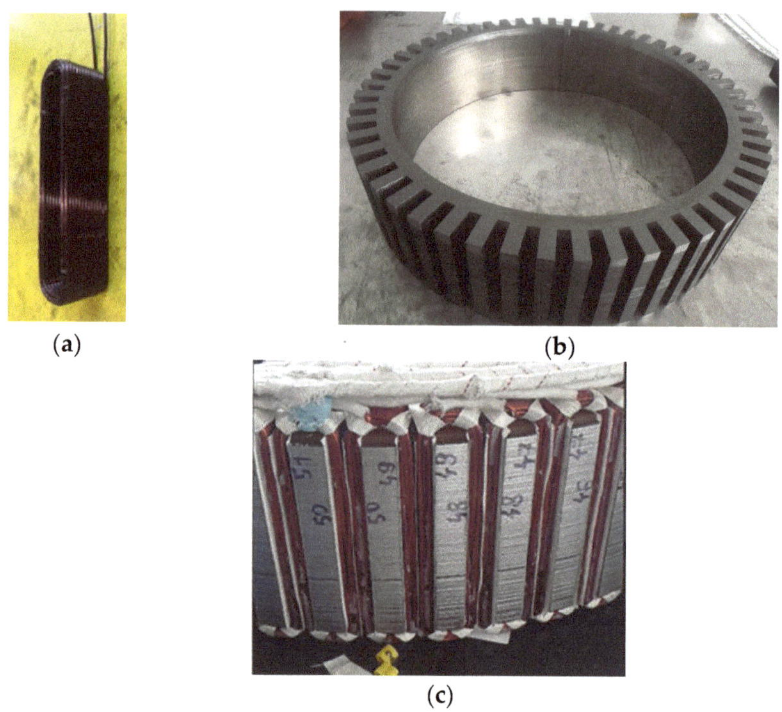

(**a**) (**b**) (**c**)

Figure 2. Physical model of: (**a**) winding coil; (**b**) magnetic core; (**c**) stator.

The drive consists of two motors installed in the rear driving axle of the car. Each of the motors will be powered by a separate inverter. The assumed battery supply voltage is U_{DC} = 400 V.

The motor was designed in the Ansys Motor-CAD (Figure 3). The dimensions of the electromagnetic and thermal circuit, such as inner and outer stator diameter, machine length, end-windings, dimensions of side heat sinks, and elements of the supporting structure and rotor diameter, result from the mechanical structure of the motor, structural elements and their strength, built-in resolver, cooling system, construction sealing, overall rims dimensions and elements of the brake system and suspension.

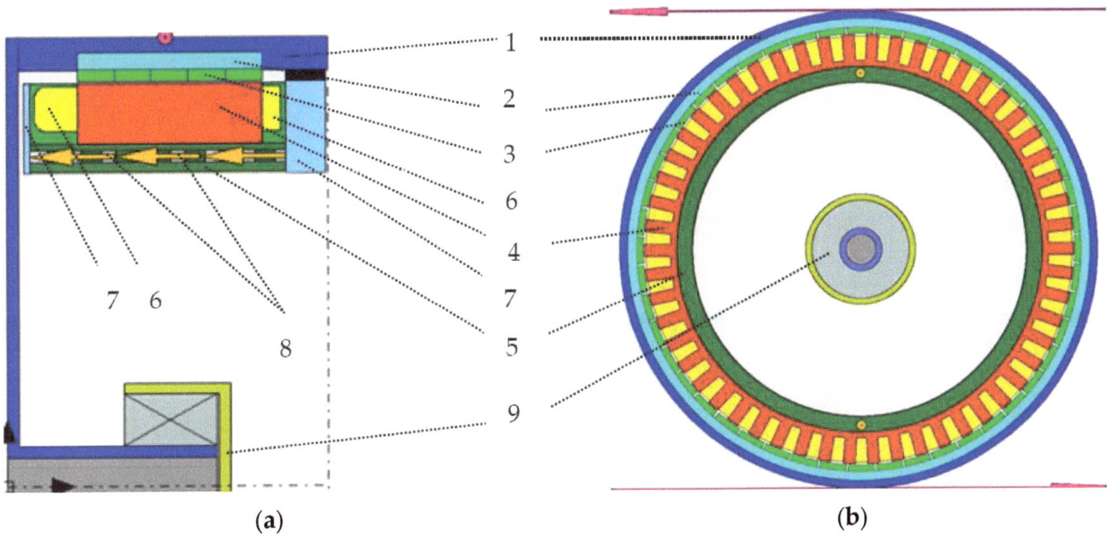

Figure 3. Motor model in Ansys Motor-CAD software (**a**) longitudinal model (**b**) transverse model; 1-rotor, 2-rotor magnetic core, 3-permanent magnets, 4-stator magnetic core, 5-supporting structure, 6–end-winding, 7-side heat sink, 8-channels with flowing coolant, 9-bearing socket.

It is important that the simulations of the electromagnetic and thermal circuit take into account the permissible space limited by various mechanical aspects, because the wheel hub motors have a compact structure, the weight, dimensions, and parameters have to be optimized. Without carrying out design work on the mechanical structure of the motor, the permissible dimensions of the electromagnetic circuit cannot be determined.

Table 1 List of selected motor design and operating parameters.

Table 1. Rated and design parameters of the motor.

Parameter	Value	Unit
$P_{max.}$	90	kW
U_{DC} (inverter)	400	V
$I_{max.}$	420	A
$T_{m\ max.}$	2300	N·m
$n_{max.}$	1200	rpm
Q (number of slots)	54	-
2p (number of pole pairs)	48	-
Copper weight	4.9	kg
Iron weight	14.7	kg
Magnets weight	3.1	kg

Figure 4 shows the FEM model of the motor with the calculated magnetic induction distribution in the magnetic core generated by permanent magnets.

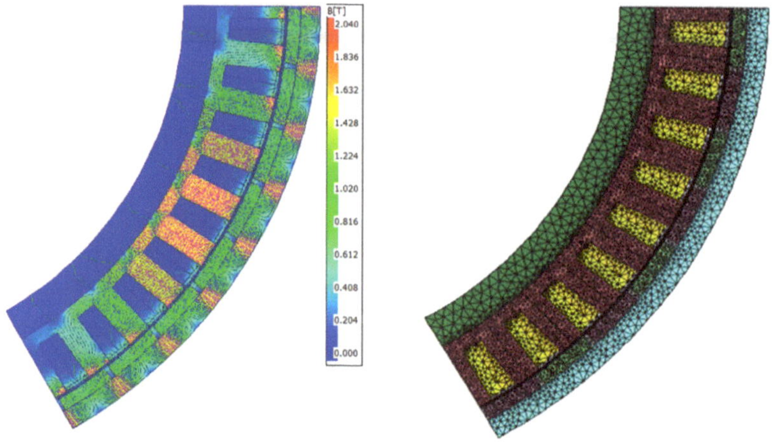

Figure 4. Calculated magnetic induction distribution from the permanent magnets.

In a traction motor, especially in a compact design with limited dimensions, it is important to calculate the power losses, because their ratio changes depending on the operating point of the electric motor.

Bertotti method has been used for the calculation of magnetic core power losses in the elaborated models [39–42]. This lets us account for hysteresis, eddy current, and excess losses:

$$\Delta P_{Fe}(t) = k_h {B_m}^2 f + \sigma \frac{b^2}{12}\left(\frac{dB}{dt}(t)\right)^2 + k_e\left(\frac{dB}{dt}(t)\right)^{\frac{3}{2}}$$

where: k_h—hysteresis factor, B_m—flux density, f—frequency, σ—conductivity, b—thickness of single electrical sheet, k_e—excess loss coefficient.

First part of the formula relates to hysteresis losses (due to the hysteresis loop of ferromagnetic material of the magnetic core). The second part of the formula covers eddy current losses (eddy currents are induced in the magnetic core). The third part relates to excess losses, caused by interaction of the external magnetic field and local magnetic fields generated by eddy currents.

In analysis, copper losses are split into constant losses and losses due to skin effect, i.e., AC winding (AC copper) losses. Copper losses are calculated on the basis of a well-known relationship:

$$\Delta P = I^2 R(v)$$

where: I—RMS-value of supply current, $R(v)$—winding resistance as a function of temperature

AC Losses on AC windings can be determined by two methods, full FEA and hybrid FEM. The full FEM method uses an accurate model where the induced eddy currents and then losses are calculated separately for each conductor. Obviously, this is the most time-consuming method. Hybrid FEM uses the flux density levels calculated by FEM for each slot, and then the losses are calculated analytically. This method is quick, but much less accurate at relatively high frequencies and relatively large conductor cross-sections. In our case, the mixed method was adopted. For the operating point with maximum speed and load, the losses were calculated using both methods, and then the correction factor was calculated according to the formula:

$$k_{\frac{Full}{Hybrid}} = \frac{\Delta P_{Cu\ Full}}{\Delta P_{Cu\ Hybrid}}$$

where AC winding losses are calculated by the Full FEA method, and AC winding losses are calculated by the Hybrid FEA method.

Since further analysis takes into account a wide load/speed range, AC winding losses are calculated with the Hybrid FEA method and each result is then multiplied by a correction factor.

The designed PMSM motor requires high torque at low rotational speeds and a wide range of rotational speeds with constant power in the area, where the DC bus voltage is the limitation. Various control strategies are used to ensure favorable motor parameters. In the area of low rotational speeds, high torque is achieved by using the maximum torque per ampere (MTPA) control strategy. The MTPA control strategy is used for minimizing winding losses. The area of a wide range of rotational speed while maintaining constant motor power is obtained using the flux weakening (FW) or maximum torque per volt (MTPV) control strategy [43–48].

Figure 5 shows the calculated motor characteristics: torque (a), mechanical power (b), efficiency (c), power losses generated in the winding (d), in the magnetic core (e) and in permanent magnets (f) in rotational speed function for seven different currents.

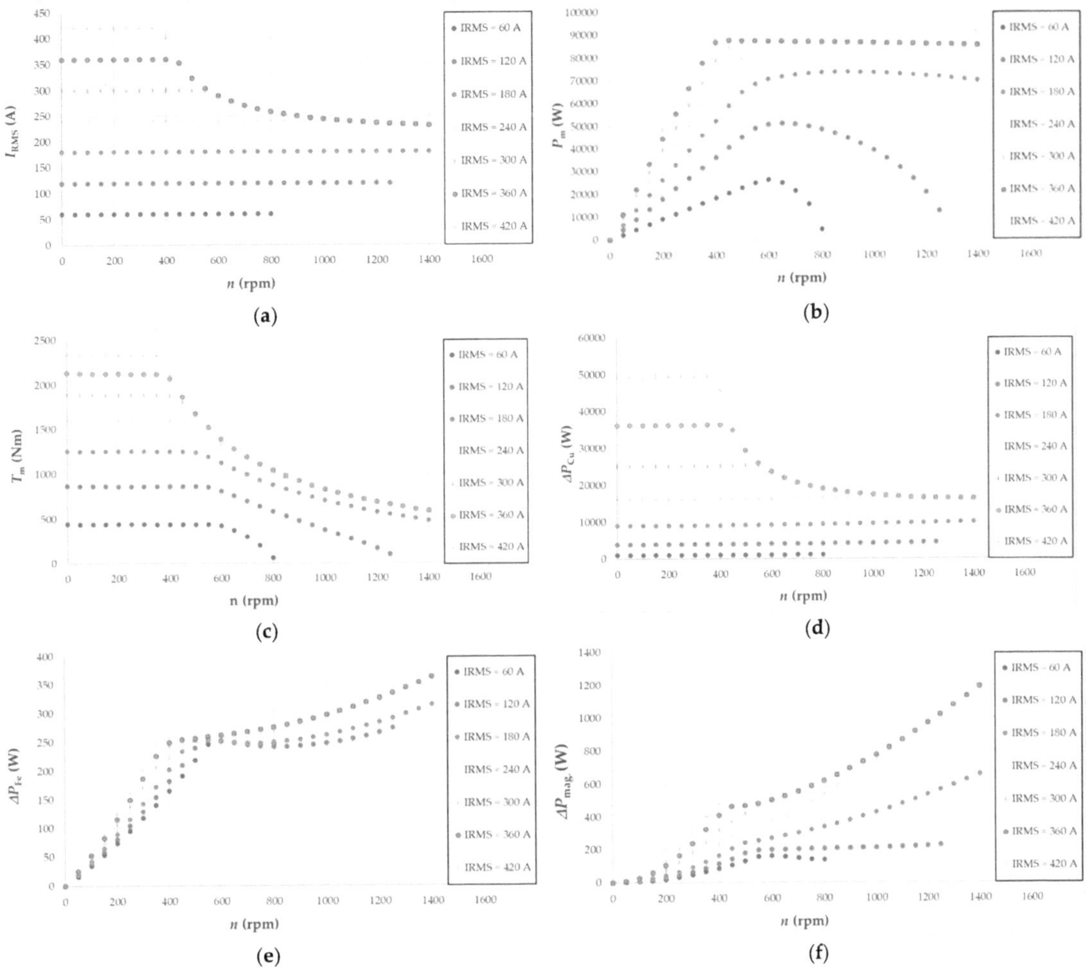

Figure 5. Calculated motor characteristics as a function of rotational speed: (**a**) supply current; (**b**) mechanical power; (**c**) torque; (**d**) power losses in the winding; (**e**) power losses in the magnetic core; (**f**) power losses in a permanent magnet.

In Figure 5a,c, it can be seen that the maximum supply current corresponds to the maximum torque only in the zone without flux weakening (MTPA control). In the zone with flux weakening, increasing the rotational speed, while maintaining the voltage limited by the DC bus, is accomplished by a space vector along the d-axis of the current, which generates a flux in the opposite direction to the flux produced by permanent magnets. In the MTPV control, the possible maximum torque is obtained at a current lower than the maximum current, which has a positive effect on the efficiency of the motor and reduces winding (Figure 5d). For currents, I_{RMS} = 240 A, I_{RMS} = 300 A, I_{RMS} = 360 A and I_{RMS} = 420 at higher rotational speed the torque and losses in the machine have the same value, because all currents are within the limits of voltage elipse and current circle defined for MTPV control (Figure 5c–f) [44]. At higher rotational speeds, the torque drops much more smoothly than in the FW control (Figure 5c). For the values of the supply currents I_{RMS} = 60 A, I_{RMS} = 120 A and I_{RMS} = 180A (Figure 5c) there is an FW control, where at a constant value of the RMS current, the share of the space vector of the current in the d axis of the motor increases, which results in an increasing weakening of the flux from permanent magnets and decreases losses in the machine. This results in an almost linear decrease of the torque, the decrease of the torque in such control, at higher rotational speeds, is much more visible than in the MTPV control. With insufficiently high RMS current values (60 A and 120 A supply current), it is impossible to achieve the maximum rotational speed and at the same time generate the appropriate torque, due to the too- high value of the current component in the d-axis in relation to the current component in the q axis of the motor.

3. Thermal Calculations

In the Ansys Motor-CAD program, a thermal motor model has been developed. Figure 6 shows a simplified equivalent thermal diagram of the motor. The thermal schema method is commonly known and commonly used in calculating electrical machines [49–51]. The program also enables the analysis of slot cross-sections on simplified FEM models in order to calibrate the network of the model.

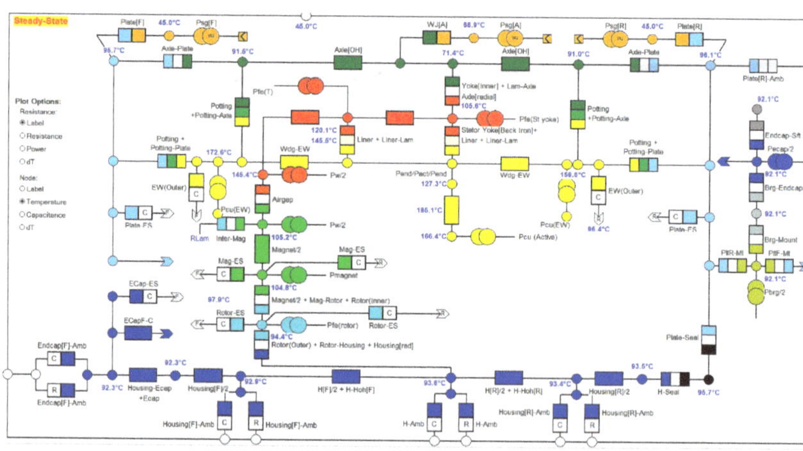

Figure 6. Simplified equivalent thermal diagram of the motor in the ANSYS Motor-CAD program.

Figure 7 shows the temperature characteristics of winding (hotspot) and magnet, calculated in the ANSYS Motr-CAD program as a function of the supply current. The calculated points of the torque correspond to the points on the calculated temperature characteristic. The calculated characteristics show that when supplied with I_{RMS} = 130 A (current density j = 19.2 A/mm^2) and the assumed operation S1, the motor produces a torque equal to T_m = 902 N m. At this operating point, the maximum temperature in the winding is set at T_{Cu} = 189 °C, while the temperature of permanent magnets is set at $T_{mag.}$ = 105 °C.

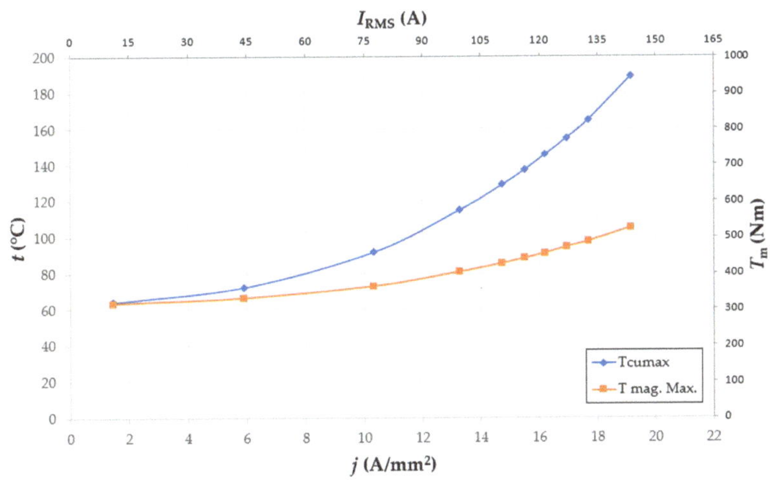

Figure 7. Calculated temperature characteristics of winding (hotspot) and magnet, calculated in the ANSYS Motr-CAD.

As the intensity and density of the current decrease, the temperature of the magnets and the winding approaches the set coolant temperature equal to $T_{cool.}$ = 60 °C. Using the insulation class allowing the winding temperature T_{Cu} = 200 °C, it can be assumed that the motor at this rotational speed works with a current density of j = 19.2 A/mm^2, while maintaining temperature safety with a margin of about 10 °C.

For this operating point, the calculated temperature distribution in individual motor components has been presented (Figure 8).

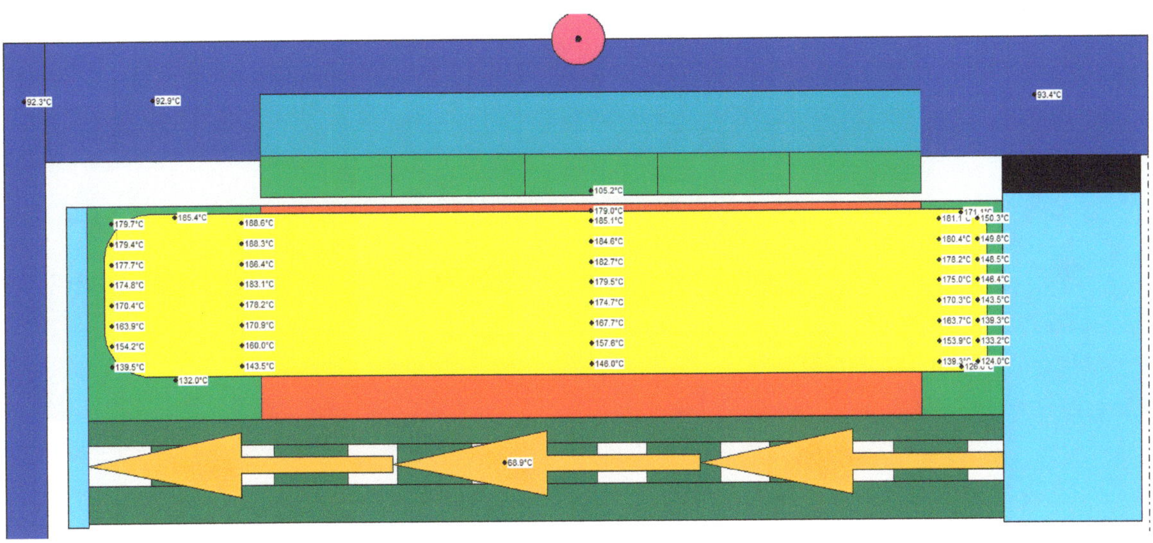

Figure 8. *Cont.*

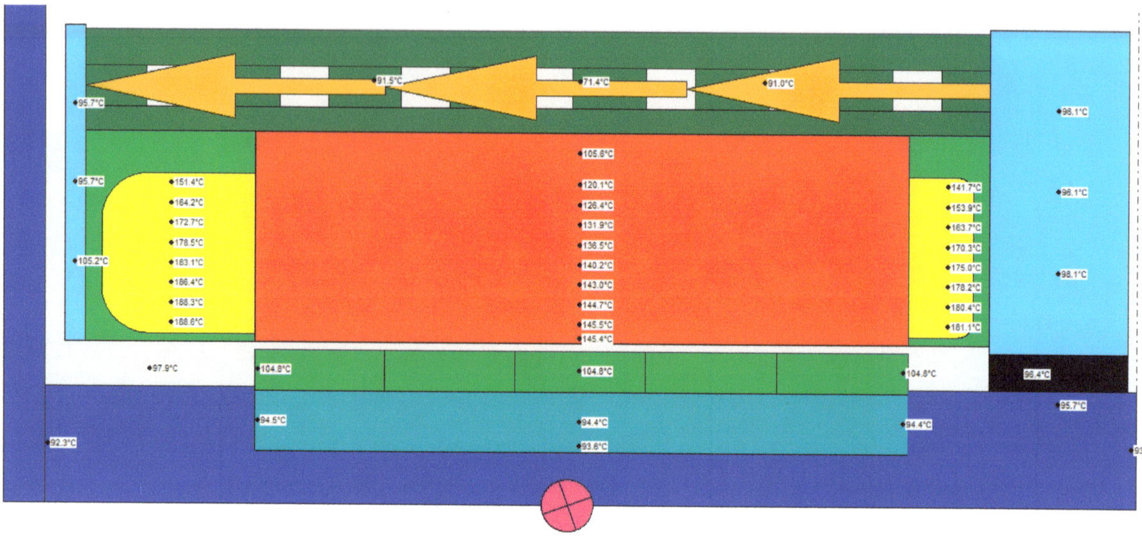

Figure 8. Calculation results of the motor temperature distribution for the operating point T_m = 902 Nm and n = 400 rpm.

The model (Figure 8) shows the calculated temperature distribution for operating point T_m = 902 Nm and n = 400 rpm.

Machine winding of the coils allows for a large filling of the slot (and a very good adjustment of the winding wire distribution in the coil). During manual winding, especially in the case where the motor is wound with a bundle of parallel wires, obtaining a large slot filling is limited, because the wires in the slot are arranged randomly, twist, and cross. It also has a negative impact on the thermal performance of the motor, because the interlacing causes the free space in the slot between the coil and the stator magnetic core, which is the path of heat dissipation from the winding to the cooling system, to be filled with air, impregnation and/or epoxy resin in places. These are spaces where thermal conductivity is much lower than copper conductivity. Figure 9 shows a cross-section of the slot model with the location of the wires in the single coil and the calculated temperature distribution in the slot for the operating point T_m = 902 Nm and speed n = 400 rpm. From the calculated temperature distribution of the winding in the slot, the maximum temperature is $T_{\text{Cu max.}}$ = 183.1 °C, mean temperature $T_{\text{Cu av.}}$ = 167.3 °C, while the minimum temperature $T_{\text{Cu min.}}$ = 133.1 °C. The calculated temperatures on the schematic model (Figure 8) are analogically $T_{\text{Cu max.}}$ = 185.4 °C, $T_{\text{Cu av.}}$ = 165.8 °C, $T_{\text{Cu min.}}$ = 127.4 °C. The results of calculations using both methods are similar with satisfactory accuracy.

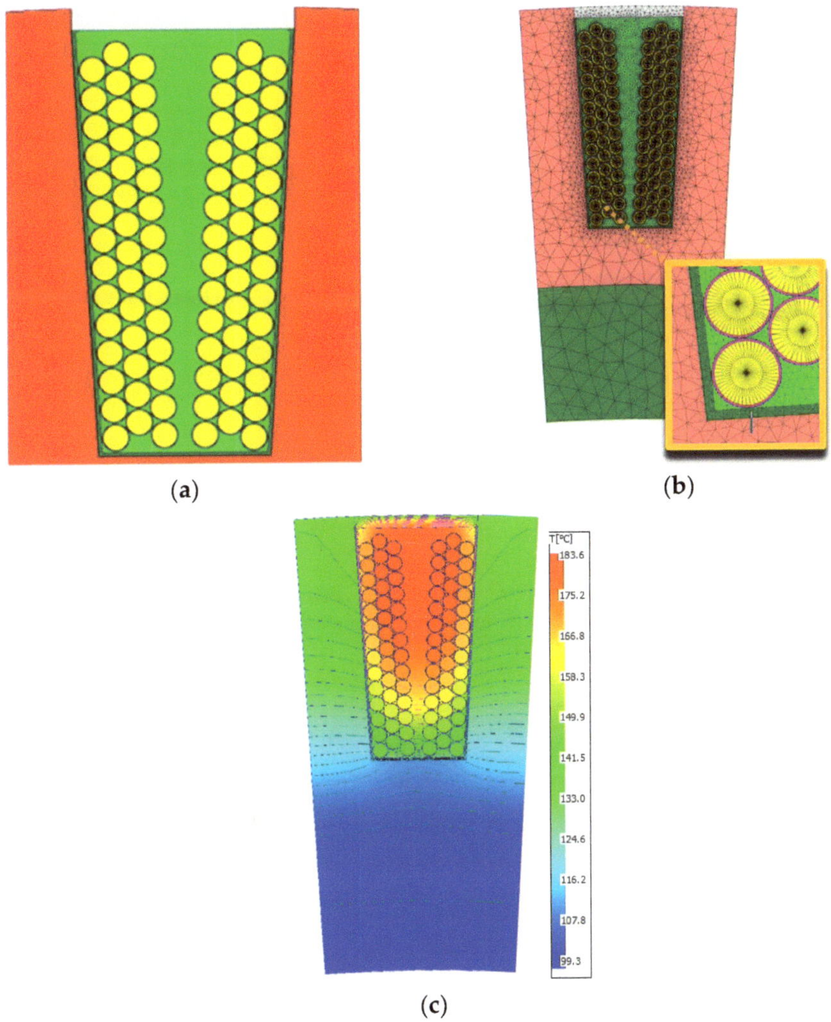

Figure 9. Longitudinal section of the circuit model of the motor with the housing: (**a**) through the stator slot; (**b**) FEM model; (**c**) simulation result (T_m = 902 Nm, n = 400 rpm).

4. Driving Cycle Simulation-Model

The Lab module of the Ansys Motor-CAD program was used to carry out the thermal analysis while driving the car, which allows for making simulations of coupled models of the electromagnetic circuit and the thermal model. During the simulation, it is possible to set the required driving cycle of the car and use a number of parameters to build a model of the vehicle. The program calculates the torque demand for the vehicle model on the basis of the vehicle speed requirements in the driving cycle. Calculations can be made taking into account the coupling with the thermal model. Such calculations take into account the influence of temperature on the operating parameters of the electromagnetic circuit.

The vehicle model is based on the analysis of the forces acting on the vehicle while driving at constant speed: rolling resistance, aerodynamic resistance, climbing force, and during acceleration and braking.

The formulas used in the model are presented below.

$$F_R = k_r mgcos\theta$$

where: F_R is rolling resistance, k_r is the rolling resistance coefficient, m is the mass of the vehicle, and g acceleration due to gravity;

$$F_D = \frac{1}{2}\rho(v + v_0)^2 C_d A_f$$

where: F_D is aerodynamic resistance, ρ is the air density, v is the velocity of the vehicle, v_0 the headwind velocity, C_d the vehicle drag coefficient, and A_f the vehicle's affective frontal area.

F_C is climbing force, calculated from the formula:

$$F_c = mgsin\theta$$

where: θ is the gradient of the slope.

F_T is the total force required for constant vehicle motion:

$$F_T = mg(k_r cos\theta + sin\theta) + \frac{1}{2}\rho(v + v_0)^2 C_d A_f$$

F_a is acceleration, calculated from the formula:

$$F_a = m.a.\delta$$

where: δ is the mass correction factor.

The torque is calculated from the formula:

$$T_{motor} = \frac{(F_a + F_T).r_w}{n_d}$$

where: r_ω is the wheel radius and n_d the final drive ratio.

The driving cycles are described as a set of points on the vehicle speed versus time characteristics. They are determined by tests carried out on real vehicles. The driving cycles are mainly used to assess the fuel consumption and emissions of a vehicle according to the standards for different types of vehicles with different drives. The driving cycle is mapped to a chassis dynamometer where tailpipe emissions are collected and analyzed to evaluate the emission factors.

Another test is the motor dynamometer test, where the evaluation is made on the basis of a set of torque and speed points.

The difference between modal and transient cycles is that modal cycles are a compilation of acceleration and constant velocity periods. They cannot represent the actual behavior of the driver. Transition cycles, on the other hand, include many speed changes typical of driving conditions on roads with average driving dynamics. Such cycles can be considered in the motor thermal calculations because they represent the average driver's ride.

Artemis cycles were used to perform a thermal analysis of the motor [52,53]. The cycle reflects driving a car around the city, taking into account the dynamics characteristic of an average driver.

Table 2 presents a summary of the main parameters of selected Artemis driving cycles: driving time, distance traveled, average speed, maximum speed, percentage of stoppage, vehicle speed up to v = 50 km/h, medium speed up to v = 90 km/h, high speed above v = 90 km/h The analysis shows that the range of vehicle speed up to v = 90 km/h is exceeded in the Artemis Road and Artemis Motorway cycle. In the Artemis Road cycle, the speed ranging from v = 90 km/h to v = 111.1 km/h represents a 2% share of the cycle. This is a speed range that is outside the intended speed range for a pure electric vehicle. On motorway cycles, the speed range above that required for electric drive is 70–71%.

Table 2. Artemis driving cycle parameters.

Parameter	Urban	Road	Motorway130	Motorway150
Driving time (s)	993	1082	1068	1068
Distance (km)	4.874	17.275	28.737	29.547
Average speed (km/h)	17.7	57.5	96.9	99.6
Maximum speed (km/h)	57.3	111.1	131.4	150.4
Percentage of the speed range over the cycle (%)				
Layover ($v = 0$ km/h)	21	2	1	1
Low speed ($0 < v \leq 50$ km/h)	77	32	15	14
Medium speed ($50 < S \leq 90$ km/h)	2	59	14	14
High speed ($v > 90$ km/h)	0	7	70	71

In the Artemis Road cycle, the speed range to $v = 90$ km/h covers 93% of the entire driving cycle, while the Artemis Urban cycle does not exceed the speed above $v = 57.3$ km/h.

Table 3 presents the vehicle parameters adopted for the calculations.

Table 3. Artemis driving cycle parameters.

Parameter	Value	Unit
Vehicle weight	3.5	t
Frontal area of the vehicle	4.4	m^2
Wheel radius	0.35	m
Vehicle rolling resistance	0.02	-
Vehicle drag coefficient	0.31	-
Mass correction factor	1.035	-
Air density	1.225	kg/m^3
Gear ratio	1	-
Motoring torque ratio	0.5	-
Generating torque ratio	0.25	-

The assumed vehicle weight is the maximum laden weight for the selected Fiat Ducato model. The frontal area and the wheel radius were determined by the characteristic dimensions of the car. Mass Correction Factor or rotational inertia factor that compensates for the increase in mass of the vehicle due to its rotating mass on board. The gear ratio is 1 because we are considering direct drive. The Motoring Torque Ratio is equal to 0.5 because we assume that each of the two motors produces half the torque needed.

Generating Torque Ratio is equal to 0.25 (in total, both motors have half the necessary braking torque in the cycle).

Figure 10 shows the vehicle speed for the three driving cycles considered in the further analysis: Artemis Urban, Artemis Motorway150, and Artemis Road.

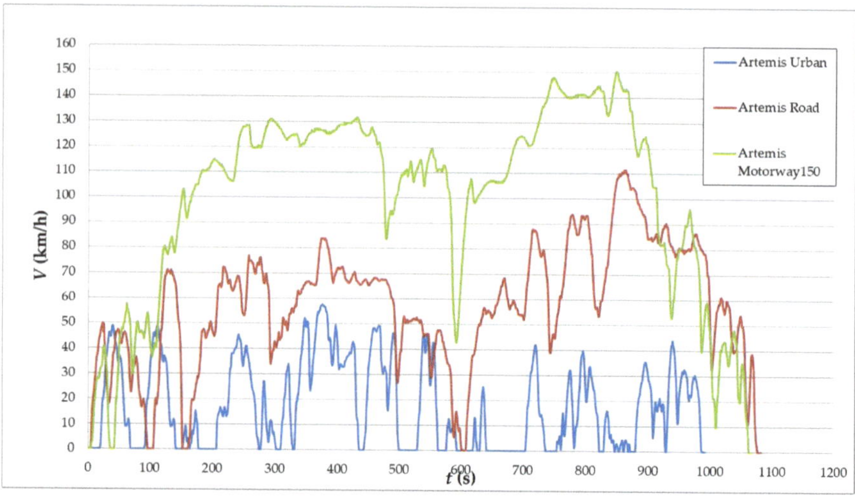

Figure 10. Vehicle speed waveforms for Artemis driving cycles.

The torque curve for the considered driving cycles was determined by the simulation method. Figure 11 shows the required waveforms of torque as a function of motor rotational speed, resulting from the demand at different vehicle speeds during the driving cycle.

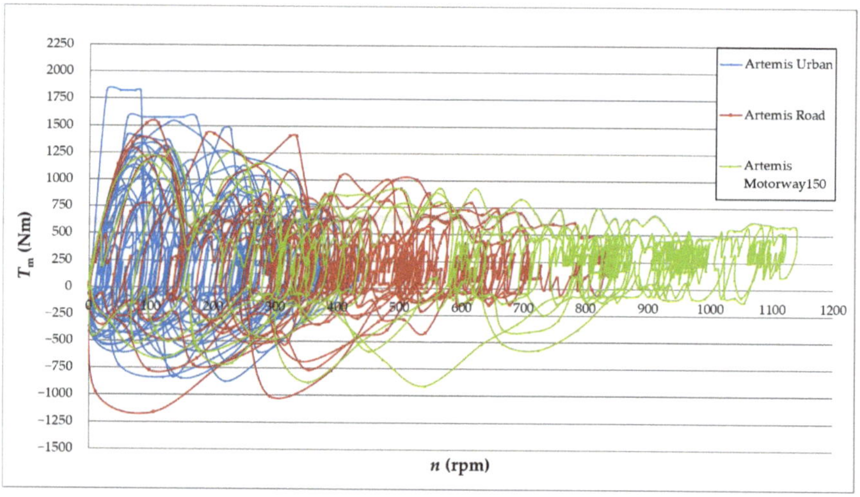

Figure 11. Curves showing the mixing of the motor operating point defined by torque and rotational speed during the Artemis Urban, Artemis Road, and Artemis Motorway150 driving cycle.

Then, the envelope of maximum torque as a function of rotational speed was determined (Figure 12). The maximum torques at various rotational speeds are the torques that the motor must generate to obtain the required acceleration in a given driving cycle

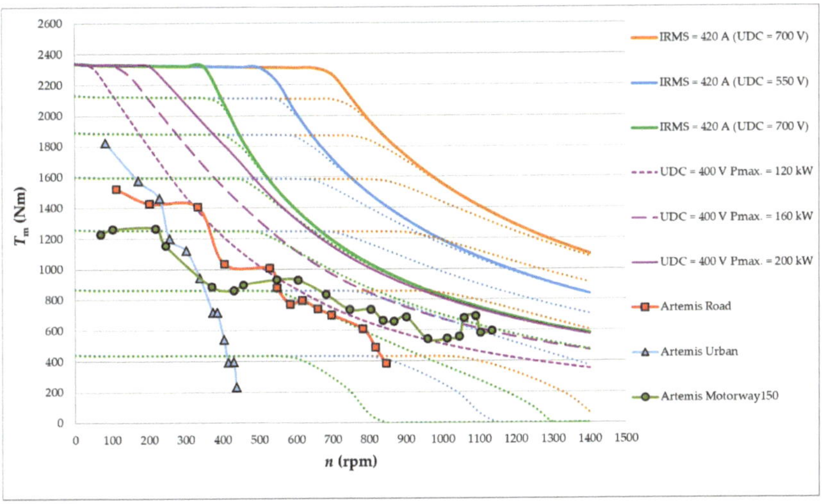

Figure 12. Calculated torque characteristics of the motor for three different supply voltages of the U_{DC} drive with the three limits of the drive input power and the maximum torque characteristics for the given rotational speed of the Artemis Urban cycle, Artemis Road and Artemis Motorway150.

Figure 12 provides the basis for the further analysis of motor performance conducted in this article. The figure shows the family of motor characteristics, calculated for the battery voltage U_N = 400 V (green color), analogous to the characteristics shown in Figure 5. The characteristics are marked with points representing the highest torque of individual driving cycles. The highest torque of approx. T_m = 1800 Nm is required for the Artemis Urban cycle at low speeds of up to approx. n = 100 rpm. The entire urban cycle is confined to the first torque control zone.

The Artemis Road driving cycle is characterized by a lower required maximum torque of approx. T_m = 1500 Nm at a rotational speed of up to approx. n = 100 rpm. At the highest motor speeds/vehicle speeds, the motor runs in the flux weakening zone.

In the Artemis Motorway150 cycle, compared to the rest of the cycles, the maximum torque required is lowest in the first control zone, while it is highest in the second control zone, and covers the range of the highest rotational speed.

Figure 12 also shows the analogous torque characteristics of the motor for higher supply voltages U_{DC} = 550 V (blue) and U_{DC} = 700 V (orange).

Increasing the supply voltage reduces the motor operating range in the flux-weakening region.

Additionally, in Figure 12 the lines limiting the operating range of the motor powered by the voltage U_{DC} = 400 V, the available battery power have been presented in purple. The maximum battery power should be selected so as not to limit the required vehicle parameters. The choice of battery power should take into account all the car components it supplies. The graph takes the battery power minus the power needed to power the rest of the components besides the motor. Power limits were presented at the level of P_{IN} = 120 kW, P_{IN} = 160 kW, P_{IN} = 200 kW. It is the power to power up both motors, i.e., if the battery is limited to P_{IN} = 120 kW, a single motor is limited by the power supply to 60 kW.

From the waveforms presented in Figure 12, it can be concluded that for the limitation of the supply power to P_{IN} = 120 kW, the urban driving cycle can be achieved, while for the on-road cycle, the power supply is insufficient at two points, which means that the assumed driving dynamics is not maintained. For such a power limitation, the Highway cycle is even less feasible. Similarly, increasing the allowable power supply of motors to P_{IN} = 160 kW will enable the achievement of the required torques for the road cycle, and increasing the power to P_{IN} = 200 kW will also cover the motorway cycle, although with a small reserve of power at the highest rotational speeds.

The following graphs, shown in Figure 13 the calculation results for the three voltages under consideration. The calculated characteristics relate to the calculations for a constant torque T_m = 750 Nm in the speed range from n = 0 rpm to n = 1400 rpm.

Figure 13a shows the torque characteristics as a function of rotational speed, calculated for the constant temperature of the winding and permanent magnets. For the voltage U_{DC} = 400 V, the torque T_m = 750 Nm cannot be achieved in the full range of rotational speeds (the torque is achieved up to a speed of n = 1050 rpm). For the voltage U_{DC} = 550 V and U_{DC} = 700 V, the considered torque is achieved in the entire rotational speed range, up to n = 1400 rpm. Figure 13b shows the I_{RMS} current waveform for various U_{DC} supply voltages. The figure shows at what rotational speed, depending on the supply voltage, it is necessary to flux weakening the motor to be able to work with the torque T_m = 750 Nm (the current value increases). Similarly, the power losses in the winding increase in Figure 13c. Figure 13d shows the calculated power losses in the stator magnetic core. While the losses in the winding increase with the increase of the supply current, the losses in the magnetic core decrease. This is due to the flux weakening process, which lowers the value of the magnetic induction in the magnetic core, thus lowering the value of power losses. Figure 13e–g shows the power losses in permanent magnets. As in the case of the losses in the stator magnetic core, also in magnets, flux weakening lowers the losses in the magnets, while the effect is limited and decreases with increasing flux weakening and current intensity, which is visible when comparing the loss waveorms for two different to request. Figure 13f,g show that for a torque smaller than T_m = 550 Nm and $T_{m'}$ = 300 Nm, the rotational speed range in which the power losses generated in the magnets are lower when powered by a lower voltage (U_{DC} = 400 V) is wider. It can be concluded that the use of a higher supply voltage will significantly reduce the power losses in the motor winding during operation in the flux weakening region, it will also reduce the power losses in the magnetic core of the motor (also only in the flux weakening region), while in the case of losses in permanent magnets, a reduction in these losses can also be expected. It should be noted that the analysis concerns the calculation of the cases of all three voltages for the same operating temperatures (T_{Cu} = 113 °C and $T_{mag.}$ = 84 °C). The losses generated in the magnets depend on the excitation, which in turn depends on the operating temperature of the magnets, which is influenced by the temperature of other elements, mainly the winding and the magnetic core.

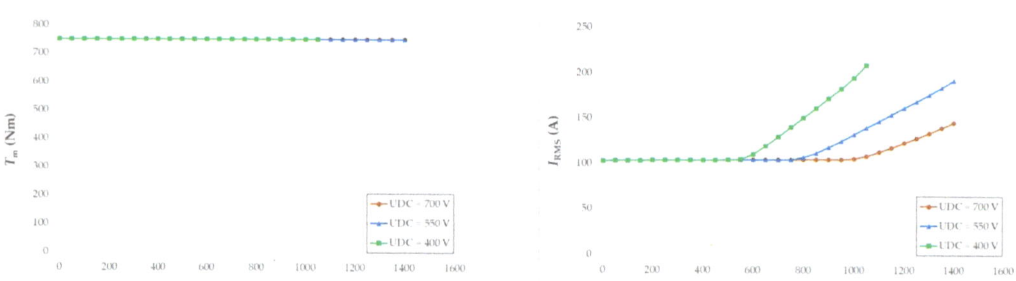

(a) (b)

Figure 13. *Cont.*

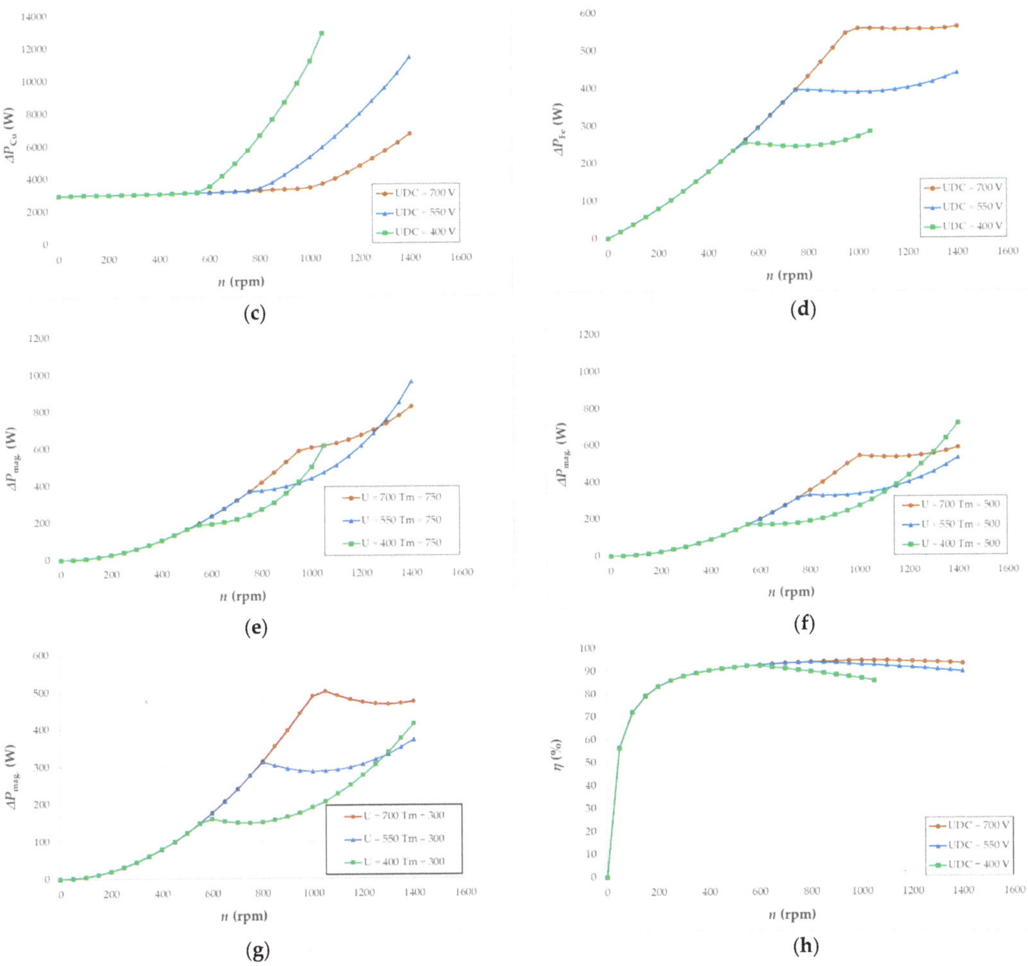

Figure 13. Calculated characteristics of motor operating parameters for three drive supply voltages as a function of rotational speed and torque T_m = 750 Nm: (**a**) torque; (**b**) supply current; (**c**) power losses in the winding; (**d**) losses in the stator magnetic core; (**e**) power losses in magnets; (**f**) power losses in magnets for Tm = 550 Nm; (**g**) power losses in magnets for T_m = 300 Nm; (**h**) motor efficiency.

Figure 13h shows the calculated performance characteristics for the same conditions as the graphs discussed. The characteristics show a large impact on the efficiency of the motor of limited voltage and the associated earlier rise of the I_{RMS} current, which is the necessity to start flux weakening at a lower rotational speed, which results in a faster increase of losses in the winding. The ratio of the winding losses to the losses generated in the magnetic core and in the permanent magnets shows that these losses determine the efficiency of the machine.

Figure 14 shows the maximum torque and power limitations, but for two different voltages, for U_{DC} = 400 V and 700 V as in Figure 12 The comparison of the operating range of the motor with different voltages is justified due to the differences in efficiency depending on the supply voltage. The comparison of the characteristics shows that with a higher supply voltage and the same power limitation, the possible range of motor operation is higher, which results in the torque reserve increasing with the increase in the permissible motor supply power. This is clearly seen in the points of the required maximum torque at the high rpm of the Artemis Motorway150 drive cycle. With the supply power limited to P_{IN} = 200 kW and the supply voltage U_{DC} = 400 V, the maximum required torque at n = 1050 rpm is barely achieved. With a voltage increased to U_{DC} = 700 V it is achieved with a torque reserve.

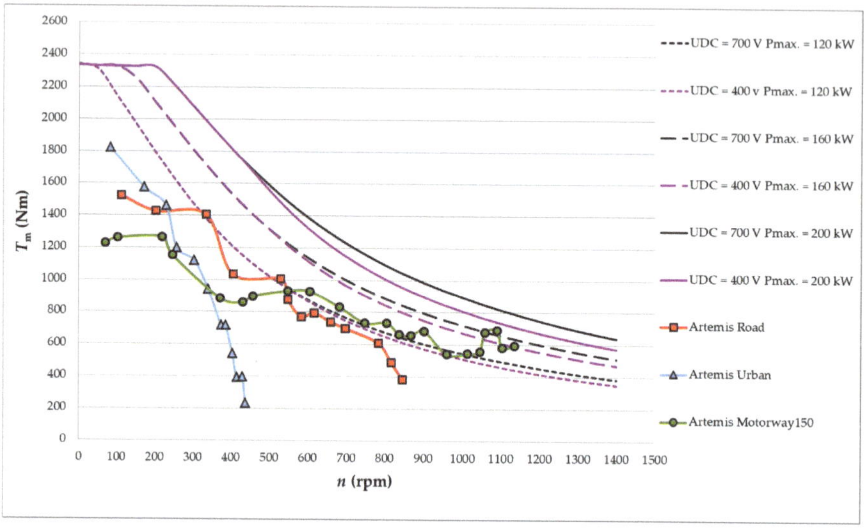

Figure 14. Maximum torque waveforms for different motor speeds in Artemis Urban, Artemis Road, and Artemis Motorway150 Cycle and plotted torque characteristics with drive input power limits for two different drive supply voltages.

The test results show the dependence of the motor operating parameters on the selected power supply parameters in various driving cycles. In the next stage of consideration, the temperature aspects, which are influenced by the presented dependencies, will be taken into account.

5. Driving Cycle Simulation-Artemis Urban Cycle

The first driving cycle under consideration is the Artemis Urban cycle, representing an average city driving. In the case of the Fiat Ducatto vehicle, this driving cycle covers its main area of work, i.e., the delivery vehicle. This sample operating range represents driving in urban conditions, such as are usually found in green zones, where only electric cars can drive.

Figure 15 shows the vehicle speed waveform for the Artemis Urban cycle. The waveform consists of certain stages that represent different driving scenarios that affect the driving dynamics of the average driver.

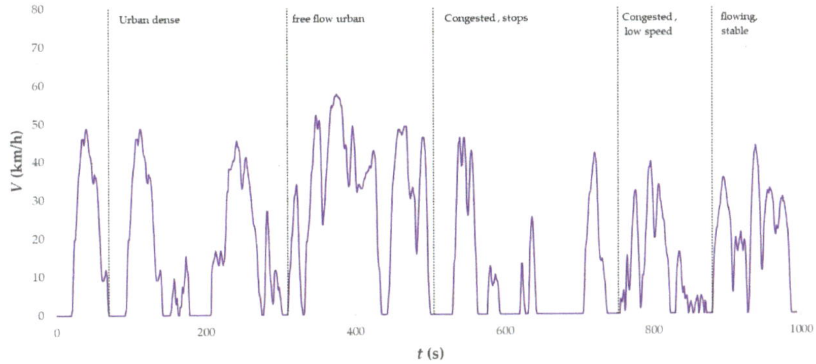

Figure 15. Vehicle speed waveform of the Artemis Urban cycle.

The first is driving in urban congestion, then driving in a relatively low urban congestion, the third is partial stops due to congested traffic then low speed in congested traffic, and free driving at a relatively constant speed. The maximum vehicle speed equal to v = 57.3 km/h occurs at the stage of driving with a relatively low density of vehicles. The speed above v = 50 km/h is 2% of the entire waveform and occurs only at this stage of the cycle. During the entire driving cycle, stops constitute 21% of the mileage, while 77% is the vehicle's movement at a speed of up to v = 50 km/h.

Figure 16 shows the calculated waveform of the required torque, which allows for the completion of the driving cycle of the considered motor, taking into account the vehicle model, operating parameters and temperature of individual motor components. The negative torque is the torque produced by the motor assuming that the motor is providing 25% of the required braking torque. The maximum braking torque is $T_{m\_generating}$ = 872 Nm.

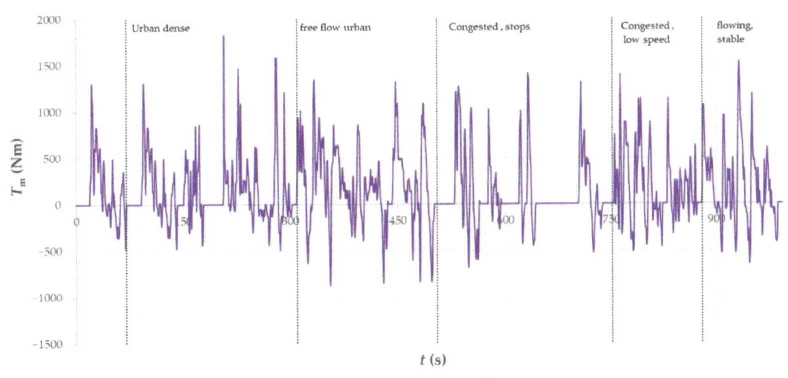

Figure 16. Vehicle torque waveform of the Artemis Urban cycle.

Figure 17 shows the calculated waveform of losses generated in the winding. The highest losses reach temporarily the value of approx. ΔP_{Cu} = 18 kW in urban dense. Figures 18 and 19 show the waveform of power losses in the stator magnetic core and in permanent magnets. Losses in the core and in magnets are very similar in nature to the waveform of vehicle speed (Figure 15). This is directly related to the dependence of these losses on the motor operating frequency. The highest losses can be seen in the free flow urban stage, where the car drives at the highest speed and usually only slows down to a slightly lower speed.

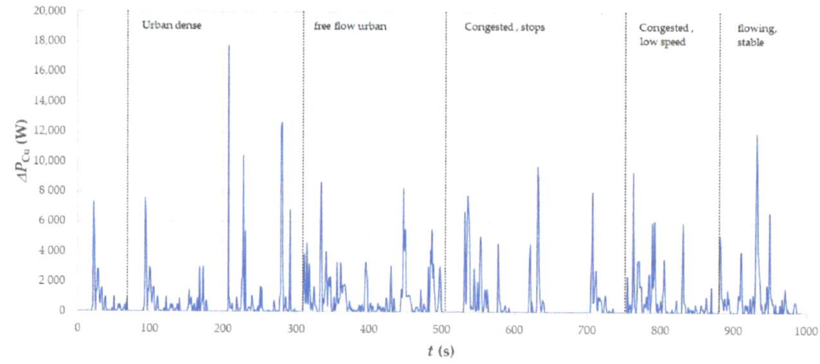

Figure 17. Calculated waveforms of loss in winding of the motor for the Artemis Urban cycle.

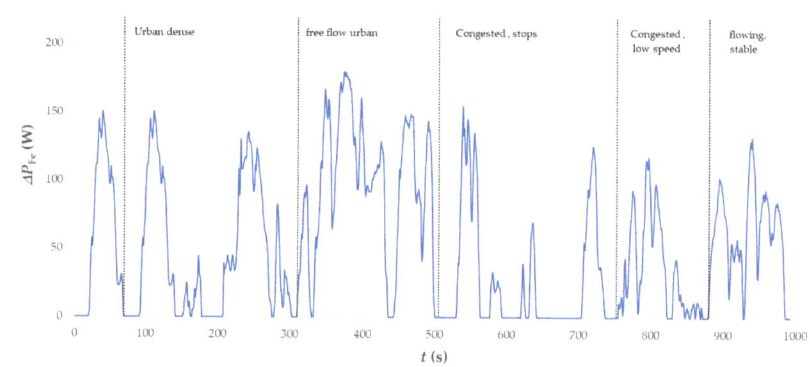

Figure 18. The calculated waveform of loss in the stator core of the motor for the Artemis Urban cycle.

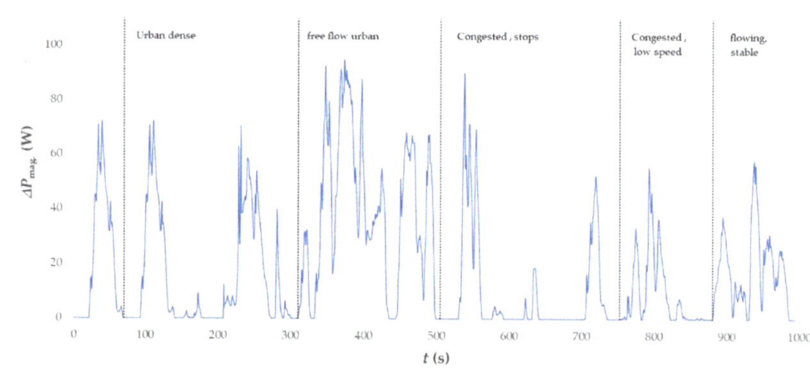

Figure 19. The calculated waveform of magnets loss in the motor for the Artemis Urban cycle.

Figure 20 shows the calculated temperature curves at the maximum point of the winding (hotspot), in permanent magnets, and in the stator yoke. The maximum temperature in the motor winding reached safe values at the level of T_{Cu} = 85 °C, as was the temperature in $T_{mag.}$ permanent magnets. = 54 °C. The stator yoke temperature has peaked at T_{stator_yoke} = 68 °C. The maximum temperature of the winding is important due to the insulation class of the winding wire and the materials to which it adheres, i.e., slot insulation and epoxy resin, in which the winding is potted. The operating temperatures of the insulation, depending on the class, may vary from tz. = 150 °C to tiz. = 200 °C. The epoxy resins used for potting the winding are also characterized by permissible operating temperatures of the order of tepoxy = 180 °C. The working temperature of permanent magnets is normally from $T_{mag.}$ = 150 °C to approx. $T_{mag.}$ = 220 °C, assuming that they do not work in an external field that would weaken their work. During flux weakening, the flux from the magnet is weakened, which results in a lowering of the permissible operating temperature. In the case of an urban cycle, the considered motor does not work in the second control zone, so the magnets are not additionally exposed to work in an external, oppositely directed magnetic field.

The temperature of the stator yoke can be important from a technological point of view.

The design of the motor assumes that the magnetic core of the stator is placed on an aluminum support structure with a cooling system. It should be analyzed whether the thermal conditions of the motor do not cause a significant temperature difference between these elements, as this may lead to a significant deterioration of the contact of the elements and deterioration of the efficiency of heat transfer from the stator to the cooling system. In the case under consideration, the temperature of the yoke is only slightly higher than the temperature of the coolant, and the temperature difference between the package and the coolant ($T_{coll.}$ = 60 °C) does not exceed the safe value of ΔT = 10 °K throughout the entire course.

The temperature of the stator yoke can be important from a technological point of view. The design of the motor assumes that the magnetic core of the stator is placed on an aluminum support structure with a cooling system. It should be analyzed whether the thermal conditions of the motor do not cause a significant temperature difference between these elements, as this may lead to a significant deterioration of the contact of the elements and deterioration of the efficiency of heat transfer from the stator to the cooling system. In this case, the temperature of the yoke is only slightly higher than the temperature of the coolant, the temperature difference between the packet and the coolant (tcoll. = 60 °C) does not exceed the safe value of Δt = 10 °K throughout the entire waveform.

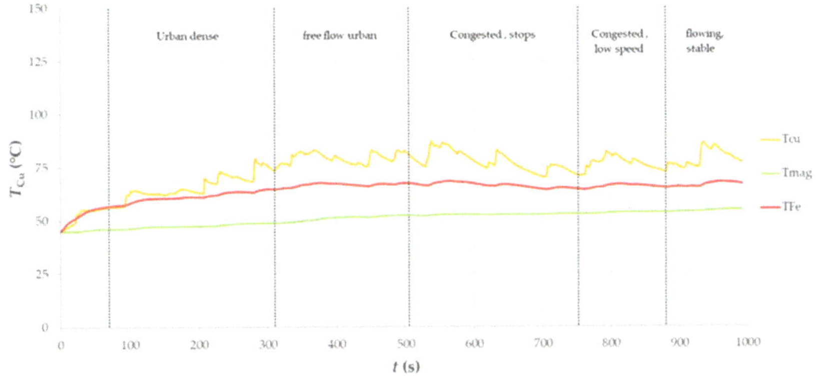

Figure 20. Calculated temperature curves in the winding (hot spot), in permanent magnets, and in the stator yoke of the motor for the considered vehicle in the Artemis Urban driving cycle.

6. Driving Cycle Simulation—Artemis Road

Another cycle under consideration is the Artemis Road cycle. Figures 21 and 22 show successively the waveform of the vehicle speed and the torque generated by the motors, necessary for its implementation with the assumed vehicle parameters. The cycle consists of successive stages divided into two parts. The first part of the stages represents secondary rural roads: urban roads before joining the traffic, driving at an inconsistent speed, driving at a constant speed. The second part of the stage represents the main roads: driving at inconsistent speed, driving at a constant speed, and urban conditions. This driving cycle is dominated by the average speed range of the vehicle (59%) from v = 50 km/h to v = 90 km/h.

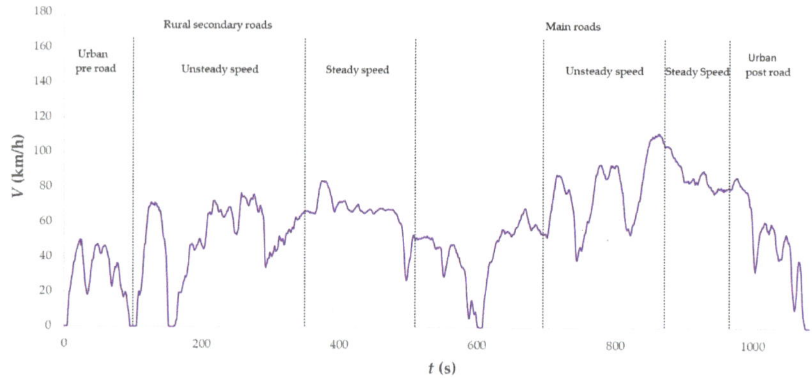

Figure 21. Vehicle speed waveform of the Artemis Road cycle.

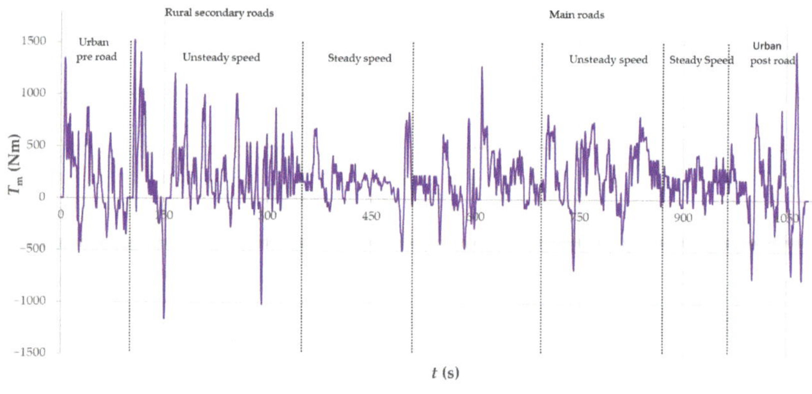

Figure 22. Vehicle torque waveform of the Artemis Road cycle.

In this driving cycle, the range of the motor's working area includes to some extent the second control zone. This is clearly visible in the diagram of the supply current (Figure 23) and the power losses generated in the winding (Figure 24) and the power losses generated in the stator magnetic core (Figure 25). In the main roads' unsteady speed and steady speed stages, the car moves at the highest speed within the work cycle. The torque required in these stages, at this speed, requires flux weakening, which causes an increase in the demand for I_{RMS} supply current at the supply voltage U_{DC} = 400 V and an increase in winding losses (Figure 24). For the voltage U_{DC} = 700 V and U_{DC} = 550 V, the current demand is almost the same. The effect of flux weakening is visible in the loss characteristics in the magnetic core (Figure 25) and in the permanent magnets (Figure 26).

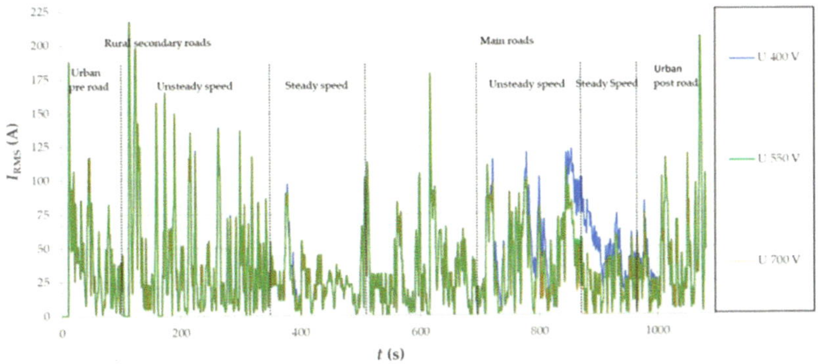

Figure 23. Calculated I_{RMS} supply current waveforms of the motor for the considered vehicle in the Artemis Road driving cycle.

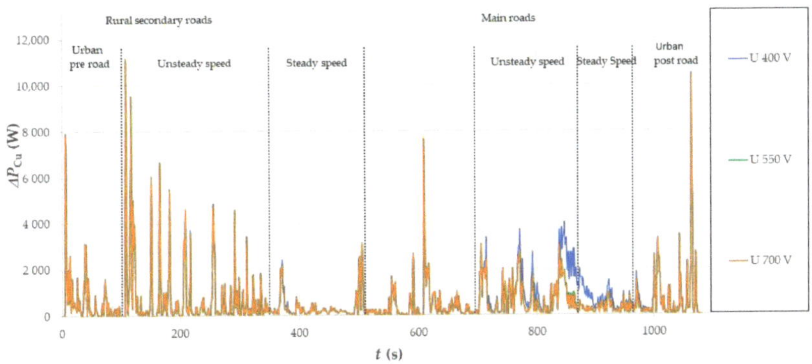

Figure 24. Calculated loss waveforms in the winding for the considered vehicle in the Artemis Road driving cycle.

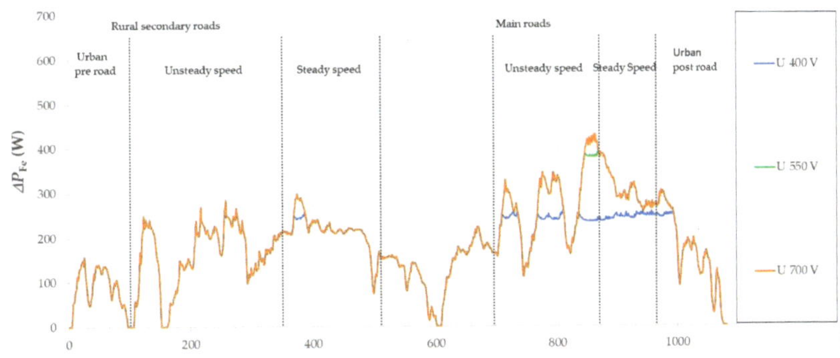

Figure 25. Calculated loss waveforms in the magnetic core of the motor for the considered vehicle in the Artemis Road driving cycle.

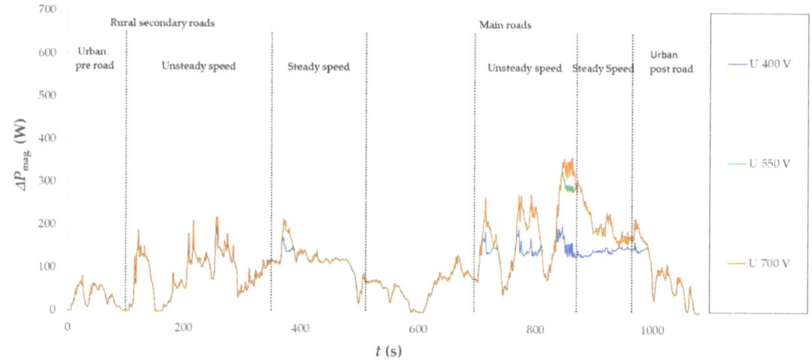

Figure 26. Calculated loss waveforms in magnets of the motor for the considered vehicle in the Artemis Road driving cycle.

Figures 27–29 show the temperature waveform in the winding (hotspot), in the permanent magnet and in the stator yoke. The increase in the supply current when the motor is supplied with the lowest voltage caused a visible increase in the winding temperature, but it is still at a safe level, below T_{Cu} = 100 °C. For the same supply voltage, the temperature of the magnets is slightly lower due to the reduced power losses by flux weakning. Despite the reduction of the power loss in the stator magnetic core, the temperature in the stator yoke increased slightly. This is due to an increase in the winding temperature, which has a significant effect on the temperature of the magnetic core, through which it gives off heat to the cooling system. For all power supply cases, the calculated winding, permanent magnet, and yoke temperatures are safe temperatures for the motor operation. The temperature in the magnet does not exceed the $T_{mag\ value\cdot}$ = 70 °C. The differences in the operating parameters between the power supply U_{DC} = 550 V and U_{DC} = 700 V are negligible, because at the higher voltage the motor does not work in the second control zone, while at the lower voltage it works to a small extent.

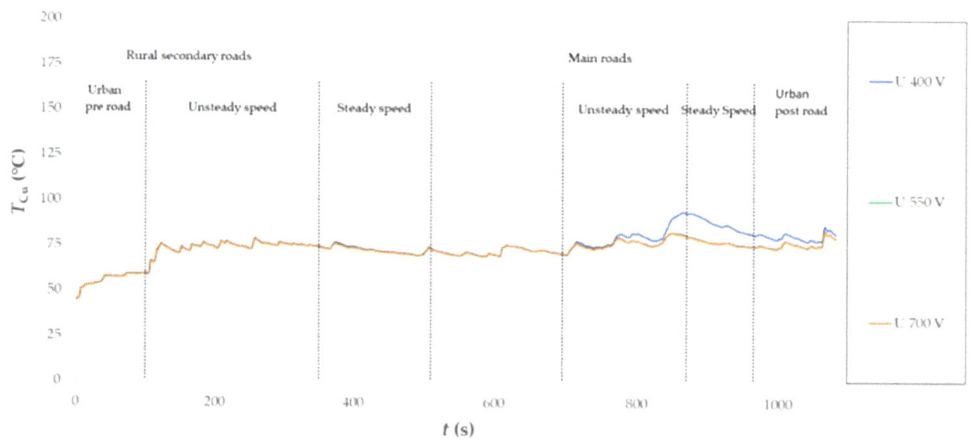

Figure 27. Calculated temperature waveforms in the motor winding (hot spot) for the considered vehicle in the Artemis Road driving cycle.

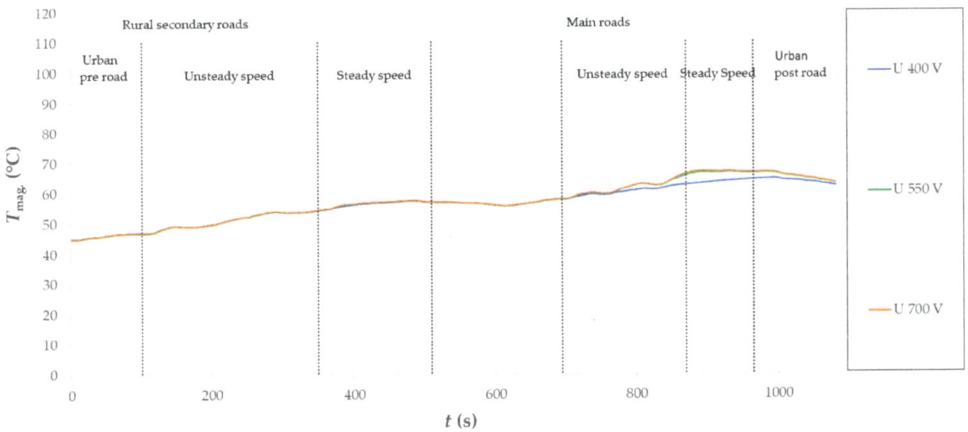

Figure 28. Calculated temperature waveforms in permanent magnets of the motor for the considered vehicle in the Artemis Road driving cycle.

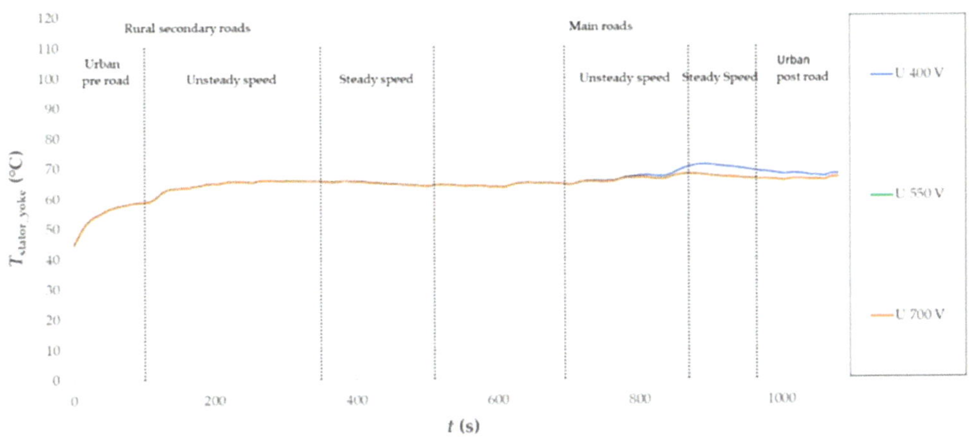

Figure 29. Calculated waveforms of temperatures in the stator yoke for the considered vehicle on the Artemis Road drive cycle.

7. Driving Cycle Simulation—Artemis Motorway150

The last cycle of work under consideration is Artemis Motorway150. The cycle represents motorway driving up to v = 150 km/h. The cycle is divided into stages before entering the motorway: urban and road, driving on the motorway: with a constant speed of up to approx. v = 130 km/h, driving at a constant speed up to v = 150 km/h and with a variable when leaving the road to urban.

The speed and torque requirements (Figures 30 and 31) make the motor work in the second control zone through all stages of driving on the highway and, similarly to the previous cycle, also to a certain extent on the road. Similar to the results presented earlier, in Figures 32 and 33, waveforms of currents and losses in the motor winding are presented, for supply with three different voltages.

The IRMS waveform shows a large difference in power demand when driving at higher speeds, depending on the supply voltage U_{DC}. The work in the second control zone is clearly visible. Similarly, as the demand for IRMS supply current increases and the supply voltage U_{DC} decreases, the losses in the winding increase, while the losses in the magnetic core and in the magnets are reduced. The differences in the considered losses are relatively large. Between the version with supply U_{DC} = 700 V and U_{DC} = 400 V, the greatest difference in winding losses was ΔP_{Cu} = 7 kW in favor of the higher supply voltage. For the same voltages, the maximum difference in the losses in permanent magnets is approx. $\Delta P_{mag.}$ = 350 W (Figure 34), and in the magnetic core of the stator ΔP_{Fe} = 350 W in favor of a lower supply voltage (Figure 35). As could be expected, the benefit of flux weakening in the context of the power or efficiency balance does not cover the losses in the winding, which is clearly visible when comparing the loss curves.

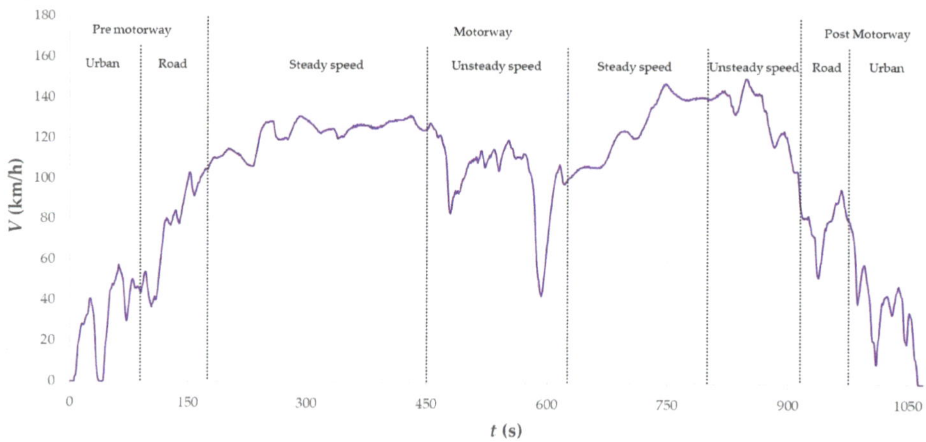

Figure 30. Vehicle speed waveform of the Artemis Motorway150 cycle.

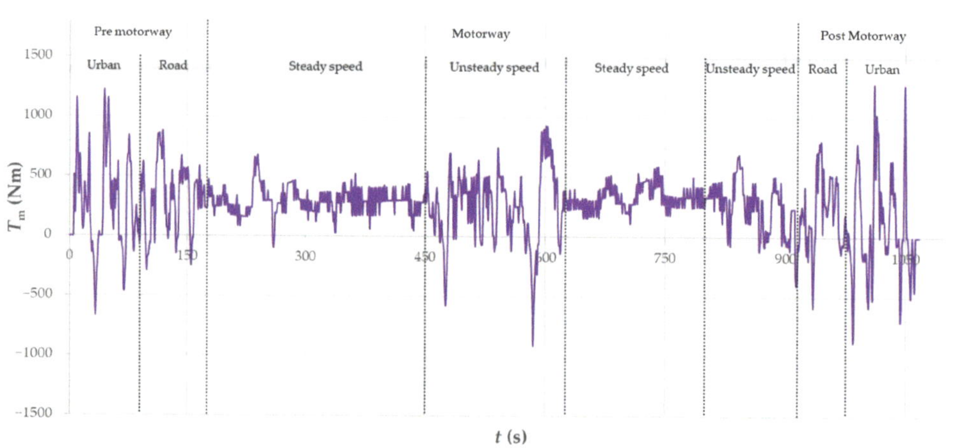

Figure 31. Calculated vehicle torque waveform of the Artemis Motorway150 cycle.

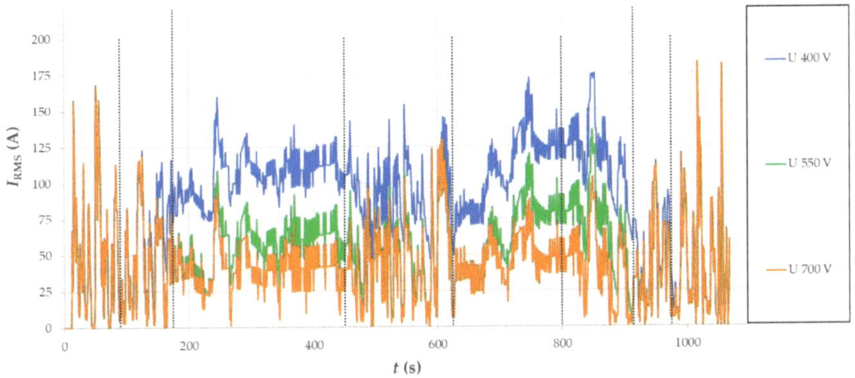

Figure 32. Calculated I_{RMS} supply current waveforms of the motor for the considered vehicle in the Artemis Motorway150 cycle.

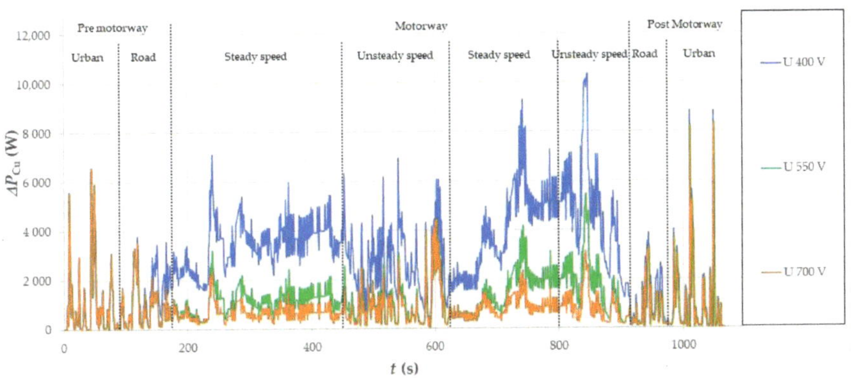

Figure 33. Calculated loss waveforms in motor windings for the considered vehicle in the Artemis Motorway150 cycle.

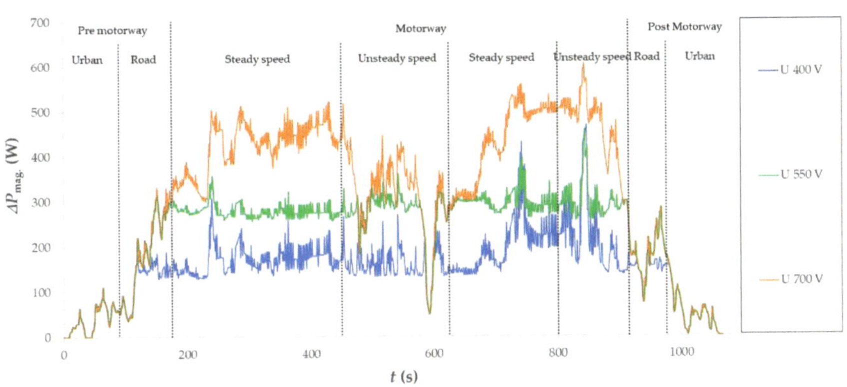

Figure 34. Calculated loss waveforms in magnets of the motor for the considered vehicle in the Artemis Motorway150 cycle.

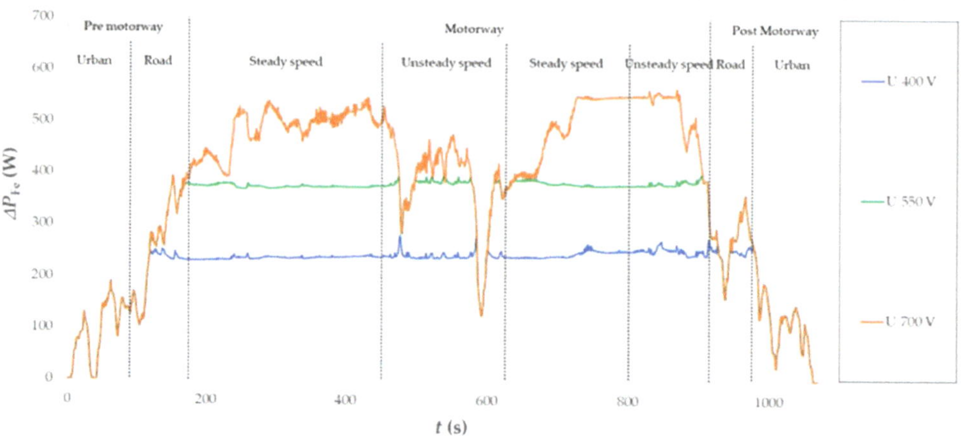

Figure 35. Calculated loss waveforms in the stator core of the motor for the considered vehicle in the Artemis Motorway150 cycle.

Figure 36 shows a comparison of the hot spot temperature curves in the winding for three supply voltages. When supplied with the voltage U_{DC} = 400 V, the maximum operating temperature of the winding reached approx. T_{Cu} = 185 °C. This temperature is relatively high, close to or exceeding (depending on the materials used) the permissible operating temperature of the insulation and epoxy resin. The high temperature of the winding also influences the temperature of the magnetic core of the stator, which, despite the reduced power losses in the second control zone (Figure 37, is high throughout the entire motorway driving range. The presented waveform of temperature and losses shows that before the car enters the motorway, the losses and temperatures of the magnetic core and the windings are very similar for all three supply voltages. The differences appear when the car enters the motorway and the vehicle moves at a higher speed (Figure 37). With increasing speed, large differences in the losses in the winding and in the magnetic core of the stator appear, depending on the supply voltage. Both in the winding and in the stator core, the temperatures increase faster when operated at the lowest voltage. Temperatures and losses are the same after the car exits the motorway. Throughout the run, the maximum stator yoke temperature was T_{stator_yoke} = 100 °C. The calculated temperature of the supporting structure (on which the stator is mounted) in the active part, i.e., under the lamination core, is ok T_{axle} = 60 °C, so the temperature difference between the elements is at the level of Δt = 40 °C. From the calculation of the thermal expansion of the materials, it follows that the equivalent air gap should increase by $l_{stator\ gap}$ = 0.127 mm. This is a rather safe value because the minimum interference between the stator and the support structure ranges from $l_{stator\ gap}$ = −0.3 to $l_{stator\ gap}$ = −0.6 mm. A big threat to the motor operation would be if the increase in the air gap thickness was close to the interference value, because then the contact resistance of the elements would increase significantly and heat dissipation to the cooling system would be much less effective.

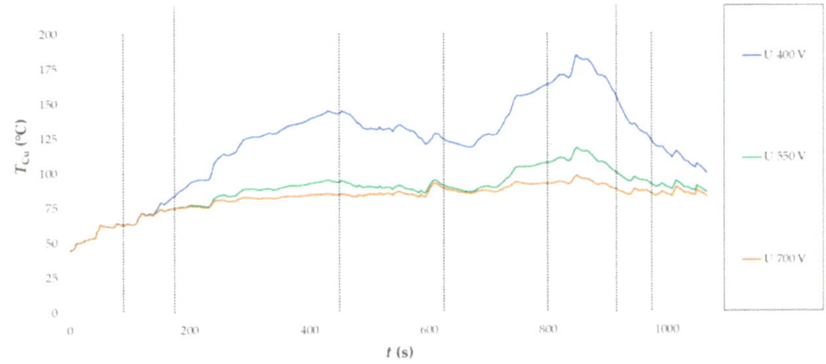

Figure 36. Calculated temperature waveforms in the motor winding (hot spot) for the considered vehicle in the Artemis Motorway150 driving cycle.

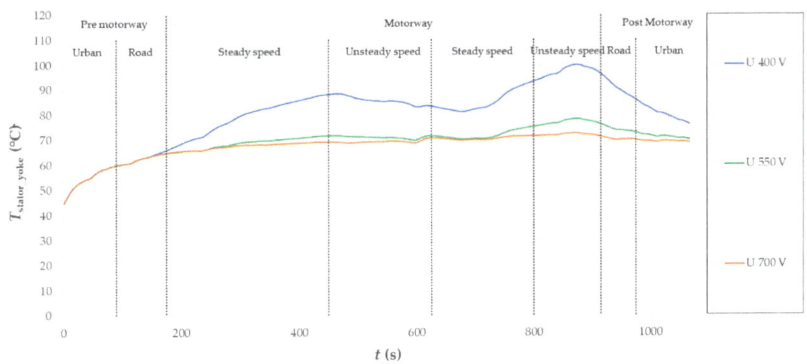

Figure 37. Calculated temperature waveforms in the motor stator yoke (hot spot) for the considered vehicle in the Artemis Motorway150 driving cycle.

The change in the length of material is assumed to be proportional to the temperature change, as expressed by the formula for linear expansion

$$x = x_0(1 + \alpha \Delta T)$$

where: x—length after temperature change, x_0—initial length, α—linear expansion coefficient, ΔT—temperature rise.

The temperature of the permanent magnets has reached the maximum level equal to $T_{mag.}$ = 90 °C in the case of the motor supplied with the highest supply voltage considered (Figure 38). This is a safe value, especially since in the case of a motor supplied with the voltage U_{DC} = 700 V, the magnet is under the influence of a lower flux weakning than in the case of lower voltage. The magnet may become partially demagnetized when exposed to a sufficiently high temperature and under the influence of an external magnetic field. Figure 39 shows the demagnetization characteristics of the applied magnets, while Figure 40 shows the distribution of induction saturation at an example position of the rotor at the operating point of the cycle for which the magnet temperature is the highest. The operating point occurs in the 847 s of cycle duration, The car moves at a speed of v = 150.4 km/h (n = 1140 rpm), the motor runs with a torque of T_m = 596 Nm, magnets temperature $T_{mag.}$ = 87.5 °C, the phase advance angle is equal to $\alpha_{advance}$ = 27.82°, current $I_{d\ RMS}$ = 42.98 A, current $I_{q\ RMS}$ = 81.46 A. The calculated distribution of induction in permanent magnets for these operating parameters showed that the minimum saturation of the induction in the areas of the magnet is $B = 0.55$ T. The calculated minimum value of the induction (due to the flux weakening) is dangerous for the magnet only at a temperature of approx. $T_{mag.} = 140$ °C (Figure 39).

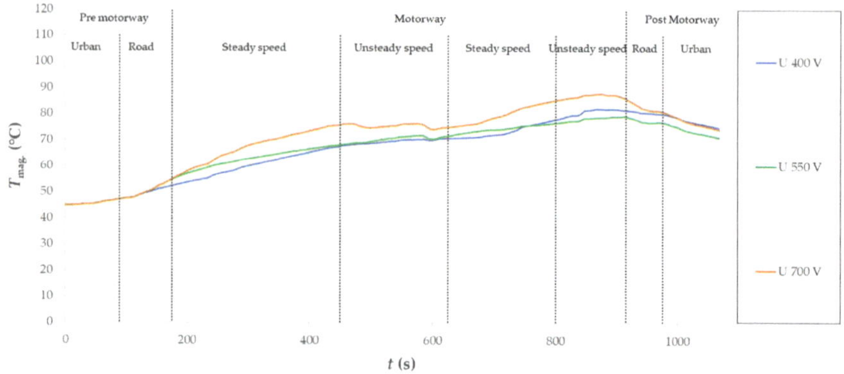

Figure 38. Calculated temperature waveforms in permanent magnets of the motor for the considered vehicle in the Motorway150 driving cycle.

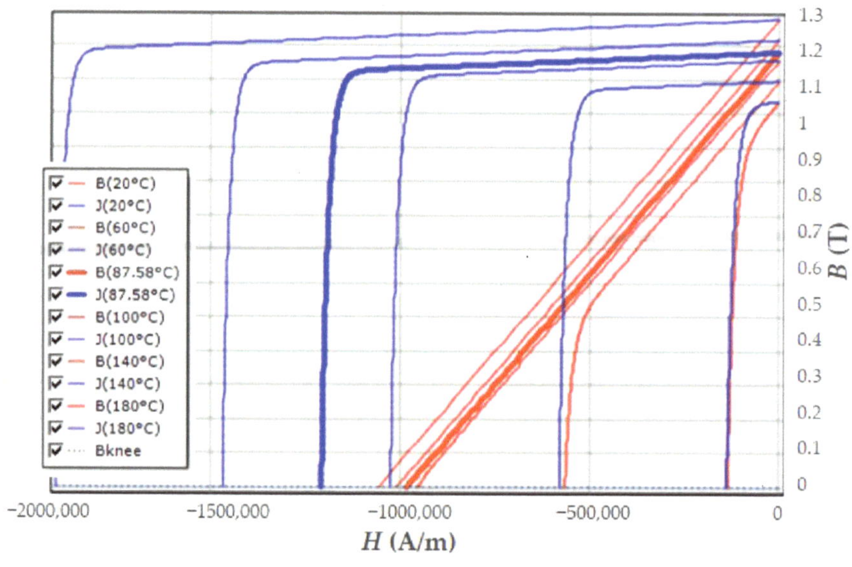

Figure 39. Demagnetization characteristics of the permanent magnets used in the model.

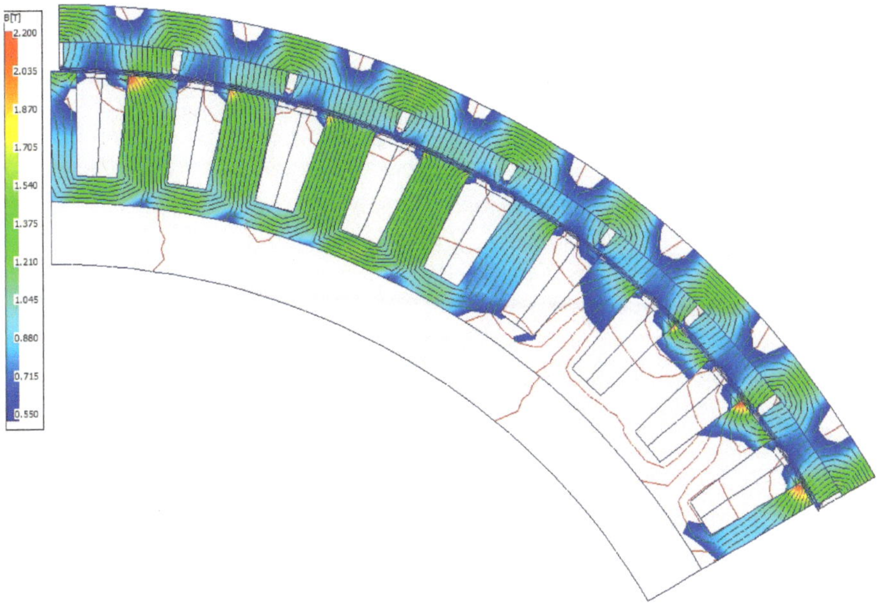

Figure 40. Calculated magnetic induction distribution of the motor with axial segmentation of magnets (x5) in the 847th second of the Artemis Motorway150 cycle duration with the operating parameters: v_{vehicle} = 150.4 km/h; n = 1140 rpm; T_{m} = 596 Nm; I_{RMS} = 94 A, $\alpha_{\text{phase advance}}$ = 27.8°).

Due to the fact that the temperatures of permanent magnets are at a safe level in all the considered supply voltages, one more analysis has been made to assess the impact of segmentation of the magnets on their operating temperature while driving the car. For this purpose, the construction of the magnet was changed to a magnet without segmentation in the computational model. A simulation of the motor operation was carried out in the Artemis Urban150 cycle, because in this cycle the achieved operating temperatures of the magnets are the highest, and, additionally, the magnets are under the influence of an external, oppositely directed magnetic field during flux weakening.

Figure 41 shows the waveforms of the maximum permanent magnet temperatures for the motor without magnet segmentation and with segmentation. As expected, the temperatures achieved are higher due to the increased power losses generated in the magnets. The calculated power losses in magnets were compared for three supply voltages and for the version of the motor with a magnet divided into five segments and without magnet segmentation (1 magnet) in Figure 42 The calculated losses in permanent magnets, in the version without segmentation, are many times higher, which is the reason for the appearance of much higher operating temperatures.

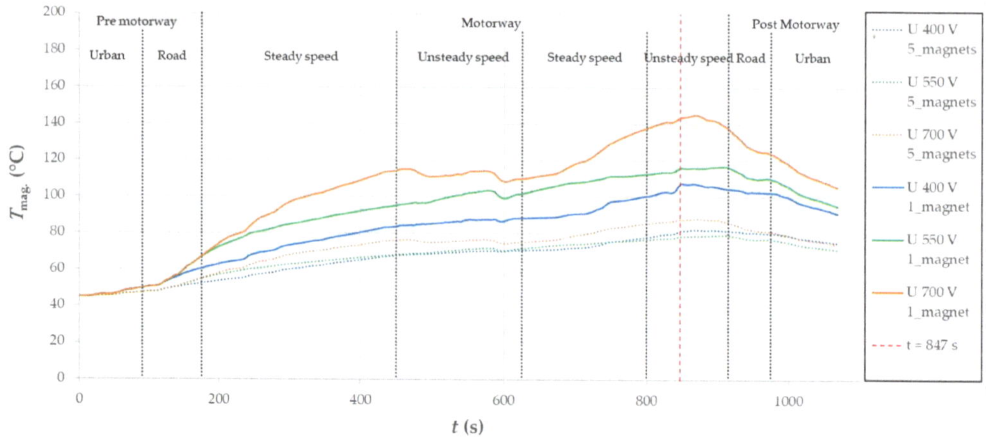

Figure 41. Calculated temperature waveforms of permanent magnets for three different supply voltages of the U_{DC} drive, for a motor with magnet segmentation (×) and a motor without magnet segmentation.

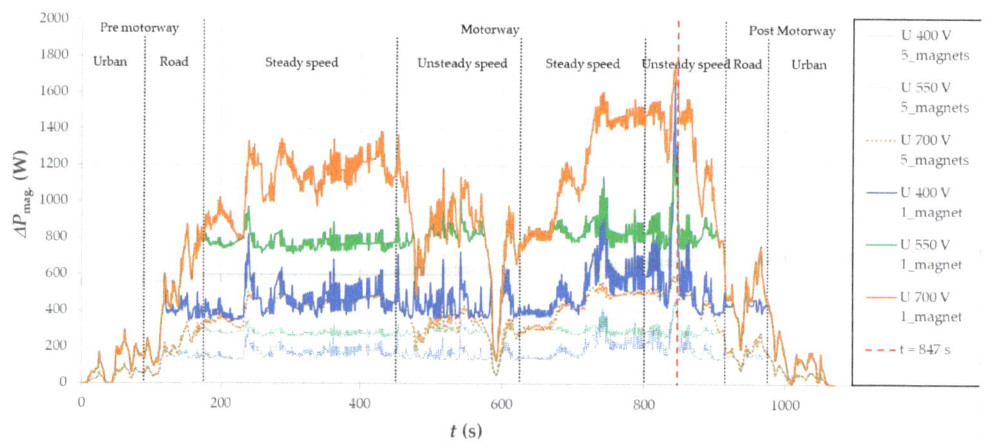

Figure 42. Calculated loss waveforms in magnets for three different supply voltages of the U_{DC} drive, for a motor with magnet segmentation (×) and a motor without magnet segmentation.

The presented calculation results show that the temperature of the magnets without segmentation when supplied with the voltage U_{DC} = 700 V reaches approx. T_{Cu} = 145 °C. This is the temperature of the magnet, reached during the operation of the motor in the second control zone. To compare the motor versions with and without magnet segmentation, the cycle point for the same time instant at high operating temperature for 847 s of cycle duration was compared. Table 4 presents the operating parameters of the considered point. Figure 43 additionally shows the distribution of the magnetic induction saturation field for the same rotor position as in Figure 40. In the case of a motor without magnet segmentation, the induction in the magnets drops below the value of B = 0.55 T in some areas. The areas are marked with red circles. Places with reduced induction are located in the slot area of the magnets. The magnet is operated at a temperature above the $T_{mag.} > 140$ °C which, in accordance with the demagnetization characteristics, indicates the possibility of partial demagnetization.

Table 4. Motor operating parameters in the 847th second of the Artemis Urban drive cycle, with magnet segmentation and without magnet segmentation.

Parameter	5 Segments	1 Segment	Unit
T_m	596	596	Nm
n	1139	1139	rpm
I_{RMS}	94	92.26	A
$I_{d\,RMS}$	42.98	32.9	A
$I_{q\,RMS}$	81.46	88	A
T_{Cu}	87.5	143.5	°C

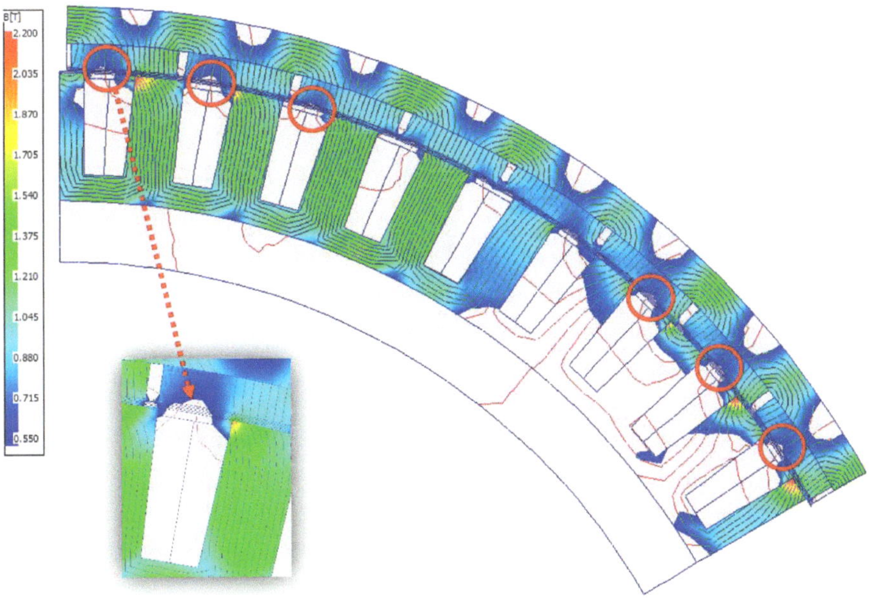

Figure 43. Calculated motor magnetic induction distribution without the axial segmentation of the magnets, in the 847th second of the Artemis Motorway150 cycle time with the operating parameters: $v_{vehicle}$ = 150.4 km/h; n = 1140 rpm; T_m = 596 Nm; I_{RMS} = 92.26 A, $\alpha_{phase\ advance}$ = 20.5°).

8. Driving Simulations—Slope

The next research that is necessary to understand the capabilities of the designed motor is uphill driving simulations. While driving, the car may be driven on roads with various types of slopes. Calculations have been made for different percentage slopes. The calculations were carried out so that at the end of the Artemis Urban cycle simulation, in 983 s of the cycle, when the car slows down (Figure 15), the drive begins with constant torque and constant speed v = 50 km/h for slopes up to 15% (Figure 44) and at a speed of v = 30 km/h for higher slopes (Figure 45). For higher slopes, a lower vehicle speed was assumed as more likely when negotiating various ramps in the city. The calculations were made similar to the previous simulation for the maximum mass of the vehicle with a load of m = 3300 kg. Table 5 presents the operating parameters set in the simulation resulting from the vehicle speed.

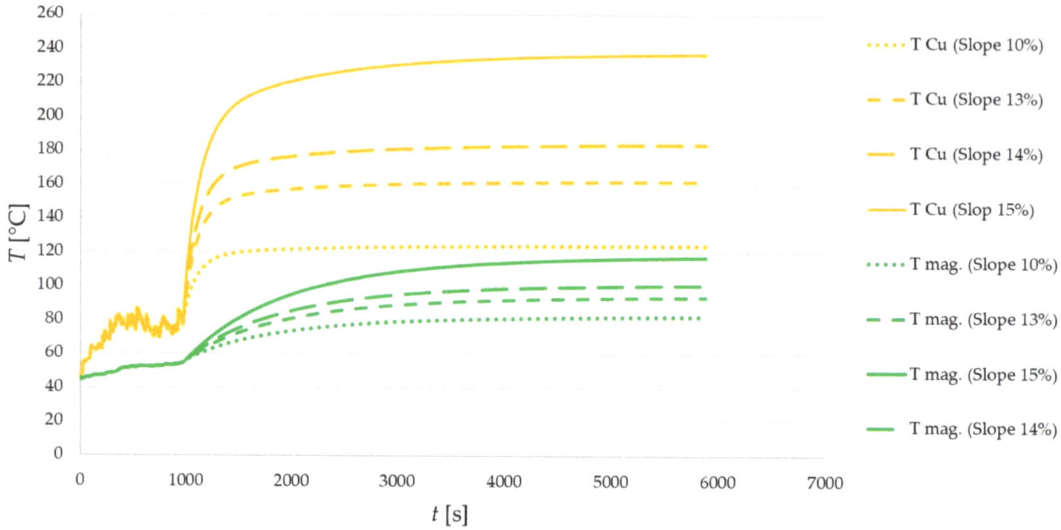

Figure 44. Calculated temperature waveforms of the winding and magnets while driving the car at a speed of v = 50 km/h, on the road with different slopes.

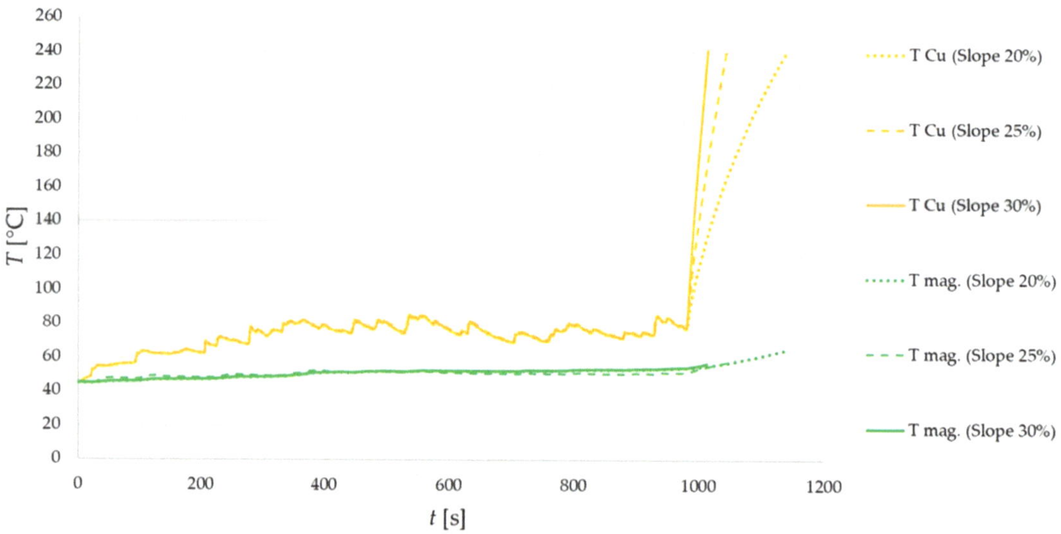

Figure 45. Calculated temperature waveforms of the winding and magnets while driving the car at a speed of v = 30 km/h, on the road with different slopes.

Table 5. Calculated motor operating parameters when driving the car on the road with different slopes.

Vehicle Speed [km/h]	Gradient [%]	Torque [Nm]	T_{cu} [°C]	T_{mag} [°C]	Work Time [s]	Distance Traveled [km]	Copper Losses [W]	Iron Losses [W]	Magnet Losses [W]
50	10	704	124	82	S1	10	2679	176.6	90.2
50	13	878	162	93,5	S1	10	4494	189.5	99.8
50	14	934	183	100	S1	10	5474	196.8	109.6
50	15	1000	200	75.5	385	5,35	7326	196.4	118.1
30	20	1259	200	60	101	0,84	9835	116.1	49.15
30	25	1542	200	56	44	0,37	14784	125.1	64.4
30	30	1825	200	56	23	0,19	19791	136.3	87

From the presented calculation results, it can be concluded that the drive system of two-wheel hub motors ensures safe, constant driving at a speed of v = 50 km/h on a road with a 13% gradient. The winding temperature is set at approx. T_{Cu} = 160 °C. The temperature of the magnets under these operating conditions is set at approx. $T_{mag.}$ = 95 °C. Both temperatures are safe for long-term operation. With a slope of the route at the level of 15%, the temperature T_{Cu} = 160 °C is reached after t = 150 s, while T_{Cu} = 200 °C was reached after t = 385 s. It can be concluded that such motor overload is permissible not permanently but until t = 385 s, which corresponds to covering the distance m = 5.34 km. Under this assumption, the temperature of the magnet will not exceed the allowable temperature.

For larger road slopes/gradients, the winding temperature exceeded T_{Cu} = 200 °C in all cases. Driving on a road with such a slope/gradient is time-limited, while slopes/gradients are relatively very large, in fact occurring in very limited sections of slopes or on climbs. Under the assumed operating conditions, the permissible maximum operating temperature T_{Cu} = 200 °C, when driving on a 30% gradient, will be achieved after driving s = 190 m, which seems to be a good result.

The calculated temperatures indicate that under the considered operating conditions, the winding temperature limits the motor operating time. The temperature of the magnets remains at an acceptable level, assuming that the operating time of the motor will be limited until the winding hotspot reaches the permissible maximum temperature. Demagnetization in such a case is not possible to appear.

Additionally, it should be remembered that the considered drive is hybrid. In cases where the vehicle runs outside the "green zone", the electric motors may be assisted by the internal combustion engine or completely unloaded.

9. Conclusions

The analysis of the motor operation, in the assumed drive, consisting of two-wheel hub motors, installed in the rear drive axle, allowed us to consider its parameters in various conditions, while driving the car. At the stage of designing the drive and the electric motor, especially those of atypical design, it is important to have a broad look at the possible scenarios of the work regimes in which the motor will work. The analysis assumed that the motor structure fills the entire available area in the rim and that the motor will be tested for three different driving cycles: city driving, suburban driving, and highway driving. The variables were the battery supply voltage and the maximum power consumed by the two electric motors.

The specification of the maximum torque requirements at a given rotational speed in a given driving cycle allowed for the presentation of the wheel hub motor's operating range in the proposed drive configuration and for the analysis of its power parameters (supply voltage, power demand).

The analysis presents the methodology of the approach to designing the power supply system and designing the power supply system as well as selecting the supply voltage and designing the motor, taking into account the operating conditions encountered in traction applications.

The simulation tests carried out have shown that for driving in the city cycle, the battery voltage U_{DC} = 400 V is a sufficient voltage. The operating temperatures obtained are much lower than the permissible for permanent magnets, windings, epoxy resin, and for the lamination core (in this case, the difference between the temperature of the lamination core and the supporting structure on which it is mounted is limited). The achieved maximum operating temperatures are so small that in the event of limiting the drive operation to the considered cycle and rotational speed range, it is possible to consider shortening the motor magnetic core lamination with a simultaneous increase in the number of turns to the value ensuring unchanged maximum torque. You can also consider reducing the supply voltage. In the urban cycle, the guarantee of maximum power for two motors, at the level of P_{IN} = 120 kW, is a sufficient value with a large margin at low rotational speeds, where the cycle requires relatively large torques and with a very large excess (unnecessary) at higher rotational speeds, where required the torque is relatively small.

In the road cycle, despite the need to work the wheel hub motor partially in the flux weakening zone, the temperatures achieved are also safe and do not come close to the maximum permissible. The demand for power increases significantly. The PIN value of 120 kW is somewhat insufficient at lower vehicle speeds and has no reserve at higher speeds.

In the case of driving on the highway, the achieved operating temperatures of the winding, when supplied with voltage U_{DC} = 400 V, are much higher, equal to or close to the maximum permissible. The temperature of the magnets remains at a safe level, as is the temperature of the magnetic core yoke (depending on the manufacturing technology and the pressing of the laminated stack/core/package onto the supporting structure). The high operating temperatures of the winding are caused by the necessity of a fairly deep flux weakening of the motor and the related increase of the motor supply current. Increasing the supply voltage causes the operating range in the second control zone to decrease so that the winding temperature is within safe limits. The higher the supply voltage, the lower the winding operating temperature in the motorway cycle. Increasing the supply voltage even to the value of U_{DC} = 700 V and a very large reduction in the flux weakening zone resulted in an increase in the power generated in the magnets, but it did not increase the operating temperature of the magnets to a dangerous value.

Due to the relatively low operating temperature of the magnets in all three driving cycles, an additional simulation of the wheel hub motor operation was performed for the highway cycle without magnet segmentation. Segmentation of the magnets involves additional production costs of the magnets themselves and the manufacture of the rotor.

The simulations of the motor operation have shown that when using a magnet without axial segmentation, when the motor is supplied with the voltage of U_{DC} = 700 V, a very large increase in power losses in the magnets will result in an increase in the operating temperature above the $T_{\text{mag temperature.}}$ > 140 °C. Additionally, according to the simulated driving cycle, the magnet is under the effect of flux weakening during this time. The analysis of the distribution of the magnetic induction field showed that at this temperature in certain areas the magnet becomes demagnetized. Lowering the operating temperature of the magnets for this operating point can in this case be achieved by using the weakening of the magnetic flux in the second motor control zone, which can be seen in the example of the supply voltage U_{DC} = 400 V, but as before it is associated with an increase in the supply current and an unacceptable increase in the winding temperature.

The use of the supply voltage U_{DC} = 550 V makes the temperature of the magnets at this operating point slightly safer, while the winding temperature is at a safe level.

In summary, simulation studies have shown that magnet segmentation is necessary to ensure the safe operation of the motor in the highway cycle.

The designed electric motor meets the requirements of the car operating cycles described in the literature. A vehicle equipped with the designed electric motors can move with a full load in urban and suburban areas without the support of the internal combustion engine. Motorway driving is also possible if the drive is configured with a higher supply voltage. If laboratory and road tests confirm the thermal load of the motor in various road conditions, then it is possible to consider reducing the size of the motor or increasing its operating range for driving on the highway.

In the next publication, the authors plan to present the results of laboratory and traction measurements of the prototype wheel hub motor. If the measurements of the motor confirm the large temperature reserves shown by simulation tests, a new wheel hub motor will be designed and built with an improved structure with improved technical parameters and reduced weight.

Author Contributions: P.D.-Development of computational models, running simulation, designing electromagnetic circuits, analysis of simulation results, preparation of the article, description and presentation of control methods, help in the analysis of the results. J.M.-Description and presentation of control methods, help in the analysis of the results, editorial article. R.K.-Analysis of simulation results, participation in the creation of proposals, editorial article. All authors have read and agreed to the published version of the manuscript.

Funding: The project "Polish electric drive solutions in wheels" ("4WHED") implemented in the Łukasiewicz Research Network is financed under specific subsidies by the Ministry of Science and Higher Education (POLAND) Number: 2/Ł-KOMEL/CŁ/2021. The research is a continuation of the project "Innovative Solutions for Direct Drive of Electric Vehicles", financed by National Centre for Research and Development under the LIDER VII program, in accordance with the agreement: LIDER/24/0082/L-7/15/NCBR/2016 (POLAND). Research on wheel hub motors, carried out by the Łukasiewicz Research Network, received the Research Award (main award) 25th Siemens Award Competition for scientists and research teams (Poland). The results of the research will be used in the work on the doctorate carried out in the Implementation Doctorate program of the Ministry of Science and Higher Education (POLAND).

Data Availability Statement: The data presented in this study are available upon request of the corresponding author. The data is not publicly available due to the implementation nature of the project.

Conflicts of Interest: The authors declare no conflict of interest.

References

1. Rahim, N.; Ping, H.; Tadjuddin, M. Design of an in-wheel axial flux brushless dc motor for electric vehicle. In Proceedings of the 2006 International Forum on Strategic Technology, Ulsan, Korea, 18–20 October 2006; pp. 16–19.
2. Zhu, J.; Cheng, K.E.; Xue, X.; Zou, Y. Design of a novel high-torque-density in-wheel switched reluctancemotor for electric vehicles. In Proceedings of the 2017 IEEE International Magnetics Conference (INTERMAG), Dublin, Ireland, 24–28 April 2017.
3. Rechkemmer, S.K.; Zhang, W.; Sawodny, O. Modeling of a Permanent Magnet Synchronous Motor of an E-Scooter for Simulation with Battery Aging Model. *IFAC-PapersOnLine* **2017**, *50*, 4769–4774. [CrossRef]
4. Zhu, X.; Shu, Z.; Quan, L.; Xiang, Z.; Pan, X. Design and Multicondition Comparison of Two Outer-RotorFlux-Switching Permanent-Magnet Motors for In-Wheel Traction Applications. *IEEE Trans. Ind. Electron.* **2017**, *64*, 6137–6148. [CrossRef]
5. Gao, P.; Gu, Y.; Wang, X. The Design of a Permanent Magnet In-Wheel Motor with Dual-Stator andDual-Field-Excitation Used in Electric Vehicles. *Energies* **2018**, *11*, 424. [CrossRef]
6. Biček, M.; Lampič, G.; Zupan, S.; Obrul, B.; Gotovac, G.; Štefe, B.; Valentinčič, J. HIGH TORQUE "IN-WHEEL" MOTORS FOR RESCUE VEHICLES. In Proceedings of the Innovative Automotive Technology—IAT 2012, DolenjskeToplice, Slovenia, 12–13 April 2012.
7. Luo, Y.; Tan, D. Lightweight design of an in-wheel motor using the hybrid optimization method. *Proc. Inst. Mech. Eng. Part D J. Automob. Eng.* **2013**, *227*, 1590–1602. [CrossRef]
8. Shin, P.S.; Kim, H.D.; Chung, G.B.; Yoon, H.S.; Park, G.S.; Koh, C.S. Shape Optimization of a Large-ScaleBLDC Motor Using an Adaptive RSM Utilizing Design Sensitivity Analysis. *IEEE Trans. Magn.* **2007**, *43*, 1653–1656. [CrossRef]
9. Seo, I.M.; Kim, H.K.; Hur, J. Design and analysis of modified spoke type BLDC motor using a ferritepermanent-magnet. In Proceedings of the 17th International Conference on Electrical Machines and Systems(ICEMS), Hangzhou, China, 22–25 October 2014; pp. 1701–1705.
10. Łebkowski, A. Design, Analysis of the Location and Materials of Neodymium Magnets on the Torque and Power of In-Wheel External Rotor PMSM for Electric Vehicles. *Energies* **2018**, *11*, 2293. [CrossRef]

11. Frajnkovic, M.; Omerovic, S.; Rozic, U.; Kern, J.; Connes, R.; Rener, K.; Biček, M. Structural Integrity of In-Wheel Motors. In Proceedings of the International Powertrains, Fuels & Lubricants Meeting, Heidelberg, Germany, 17 September 2018. [CrossRef]
12. Biček, M.; Connes, R.; Omerović, S.; Gündüz, A.; Kunc, R.; Zupan, S. The Bearing Stiffness Effect on In-Wheel Motors. *Sustainability* **2020**, *12*, 4070. [CrossRef]
13. Li, G.; Wang, Y.; Zong, C. Driving State Estimation of Electric Vehicle with Four-wheel-hub-motors. *Qiche Gongcheng/Automot. Eng.* **2018**, *40*, 150–155.
14. Wanner, D.; Kreußlein, M.; Augusto, B.; Drugge, L.; Stensson Trigell, A. Single wheel hub motor failures and their impact on vehicle and driver behavior. *Veh. Syst. Dyn.* **2016**, *54*, 1345–1361. [CrossRef]
15. Lampič, G.; Detela, A.; Valentinčič, J. Management of innovative technology of Elaphe* in-wheel. Electric motors—A case study. In Proceedings of the 9th International Conference on Management of Innovative Technologies MIT'2007, Fiesa, Slovenia, 9–10 October 2007.
16. Kostic Perovic, D. Making the Impossible, Possible—Overcoming the Design Challenges of In Wheel Motors. *World Electr. Veh. J.* **2012**, *5*, 514–519. [CrossRef]
17. Ślaski, G.; Gudra, A.; Borowicz, A. Analysis of the influence of additional unsprung mass of in-wheel motors on the comfort and safety of a passenger car. *Arch. Automot. Eng.—Arch. Motoryz.* **2014**, *65*, 51–64.
18. Parczewski, K.; Romaniszyn, K.; Wnęk, H. Influence of electric motors assembly in hubs of vehicle wheels on the dynamics of movement, especially on surfaces with different adhesion coefficient. *Combust. Engines* **2019**, *179*, 58–64. [CrossRef]
19. Dukalski, P.; Będkowski, B.; Parczewski, K.; Wnęk, H.; Urbaś, A.; Augustynek, K. Analysis of the influence of assembly electric motors in wheels on behaviour of vehicle rear suspension system. In *IOP Conference Series: Materials Science and Engineering*; IOP Publishing: Bristol, UK, 2018; Volume 421. [CrossRef]
20. Parczewski, K.; Wnek, H. Comparison of overcoming inequalities of the road by a vehicle with a conventional drive system and electric motors placed in the wheels. In Proceedings of the Transport Means 2020, Palanga, Lithuania, 30 September–2 October 2020.
21. Dukalski, P.; Będkowski, B.; Parczewski, K.; Wnęk, H.; Urbaś, A.; Augustynek, K. Analysis of the Influence of Motors Installed in Passenger Car Wheels on the Torsion Beam of the Rear Axle Suspension. *Energies* **2022**, *15*, 222. [CrossRef]
22. Szewczyk, P.; Łebkowski, A. Studies on Energy Consumption of Electric Light Commercial Vehicle Powered by In-Wheel Drive Modules. *Energies* **2021**, *14*, 7524. [CrossRef]
23. Available online: http://in-wheel.com/en/ (accessed on 5 June 2022).
24. Available online: https://www.ziehl-abegg.com/de/en/product-range/automotive/in-wheel-hub-motors/ (accessed on 5 June 2022).
25. Available online: https://www.proteanelectric.com/ (accessed on 5 June 2022).
26. Available online: https://www.lordstownmotors.com/pages/tech (accessed on 5 June 2022).
27. Bedkowski, B.; Dukalski, P.; Jarek, T.; Wolnik, T. Tests of Electrical Motor for Installation in the Wheel Hub of an Electric Car. In *IOP Conference Series Materials Science and Engineering*; IOP Publishing: Bristol, UK, 2020; Volume 841, p. 012003.
28. Dukalski, P.; Krok, R. Selected Aspects of Decreasing Weight of Motor Dedicated to Wheel Hub Assembly by Increasing Number of Magnetic Poles. *Energies* **2021**, *14*, 917. [CrossRef]
29. Tomasz, W.; Szczepan, O.; Łukasz, C.; Tomasz, J.; Vojtech, S. Design methods for limiting rotor losses in a fractional slot PMSM motor with high power density. *Arch. Electr. Eng.* **2022**, *71*, 963–979. [CrossRef]
30. Yamazaki, K.; Shina, M.; Kanou, Y.; Miwa, M.; Hagiwara, J. Effect of Eddy Current Loss Reduction by Segmentation of Magnets in Synchronous Motors: Difference between Interior and Surface Types. *IEEE Trans. Magn.* **2009**, *45*, 4756–4759. [CrossRef]
31. Martin, F.; Zaim ME, H.; Tounzi, A.; Bernard, N. Improved Analytical Determination of Eddy Current Losses in Surface Mounted Permanent Magnets of Synchronous Machine. *IEEE Trans. Magn.* **2014**, *50*, 1–9.
32. Huang, W.Y.; Bettayeb, A.; Kaczmarek, R.; Vannier, J.C. Optimization of Magnet Segmentation for Reduction of Eddy-Current Losses in Permanent Magnet Synchronous Machine. *IEEE Trans. Energy Convers.* **2010**, *25*, 381–387. [CrossRef]
33. Upadhayay, P.; Patwardhan, V. Magnet Eddy-Current Losses in External Rotor Permanent Magnet Generator. In Proceedings of the International Conference on Renewable Energy Research and Applications, Madrid, Spain, 20–23 October 201.
34. Ouamara, D.; Dubas, F. Permanent-Magnet Eddy-Current Losses: A Global Revision of Calculation and Analysis. *Math. Comput. Appl.* **2019**, *24*, 67. [CrossRef]
35. Gao, R.-Z.; Zhai, L.; Su, L.-W. The Research on Full-speed Field Weakening Control Method of Electric Vehicle Interior Permanent Magnet Synchronous Motor. *ITM Web Conf.* **2017**, *11*, 09001. [CrossRef]
36. Ustun, O.; Kivanc, O.C.; Senol, S.; Fincan, B. On Field Weakening Performance of a Brushless Direct Current Motor with Higher Winding Inductance: Why Does Design Matter? *Energies* **2018**, *11*, 3119. [CrossRef]
37. Rivera, C.A.; Poza, J.; Ugalde, G.; Almandoz, G. Field Weakening Characteristics Computed with FEM-Coupled Algorithms for Brushless AC Motors. *Energies* **2018**, *11*, 1288. [CrossRef]
38. Lee, H.-J.; Shon, J.-G. Improved Voltage Flux-Weakening Strategy of Permanent Magnet Synchronous Motor in High-Speed Operation. *Energies* **2021**, *14*, 7464. [CrossRef]
39. Kowal, D.; Sergeant, P.; Dupré, L.; Vandenbossche, L. Comparison of Iron Loss Models for Electrical Machines WithDifferent Frequency Domain and Time Domain Methodsfor Excess Loss Prediction. *IEEE Trans. Magn.* **2015**, *51*, 1–10. [CrossRef]

40. Yamazaki, K.; Fukushima, N. Iron-Loss Modeling for Rotating Machines: Comparison between Bertotti's Three-Term Expression and 3-D Eddy-Current Analysis. *IEEE Trans. Magn.* **2010**, *46*, 3121–3124. [CrossRef]
41. Zhao, H.; Luo, Y.; Ren, H.W.; Peter, B. A Complete Model for Iron Losses Prediction in Electric Machines Including Material Measurement, Data Fitting, FE Computation and Experimental Validation. *Electr. Rev.* **2012**, *88*, 52–55.
42. Bertotti, G. General properties of power losses in soft ferromagnetic material. *IEEE Trans. Magn.* **1988**, *24*, 621–630. [CrossRef]
43. Fadel, M.; Sepulchre, L.; Pietrzak-David, M. Deep Flux-Weakening Strategy with MTPV for High-Speed IPMSM for Vehicle Application. *IFAC-PapersOnLine* **2018**, *51*, 616–621. [CrossRef]
44. Hu, D.; Zhu, L.; Xu, L. Maximum Torque per Volt operation and stability improvement of PMSM in deep flux-weakening Region. In Proceedings of the 2012 IEEE Energy Conversion Congress and Exposition (ECCE), Raleigh, NC, USA, 15–20 September 2012; pp. 1233–1237. [CrossRef]
45. Ji, X.; Lv, Y. Field weakening control of PMSM used in an electric power steering system. In Proceedings of the 2011 International Conference on Electric Information and Control Engineering, ICEICE 2011, Wuhan, China, 15–17 April 2011; pp. 2194–2199. [CrossRef]
46. Neapolitan, R.E.; Hee Nam, K. PMSM Control Methods. In *AC Motor Control and Electrical Vehicle Applications*; CRC Press: Boca Raton, FL, USA, 2019; Available online: https://www.taylorfrancis.com/chapters/mono/10.1201/9781315200149-7/pmsm-control-methods-richard-neapolitan-kwang-hee-nam (accessed on 5 November 2020).
47. Nguyen, Q.K.; Petrich, M.; Jörg, R.-S. Implementation of the MTPA and MTPV control with online parameter identification for a high speed IPMSM used as traction drive. In Proceedings of the 2014 International Power Electronics Conference (IPEC-Hiroshima 2014—ECCE ASIA), Hiroshima, Japan, 18–21 May 2014; Volume 4, pp. 318–323. [CrossRef]
48. Sepulchre, L.; Fadel, M.; Pietrzak-David, M.; Porte, G. MTPV Flux-Weakening Strategy for PMSM High Speed Drive. *IEEE Trans. Ind. Appl.* **2018**, *54*, 6081–6089. [CrossRef]
49. Będkowski, B.; Madej, J. The innovative design concept of thermal model for the calculation of the electromagnetic circuit of rotating electrical machines. *Eksploat. i Niezawodn.—Maint. Reliab.* **2015**, *17*, 481–486. [CrossRef]
50. Krok, R. *Sieci Cieplne w Modelowaniu Pola Temperatury w Maszynach Elektrycznych prądu Przemiennego*; Wydawnictwo Politechniki Śląskiej, Monograph: Gliwice, Poland, 2010.
51. Motor-CAD v14.1 Manual (11/01/2021).
52. André, M. The ARTEMIS European driving cycles for measuring car pollutant emissions. *Sci. Total Environ.* **2005**, *334–335*, 73–84. [CrossRef]
53. Nyberg, P.; Frisk, E.; Nielsen, L. Using Real-World Driving Databases to Generate Driving Cycles with Equivalence Properties. *IEEE Trans. Veh. Technol.* **2015**, *65*, 4095–4105. [CrossRef]

designs

MDPI

Article

Design of an Integrated, Six-Phase, Interleaved, Synchronous DC/DC Boost Converter on a Fuel-Cell-Powered Sport Catamaran

Michael Rimondi †, Riccardo Mandrioli, Vincenzo Cirimele, Lohith Kumar Pittala, Mattia Ricco * and Gabriele Grandi

Department of Electrical, Electronic and Information Engineering, University of Bologna, 40136 Bologna, Italy
* Correspondence: mattia.ricco@unibo.it; Tel.: +39-051-209-3591
† Current Address: Ducati Motor Holding S.p.A., 40132 Bologna, Italy.

Abstract: This paper describes the preliminary analysis, design and implementation phases of a DC/DC boost converter dedicated to the Futura catamaran propulsion chain developed by the UniBoAT team at the University of Bologna. The main goal of the project was the reduction of the converter's weight by eliminating the use of heat sinks and by reducing the component size, especially inductors and capacitors. The obtained converter is directly integrated into the structure containing the fuel-cell stack. The realized converter was based on an interleaved architecture with six phases controlled through the average current mode control. The design was validated through simulations carried out using the LT-Spice software, whereas experimental validations were performed by means of both bench tests and on-field tests. Detailed thermal and efficiency analyses were provided with the bench tests under the two synchronous and non-synchronous operating modes and with the adoption of the phase-shedding technique. Prototype implementation and performance in real operating conditions are discussed in relation to on-field tests. The designed converter can be used in other applications requiring a voltage-controlled boost converter.

Keywords: fuel cell; DC-DC converter; digital SMPS; interleaved converter; maritime propulsion

Citation: Rimondi, M.; Mandrioli, R.; Cirimele, V.; Pittala, L.K.; Ricco, M.; Grandi, G. Design of an Integrated, Six-Phase, Interleaved, Synchronous DC/DC Boost Converter on a Fuel-Cell-Powered Sport Catamaran. *Designs* **2022**, *6*, 113. https://doi.org/10.3390/designs6060113

Academic Editor: Loránd Szabó and Feng Chai

Received: 23 September 2022
Accepted: 10 November 2022
Published: 14 November 2022

1. Introduction

In recent years, there has been strong renewed interest in hydrogen production technologies and fuel cells for powering electric propulsion systems. The interest in hydrogen is all the greater in those contexts where high autonomy is required, and the exclusive use of battery packs as power sources introduces strong volume and weight issues. One such context is undoubtedly the maritime transport sector, where all-electric ships are garnering increasing interest [1].

The Argonauts team of the University of Bologna-UniBoAT is a research project involving students and researchers that moves, with these considerations, toward expanding the knowledge about the use of sustainable propulsion systems in the context of mobility.

In recent times, the UniBoAT team developed Futura, a racing catamaran prototype powered by CO_2-free energy sources supplying an outboard electric motor. The Futura's propulsion system is based on a mix of currently available sustainable technologies: high-efficiency photovoltaic panels, a self-built Li-Ion battery and a high-pressure hydrogen tank supplying a proton-exchange membrane fuel cell (PEMFC).

In July 2021, the Argonauts team participated for the first time in the "Monaco Energy Boat Challenge 2021", an event hosted by Yacht Club de Monaco [2]. The boat developed by the team competed in the Energy class, which consists of designing the chassis and the propulsion system to be installed on the catamaran hulls. The hulls were provided by the organization, and they were the same for all participants of the class. The Monaco Energy Boat Challenge includes four challenges: an endurance race of 3 h, a slalom track, a speed record and a one-by-one elimination race. The Monaco Energy Boat Challenge rules for the endurance race stipulate a maximum on-board stored energy of 10 kWh and a maximum boat weight, without a pilot, of 250 kg. In the 2021 prototype, the energy was managed 60% by a battery and 40% by a fuel cell. The fuel cell was operating through its own OEM controller and a DC-DC converter with a total weight of 6 kg (excluding stack weight). Details of the Futura catamaran are visible in Figure 1.

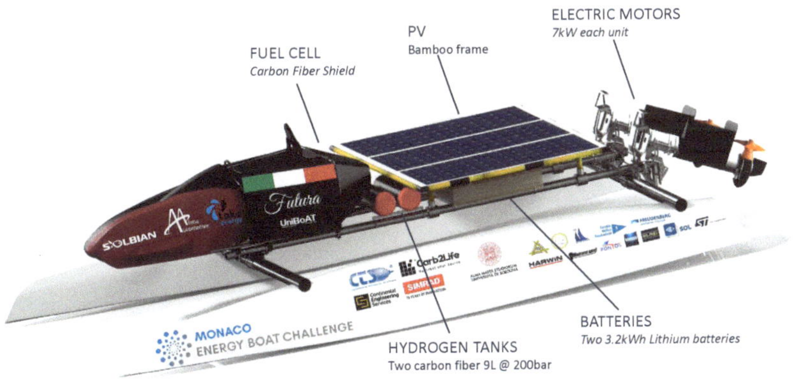

Figure 1. Layout of the racing catamaran Futura developed by UniBoAT.

This work moves from this first architecture with the aim of improving the converter stage within the development of the new 2022 Futura prototype. The main idea behind the new concept was to obtain a completely novel design, making the fuel cell and converter assembly a single, integrated, stand-alone device based on automotive-grade components. This process, while starting from well-established practices regarding the structure and control of converters, required original solutions in both mechanical integration and design issues for power converters.

One of the most sensitive aspects of the project relates to the peculiar operation of fuel cells. Although FCs demonstrate good output power capability during steady-state operation, the dynamic response during transients, especially at peak power demand, is relatively slow [3]. This is why FCs are often combined with energy storage systems such as batteries or supercapacitors to improve the capabilities of supplying energy during fast power transient and to recover energy during regenerative breaking [4]. Therefore, a DC/DC boost converter is essential to match the FCs' low operating voltage with the generally higher voltage of the lithium battery packs [5]. In the studied application, and for other applications for electric propulsion, the DC/DC converter has to meet several requirements. They are here briefly summarized:

1. Low weight and small volume (i.e., high gravimetric and volumetric power density).
2. High reliability.
3. High efficiency.
4. Low current ripple to extend fuel cells' lifetime [6].

Several DC/DC boost converter topologies have been documented in the literature [7,8]. Their advantages and disadvantages are effectively summarized and compared in [9,10].

Standard boost converter topology, despite being extensively used in renewable energy applications [11], is not able to meet all the mentioned requirements. This is because the use of massive inductors and capacitors is necessary to keep the current ripple at the required low level [12].

Isolated resonant LLC converters are valid alternatives, especially in such applications with high-voltage levels where galvanic insulation between FC and battery is required [13,14]. Resonant DC/DC converters provide higher voltage gain and allow one to increase the conversion efficiency by using soft-switching techniques [15]. However, LLC converters require more complex control and the use of additional components, such as the compensation capacitors and the (often custom-made) transformer. Eventually, the efficiency of these converters is greatly variable with changes in the working point. This makes this architecture not optimal for a wide operating range, especially in the case of a light load [16].

Multi-phase interleaved boost converters (IBC) are promising candidates. This topology family consists of several standard boost stages (phases) connected in parallel and independently controlled [17–19]. The interleaving technique, i.e., the control of the different phases with different relative phase shifts, is the most preferred way for parallelizing because it allows obtaining relevant reductions in input current ripple and output voltage ripple while increasing their frequency, together with low switching loss and a usually faster transient response [12,17,20]. Despite this architecture requiring an increase in the number of inductors (one for each phase), an increase in the ripple frequency allows a strong reduction in their size [21,22]. Moreover, the efficiency of interleaved converters can be kept high across the whole operating range by adopting phase shedding [23]. It is for these reasons that this topology was the one adopted in the application that is the subject of this paper.

This work thus starts by choosing architectures and methods validated in the literature and describes their adaptation to create a novel conversion stage. In particular, we intend to present all those design choices that allowed us to integrate the DC/DC converter directly into the metal frame containing the FC stack, eliminating the use of a dedicated heat sink without altering the structure and performance of the FC. As better described in the dedicated Section 2, this converter was also developed to operate in a manner compatible with the short circuit unit. The IBC here proposed has been designed to work at a high switching frequency of 400 kHz in order to reduce as much as possible the inductors' size and weight. This choice has a great impact on PCB design and the selection of the components. The design of the electronic boards and traces and ground-plane placing are fundamental to ensuring low switching losses and limiting electromagnetic emissions. Some unconventional choices in PCB design are also discussed. The converter and the whole system of FC+converter have been tested in both laboratory and real conditions.

The remainder of the paper is organized as follows. Section 2 provides the description of the application. Sections 3 and 4 discuss the design of the multi-phase interleaved converter and its firmware, respectively. The simulation results and experimental validation details are provided in Sections 5 and 6. Finally, the conclusions are presented in Section 7.

2. Description of the Application

Along with other technologies that make use of renewable sources, fuel cells (FC) are gaining much attention as practical solutions to respond to the environmental issues related to vehicular propulsion. PEMFC is the most promising FC technology because of its higher efficiency, low operating temperature and pressure and easy maintenance [24]. The application referred to in this work has been developed around the air-cooled PEMFC Horizon H-2000 shown in Figure 2. This fuel cell has a rated power of 2 kW for a size of 256 mm × 350 mm × 183 mm and a weight of 10 kg. This FC is originally sold together with its OEM controller (Figure 2b) weighing 2.5 kg and without any power converter device.

The FC is composed of a stack of 48 PEMFCs which are enclosed in an aluminum frame that supports and protects them. The two fans visible on the front side are devoted to airflow circulation with the aims of providing oxygen for the FC reaction and cooling down the device.

Each cell of the stack has an open circuit voltage of 0.95 V, resulting in a no-load voltage of about 45.6 V. The electrical behavior of the whole stack is shown in Figure 3.

(a) (b)

Figure 2. Horizon H-2000 PEMFC Stack (**a**) and its OEM controller (**b**). Images retrieved [25].

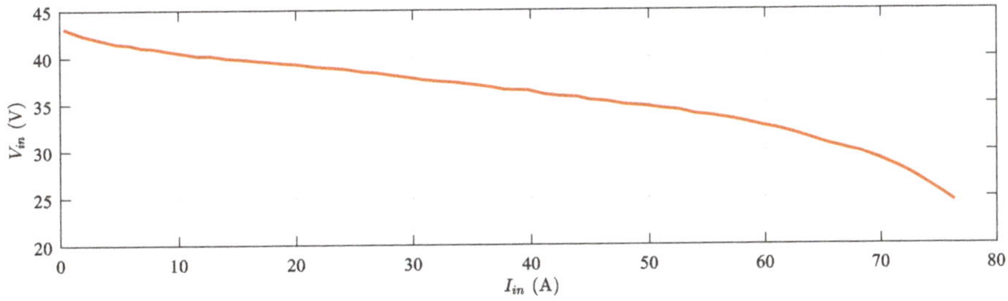

Figure 3. H-2000 PEMFC voltage–current characteristic.

The most critical issue owing to PEMFC is that its efficiency is greatly influenced by its operating point. For this reason, its output current must be accurately regulated by means of a dedicated power converter [26]. The regulation provided by the DC-DC converter is fundamental also because the PEMFCs' output voltage varies based on the condition of its stack's temperature and hydrogen pressure [27].

It is worth mentioning that the adopted PEMFC is equipped with the so-called short circuit unit (SCU). The SCU is a device that short-circuits the PEMFC output terminals for an interval of 100 ms every 10 s [28]. As indicated by different scientific works [29], this short circuiting procedure improves the performance of the stack over time, aiding also the moistening of the proton exchange membranes of the cells.

For the FC output regulation, many works, specifically those in the automotive field, proposed topologies such as the buck-boost converter, the interleaved boost converter, and other isolated DC-DC converters [30,31]. Among these architectures, the DC-DC boost converter is the one that presents high efficiency, together with a reduced component count [9,32]. According to this, the present work considers a synchronous interleaved boost converter (SIBC) specifically designed to work with the PEMFC and a contemporary acting as a charger for a Li-ion battery pack [20]. The coupling with the battery is necessary because the principle of operation on which the fuel cell is based does not allow it to operate under conditions where high dynamics are required unless its performance is compromised. In the application under consideration, the power demands during the competition would instead require sudden power variations. Therefore, the battery is used to supply the rapid variations and power peaks that the fuel cell alone would not be able to handle [33]. The resulting traction chain scheme and the related energy flow diagram are reported in Figure 4.

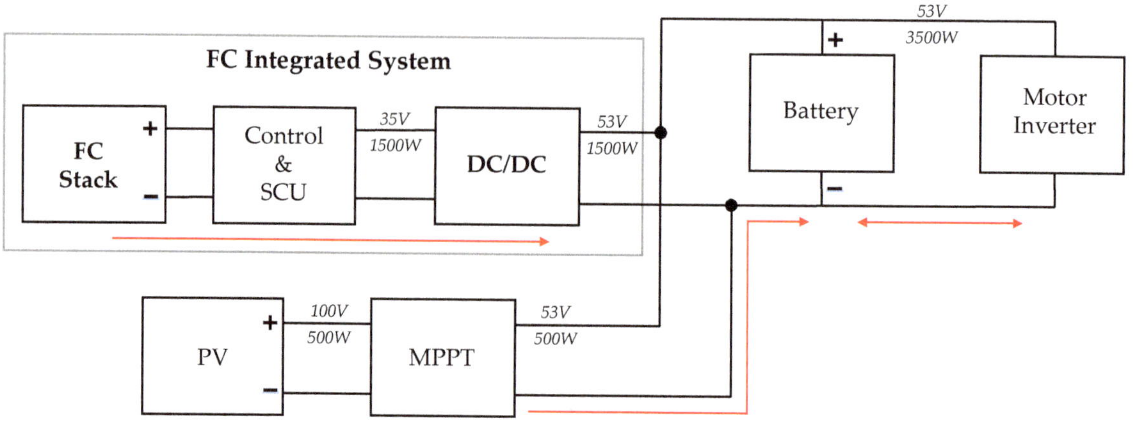

Figure 4. Traction chain scheme of the Futura catamaran and related energy flow diagram.

Along the whole traction chain of the catamaran, the voltage level is limited to a threshold value of 60 V. This choice allowed us to increase the safety level, avoiding at the same time the use of isolated power converter topologies, providing a strong benefit in terms of overall weight reduction.

As the converter has been specifically designed for sports competition, its design had as its key targets the maximization of efficiency, the highest possible weight reduction and mandatory high reliability. The adopted PEMFC offered approximately 60% efficiency when working in a power generation range lower than (10 kW). According to the required accuracy in the regulation of output voltage and currents, the power electronic converter is controlled through the average current mode control (ACMC) technique. This control technique in fact, by directly regulating the average current flowing in the inductor, ensures enhanced accuracy in the input current control [34]. Other common switching-mode power supply (SMPS) control strategies, e.g., peak-current mode, suffer from poor noise immunity, a need for slope compensation, and most of the possible peak-to-average current errors [35,36].

3. Design of the Converter

This section discusses the SIBC's hardware (HW) and firmware (FW) design choices. The basic technical requirements for the design of the converter are reported in Table 1.

Table 1. The converter's technical requirements.

Parameter	Symbol	Nominal Value	Min.	Max.	Unit
Input voltage	V_{in}	35	25	40	V
Input current	I_{in}	45	10	60	A
Input current ripple	ΔI_{in}	<1%	-	-	-
Output voltage	V_{out}	53	38	58.8	V
Rated efficiency	η	>95%	-	-	-
Total weight	-	<1.5	-	-	kg

As mentioned in Section 2, the design of the converter started from the fundamental requirements of reliability under the goals of weight minimization and efficiency maximization. Among the different possible architectures of the converter, the interleaved boost topology has been considered the most suitable one. This topology has numerous features that make it particularly suitable for the application under consideration. The interleaved structure allowed the current to be distributed over multiple switches, reducing both conduction and switching losses. By acting on the control of the different interleaved legs so as to operate them with appropriate phase shifting, it is possible to reduce the ripple amplitude of the output current and increase its effective frequency, thereby reducing the current operated by each switch, making it possible to choose components with a lower current value and allowing higher switching frequencies. All this results in a strong reduction in all filter elements, i.e., inductors and capacitors, resulting in strong reductions of weights. Furthermore, the reduction of the ripples in input current and voltage has been proved to significantly improve the PEMFC lifetime, hence increasing the reliability of the whole power conversion stage [6,37–39].

The interleaved architecture also meets the need for reliability by strongly improving the fault tolerance of the converter. In fact, even if one of the switches fails, the remaining phases can continue to operate by accepting a derating of the current [40].

Finally, the interleaved configuration opens up the possibility to perform phase shedding at the low-load level to improve converter efficiency [23].

3.1. Converter Architecture

Following a comparative evaluation that considered the usable space constraints; voltage and current levels; and possible micro-controllers that could be employed to implement the converter control, the architecture shown in Figure 5 was adopted.

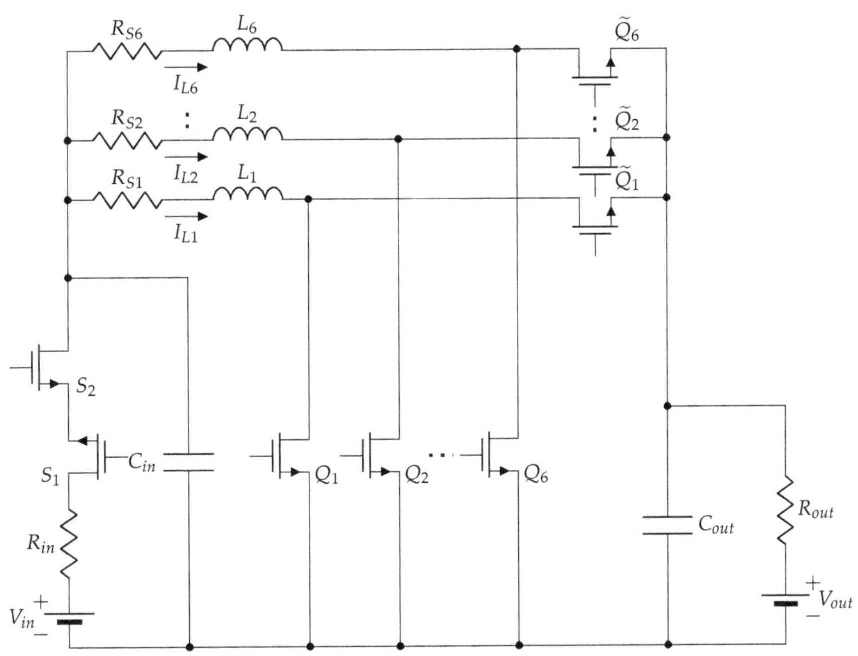

Figure 5. Power stage schematic.

Hence, the implemented converter resulted in six interleaved legs in the SIBC. The two mosfets, S_1 and S_2, connected in the so-called back-to-back configuration, are operated as electronic relays to guarantee the possibility of decoupling the source and the load voltages to prevent the circulation of currents through the body diodes of the top side mosfets $\tilde{Q}_{1,\ldots,6}$. This decoupling action is also essential to prevent the discharge of the SIBC's input bus capacitors during the PEMFC short-circuiting, allowing one to disconnect the PEMFC output from the converter's input side in case of occurrences of system faults. Hence, these mosfets have been selected with the lowest possible drain-to-source on-state resistance $R_{DS(on)}$ with the aim of reducing the conduction losses. In fact, these switches are mainly operated in static conditions, for which the dynamic performances assume secondary importance. The actual SIBC is based purely on mosfets avoiding diodes. This is again with the aim of taking advantage of the lower conduction resistance that these devices present with respect to diodes. According to the adopted six-leg architecture, the control of each leg is phase-shifted of $\pi/3$ rad, and each mosfet has to conduct only 1/6 of the whole current managed by the whole converter. The adopted mosfets can be operated at a switching frequency f_s of 400 kHz. Therefore, the output current ripple will present a frequency six times higher. This has allowed us to reduce the size of the filter inductors $L_{1,\ldots,6}$ on each leg and the size of the electrolytic capacitor C_{out}.

The resistors $R_{s1,\ldots,s6}$ are shunt resistors that have been chosen as current sensing devices thanks to their simplicity, extremely reduced power losses and good performances offered with respect to other current sensing devices [41].

3.2. Inductor Design

The sizing of the inductors has moved from the evaluation of the expected duty cycle operating range through Equations (1)–(3) that refer to minimum, maximum and nominal duty cycles, respectively, [18,42].

$$D_{\text{min}} = \frac{V_{\text{out,min}} - V_{\text{in,max}}}{V_{\text{out,min}}} \tag{1}$$

$$D_{\text{max}} = \frac{V_{\text{out,max}} - V_{\text{in,min}}}{V_{\text{out,max}}} \tag{2}$$

$$D_{\text{n}} = \frac{V_{\text{out,n}} - V_{\text{in,n}}}{V_{\text{out,n}}} \tag{3}$$

The value of the inductor is put in correlation with the duty cycle and the input voltage through Equation (4) by imposing the maximum allowed ripple current ΔI_{L}:

$$\Delta I_{\text{L}} = \frac{V_{\text{in}}}{L} D\, T \tag{4}$$

where T is the switching period.

The minimum value of the inductance is evaluated through Equation (5) with respect to the nominal average inductor current $I_{\text{L,avg}}$. For the sake of simplicity, the maximum allowed inductor ripple current ΔI_{L} has been set as equal to 60% of $I_{\text{L,avg}}$.

$$L_{\text{min}} = D_{\text{max}} \frac{V_{\text{in,min}}}{\Delta I_{\text{L}} f_{\text{s}}} \tag{5}$$

The obtained value needs an additional assessment with the minimum value of inductance $L_{\text{min,ccm}}$ that guarantees that the converter will operate in continuous conduction mode (CCM) for values of the input current higher than 30 A.

This is particularly important because synchronous switching is possible only if CCM is guaranteed (this aspect is detailed in Section 4.3). $L_{\text{min,ccm}}$ is evaluated as:

$$L_{\text{min,ccm}} = D_{\text{n}} \frac{V_{\text{in,n}}}{2 I_{\text{L,avg}} f_{\text{s}}} \tag{6}$$

The actual value of the adopted inductor is related to the most stringent constraint value among L_{min} and $L_{\text{min,ccm}}$, i.e., the one able to guarantee the respect of both conditions [43]. The values obtained in the design phase are reported in Table 2.

3.3. Other Components of the Power Stage

In this specific application, the choice of the output bus capacitors is not critical because of the adopted interleaved architecture and the contemporary direct connection to the battery pack that helps to smooth the output voltage. However, a 47 µF electrolytic capacitor was still inserted to filter out some of the harmonic content of the residual ripple. Conversely, the choice of the proper input capacitance was carried out with the help of circuit simulation to guarantee the respect of the desired ripple specifications while verifying compliance with the physical space limits.

Table 2. Design specifications of the inductors and characteristics of the chosen [43].

Parameter	Value	Unit
	Design specifications	
L_{min}	6.89	µH
$L_{min,ccm}$	3.589	µH
$I_{L,avg}$	10	A
	Selected component characteristics	
L_n	6.8	µH
$I_{\Delta 25^\circ C}$	36	A
I_{sat}	36	A
DCR_{typ}	2.84	Ω

The evaluation led to the choice of a solution based on the parallel connection of six 1 µF multi-layer ceramic capacitors (MLLCs) and one 47 µF electrolytic SMD capacitor for each phase. These hybrid solutions based on two different technologies allowed us to realize efficient filtering on the whole harmonic spectrum of the current ripple [44]. The power mosfets were selected from the Infineon Technologies AG SMD bottom-cooled portfolio [45] to achieve a reasonable balance between expected conduction losses (i.e., low on resistance) and switching losses (i.e., low gate capacitance). In each phase, a fast, dual-channel mosfet gate-driver was used [46]. This device provides input-to-output isolation, protecting the MCU, and thanks to a high-output current rating, it is able to drive the mosfet at a high switching frequency. The high-side mosfet gate-to-source voltage is obtained by means of bootstrap circuitry consisting of a 100 nF ceramic capacitor and a Schottky diode. Chosen components are listed in Table 3.

Table 3. Main electronic components.

Description	Manufacturer	Model Number
Power inductor	Vishay	IHLP8787MZER6R8M5A [43]
Electrolitic input capacitor	Vishay	MAL214699803E3 [44]
Power mosfet	Infineon Technologies	BSC072N08NS5 [45]
Mosfet gate-driver	Infineon Technologies	2EDF7275K [46]
Shunt resistance	Bourns	CSS2H-3920R-1L00F [47]
Current sense amplifier	Texas Instruments	INA290A3 [48]

The datasheet values of the selected components were put in the circuit simulation model built in the LTspice (by Analog Devices Inc., Wilmington, MA, USA) environment to assess the power stage performance. Additional details are provided in Section 5.

3.4. PCB

Downstream of the selection of the components and the verification of the main converter performances through circuit simulation, the design of the PCB followed. Its manufacturing specifications are summarized in Table 4.

The thickness of the copper layer was chosen to ensure the proper current density ratings. Sizing was done using Saturn PCB Design Toolkit [49]. The PCB was designed while making reference to the FR4 dielectric material and taking the input power supply line as a reference with the most stressed trace. As the FR-4 material presents a high-glass transition temperature T_g, the maximum working temperature was assumed to equal 100 °C. In addition, due to the quite high switching frequency and the relevant number of analog signal traces on the board, a four-layer PCB was chosen as a suitable layout able to provide appropriate shielding to possible electromagnetic interferences (EMI) [50].

Table 4. PCB specifications.

Parameter	Value	Unit
Material	FR-4	-
Board Thickness	1.6	mm
Number of layers	4	-
Copper trace width	70	µm
Board width	90	mm
Board length	310	mm
T_g	155	°C

Figure 6 shows the obtained PCB with the second-phase components concealed to make visible the circuits underneath. These parts are also detailed through the magnification on the left side.

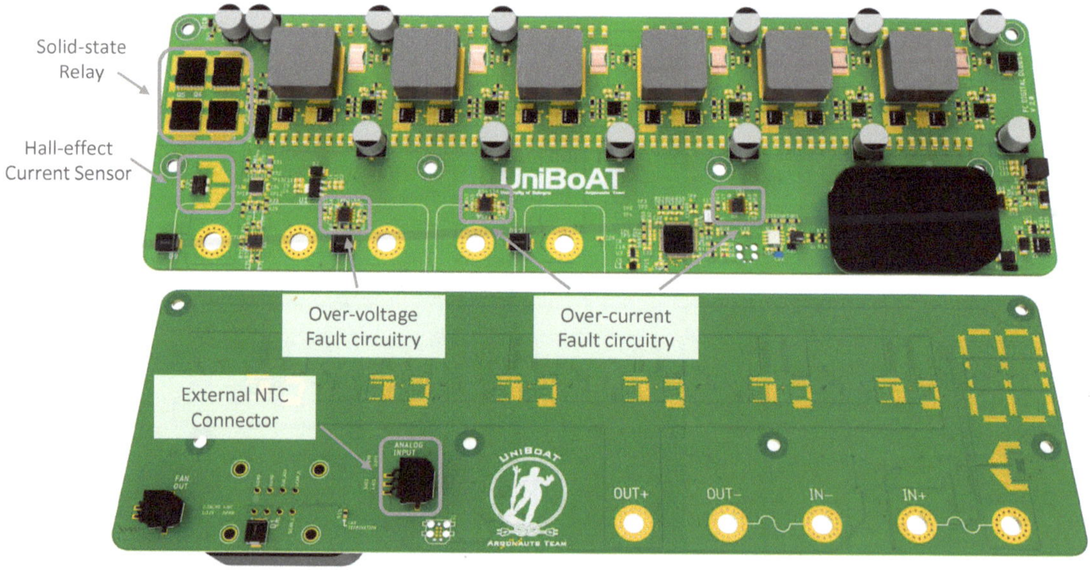

Figure 6. Top and bottom sides of the designed PCB with components.

For what concerns the placement of the components, particular effort was made to minimize the commutation loop of the mosfets. According to the high operation frequency adopted for the switches, the reduction of these loops is essential for the reduction of parasitic inductances resulting from the circuit traces that could introduce intolerable oscillations and overshoots during mosfet transitions [51]. In order to minimize the number of components adopted with a view to increasing the reliability of the converter, capacitors were not inserted near the rheophores of the mosfets, but the reduction in the switching loop area was obtained by building the loop traces vertically from top to bottom, utilizing inner layers. This allowed to restrict the loop area to the only PCB core thickness [50].

The bottom side of the PCB was kept empty to aid the heat transfer towards the PEMFC aluminum frame. In fact, in order to push weight reduction as far as possible, no specific heat sink was provided for the converter, but the external structure of the fuel cell was exploited instead. The thermal interface was further improved through a thermal gap pad and by placing 0.4 mm vias close to most stressed power components (make reference to Figure 7 magnification).

Figure 7. Detail of one phase on the PCB.

3.5. Safety and Fault Protection

Safety was taken into account during the design phase. Despite the conventional peak-current mode control scheme, where the overcurrent can be easily identified and reduced on a cycle-by-cycle basis, this approach is not possible under ACMC due to the high switching frequency [52]. However, thanks to the advanced features of the adopted microcontroller that provide dedicated fault input lines, all gate drive signals are immediately turned off once the fault condition is detected. In this way, low latency fault protections can be implemented.

The two substantial faults considered during the design phase were:

- Inductor overcurrent: Turn off the power stage if any of the inductor currents reach the fault threshold. This avoids a mosfet fault caused by controller errors and at the same time protects from output short-circuits. The threshold is made software-configurable via a CAN bus and must be 8–10 A higher than the average phase current taking the ripple into account.
- Input and output overvoltage: Turn off the power stage as soon as input or output voltages exceed a critical threshold level. This allows one to safeguard the electrolytic capacitors and prevents battery overvoltage.

The comparable logical network for overcurrent fault prevention is shown in Figure 8. The current measured at the shunt resistor is amplified and fed directly into each phase's open-drain operational amplifier (Op-Amp). Using the Op-Amp, each stage is configured to compare the signal to a voltage generated by a digital to analog converter (DAC). Then, the output of all the comparators is linked to form an OR gate, which is further connected to a high-resolution timer (HRTIMER) that generates the switching signals. Overvoltage protection uses a similar concept, except that the threshold is determined by a voltage divider; thus, it is fixed on the HW side to $V_{in,OV} = 50$ V and $V_{out,OV} = 59$ V. Finally, hardware protection against overcurrents is provided by means of fuses installed on both the input and output stages in the points indicated by the fuse symbol on the bottom side silk screen in Figure 6.

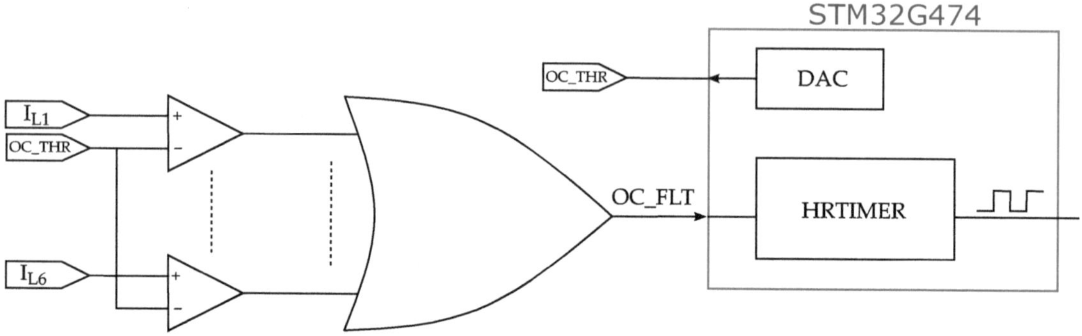

Figure 8. Overcurrent fault protection equivalent logical network.

4. Firmware Design

SIBCs require precise digital control, which needs the use of a microcontroller unit (MCU) featured with specific peripherals. The STM32G474 MCU by STMicroelectronics was chosen because this MCU is designed to address specific digital power conversion applications, such as Digital Switched-Mode Power Supply (D-SMPS) and motor-inverters.

The STM32G474 MCU has dedicated peripherals called HRTIMER that can achieve 184 ps resolution and generate complementary six-pair PWM output signals using a delay-locked loop (DLL). In addition, a dead-time generation unit is added to govern the signal's rising and falling edge without the need for a specific software implementation. The main computation unit is an Arm® 32-bit Cortex®-M4 CPU with a floating-point unit (FPU) operating at 170 MHz, ensuring adequate calculation performance to handle the controller loops and CAN communication interfaces [53].

4.1. Architecture

The firmware architecture is distinguished into two levels. One level is devoted to the soft-periodic functionalities, such as the external communication interface and the main finite state machine (FSM). These can be implemented on a real-time operating system, i.e., FreeRTOS developed for STM32 MCUs. A second level of the firmware manages the so-called hard-periodic functionalities, i.e., current control loops and duty cycle synchronous updates. These hard-period functionalities are executed inside the time-driven interrupt services routines.

Figure 9 depicts one of the six parallel boost phases' control logic. A proportional-integral (PI) controller is used to regulate the current. The amplified signal taken from the shunt resistors is used to implement overcurrent protection before being averaged by a low-pass filter (made by a simple RC network) and monitored by the analog to digital converter (ADC) integrated into the MCU. The microcontroller keeps the control loop at 20 kHz and writes the updated duty cycle set-point to the HRTIMER control register.

4.2. PWM Signal Generation

The generation and updating of PWM signals are some of the most critical and challenging tasks for the management of the interleaved architecture. In fact, the parallel nature of the selected topology necessitates accurate duty cycle synchronization.

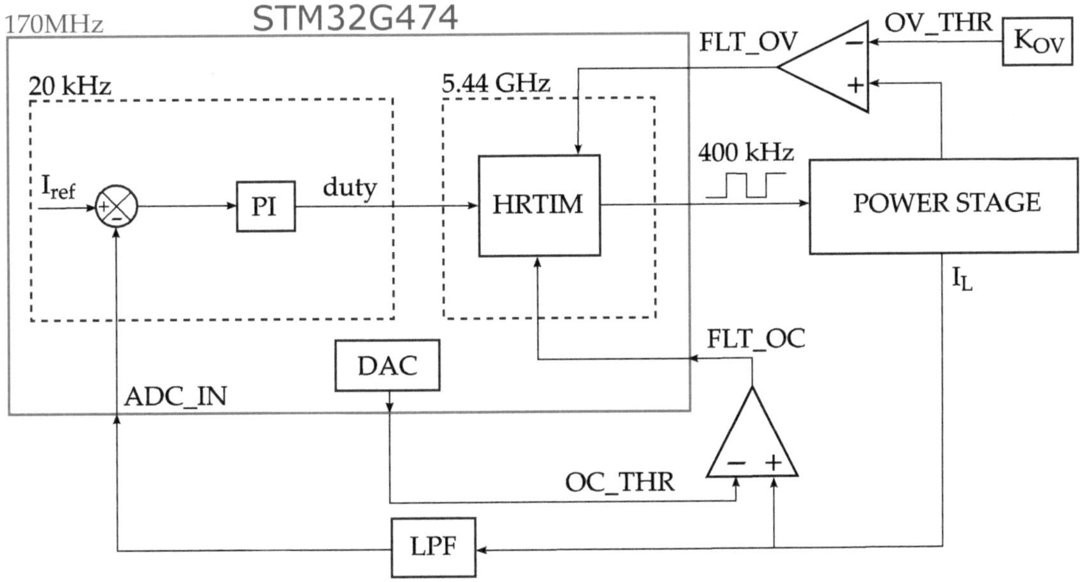

Figure 9. Single-phase controller block diagram with fault detection lines.

Among the specialized peripherals integrated into the MCU, the HRTIMER was designed specifically for this purpose. It comprises seven counting units, including a master and six slave timers, each of them with two output lines. For each timer unit, there are also four comparators. The master timer oversees phase management: while regularly resetting the slave clocks, it establishes a phase connection between the different legs of the boost converter. The relative phase shift is calculated by dividing 360° by the number of phases, which is maintained at 60° in this application. The master timer is set to continuous mode; thus, it counts the switching period roll-back till zero and starts again. The slave timers, instead, are set to re-triggerable single-shot mode and wait for master timer comparator events before start counting after their period is reached. The duty is imposed separately from the phase shift into each of the timers. This allows for phase shedding, which helps modify the number of active phases in real-time depending on the operating conditions (especially for light load [23]).

What has just been described is detailed graphically in Figure 10.

Analog conversions are initiated by a slave timer's second comparator, which is set to create a trigger at half the switching-off time. The ADC channel corresponding to the proper current signal is sampled when the event occurs. Switching disturbances are minimized as much as is feasible by selecting the sample point [54].

4.3. *Reverse Current Control*

Another fundamental aspect that needs to be considered while controlling a synchronous DC-DC converter connected to a battery is the occurrence and management of the DCM operation condition. In fact, when the inductor current falls to zero, the battery starts to supply current toward the input through the high-side mosfet, which is, at this time, conducting. DCM operation occurs under a light load, i.e., when the average inductor current is lower in comparison to the ripple. This phenomenon can be disruptive for the converter and cause degradation of the FC; therefore, it must be avoided. The easiest solution is to adopt the classical non-synchronous boost topology which intrinsically prevents negative current flow by replacing top side mosfet with a diode, but it has the drawback of efficiency loss (2–3%).

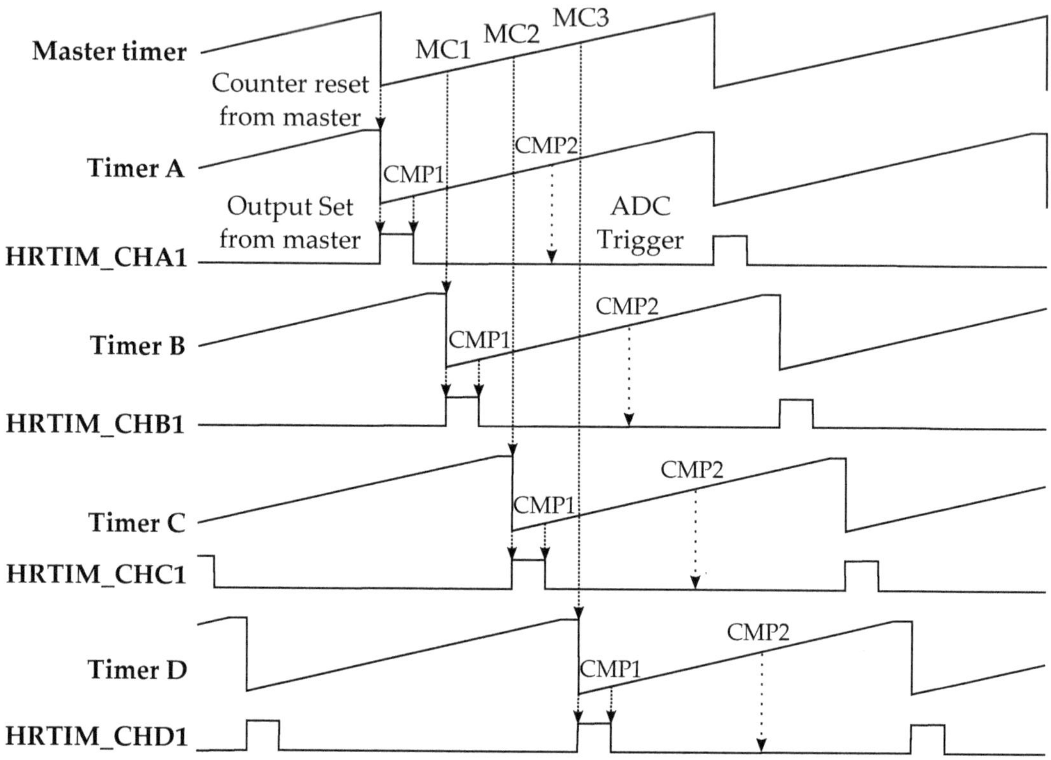

Figure 10. Timing diagram of PWM generation for 4 phases using high resolution timers.

The solution implemented here follows the mode boundary detector approach presented in [55]. This method identifies the two operating conditions on the basis of the parameter k, defined in Equation (7).

$$k = I_L - \frac{\Delta I_L}{2} \tag{7}$$

The converter operates in CCM for all values of $k > 0$; conversely, it operates in DCM for $k < 0$. At every control loop iteration, according to the k sign, the topside mosfet is kept off in case of DCM or enabled for CCM. A safe margin is also added, and non-synchronous mode is kept at every set point change and during the start-up phase.

5. Simulation Results

As introduced in Section 3.3, the validation of the design was performed by simulating the main components of the power stage on LTSpice.

Infineon Technologies AG provides the SPICE model for the selected device, allowing for a quite accurate simulation of its dynamic behavior [56]. The circuit model of one phase of the power stage is shown in Figure 11. It includes three circuit blocks: dead-time generation logic, gate drive, and bootstrap circuitry; and one phase of the synchronous boost converter. Battery and FC are simply modeled as non-ideal voltage sources with series resistance to approximate their dynamics under load. The adopted value of the battery ESR was obtained by starting from the available information referred to the single cell ESR and considering their parallel–series arrangement inside the whole battery pack. The fuel cell ESR was extracted as a first-order approximation of the voltage–current characteristic curve reported in Figure 3.

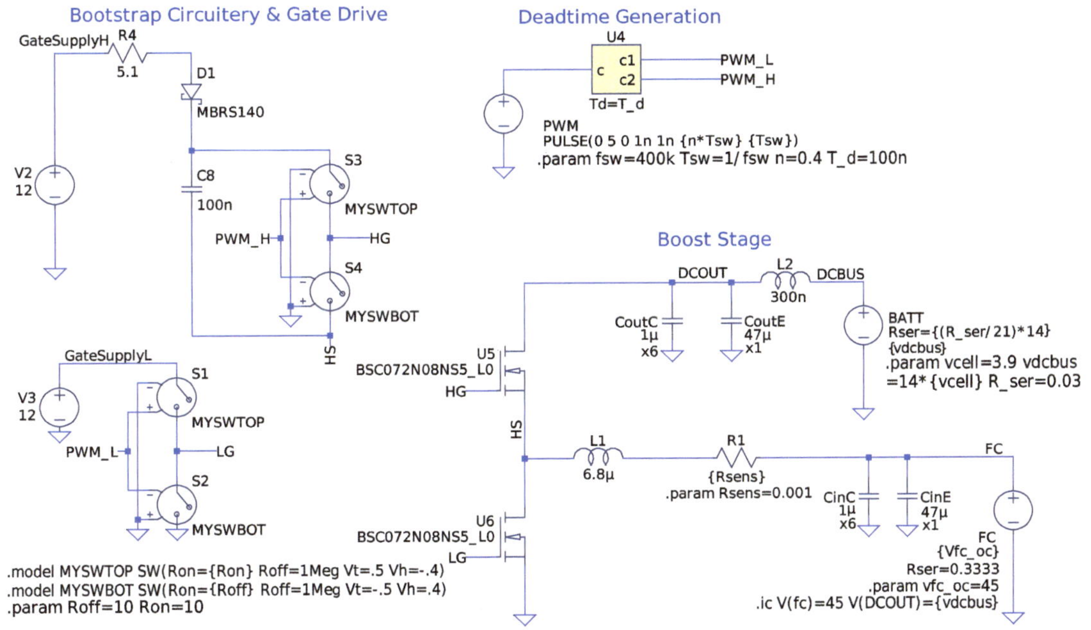

Figure 11. LTspice model of one phase of the boost converter.

Several simulations under different working conditions have been carried out to verify that the power components were working within tolerable voltage and current ranges and to validate the expected efficiency of the whole converter. The maximum values of critical electrical quantities for the main components of the power circuit have been given in Table 5. These values, extracted from the respective datasheets, are compared with the maximum values obtained in the simulation.

It appears clear that all the components are working well below their maximum rated limits. Indeed, an adequate safety margin was taken in order to to ensure the highest possible reliability by considering simulation uncertainties and accounting for possible variability due to variation in ambient working temperature. Table 5 reports also the information related to the bootstrap diode used for the powering of the high mosfet drive.

Finally, the whole six-phase power stage was simulated to assess the complete design with respect to the initial input–output ripple requirements. Figure 12 shows the waveform of the switching node voltage and the inductor current while the converter is operating at nominal conditions (i.e., $V_{in} = 34$ V, $V_{out} = 53.5$ V and $I_{in} = 40$ A).

Table 5. Main components adopted for the power stage. Considered electrical limit parameters extracted from the datasheets and related maximum values obtained in simulation.

Selected Device	Assessed Parameter	Max. Value from Simulation	Max. Rated Value	Unit
Mosfet	$I_{D,max}$	17.4	74	A
	$I_{D,rms}$	9.3	40	A
	$V_{DS,max}$	64	80	V
Inductor	$I_{L,rms}$	10.2	36	A
Shunt Resistor	P_{rms}	0.1	5	W
Input Capacitor	$I_{Cout,rms}$	0.45	0.47	A
Bootstrap Diode	$I_{bs,max}$	2.1	10	A
	$I_{bs,rms}$	0.1	2	A

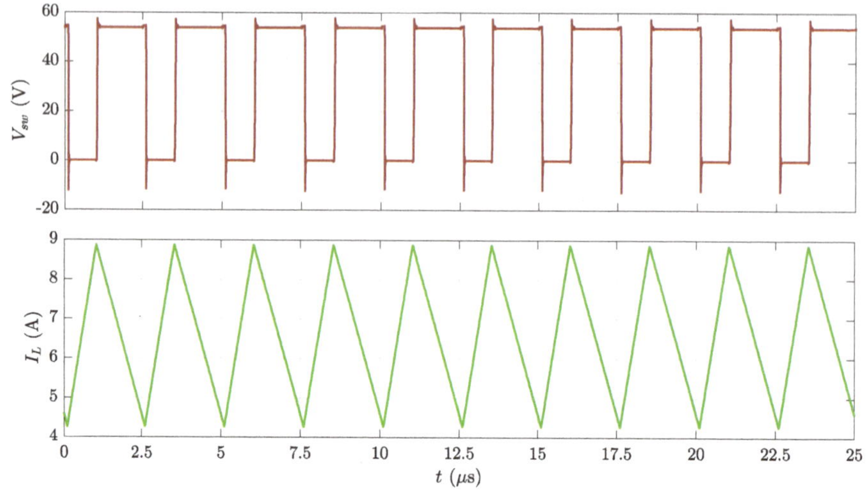

Figure 12. Switching node voltage (**top**). Inductor current (**bottom**).

Even if the inductor current ripple related to each phase presents a relatively high value of 4.5 A, the combination of the input capacitor and the operation of the six-phase interleaved arrangement has to lead to a low-input-current oscillation with amplitude slightly lower than 0.3 A. The resulting input current ripple shown in Figure 13 results in 0.7% of I_{in}, satisfying the converter's requirement.

The obtained power conversion efficiency was within the range 95.5% $< \eta <$ 97.5% when the converter worked in synchronous mode and 94% $< \eta <$ 95% under non-synchronous-mode operation.

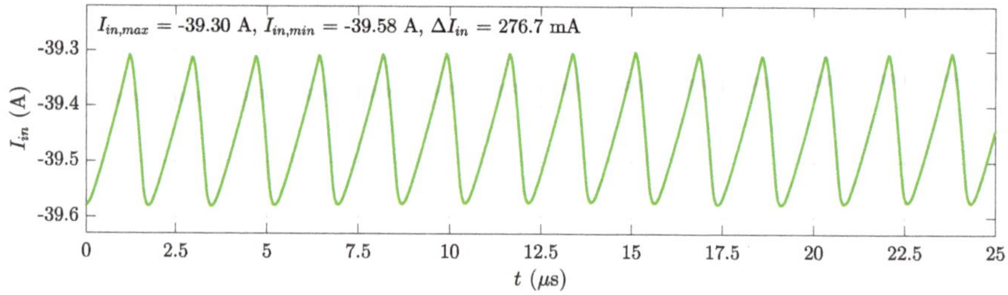

Figure 13. Input current ripple.

6. Experimental Validation

This section presents the results regarding the tests carried out on the prototype built on the basis of the evaluations and analyses presented in the previous sections. The results presented refer to both the tests carried out in the laboratory and those carried out in the real field.

6.1. Bench Test

The laboratory tests were carried out by using a bench power supply and an electronic active load to validate the converter functionalities and its performances in safe conditions.

One of the most important electrical parameters assessed during the laboratory test was the inductor current. This quantity was acquired by connecting an oscilloscope probe to the output of the current-sense amplifier on the converter board. Figure 14 shows the measured waveform of the inductor current when the converter operates under the conditions reported in Table 6.

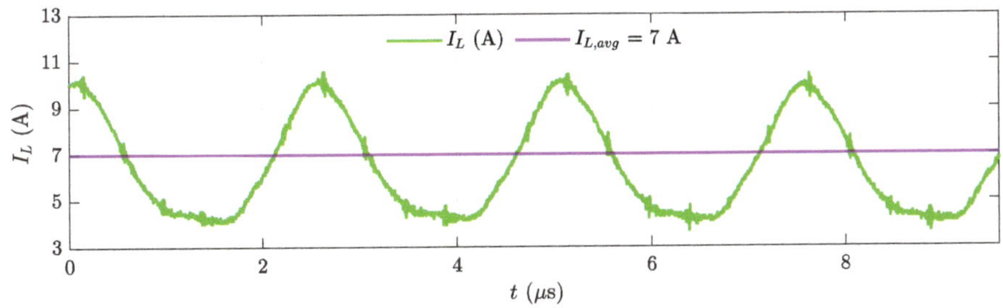

Figure 14. Inductor current waveform for $I_{\text{in}} = 42$ A.

Table 6. Bench test operating condition.

Parameter	Value	Unit
V_{in}	34	V
V_{out}	53.8	V
I_{in}	42	A
Duty cycle D	0.41	-
I_{ph}	7	A
P_{in}	1430	W

It can be seen that the comparison of the data obtained in simulation versus the measurements indicates a 1.28 A (+28%) increment in the amplitude of the ripple of the inductor current due to real inductor uncertainty and board non-idealities.

Another series of tests has concerned the verification of the cooling performance and the respecting of the temperature limits under the two synchronous and non-synchronous operating modes. The thermal camera images shown in Figure 15 were taken in steady-state conditions and with the fuel-cell case at 45 °C.

In non-synchronous mode, the high-side mosfet, which practically works as a diode, is the most stressed component. In said condition, this component reaches 100 °C. In synchronous mode, instead, the loss, and thus the heat, are distributed evenly over both mosfets of each phase. This operating mode allows the temperature to be lowered by 20 °C compared to the non-synchronous mode.

The results related to the converter efficiency in the different operating modes are summarized in the graphs of Figure 16.

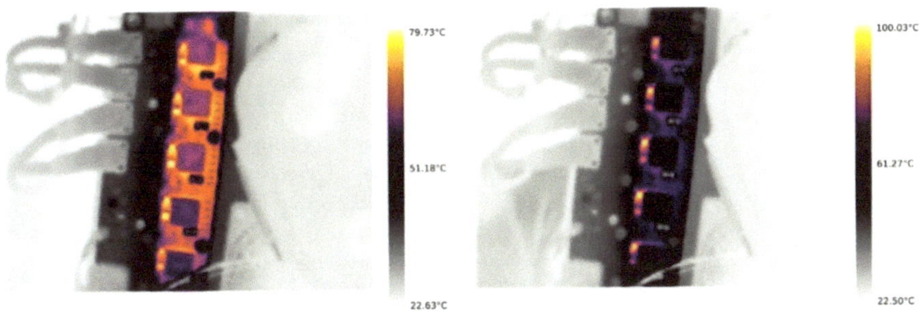

Figure 15. Temperatures of the mosfets at $I_{\text{in}} = 45$ A for the non-synchronous operation mode (**right**) and for the synchronous operation mode (**left**) obtained with a thermal camera.

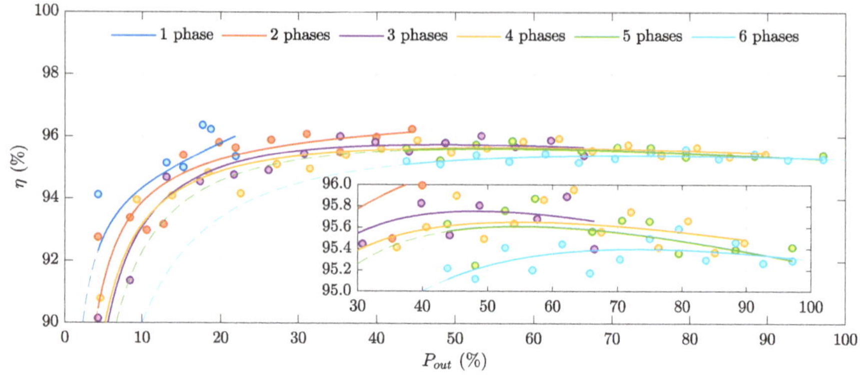

(a)

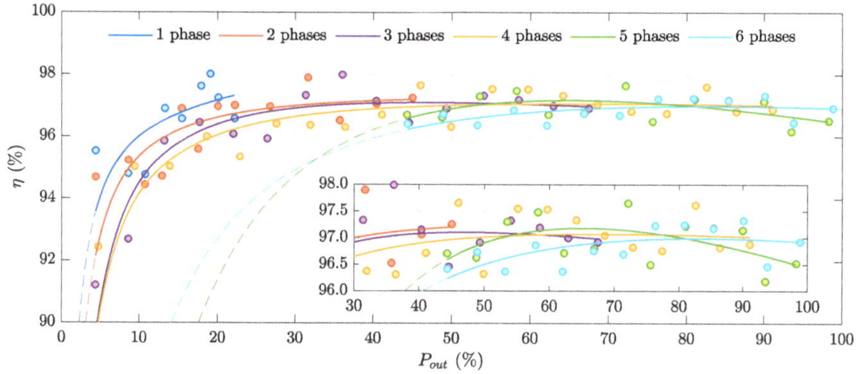

(b)

Figure 16. Converter efficiencyw with various numbers of phases operating versus power increase normalized to the rated power of 1.5 kW in non-synchronous (**a**) and synchronous (**b**) modes of operation. The experimental data are represented with colored circles, and the traces represent the interpolations of the datasets.

Tests were conducted by increasing the number of active phases with the increase in the output power of the converter. In this way, multiple efficiency traces were achieved, and the effectiveness of the phase shedding technique was readily validated. Each phase was rated 250 W (i.e., 1/6 of the total rated power of the converter of 1.5 kW). Nevertheless, the tests were performed for a short time above the rated value to increase the experimental dataset's range. For Figure 16, the experimental data, indicated with the colored circles, were interpolated to provide a continuous indication of the trend in efficiency as the number of active phases increases. These validations were carried out for both non-synchronous (see Figure 16a) and synchronous modes of operation (see Figure 16b). The observations indicate that during low-power operation, the use of a limited number of phase converters (one or two phases) provides better efficiency than operating with all six phases. In contrast, the use of more phases for a power closer to the rated level allows for an almost flat efficiency profile.

For a more immediate and clear visualization of the results, Figure 17 shows the efficiency trend in the case of permanent use of all six phases and in the case of the use of the phase shedding technique.

The interpolation for the phase shedding was computed bearing in mind that each leg can withstand at most 1/6 of the rated power; this means that only points displayed as red dots were eligible for the calculation. This assumption was made to preserve each active phase within the range of its maximum local efficiency [57]. The efficiency trace related to all six phases is identical to the one visible in Figure 16. In this case, a 1% difference between the non-synchronous (Figure 17a) and synchronous (Figure 17b) modes of operation can be observed. During low-power operation, the efficiency improved by a few percentage points between phase shedding and six phases used in both operating modes.

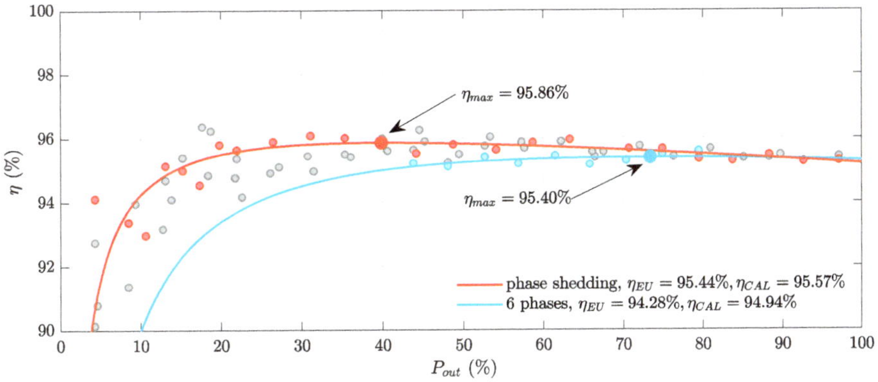

(**a**)

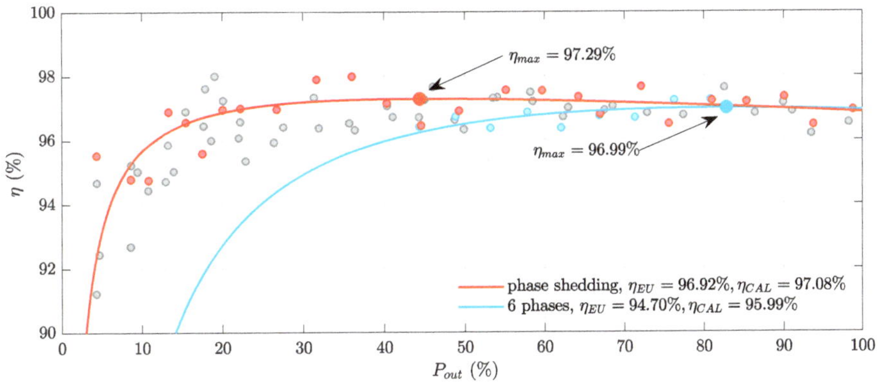

(**b**)

Figure 17. Efficiency comparing interpolated phase shedding and six phases of non-synchronous operation (**a**) and synchronous operation (**b**). Experimental data are represented with colored circles, and continuous traces represent the interpolation.

Table 7 provides European (η_{EU}) and Californian (η_{CAL}) efficiencies [58], together with the maximum experimental and interpolated points. In this way, multiple figures of merit are provided, ensuring overall efficiency assessment in both modes of operation.

Table 7. European, Californian, maximum experimental, and maximum interpolated efficiencies computed for both non-synchronous and synchronous operating modes.

Mode		η_{EU}	η_{CAL}	$\eta_{max,EXP}$	$\eta_{max,INT}$
Non-Synchronous	Phase shedding	95.44%	95.57%	96.08%	95.86%
	Six phases	94.28%	94.94%	95.59%	95.40%
Synchronous	Phase shedding	96.92%	97.08%	97.99%	97.29%
	Six phases	94.70%	95.99%	97.33%	96.99%

In the case of non-synchronous mode, the η_{EU} is 1% higher with the use of phase shedding. Similarly, a 0.63% difference can be seen in the case of η_{CAL}. Moreover, a difference lower than 0.5% can be seen for the maximum efficiency value (η_{max}) calculated from experimental data ($\eta_{max,EXP}$) and from interpolated data ($\eta_{max,INT}$). On the other hand, phase shedding in synchronous mode resulted in η_{EU} and η_{CAL} being 2% and 1% higher, respectively, than the continuous operation with all six phases. This demonstrates the effectiveness of the phase-shedding technique.

In conclusion, phase shedding's effectiveness was assessed for both synchronous and non-synchronous modes. In particular, synchronous mode proved to be more efficient at all working points and in all operating modes.

6.2. In-Field Test

The resulting complete prototype composed of the fuel-cell stack, the controller (with the SCU inside) and the DC-DC converter is shown in Figure 18.

Figure 18. Photo of the complete system prototype comprising the H-2000 fuel cell, the controller (**left**-side board) and the DC-DC converter (**right**-side board) with and without enclosures.

The electrical and electronic parts were equipped with automotive-grade connectors and enclosed inside 3D printed covers specifically designed to guarantee equivalent IP66 protection. The data during the boat cruising were collected by means of a vehicle control unit (VCU). The VCU recorded all operating variables at 10 ms and provided different log files that were imported in Matlab for the offline analysis of the performance in the real case scenario.

Figure 19 shows the input current, input power and output power waveforms measured by the DC-DC MCU and transmitted to the VCU through the CAN bus.

The input-current behavior highlights the intervention of SCU. When SCU is about to take action, the FC controller notifies the DC-DC converter that a short circuit is going to occur. Then, the DC-DC switches are turned off and the back-to-back input mosfets open. Once the SCU intervention period is over, the DC-DC automatically returns to operating normally. Naturally, this cyclic operation causes a decrease in the average power delivered by the fuel cell, as can be noticed by observing the power waveforms.

The operating temperatures were continuously monitored by means of two NTC sensors glued with thermal compound to the first phase mosfets and other three SMD NTCs installed directly on the PCB, as shown in Figure 20.

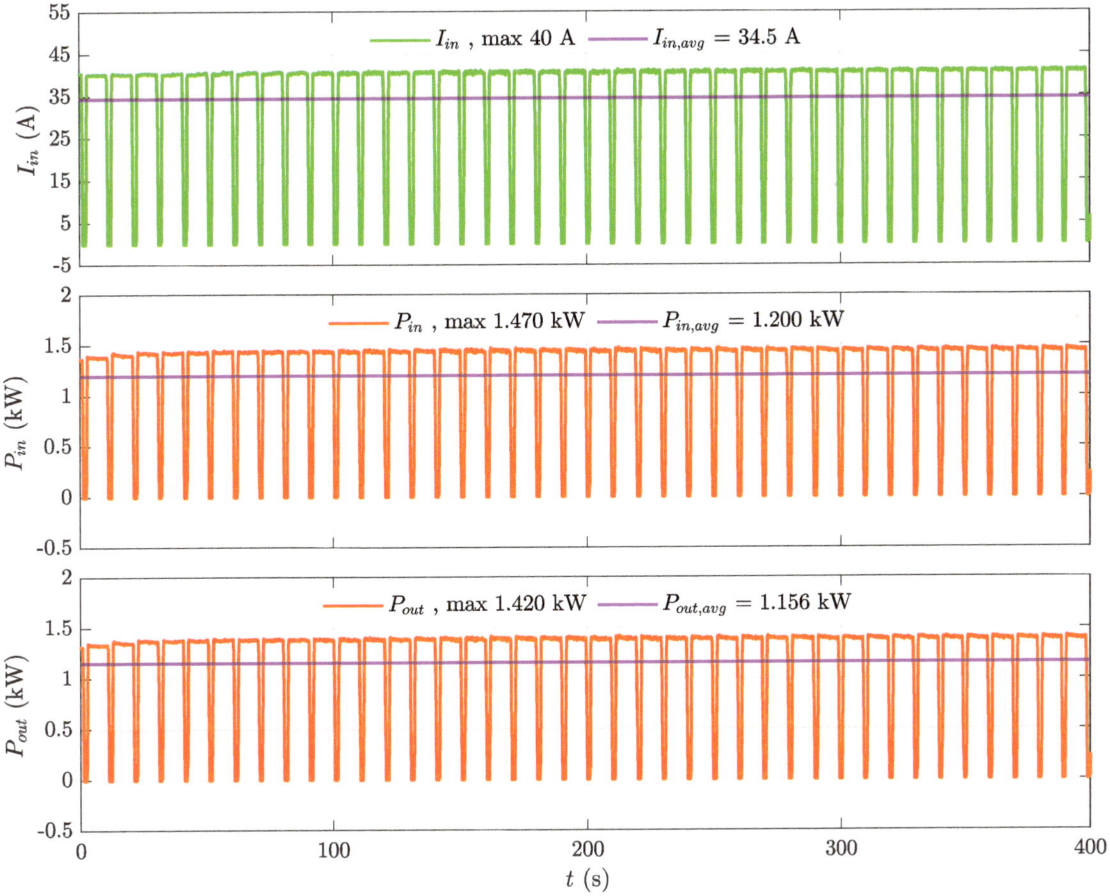

Figure 19. Data acquired during the sea trial with the reference input current target of 40 A (**top**): input power (**middle**) and output power (**bottom**).

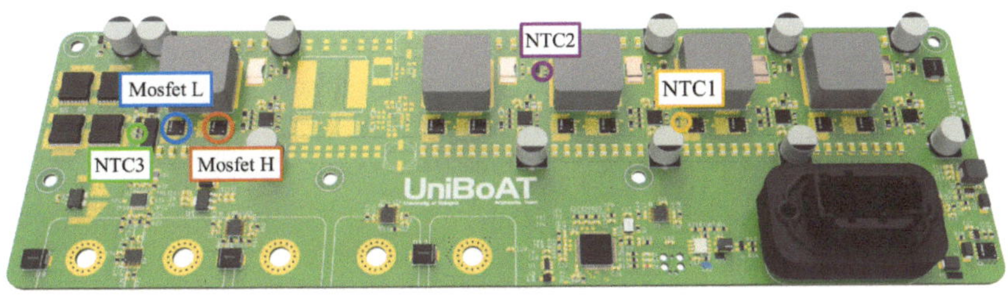

Figure 20. Placement of the temperature sensors on the PCB and mosfets.

The measured temperatures are shown in Figure 21. At steady-state, all the temperatures were verified to be below the ratings of the components with an adequate margin. These results demonstrated good performance of thermal dissipation.

The results obtained from the two tests and the different set of working conditions are summarized in Table 8.

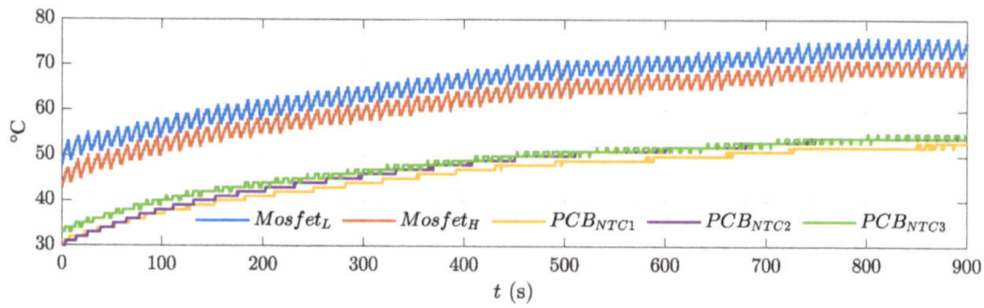

Figure 21. PCB and mosfet temperatures for rated $P_{in} = 1.5$ kW.

Table 8. In-field test results.

Parameter	Test 1	Test 2	Unit
V_{in}	35.3	34.7	V
V_{out}	50.5	53.0	V
I_{in}	40	46	A
D	0.33	0.37	-
$P_{avg,in}$	1200	1382	W
$P_{avg,out}$	1156	1342	W
T_{mosfet}	76	83	°C
η_{avg}	96.3%	97.1%	-

Figure 22 shows the motor current, the boat speed, the battery voltage, and the converter's input current obtained during the sea trial cycle used to test the whole boat power and traction chain.

The rapid changes in the speed that mirror fast transients in the motor current are due to the boat's acceleration, deceleration, and veering. The motor current goes to zero during the period of stopping of a few tens of seconds. It can be noticed that a slight variation in battery voltage (V_{bat}) occurs in correspondence with the motor current variation. Despite the input/output voltage variation or load current variation, the converter input current ($I_{in,DC\text{-}DC}$) is always kept constant at the rated value. Even in the real data shown in Figure 23, the fuel cell power is clearly kept constant while the battery power is changing in accordance with the motor's behavior. This proves the better performance of the regulator during the load transients.

Figure 23 is reported to provide a comprehensive view of the power chain power flows. The shown data were collected during the endurance race held for 3 h. In the first half-hour phase of the race, the fuel cell was not used in order to preserve the limited amount of hydrogen stored in the tanks onboard that was instead used in the subsequent two and half hours. Consequently, the power associated with the FC ($P_{DC\text{-}DC}$) was zero. Therefore, the power provided to the motor (P_{motor}) during the first phase was a combination of battery (P_{bat}) and PV (P_{pv}) power. The majority of the power of the first phase was provided by the battery. This allows one to keep the DC bus voltage as high as possible to avoid absorbing a high current from the battery once a lower state of charge is reached. This helps to preserve the battery's health. When the FC starts contributing, a reduction in power provided by the battery can be observed. It is more clearly visible from the moving average plot ($P_{bat,avg}$) calculated by considering an averaging window width of 10 s. This assumption was made based on the interval of SCU action. A magnification of the plots was provided to show how the full power for the traction is provided by the battery during each SCU activation.

About half way through the race, the motor power was reduced to preserve the total energy for the next races mentioned in the introduction. The fuel cell was operated at the rated value in the entire race; therefore, a reduction in battery power was observed during this period. The sharp load transients visible in Figure 23 occurred for various reasons, such as avoiding dangerous situations while sealing and fast overtakes.

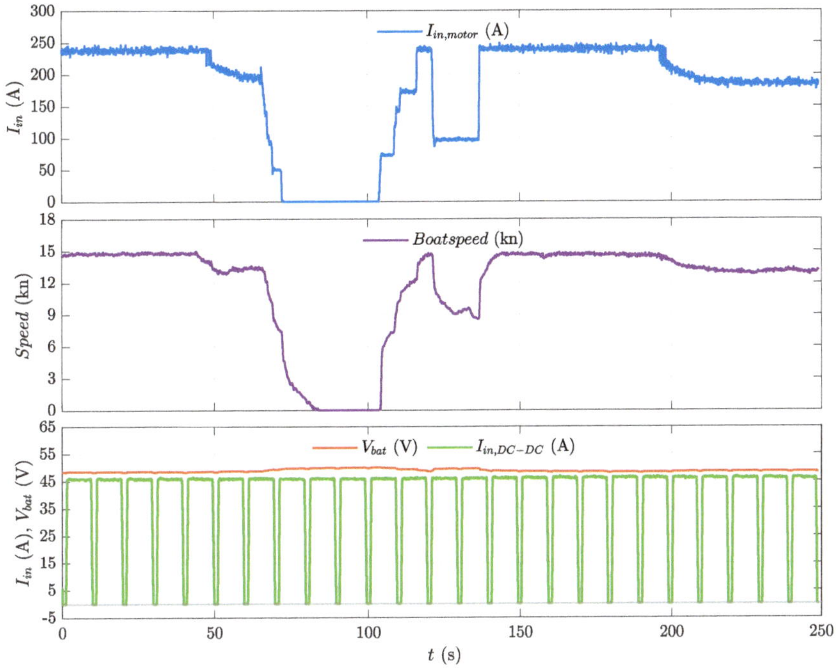

Figure 22. Data acquired during the sea trail depicting the motor current transients (**top**), boat speed in knots (**middle**), constant DC-DC converter input current, and battery voltage (**bottom**). One knot equals 1.85 km/h.

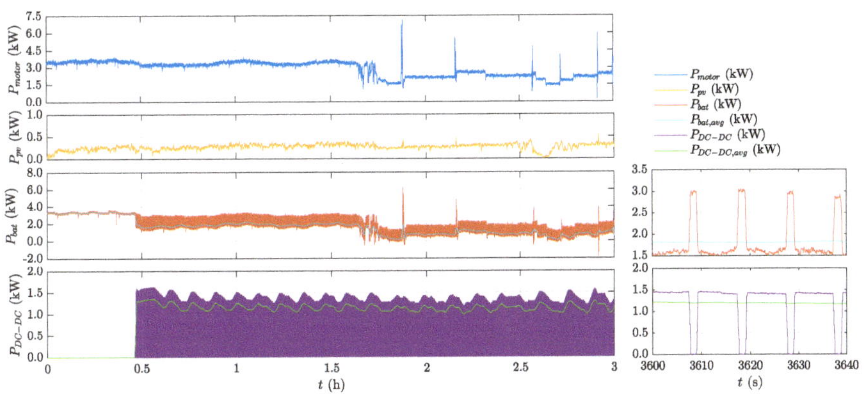

Figure 23. Data acquired during endurance race depicting the motor power, PV power, battery power, and DC-DC converter input power, respectively; and the battery and DC-DC power moving average tested under 10 s of window length. Right-hand plots are the zoomed-in versions of battery and DC-DC power.

7. Conclusions

The boost DC-DC converter discussed in the present work was demonstrated to fulfill the design specifications and proved to be reliable when working both in the laboratory and in real operating conditions during navigation and competition.

Thanks to the choice of high-performing electrical components and the extensive analysis of the PCB layout, the switching frequency could be kept high without compromising the whole converter's performance. This, along with the optimized FW and control strategies, made it possible to realize a lightweight (700 g) and efficient (94.5% $< \eta <$ 97.5%) converter.

The digital implementation of the ACMC precisely regulated the FC current, which followed the reference set-point even in the presence of the fast variation caused by the presence of the SCU operation. The common drawback of ACMC's inability to perform cycle-by-cycle overcurrent protection was completely overcome by using specifically designed fault circuitry and low-latency MCU input lines.

The converter's modularity is now under study to achieve a higher FC power rating employing the same base structure discussed in this paper. In this way, technological transfer and industrialization for applications not strictly related to naval racing that could take advantage of the great power density ensured by using FC case as a converter heat sink could be enabled by taking advantage of the scale economy.

Author Contributions: Conceptualization, M.R. (Michael Rimondi), R.M., V.C. and M.R. (Mattia Ricco); methodology, M.R. (Michael Rimondi), R.M. and M.R. (Mattia Ricco); software, M.R. (Michael Rimondi), R.M. and L.K.P.; validation, R.M., V.C., M.R. (Mattia Ricco) and G.G.; formal analysis, R.M., V.C. and M.R. (Mattia Ricco); investigation, M.R. (Michael Rimondi), R.M. and L.K.P.; resources, M.R. (Mattia Ricco) and G.G.; data curation, M.R. (Michael Rimondi), R.M., V.C. and L.K.P.; writing—original draft preparation, M.R. (Michael Rimondi), R.M., V.C. and L.K.P.; writing—review and editing, M.R. (Michael Rimondi), R.M., V.C., M.R. (Mattia Ricco) and G.G.; visualization, M.R. (Michael Rimondi), R.M., V.C. and L.K.P.; supervision, R.M., V.C., M.R. (Mattia Ricco) and G.G. All authors have read and agreed to the published version of the manuscript.

Funding: This research received no external funding.

Data Availability Statement: Not applicable.

Acknowledgments: The authors would like to thank all UniBoAT team members for the work done together and for the important experience of human and professional growth.

Conflicts of Interest: The authors declare no conflict of interest.

Abbreviations

The following abbreviations are used in this manuscript:

ACMC	Average Current Mode Control
ADC	Analogue to Digital Converter
CAN	Controller Area Network
CCM	Continuous Conduction Mode
DAC	Digital to Analogue Converter
DC-DC	Direct current to direct current
SMPS	Switching-Mode Power Supply
DCM	Discontinuous Conduction Mode
DLL	Delay Locked Loop
EMI	Electromagnetic interference
FC	Fuel Cell
FPU	Floating Point Unit
FSM	Finite State Machine
FW	Firmware
HW	Hardware
DCR	DC Resistance
MCU	MicroController Unit
MPPT	Maximum Power Point Tracking
OEM	Original Equipment Manufacturer
PCB	Printed Circuit Board
PEM	Proton Exchange Membrane
PEMFC	Proton Exchange Membrane Fuel Cell

PI	Proportional integrative
PWM	Pulse Width Modulation
SCU	Short Circuit Unit
SIBC	Syncronous Interleaved Boost Converter
VCU	Vehicle Control Unit

References

1. Han, J.; Charpentier, J.F.; Tang, T. An Energy Management System of a Fuel Cell/Battery Hybrid Boat. *Energies* **2014**, *7*, 2799–2820. [CrossRef]
2. Monaco Energy Boat Challenge. Available online: https://energyboatchallenge.com/en/home-english/ (accessed on 9 November 2022).
3. Tang, Y.; Yuan, W.; Pan, M.; Wan, Z. Experimental investigation on the dynamic performance of a hybrid PEM fuel cell/battery system for lightweight electric vehicle application. *Appl. Energy* **2011**, *88*, 68–76. [CrossRef]
4. Jin, K.; Ruan, X.; Yang, M.; Xu, M. A Hybrid Fuel Cell Power System. *IEEE Trans. Ind. Electron.* **2009**, *56*, 1212–1222. [CrossRef]
5. Warncke, M.; Fahlbusch, S.; Hoffmann, K.F. DC/DC-converter for fuel cell integration in more electric aircraft applications. In Proceedings of the 2017 19th European Conference on Power Electronics and Applications (EPE'17 ECCE Europe), Warsaw, Poland, 11–14 September 2017; pp. 1–10. [CrossRef]
6. Guilbert, D.; Gaillard, A.; Mohammadi, A.; N'Diaye, A.; Djerdir, A. Investigation of the interactions between proton exchange membrane fuel cell and interleaved DC/DC boost converter in case of power switch faults. *Int. J. Hydrog. Energy* **2015**, *40*, 519–537. [CrossRef]
7. Shojaeian, H.; Hasanzadeh, S.; Salehi, S.M. A Single Switch High Voltage Gain DC-DC Converter Based on Coupled Inductor and Switched-Capacitor for Renewable Energy Systems. In Proceedings of the 2021 12th Power Electronics, Drive Systems, and Technologies Conference (PEDSTC), Tabriz, Iran, 2–4 February 2021; pp. 1–6. [CrossRef]
8. Cao, G.; Zhang, J.; Wang, Y. Non-isolated Ultrahigh Boost Ratio DCDC Converter with Coupled Inductor and Charge Pump. In Proceedings of the 2021 IEEE 1st International Power Electronics and Application Symposium (PEAS), Shanghai, China, 13–15 November 2021; pp. 1–4. [CrossRef]
9. Forouzesh, M.; Siwakoti, Y.P.; Gorji, S.A.; Blaabjerg, F.; Lehman, B. Step-Up DC–DC Converters: A Comprehensive Review of Voltage-Boosting Techniques, Topologies, and Applications. *IEEE Trans. Power Electron.* **2017**, *32*, 9143–9178. [CrossRef]
10. Yu, X.; Starke, M.; Tolbert, L.; Ozpineci, B. Fuel cell power conditioning for electric power applications: A summary. *Electr. Power Appl. IET* **2007**, *1*, 643–656. [CrossRef]
11. Ellabban, O.; Hegazy, O.; Van Mierlo, J.; Lataire, P. Dual loop digital control design and implementation of a DSP based high power boost converter in fuel cell electric vehicle. In Proceedings of the 2010 12th International Conference on Optimization of Electrical and Electronic Equipment, Brasov, Romania, 20–22 May 2010; pp. 610–617. [CrossRef]
12. Thiyagarajan, A.; Praveen Kumar, S.G.; Nandini, A. Analysis and comparison of conventional and interleaved DC/DC boost converter. In Proceedings of the Second International Conference on Current Trends In Engineering and Technology-ICCTET 2014, Coimbatore, India, 8 July 2014; pp. 198–205. [CrossRef]
13. Lee, J.Y.; Jeong, Y.S.; Han, B.M. An Isolated DC/DC Converter Using High-Frequency Unregulated *LLC* Resonant Converter for Fuel Cell Applications. *IEEE Trans. Ind. Electron.* **2011**, *58*, 2926–2934.[CrossRef]
14. Boscaino, V.; Miceli, R.; Buccella, C.; Cecati, C.; Latafat, H.; Razi, K. Fuel Cell power system with LLC resonant DC/DC converter. In Proceedings of the 2014 IEEE International Electric Vehicle Conference (IEVC), Florence, Italy, 17–19 December 2014; pp. 1–6. [CrossRef]
15. Ashique, R.H.; Salam, Z.; Maruf, M.H.; Shihavuddin, A.; Islam, M.T.; Rahman, M.F.; Kotsampopoulos, P.; Fayek, H.H. A Comparative Analysis of Soft Switching Techniques in Reducing the Energy Loss and Improving the Soft Switching Range in Power Converters. *Electronics* **2022**, *11*, 1062. [CrossRef]
16. Bhuvaneswari, C.; Babu, R.S.R. A review on LLC Resonant Converter. In Proceedings of the 2016 International Conference on Computation of Power, Energy Information and Commuincation (ICCPEIC), Chennai, India, 20–21 April 2016; pp. 620–623. [CrossRef]
17. Khanipah, N.H.A.; Azri, M.; Talib, M.H.N.; Ibrahim, Z.; Rahim, N.A. Interleaved boost converter for fuel cell application with constant voltage technique. In Proceedings of the 2017 IEEE Conference on Energy Conversion (CENCON), Kuala Lumpur, Malaysia, 30–31 October 2017; pp. 55–60. [CrossRef]
18. Rahavi, J.A.; Kanagapriya, T.; Seyezhai, R. Design and analysis of Interleaved Boost Converter for renewable energy source. In Proceedings of the 2012 International Conference on Computing, Electronics and Electrical Technologies (ICCEET), Tamil Nadu, India, 21–22 March 2012; pp. 447–451. [CrossRef]
19. Salehi, S.M.; Dehghan, S.M.; Hasanzadeh, S. Interleaved-Input Series-Output Ultra-High Voltage Gain DC–DC Converter. *IEEE Trans. Power Electron.* **2019**, *34*, 3397–3406. [CrossRef]
20. Drobnic, K.; Grandi, G.; Hammami, M.; Mandrioli, R.; Ricco, M.; Viatkin, A.; Vujacic, M. An Output Ripple-Free Fast Charger for Electric Vehicles Based on Grid-Tied Modular Three-Phase Interleaved Converters. *IEEE Trans. Ind. Appl.* **2019**, *55*, 6102–6114. [CrossRef]

21. Wang, H. Design and Control of a 6-Phase Interleaved Boost Converter Based on SiC Semiconductors with EIS Functionality for Fuel Cell Electric Vehicle. Ph.D. Thesis, Université Bourgogne Franche-Comté, Besancon, France, 2019.
22. Velázquez-Elizondo, P.E.; Araujo-Vargas, I.; Villarruel-Parra, A.; Gómez-Olguín, F.J.; Cano-Pulido, K.; Granados-Luna, T.R. Six-phase dual-interleaved buck-boost converter for high power-density automotive applications. *J. Eng.* **2019**, *2019*, 3592–3597. [CrossRef]
23. Alharbi, M.A.; Dahidah, M.; Ali, S.; Ethni, S.; Pickert, V. Current Ripple Minimisation Based on Phase-Shedding of DC-DC Interleaved Converters for EV Charging System. In Proceedings of the IECON 2019-45th Annual Conference of the IEEE Industrial Electronics Society, Lisbon, Portugal, 14–17 October 2019; Volume 1, pp. 3456–3462. [CrossRef]
24. Andújar, J.; Segura, F.; Durán, E.; Rentería, L. Optimal interface based on power electronics in distributed generation systems for fuel cells. *Renew. Energy* **2011**, *36*, 2759–2770. [CrossRef]
25. Horizon. H2000 PEM Fuel Cell. 2020. Available online: https://www.horizoneducational.com/h-2000-pem-fuel-cell-2kw/p1357 (accessed on 9 November 2022).
26. Jarry, T.; Lacressonnière, F.; Jaafar, A.; Turpin, C.; Scohy, M. Modeling and Sizing of a Fuel Cell—Lithium-Ion Battery Direct Hybridization System for Aeronautical Application. *Energies* **2021**, *14*, 7655. [CrossRef]
27. Barragán, A.J.; Enrique, J.M.; Segura, F.; Andújar, J.M. Iterative Fuzzy Modeling of Hydrogen Fuel Cells by the Extended Kalman Filter. *IEEE Access* **2020**, *8*, 180280–180294. [CrossRef]
28. Gupta, G.; Wu, B.; Mylius, S.; Offer, G. A systematic study on the use of short circuiting for the improvement of proton exchange membrane fuel cell performance. *Int. J. Hydrogen Energy* **2016**, *42*, 4320–4327. [CrossRef]
29. Grumm, F.; Schumann, M.; Cosse, C.; Plenz, M.; Lücken, A.; Schulz, D. Short Circuit Characteristics of PEM Fuel Cells for Grid Integration Applications. *Electronics* **2020**, *9*, 602. [CrossRef]
30. Kabalo, M.; Blunier, B.; Bouquain, D.; Miraoui, A. State-of-the-art of DC-DC converters for fuel cell vehicles. In Proceedings of the 2010 IEEE Vehicle Power and Propulsion Conference, Lille, France, 1–3 September 2010; pp. 1–6. [CrossRef]
31. Alavi, O.; Rajabloo, T.; De Ceuninck, W.; Daenen, M. Non-Isolated DC-DC Converters in Fuel Cell Applications: Thermal Analysis and Reliability Comparison. *Appl. Sci.* **2022**, *12*, 5026.[CrossRef]
32. Hasaneen, B.M.; Elbaset Mohammed, A.A. Design and simulation of DC/DC boost converter. In Proceedings of the 2008 12th International Middle-East Power System Conference, Aswan, Egypt, 12–15 March 2008; pp. 335–340. [CrossRef]
33. Song, K.T.; Song, S.H.; Wang, S.S. A Fuel Cell Power Supply System Based on Charging Scheduling of Multiple Lithium Batteries. In Proceedings of the 2018 International Conference on System Science and Engineering (ICSSE), New Taipei City, Taiwan, 28–30 June 2018; pp. 1–6. [CrossRef]
34. Latha, K.; Maadhuri, M.; Maheswari, B.U.; Rajalakshmi, N.; Dhathathreyan, K.S. Current Mode Control of Fuel Cell Fed Interleaved Boost Converter for Ripple Current Minimization. In Proceedings of the ASME 2011 9th International Conference on Fuel Cell Science, Engineering and Technology, Washington, DC, USA, 7–10 August 2011.
35. Dixon, L. Average current mode control of switching power supplies. In *Unitrode Power Supply Design Seminar Handbook* ; Unitrode Corporation: Merrimack, NH, USA, 1990; pp. 1–5.
36. Saini, D.K.; Reatti, A.; Kazimierczuk, M.K. Average current-mode control of buck dc-dc converter with reduced control voltage ripple. In Proceedings of the IECON 2016-42nd Annual Conference of the IEEE Industrial Electronics Society, Florence, Italy, 23–26 October 2016; pp. 3270–3275. [CrossRef]
37. Kim, J.; Jang, M.; Cho, B.; Tak, Y. Characteristic analysis and modeling on PEMFC degradation associated with low frequency ripple current effects. In Proceedings of the 2011 IEEE Energy Conversion Congress and Exposition, Phoenix, AZ, USA, 17–22 September 2011; pp. 3336–3342. [CrossRef]
38. Barhoumi, E.M.; Farhani, S.; Bacha, F. High efficiency power electronic converter for fuel cell system application. *Ain Shams Eng. J.* **2021**, *12*, 2655–2664. [CrossRef]
39. Sampath, S.; Rahiman, Z.; Chenniappan, S.; Sundaram, E.; Subramaniam, U.; Padmanaban, S. Efficient Multi-Phase Converter for E-Mobility. *World Electr. Veh. J.* **2022**, *13*, 67. [CrossRef]
40. Ruffo, R.; Cirimele, V.; Diana, M.; Khalilian, M.; Ganga, A.L.; Guglielmi, P. Sensorless Control of the Charging Process of a Dynamic Inductive Power Transfer System With an Interleaved Nine-Phase Boost Converter. *IEEE Trans. Ind. Electron.* **2018**, *65*, 7630–7639. [CrossRef]
41. Patel, A.M.; Ferdowsi, M. Advanced Current Sensing Techniques for Power Electronic Converters. In Proceedings of the 2007 IEEE Vehicle Power and Propulsion Conference, Arlington, TX, USA, 9–12 September 2007; pp. 524–530. [CrossRef]
42. Texas Instruments. *AN-1820 LM5032 Interleaved Boost Converter*; Application Report; Texas Instrument: Dallas, TX, USA, 2013. Available online: https://www.ti.com/lit/an/snva335a/snva335a.pdf (accessed on 9 November 2022).
43. Vishay Dale. IHLP8787 Automotive Inductors. 2020. Available online: https://www.vishay.com/docs/34350/ihlp8787mz5a.pdf (accessed on 9 November 2022).
44. Vishay BCcomponents. High Temperature SMD Aluminum Electrolytic Capacitors. 2016. Available online: https://www.vishay.com/docs/28403/146cti.pdf (accessed on 9 November 2022).
45. Infineon Technologies. OptiMOS^{TM}5 Power-Transistor. 2019. Available online: https://www.infineon.com/dgdl/Infineon-BSC072N08NS5-DataSheet-v02_01-EN.pdf?fileId=5546d4624ad04ef9014ae3d8e8192c7c (accessed on 9 November 2022).
46. Infineon Technologies. Dual-Channel, Isolated Mosfet Gate-Driver. 2021. Available online: https://www.infineon.com/dgdl/Infineon-2EDF7175F-DataSheet-v02_07-EN.pdf?fileId=5546d462636cc8fb0163b09026be3060 (accessed on 9 November 2022).

47. Bourns. Current Sense Resistor. 2016. Available online: https://www.bourns.com/docs/Product-Datasheets/css2h-3920.pdf (accessed on 9 November 2022).
48. Texas Instruments. Ultra-Precise Current Sense Amplifier. 2021. Available online: https://www.ti.com/lit/gpn/ina290-q1 (accessed on 9 November 2022).
49. Saturn PCB Design Toolkit Version 8.20 Website. Available online: https://saturnpcb.com/saturn-pcb-toolkit/ (accessed on 10 November 2022).
50. Pace, L.; Idir, N.; Duquesne, T.; De Jaeger, J.C. Parasitic Loop Inductances Reduction in the PCB Layout in GaN-Based Power Converters Using S-Parameters and EM Simulations. *Energies* **2021**, *14*, 1495. [CrossRef]
51. Sochor, P.; Huerner, A.; Elpelt, R. Commutation loop design for optimized switching behavior of CoolSiC(exp TM) MOSFETs using compact models. In Proceedings of the PCIM Europe Digital Days 2020—International Exhibition and Conference for Power Electronics, Intelligent Motion, Renewable Energy and Energy Management, Nuremberg, Germany, 5–7 May 2020; pp. 1–8.
52. Sheehan, R. Understanding and Applying Current-Mode Control Theory. In Proceedings of the Power Electronics Technology Exhibition and Conference, Dallas, TX, USA, 30 October–1 November 2007. Available online: https://www.ti.com/lit/an/snva555/snva555.pdf?ts=1668164086792&ref_url=https
53. STMicroelectronics. STM32G474 MCU. 2021. Available online: https://www.st.com/resource/en/datasheet/stm32g474qe.pdf (accessed on 9 November 2022).
54. Cuoghi, S.; Mandrioli, R.; Ntogramatzidis, L.; Gabriele, G. Multileg Interleaved Buck Converter for EV Charging: Discrete-Time Model and Direct Control Design. *Energies* **2020**, *13*, 466. [CrossRef]
55. Kim, M.J.; Shin, M.H.; Choi, S.C.; Bae, K.C.; Won, C.Y.; Jung, Y.C. Reverse current control method of synchronous boost converter for fuel cell using a mode boundary detector. In Proceedings of the 2014 IEEE International Conference on Industrial Technology (ICIT), Busan, Korea, 26 February–1 March 2014; pp. 359–364. [CrossRef]
56. SPICE Model of the BSC072N08NS Mosfet by Infineon Technologies. Available online: https://www.infineon.com/cms/en/product/power/mosfet/n-channel/bsc072n08ns5/#!simulation (accessed on 9 November 2022).
57. Cousineau, M.; Monroy, M.; Lorenzi Pol, W.; Hureau, L.; Aulagnier, G.; Goyhenetche, P.; Rolland, E.; Flumian, D. Decentralized Phase Shedding with Low Power Mode for Multiphase Converter. *Energies* **2021**, *14*, 6748. [CrossRef]
58. Rampinelli, G.; Krenzinger, A.; Chenlo Romero, F. Mathematical models for efficiency of inverters used in grid connected photovoltaic systems. *Renew. Sustain. Energy Rev.* **2014**, *34*, 578–587. [CrossRef]

applied sciences

MDPI

Article

Analytical Calculation of Air-Gap Magnetic Field Distribution in IPMSMs with Mixed Eccentricity Accounting for Bridge Saturation

Yu Sun [1,2], Lichen Gu [2], Peijin Liu [2,*], Jiangcheng Chen [3] and Donghong Cheng [2]

1 School of Science, Xi'an University of Architecture and Technology, Xi'an 710055, China
2 School of Mechanical and Electrical Engineering, Xi'an University of Architecture and Technology, Xi'an 710055, China
3 Shenzhen Academy of Robotics, Shenzhen 518057, China
* Correspondence: liuxpj@xauat.edu.cn

Abstract: This paper presents an analytical model for calculating the detailed air-gap magnetic field distribution in interior permanent magnet synchronous motors (IPMSMs) with mixed eccentricity. In order to improve the efficiency of model solving, a modeling strategy combining the equivalent magnetic circuit network method and the subdomain method was adopted. Specifically, the magnetic field distribution of the rotor was modeled by using the magnetic circuit analysis method, and the magnetic field distributions of the air-gap, slot opening and stator slot along the radial direction were modeled in different regions according to their structure ruler. Then, the influence of bridge saturation was considered. Moreover, based on the analysis of the air-gap geometric structure with mixed eccentricity, a detailed spatiotemporal analytical model of the air-gap magnetic field was established, which provides a more accurate description of the mixed eccentricity composed of static and dynamic rotor eccentricities of different severity. The analytical models were compared with the corresponding models established by the finite element method, which proved the accuracy and validity of the models established in this paper. Finally, some key features related to radial and tangential air-gap flux density were extracted, which can significantly reflect the characteristics of eccentricities. The main findings reported in this paper will be of benefit for developing methods for early identification and diagnosis of eccentricity faults in IPMSMs.

Keywords: IPMSMs; mixed eccentricity; bridge saturation; air-gap magnetic field; analytic calculation

Citation: Sun, Y.; Gu, L.; Liu, P.; Chen, J.; Cheng, D. Analytical Calculation of Air-Gap Magnetic Field Distribution in IPMSMs with Mixed Eccentricity Accounting for Bridge Saturation. *Appl. Sci.* **2022**, *12*, 11956. https://doi.org/10.3390/app122311956

Academic Editors: Loránd Szabó and Feng Chai

Received: 7 November 2022
Accepted: 22 November 2022
Published: 23 November 2022

1. Introduction

Recently, interior permanent magnet synchronous motors (IPMSMs) have been widely used in electric vehicles and other industrial fields due to their characteristics of high efficiency, high power density, carbon free and zero emission [1–4]. As a result, the requirements for safe and healthy operation with high performance are becoming higher and higher. In actual application, due to the manufacturing, installation and variation in support stiffness, rotor mixed eccentricity is a widespread phenomenon in the operation of the IPMSM, which leads to the distortion of the air-gap magnetic field of the motor and affects the performance and efficiency of the motor, and even affects the service life of the motor [5]. Therefore, it is important to clarify the variation rule and characteristics of the air-gap magnetic field in IPMSMs with mixed eccentricity, so as to further realize the early diagnosis of mixed eccentricity faults and the real-time control, which is of great significance for the healthy and efficient operation of the motor [5–7].

In order to build the model of air-gap magnetic field, two types of methods are usually used: finite element methods and analytical methods. The finite element method is a relatively mature method for numerical modeling. Through the refined solid models and the configuration of constraints, the air-gap magnetic field can be accurately calculated under different operation conditions. However, the real-time performance is not good, and the consumption of computing resources is large. Therefore, it has been mostly used in the design optimization and structural verification of the motor [8,9]. In contrast, the analytical methods can reveal the coupling law between mechanical/electrical parameters and magnetic field through mathematical modeling, the motor and has a fast running speed [3,10]. Therefore, analytical method has been widely used in the research of motor fault diagnosis mechanisms. However, there are many factors affecting the spatial and temporal distribution of the air-gap magnetic field in IPMSMs. The main factors include the load characteristics, the cogging effect formed between the permanent magnet and stator teeth, and the magnetic saturation in the bridge region. In addition, in the study of the air-gap magnetic field under rotor eccentricity, single static eccentricity, or dynamic eccentricity, has been considered in previous research [11–13], which is different from the mixed eccentricity in actual operation. Therefore, the establishment of the analytical model of air-gap magnetic field accounts for the above factors is a key issue to accurately calculate the spatial and temporal distribution of the air-gap magnetic field under the general operation of IPMSMs.

At present, the analytical methods for modeling the air-gap magnetic field of permanent magnet synchronous motors (PMSMs) include the equivalent magnetic circuit network method [14–17], the subdomain model method [18–21], and the magneto-motive force-magnetic permeability method [22–26]. In [14,15], the equivalent magnetic circuit network method was used to model and analyze the rotor magnetic field of PMSM to verify the effect of the optimal design. The same method was used to model the armature magnetic field of the permanent magnet linear synchronous motor analytically in [16]. By combining the equivalent magnetic circuit network method and the Carter coefficient method, the influence of the cogging torque on the distribution of the air-gap magnetic field in the IPMSM was analyzed in [17]. However, the equivalent magnetic circuit network method can only calculate the radial component of the air-gap magnetic field and cannot calculate the tangential component of the air-gap magnetic field. What is more, the subcircuit division needs to be more detailed in order to improve the calculation accuracy, and the amount of computation would be greatly increased [17]. For the subdomain modeling method, this was adopted to model the rotor magnetic field under no-load and under-load conditions in [18], in which the effect of saturation in iron core was taken into account. In [19], the subdomain model method was used to analyze the air-gap magnetic fields of the permanent magnet motor under no-load and load conditions. In this study, in order to calculate the influence of the saturation effect of the rotor iron core, the saturation effect is equivalent to the sector-shaped saturation region. In [20], the subdomain modeling method was used to analyze the distribution of the air-gap magnetic field of surface-mounted permanent magnet synchronous motors with different slot structures. In [21], the subdomain modeling method was used to model the armature reaction magnetic field of the PMSM, and the inductance parameters were identified. In this study, the effect of saturation in iron core was ignored. For the subdomain method, the mutual influence of the slots can be considered, but it is required to solve the partial differential equation and solve the undetermined coefficient through boundary conditions, which lead to a large amount of calculation. The most important thing for the subdomain method is that the effect of saturation in iron core on the air-gap magnetic field of the motor cannot be calculated directly. The magneto-motive force-magnetic permeability method [22–26] generally assumes that the magnetic permeability of the rotor core is infinite, so the nonlinear effect of saturation in the iron core on the air-gap magnetic field cannot be considered. At the same time, although the spatial distribution of the armature reaction magnetic field can be calculated by vector superposition, the effect of magnetic saturation is ignored [27,28]. We can conclude from the above analysis that, in the analytical modeling of air-gap magnetic field of the PMSM, a high-dimensional model and complex boundary conditions will be introduced due to the magnetic saturation in iron core. As a result, the influence of magnetic saturation in iron core has been ignored in the single analytical modeling method, which makes it difficult to accurately reflect the magnetic field distribution.

In view of the existing problems in the current analytical methods and in order to accurately calculate the air-gap magnetic field distribution in IPMSMs with mixed eccentricity, an analytical model that accounts for the influence of magnetic saturation in the bridge regions was established by combining the equivalent magnetic circuit network method and the subdomain method. This analytical model shows high accuracy, which has been verified by finite element simulation. Then, using the established model, the distribution of the air-gap magnetic field of the IPMSM with mixed eccentricity values of different severity was calculated, and features reflecting the eccentricity were extracted. The results of this study lay a foundation for developing early identification and diagnosis methods for eccentricity faults in IPMSMs.

2. Methods

The analytical model of air-gap magnetic field distribution in the IPMSM with mixed eccentricity can be established based on the normal model whose air-gap length is replaced by the air-gap geometric model with mixed eccentricity. In this section, the analytical model of the air-gap magnetic field in the IPMSM is established with the following assumptions: (1) The end effect, the rotor core conductivity and the eddy current effect are ignored. (2) The magnetic permeability of the rest of the iron core is infinite except for magnetic saturation in the bridge. (3) The stator slots and slot openings are approximately fan-shaped. The normal air-gap magnetic field considering bridge saturation is composed of the no-loaded air-gap magnetic field of the permanent magnet, the armature reaction air-gap magnetic field and the equivalent magnetic field of the effect of bridge saturation. During the establishment of each magnetic field model, the motor is divided into four subdomains along the radial direction: rotor, air-gap, slot opening and stator slot. The air-gap flux density contains temporal and spatial variables and is described with a radial component and a tangential component. The analytical method for efficient calculation of air-gap magnetic field is formed by combining the equivalent magnetic circuit network method and the subdomain method.

2.1. *Analytical Model of the Normal Air-Gap Magnetic Field*

2.1.1. Model of the Equivalent Magnetic Circuit of the Rotor

The flux is produced by the scalar magnetic potential on the rotor surface, and the part entering the air-gap is solved by using the equivalent magnetic circuit network method. The equivalent magnetic circuit is shown in Figure 1.

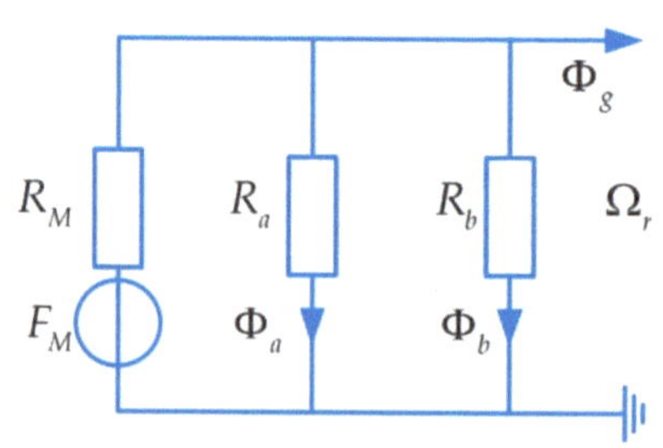

Figure 1. The equivalent magnetic circuit.

The permanent magnet magnetomotive force F_M, equivalent reluctance R_M, extreme flux leakage Φ_b and interpolar flux leakage Φ_i can be expressed by the motor parameters as

$$F_M = \frac{B_M \cdot h_M}{\mu_0 \mu_r} \tag{1}$$

$$R_M = \frac{h_M}{\mu_0 \mu_r L_M L_Z} \tag{2}$$

$$\Phi_b = B_s h_b L_Z \tag{3}$$

$$\Phi_i = B_s \frac{h_i}{2} L_Z \tag{4}$$

where B_M is the residual magnetic induction intensity of the permanent magnet, h_M is the thickness of the permanent magnet, μ_0 is the vacuum permeability, μ_r is the relative permeability of the permanent magnet, L_M is the width of the permanent magnet, L_Z is the axial length of the iron core, B_S is the magnetic induction intensity of bridge saturation, h_b is the thickness of the end of bridge, and h_i is the thickness of the interpolar magnetic in the bridge. The scalar magnetic potential of the rotor surface can be expressed as

$$\Omega_r = F_M - (\Phi_b + \Phi_i + \Phi_{gr}) R_M \tag{5}$$

where Φ_{gr} is the magnetic flux entering the air-gap. It can be calculated as

$$\Phi_{gr} = \frac{F_M}{R_M} - \frac{\Omega_r}{R_M} - \Phi_b - \Phi_i \tag{6}$$

2.1.2. Model of the Air-Gap Magnetic Field of Permanent Magnet under No-Load

According to the assumptions, when the stator winding is an open circuit, the vector magnetic potential of each subdomain satisfies the Laplace Equation [1]:

$$\frac{\partial^2 \varphi}{\partial r^2} + \frac{1}{r}\frac{\partial \varphi}{\partial r} + \frac{1}{r^2}\frac{\partial^2 \varphi}{\partial \alpha^2} = 0 \tag{7}$$

The general solution of vector magnetic potential in air-gap is

$$\mathbf{A}_{gp} = \sum_{k=1}^{\infty}\left[\mathbf{A}_{gpk}\left(\frac{r}{R_s}\right)^{k} + \mathbf{B}_{gpk}\left(\frac{r}{R_s}\right)^{-k}\right]\cos(k\alpha) + \sum_{k=1}^{\infty}\left[\mathbf{C}_{gpk}\left(\frac{r}{R_s}\right)^{k} + \mathbf{D}_{gpk}\left(\frac{r}{R_s}\right)^{-k}\right]\sin(k\alpha) \tag{8}$$

where $\mathbf{A}_{gp}$ is the vector magnetic potential of the permanent magnet with no-load, and $\mathbf{A}_{gpk}$, $\mathbf{B}_{gpk}$, $\mathbf{C}_{gpk}$ and $\mathbf{D}_{gpk}$ are the Fourier series undetermined coefficients of the air-gap magnetic field of the permanent magnet, R_S is the inner radius of the stator, k is the harmonic order of the air-gap magnetic field, α is the circumferential angle of the air-gap space, and r is the length from the center of the radial rotor to the inner boundary of the stator.

Then, the radial component and tangential component of the flux density are described by Equation (9) and Equation (10), respectively

$$B_r = \frac{1}{r}\frac{\partial \mathbf{A}_{gp}}{\partial \alpha} \tag{9}$$

$$B_t = -\frac{\partial \mathbf{A}_{gp}}{\partial r} \tag{10}$$

by substituting Equation (8) into Equations (9) and (10)

$$B_{rpg} = -\sum_{k=1}^{\infty}\frac{k}{r}\left[\mathbf{A}_{gpk}\left(\frac{r}{R_s}\right)^{k} + \mathbf{B}_{gpk}\left(\frac{r}{R_s}\right)^{-k}\right]\cos(k\alpha) + \sum_{k=1}^{\infty}\frac{k}{r}\left[\mathbf{C}_{gpk}\left(\frac{r}{R_s}\right)^{k} + \mathbf{D}_{gpk}\left(\frac{r}{R_s}\right)^{-k}\right]\sin(k\alpha) \tag{11}$$

$$B_{tpg} = -\sum_{k=1}^{\infty}\frac{k}{r}\left[\mathbf{A}_{gpk}\left(\frac{r}{R_s}\right)^{k} - \mathbf{B}_{gpk}\left(\frac{r}{R_s}\right)^{-k}\right]\cos(k\alpha) - \sum_{k=1}^{\infty}\frac{k}{r}\left[\mathbf{C}_{gpk}\left(\frac{r}{R_s}\right)^{k} - \mathbf{D}_{gpk}\left(\frac{r}{R_s}\right)^{-k}\right]\sin(k\alpha) \tag{12}$$

where B_{rpg} and B_{tpg} are the radial and tangential components of air-gap flux density of permanent magnet with no-load, respectively, and k is the harmonic frequency of air-gap flux density.

The Fourier undetermined coefficients can be solved by the interface conditions of each subdomain of the IPMSM under no-load conditions. The interface conditions of each subdomain satisfy the conditions that: the vector magnetic potential and tangential flux density are both continuous on the interface between the stator slot and slot opening; the vector magnetic potential and tangential flux density are also continuous on the interface between the slot opening and air-gap; the scalar magnetic potential and air-gap magnetic flux are continuous on the interface between rotor and air-gap. According to the above interface conditions, the constraint Equations of the undetermined coefficients can be derived as

$$\begin{bmatrix} 0 & 0 & 0 & \mathbf{K}_{14} & \mathbf{K}_{15} & \mathbf{K}_{16} & \mathbf{K}_{17} & \mathbf{K}_{18} \\ 0 & \mathbf{K}_{22} & \mathbf{K}_{23} & 0 & 0 & \mathbf{K}_{26} & \mathbf{K}_{27} & 0 \\ 0 & \mathbf{K}_{32} & \mathbf{K}_{33} & \mathbf{K}_{34} & \mathbf{K}_{35} & \mathbf{K}_{36} & \mathbf{K}_{37} & 0 \\ 0 & 0 & 0 & \mathbf{K}_{44} & \mathbf{K}_{45} & 0 & 0 & \mathbf{K}_{48} \\ 0 & 0 & 0 & 0 & 0 & \mathbf{K}_{56} & \mathbf{K}_{57} & \mathbf{K}_{58} \\ \mathbf{K}_{61} & \mathbf{K}_{62} & \mathbf{K}_{63} & 0 & 0 & 0 & 0 & 0 \\ 0 & \mathbf{K}_{72} & \mathbf{K}_{73} & \mathbf{K}_{74} & \mathbf{K}_{75} & 0 & 0 & 0 \\ \mathbf{K}_{81} & \mathbf{K}_{82} & \mathbf{K}_{83} & 0 & 0 & 0 & 0 & 0 \end{bmatrix} \begin{bmatrix} \Omega_r \\ \mathbf{A}_{gpk} \\ \mathbf{B}_{gpk} \\ \mathbf{C}_{gpk} \\ \mathbf{D}_{gpk} \\ \mathbf{A}_{sopi} \\ \mathbf{B}_{sopi} \\ \mathbf{A}_{spi} \end{bmatrix} = \begin{bmatrix} \mathbf{Y} \\ 0 \\ 0 \\ 0 \\ 0 \\ 0 \\ 0 \\ 0 \end{bmatrix} \tag{13}$$

where $\mathbf{K}_{11}$~$\mathbf{K}_{88}$ are the coefficient matrixes under each boundary condition, respectively, which depend on the motor structure parameters, and $\mathbf{Y}$ is determined by the rotor magnetic field. $\mathbf{A}_{gpk}$, $\mathbf{B}_{gpk}$, $\mathbf{C}_{gpk}$ and $\mathbf{D}_{gpk}$ are the undetermined coefficients of Fourier series of air-gap flux density general solution, respectively. $\mathbf{A}_{sopi}$ and $\mathbf{B}_{sopi}$ are the undetermined coefficients of flux density Fourier general solution in the slot opening domain, respectively, and i is the ith stator slot. $\mathbf{A}_{spi}$ is the undetermined coefficient of the $\mathbf{Y}$ Fourier solution in the stator slot region [19,20].

By solving the matrix Equation (13), each undetermined coefficient can be obtained, and substituting them into Equations (11) and (12), the radial and tangential components of the air-gap flux density of the permanent magnet of the IPMSM can be obtained.

2.1.3. Model of the Armature Reaction Magnetic Field

When analyzing the armature reaction magnetic field, there is current excitation in the stator slot winding, and the vector magnetic potential in the stator slot region satisfies the Poisson Equation:

$$\frac{\partial^2 A}{\partial r^2} + \frac{1}{r}\frac{\partial A}{\partial r} + \frac{1}{r^2}\frac{\partial^2 A}{\partial \alpha^2} = -\mu_0 J \tag{14}$$

The form of general solutions of armature reaction air-gap flux density is the same as that of air-gap flux density under no-load, but the value of undetermined coefficient is different. The interface conditions of each subdomain of armature reaction under load of the IPMSM are as follows: The vector magnetic potential and tangential flux density are continuous on the interface between stator slot domain and slot opening domain; the vector magnetic potential and tangential flux density are continuous on the interface between slot opening domain and air-gap domain; the tangential air-gap flux density is 0 on the interface between air-gap domain and rotor domain. According to the interface conditions, the constraint equations of the undetermined coefficients can be derived and described as

$$\begin{bmatrix} 0 & \mathbf{K}_{12} & \mathbf{K}_{13} & 0 & \mathbf{K}_{15} & \mathbf{K}_{16} & 0 \\ 0 & \mathbf{K}_{22} & \mathbf{K}_{23} & \mathbf{K}_{24} & \mathbf{K}_{25} & \mathbf{K}_{26} & \mathbf{K}_{27} \\ 0 & 0 & 0 & \mathbf{K}_{34} & \mathbf{K}_{35} & 0 & 0 \\ 0 & 0 & 0 & 0 & 0 & \mathbf{K}_{46} & \mathbf{K}_{47} \\ \mathbf{K}_{51} & \mathbf{K}_{52} & \mathbf{K}_{53} & 0 & 0 & 0 & 0 \\ 0 & \mathbf{K}_{62} & \mathbf{K}_{63} & \mathbf{K}_{64} & \mathbf{K}_{65} & 0 & 0 \\ \mathbf{K}_{71} & \mathbf{K}_{72} & \mathbf{K}_{73} & 0 & 0 & 0 & 0 \end{bmatrix} \begin{bmatrix} \mathbf{A}_{gak} \\ \mathbf{B}_{gak} \\ \mathbf{C}_{gak} \\ \mathbf{D}_{gak} \\ \mathbf{A}_{soai} \\ \mathbf{B}_{soai} \\ \mathbf{A}_{sai} \end{bmatrix} = \begin{bmatrix} \mathbf{Y}_1 \\ 0 \\ 0 \\ 0 \\ \mathbf{Y}_2 \\ \mathbf{Y}_3 \\ \mathbf{Y}_4 \end{bmatrix} \tag{15}$$

where $\mathbf{K}_{12}$~$\mathbf{K}_{77}$ are the equation coefficient matrixes under each boundary condition, depending on the motor structure parameters, and $\mathbf{Y}_1$~$\mathbf{Y}_4$ are determined by the armature current density under load conditions. $\mathbf{A}_{gak}$, $\mathbf{B}_{gak}$, $\mathbf{C}_{gak}$ and $\mathbf{D}_{gak}$ are the undetermined Fourier series coefficients of the air-gap flux density general solution, respectively. $\mathbf{A}_{soai}$ and $\mathbf{B}_{soai}$ are the undetermined coefficients of the flux density Fourier general solution in the stator slot opening domain, respectively, and i is the ith stator slot. $\mathbf{A}_{sai}$ is the undetermined coefficient of the flux density Fourier general solution in the stator slot region [19,20]. By solving Equation (15), each undetermined coefficient can be obtained. Furthermore, the radial and tangential components of the air-gap flux density of the armature reactive of the IPMSM can be obtained.

2.1.4. Model of the Magnetic Field in the Bridge Regions

The effect of the bridge saturation will affect the armature reactive magnetic field, which is essentially due to the magnetic voltage drop caused by the armature reactive magnetic field passing through the bridge of the rotor. The subdomain method combined with the equivalent magnetic circuit method is used to calculate the magnetic field in the bridge regions. The motor is divided into three subdomains: stator, air-gap and rotor. The interface conditions of each subdomain are as follows: the tangential air-gap flux density at the interface between the air-gap and stator is 0; the scalar magnetic potential at the interface between the air-gap and rotor is continuous; the air-gap flux density at the interface between the rotor domain and air-gap domain is continuous. The constraint equations of the undetermined coefficient are obtained as

$$\begin{bmatrix} \mathbf{K}_{11} & \mathbf{K}_{12} & \mathbf{K}_{13} & \mathbf{K}_{14} & \mathbf{K}_{15} \\ \mathbf{K}_{21} & \mathbf{K}_{22} & 0 & 0 & 0 \\ 0 & 0 & \mathbf{K}_{33} & \mathbf{K}_{34} & 0 \\ \mathbf{K}_{41} & \mathbf{K}_{42} & 0 & 0 & \mathbf{K}_{45} \\ 0 & 0 & \mathbf{K}_{53} & \mathbf{K}_{54} & \mathbf{K}_{55} \end{bmatrix} \begin{bmatrix} \Omega_r \\ \mathbf{A}_{gbk} \\ \mathbf{B}_{gbk} \\ \mathbf{C}_{gbk} \\ \mathbf{D}_{gbk} \end{bmatrix} = \begin{bmatrix} \mathbf{Y} \\ 0 \\ 0 \\ 0 \\ 0 \end{bmatrix} \tag{16}$$

where $\mathbf{K}_{11}$~$\mathbf{K}_{55}$ depend on the structural parameters of the motor, and $\mathbf{Y}$ depends on the flux leakage in the bridge regions. $\mathbf{A}_{gbk}$, $\mathbf{B}_{gbk}$, $\mathbf{C}_{gbk}$ and $\mathbf{D}_{gbk}$ are the undetermined Fourier series coefficients of the air-gap magnetic general solution, respectively [19,20]. By solving Equation (16), the undetermined coefficients can be calculated. Further, the radial component and tangential component of the air-gap flux density of the rotor magnetic field in the bridge regions of the IPMSM can be found. B_{rbg} and B_{tbg} can be obtained, whose Fourier expressions are the same as those of Equations (11) and (12).

By superimposing the air-gap magnetic field under no-load conditions, armature reaction magnetic field and rotor magnetic field in the bridge regions, the analytical model of the normal air-gap magnetic field can be described as

$$B_r = B_{rpg} + B_{rag} + (B_{rbg} + B_{rpg}) \tag{17}$$

$$B_t = B_{tpg} + B_{tag} + (B_{tbg} + B_{tpg}) \tag{18}$$

2.2. Geometric Model of the Air-Gap with Mixed Eccentricity

The eccentricity caused in the process of motor manufacturing and assembly is an objective factor that cannot be ignored in motor design optimization and performance analysis. Similarly, both static eccentricity and dynamic eccentricity co-exist at the same time during motor operation. Thus, the accurate geometric model of the air-gap with mixed eccentricity is established in this paper. It is assumed that the boundary line of the rotor cross-section is almost an ideal circle, and the shape and position of the stator and rotor do not change in the axial direction. Moreover, the position of the minimum air-gap relative to the stator during static eccentricity remains unchanged [29], and the position of the minimum air-gap relative to the rotor during dynamic eccentricity remains unchanged too [30].

As shown in Figure 2, when there is mixed eccentricity in the motor, the air-gap length $g(\alpha,t)$ is a function of spatial angle α and time t. R_s and R_r are motor stator inner diameter and rotor outer diameter, respectively, δ is the eccentric angle, O_S is the center of the stator circle, O_R is the center of the rotor circle, and $O_{R'}$ is the center of the motion track of the rotor center. $d_{SE} = ||O_S O_R||$ is the static eccentricity distance, while $d_{DE} = ||O_S O_{R'}||$ is the dynamic eccentricity distance. We draw a straight line through point O_R to intersect the circle O_S and the circle O_R at points A and B, respectively. The length of line segment AB is $g(\alpha,t)$, and the angle between the line segment O_RA and the polar axis is α, namely, the angle of the air-gap space. The angle between the line segment $O_R O_{R'}$ and the pole axis is ωt, namely, the rotor rotation angle. According to the trigonometric relationship, Equation (19) can be yielded:

$$\angle O_R O_{R'} O_S = \begin{cases} \pi - \delta + \omega t & \omega t - \delta < 2k\pi \\ \pi - \omega t + \delta & 2k\pi \le \omega t - \delta < (2k+1)\pi \\ \omega t - \pi - \delta & (2k+1)\pi \le \omega t - \delta < 2(k+1)\pi \end{cases} \quad k = 0,1,2\cdots \tag{19}$$

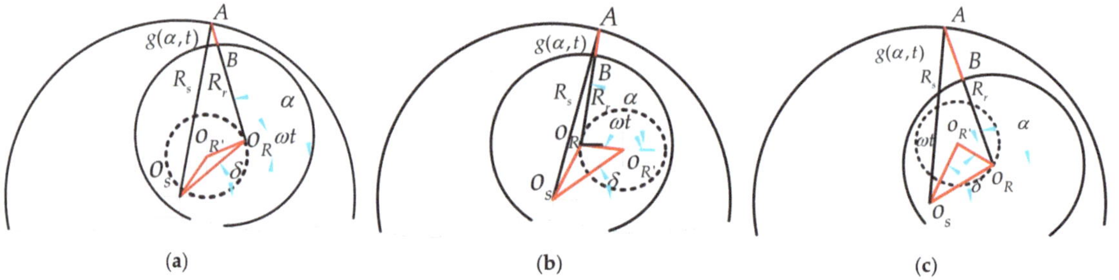

Figure 2. The geometric model of the air-gap with mixed eccentricity. (**a**) $\omega t - \delta < 2k\pi$; (**b**) $2k\pi \le \omega t - \delta < (2k+1)\pi$; (**c**) $(2k+1)\pi \le \omega t - \delta < 2(k+1)\pi$.

Further, the length of O_sO_R can be yielded as

$$\begin{aligned} \|O_S O_R\| &= \sqrt{d_{SE}{}^2 + d_{DE}{}^2 - 2d_{SE} \cdot d_{DE} \cdot \cos\angle O_R O_{R'} O_S} \\ &= \sqrt{d_{SE}{}^2 + d_{DE}{}^2 + 2d_{SE} \cdot d_{DE} \cdot \cos\angle(\delta - \omega t)} \end{aligned} \tag{20}$$

Therefore, in $\Delta O_s O_R O_{R'}$

$$\angle O_S O_R O_{R'} = \arccos\frac{\|O_S O_R\|^2 + d_{DE}{}^2 - d_{SE}{}^2}{2\|O_S O_R\| \cdot d_{DE}} \tag{21}$$

and in $\Delta A\, O_S O_{R'}$

$$\angle A O_R O_S = \begin{cases} \pi - \alpha + \omega t + \angle O_S O_R O_{R'} & \omega t - \delta < 2k\pi \\ \pi - \alpha + \omega t - \angle O_S O_R O_{R'} & 2k\pi \le \omega t - \delta < (2k+1)\pi \\ \angle O_S O_R O_{R'} - \alpha + \omega t - \pi & (2k+1\ \pi \le \omega t - \delta < 2(k+1)\pi \end{cases} \quad k = 0,1,2\cdots \tag{22}$$

$$R_S{}^2 = \|O_S O_R\|^2 + \|AO_R\|^2 - 2\|O_S O_R\| \cdot \|AO_R\| \cdot \cos\angle AO_R O_S \tag{23}$$

According to Equation (23), it can be obtained that

$$\begin{aligned} \|AO_R\| &= \|O_S O_R\| \cdot \cos\angle AO_R O_S + \sqrt{(\|O_S O_R\| \cdot \cos\angle AO_R O_S)^2 - (\|O_S O_R\|^2 - R_S{}^2)} \\ &= \|O_S O_R\| \cdot \cos\angle AO_R O_S + \sqrt{R_S{}^2 - (\|O_S O_R\| \cdot \sin\angle AO_R O_S)^2} \end{aligned} \tag{24}$$

It can be found in Figure 2 that the air-gap length can be described as

$$g(\alpha, t) = \|AO_R\| - R_r \tag{25}$$

and the ratio of the static eccentricity can be described as

$$\varepsilon_S = \frac{d_{SE}}{g_0} \tag{26}$$

while the ratio of dynamic eccentricity can be described as

$$\varepsilon_D = \frac{d_{DE}}{g_0} \tag{27}$$

where g_0 is the length of air-gap in normal state.

The air-gap length of mixed eccentricity can be obtained by combining Equations (19)–(27) and can be described as

$$g(\alpha,t) = \begin{cases} \sqrt{R_s{}^2 - \sin^2\left[\alpha - \omega t - \arccos\left(\frac{\varepsilon_D + \varepsilon_S\cos(\delta-\omega t)}{\sqrt{\varepsilon_S{}^2+\varepsilon_D{}^2+2\varepsilon_S\cdot\varepsilon_D\cos(\delta-\omega t)}}\right)\right] \cdot g_0{}^2\left[\varepsilon_S{}^2+\varepsilon_D{}^2+2\varepsilon_S\cdot\varepsilon_D\cos(\delta-\omega t)\right]} \\ -R_r - g_0\sqrt{\varepsilon_S{}^2+\varepsilon_D{}^2+2\varepsilon_S\cdot\varepsilon_D\cos(\delta-\omega t)} \\ \cdot\cos\left[\alpha - \omega t - \arccos\left(\frac{\varepsilon_D + \varepsilon_S\cos(\delta-\omega t)}{\sqrt{\varepsilon_S{}^2+\varepsilon_D{}^2+2\varepsilon_S\cdot\varepsilon_D\cos(\delta-\omega t)}}\right)\right] & ,\ \omega t - \delta < 2k\pi \\ \sqrt{R_s{}^2 - \sin^2\left[\omega t - \alpha - \arccos\left(\frac{\varepsilon_D + \varepsilon_S\cos(\delta-\omega t)}{\sqrt{\varepsilon_S{}^2+\varepsilon_D{}^2+2\varepsilon_S\cdot\varepsilon_D\cos(\delta-\omega t)}}\right)\right] \cdot g_0{}^2\left[\varepsilon_S{}^2+\varepsilon_D{}^2+2\varepsilon_S\cdot\varepsilon_D\cos(\delta-\omega t)\right]} \\ -R_r - g_0\sqrt{\varepsilon_S{}^2+\varepsilon_D{}^2+2\varepsilon_S\cdot\varepsilon_D\cos(\delta-\omega t)} \\ \cdot\cos\left[\omega t - \alpha - \arccos\left(\frac{\varepsilon_D + \varepsilon_S\cos(\delta-\omega t)}{\sqrt{\varepsilon_S{}^2+\varepsilon_D{}^2+2\varepsilon_S\cdot\varepsilon_D\cos(\delta-\omega t)}}\right)\right] & ,\ 2k\pi \le \omega t - \delta < (2k+1)\pi \\ \sqrt{R_s{}^2 - \sin^2\left[\alpha - \omega t - \arccos\left(\frac{\varepsilon_D + \varepsilon_S\cos(\delta-\omega t)}{\sqrt{\varepsilon_S{}^2+\varepsilon_D{}^2+2\varepsilon_S\cdot\varepsilon_D\cos(\delta-\omega t)}}\right)\right] \cdot g_0{}^2\left[\varepsilon_S{}^2+\varepsilon_D{}^2+2\varepsilon_S\cdot\varepsilon_D\cos(\delta-\omega t)\right]} \\ -R_r - g_0\sqrt{\varepsilon_S{}^2+\varepsilon_D{}^2+2\varepsilon_S\cdot\varepsilon_D\cos(\delta-\omega t)} \\ \cdot\cos\left[\alpha - \omega t - \arccos\left(\frac{\varepsilon_D + \varepsilon_S\cos(\delta-\omega t)}{\sqrt{\varepsilon_S{}^2+\varepsilon_D{}^2+2\varepsilon_S\cdot\varepsilon_D\cos(\delta-\omega t)}}\right)\right] & ,\ (2k+1)\pi \le \omega t - \delta < 2(k+1)\pi \end{cases} \tag{28}$$

Obviously, Equation (28) is a general expression suitable for static eccentricity, dynamic eccentricity, and mixed eccentricity. By substituting $g(\alpha,t)$ calculated from Equation (28) into Equations (16) and (17), the air-gap length becomes a function of α and t. The analytical model reflecting the mixed eccentric air-gap magnetic field is finally obtained.

2.3. Model Validation Method

In order to verify the accuracy of the analytical model established in this paper, the finite element model of the IPMSM was established as the benchmark model for comparison, which is shown in Figure 3. In the modeling, the assumptions are as follows: (1) The material is isotropic; (2) the permeability of the material is uniform and does not vary with temperature. According to the actual structure of the IPMSM and the division of permeance units, the flux density distribution of the air-gap is calculated by equivalent magnetic network modeling, and the schematic diagram of the basic modeling process is shown in Figure 3c. Please refer to [31] for more detail on the modeling process.

Several typical geometric structures of the permeance units and the corresponding calculation formula of the permeance are shown in Table 1. In consideration of computational accuracy and speed, two-layer air-gap permeance units are adopted and the mesh thickness of each layer is 0.35 mm.

The main parameters of the IPMSM analyzed in this paper are shown in Table 2. The air-gap magnetic field distributions with and without mixed eccentricity are calculated using established analytical models and are both compared with corresponding results calculated with the finite element models. The radial and tangential components of the air-gap magnetic field are compared separately. The accuracy of the analytical model is characterized by the root mean square percentage error (*RMSPE*), described as

$$RMSPE = \sqrt{\frac{1}{N_d}\sum_{i=1}^{N_d}\frac{(S_{Ai} - S_{Ei})^2}{S_{Ei}{}^2}} \times 100\% \tag{29}$$

where N_d is the number of samples of the radial or tangential component of air-gap magnetic field in one period, S_{Ai} is the ith value of the radial or tangential component sequence calculated with the analytical model, and S_{Ei} is the ith value of the radial or tangential component sequence calculated with the finite element model.

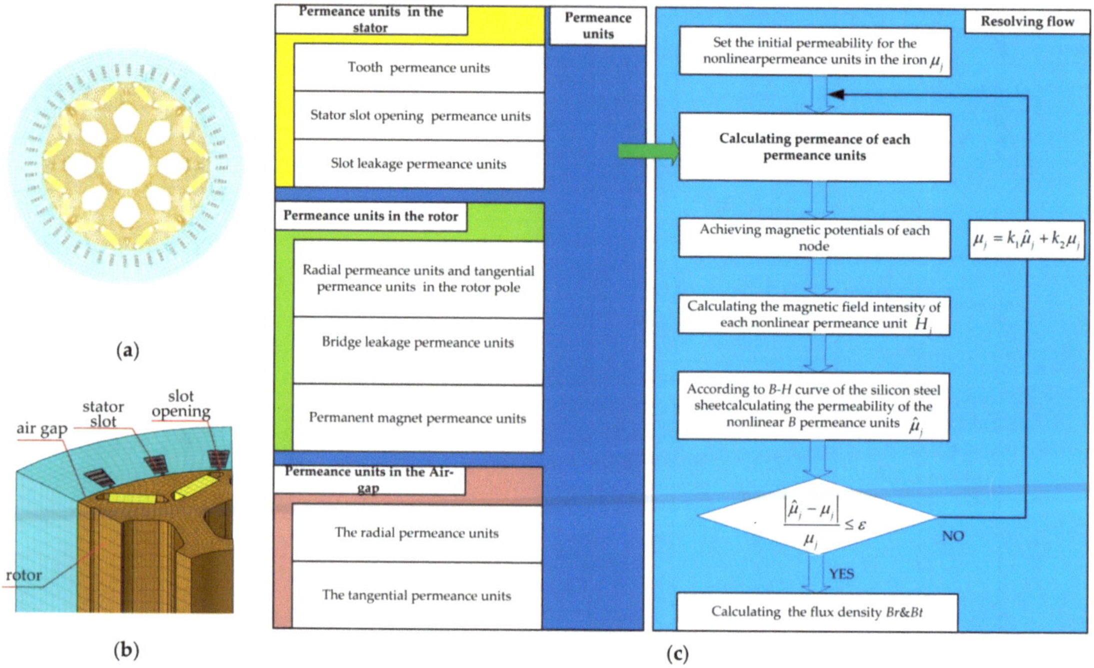

Figure 3. The finite element model of the IPMSM. (**a**) End view drawing; (**b**) grid diagram; (**c**) the basic modeling process.

Table 1. The types of finite elements.

Permeance Units	Permeance Formulas	Geometric Structure
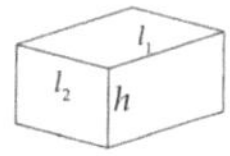 Rectangular	$G = \mu l_1 l_2 / h$	stator slot rotor pole air-gap magnetic bridge
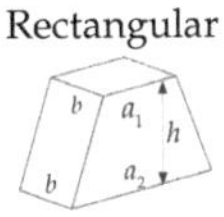 Trapezoidal	$G = \mu(a_1 + a_2)b/2h$	stator tooth
 Fan-shaped	$G = 4\mu l_3/\pi$	stator slot opening

Table 2. The main parameters of IPMSM.

Parameters	Numerical Values	Parameters	Numerical Values
Rated power P_n	160 kW	Axial length of the iron core L_r	130 mm
Rated speed n_0	5100 rpm	Number of wire layers per groove N_t	4
Number of pole-pairs p	4	Number of parallel branches N_p	2
Number of slots N_s	48	Magnet thickness h_M	6 mm
Radius of the slot bottom R_{sb}	93.6 mm	Width of the ermanent magnet L_M	17 mm
Radius of the slot top R_{st}	80.6 mm	Permeability of the vacuum μ_0	$4\pi \times 10^{-7}$ H/m
Groove width b_{sa}	0.05 rad	Relative permeability of the permanent magnet μ_r	1.059
Rabbet width b_{oa}	0.011 rad	Width of the bridge α_b	0.035 rad
Inner radius of the stator R_S	79.8 mm	End thickness of the bridge h_b	2.1 mm
External radius of the rotor R_r	79.1 mm	Interpolar thickness of the bridge h_i	2.8 mm
Basic frequency f_0	340 Hz		

3. Results and Discussion

3.1. Model Validation

3.1.1. The Normal Case

In this section, the accuracy of calculation results of air-gap magnetic field without eccentricity was examined. The spatiotemporal distributions of the radial flux density and tangential flux density of the air-gap magnetic field obtained by analytical method and FEA are shown in Figure 4, where the initial spatial angle of the temporal distribution calculation is 3.7° in this paper.

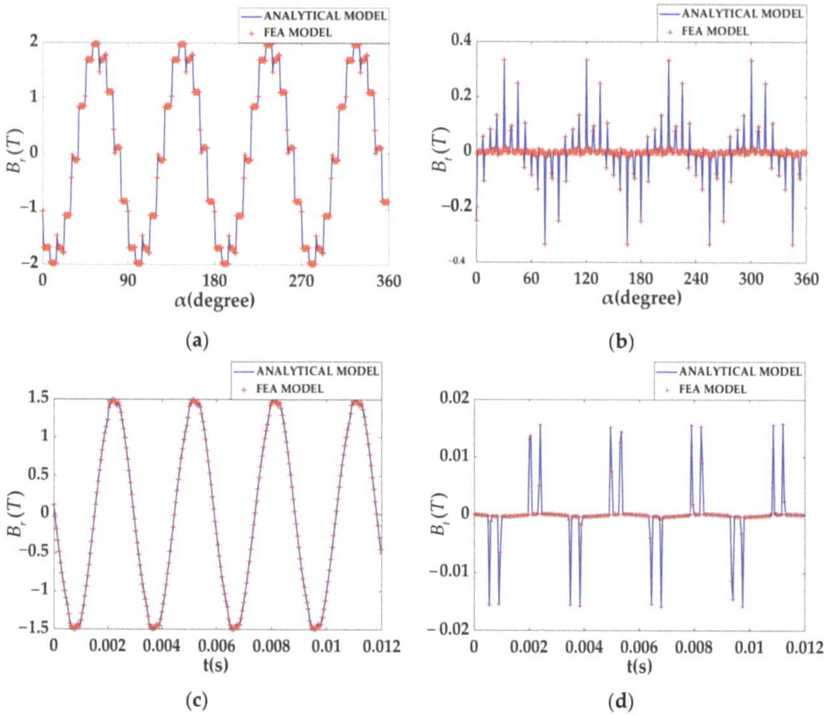

Figure 4. The spatial and temporal distributions of the normal air-gap flux density. (**a**) The spatial distribution of radial flux density; (**b**) The spatial distribution of tangential flux density; (**c**) The temporal distribution of radial flux density; (**d**) The temporal distribution of tangential flux density.

According to Equation (29), the *RMSPEs* of the spatial radial component and the spatial tangential component are 0.9% and 0.89%, respectively, while the *RMSPEs* of the temporal radial component and the temporal tangential component are 1.01% and 1.1%, respectively. This shows that the results calculated by the proposed analytical model are consistent with the results of FEA, which demonstrates the accuracy of the proposed analytical model.

3.1.2. The Eccentric Case

In order to show that the analytical model in this paper can accurately and effectively reflect the phenomenon of eccentricity, the spatiotemporal radial and tangential distributions of the air-gap flux density under the conditions of ε_S = 50%, and ε_D = 50% were calculated. Figure 5 shows the spatiotemporal distributions of radial and tangential components of the air-gap flux density. Equation (29) was also used to analyze the model error. The calculation results show that the *RMSPEs* of the spatial radial and tangential components are 1.02% and 0.97%, respectively. Additionally, *RMSPEs* of the temporal radial and tangential components are 1.11% and 1.12%, respectively. Therefore, the analytical model of the air-gap magnetic field distribution with mixed eccentricity is consistent with the FEA model, which verifies the accuracy and effectiveness of the proposed modelling method.

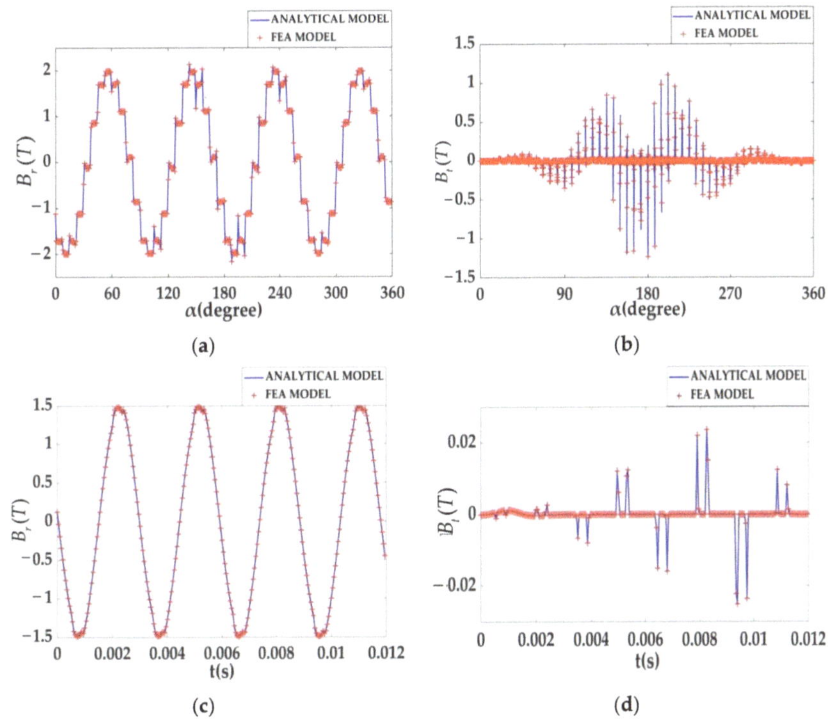

Figure 5. The spatial and temporal distributions of air-gap flux density with mixed eccentricity. (**a**) The spatial distribution of radial flux density; (**b**) The spatial distribution of tangential flux density; (**c**) The temporal distribution of radial flux density; (**d**) The temporal distribution of tangential flux density.

In addition, from the spatial diagram of the radial components, the peak values of the magnetic flux density of the air-gap are mainly affected by the magnetic bridge of the rotor and are close to 2T. Especially in the case of mixed eccentricity, the tightest of the air-gap can reach a saturation state. Once saturated, the maximum magnetic density reaches 2.07T, which is consistent with the results of finite element analysis.

3.2. Influence of the Mixed Eccentricity on the Air-Gap Magnetic Field

In order to investigate the influence of mixed eccentricity on the spatiotemporal distribution of the air-gap magnetic field, Figure 6 shows the three-dimensional spatiotemporal distribution of air-gap flux density under the conditions of $\varepsilon_S = 50\%$ and $\varepsilon_D = 50\%$. Figure 6a shows the three-dimensional spatiotemporal distribution of radial air-gap flux density, and Figure 6b shows the three-dimensional spatiotemporal distribution of tangential air-gap flux density. The blue curves along the x axis (times) are the curves of the variation in radial and tangential flux density with time at a given space angle, while the blue curves along the y axis (space angle) are the curves of the variation of radial and tangential components of air-gap flux density with space angle at a given moment. It can be seen from the figures that, under different space angles, the temporal distributions of the radial and tangential flux density are completely different, while the spatial distributions of the radial and tangential flux density at different times are relatively consistent. Therefore, in the following parts of the paper, only the spatial distribution of the air-gap magnetic field is analyzed.

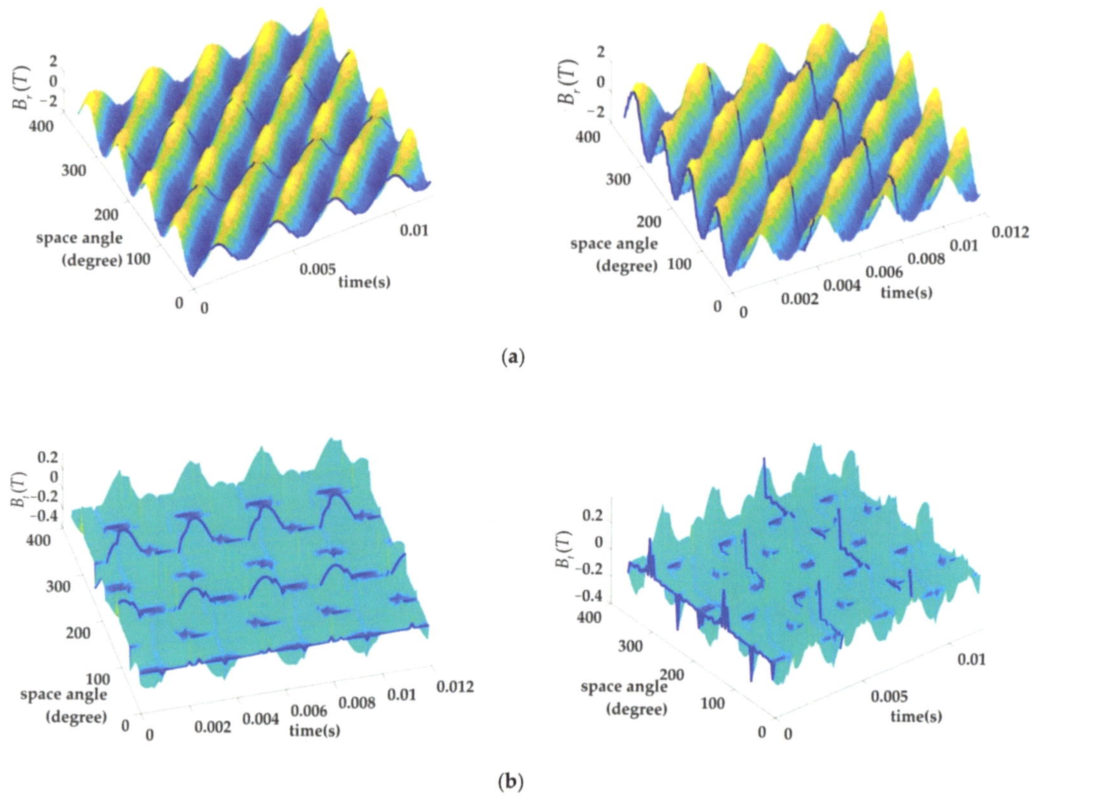

Figure 6. The three-dimensional spatiotemporal characteristics of air-gap magnetic field with mixed eccentricity. (**a**) The spatiotemporal distribution of radial component; (**b**) the spatiotemporal distribution of tangential component.

According to the above calculation, it can be found that mixed eccentricity leads to significant changes in the distribution of the air-gap magnetic field. In order to characterize this change, some features representing the radial and tangential components of the magnetic field were extracted, and their variation between different degrees of mixed eccentricity was analyzed.

3.2.1. Analysis of the Radial Component of the Air-Gap Flux Density

The radial un-symmetry ($USYr$) and the radial total harmonic distortion rate ($THDr$) are designed for the radial component of the air-gap flux density, which can be described as

$$USYr = \frac{B_{rmax} - B_{rmin}}{B_{rn}} \times 100\% \tag{30}$$

$$THDr = \sqrt{\sum_{k}^{\infty}\left[\frac{B_r(k)}{B_{r1}}\right]^2} \times 100\% \tag{31}$$

where B_r is the spatial distribution of radial component of air-gap flux density, and B_{rmax} and B_{rmin} are the maximum and minimum root mean square values of all periods of B_r. B_{rn} is the root mean square value without eccentricity. B_{r1} is the amplitude of fundamental component of the radial air-gap flux density, and $B_r(k)$ is the amplitude of kth harmonic component of radial air-gap flux density.

In order to study the influence of different degrees of static and dynamic eccentricities on radial air-gap flux density, the results of two situations are shown in Table 3; ε_S = 20% and tε_S = 50% for the first situation, while ε_D = 50% and ε_D = 20% for the second situation. From the table, we can find that the spatial distribution of the air-gap radial flux density is distorted by mixed eccentricity. Compared with the case without mixed eccentricity, the values of *USYr* and *THDr* increase when the mixed eccentricity occurs. In the situation of ε_S = 50% and ε_D = 20%, the *USYr* increases by 0.19%, the amplitude of spatial harmonics increases by 0.09% averagely, and the *THDr* also increases by 0.02%. It can be seen that the influence of static eccentricity of the rotor on air-gap radial flux density is greater than that of dynamic eccentricity.

Table 3. The spatial harmonic amplitudes, *USYr* and *THDr* with different mixed eccentricities.

Mixed Eccentric Conditions	Spatial Harmonic Amplitudes										*USYr*	*THDr*
	11	13	23	25	35	47	59	71	83	85	(%)	(%)
Normal	0.2084	0.1309	0.114	0.0519	0.07635	0.05658	0.04392	0.03379	0.02624	0.01964	0	15.52
ε_S = 20% & ε_D = 50%	0.2128	0.1355	0.1246	0.05802	0.09288	0.07738	0.06674	0.05448	0.04471	0.04344	5.21	17.12
ε_S = 50% & ε_D = 20%	0.2128	0.1356	0.1248	0.05809	0.09315	0.07774	0.06717	0.05488	0.04508	0.04347	5.40	17.14

In order to further highlight the influence of mixed eccentricity on the features of radial air-gap flux density, the degrees and steps of static and dynamic eccentricities were set to be the same, so the degree of mixed eccentricity is the same as the degree of static eccentricity or dynamic eccentricity for the analysis of the variation characteristic of Br with a mixed-eccentric fault. The air-gap flux density was calculated when the static eccentricity angle was 2°, and the result is shown in Figure 7. The values of *USYr* and *THDr* of the radial air-gap flux density also increase with the increase in the degree of mixed eccentricity. When the degree of the mixed eccentricity is low (ε_m < 40%), the values of *USYr* are low and change smoothly, while when the ε_m > 40%, the values of *USYr* increase significantly.

3.2.2. Analysis of the Tangential Air-Gap Flux Density Component

We define the tangential un-symmetry (*USYt*) of the tangential air-gap flux density component as

$$USYt = \frac{B_{tmax} - B_{tmin}}{B_{tn}} \times 100\% \tag{32}$$

where B_t is the spatial distribution of tangential component of air-gap flux density, and B_{tmax} and B_{tmin} are the maximum and minimum root mean square values of all periods of B_t. B_{tn} is the root mean square value without eccentricity.

When mixed eccentricity occurs, the tangential air-gap flux density curve will be distorted and become a non-stationary curve. Thus, spectrum analysis based on Fourier transform cannot accurately describe the frequency and amplitude characteristics of tangential air-gap flux density under different spatial angles. In this paper, the wavelet transform method was used to decompose the original tangential air-gap flux density, and the number of decomposition layers was 7. Among them, the low-frequency components a6 and a7 were closer to the envelope of the original signal. As the eccentricity increases, the distortion becomes serious, that is, the asymmetry tends to be serious. The high-frequency components d1, d2, and d3 are obvious periodic signals, and as the degree of eccentricity increases, the signal energy distributions converge toward the center of the samples, as shown in Figure 8.

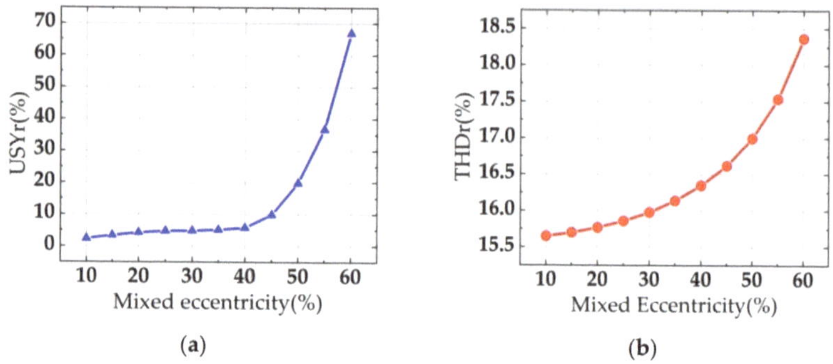

Figure 7. The values of *USYr* and *THDr*. (**a**) *USYr*; (**b**) *THDr*.

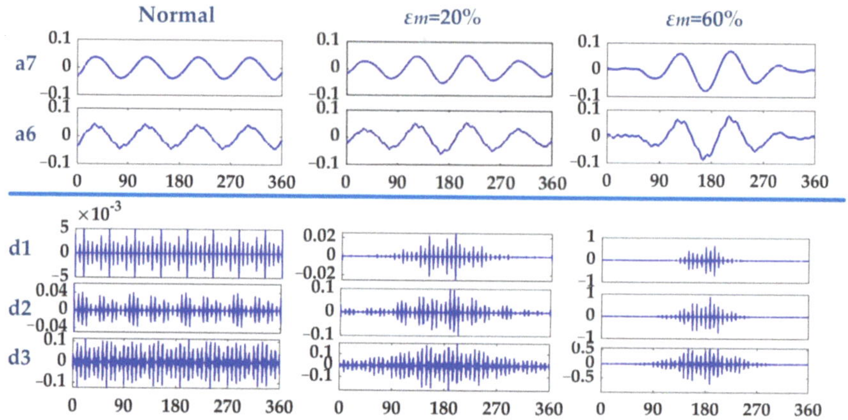

Figure 8. The results of wavelet decomposition of tangential air-gap flux density.

Therefore, the feature of energy proportion is defined to describe the degree of aggregation of signal energy, which is expressed as

$$E_{oi} = \frac{E_{Mi}}{E_T} \times 100\% \tag{33}$$

where i = 1,2 are the coefficients of the first layer and the second layer of the high-frequency components of the signal, respectively, signal samples are equally divided into 10 intervals, and E_{Mi} is the signal energy of the fifth and sixth intervals in the middle. E_T is the total energy of signal samples. The signal energy is calculated by

$$E = \sum_{m=1}^{N_d} s(m)^2 \tag{34}$$

where N_d is the calculated sample length and s is the calculated sample values.

The samples of low-frequency component a7 were used to calculate the *USYt*, and the samples of high-frequency components d1, d2, and d3 were used to calculate the E_{oi}. The results are shown in Figure 9. In Figure 9a, *USYt* increases significantly with the increase in degree of mixed eccentricity, indicating that mixed eccentricity has a serious impact on the low-frequency component of tangential air-gas flux density. In Figure 9b, E_{oi} also increases correspondingly. The E_{oi} of the d1 component is most sensitive to the severity of mixed eccentricity.

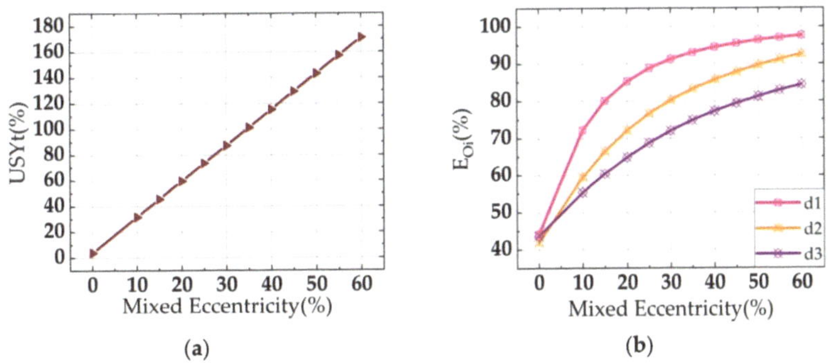

Figure 9. The influence of mixed eccentricity on the values of *USYt* and E_{Oi}: (**a**) *USYt*; (**b**) E_{Oi}.

3.3. Influence of Bridge Saturation on Air-Gap Magnetic Field with Mixed Eccentricity

In order to analyze the influence of bridge saturation on the air-gap magnetic field distribution with mixed eccentricity, the radial spatial component of air-gap flux density was calculated with the analytical model without considering saturation. Specifically, the calculation was performed by ignoring Equation (16) while solving Equations (17) and (18). Then, the radial un-symmetry *USYnr* and the radial total harmonic distortion rate *THDnr* were further calculated for the case without considering bridge saturation and were then compared with the results calculated in Section 3.2.1 in which the bridge saturation was considered, as shown in Figure 10.

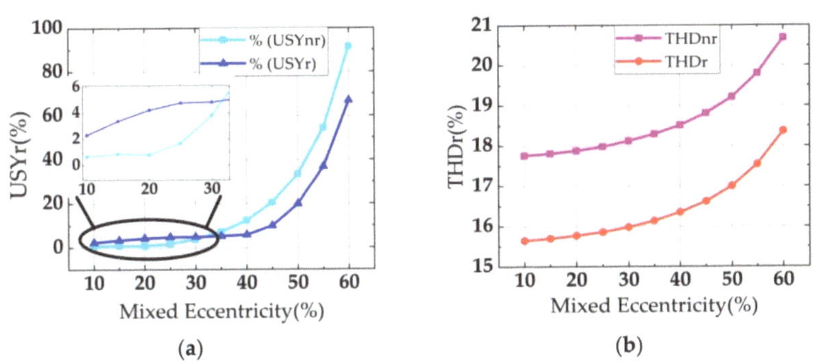

Figure 10. The influence of the bridge saturation on *USYnr* and *THDnr*: (**a**) *USYn*; (**b**) *THDnr*.

It can be seen from Figure 10 that the values of *USYnr* are lower than those of *USYr* when the degree of mixed eccentricities are low. In other words, when the analytical model does not consider the bridge saturation, the calculation results show that the eccentricity has little effect on the magnetic field. So, it is difficult to identify the characteristics of the early stage of eccentricity fault. When the degree of mixed eccentricity is high, the increase in *USYnr* is greater than that of *USYr*. That is, it seems that the impact on the magnetic field is greater than in reality when the model without considering the bridge saturation, which will cause misjudgment of the degree of eccentricity. In addition, we found that the bridge saturation has little effect on *THDnr*.

4. Conclusions

Mixed eccentricity is a common fault of IPMSMs, which will distort the air-gap magnetic field and affect the motor performance. In this paper, an analytical model of the air-gap magnetic field distribution in IPMSMs with mixed eccentricity considering the effect of bridge saturation is established, and the accuracy is verified by comparison with finite element analysis. On this basis, features of the air-gap magnetic field distribution that can effectively reflect the mixed eccentricity fault are extracted. The calculation results show that features change significantly with the variation in the degree of mixed eccentricity, which can accurately reflect the evolution process and characteristics of the mixed eccentricity. In the future, we will further study the relationship between the features extracted in this paper and the monitored quantities such as electrical parameters and mechanical parameters and achieve early detection and diagnosis of mixed eccentricity.

Author Contributions: Conceptualization, Y.S. and P.L.; methodology, Y.S.; software, Y.S. and P.L.; validation, Y.S. and J.C.; formal analysis, P.L.; investigation, Y.S. and J.C.; resources, L.G.; data curation, Y.S. and J.C.; writing—original draft preparation, Y.S.; writing—review and editing, P.L.; supervision, L.G.; project administration, L.G. and D.C. All authors have read and agreed to the published version of the manuscript.

Funding: This research was funded by the key research and development program of Shaanxi Province, China (Grant no. 2022GY-134), Science and Technology Foundation of Xi'an University of Architecture and Technology, China (Grant no. ZR19059), Natural Science Foundation of Guangdong Province, China (Grant no. 2021A1515011423), and Shenzhen basic research project (Grant no. JCYJ20190806172007629).

Institutional Review Board Statement: Not applicable.

Informed Consent Statement: Not applicable.

Data Availability Statement: Not applicable.

Conflicts of Interest: The authors declare no conflict of interest.

References

1. Čorović, S.; Miljavec, D. Modal analysis and rotor-dynamics of an interior permanent magnet synchronous motor: An experimental and theoretical study. *Appl. Sci.* **2020**, *10*, 5881. [CrossRef]
2. Zhang, G.; Yu, W.; Hua, W.; Cao, R.; Qiu, H.; Guo, A. The Design and Optimization of an Interior, Permanent Magnet Synchronous Machine Applied in an Electric Traction Vehicle Requiring a Low Torque Ripple. *Appl. Sci.* **2019**, *9*, 3634. [CrossRef]
3. Zhu, L.; Jiang, S.; Zhu, Z.; Chan, C. Analytical modeling of open-circuit air-gap field distributions in multisegment and multilayer interior permanent-magnet machines. *IEEE Trans. Magn.* **2009**, *45*, 3121–3130. [CrossRef]
4. Zhang, Z.; Xia, C.; Yan, Y.; Geng, Q.; Shi, T. A hybrid analytical model for open-circuit field calculation of multilayer interior permanent magnet machines. *J. Magn. Magn. Mater.* **2017**, *435*, 136–145. [CrossRef]
5. Tian, K.; Zhang, T.; Ai, Y.; Zhang, W. Induction motors dynamic eccentricity fault diagnosis based on the combined use of WPD and EMD-simulation study. *Appl. Sci.* **2018**, *8*, 1709. [CrossRef]
6. Liu, Z.; Huang, J.; He, S. Diagnosis of air-gap eccentricity and partial demagnetisation of an interior permanent magnet synchronous motor based on inverse transient complex inductance vector theory. *IET Electr. Power Appl.* **2018**, *12*, 1166–1175. [CrossRef]
7. Liu, Z.; Huang, J.; Li, B. Diagnosing and distinguishing rotor eccentricity from partial demagnetisation of interior PMSM based on fluctuation of high-frequency d-axis inductance and rotor flux. *IET Electr. Power Appl.* **2017**, *11*, 1265–1275. [CrossRef]
8. Masmoudi, A. Fractional-Slot Concentrated Windings: Design and Analysis. In *Design and Electromagnetic Feature Analysis of AC Rotating Machines*; Springer: Berlin/Heidelberg, Germany, 2019; pp. 55–85.
9. Li, L.; Li, W.; Li, D.; Zhang, X.; Fan, Y. Influence of sleeve thickness and various structures on eddy current losses of rotor parts and temperature field in surface mounted permanent-magnet synchronous motor. *IET Electr. Power Appl.* **2018**, *12*, 1183–1191. [CrossRef]
10. An, Y.; Ma, C.; Zhang, N.; Guo, Y.; Degano, M.; Gerada, C.; Bu, F.; Yin, X.; Li, Q.; Zhou, S. Calculation model of armature reaction magnetic field of interior permanent magnet synchronous motor with segmented skewed poles. *IEEE Trans. Energy Convers.* **2021**, *37*, 1115–1123. [CrossRef]
11. Caruso, M.; Di Tommaso, A.O.; Miceli, R.; Viola, F. A Cogging Torque Minimization Procedure for Interior Permanent Magnet Synchronous Motors Based on a Progressive Modification of the Rotor Lamination Geometry. *Energies* **2022**, *15*, 4956. [CrossRef]
12. Park, Y.; Fernandez, D.; Lee, S.B.; Hyun, D.; Jeong, M.; Kommuri, S.K.; Cho, C.; Reigosa, D.D.; Briz, F. Online detection of rotor eccentricity and demagnetization faults in PMSMs based on hall-effect field sensor measurements. *IEEE Trans. Ind. Appl.* **2018**, *55*, 2499–2509. [CrossRef]
13. Li, Y.; Lu, Q.; Zhu, Z.Q. Unbalanced magnetic force prediction in permanent magnet machines with rotor eccentricity by improved superposition method. *IET Electr. Power Appl.* **2017**, *11*, 1095–1104. [CrossRef]
14. Kim, H.J.; Moon, J.W. Improved rotor structures for increasing flux per pole of permanent magnet synchronous motor. *IET Electr. Power Appl.* **2018**, *12*, 415–422. [CrossRef]

15. Azarinfar, H.; Aghaebrahimi, M.R. Design, analysis and fabrication of a novel transverse flux permanent magnet machine with disk rotor. *Appl. Sci.* **2017**, *7*, 860. [CrossRef]
16. Krämer, C.; Kugi, A.; Kemmetmüller, W. Modeling of a permanent magnet linear synchronous motor using magnetic equivalent circuits. *Mechatronics* **2021**, *76*, 102558. [CrossRef]
17. Song, I.-S.; Jo, B.-W.; Kim, K.-C. Analysis of an IPMSM Hybrid Magnetic Equivalent Circuit. *Energies* **2021**, *14*, 5011. [CrossRef]
18. Yu, Z.; Li, Y.; Jing, Y.; Du, J.; Wang, Z. Analytical model for magnetic field calculation of SPMSM with chamfered pole considering iron core saturation. *IET Electr. Power Appl.* **2020**, *14*, 1856–1864. [CrossRef]
19. Liang, P.; Chai, F.; Li, Y.; Pei, Y. Analytical prediction of magnetic field distribution in spoke-type permanent-magnet synchronous machines accounting for bridge saturation and magnet shape. *IEEE Trans. Ind. Electron.* **2016**, *64*, 3479–3488. [CrossRef]
20. Yang, J.; Chen, J.; Yang, G.; Zhao, S.; Zhang, C. Accurate Calculation of Magnetic Field of Same-Pole and Same-Slot Surface-Mounted Three-Phase Permanent Magnet Synchronous Motor. *IEEE Trans. Magn.* **2022**, *58*, 8204910. [CrossRef]
21. Shin, K.-H.; Cho, H.-W.; Lee, S.-H.; Choi, J.-Y. Armature reaction field and inductance calculations for a permanent magnet linear synchronous machine based on subdomain model. *IEEE Trans. Magn.* **2017**, *53*, 8105804. [CrossRef]
22. Ma, F.; Yin, H.; Wei, L.; Wu, L.; Gu, C. Analytical calculation of armature reaction field of the interior permanent magnet motor. *Energies* **2018**, *11*, 2375. [CrossRef]
23. Su, P.; Hua, W.; Wu, Z.; Han, P.; Cheng, M. Analysis of the operation principle for rotor-permanent-magnet flux-switching machines. *IEEE Trans. Ind. Electron.* **2017**, *65*, 1062–1073. [CrossRef]
24. Zheng, J.; Zhao, W.; Ji, J.; Zhu, J.; Gu, C.; Zhu, S. Design to Reduce Rotor Losses in Fault-Tolerant Permanent-Magnet Machines. *IEEE Trans. Ind. Electron.* **2018**, *65*, 8476–8487. [CrossRef]
25. Zhu, X.; Hua, W.; Wang, B.; Dai, N. Comparison of stator-and rotor-surface-mounted PM brushless machines. *IET Electr. Power Appl.* **2020**, *14*, 62–70. [CrossRef]
26. Cheng, M.; Zhu, X.; Wang, Y.; Wang, R.; Wang, W. Effect and inhibition method of armature-reaction field on superconducting coil in field-modulation superconducting electrical machine. *IEEE Trans. Energy Convers.* **2019**, *35*, 279–291. [CrossRef]
27. Ma, C.; Li, Q.; Lu, H.; Liu, Y.; Gao, H. Analytical model for armature reaction of outer rotor brushless permanent magnet DC motor. *IET Electr. Power Appl.* **2018**, *12*, 651–657. [CrossRef]
28. Deng, W.; Zuo, S. Comparative study of sideband electromagnetic force in internal and external rotor PMSMs with SVPWM technique. *IEEE Trans. Ind. Electron.* **2018**, *66*, 956–966. [CrossRef]
29. Ebrahimi, B.M.; Faiz, J. Configuration impacts on eccentricity fault detection in permanent magnet synchronous motors. *IEEE Trans. Magn.* **2012**, *48*, 903–906. [CrossRef]
30. Rezaee-Alam, F.; Rezaeealam, B.; Faiz, J. Unbalanced magnetic force analysis in eccentric surface permanent-magnet motors using an improved conformal mapping method. *IEEE Trans. Energy Convers.* **2016**, *32*, 146–154. [CrossRef]
31. Guo, L.; Xia, C.; Wang, H.; Wang, Z.; Shi, T. Improved equivalent magnetic network modeling for analyzing working points of PMs in interior permanent magnet machine. *J. Magn. Magn. Mater.* **2018**, *454*, 39–50. [CrossRef]

energies

MDPI

Article

A Novel High-Speed Permanent Magnet Synchronous Motor for Hydrogen Recirculation Side Channel Pumps in Fuel Cell Systems

Kang Ma, Ye Liu *, Ziqiang Wei, Jianfei Yang and Baocheng Guo

College of Electrical and Automation Engineering, Nanjing Normal University, Nanjing 210023, China
* Correspondence: liuye@njnu.edu.cn

Abstract: In hydrogen recirculation side channel pumps, the motor rotor is exposed to a high-pressure mixture of steam and hydrogen, which makes hydrogen embrittlement occur in permanent magnets (PMs). A protective coating is necessary for the PMs in high-pressure hydrogen. However, in the process of sleeve interference installation, the protective coating of the PMs is easily damaged. This paper proposes two surface-mounted insert permanent magnet (SIPM) synchronous motor topologies, SIPM1 and SIPM2, in which the retaining sleeves can be eliminated and the PM protective coating is safe in the assembling process. A dovetail PM and rotor core structure is used to protect the PM with higher rotor strength without retaining the sleeve. The electromagnetic performance of the motors with different rotors, including airgap flux density, output torque, torque ripple, and energy efficiency is compared and optimized. It is concluded that the output torque of the SIPM motor can be promoted by 22.4% and torque ripple can be reduced by 2.9%, while the PM volume remains the same as that of the conventional SPM motor. At the same time, the SIPM motor can have lower harmonic contents of back electromotive force (EMF) and rotor loss compared to the SPM motor with a retaining sleeve. Furthermore, the stress of the PM is analyzed under conditions of PM glue action and failure. The proposed SIPM2 has the ability to operate safely at high-speed and high-temperature operating conditions when the PM glue fails.

Keywords: surface-mounted insert permanent magnet; dovetail permanent magnets; retaining sleeve; rotor stress; torque improvement

Citation: Ma, K.; Liu, Y.; Wei, Z.; Yang, J.; Guo, B. A Novel High-Speed Permanent Magnet Synchronous Motor for Hydrogen Recirculation Side Channel Pumps in Fuel Cell Systems. *Energies* **2022**, *15*, 8913. https://doi.org/10.3390/en15238913

Academic Editors: Loránd Szabó and Feng Chai

Received: 21 October 2022
Accepted: 23 November 2022
Published: 25 November 2022

1. Introduction

In recent years, environmentally friendly transportation and clean energy are popular topics due to depleting fossil fuel energy and concerns of environmental protection [1,2]. The availability of power-intensive and efficient electric motors plays an important role in the development and rapid implementation of hybrid electric vehicles (HEVs) [3,4]. High-speed brushless permanent magnet synchronous motors (PMSMs) have been attracting wide attention due to their high efficiency and high-power density and have been extensively applied in industrial applications such as compressors, pumps, electric vehicles, and motor tool spindle drives [5–7]. Hydrogen fuel cells (HFC) have been widely used in HEVs as driving motors due to their high efficiency, low environmental influence, and good renewability. It is inevitable that excess hydrogen is wasted when the HFC is working. Fortunately, it has been concluded that this problem can be optimized by adding a recycling system to the outlet side of the stack in order to collect excess hydrogen and recirculate it back to the HFC stack by the recirculation pumps (RPs) and ejector to achieve hydrogen secondary utilization [8,9]. The hydrogen system of the HFC is shown in Figure 1. Hydrogen mixed with water, which exhaust from the anode outlet of the stack, is dewatered by the water separator. Dried hydrogen then flows into the RPs and ejector for secondary electricity generation. The ejector, combined with RPs, has the following enhancements compared to ejector-only systems:

- Wider operation range: pressure difference and flow rate;
- Cover low power system load working condition;
- Increased flexibility of shut-down procedure.

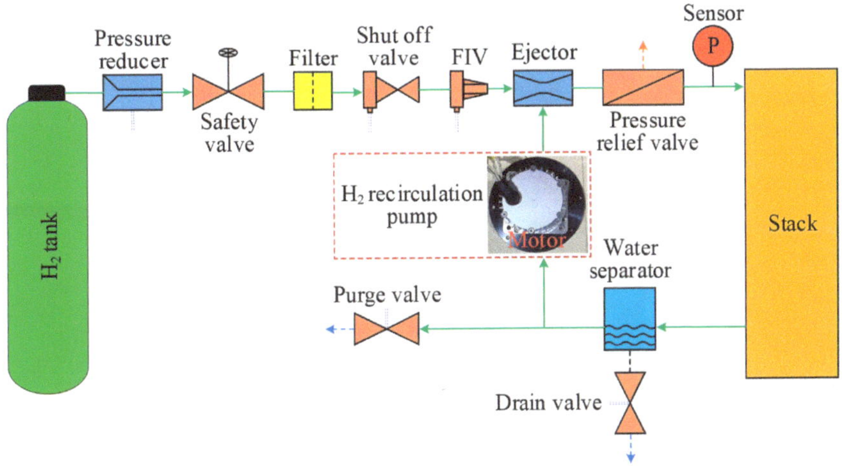

Figure 1. Hydrogen system.

RPs need to be driven with a motor, and high-speed PMSMs can meet the requirements of RPs due to their advantages. In the case of special operating conditions, such as high-pressure hydrogen, the rare-earth PM material is facilely pulverized due to the fact that hydrogen atoms may enter the interior along the grain boundary of the PMs. A protective coating is necessary for the PMs in the hydrogen recirculation side channel pumps (HRSCPs). It is worth noting that PM materials are generally very weak in tension and cannot afford large centrifugal forces at high-speed operation. In order to protect the PM materials at high-speed operation, a retaining sleeve is proposed to be fixed on the outer surface of the PM materials by interference installation.

There are two existing retaining sleeve materials to protect PMs: one is to use a high-strength non-magnetic alloy sleeve, such as stainless steel, and the other is to bundle rotors with high-strength fibers, such as carbon fiber [10,11]. However, the motor's manufacturing cost and complexity can be increased due to the introduction of a retaining sleeve. Meanwhile, the eddy current loss and poor heat dissipation capability can be introduced by non-magnetic alloy and carbon fiber, respectively. It is worth noting that the retaining sleeve relatively increases the physical airgap length, and the additional resistance can be crossed by flux lines, which leads to the reduction of output torque and efficiency [12]. Meanwhile, the protective coating of the PMs can be easily damaged in the process of sleeve interference installation. For these reasons, a sleeve-free high-speed motor is now in demand.

At present, sleeve-free is used mainly in interior permanent magnet (IPM) synchronous motors [13,14]. Reference [15] proposes to use an amorphous material for the rotor core to eliminate the need for a sleeve. The low core losses and high-strength of amorphous material are interesting properties for high-speed motors. However, there is a well-known problem of slotting that affects amorphous material during the manufacturing process. Laser cutting can cause short circuits, and the wire electrical discharge machining method is more time-consuming. A novel retaining sleeve is proposed in reference [16] to reduce eddy current loss. The sleeve opening is imposed to the surface within a range that meets mechanical safety. The losses are further reduced by adjusting the skew angle of the opening area. The sleeve eddy current losses in the optimized sleeve decreased by 72.6% compared with the initially produced model. However, the rotor retaining sleeve is not completely eliminated.

Zhao proposes an optimal design of a novel asymmetrical rotor to obtain performance improvement [17]. The proposed design of the asymmetrical rotor is employed to improve the torque characteristics by establishing rotor asymmetry to allow the reluctance torque and the magnetic torque to reach maximum. It is regrettable that Zhao only investigated the design at low speed. Reference [18] compares rotor stress in different PM installation methods, including where PMs are fixed with glue to the surface of the rotor core and installed without glue by using a retaining sleeve. Nevertheless, the glue failure condition is not taken into account when a retaining sleeve is not used to protect the PM. It is desirable to investigate whether or not the PM falls off the rotor core when the glue fails with sleeve-free.

In this paper, two novel surface-mounted insert permanent magnet (SIPM) motors are proposed to reduce the expense of a retaining sleeve and prevent the PM from falling out of the rotor core. Assessment is also conducted on electromagnetic performance, such as output torque, which can be promoted due to the elimination of the sleeve resistance. This paper is organized as follows: the topologies and main parameters of the conventional SPM motor and proposed SIPM motors are described in Section 2. In Section 3, the influence of rotor design parameters on electromagnetic performance, including back electromotive-force (EMF), output torque, and torque ripple, is investigated. Meanwhile, the rotor stress of SIPM is analyzed to help determine parameters. In Section 4, the electromagnetic and mechanical performance of the SIPM and conventional SPM motors are compared to prove the superiority of the proposed SIPM motors. Finally, Section 5 gives the conclusion. From the literature survey, there have been few technical articles that investigate the sleeve-free applications in SPM motors at high-speed rotation and special working environments, and the main contribution of this paper is to propose a promising solution to enhance the rotor strength of a sleeve-free SPM motor in HRSCPs.

2. Motors with Different Rotors

Motor windings is a key part of achieving electromechanical energy conversion, and includes the distribution windings and concentrated windings in the motor. Compared to the more complex end forms of distributed windings, concentrated windings have the advantage of simple winding ends and low resistance. In this paper, a low power motor is investigated for HRSCP, and concentrated windings with low loss can meet HRSCP's high efficiency requirement. The SPM rotor is chosen as the rotor type instead of the IPM rotor due to the fact that SPM rotors are easily designed to optimize the air-gap flux density waveform of the motor to a sinusoidal distribution, thus improving the performance of the motor.

In order to compare the performance of the motors among the SPM and SIPM, the 6-slot/4-pole PMSM with fractional-slot concentrated-winding (FSCW) is adopted, as shown in Figure 2a. Figure 2b shows the conventional SPM rotor with the retaining sleeve. The detailed parametric models of SIPM rotors are shown in Figure 2c,d. It is worth noting that the motors have the same stator, axial length, and airgap length, and the main design parameters are given in Table 1. The PMs thickness of SIPM rotors is not given in Table 1 due because the shape needs to be adjusted to optimize performance. Meanwhile, the volume of PMs in the SIPM rotors is the same as that in the SPM rotor. The materials of stator, rotor, and PM are T-100, Steel-1010, and NdFeB, respectively. The thickness of the silicon steel sheet is 0.1 mm.

For high-speed range operation, the SPM rotor employs a sleeve to protect the PMs, while the SIPM rotors possess the dovetailed rotor core to implement it; the PMs are locked inside the rotor core by the slotted core. There will be additional closed leakage flux lines generated in the contact between the dovetailed PM and rotor core in the SIPM rotors, which leads to increasing harmonics order. To minimize the harmonics order, there are many alternative methods available for tackling this problem. In particular, skewing of the stator slots has been widely used to reduce cogging torque. However, the skewing of stator slots is difficult and can therefore increase production costs. In this paper, the harmonics order is weakened by changing the shape of the magnet poles [19,20].

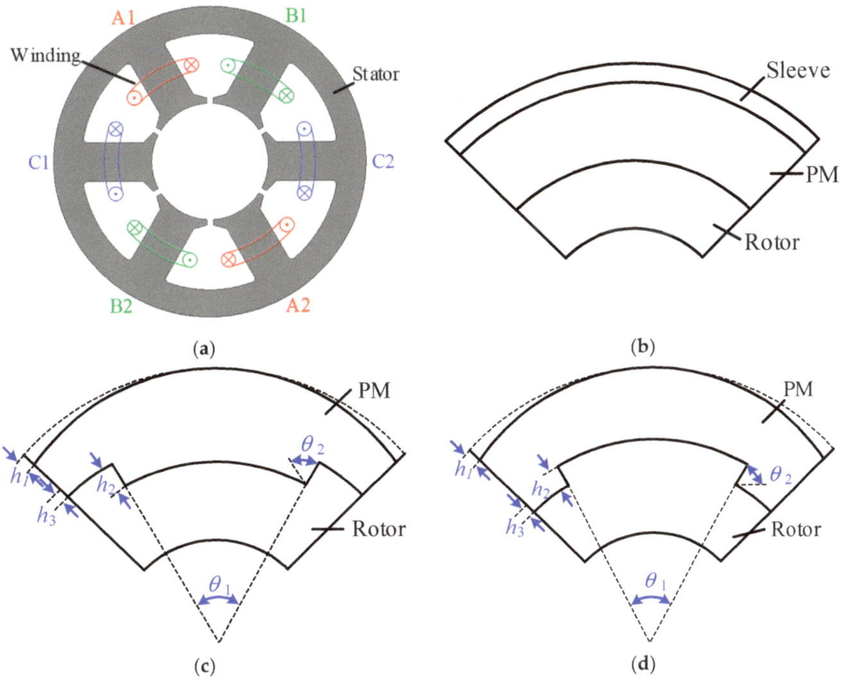

Figure 2. Cross-sectional view of 6-slot/4-pole motor with different rotors. (**a**) 6-slot stator and windings. (**b**) Conventional SPM rotor. (**c**) SIPM1 rotor. (**d**) SIPM2 rotor.

Table 1. Main design parameters of 6-slot/4-pole motor.

<table>
<tr><th rowspan="2">Parameter</th><th>SPM</th><th>SIPM1</th><th>SIPM2</th></tr>
<tr><th colspan="3">Value</th></tr>
<tr><td>Number of slots</td><td colspan="3">6</td></tr>
<tr><td>Number of poles</td><td colspan="3">4</td></tr>
<tr><td>Stator outer diameter (mm)</td><td colspan="3">80</td></tr>
<tr><td>Stator inner diameter (mm)</td><td colspan="3">30</td></tr>
<tr><td>Air-gap length (mm)</td><td colspan="3">1.5</td></tr>
<tr><td>Rotor outer diameter (mm)</td><td colspan="3">28.5</td></tr>
<tr><td>Stack length (mm)</td><td colspan="3">30</td></tr>
<tr><td>Sleeve thickness (mm)</td><td>1</td><td colspan="2">0</td></tr>
<tr><td>PM thickness (mm)</td><td>4</td><td colspan="2">-</td></tr>
<tr><td>DC Voltage (V)</td><td colspan="3">24</td></tr>
<tr><td>Rated phase current (Arms)</td><td colspan="3">35</td></tr>
<tr><td>Rated speed (r/min)</td><td colspan="3">25,000</td></tr>
</table>

3. Parametric Optimization of SIPM Motors

In order to optimize the electromagnetic performance and reliability of the mechanical structure and avoid the geometric conflict, four key design parameters of the SIPM rotors are chosen for the performance optimizations, and their range of variations are listed in (1).

$$\begin{cases} 20^{\circ} \leq \theta_1 \leq 70^{\circ} \\ 0^{\circ} \leq \theta_2 \leq 60^{\circ} \\ 0\text{ mm} \leq h_1 \leq 1.5\text{ mm} \\ 0.5\text{ mm} \leq h_2 \leq 1.5\text{ mm} \end{cases} . \quad (1)$$

The volume of PMs in three rotors keep the same by adjusting h_3, thus, h_3 is not given a range of variations.

3.1. Electromagnetic Optimization of SIPM

Maxwell software is adopted to analyze the electromagnetic performance of the conventional SPM and the proposed SIPM motors, including field distribution, back-EMF, torque characteristics, and loss. The no-load back-EMF and torque characteristics are selected as the design objectives of SIPM motors in this section.

Figure 3 shows the variations of output torque and torque ripple, with trimming thickness h_1 of the PMs edge in the SPM motor without sleeve. Apparently, both output torque and torque ripple decrease with h_1. However, the decrease in torque ripple is much smaller than that of output torque, which demonstrates that the adopted shape of the magnet poles can effectively weaken the torque ripple with a slight sacrifice of output torque.

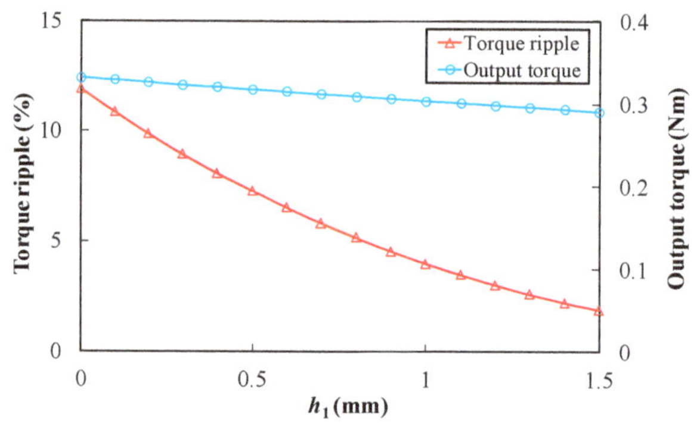

Figure 3. Torque characteristics vs. h_1.

Figure 4 shows that FE predicted the three-dimensional (3-D) view of the output torque, torque ripple, and total harmonic distortion (THD) of no-load EMF with h_2 and θ_1 at the rated current and speed. THD is defined as

$$THD = \frac{\sqrt{\sum_{n=2}^{\infty} E_n^2}}{E_1} \quad (2)$$

where E_n is the amplitude of the no-load EMF harmonic. E_1 is the fundamental no-load EMF. It can be observed that the output torque increases with the increase of θ_1 and the output torque decreases with the increase of h_2 in the SIPM1. The torque ripple reaches maximum when θ_1 is 35°, whilst the torque ripple decreases with the increase of h_2 when θ_1 is more than 50°. The no-load EMF has the largest THD when θ_1 is 40°, and the THD increases with the increase of h_2 when θ_1 is less than 60°. The torque characteristics and THD of the SIPM2 with both h_2 and θ_1 are investigated, and the results are shown in Figure 4b. It is evident that the output torque decreases with the increase of h_2 and θ_1 due to the fact that the volume of PMs is reduced. It has the maximum torque ripple when θ_1 is 40°, whilst the torque ripple increases with the increase of h_2 in the full range of θ_1. The no-load EMF has the largest THD when θ_1 reaches 50°.

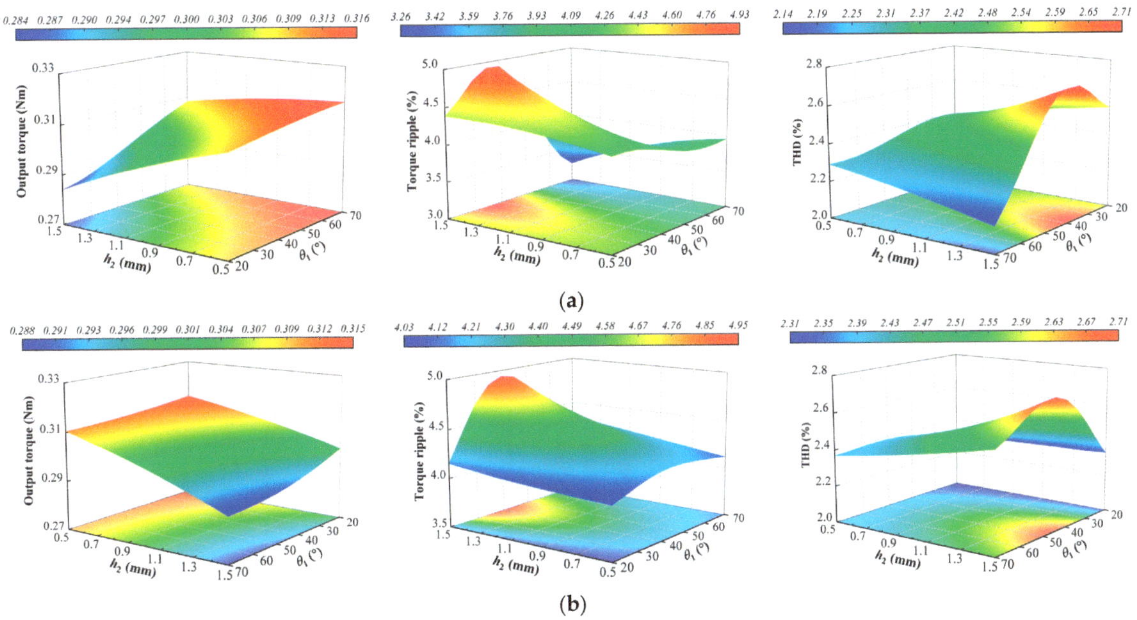

Figure 4. Torque, torque ripple, and THD vs. h_2 and θ_1. (**a**) SIPM1. (**b**) SIPM2.

3.2. Mechanical Strength Optimization of SIPM

In order to calculate the mechanical stress generated in the rotor modal, the 3-D FEA model is established. The mechanical properties of the materials used in the rotor are listed in Table 2. According to the first strength theory (maximum tensile stress theory), the PM tangential stress of brittle material should be less than its maximum tensile strength. The stress ranges of PM are listed (3).

$$\sigma_t = \sigma_{t1} + \sigma_{t2} < \sigma_{t,max} \tag{3}$$

where σ_t, σ_{t1}, σ_{t2}, and $\sigma_{t,max}$ are the tangential stress, mechanical tangential stress, thermal stress, and maximum tensile strength of the PM, respectively. The tangential stress of the PM with the SIPM1 and SIPM2 is calculated in the ANSYS Workbench at rated operating conditions. In order to protect these PMs embedded in the surface of the rotor, θ_2 should be given an angle, i.e., the PM or rotor core should be designed to become a dovetailed one, as given in Figure 2c,d. θ_2 is determined by simulating the failure state of the PM glue. The θ_2 has a small effect on the rotor strength when the PMs are not separated from the rotor core in FEA; thus, θ_2 is not analyzed separately in this section.

Table 2. Material properties.

Parameter	Rotor Core	PM
Density (kg/m^3)	7650	7650
Young's modulus (GPa)	210	160
Poisson ratio	0.27	0.24
CTE(µm/m/°C)	12	12
Tensile strength (MPa)	450	80

Figure 5 shows stress results with h_2 and θ_1, and the stresses are maximum value. The tangential stress of PM is much lower than the tensile strength due to the fact that the contact relationship is set to a fixed constraint for simulating the PM glue action state. It can be seen that the stress increases with the increase of θ_1 in the SIPM; the result is the same for h_2.

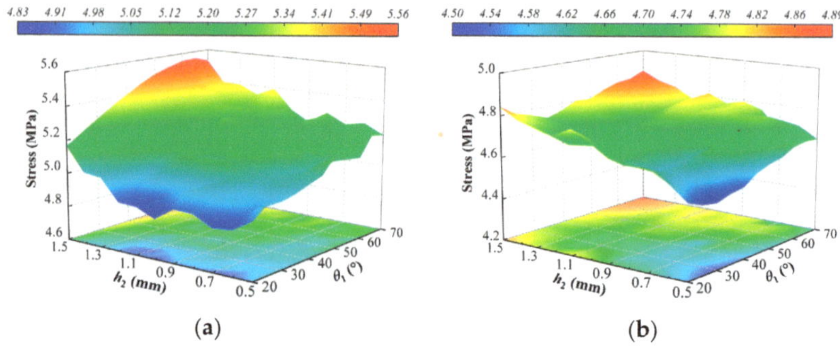

Figure 5. The tangential stress of PM. (**a**) SIPM1. (**b**) SIPM2.

3.3. Optimization Result

In this paper, the three different rotors (SPM, SIPM1, and SIPM2) have the same PM volume by adjusting h_3. Based on the tradeoff between the output torque, torque ripple, the THD of EMF, and stress, the key parameters of SIPM1 and SIPM2 are listed in Table 3. The optimized SIPM1 and SIPM2 rotor structures are shown in Figure 6.

Table 3. Optimization results of key parameters for SIPM.

Parameter	SIPM1	SIPM2
	Value	
θ_1 (°)	55	60
θ_2 (°)	40	10
h_1 (mm)	1	1
h_2 (mm)	1	1.2
h_3 (mm)	0.5	0.9

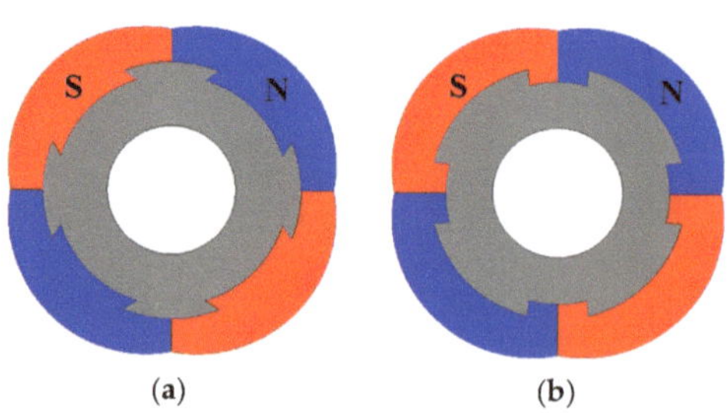

Figure 6. Rotor structures. (**a**) SIPM1. (**b**) SIPM2.

4. Performance Analysis and Comparison

In the previous section, the investigated motors are optimized by the FEA. In this section, the electromagnetic performances and mechanical strength of the SIPM motors are investigated and compared with the SPM motor.

4.1. No-Load Airgap Flux Density and Back-EMF

The no-load flux density distribution of these motors is shown in Figure 7. It can be seen that the flux density in the stator yoke of SIPM is larger than that of the SPM, which suggests that the SIPM can obtain larger PM flux-linkage. The no-load airgap flux density and harmonics order of the SPM and SIPM motors are shown in Figure 8. It can be seen that the SIPM1 motor has a slightly larger fundamental airgap flux density than the SIPM2 motor because it has a smaller rotor core volume to weaken the leakage flux effect. Meanwhile, it was found that the SIPM motors have higher fundamental airgap flux density than conventional SPM motors due to fact that the proposed SIPM motors eliminate the retaining sleeve, which reduces airgap resistance.

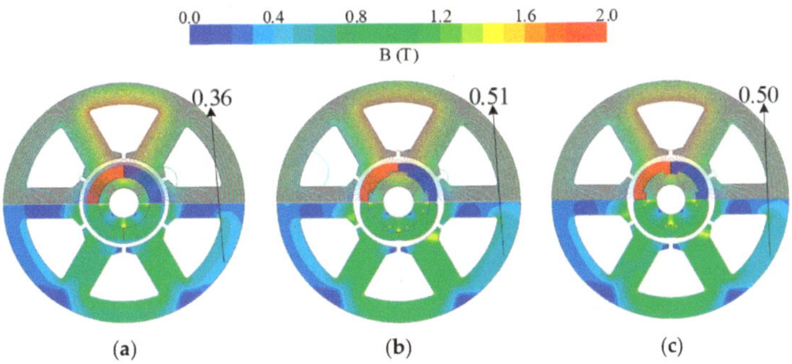

Figure 7. No-load flux density distribution of SPM and SIPM motors. (**a**) SPM. (**b**) SIPM1. (**c**) SIPM2.

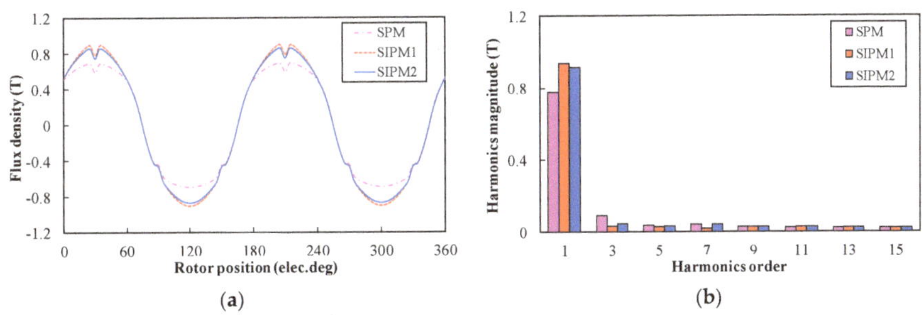

Figure 8. No-load airgap flux density. (**a**) Waveforms. (**b**) Harmonics order.

Figure 9 shows the no-load phase back-EMF and harmonics order of the SPM and SIPM motors at rated operating conditions. It can be seen that the amplitudes of these back-EMF agree well with the airgap flux density characteristics. Apparently, both SIPM1 and SIPM2 motors have a slightly lower harmonics order in fifth and seventh than the SPM motor, and they have lower THD due to the larger fundamental back-EMF.

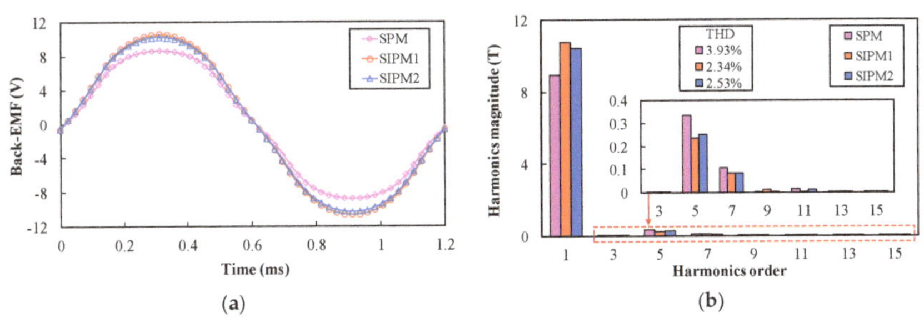

Figure 9. No-load EMF. (**a**) Waveforms. (**b**) Harmonics order.

4.2. Output Torque Characteristics

The cogging torque cycles per mechanical period, N_c, is defined as follows:

$$N_c = LCM(N_s, 2p) \quad (4)$$

where LCM is the least common multiple, N_S is the stator slot number, and p is the rotor pole-pair number. The period of cogging torque is the same due to the fact that the stator slot number and pole number have same LCM in these motors. As can be seen in Figure 10, three different motors have small amplitude of cogging torque, and the SIPM2 has the lowest value. Figure 11 shows the output torque of the SPM and SIPM motors under i_d = 0 control. The torque characteristics of these motors are displayed in Table 4. Where T_{out} is the output torque, $\Delta T_{pk\text{-}pk}$ is the peak-to-peak value of output torque, T_c is the cogging torque, T_{ripple} is the torque ripple, V_{pm} is the volume of the PM, and η_{pm} is the output torque per PM volume. As can be seen in Table 4, the presented SIPM1 motor has 1.22 times the output torque when the SPM motor is taken as the reference value. The SPM and SIPM1 motors have the largest and lowest torque ripple of the three motors, respectively. Meanwhile, the SIPM1 motor has the highest PM utilization ratio when PM usage remains the same.

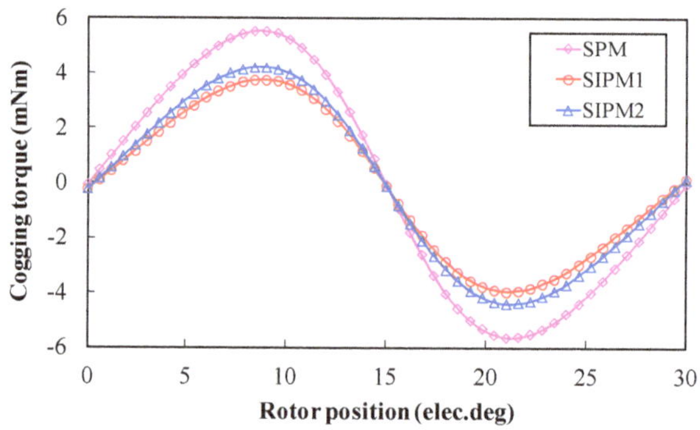

Figure 10. Cogging torque.

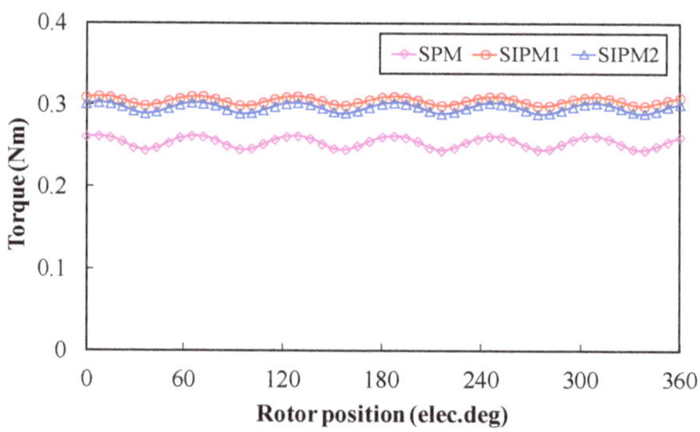

Figure 11. Output torque.

Figure 12 shows the waveforms in these motors of the output torque versus phase current. From Figure 12, it can be seen that the SPM and two SIPM motors at the output torque varies almost linearly with the phase current over a wide range of phase currents. At the same time, the SIPM motors still achieves a higher output torque and PM utilization ratio than the conventional SPM motor.

Table 4. Torque characteristic of motors.

Characteristic	SPM	SIPM1	SIPM2
T_{out} (Nm)	0.254	0.311	0.297
T_{out} (%)	100	122.4	116.9
$\Delta T_{pk\text{-}pk}$ (Nm)	0.017	0.012	0.013
T_c (mNm)	5.54	3.75	4.21
T_{ripple} (%)	6.7	3.8	4.4
V_{pm} (cm^3)	7.9	7.9	7.9
η_{pm} (Nm/cm^3)	0.032	0.039	0.038

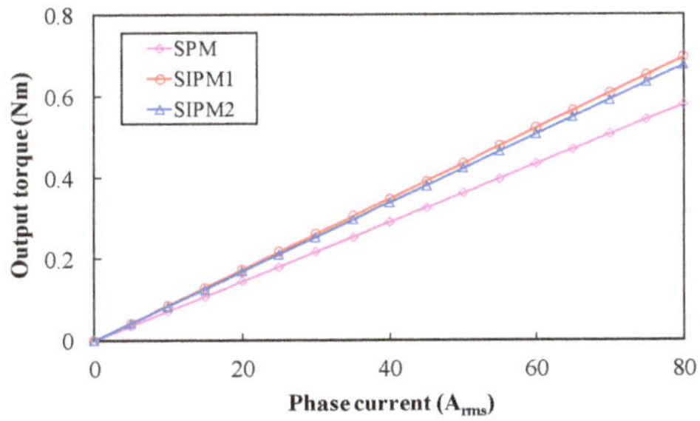

Figure 12. Output torque versus phase current.

4.3. Energy Loss and Efficiency

Conduction losses in the stator winding are typically the sum of two contributions: the dc loss and the ac loss. The total copper loss is calculated to be 6.3 W at the rated current. The distributions of loss density in the SPM and SIPM motors at rated operating conditions are shown in Figure 13. It can be observed that the SIPM motors have larger core loss density than the SPM motor due to the fact that they have a higher stator core flux density. Figure 13b shows eddy current loss between the sleeve and the PM. There is similar eddy current loss in the PM between the SIPM1 and SIPM2 motors. It is worth noting that the eddy current loss of the SPM motor is much higher than those of the SIPM motors, as shown in Figure 14. The main reason is that the armature field of the SPM motor has larger airgap resistance due to the existing stainless-steel sleeve.

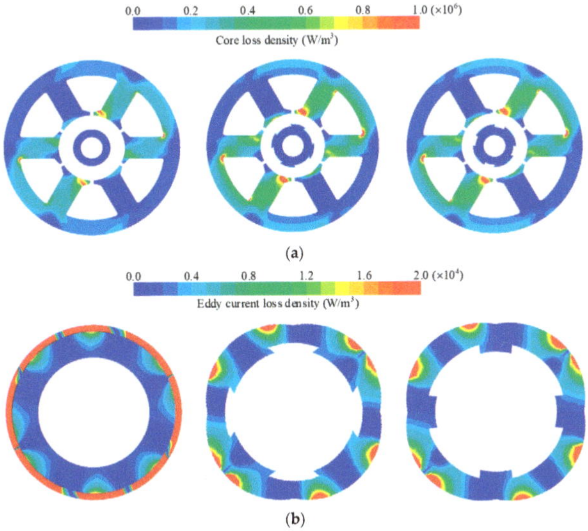

Figure 13. Distribution of loss density. (**a**) Core loss density. (**b**) Eddy current loss density.

The loss, including PM and sleeve eddy current loss, stator and rotor core loss, and copper loss, and the motor's energy efficiency are shown in Figure 14. It can be seen that the SPM motor has less stator core loss than the SIPM motors, which results in higher efficiency than the SIPM motors due to slightly lower output power. Although the SIPM motors have larger loss than the conventional SPM motor, they can still acquire similar efficiency due to the improved output power.

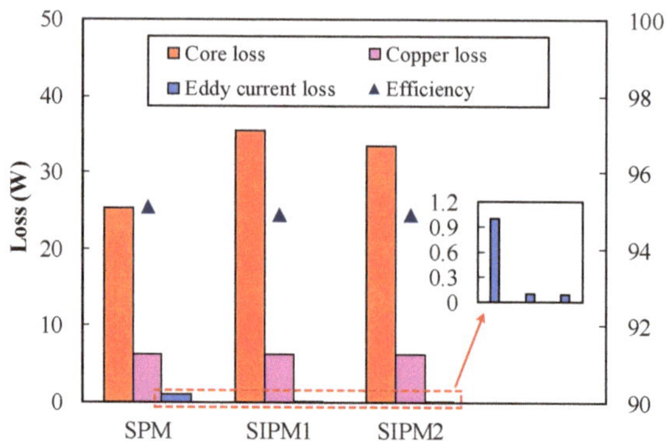

Figure 14. Energy loss and efficiency.

4.4. Mechanical Strength

The rotor stress and deformation, as two significant mechanical factors, are analyzed and compared in the SIPM motors at the rated speed of 25,000 r/min. According to (3), thermal stress was simulated by increasing the rotor temperature to 80 °C. Meanwhile, the contact relationship of PM is set to frictional contact for simulating the PM glue failure state, and the friction factor is 0.15. The distribution of PM stress between the SIPM1 and SIPM2 motors, including radial, tangential, and equivalent, are shown in Figures 15 and 16, respectively. It is shown that the radical stress of PM in the SIPM1 is 150.81 MPa, which is much higher than this in the SIPM2, about 34.46 MPa. However, there is slight difference in the PM's tangential stress between the two configurations. The equivalent stress of the PM in the two configurations is the tensile stress. It is worth noting that the equivalent stress of the PM in the SIPM1 is 155.59 MPa, when the speed and temperature are 25,000 r/min and 80 °C, contrasting that of the ratio of the SIPM2, which is only 71.86 MPa. It can be seen that the equivalent stress of the PM in the SIPM1 motor has exceeded the maximum tensile stress, which is only 80 MPa, and the SIPM2motor is lower than the safe boundary.

Figure 17 shows the PM deformation between the SIPM1 and SIPM2 motors; the position of maximum deformation is at the dovetail slot. It can be seen that the maximum deformation in the SIPM1 motor is obviously higher than that in the SIPM2 motor, and the difference between the two configurations is nearly double. This is due to the fact that the SIPM1 motor has higher equivalent stress than the SIPM2 motor at the dovetail slot.

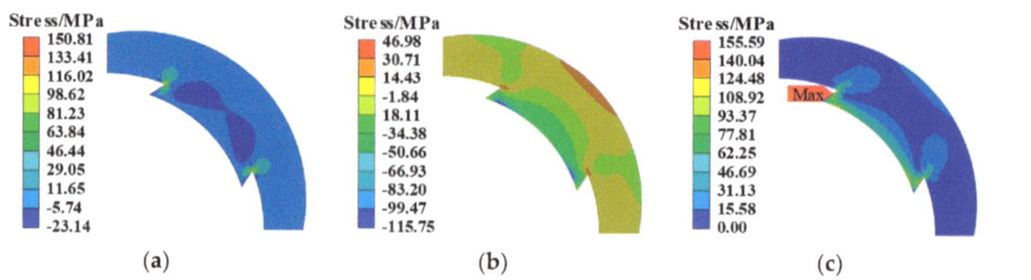

Figure 15. Distribution of PM stress with SIPM1. (**a**) Radial stress. (**b**) Tangential stress. (**c**) Equivalent stress.

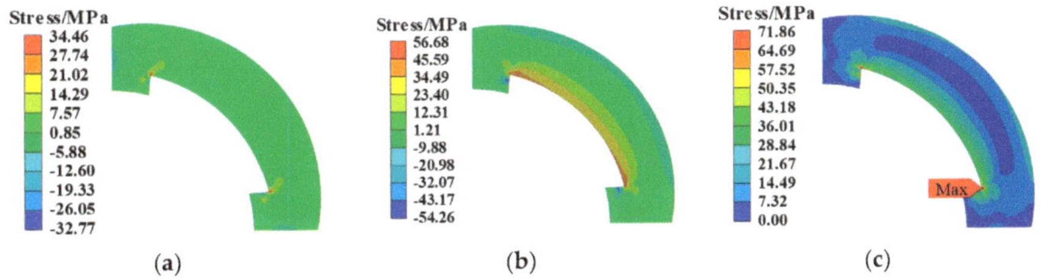

Figure 16. Distribution of PM stress with SIPM2. (**a**) Radial stress. (**b**) Tangential stress. (**c**) Equivalent stress.

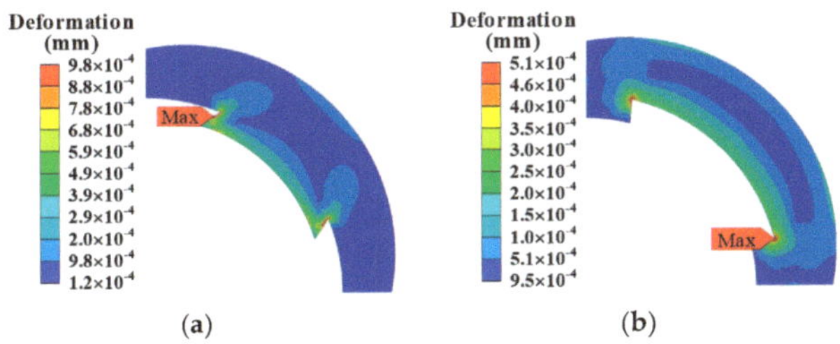

Figure 17. PM deformation with SIPM. (**a**) SIPM1. (**b**) SIPM2.

5. Conclusions

In this paper, PM motors without retaining sleeves, which adopt novel motor rotor topology, are presented for HRSCPs. The performance of the SIPM motors is optimized by the established FE analysis method. Meanwhile, the electromagnetic and mechanical strength performance of the proposed SIPM motors is compared and analyzed with the conventional SPM motor. The detailed conclusions are as follows:

The proposed SIPM motors adopt the dovetail PM and rotor core to protect the PM with higher rotor strength without a retaining sleeve, and the risk of damage to the protective coating will be weakened. Meanwhile, it achieves up to 22% higher output torque than the SPM motor due to reduced airgap resistance. Meanwhile, the proposed PM structure is slightly more complex than the traditional tile-type structure. However, the manufacturing cost of the PM increase is limited, while the PMs volume remains the same among the three motors. The retaining sleeve is eliminated to simplify the process, and the cost of the whole motor is reduced.

By trimming the thickness of the PMs edge, the torque ripple of the proposed SIPM motors can be reduced compared to the SPM motor. Therefore, it has a single parameter optimization for the torque ripple.

Due to the eliminated retaining sleeve, SIPM motors have a lower eddy current loss than the SPM motor. However, the core loss of the SIPM motors is higher than the conventional SPM motor due to the fact that the stator core has higher flux density in the SIPM motors. In addition, the enhanced output torque leads to SIPM motors obtaining slightly lower energy efficiency, while the SIPM motors have higher total loss.

Finally, the SIPM2 motor has better mechanical performance when PM glue is not considered, including stress and deformation. The SIPM2 motor has a lower equivalent stress of the PM, which is lower than the maximum tensile stress, contrasting that of the ratio of the SIPM1, which has twice the maximum tensile stress.

Author Contributions: Investigation, K.M. and Y.L.; writing—original draft, K.M. and Y.L.; software, Z.W. and J.Y.; writing—review and editing, B.G. All authors have read and agreed to the published version of the manuscript.

Funding: This work was supported by National Natural Science Foundation of China under Award 52107051 and the Natural Science Foundation of Jiangsu Province under Award BK20210570.

Conflicts of Interest: The authors declare no conflict of interest.

References

1. Luo, X.; Pan, L.; Yang, J. Mineral Resource Constraints for China's Clean Energy Development under Carbon Peaking and Carbon Neutrality Targets: Quantitative Evaluation and Scenario Analysis. *Energies* **2022**, *15*, 7029. [CrossRef]
2. Changmai, P.; Deka, S.; Kumar, S.; Babu, T.S.; Aljafari, B.; Nastasi, B. A Critical Review on the Estimation Techniques of the Solar PV Cell's Unknown Parameters. *Energies* **2022**, *15*, 7212. [CrossRef]
3. Wei, D.; He, H.; Cao, J. Hybrid electric vehicle electric motors for optimum energy efficiency: A computationally efficient design. *Energy* **2020**, *203*, 117779. [CrossRef]
4. Li, Z.; Khajepour, A.; Song, J. A comprehensive review of the key technologies for pure electric vehicles. *Energy* **2020**, *182*, 824–839. [CrossRef]
5. Beik, O.; Schofield, N. High-Voltage Hybrid Generator and Conversion System for Wind Turbine Applications. *IEEE Trans. Ind. Electron.* **2017**, *65*, 3220–3229. [CrossRef]
6. Ismagilov, F.R.; Vavilov, V.E.; Miniyarov, A.H.; Urazbakhtin, R.R. Super high-speed electric motor with amorphous magnetic circuit for the hydrogen fuel cell air supply system. *Int. J. Hydrogen Energy* **2018**, *43*, 11180–11189. [CrossRef]
7. Golovanov, D.; Gerada, D.; Sala, G.; Degano, M.; Trentin, A.; Connor, P.H.; Xu, Z.; La Rocca, A.; Galassini, A.; Tarisciotti, L.; et al. 4-MW Class High-Power-Density Generator for Future Hybrid-Electric Aircraft. *IEEE Trans. Transp. Electrif.* **2021**, *7*, 2952–2964. [CrossRef]
8. Dadvar, M.; Afshari, E. Analysis of design parameters in anodic recirculation system based on ejector technology for PEM fuel cells: A new approach in designing. *Int. J. Hydrogen Energy* **2014**, *39*, 12061–12073. [CrossRef]
9. Hong, Z.; Li, Q.; Han, Y.; Shang, W.; Zhu, Y.; Chen, W. An energy management strategy based on dynamic power factor for fuel cell/battery hybrid locomotive. *Int. J. Hydrogen Energy* **2018**, *43*, 3261–3272. [CrossRef]
10. Shen, J.; Hao, H.; Yuan, C. FEA study on circumferential grooves on rotor retaining sleeve of high-speed pm brushless motors. *Proc. Chin. Soc. Electr. Eng.* **2012**, *12*, 53–60.
11. Binder, A.; Schneider, T.; Klohr, M. Fixation of buried and surface mounted magnets in high-speed permanent magnet synchronous motors. In Proceedings of the Fourtieth IAS Annual Meeting. Conference Record of the 2005 Industry Applications Conference, Hong Kong, China, 2–6 October 2005; Volume 42, pp. 1031–1037. [CrossRef]
12. Li, C.; Wang, K.; Liu, C. Torque Improvement and Cost Reduction of Permanent Magnet Motors With a Dovetailed Consequent-Pole Rotor. *IEEE Trans. Energy Convers.* **2018**, *33*, 1628–1640. [CrossRef]
13. Rao, J.; Qu, R.; Ma, J.; Xu, W. Investigate the influence of magnetic bridge design on mechanical strength and electromagnetic characteristics in high speed IPM motors. In Proceedings of the 2014 17th International Conference on Electrical Machines and Systems (ICEMS), Hangzhou, China, 22–25 October 2014; pp. 22–27. [CrossRef]
14. Arumugam, P.; Xu, Z.; La Rocca, A.; Vakil, G.; Dickinson, M.; Amankwah, E.; Hamiti, T.; Bozhko, S.; Gerada, C.; Pickering, S.J.; et al. High-Speed Solid Rotor Permanent Magnet Motors: Concept and Design. *IEEE Trans. Transp. Electrif.* **2016**, *2*, 391–400. [CrossRef]
15. Liu, Y.; Ou, J.; Schiefer, M.; Breining, P.; Grilli, F.; Doppelbauer, M. Application of an Amorphous Core to an Ultra-High-Speed Sleeve-Free Interior Permanent-Magnet Rotor. *IEEE Trans. Ind. Electron.* **2018**, *65*, 8498–8509. [CrossRef]
16. Jun, H.-W.; Lee, J.; Lee, H.-W.; Kim, W.-H. Study on the Optimal Rotor Retaining Sleeve Structure for the Reduction of Eddy-Current Loss in High-Speed SPMSM. *IEEE Trans. Magn.* **2015**, *51*, 1–4. [CrossRef]
17. Zhao, W.; Lipo, T.A.; Kwon, B.-I. Optimal Design of a Novel Asymmetrical Rotor Structure to Obtain Torque and Efficiency Improvement in Surface Inset PM Motors. *IEEE Trans. Magn.* **2015**, *51*, 1–4. [CrossRef]
18. Zubkov, Y.V.; Vereshagin, V.E. Designing of High Speed Permanent Magnet Electric Motors with Rotor Strength Verification. In Proceedings of the 2021 International Ural Conference on Electrical Power Engineering (UralCon), Magnitogorsk, Russia, 24–26 September 2021; pp. 414–418. [CrossRef]
19. Wang, K.; Zhu, Z.Q.; Ombach, G. Torque Enhancement of Surface-Mounted Permanent Magnet Motor Using Third-Order Harmonic. *IEEE Trans. Magn.* **2014**, *50*, 104–113. [CrossRef]
20. Du, Z.S.; Lipo, T.A. High Torque Density and Low Torque Ripple Shaped-Magnet Motors Using Sinusoidal Plus Third Harmonic Shaped Magnets. *IEEE Trans. Ind. Appl.* **2019**, *55*, 2601–2610. [CrossRef]

machines

Article

Modeling and Simulation of a Novel Low-Speed High-Torque Permanent Magnet Synchronous Motor with Asymmetric Stator Slots

Shaokai Kou [1,2,3], Ziming Kou [1,2,3,4,*], Juan Wu [1,2,3] and Yandong Wang [1,2,3]

1 College of Mechanical and Vehicle Engineering, Taiyuan University of Technology, Taiyuan 030024, China
2 National-Local Joint Engineering Laboratory of Mining Fluid Control, Taiyuan 030024, China
3 Shanxi Province Engineering Technology Research Center for Mine Fluid Control, Taiyuan 030024, China
4 School of Mechanical Engineering, Anhui University of Science and Technology, Huainan 232001, China
* Correspondence: zmkou@163.com

Abstract: Focusing on the unstable electromagnetic performance of an air gap magnetic field caused by torque ripple and harmonic interference of a multi-slot and multi-pole low-speed high-torque permanent magnet synchronous motor (LHPMSM), an asymmetric stator slot is proposed to improve the comprehensive electromagnetic performance of the motor. Moreover, this paper develops an exact analytical model which predicts the magnetic field distribution based on Laplace's and Poisson's equations. The stator slot asymmetry is introduced into the model and solved by the method of separating variables. Taking a 40p168s LHPMSM as an example, numerical results of the no-load flux density field distributions are obtained by the finite element method (FEM) and employed to validate the analytical model. The influence of stator slot asymmetric structure on electromagnetic characteristics is subsequently analyzed. The results show that, compared with the semi-closed slot model, the asymmetric slot has better torque characteristics, and the electromagnetic characteristics of the motor can be significantly improved by optimizing the stator slot asymmetry. Finally, a prototype is manufactured and tested to validate the theoretical analysis.

Keywords: low-speed high-torque permanent magnet synchronous motor (LHPMSM); asymmetric slot; analytical model; slot asymmetry; electromagnetic performance

Citation: Kou, S.; Kou, Z.; Wu, J.; Wang, Y. Modeling and Simulation of a Novel Low-Speed High-Torque Permanent Magnet Synchronous Motor with Asymmetric Stator Slots. *Machines* **2022**, *10*, 1143. https://doi.org/10.3390/machines10121143

Academic Editors: Loránd Szabó and Feng Chai

Received: 24 October 2022
Accepted: 26 November 2022
Published: 1 December 2022

1. Introduction

The low-speed, high-torque permanent magnet synchronous motor (LHPMSM) has the advantage of low energy consumption, high efficiency, high power density, etc. It has been widely used in industrial applications such as mine transportation and mining, wind power generation, and many other fields [1–5]. However, the LHPMSM usually has the characteristics of multiple slots and multiple poles, large volume, and low installation accuracy. Moreover, the drive system employs the external motor and transmits torque through the output shaft, aggravating friction loss and increasing space volume. The commonly used transmission type of the motor is the external direct-drive permanent magnet motor or magnetic gear, with sufficient installation space. However, it is difficult to employ for some applications in limited space, such as fully mechanized mining faces and roadways in the underground mine.

This paper proposes a novel integrated LHPMSM which employs the transmission structure without a reducer, coupling, or high-speed shaft. Thus, the structure substantially reduces the total volume and assembly cost of the mechanical transmission system. However, its rotor bears the external load, which causes torque ripples. Therefore, it is necessary to improve the electromagnetic performance of the motor through structural design, especially for the stator and rotor.

Some research on improving the electromagnetic performance of the PMSM has been carried out. From the perspective of optimizing the structure of PMs, they mainly employ the segmented inclined magnetic pole [6,7], the segmented permanent magnet [8], the eccentric magnetic pole [9], the unequal-thickness magnetic pole [10,11], the combined magnetic pole [12], different pole arc coefficients [13,14], etc. However, for the multi-pole LHPMSM, the complex shape of permanent magnets greatly increases the production cost, with a large amount of material waste. Therefore, the inner arc eccentric PM described in [9] is adopted. The eccentricity of the PM is equal to the maximum magnetization thickness, which not only has low manufacturing cost, but also avoids material waste.

Optimizing the structure of stator slots is another effective technique to improve the electromagnetic performance of the LHPMSM. In [15], an eccentric stator tooth was proposed, and the influence of eccentricity on the cogging torque was analyzed; the results showed that compared with the non-eccentric structure, the eccentric structure had a more significant weakening effect on cogging torque. In [16], a novel arc-shaped tooth top stator slot structure for a multi-slot permanent magnet synchronous motor is proposed, the coupling relationship between tooth height and tooth top arc radius is established, and the tooth top circular arc radius value with the best electromagnetic characteristics (including cogging torque, torque ripple, air gap magnetic flux density harmonic, average electromagnetic torque) is obtained through a multi-objective optimization design method. In [17], a novel multi-tooth structure of two tooth permanent magnet brushless DC motor was proposed and investigated by 2D and 3D FEM models; the results revealed that the cogging torque of this structure accounts for only about 5% of the electromagnetic torque. In [18], a single-layer slot and double-layer slot combined with the stator core model for a multi-slot permanent magnet synchronous motor was proposed, in which the single-layer slot is composed of two asymmetric stator slots and three semi-closed slots. The influence of the winding factor on the harmonic of the magnetomotive force (MMF) was analyzed; the result shows that the MMF harmonics can be effectively suppressed and the output torque of the motor can be improved by properly selecting the winding factor ratio of the two slots. In [19], three optimization methods of stator slot parameters against a semi-closed slot, including changing the stator tooth width, unequal tooth width, and unequal slot width, are adopted to analyze the harmonic content of the torque ripple and vibration amplitude of electromagnetic force. The results show that the method of unequal tooth widths has the best reduction effect on the electromagnetic vibration and torque ripple. In [20], a slot-opening shift method was presented to reduce the cogging torque, and the calculation method for the shifting angle was expressed. In [21], a skewed unequal width adjacent stator tooth structure was proposed to weaken the 6th torque ripple harmonic of stator teeth caused by synchronous inductance difference; the axial skewed segment technology of PM was combined to point out that the 6th, 12th, and 24th cogging torque harmonics can be eliminated when the electrical angle of the half harmonic period is an integer multiple of the skew angle of adjacent PMs. Moreover, in [22,23], the Fourier decomposition coefficient of the cogging torque with an unequal tooth width and slot size of the stator was deduced, and its specific relationship with the adjacent stator tooth width and slot size was obtained. In [24], the effects of the slotted tooth, step tooth, and eccentric tooth on improving torque characteristics were compared by optimizing the slotting depth, step width, and eccentricity, respectively. Although the stepped tooth had better performance in reducing torque ripple and cogging torque than the slotted tooth, its average torque was the lowest. The average torque will be improved by increasing the number of steps. However, the harmonic content of the flux density was more severe. Therefore, the slotted tooth was more suitable for improving the torque density.

In this paper, an asymmetric slot structure is proposed, in which a V-shaped slot is opened on one side of the stator slot to reduce the edge flux linkage. On the other side, the semi-closed slot is retained to facilitate improving the average torque by optimizing the slot width. However, the complexity of the stator slot structure will inevitably lead to difficulty in magnetic field modeling, and it is challenging to obtain an accurate analytical model by using the traditional subdomain (SD) model method. The exact SD model is an elegant way to analytically determine magnetic fields in electrical machines. This paper simplifies irregular areas in the stator slot as a sector annular. The asymmetry of the slot opening and slot body is defined to derive the asymmetric exact SD model.

The rest of this paper is organized as follows. Section 2 introduces the structure of the LHPMSM, as well as the asymmetric slot structure and its simplified model. Sections 3 and 4 describe the process of asymmetric SD modeling, including slot-opening asymmetric SD and slot body asymmetric SD. In Section 5, the FEA models of the semi-closed slot and asymmetric slot are introduced and simulated. Moreover, the effect of the slot opening asymmetry and slot body asymmetry on the electromagnetic performance is analyzed in Section 6. An LHPMSM prototype with a 40-pole/168-slot is manufactured and tested in Section 7. Conclusions are drawn in Section 8.

2. Structure and Analysis Model

2.1. Structure of LHPMSM

In this research, a 168-slot/40-pole LHPMSM is introduced to verify the asymmetric exact SD model. Figure 1 shows the 3D structure of the LHPMSM, which is a permanent magnet external-rotor mine hoist. In the whole configuration, the main shaft and stator are fixed by a flange; a hexagonal anti-torsion structure is adopted on both sides of the main shaft. Figure 2 shows the PMs adhered to the inner surface of the drum and fixed with circumferential and axial non-magnetic blocks. Moreover, the steel wire ropes (load) are wound on the outer surface of the drum. Figure 3 shows the configuration of the asymmetric stator slot.

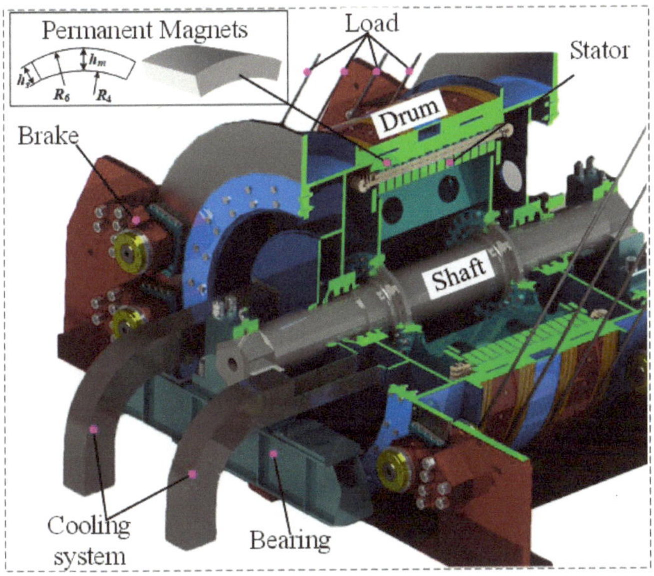

Figure 1. Three-dimensional structure of the LHPMSM.

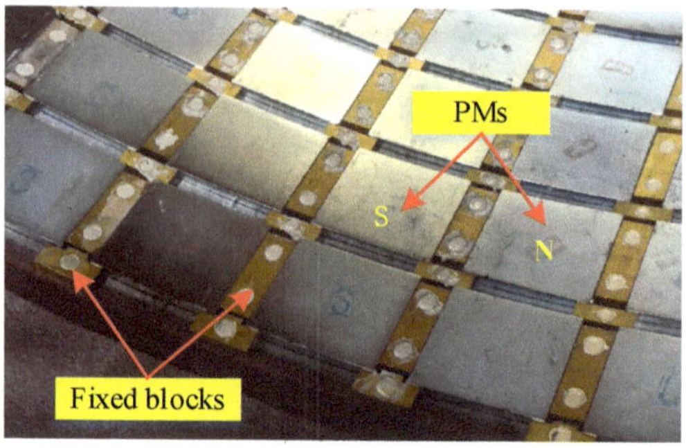

Figure 2. Configuration of the rotor.

Figure 3. Configuration of the stator slot.

2.2. Analysis Model

Figure 4a shows the single asymmetric stator slot parametric model. The slot opening is the asymmetric structure in which the arc length from the center line of the slot opening to the slot body within the slot is unequal, and the slot opening angle is β. Moreover, one side of the tooth tip is collinear with the slot body on the same side. The angles from the center line of the slot opening to the slot body on both sides are $\beta/2$ and $\beta 1$, respectively. On one side of the slot body, the structure can be equivalent to opening a U-shaped groove, where l is the groove depth. The other side retains the unilateral structure of the semi-closed slot. To facilitate modeling, we simplified the original model so that the irregular region in the slot is simplified as the sector annular area. The angle from the bottom edge line on one side of the area to the slot body is α, and the slot-pitch angle is δ. Then, we can define the following geometric relationships:

(1) The asymmetry of the slot opening: the deviation degree of the center line of slot opening relative to the center line of stator slot, which is defined as λ, as shown in Equation (1). It should be noted that when $\beta 1 < \delta/2$, the slotting effect has a serious impact on the magnetic flux density. Therefore, this situation is ignored during the analysis.

$$\lambda = \frac{2\beta_1}{\delta} = \frac{2\delta - \beta}{\delta}\left(\frac{\delta}{2} \leq \beta < \delta,\ 1.5 \leq \lambda < 2\right) \tag{1}$$

(2) The asymmetry of the inside slot: the deviation degree of the center line of the asymmetric region within the slot relative to the center line of the stator slot, which is defined as

$$\xi = \frac{\alpha + \delta}{\delta} \quad (0 < a < \delta,\ 1 < \xi < 2) \tag{2}$$

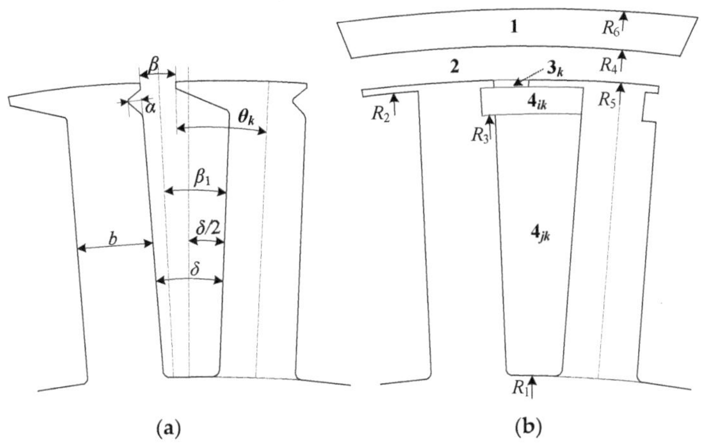

Figure 4. Single asymmetric stator slot model. (**a**) Original parametric model; (**b**) simplified model.

3. Analytical Solution of Magnetic Field

In this paper, the following assumptions are made to enable and simplify the analytical solution: (1) the PMs adopt radial magnetization; (2) the end effect is neglected; (3) the permeability of stator iron is infinite; (4) the relative permeability of PM is equal to 1.

Figure 4a shows the 1/8 geometry of LHPMSM with an asymmetric slot. Figure 4b shows the solution domain of the motor, which is divided into five subdomains: PMs (Region 1), air gap (Region 2), asymmetric slot opening (Region 3*k*), asymmetric slot body (Region 4*ik*), and symmetrical slot body (Region 4*jk*), with $k = 1, 2, \ldots, Q$, and Q being the number of slots.

The parameters in the model are radius of slot bottom, R_1, inner radius of slot opening, R_2, outer radius of symmetrical slot body, R_3, inner radius of PMs, R_4, outer radius of slot opening, R_5, outer radius of PMs, R_6, and the initial position of kth slot, θ_k. θ_k and δ can be expressed by λ as

$$\beta = (2-\lambda)\delta \tag{3}$$

$$\theta_k = \frac{k\pi}{Q} + \left(\frac{\delta}{2} - \beta\right) = \frac{k\pi}{Q} + \left(\lambda - \frac{3}{2}\right)\delta \tag{4}$$

The sth subdomain magnetic vector potential A_s satisfies the governing function as

$$\nabla^2 A_s = \begin{cases} 0 & s = 2, 3k, 4ik \\ \frac{\mu_0}{r} \cdot \frac{\partial M_r}{\partial \theta} & s = 1 \\ -\mu_0 J & s = 4jk1, 4jk2 \end{cases} \tag{5}$$

where μ_0, M_r and J are the permeability of vacuum, radial component of the magnetization vector in the PM subdomain, and current density in the slot, respectively.

The general solutions of the vector potential A_s can be expressed as follows [25]:

$$\begin{aligned} A_s(r,\theta) &= A_{s0} + B_{s0}\ln r \\ &+ \sum_{n=1}^{\infty} \left(A_{sn} f_{sa}(r) + B_{sn} f_{sb}(r)\right) \cdot \cos(p_s\theta_s) \\ &+ \sum_{n=1}^{\infty} \left(C_{sn} f_{sa}(r) + D_{sn} f_{sb}(r)\right) \cdot \sin(p_s\theta_s) + A_{sp} \end{aligned} \tag{6}$$

where A_{s0}, B_{s0}, A_{sn}, B_{sn}, C_{sn} and D_{sn} are integral constants; f_{sa} and f_{sb} are functions of the r; θ_s is the angle of sth subdomain, $\theta_s = \theta - \theta_{s0}$, θ_{s0} is the initial phase angle of sth subdomain; η_s is the Fourier series period. The expressions of f_{sa}, f_{sb} and p_s are

$$f_{sa}(r) = \left(\frac{r}{R_{sO}}\right)^{\frac{2n\pi}{\eta_s}}, f_{sb}(r) = \left(\frac{r}{R_{sI}}\right)^{-\frac{2n\pi}{\eta_s}} \text{ and } p_s = \frac{2n\pi}{\eta_s}$$

where R_{sO} and R_{sI} are the outer radius and inner radius of sth subdomains, respectively.

3.1. Field Solutions in Asymmetric Slot Opening

As shown in Figure 4a, the span angle of kth asymmetric slot opening ranges from θ_k to $\theta_k + \beta$. The Laplace equation in polar coordinates of the asymmetric slot opening subdomain is

$$\begin{cases} \frac{\partial^2 A_{3k}}{\partial r^2} + \frac{1}{r}\frac{\partial A_{3k}}{\partial r} + \frac{1}{r^2}\frac{\partial^2 A_{3k}}{\partial \theta^2} = 0 \\ R_2 \le r \le R_5\,,\ \theta_k \le \theta \le \theta_k + \beta \end{cases} \tag{7}$$

As can be seen from Figure 5c,d, the boundary conditions of the kth slot opening subdomain write

$$\left.\frac{\partial A_{3k}}{\partial \theta}\right|_{\theta=\theta_k} = 0,\ \left.\frac{\partial A_{3k}}{\partial \theta}\right|_{\theta=\theta_k+\beta} = 0 \tag{8}$$

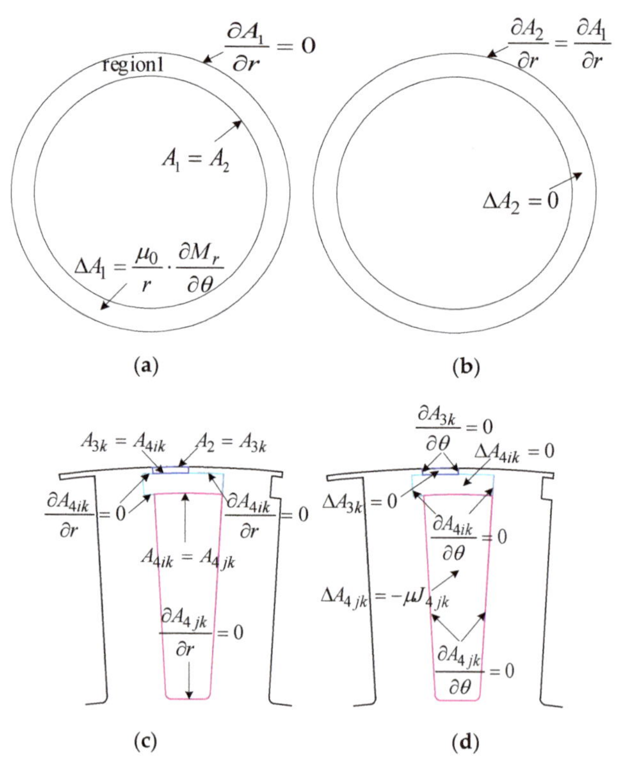

Figure 5. Boundary conditions of each subdomain. (**a**) PMs; (**b**) air gap; (**c**) radial asymmetric slot opening and slot body; (**d**) circumferential asymmetric slot opening and slot body.

The continuity of the radial component of the flux density leads to

$$A_{3k}(r,\theta)|_{r=R_2} = A_{4ik}(r,\ \theta)|_{r=R_2} \tag{9}$$

$$A_{3k}(r,\theta)|_{r=R_5} = A_2(r,\theta)|_{r=R_5} \tag{10}$$

Taking into account (6) and (7) and the boundary conditions shown in Equation (8), the general solution of vector potential A_{3k} in the asymmetric slot opening subdomain can be derived as

$$A_{3k}(r,\theta) = A_{3k} + B_{3k}\ln r + \sum_{m=1}^{\infty}(A_{3km}f_{3ka}(r) + B_{3km}f_{3kb}(r))\cdot\cos(p_{3k}\theta_{3k}) \tag{11}$$

where

$$f_{3ka}(r) = \left(\frac{r}{R_5}\right)^{\frac{m\pi}{(2-\lambda)\delta}}, f_{3kb}(r) = \left(\frac{r}{R_2}\right)^{-\frac{m\pi}{(2-\lambda)\delta}}, \theta_{3k} = \theta - \theta_k \text{ and } p_{3k} = \frac{m\pi}{(2-\lambda)\delta}$$

The constants A_{3k}, B_{3k}, A_{3kn}, and B_{3kn} are determined using Fourier series expansion as

$$A_{3k} + B_{3k}\ln R_2 = \frac{1}{(2-\lambda)\delta}\int_{\theta_k}^{\theta_k+(2-\lambda)\delta} A_{4ik}(R_2,\theta)d\theta \tag{12}$$

$$A_{3k} + B_{3k}\ln R_5 = \frac{1}{(2-\lambda)\delta}\int_{\theta_k}^{\theta_k+(2-\lambda)\delta} A_2(R_5,\theta)d\theta \tag{13}$$

$$A_{3km} = \frac{2}{(2-\lambda)\delta}\int_{\theta_k}^{\theta_k+(2-\lambda)\delta} A_{4ik}(R_2,\theta)\cos\frac{m\pi}{(2-\lambda)\delta}(\theta-\theta_k)d\theta \tag{14}$$

$$B_{3km} = \frac{2}{(2-\lambda)\delta}\int_{\theta_k}^{\theta_k+(2-\lambda)\delta} A_2(R_5,\theta)\cos\frac{m\pi}{(2-\lambda)\delta}(\theta-\theta_k)d\theta \tag{15}$$

3.2. Field Solutions in Asymmetric Slot Body

As shown in Figure 4b, it can be seen that the span angle of the kth asymmetric slot body ranges from $\theta_k - (\lambda - 1)\delta$ to $\theta_k + \delta\xi - (\lambda - 1)\delta$. The Laplace equation in polar coordinates of the asymmetric slot body subdomain is

$$\begin{cases} \frac{\partial^2 A_{4ik}}{\partial r^2} + \frac{1}{r}\frac{\partial A_{4ik}}{\partial r} + \frac{1}{r^2}\frac{\partial^2 A_{4ik}}{\partial \theta^2} = 0 \\ R_3 \le r \le R_2, \theta_k - (\lambda-1)\delta \le \theta \le \theta_k + \delta\xi - (\lambda-1)\delta \end{cases} \tag{16}$$

As can be seen in Figure 5c,d, the boundary conditions of the kth asymmetric slot body subdomain are

$$\left.\frac{\partial A_{4ik}}{\partial \theta}\right|_{\theta=\theta_k-(\lambda-1)\delta} = 0, \quad \left.\frac{\partial A_{4ik}}{\partial \theta}\right|_{\theta=\theta_k+\delta\xi-(\lambda-1)\delta} = 0 \tag{17}$$

The continuity of the radial component of the flux density leads to

$$A_{4ik}(r,\theta)|_{r=R_3} = A_{4jk1}(r,\theta)\Big|_{r=R_3} \tag{18}$$

Similarly, the general solution of vector potential A_{3k} in the asymmetric slot body subdomain can be derived as

$$A_{4ik}(r,\theta) = A_{4ik} + B_{4ik}\ln r + \sum_{u=1}^{\infty}(A_{4iku}f_{4ikua}(r) + B_{4iku}f_{4ikub}(r))\cdot\cos(p_{4iku}\theta_{4ik}) \tag{19}$$

where

$$f_{4ikva}(r) = \left(\frac{r}{R_2}\right)^{\frac{u\pi}{\delta\xi}}, f_{4ikb}(r) = \left(\frac{r}{R_3}\right)^{-\frac{u\pi}{\delta\xi}}, \theta_{4ik} = \theta - \theta_k + (\lambda-1)\delta \text{ and } p_{4iku} = \frac{u\pi}{\delta\xi}$$

The constants A_{4ik}, B_{4ik}, A_{4ikn}, and B_{4ikn} are determined using Fourier series expansion:

$$A_{4ik} + B_{4ik} \ln R_2 = \frac{1}{\delta\xi} \int_{\theta_k-(\lambda-1)\delta}^{\theta_k+\delta\xi-(\lambda-1)\delta} A_{3k}(R_2,\theta) d\theta \tag{20}$$

$$A_{4ik} + B_{4ik} \ln R_3 = \frac{1}{\delta\xi} \int_{\theta_k-(\lambda-1)\delta}^{\theta_k+\delta\xi-(\lambda-1)\delta} A_{4jk1}(R_3,\theta) d\theta \tag{21}$$

$$A_{4iku} = \frac{2}{\delta\xi} \int_{\theta_k-(\lambda-1)\delta}^{\theta_k+\delta\xi-(\lambda-1)\delta} A_{3k}(R_2,\theta) \cos \frac{u\pi}{\delta\xi}(\theta - \theta_k + (\lambda-1)\delta) d\theta \tag{22}$$

$$B_{4iku} = \frac{2}{\delta\xi} \int_{\theta_k-(\lambda-1)\delta}^{\theta_k+\delta\xi-(\lambda-1)\delta} A_{4jk1}(R_3,\theta) \cos \frac{u\pi}{\delta\xi}(\theta - \theta_k + (\lambda-1)\delta) d\theta \tag{23}$$

3.3. Cogging Torque Calculation

In the polar coordinate system, the radial and tangential flux density distribution in the air gap is given by

$$B_{r2}(r,\theta) = \frac{1}{r}\frac{\partial A_2(r,\theta)}{\partial \theta}, B_{\theta 2}(r,\theta) = -\frac{\partial A_2(r,\theta)}{\partial r} \tag{24}$$

where B_{r2} and $B_{\theta 2}$ are the radial and tangential components of the air gap flux density at r, respectively.

According to the Maxwell stress tensor, a circle of radius rag in the air-gap subdomain is taken as the integration path. The electromagnetic torque is expressed as follows:

$$T_c = \frac{L_{ef} {r_{ag}}^2}{\mu_0} \int_0^{2\pi} B_{r2}(r_{ag},\theta) \cdot B_{\theta 2}(r_{ag},\theta) d\theta \tag{25}$$

where L_{ef} is the stack length and r_{ag} is the radius of the air-gap subdomain.

3.4. Field Solutions in the Rest Subdomains

According to Equation (5), the magnetic vector potential expressions of the permanent magnet subdomain, the air gap subdomain, and the symmetrical slot body subdomain are given by

$$\text{region1}: \begin{cases} \frac{\partial^2 A_1(r,\theta)}{\partial r^2} + \frac{1}{r}\frac{\partial A_1(r,\theta)}{\partial r} + \frac{1}{r^2}\frac{\partial^2 A_1(r,\theta)}{\partial \theta^2} = \frac{\mu_0}{r} \cdot \frac{\partial M_r}{\partial \theta} \\ R_4 \le r \le R_6,\ 0 \le \theta \le 2\pi \end{cases} \tag{26}$$

$$\text{region2}: \begin{cases} \frac{\partial^2 A_2(r,\theta)}{\partial r^2} + \frac{1}{r}\frac{\partial A_2(r,\theta)}{\partial r} + \frac{1}{r^2}\frac{\partial^2 A_2(r,\theta)}{\partial \theta^2} = 0 \\ R_5 \le r \le R_4,\ 0 \le \theta \le 2\pi \end{cases} \tag{27}$$

$$\text{region}4jk1: \begin{cases} \frac{\partial^2 A_{4jk1}(r,\theta)}{\partial r^2} + \frac{1}{r}\frac{\partial A_{4jk1}(r,\theta)}{\partial r} + \frac{1}{r^2}\frac{\partial^2 A_{4jk2}(r,\theta)}{\partial \theta^2} = -\mu_0 J_{4jk1} \\ R_{4jk12} \le r \le R_3,\ \theta_k - (\lambda-1)\delta \le \theta \le \theta_k + (2-\lambda)\delta \end{cases} \tag{28}$$

$$\text{region}4jk2: \begin{cases} \frac{\partial^2 A_{4jk2}(r,\theta)}{\partial r^2} + \frac{1}{r}\frac{\partial A_{4jk2}(r,\theta)}{\partial r} + \frac{1}{r^2}\frac{\partial^2 A_{4jk2}(r,\theta)}{\partial \theta^2} = -\mu_0 J_{4jk2} \\ R_1 \le r \le R_{4jk12},\ \theta_k - (\lambda-1)\delta \le \theta \le \theta_k + (2-\lambda)\delta \end{cases} \tag{29}$$

where J_{4jk1} and J_{4jk2} is the current density.

As shown in Figure 5, the associated boundary conditions in PMs, air-gap, and slot subdomains are

$$\text{region1}: \left.\frac{\partial A_1}{\partial r}\right|_{r=R_6} = 0 \tag{30}$$

$$A_1(r,\theta)|_{r=R_4} = A_2(r,\theta)|_{r=R_4} \tag{31}$$

$$H_1(r,\theta)|_{r=R_4} = H_2(r,\theta)|_{r=R_4} \tag{32}$$

$$\text{region2}: \left.\frac{\partial A_2}{\partial r}\right|_{r=R_5} = \begin{cases} \left.\frac{\partial A_{3k}}{\partial r}\right|_{r=R_5} & \theta_k \le \theta \le \theta_k + \beta \\ 0 & \text{elsewhere} \end{cases} \tag{33}$$

$$A_2(r,\theta)|_{r=R_5} = \begin{cases} A_{3k}(r,\theta)|_{r=R_5} & \theta_k \le \theta \le \theta_k + \beta \\ 0 & \text{elsewhere} \end{cases} \tag{34}$$

$$\text{region}4jk1: \left.\frac{\partial A_{4jk1}}{\partial r}\right|_{r=R_3} = \begin{cases} \left.\frac{\partial A_{4ik}}{\partial r}\right|_{r=R_3} & \theta_k - (\lambda-1)\delta \le \theta \le \theta_k + (2-\lambda)\delta \\ 0 & \text{elsewhere} \end{cases} \tag{35}$$

$$A_{4jk1}(r,\theta)\Big|_{r=R_3} = A_{4ik}(r,\theta)\ |_{r=R_3}, \theta_k - (\lambda-1)\delta \le \theta \le \theta_k + (2-\lambda)\delta \tag{36}$$

$$\left.\frac{\partial A_{4jk1}}{\partial r}\right|_{r=R_{4jk13}} = \left.\frac{\partial A_{4jk2}}{\partial r}\right|_{r=R_{4jk13}} \tag{37}$$

$$A_{4jk1}(r,\theta)\Big|_{r=R_{4jk13}} = A_{4jk2}(r,\theta)\Big|_{r=R_{4jk13}} \tag{38}$$

$$\left.\frac{\partial A_{4jk1}}{\partial \theta}\right|_{\theta=\theta_k-(\lambda-1)\delta} = 0, \quad \left.\frac{\partial A_{4jk1}}{\partial \theta}\right|_{\theta=\theta_k+\delta\xi-(\lambda-1)\delta} = 0 \tag{39}$$

$$\text{region}4jk2: \left.\frac{\partial A_{4jk1}}{\partial r}\right|_{r=R_1} = 0 \tag{40}$$

$$\left.\frac{\partial A_{4jk2}}{\partial \theta}\right|_{\theta=\theta_k-(\lambda-1)\delta} = 0, \quad \left.\frac{\partial A_{4jk2}}{\partial \theta}\right|_{\theta=\theta_k+\delta\xi-(\lambda-1)\delta} = 0 \tag{41}$$

where $R_{4jk13} = \sqrt{(R_1^2 + R_3^2)/2}$.

The general solution of vector potential A_1, can be derived as

$$\begin{aligned} A_1(r,\theta) &= \sum_{n=1}^{\infty}\left(A_{1n}\left(\tfrac{r}{R_6}\right)^n + B_{1n}\left(\tfrac{r}{R_4}\right)^{-n}\right)\cdot \cos n\theta \\ &+ \sum_{n=1}^{\infty}\left(C_{1n}\left(\tfrac{r}{R_6}\right)^n + D_{1n}\left(\tfrac{r}{R_4}\right)^{-n}\right)\cdot \sin n\theta + A_{1p} \\ A_{1p}(r,\theta) &= \begin{cases} \sum_{n=1}^{\infty}\left[\frac{\mu_0 n M_{rn} r}{n^2-1}(-\cos(n\theta)+\sin(n\theta))\right] n \ne 1 \\ \sum_{n=1}^{\infty}\left[\frac{\mu_0 M_{rn} r \ln r}{2}(\cos(n\theta)-\sin(n\theta))\right] n = 1 \end{cases} \end{aligned} \tag{42}$$

where M_{rn} is the radial component for remanent magnetization of PMs.

Taking into account the boundary conditions (30), the following equations can be obtained:

$$B_{1n} = \begin{cases} \left(A_{1n} - M_{rn}\cdot\frac{R_6\mu_0}{n^2-1}\right)\cdot\left(\frac{R_4}{R_6}\right)^n & n \ne 1 \\ \left(A_{1n} + M_{rn}\frac{R_6(1+\ln R_6)\mu_0}{n(n^2-1)}\right)\cdot\left(\frac{R_4}{R_6}\right)^n & n = 1 \end{cases} \tag{43}$$

$$D_{1n} = \begin{cases} \left(C_{1n} + M_{rn}\cdot\frac{R_6\mu_0}{n^2-1}\right)\cdot\left(\frac{R_4}{R_6}\right)^n & n \ne 1 \\ \left(C_{1n} - M_{rn}\frac{R_6(1+\ln R_6)\mu_0}{n(n^2-1)}\right)\cdot\left(\frac{R_4}{R_6}\right)^n & n = 1 \end{cases} \tag{44}$$

Then, according to Equations (42)–(44), the vector potential $A_1(r,\theta)$ can be rewritten as

$$\begin{aligned} A_1(r,\theta) &= \sum_{n=1}^{\infty}[A_{1n}f_1(r) - M_{rn}f_{1rn}(r)]\cdot\cos n\theta \\ &+ \sum_{n=1}^{\infty}[C_{1n}f_1(r) + M_{rn}f_{1rn}(r)]\cdot\sin n\theta \end{aligned} \tag{45}$$

where

$$f_1(r) = \left[\left(\frac{r}{R_6}\right)^n + \left(\frac{R_4}{R_6}\right)^n\left(\frac{r}{R_4}\right)^{-n}\right]$$

$$f_{1rn}(r) = \begin{cases} \frac{\mu_0}{n^2-1}\left(R_6\cdot\left(\frac{R_4}{R_6}\right)^n\left(\frac{r}{R_4}\right)^{-n}\right)+nr & n\neq 1 \\ -\frac{\mu_0}{2}\left(\frac{R_6(1+\ln R_6)}{n}\left(\frac{R_4}{R_6}\right)^n\left(\frac{r}{R_4}\right)^{-n}+r\ln r\right) & n=1 \end{cases}$$

The general solution of vector potential A_2 can be derived as

$$\begin{aligned} A_2(r,\theta) &= \sum_{n=1}^{\infty}\left(A_{2n}\left(\frac{r}{R_4}\right)^n + B_{2n}\left(\frac{r}{R_5}\right)^{-n}\right)\cdot\cos n\theta \\ &+\sum_{n=1}^{\infty}\left(C_{2n}\left(\frac{r}{R_4}\right)^n + D_{2n}\left(\frac{r}{R_5}\right)^{-n}\right)\cdot\sin n\theta \end{aligned} \quad (46)$$

As shown in Figure 6, the region $4jk$ adopts double-layer overlapping short-distance winding, with the upper and lower slot areas being equal. The general solution of vector potential A_{4jk1} and A_{4jk2} can be derived as

$$\begin{aligned} A_{4jk1}(r,\theta) = A_{4jk1} + \sum_{u=1}^{\infty}\left(A_{4jku1}f_{4jkua1}(r)+B_{4jku1}f_{4jkub1}(r)\right)\cdot\cos(p_{4jku1}\theta_{4jk1}) + A_{4jk1p} \\ f_{4jkua1}(r) = \left(\frac{r}{R_3}\right)^{\frac{u\pi}{\delta}}, f_{4jkub1}(r) = \left(\frac{r}{R_{4jk13}}\right)^{-\frac{u\pi}{\delta}} \\ \theta_{4jk1} = (\theta-\theta_k+(\lambda-1)\delta) \\ p_{4jku1} = \frac{u\pi}{\delta} \end{aligned} \quad (47)$$

$$\begin{aligned} lA_{4jk2}(r,\theta) = A_{4jk2} + \sum_{u=1}^{\infty}\left(A_{4jku2}f_{4jkua2}(r)+B_{4jku2}f_{4jkub2}(r)\right)\cdot\cos(p_{4jku2}\theta_{4jk2}) + A_{4jku2p} \\ f_{4jkua2}(r) = \left(\frac{r}{R_{4jk13}}\right)^{\frac{u\pi}{\delta}}, f_{4jkub2}(r) = \left(\frac{r}{R_1}\right)^{-\frac{u\pi}{\delta}} \\ \theta_{4jk2} = (\theta-\theta_k+(\lambda-1)\delta) \\ p_{4jku2} = \frac{u\pi}{\delta} \end{aligned} \quad (48)$$

$$A_{4jku1p} = \frac{1}{2}\mu_0 J_{4jk1}(R_{4jk13}^2\ \ln r - \frac{1}{2}r^2) + \frac{1}{2}\mu_0 J_{4jk2}\left(R_1^2 - R_{4jk13}^2\right)\ln r \quad (49)$$

$$A_{4jku2p} = \frac{1}{2}\mu_0 J_{4jk2}(R_1^2\ln r - \frac{1}{2}r^2) \quad (50)$$

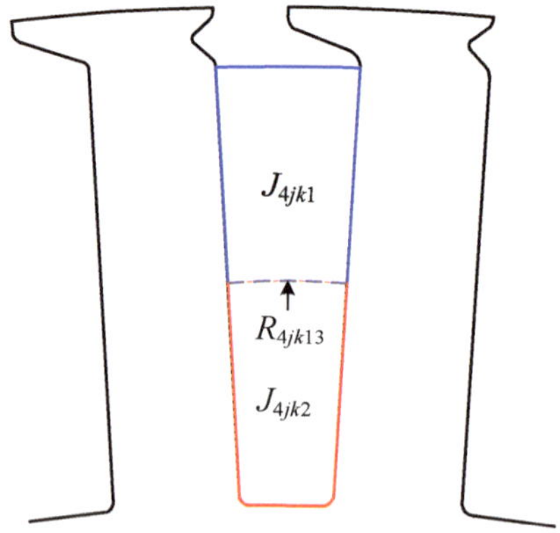

Figure 6. Winding configurations.

4. Integration Constants

In order to obtain the integration constants in the expressions of the vector potential, the interface conditions between subdomains need to be applied.

4.1. Interface between Region 1 and Region 2

According to Equations (31), (45) and (46), the following equations can be obtained:

$$2A_{1n}\left(\frac{R_4}{R_6}\right)^n - A_{2n} - B_{2n}\left(\frac{R_4}{R_5}\right)^{-n} = M_{rn}f_{1rn}(R_4) \tag{51}$$

$$2C_{1n}\left(\frac{R_4}{R_6}\right)^n - C_{2n} - D_{2n}\left(\frac{R_4}{R_5}\right)^{-n} = -M_{rn}f_{1rn}(R_4) \tag{52}$$

The interface condition (32) can be rewritten as

$$H_1(r,\theta)|_{r=R_4} = H_2(r,\theta)|_{r=R_4} \rightarrow \frac{1}{\mu_r}\cdot\frac{\partial A_1(r,\theta)}{\partial r}\bigg|_{r=R_4} = \frac{\partial A_2(r,\theta)}{\partial r}\bigg|_{r=R_4} \tag{53}$$

According to Equations (45), (46) and (53), the following equations can be obtained:

$$A_{2n} - B_{2n}\left(\frac{R_4}{R_5}\right)^{-n} = M_{rn}f_{2rn}(R_4) \tag{54}$$

$$C_{2n} - D_{2n}\left(\frac{R_4}{R_5}\right)^{-n} = -M_{rn}f_{2rn}(R_4) \tag{55}$$

where

$$f_{2rn}(R_4) = \begin{cases} \left[\frac{n}{R_4}\frac{\mu_0}{n^2-1}\left(R_6\cdot\left(\frac{R_4}{R_6}\right)^n\right) - n\right] & n \neq 1 \\ \left[-\frac{n}{R_4}\frac{\mu_0}{2}\frac{R_6(1+\ln R_6)}{n}\left(\frac{R_4}{R_6}\right)^n + \frac{\mu_0}{2}(1+\ln R_4)\right] & n = 1 \end{cases}$$

4.2. Interface between Region 2 and Region 3k

According to Equations (13), (15), (33) and (34), the following equation can be obtained:

$$\begin{aligned} A_{3k} + B_{3k}\ln R_5 &= \frac{1}{(2-\lambda)\delta}\left[\sum_{n=1}^{\infty}\left(A_{2n}\left(\frac{R_5}{R_4}\right)^n + B_{2n}\right)\cdot l(n,k)\right. \\ &\quad \left. + \sum_{n=1}^{\infty}\left(C_{2n}\left(\frac{R_5}{R_4}\right)^n + D_{2n}\right)\cdot s(n,k)\right] \end{aligned} \tag{56}$$

$$\begin{aligned} B_{3km} = \frac{2}{(2-\lambda)\delta}&\left\{\sum_{n=1}^{\infty}\left(A_{2n}\left(\frac{R_5}{R_4}\right)^n + B_{2n}\right)f(m,n,k)\right. \\ &\left. + \sum_{n=1}^{\infty}\left(C_{2n}\left(\frac{R_5}{R_4}\right)^n + D_{2n}\right)g(m,n,k)\right\} \end{aligned} \tag{57}$$

where

$$l_{2,3k}(n,k) = \int_{\theta_k}^{\theta_k+(2-\lambda)\delta} \cos n\theta d\theta$$

$$s_{2,3k}(n,k) = \int_{\theta_k}^{\theta_k+(2-\lambda)\delta} \text{sins} n\theta d\theta$$

$$\gamma_{2,3k}(n,k) = \int_{\theta_k}^{\theta_k+(2-\lambda)\delta} \sin\left(\frac{m\pi}{(2-\lambda)\delta}(\theta-\theta_k)\right)d\theta$$

$$f_{2,3k}(m,n,k) = \int_{\theta_k}^{\theta_k+(2-\lambda)\delta} \cos n\theta\cdot\cos\left(\frac{m\pi}{(2-\lambda)\delta}(\theta-\theta_k)\right)d\theta$$

$$g_{2,3k}(m,n,k) = \int_{\theta_k}^{\theta_k+(2-\lambda)\delta} \sin n\theta\cdot\cos\left(\frac{m\pi}{(2-\lambda)\delta}(\theta-\theta_k)\right)d\theta$$

Then, according to Equation (10) or Equation (34), the following equation can be obtained:

$$A_{2n}\left(\frac{R_5}{R_4}\right)^n + B_{2n} = \sum_{m=1}^{\infty}\left[A_{3km} + B_{3km}\left(\frac{R_5}{R_2}\right)^{-\frac{m\pi}{(2-\lambda)\delta}}\right]\cdot f(m,n,k) \tag{58}$$

$$C_{2n}\left(\frac{R_5}{R_4}\right)^n + D_{2n} = \sum_{m=1}^{\infty}\left[A_{3km} + B_{3km}\left(\frac{R_5}{R_2}\right)^{-\frac{m\pi}{(2-\lambda)\delta}}\right]\cdot g(m,n,k) \tag{59}$$

4.3. Interface between Region 3k and Region 4ik

According to Equations (9), (11), (12) (14), (19), (20) and (22), the following equation can be obtained:

$$\begin{aligned} A_{3km} &= \tfrac{2}{(2-\lambda)\delta}\Big[(A_{4ik} + B_{4ik}\ln R_2) l_{3k,4ik}(m,k) \\ &+ \sum_{u=1}^{\infty}\left(A_{4iku} + B_{4iku}\left(\tfrac{r}{R_2}\right)^{-\frac{u\pi}{\delta\xi}}\right) f_{3k,4ik}(u,m,k)\Big] \end{aligned} \tag{60}$$

$$A_{4iku} = \frac{2}{\delta\xi}\cdot\left[A_{3k} + B_{3k}\ln R_2 + \sum_{m=1}^{\infty}\left(A_{3km}\left(\frac{R_2}{R_5}\right)^{\frac{m\pi}{(2-\lambda)\delta}} + B_{3km}\right)\right]\cdot g_{3k,4ik}(u,m,k) \tag{61}$$

$$\begin{aligned} A_{3k} + B_{3k}\ln R_2 &= \tfrac{1}{(2-\lambda)\delta}\Big[A_{4ik} + B_{4ik}\ln R_2 \\ &+ \sum_{v=1}^{\infty}\left(A_{4iku} + B_{4iku}\left(\tfrac{r}{R_2}\right)^{-\frac{u\pi}{\delta\xi}}\right)\cdot s(u,k)\Big] \end{aligned} \tag{62}$$

$$\begin{aligned} A_{4ik} + B_{4ik}\ln R_2 &= \tfrac{1}{\delta\xi}\Big[A_{3k} + B_{3k}\ln R_2 \\ &+ \sum_{m=1}^{\infty}\left(A_{3km}\left(\tfrac{R_2}{R_5}\right)^{\frac{m\pi}{(2-\lambda)\delta}} + B_{3km}\right)\cdot r(m,k)\Big] \end{aligned} \tag{63}$$

where

$$l_{3k,4ik}(m,k) = \int_{\theta_k-(\lambda-1)\delta}^{\theta_k+\delta\xi-(\lambda-1)\delta}\cos\left(\frac{m\pi}{(2-\lambda)\delta}(\theta-\theta_k)\right)d\theta$$

$$s_{3k,4ik}(u,k) = \int_{\theta_k}^{\theta_k+(2-\lambda)\delta}\cos\left(\frac{u\pi}{\delta\xi}\cdot(\theta-\theta_k+(\lambda-1)\delta\right)d\theta$$

$$\begin{aligned} f_{3k,4ik}(u,m,k) &= \int_{\theta_k}^{\theta_k+(2-\lambda)\delta}\cos\left(\tfrac{m\pi}{(2-\lambda)\delta}(\theta-\theta_k)\right) \\ &\times\cos\tfrac{u\pi}{\delta\xi}(\theta-\theta_k+(\lambda-1)\delta)d\theta \end{aligned}$$

$$\begin{aligned} g_{3k,4ik}(u,m,k) &= \int_{\theta_k-(\lambda-1)\delta}^{\theta_k+\delta\xi-(\lambda-1)\delta}\cos\left(\tfrac{m\pi}{(2-\lambda)\delta}(\theta-\theta_k)\right) \\ &\times\cos\tfrac{u\pi}{\delta\xi}(\theta-\theta_k+(\lambda-1)\delta)d\theta \end{aligned}$$

Then, according to Equation (9), the following equation can be obtained:

$$\begin{aligned} &\sum_{m=1}^{\infty}\left(A_{3km}\left(\tfrac{R_2}{R_5}\right)^{\frac{m\pi}{(2-\lambda)\delta}} + B_{3km}\right)\cdot l_{3k,4ik}(m,k) \\ &= \sum_{u=1}^{\infty}\left(A_{4iku} + B_{4iku}\left(\tfrac{R_2}{R_3}\right)^{-\frac{u\pi}{\delta\xi}}\right)\cdot s_{3k.4ik}(u,k) \end{aligned} \tag{64}$$

4.4. Interface between Region 4ik and Region 4jk1

According to Equations (18), (19), (21), (23), (35) and (36), the following equation can be obtained:

$$\begin{aligned} B_{4iku} &= 2\left(A_{4jk1} + A_{4jku1p}\right) \\ &+ \tfrac{1}{\delta\xi}\sum_{u=1}^{\infty}\left(A_{4jku1} + B_{4jku1}\left(\tfrac{R_3}{R_{4jk12}}\right)^{-\frac{u\pi}{\delta}}\right)\cdot l_{4ik,4jk1}(u,k) \end{aligned} \tag{65}$$

$$\begin{aligned} A_{4jku1} &= \tfrac{2}{\delta\xi}\cdot\int_{\theta_k}^{\theta_k+(2-\lambda)\delta} \tfrac{\partial A_{4ik}(r,\theta)}{\partial r}\Big|_{r=R_3} \cos\left(\tfrac{u\pi}{\delta\xi}(\theta-\theta_k+(\lambda-1)\delta\right)d\theta \\ &= \tfrac{2}{\delta\xi R_3}\cdot\left\{B_{4ik}+\tfrac{u\pi}{\delta\xi}\cdot\sum_{u=1}^{\infty}\left[\left(A_{4iku}\left(\tfrac{R_3}{R_2}\right)^{\frac{u\pi}{\delta\xi}}\right)-B_{4iku}\right]\cdot s_{4ik,4jk1}(u,k)\right\} \end{aligned} \tag{66}$$

$$\begin{aligned} A_{4ik}+B_{4ik}\ln R_3 &= A_{4jk1}+A_{4jku1p} \\ &+\tfrac{1}{\delta\xi}\cdot\sum_{u=1}^{\infty}\left(A_{4jku1}+B_{4jku1}\left(\tfrac{R_3}{R_{4jk12}}\right)^{-\frac{u\pi}{\delta}}\right)\cdot l_{4ik,4jk1}(u,k) \end{aligned} \tag{67}$$

$$B_{4ik}=\frac{1}{2}\mu_0\left[J_{4jk1}(R_{4jk13}^2-R_3^2)+J_{4jk2}\left(R_1^2-R_{4jk13}^2\right)\right] \tag{68}$$

$$\sum_{u=1}^{\infty}\left(A_{4jku1}+B_{4jku1}\left(\frac{R_3}{R_{4jk13}}\right)^{-\frac{u\pi}{\delta}}\right)=\sum_{u=1}^{\infty}\left(A_{4iku}\left(\frac{R_3}{R_2}\right)^{\frac{u\pi}{\delta\xi}}+B_{4iku}\right) \tag{69}$$

where

$$l_{4ik,4jk1}(u,k)=\int_{\theta_k-(\lambda-1)\delta}^{\theta_k+\delta\xi-(\lambda-1)\delta}\left(1+\cos\frac{2u\pi}{\delta\xi}(\theta-\theta_k+(\lambda-1)\delta)\right)d\theta$$

$$s_{4ik,4jk1}(u,k)=\int_{\theta_k}^{\theta_k+(2-\lambda)\delta}\left(1+\cos\frac{2u\pi}{\delta\xi}(\theta-\theta_k+(\lambda-1)\delta)\right)d\theta$$

4.5. Interface between Region 4jk1 and Region 4jk2

According to Equations (37)–(41) and (47)–(50), the following equation can be obtained:

$$\begin{aligned} A_{4jk2}+\tfrac{1}{2}\mu_0 J_{4jk2}(R_{4jk12}^2\ln R_{4jk12}-\tfrac{1}{2}R_{4jk12}{}^2) &= \\ A_{4jk1}+\tfrac{1}{2}\mu_0 J_{4jk1}(R_{4jk12}^2\ln R_{4jk12}-\tfrac{1}{2}R_{4jk12}{}^2) \\ +\tfrac{1}{2}\mu_0 J_{4jk2}\left(R_1^2-R_{4jk12}^2\right)\ln R_{4jk12} &= A_{4ik}+B_{4ik}\ln R_3 \end{aligned} \tag{70}$$

$$B_{4jku2}=A_{4jku2}\left(\frac{R_1}{R_{4jk13}}\right)^{\frac{u\pi}{\delta}} \tag{71}$$

$$\sum_{u=1}^{\infty}\left(A_{4jku1}\left(\frac{R_{4jk13}}{R_3}\right)^{\frac{u\pi}{\delta}}+B_{4jku1}\right)=\sum_{u=1}^{\infty}\left[A_{4jku2}\left(1+\left(\frac{R_1}{R_{4jk13}}\right)^{\frac{2u\pi}{\delta}}\right)\right] \tag{72}$$

by rewriting Equations (51), (52), and (54)–(72) in matrix and vector form. Then, the integral constants of each region can be obtained through matrix operation. In addition, the detailed derivation of $l_{2,3k}$, $s_{2,3k}$, $\gamma_{2,3k}$, $f_{2,3k}$, $g_{2,3k}$, $l_{3k,4ik}$, $s_{3k,4ik}$, $f_{3k,4ik}$, $g_{3k,4ik}$, $l_{4ik,4jk1}$, $s_{4ik,4jk1}$ and the method of matrix operation can be seen in [26].

5. FEM Results and Analysis

To verify the proposed analytical model, two LHPMSM prototype machines are manufactured with a semi-closed slot and asymmetric slot, respectively. The FEA models of the two prototypes are shown in Figure 7. The main geometric dimensions are given in Figure 4 and Table 1 and machine parameters are shown in Table 2. They have the same structural design parameters except for the stator teeth. The PMs and stator materials are N38SH and DW470-50, respectively.

Figure 7. The FEM analysis of two LHPMSM prototype machines: (**a**) semi-closed slot; (**b**) asymmetric slot.

Table 1. Main Geometric Dimensions of Prototype Machines.

Symbol	Quantity	Title 3	
		Semi-Closed Slot	**Asymmetric Slot**
p	Pole pairs number	20	20
Q	Slot number	168	168
R_1	Radius of slot bottom	764 mm	764 mm
R_2	Inner radius of slot opening	824 mm	824 mm
R_3	Outer radius of slot body	821 mm	\
R_4	Inner radius of PMs	828 mm	828 mm
R_5	Outer radius of slot opening	825 mm	825 mm
R_6	Outer radius of PMs	840 mm	840 mm
b	Tooth width of stator	20 mm	20 mm
h_0	Toothed boots height	1 mm	1 mm
l	The depth of U-shaped groove	1.29 mm	\
hs_1	The width of U-shaped groove	3 mm	\
hs_2	Stator tooth height	57 mm	57 mm
g	Air-gap length	3 mm	3 mm
α_p	Pole-arc coefficient	0.94	0.94
λ	The asymmetry of slot opening	variable	1
ξ	The asymmetry of inside slot	variable	1
δ	Slot pitch angle of stator	2.14°	2.14°

Table 2. Parameters of Prototype Machines.

Symbol	Quantity	Value
U_N	Rated voltage	380 V
I_N	Rated current	20 A
P_N	Rated power	12 kW
n_N	Rated rotating speed	50 rpm
R_N	Resistance per phase	1.05 Ω
\	Number of conductors	28
\	Number of parallel branches	8

5.1. Air-Gap Magnetic Flux Density

Figure 8 shows the comparison between the AM and FEM results of the radial and tangential flux density with the semi-closed slot and asymmetric slot at the middle of the air-gap subdomain (r = 826.5 mm). From the figures, although the end effect is ignored, the air-gap flux density waveforms obtained by the two methods are still in good agreement. Moreover, the slotting effect leads to the distortion of the flux density waveforms, while the waveform with the asymmetric slot has higher peaks than that of the semi-closed slot and may have a higher harmonic distortion rate at these locations. Consequently, the harmonic analyses of the air-gap flux density for the two models are shown in Figure 9. The fundamental wave amplitudes of air-gap flux density between the semi-closed slot model and asymmetric slot model are 1.118T and 1.146T, respectively. In addition, the total harmonic distortion (THD) rate of the air-gap flux density for the two models can be calculated by (73), which are 26.95% and 29.56%, respectively.

$$THD_{ag} = \frac{\sqrt{\sum_{i=1}^{\infty} B_{2i+1}^2}}{B_1} \times 100\% \tag{73}$$

where B_1 is the fundamental value of air gap flux density and B_{2i+1} is the $(2i + 1)$th harmonic contents of air gap flux density.

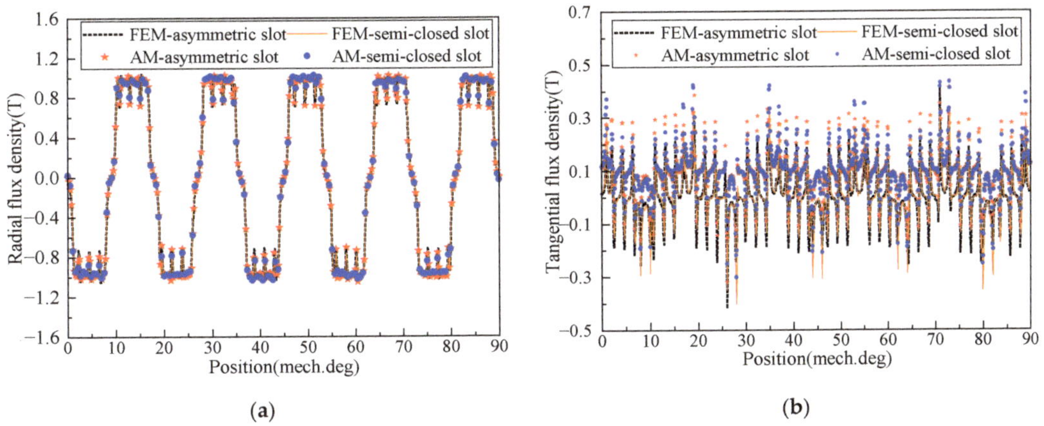

Figure 8. Air gap flux density of semi-closed slot model and asymmetric slot model: (**a**) radial; (**b**) tangential.

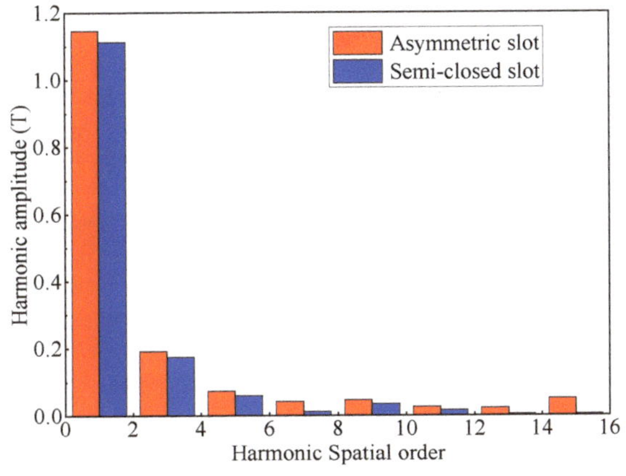

Figure 9. Comparison of fundamental wave amplitude of air gap flux density between semi-closed slot model and asymmetric slot model.

As shown in Figure 9, the asymmetric slot model has a higher fundamental wave amplitude of air-gap magnetic flux density than the semi-closed slot model, which indicates that it can withstand a greater magnetic load. This will reduce the size of the motor and, thus, the manufacturing cost. However, the position of the asymmetric slot is slightly different from that of the semi-closed slot in spatial location, which cannot avoid the influence of the slotting effect. Therefore, the asymmetric slot model has a slightly higher

THD of air-gap flux density than the semi-closed slot model, which is mainly caused by the third harmonic content. The harmonic amplitude of the air-gap flux density can be effectively suppressed by optimizing the structure of the stator slot, which will be discussed in Section 6.

Figure 10a shows the 2D-FEM results of no-load back EMF for comparison between the semi-closed slot model and the asymmetric slot model. It can be seen that the amplitudes of the two models—the former is 42.717 V and the latter is 41.677 V—are very close, but the waveform of the latter tends to be a sinusoidal wave. In addition, the harmonic analysis results are shown in Figure 10b. The THD of no-load back EMF for the asymmetric slot model and semi-closed slot model are 11.82% and 10.86%, respectively.

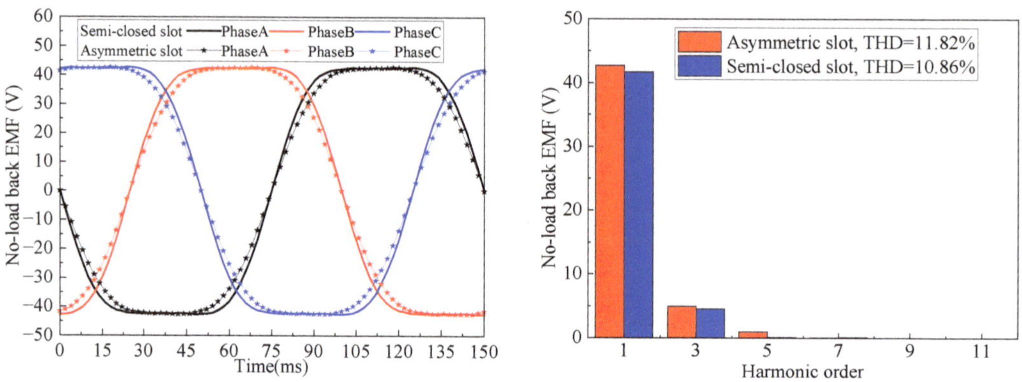

(**a**) (**b**)

Figure 10. Comparison of no-load back EMF between semi-closed slot model and asymmetric slot model: (**a**) waveform; (**b**) harmonic analysis.

5.2. Electromagnetic Torque and Cogging Torque

The LHPMSM constantly suffers heavy load and occasionally shock load during operation. Therefore, the driving system requires a high output capacity and stability, which depends on the amplitude of the electromagnetic torque and the torque ripple ratio. Moreover, the cogging torque can cause torque ripple, particularly at low speeds. Consequently, the electromagnetic torque performance and cogging torque performance are analyzed, respectively.

Figure 11a shows the FEM results of the electromagnetic torque of the asymmetric slot model and semi-closed slot model. The average electromagnetic torques of the two models are 2.25 kN·m and 2.19 kN·m, respectively. The corresponding torque ripples of the two models are 2.18% and 2.43%, respectively.

Figure 11b shows the AM and FEM analysis of the cogging torque of the asymmetric slot model and semi-closed slot model. The peak values of the cogging torque are 10.67 N·m and 12.65 N·m, respectively.

Comparisons of electromagnetic characteristics between the asymmetric slot model and traditional semi-closed slot model are shown in Table 3. Note that the '↑' in the table represents the increase rate and the '↓' represents the decrease rate. In summary, the asymmetric slot model reveals better torque performance than the semi-closed slot model, which is mainly because the former can be equivalent to the partially unequal width teeth. This structure can revise the difference of synchronous inductance between two adjacent stator teeth, but the actual effect is affected by the stator tooth structure [21]. Therefore, the cogging torque and torque ripple can be weakened but cannot be eliminated. The torque characteristics will be significantly improved by optimizing the U-shaped groove.

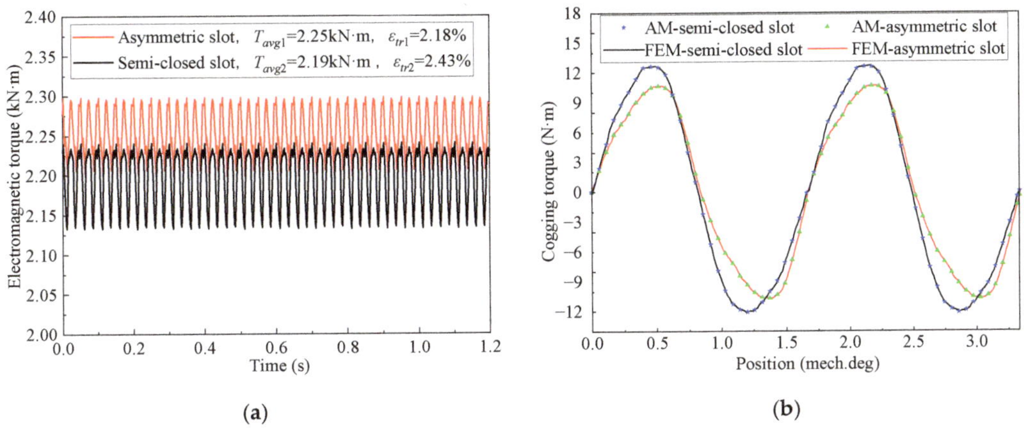

Figure 11. Comparison of electromagnetic characteristics between semi-closed slot model and asymmetric slot model: (**a**) electromagnetic torque; (**b**) cogging torque.

Table 3. Comparison of Electromagnetic Characteristics Between Asymmetric Slot Model and Traditional Semi-closed Slot Model.

Comparison	Air-Gap Flux Density (T)		Air-Gap Flux Density THD (%)	Average Value of Electromagnetic Torque T_{avg} (kN·m)	Torque Ripple (%)	Cogging Torque (N·m)	No-Load Back EMF THD (%)
	B_r	B_t					
semi-closed slot	1.118	0.076	26.95%	2.19	2.43%	12.65	10.86%
Asymmetric slot	1.146	0.082	29.56%	2.25	2.18%	10.67	11.82%
change	2.4%↑	7.3% ↑	8.8% ↓	2.6% ↑	10.2% ↓	15.65% ↓	8.12% ↑

6. Influence of Key Parameters

Many motor design parameters, such as the air-gap length, pole–slot ratio, pole-arc coefficient, etc., have a significant impact on electromagnetic characteristics, especially for the air-gap flux density, electromagnetic torque, back EMF, etc. As shown in Section 2, the critical design parameters are λ and ξ against the proposed asymmetric slot model. Therefore, the influence of key parameters on electromagnetic performances are investigated in this section.

Based on the relationship between the λ and β given in Equation (1), the asymmetric slot model with different values of λ is modeled by 2-D FEA parametric modeling technology. It should be noted that the feasibility of assembly needs to be taken into consideration. The actual width of the slot opening is only 3 mm for λ = 1.75. It is difficult to complete wire-wrapping for λ > 1.75. Thus, the range of λ is defined as 1.5 to 1.75, and the step size is 0.05.

Figure 12a shows the influence of λ and ξ on the air-gap flux density of the motor. The amplitude of the air-gap flux density increases first then decreases significantly with the increase in λ and ξ. The air-gap flux density reaches the maximum value of 1.181T when λ = 1.6 and ξ = 1.5. According to Equations (1) and (2), and Table 1, the values of β and α are 0.856° and 1.07°, respectively.

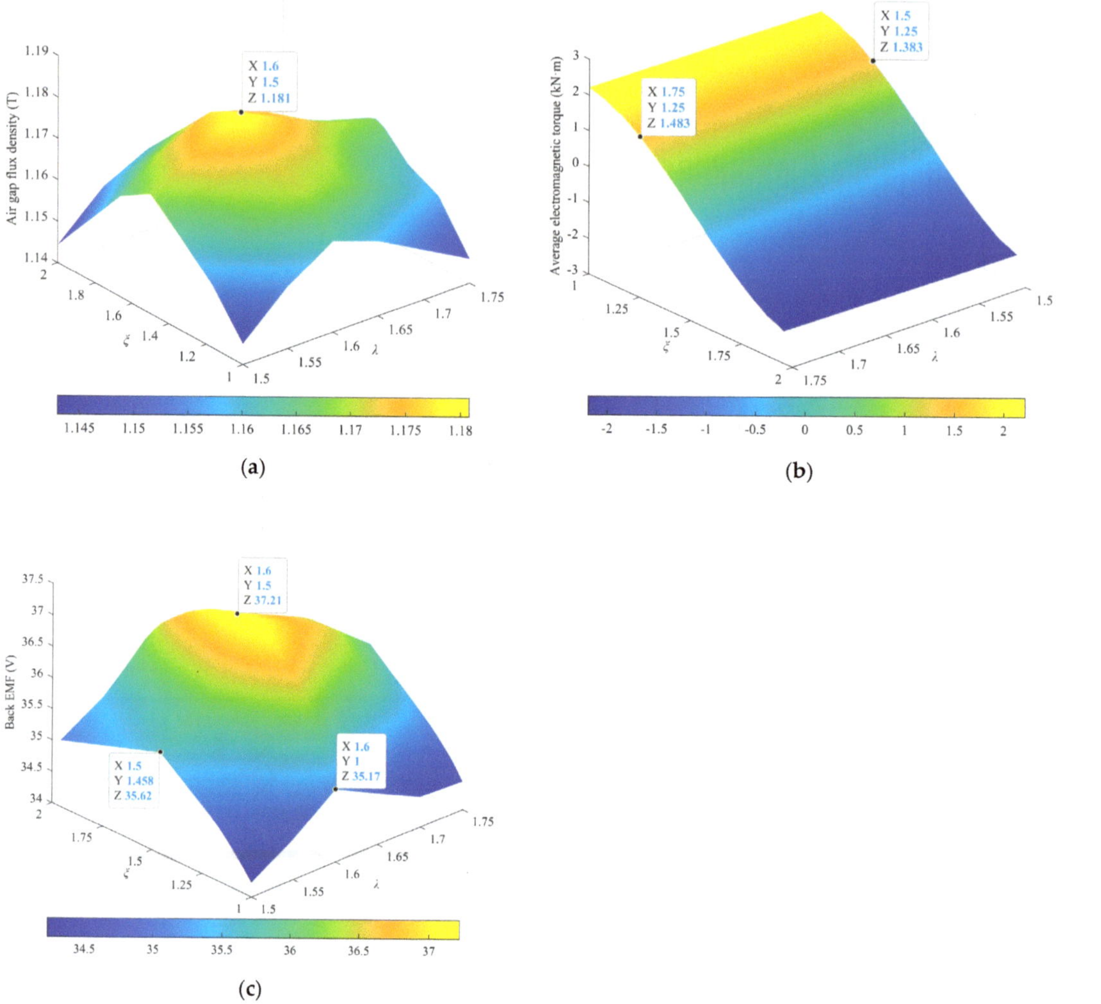

Figure 12. Surface chart of electromagnetic characteristics with λ and ξ: (**a**) fundamental wave amplitude of air-gap flux density; (**b**) electromagnetic torque; (**c**) fundamental wave amplitude of no-load back EMF.

In Figure 12b, as the value of λ continues to increase, the average electromagnetic torque is decreased monotonically. Moreover, the direction of electromagnetic torque will be changed when the λ reaches a specific value, which is because λ is related to the initial position of the slot. On the other hand, as the value of ξ continues to increase, the electromagnetic torque changes slightly.

As shown in Figure 12c, the fundamental wave amplitude of no-load back EMF increases first then decreases significantly with the increase in λ and ξ. The amplitude of no-load back EMF reaches the maximum value of 37.21 V when $\lambda = 1.6$ and $\xi = 1.5$.

7. Experimental Result

To verify the motor performance, an LHPMSM with a 40-pole rotor and 168-slot stator is designed based on the design parameters in Table 1. The values of λ and ξ are 1.65 and 1.5, respectively. The experimental platform is shown in Figure 13, and it mainly includes the experimental prototype, console, hydraulic station, power cabinet, three-level frequency converter, power analyzer, current transformer, and encoder. The frequency converter is set to the LOCAL model, and the data are transmitted to the industrial control computer through the Bluetooth interface.

Figure 13. The testing platform of the LHPMSM prototype.

As seen from Figure 14, the external surface of the drum is not wrapped with steel wire rope, and its normal operation can be equivalent to the no-load working condition. The experimental prototype rated parameters are set by the three-level frequency converter, as given in Table 2.

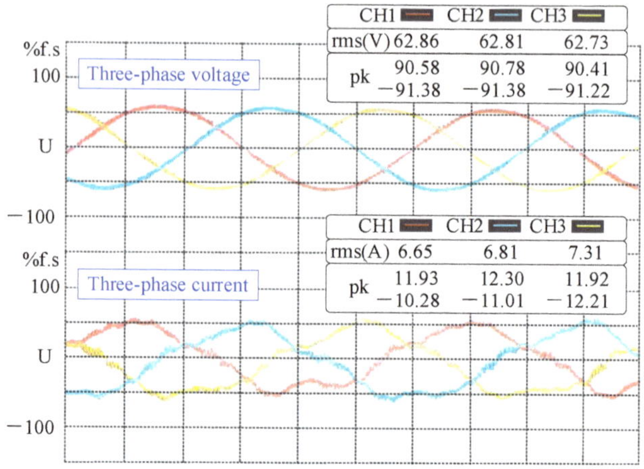

Figure 14. Measured three-phase currents and voltages at the rated condition.

The input current and input voltage are measured at 17.2 Hz. The measured waveforms are shown in Figure 15a. The RMS values of the three-phase input voltage are 62.86 V, 62.81 V, and 62.73 V, respectively. The RMS values of the three-phase input current are 6.65 A, 6.81 A, and 7.31 A, respectively.

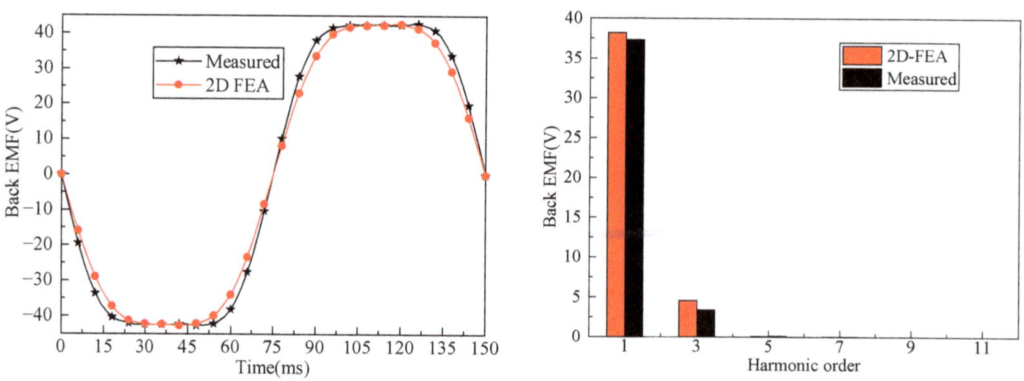

(a) (b)

Figure 15. Two-dimensional FEM calculated and measured back EMF waveforms at 50 rpm: (**a**) waveforms; (**b**) harmonic analysis.

Figure 15b shows that the measured back EMF is around 2.4% lower than that of the 2D FEM result. It can be calculated that the THD of the 2D FEM and the measured back EMF are 12.09% and 9.16%, respectively.

8. Conclusions

In this paper, a novel design of an integrated LHPMSM that includes asymmetric stator slots is proposed. To predict the magnetic field distribution of the complex structures, an analysis model with two correction factors is established and solved by the method of separating variables. The analysis results are in good agreement with the finite element calculation results. In addition, the influential factors of the motor electromagnetic performance are further analyzed. To validate the effectiveness of theoretical analysis, a 168-slot/40-pole LHPMSM with an asymmetric stator slot is manufactured and measured. The following conclusions can be obtained:

(1) For LHPMSM with multiple slots, the effective amplitude of the radial air gap flux density is increased by 2.4%, the harmonic of the air gap flux density is weakened by 8.8%, the average torque of the motor is increased by 2.6%, the torque ripple is weakened by 10.2%, and the cogging torque is weakened by 15.65% compared with the traditional semi-closed stator slot.

(2) The two correction factors, i.e., the asymmetry of the slot opening λ and the asymmetry of the inside slot ξ have significant effects on the electromagnetic characteristic of the motor. Specifically, the 168-slot/40-pole LHPMSM has the best electromagnetic characteristics when $\lambda = 1.6$ and $\xi = 1.5$. The values of λ and ξ are related to the slot pitch angle of the stator. In addition, we found that the electromagnetic torque is slightly affected by ξ. Therefore, it is necessary to properly select the two factors to obtain the optimum electromagnetic characteristic.

Author Contributions: S.K.: Simulation, Writing review and editing, Visualization. Z.K.: Conceptualization, Supervision, Methodology, Funding acquisition, Project administration; J.W.: Funding acquisition, Project administration; Y.W.: Formal analysis. All authors have read and agreed to the published version of the manuscript.

Funding: National key R&D program-funded projects (grant no. 2020YFB1314103), the National Natural Science Foundation of China (grant no. 52274156), Guizhou science and technology support plan projects (grant no. Guizhou science and technology cooperation support [2022] general 004), Shanxi province raised funds to support the scientific research projects of the returned overseas person (grant no. 2021-049).

Data Availability Statement: Not applicable.

Conflicts of Interest: The authors declare that they have no known competing financial interest or personal relationships that could have appeared to influence the work reported in this paper.

References

1. Lu, E.; Li, W.; Yang, X.F.; Xu, S.Y. Composite Sliding Mode Control of a Permanent Magnet Direct-Driven System For a Mining Scraper Conveyor. *IEEE Access* **2017**, *5*, 22399–22408. [CrossRef]
2. Tlali, P.M.; Wang, R.J.; Gerber, S.; Botha, C.D.; Kamper, M.J. Design and Performance Comparison of Vernier and Conventional PM Synchronous Wind Generators. *IEEE Trans. Ind. Appl.* **2020**, *56*, 2570–2579. [CrossRef]
3. Sheng, L.C.; Li, W.; Wang, Y.Q.; Fan, M.B.; Yang, X.F. Sensorless Control of a Shearer Short-Range Cutting Interior Permanent Magnet Synchronous Motor Based on a New Sliding Mode Observer. *IEEE Access* **2017**, *5*, 18439–18450. [CrossRef]
4. Xie, B.C.; Zhang, Y.; Wang, J.Q.; Liang, B.X.; Zhang, F.G. An efficient multidisciplinary design research for the integrated low speed permanent magnet motor system based on analytical and numerical hybrid analysis. *Energy Rep.* **2022**, *8*, 199–208. [CrossRef]
5. Gan, B.P.; Zhang, B.Y.; Li, Q.S.; Feng, G.H. Investigation into fault tolerant capability of new modular low-speed and high-torque direct-drive permanent magnet motor based on unequal span winding. *IET Electr. Power Appl.* **2021**, *15*, 1358–1383. [CrossRef]
6. Chu, W.Q.; Zhu, Z.Q. Reduction of On-Load Torque Ripples in Permanent Magnet Synchronous Machines by Improved Skewing. *IEEE Trans. Magn.* **2013**, *49*, 3822–3825. [CrossRef]
7. Wang, H.M.; Guan, P.J.; Liu, S.; Wu, S.; Shi, T.N.; Guo, L.Y. Modeling and Analyzing for Magnetic Field of Segmented Surface-Mounted PM Motors with Skewed Poles. *J. Electr. Eng. Technol.* **2022**, *17*, 1097–1109. [CrossRef]
8. Pang, Y.; Zhu, Z.Q.; Feng, Z.J. Cogging Torque in Cost-Effective Surface-Mounted Permanent-Magnet Machines. *IEEE Trans. Magn.* **2011**, *47*, 2269–2276. [CrossRef]
9. Ashabani, M.; Mohamed, Y. Multiobjective Shape Optimization of Segmented Pole Permanent-Magnet Synchronous Machines With Improved Torque Characteristics. *IEEE Trans. Magn.* **2011**, *47*, 795–804. [CrossRef]
10. Zhao, J.; Hu, H.; Li, B. *Influence of Edge Permanent-Magnet Shape on Performance of an Arc-Linear Permanent-Magnet Synchronous Machine*; IEEE: New York, NY, USA, 2015.
11. Wang, H.M.; Liu, S.; Wu, S.; Guo, L.Y.; Shi, T.N. Optimal Design of Permanent Magnet Structure to Reduce Unbalanced Magnetic Pull in Surface-Mounted Permanent-Magnet Motors. *IEEE Access* **2020**, *8*, 77811–77819. [CrossRef]
12. Peng, C.; Wang, D.H.; Feng, Z.K.; Wang, B.D. A New Segmented Rotor to Mitigate Torque Ripple and Electromagnetic Vibration of Interior Permanent Magnet Machine. *IEEE Trans. Ind. Electron.* **2022**, *69*, 1367–1377. [CrossRef]
13. Xing, Z.Z.; Wang, X.H.; Zhao, W.L. Cogging torque reduction based on segmented skewing magnetic poles with different combinations of pole-arc coefficients in surface-mounted permanent magnet synchronous motors. *IET Electr. Power Appl.* **2021**, *15*, 200–213. [CrossRef]

14. Peng, C.; Wang, D.H.; Wang, B.D.; Li, J.C.; Xu, C.; Wang, X.H. Torque Ripple and Electromagnetic Vibration Suppression in Permanent Magnet Synchronous Motor Using Segmented Rotor With Different Pole Widths. *IEEE Trans. Magn.* **2022**, *58*, 5. [CrossRef]
15. Wang, K.; Liang, Y.P.; Wang, D.M.; Wang, C.G. Cogging torque reduction by eccentric structure of teeth in external rotor permanent magnet synchronous motors. *IET Electr. Power Appl.* **2019**, *13*, 57–63. [CrossRef]
16. Ahn, J.M.; Son, J.C.; Lim, D.K. Optimal Design of Outer-Rotor Surface Mounted Permanent Magnet Synchronous Motor for Cogging Torque Reduction Using Territory Particle Swarm Optimization. *J. Electr. Eng. Technol.* **2021**, *16*, 429–436. [CrossRef]
17. Qu, H.; Zhu, Z.Q. Analysis of Split-Tooth Stator Slot PM Machine. *IEEE Trans. Ind. Electron.* **2021**, *68*, 10580–10591. [CrossRef]
18. Huang, J.X.; Sui, Y.; Yin, Z.S.; Yuan, Z.H.; Zheng, P. A Novel High Torque Density Dual Three-Phase PMSM With Low Space Harmonic Content. *IEEE Trans. Magn.* **2022**, *58*, 7. [CrossRef]
19. Xing, Z.Z.; Wang, X.H.; Zhao, W.L. Optimization of Stator Slot Parameters for Electromagnetic Vibration Reduction of Permanent Magnet Synchronous Motors. *IEEE Trans. Transp. Electrif.* **2022**, *8*, 4337–4347. [CrossRef]
20. Liu, T.; Huang, S.D.; Gao, J.; Lu, K.Y. Cogging Torque Reduction by Slot-Opening Shift for Permanent Magnet Machines. *IEEE Trans. Magn.* **2013**, *49*, 4028–4031. [CrossRef]
21. Petrov, I.; Ponomarev, P.; Alexandrova, Y.; Pyrhonen, J. Unequal Teeth Widths for Torque Ripple Reduction in Permanent Magnet Synchronous Machines With Fractional-Slot Non-Overlapping Windings. *IEEE Trans. Magn.* **2015**, *51*, 9. [CrossRef]
22. Tang, X.; Wang, X.; Sun, S.; Zhao, J. Analytical analysis and study of reduction methods of cogging torque in line-start permanent magnet synchronous motors. *Proc. CSEE* **2016**, *36*, 1395–1403. [CrossRef]
23. Tang, X.; Wang, X.; Tian, M.; Meng, X. Study of reduction methods of cogging torque in line-start permanent magnet synchronous motor by changing the parameters of stator teeth and slots. *Trans. China Electrotech. Soc.* **2016**, *31*, 1–8.
24. Zhao, J.; Yan, Y.-S.; Chen, H.; Liu, X.-D. Effect of different rotor configurations on torque characteristics of a 10-pole/12-slot flux switching permanent magnet machine. *Electr. Mach. Control* **2016**, *20*, 51–56. [CrossRef]
25. Hu, H.Z.; Zhao, J.; Liu, X.D.; Guo, Y.G. Magnetic Field and Force Calculation in Linear Permanent-Magnet Synchronous Machines Accounting for Longitudinal End Effect. *IEEE Trans. Ind. Electron.* **2016**, *63*, 7632–7643. [CrossRef]
26. Gao, L.L.; Cai, Z.Q.; Liang, Y.P.; Wang, D.M.; Niu, Q.; Li, J. An Improved Analytical Model of Magnetic Field in Surface-Mounted Permanent Magnet Synchronous Motor With Magnetic Pole Cutting. *IEEE Access* **2021**, *9*, 142804–142814. [CrossRef]

applied sciences

MDPI

Article

Design Optimization of the Magnet-Free Synchronous Homopolar Motor of a Subway Train

Vladimir Dmitrievskii [1], Vladimir Prakht [1,*], Vadim Kazakbaev [1] and Alecksey Anuchin [2]

[1] Department of Electrical Engineering, Ural Federal University, 620002 Yekaterinburg, Russia
[2] Department of Electric Drives, Moscow Power Engineering Institute, 111250 Moscow, Russia
* Correspondence: va.prakht@urfu.ru; Tel.: +7-909-028-49-25

Featured Application: The presented results can be used in designing of synchronous homopolar machines and traction electrical drives.

Abstract: Synchronous homopolar motors have no permanent magnets and their excitation winding is fixed at their stator. However, they can be a good alternative to induction and permanent magnet motors in traction applications requiring a wide constant power speed range. They provide an excitation flux control and a highly reliable brushless rotor design. This article presents the procedure and results of optimizing a 370 kW synchronous homopolar motor for driving subway train. The optimization procedure was developed to take into account the subway train moving trajectory. The analysis considers only a limited number of steady-state operating points of the motor to reduce computation time. The optimization results show a significant improvement of the target parameters of the traction drive. The optimization makes it possible to significantly reduce the losses in the operating cycle, as well as the torque ripple of the motor and the current rating of the traction inverter.

Keywords: synchronous homopolar motor; electrically excited synchronous motor; Nelder–Mead method; optimal design of electric machines; subway train; constant power speed range; traction drive

Citation: Dmitrievskii, V.; Prakht, V.; Kazakbaev, V.; Anuchin, A. Design Optimization of the Magnet-Free Synchronous Homopolar Motor of a Subway Train. *Appl. Sci.* **2022**, *12*, 12647. https://doi.org/10.3390/app122412647

Academic Editors: Loránd Szabó and Feng Chai

Received: 16 November 2022
Accepted: 8 December 2022
Published: 9 December 2022

1. Introduction

The interior permanent magnet synchronous motors (PMSM) are widespread in electric traction. The excitation of such motors is made by means of rear-earth magnets. Thus, these motors have high power-to-weight ratio and high efficiency [1,2]. However, PMSMs have the following disadvantages. (1) PMSMs have high cost due to presence of expensive rear-earth magnets. (2) The use of rear-earth magnets causes the dependence on the limited number of the magnet suppliers in the production chain [3]. The prices of the rear-earth magnets are unstable and can seriously vary over time in a few years [4,5]. (3) The extraction of rear-earth elements, which are used in production of the magnets, harms the environment [6]. (4) Permanent magnets are prone to irreversible demagnetization due to the high operating temperature of more than 120 °C and strong demagnetization field, inherent in a traction motor with limited dimensions. (5) PMSMs have high power loss at a rotational speed close to or equal to the maximum in the field weakening region due to a large stator winding current required to compensate the uncontrolled magnetomotive force (MMF) of permanent magnets [7–9]. (6) The presence of high electromotive force (EMF) in the windings during rotation can results in fire in case of short-circuiting. This is very dangerous particularly in subway and trains, those have huge kinetic energy and cannot stop immediately. In case of subway train, the probability of fire should be excluded at all due to the limited volume in the tunnel and difficultness of evacuation. Therefore, utilization of the PMSMs in such applications is undesirable.

The highlighted disadvantages of PMSMs with rear-earth magnets limit their applications in subway, trams, mining trucks, etc. To overcome these disadvantages of PMSMs the leading car manufactures such as BMW (models BMW iX3) and Renault (models Renault Zoe, Fluence, Megane E-TECH) have developed DC-excited synchronous motors (DCESM) with the excitation winding on the rotor (BMW [10,11], Renault [12]). The constant power speed range is provided by means of the flux regulation of DCESM [13,14]. However, the electrical machines of BMW iX3 and Renault Zoe have the slip rings and a brushed contact to feed the excitation winding on the rotor, which limits the motor speed and reduces its reliability [14].

An alternative could be high-reliable and brushless synchronous homopolar machines (SHM) having both excitation and power winding on stator, and a salient-pole rotor. The main advantages of SHMs compared to DCESMs are highly reliable brushless rotor design and concentrated excitation winding, the number of coils of which is small and does not depend on the number of motor poles, while in a DCESM the number of coils (and, therefore, the length of the wire and losses) increase with the number of poles. Such SHMs are widely used as the high-reliable generators in passenger railway cars, on-board generators for ships and aircraft [15,16], automotive generators [17], and in welding units [18].

In papers [7,19–23], the SHM is used as a traction motor of a mining truck. In [19], the simplified computation method of the traction SHM was described, and the experimental verification of the proposed method was performed. In [20], the control strategy of such traction motor was considered. In [21], this machine was optimized by means of the developed optimization procedure, which is now used at the design stage of the new motors. In [22], the optimization was made in order to minimize the installed power of the traction inverter of the mining truck. In [7], the parameters of the SHM are compared with the interior permanent magnet synchronous motor of the same power driving the same mining truck. In [23], the parameters of the SHM are compared with the parameters of the DCESM in the same application.

However, the review shows absence of use of SHM in other traction applications. This paper considers an optimization process for the synchronous homopolar motor for a traction electric drive of a subway train. The optimization was performed with respect to the typical speed and torque profile during transition between station considering acceleration and deceleration phases. The following optimization targets were selected: decrease in average losses along the way from station to station, decrease in installed power capacity of the traction inverter, and reduction of the torque ripple.

The one-criterion Nelder–Mead method is applied in this study to optimize the SHM design. An important advantage of the Nelder–Mead method over other methods that are often used to optimize electrical machines [24–28] is the significant savings in computational time, which makes it possible to increase the number of parameters to optimize, as well as to apply more complex optimization criteria, for the calculation of which it is necessary to calculate several load points of the machine [22]. This advantage is important for optimizing traction machines.

The optimization procedure was developed that takes into account the subway train moving trajectory. This analysis considers only a limited number of operating points making it possible to optimize the motor according to the specified cost function. The optimization results show a significant improvement in the target parameters of the traction drive.

2. Main Design Parameters of the SHM

Figure 1 shows the considered SHM design. The SHM has two stacks on the stator and two stacks on the rotor made of laminated steel. The stator stack and the rotor stack located in the front of the stator stack form the stator-rotor stack combination (SRSC). The stator lamination has 60 slots and 8-pole winding with a number of slots per pole and phase $q = 60/(8{\cdot}3) = 2.5$. The laminations are assembled from sheets of electrical steel 0.5 mm thick.

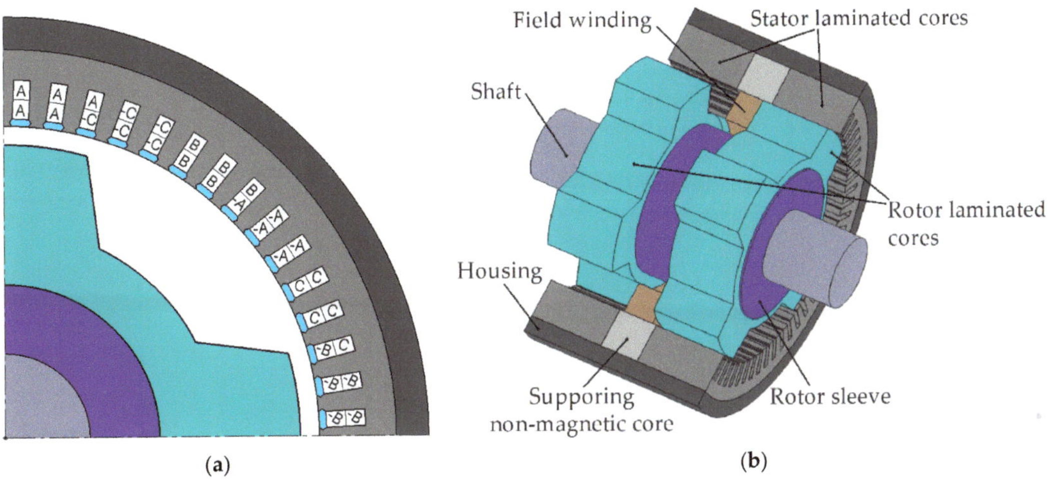

Figure 1. Sketch of the SHM geometry: (**a**) Cross-section and stator power winding configuration (1/4 of the motor cross-section is shown, the other parts are symmetrical). (**b**) General view. The stator winding is not shown to avoid cluttering up the figure.

An excitation winding is located between the SRSCs. The rotor has no windings. Each rotor stack has 4 teeth, and the teeth of the rotor stacks are shifted by 45 mechanical degrees with respect to each other. The SHM housing and the rotor sleeve carry the stator lamination and the rotor lamination, respectively. They are made of a solid (not laminated) steel and provide the link to the flux produced by the excitation winding. No standard regulates the magnetic properties of solid steels, and, therefore, they can have significant dispersion. To guarantee that the housing and the sleeve are able to conduct the required flux, the flux density in these parts is restricted with the value of 1.6 T.

Figure 2 shows the inverter circuit diagram for powering the considered traction motor. To control the three-phase winding, field-oriented control is used, in which the stator current vector is formed in a coordinate system rotating synchronously with the rotor. The excitation winding current is set using PWM proportional to the armature current. The offset angle of the armature current vector from the direction of the greatest magnetic conductivity of the rotor pole for various operating points and the ratio between the armature current and the excitation current are reported in Section 6.

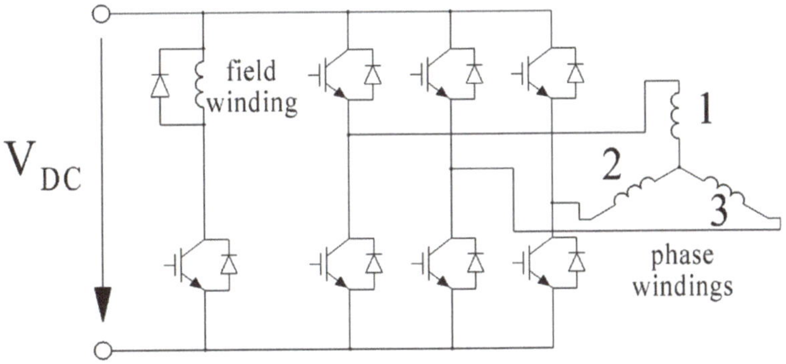

Figure 2. Schematics of the three-phase inverter with a DC-chopper for the excitation winding.

3. Operating Cycle and Operating Points of the Subway Train Motor

The transition of the train from one station to another has the following stages shown in Figure 3 [29]:

- Acceleration with the constant torque T_0;
- Acceleration with the constant power to maximum speed n_{max};
- Coasting with slow deceleration due to friction or movement with constant speed (cruising);
- Braking with the constant power;
- Braking with the constant torque T_0.

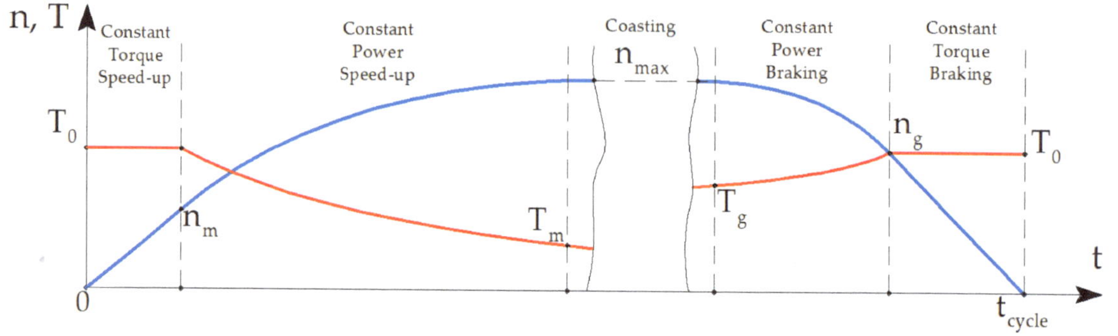

Figure 3. Sketch of the speed (blue) and torque (red) profiles in time of the traction motor of the subway train.

The considered powertrain is designed with the supercapacitors to store the energy of the braking. The rated voltage of the DC catenary is 750 V ($V_{DC\ rated}$). It was chosen as the minimal voltage for supercapacitors, whereas the maximum voltage of supercapacitors is chosen to be 1050 V. The supercapacitors are feeding traction inverter directly. During the start of the drive, the supercapacitors can be fully charged or partially discharged but the control system tries to discharge them to 750 V at the end of the acceleration phase in order to be ready for braking. These circumstances define the limit the maximum phase-to-phase voltage of the motor during acceleration stage of 750 V. A higher voltage limit can be applied for the braking modes of the motor to perform more efficient braking.

The demanded speed-torque curve of the traction motor is depicted in Figure 4. The right part of the plot corresponds to the driving mode of operation, whereas the left part of the plot represents generating mode. For the sake of simplicity, the negative torque in the generator mode is shown in absolute values above zero. The motor must produce maximum torque T_0 = 1240 N·m and the maximum speed n_{max} = 4280 rpm.

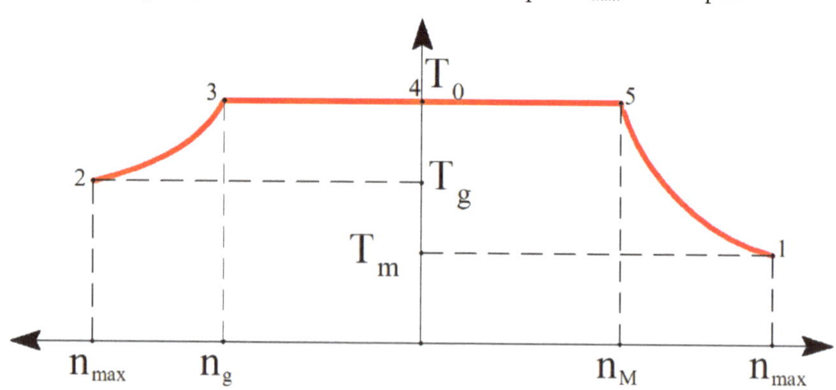

Figure 4. Demanded speed-torque curve of the subway traction drive.

The operation points in Figure 4 are enumerated in order of increasing torque:

1. Driving mode at the maximum speed n_{max};
2. Generating mode at the maximum speed n_{max};

3. Generating mode in the point of transition from constant power to constant torque at the speed n_g = 2854 rpm;
4. Zero speed mode with the maximum torque T_0 = 1240 N·m;
5. Driving mode at the point of transition from constant torque to constant power at the speed n_m = 1427 rpm.

4. Average Loss Calculation

Usually, the subway stations are located near of each other. Therefore, the duration of the cruising is not too much. Moreover, the required power in the cruising mode is much smaller than the one needed in other operation modes excluding stops. Therefore, the cruising mode and coasting were excluded from consideration.

Average value of some variable A during operation cycle depicted in Figure 4 is defined by equation:

$$< A >= \frac{\int_{0}^{t_{cycle}} Adt}{t_{cycle}} = \frac{\int_{cycle} Adt}{\int_{cycle} dt} \tag{1}$$

where t_{cycle} is the duration of the cycle; t is time.

The mechanical subsystem of the train can be represented by:

$$J\frac{d\Omega}{dt} = \frac{2\pi}{60}J\frac{dn}{dt} = T \tag{2}$$

where Ω is the motor mechanical angular frequency in radians per second; $n = 60{\cdot}\Omega/(2{\cdot}\pi)$ is the motor rotational speed in revolutions per minute; J is the total equivalent inertia of the train at the motor shaft; T is the equivalent torque at the motor shaft. Neglecting the friction and the slopes, the equivalent torque T is equal to the torque produced by the motor.

By expressing dt from (2) and substituting it into (1) the average value can be expressed as:

$$< A > = \frac{\int_{cycle} \frac{A}{T}dn}{\int_{cycle} \frac{dn}{T}} \tag{3}$$

Let's estimate the value of integral in the numerator of (3) with a help of quadrature formulas. The operation mode of the motor changes in operation points 1–5 (see Figure 4), and there is no derivative of the torque with respect to speed. Therefore, it is reasonable to use the equation of first algebraic order (i.e., trapezoidal rule). Summing up the products of arithmetic means of the integrands A/T and the speed differences at the separate segments from Figure 4 (the segments from the point 2 to point 3, from 3 to 4, from 4 to 5, and from 5 to 1) results in the following:

$$\begin{gathered}\int_{cycle} \frac{A}{T}dn \approx \left(\frac{A_2}{T_2} + \frac{A_3}{T_0}\right)\frac{n_{\max}-n_g}{2} + \left(\frac{A_3}{T_0} + \frac{A_4}{T_0}\right)\frac{n_g}{2} + \dots \\ \dots + \left(\frac{A_4}{T_0} + \frac{A_5}{T_0}\right)\frac{n_m}{2} + \left(\frac{A_5}{T_0} + \frac{A_1}{T_1}\right)\frac{n_{\max}-n_m}{2},\end{gathered} \tag{4}$$

where A_1–A_5 are the values of the calculated function A at operating points 1–5 (see Table 1), respectively.

Table 1. Operating points of the traction motor considered during optimization.

Operating Point, i	Operating Point Name	Speed, rpm	Torque, N·m	w_i	k_i
1	Driving mode; maximum speed	4380	413.4	0.363	0.97
2	Generating mode; maximum speed	4380	826.9	0.091	0.99
3	Generating mode; transition from constant power to constant torque operation modes	2854	1240	0.182	1.1
4	Zero speed	0	1240	0.182	0.97
5	Driving mode; transition from constant torque to constant power operation modes	1427	1240	0.182	0.97

Approximation of the denominator of (3) can be expressed from (4) by substituting $A \equiv 1$. Substituting (4) into (3) as well as $T_2 = T_0 \cdot n_g/n_{max}$, $T_1 = T_0 \cdot n_m/n_{max}$, it is possible to find similarly summand in each product with A_i. Then the average loss estimation during operation cycle can be obtained as a weighted-average losses in points 1–5 of the operation cycle as:

$$< A > \approx \sum_{i=1,2,3,4,5} w_i A_i \tag{5}$$

In (5), the weight coefficients w_i normalized on the basis unity sum of all coefficients can be evaluated by:

$$w_i = \frac{W_i}{\sum\limits_{1}^{5} W_i} \tag{6}$$

In (6), the coefficients *Wi* are calculated as according to (7):

$$W_1 = \frac{n_{\max}(n_{\max} - n_m)}{n_m};\ W_2 = \frac{n_{\max}(n_{\max} - n_g)}{n_g};\ W_3 = n_{\max};\ W_4 = (n_g + n_m);\ W_5 = n_{\max}. \tag{7}$$

Average value estimate (5) is relevant for any function $A(t)$ in the considered cycle from Figure 3. Here, this Equation (5) is used to evaluate average electric losses in the electric machine $<P_{loss\ el}>$. The losses during coasting are not taken into account due to the fact that they are small. Only the acceleration and deceleration phases are considered.

5. Voltage Limit

Kinetic energies of the train in operation points 2 and 3 (see Figure 4) equal to $J \cdot {\Omega_{\max}}^2/2$ and $J \cdot {\Omega_g}^2/2$, respectively. That is the train loses $1 - {\Omega_g}^2/{\Omega_{\max}}^2 = 1 - {n_g}^2/{n_{max}}^2$ part of its kinetic energy when braking from point 2 to 3. This energy is converted from the motion to the charge of the supercapacitor. The ratio between the energy accumulated in the supercapacitor during segment 2–3 and the total energy being accumulated at both segments 2–3 and 3–4 is $({V_3}^2 - {V_{DC\ rated}}^2)/({V_{max}}^2 - {V_{DC\ rated}}^2)$, where V_3 is the DC voltage in operating point 3; $V_{DC\ rated}$ is the rated voltage of the DC catenary; V_{max} is the maximum DC voltage. Therefore, the expected level of the DC link or supercapacitor voltage in the point 3 can be evaluated according to (8):

$$V_3 = \sqrt{V_{DC\ rated}^2 + (V_3^2 - V_{DC\ rated}^2)(1 - \frac{n_g^2}{n_{max}^2})} = 929\ \text{V}. \tag{8}$$

Thus, the desired motor phase-to-phase voltage amplitude V_i at operating points i = 1,2,3,4,5 can distinguish from $V_{DC\ rated}$ = 750 V. Table 1 shows the main parameters of the motor in the operation points i = 1,2,3,4,5, including weight coefficients being evaluated using (6) and normalized in addition to coefficient $k_i = V_i/V_{DC\ rated}$.

The expected DC voltage in the operation point 3 is 100%·(929 − 750)/750 = 24% higher than $V_{DC\ rated}$. Coefficient k_3 = 1.1 is selected with a large margin, and it is expected that the magnitude of the phase-to-phase voltage is higher than the rated voltage of the grid by 10% in the operating point 3. Moreover, a 3% margin for $V_{DC\ rated}$ is selected as the expected voltage drop across the power semiconducting switches in the driving mode, and a 1% margin for $V_{DC\ rated}$ is selected at operating point 2.

6. SHM Optimization Parameters and Procedure

Optimization of the motor parameters is performed using Nelder–Mead method. The Nelder–Mead algorithm used for the optimization is well known [30] and is included in the basic MATLAB software package (function "fminsearch"). This method is a one-criterion optimization method that uses only one objective function. The conventional approach for implementing multi-criteria optimization using a one-criterion method is to include in the optimized function all optimization objectives as a sum or product with certain weight coefficients [22].

The cost function is composed taking into account the following targets listed in descending order of importance:

1. Minimization of average losses $<P_{losses}>$;
2. Minimization of the maximum stator current magnitude during the entire operation cycle taking into account available voltage limit;
3. Minimization of the maximum symmetrized torque ripple along the entire operation cycle max $(TRsym)$;
4. Minimization of the maximum non-symmetrized torque ripple along the entire operation cycle max(TR).

A non-symmetrized torque ripple is produced by a single SRSC. A symmetrized torque ripple is produced by a SHM as a whole. Details of *TR* and *TRsym* are given in [19]. According to the targets of optimization the cost function can be written as their product. The weights are the powers to which the multipliers are raised:

$$F = < P_{loss\_el} > \max(I_i)^{0.7} \max(TRsym_i)^{0.025} \max(TR_i)^{0.01} f\left(\frac{\max(B_i)}{1.6[T]}\right), \quad f(x) = \begin{cases} 1, x < 1.6[T] \\ x^3, otherwise \end{cases}, \tag{9}$$

where $<P_{loss\_el}>$ are the electric losses; I_i is the value of current in the *i*th operation point; $TRsym_i$ is the value of the symmetrized torque ripple in the *i*th operation point; *TR* is the value of the non-symmetrized torque ripple in the *i*th operation point, B_i the magnetic flux density in the housing and the sleeve.

The one-criterion Nelder–Mead method is applied in this study to optimize the SHM design. The Nelder–Mead method belongs to unconstrained optimization methods. The optimization constraints could be specified simply by assigning infinite values to the objective function when the constraint conditions are not met. However, this would lead to a rapid decrease in the volume of the simplex. For this reason, objective function (9) is modified by using the 'soft constraints' with the penalty growing rapidly in the forbidden area (if $B_i \geq 1.6$ T).

Optimization was performed neglecting the discrete nature of some parameters. The number of turns in the section of power winding N_{sec} is assumed to be a real number; it is not checked that the width and height was selected from the standard [31]. The number of turns is selected according to $V_{DC\ rated} = \max(V_i/k_i)$ [23]. The number of parallel branches is assumed to be equal to 4. Figure 5 shows the geometric parameters of the SHM.

The parameters fixed during the optimization and varied ones are given in Tables 2 and 3. Additionally, the areas of the cross-sections of the stator housing and the rotor sleeve are assumed to be equal since they transfer the same excitation flux. The shaft is assumed to be nonmagnetic or to have poor magnetic conductivity.

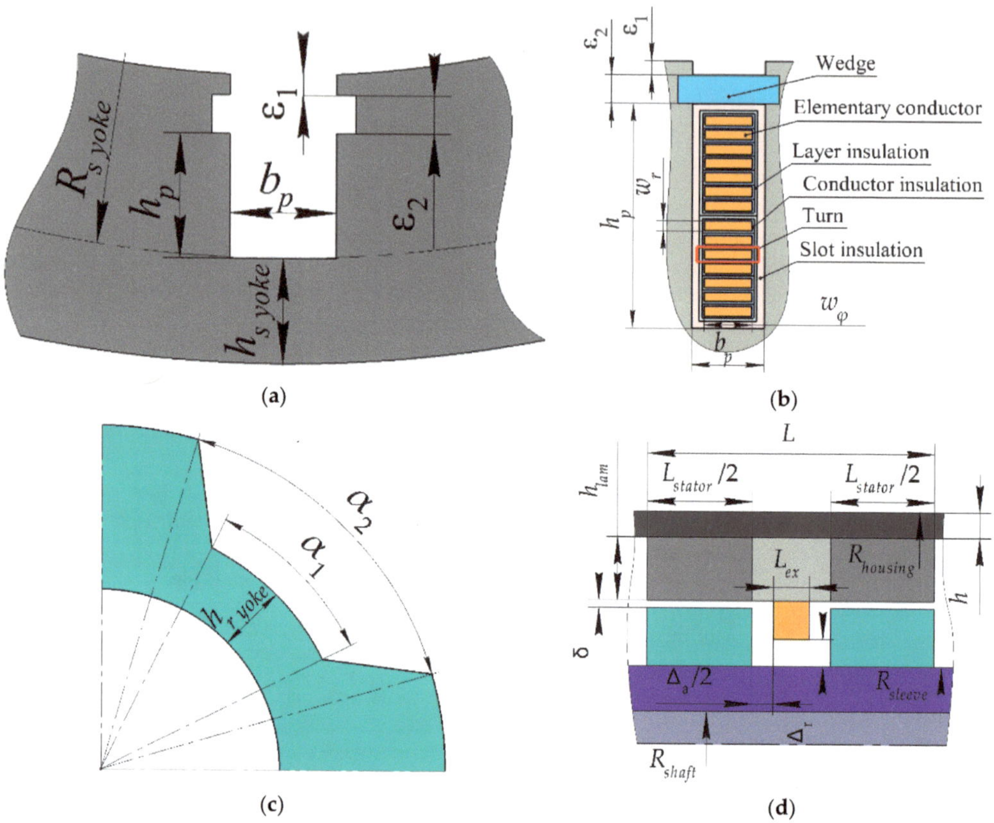

Figure 5. SHM geometric parameters. (**a**) Stator slot; (**b**) stator winding; (**c**) rotor core; (**d**) other dimensions.

Table 2. Some geometric parameters fixed during the optimization.

Parameter	Value
Machine length without end winding parts L, mm	302.5
Length of one stator stack	151.25
Stator housing radius, mm	267
Axial clearance between excitation winding and rotor, Δ_a, mm	29
Radial clearance between field winding and rotor Δ_r, mm	22
Shaft radius R_{shaft}, mm	40
Stator lamination yoke $h_{s\ yoke}$, mm	21
Rotor lamination yoke $h_{r\ yoke}$, mm	17
Stator wedge thickness, ε_2, mm	2
Stator unfilled area thickness, ε_1, mm	3

All of these determine other geometric parameters of the rotor and stator lamination including the inner rotor lamination radius that is the radius of the rotor sleeve. For example, the outer radius of stator lamination is determined by the outer housing radius and the housing thickness.

Table 3. Geometric parameters varied during the optimization.

Parameter	Initial Design	Optimized Design
Housing thickness h, mm	20	25.7
Total stator stacks length L_{stator}, mm	210	228
Stator slot depth, h_p, mm	36	30.6
Stator slot width, b_p, mm	7	7.5
Airgap width δ, mm	2	3.0
Rotor slot thickness, α_1, deg	36	36.7
Rotor slot thickness, α_2, deg	63	64.3
Angles of field weakening at operating points 1,2,3,4 electrical radians	0.5; 0.8; 0.4; 0.3	0.64; 0.91; 0.34; 0.17
Current ratio	5	5.95

The sizes of the armature wire w_x and w_y (see Figure 5b) necessary to compute the DC and eddy current loses in it [23] are determined according to Equation (10):

$$b_p = w_x + a_x;\ h_p = 2\cdot(w_y + \Delta w)\cdot N_{sec} + a_y, \tag{10}$$

where a_x = 1.51 mm, a_y = 1.8 mm, Δw = 0.31 mm are the constants taking into account the width of the slot insulation, layer insulation, impregnation, etc.

Only DC losses are assumed to be in the excitation winding. To calculate them, the copper filling factor 0.8 is adopted. Among parameters varied during the optimization, there are current angles measured from the middle of the rotor tooth in modes 1,2,3,4.

To reduce the number of the optimized parameters, the angle in operating point 5 is assumed to be the same as in operating point 4, the ratio of currents in in the excitation winding section, and the armature winding layer is assumed to be the same in all modes. Additionally, the ratio of the rotor slot widths α_2/α_1 (see Figure 5c) is assumed to be constant during the optimization. The mechanical losses are assumed to be mainly windage losses and proportional to the speed cubed. The mechanical losses at n_{max} are assumed to be 3.55 kW.

7. Optimization Results

Figure 6 shows the change of the electrical losses and current magnitude during optimization stage. Figure 7 depicts the simultaneous change of the symmetrized and non-symmetrized torque ripple. Figure 8 illustrates the change of the cost function value along the optimization.

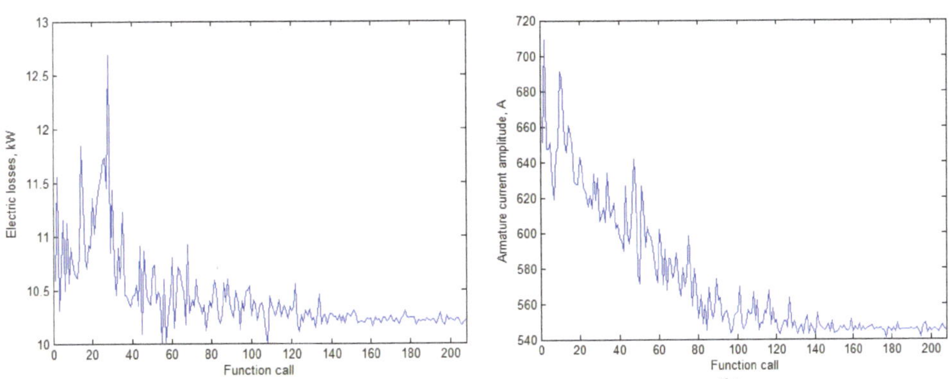

Figure 6. Change of the motor parameters during optimization. (**a**) Average losses; (**b**) maximum magnitude of the power winding current.

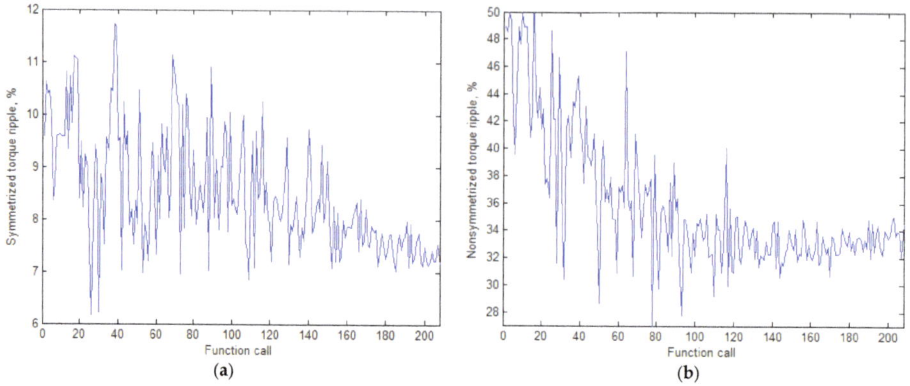

Figure 7. Change of the motor parameters during optimization. (**a**) Symmetrized torque ripple; (**b**) nonsymmetrized torque ripple.

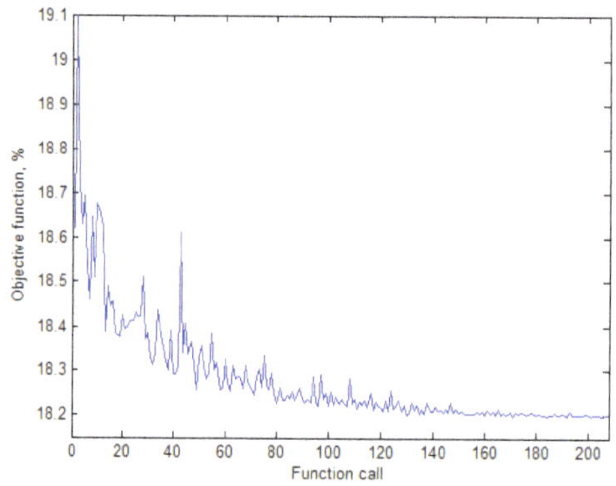

Figure 8. Variation of the cost function value according to (9) during optimization.

Figures 9 and 10 depict the geometry and induction value in the SHM before and after optimization. Table 4 shows the results of optimization.

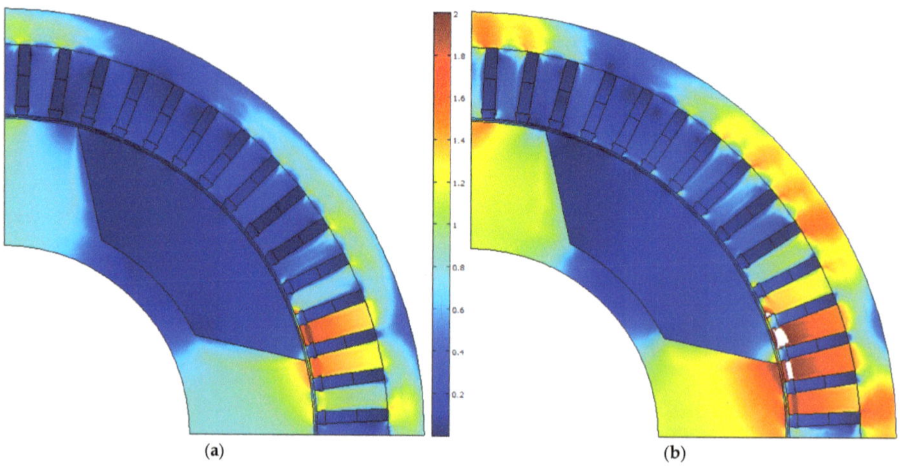

Figure 9. The cross-section of the initial design of the synchronous homopolar motor and the magnitude plot of flux density (T); white areas mark the extreme saturation level (>2 T). (**a**) Operating point 1; (**b**) operating point 4.

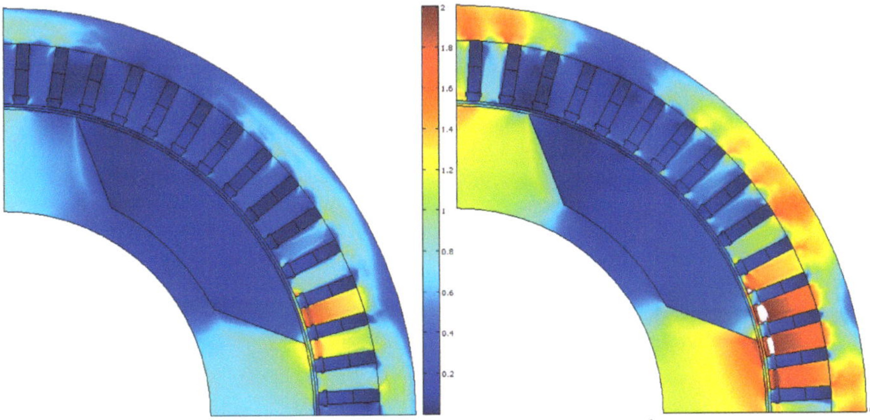

(**a**) (**b**)

Figure 10. The cross-section of the optimized design of the synchronous homopolar motor and the magnitude plot of flux density (T); white areas mark the extreme saturation level (>2 T). (**a**) Operating point 1; (**b**) operating point 4.

Table 4. Optimization results.

Parameter	Initial Design					Optimized Design				
Operating Point, *i*	1	2	3	4	5	1	2	3	4	5
Rotational speed n, rpm	4280	4280	2854	0	1427	4280	4280	2854	0	1427
Amplitude of the armature phase current I_{arm}, A	316	554	642	647	651	311	541	547	542	545
Efficiency, %	94.3	94.9	95.4	0.0	94.3	94.8	95.0	95.6	0.0	94.2
Output mechanical power P_{mech}, kW	185.3	−370.6	−370.6	0.0	185.3	185.3	−370.6	−370.6	0.0	185.3
Input electrical power, kW	196.6	−351.7	−353.7	8.3	196.4	195.4	−352.2	−354.4	8.9	196.8
Mechanical loss, kW	3.55	3.55	1.06	0.00	0.14	3.55	3.55	1.06	0.00	0.14
Armature DC copper loss, kW	1.79	5.51	7.39	7.50	7.60	2.48	7.53	7.67	7.55	7.62
Armature eddy-current copper loss, kW	1.15	3.96	3.25	0.00	0.83	0.59	2.25	2.16	0.00	0.60
Stator lamination loss, kW	4.14	4.63	3.99	0.00	1.68	2.87	3.39	3.64	0.00	1.65
Rotor lamination loss, kW	0.48	0.62	0.39	0.00	0.10	0.19	0.30	0.22	0.00	0.06
Excitation copper loss, kW	0.18	0.59	0.79	0.79	0.79	0.46	1.41	1.44	1.40	1.40
Total loss, kW	11.30	18.86	16.88	8.29	11.14	10.14	18.43	16.19	8.95	11.47
Average losses according to Formula (5)			12.43					12.02		
Number of turns in armature winding			5.93					6.68		
Power factor	0.986	−0.996	−0.924	1.000	0.920	0.994	−1.000	−0.927	1.000	0.905
Line-to-line voltage amplitude V_a, V.	728	749	699	14	383	718	755	825	16	468
Nonsymmetrized Torque ripple, %	48.9	33.7	23.4	23.4	23.4	33.9	25.9	18.8	19.0	19.0
Symmetrized torque ripple, %	9.59	7.40	4.90	5.04	5.03	7.21	5.47	3.77	3.91	3.91
Magnetic flux density in the housing and the sleeve, T	1.13	1.43	1.70	1.73	1.73	0.95	1.33	1.56	1.60	1.60

The following conclusions can be made by comparing motor parameters from Table 4 before and after optimization:

1. In the initial design, the magnetic flux density in the housing and the sleeve exceeds the chosen level of 1.6 T. After the optimization, the restriction of this value is fulfilled.
2. Average losses were reduced by $(12.43 - 12.03)/12.43 = 3.2\%$;

3. The maximum current magnitude before optimization was 647 A at operation point 4, at zero speed and maximum torque. After optimization, the maximum current magnitude becomes at the operation point 3 at the speed n_g when transiting from the constant power to constant torque in generating mode. Thus, the maximum inverter current magnitude was reduced by (647 − 547)/547 = 15.5%.
4. Maximum symmetrized torque ripple (ripple at the shaft) occurs in the operation point 1 in the driving mode at maximum speed. After optimization its level was reduced by (9.59 − 7.21)/7.21 = 24.8%.

8. Conclusions

This article discusses the procedure and results of optimizing a 370 kW synchronous homopolar motor for driving subway train. The motor characteristics are optimized taking into account the subway train moving trajectory, namely acceleration and braking stages. A single-objective Nelder-Mead algorithm was used during our procedure. The number of motor operating points necessary to compute at one cost function call in the proposed approach is only five, which makes the computational efforts acceptable for the optimization.

The following optimization targets were selected: decreasing the average losses along the way from station to station, decreasing the installed power capacity of the traction inverter, and the reduction of the torque ripple. The optimization results show a significant improvement of the target parameters of the traction drive. Average motor losses were reduced by 3.2%. The required current of the solid-state inverter was reduced by 15.5%. The motor torque ripple was reduced by 24.8%.

Author Contributions: Conceptual approach, A.A., V.D. and V.P.; data curation, V.D. and V.K.; software, V.D. and V.P.; calculations and modeling, A.A., V.D., V.K. and V.P.; writing—original draft, A.A., V.D., V.K. and V.P.; visualization, V.D. and V.K.; review and editing, A.A., V.D., V.K. and V.P. All authors have read and agreed to the published version of the manuscript.

Funding: The research was performed with the support of the Russian Science Foundation grant (Project No 21-19-00696).

Institutional Review Board Statement: Not applicable.

Informed Consent Statement: Not applicable.

Data Availability Statement: Data are contained within the article.

Acknowledgments: The authors thank the editors and reviewers for careful reading and constructive comments.

Conflicts of Interest: The authors declare no conflict of interest.

References

1. Gundogdu, T.; Zhu, Z.-Q.; Chan, C.C. Comparative Study of Permanent Magnet, Conventional, and Advanced Induction Machines for Traction Applications. *World Electr. Veh. J.* **2022**, *13*, 137. [CrossRef]
2. Dianov, A. Highly Efficient Sensorless Multicontrol Mode Compressor IPMSM Drive with Seamless Transitions. *IEEE Trans. Power Electron.* **2022**, *37*, 9129–9137. [CrossRef]
3. Schulze, R. Reducing Environmental Impacts of the Global Rare Earth Production for Use in Nd-Fe-B Magnets, How Magnetic Technologies Much Can Recycling Contribute? Ph.D. Thesis, Technical University of Darmstadt, Darmstadt, Germany, December 2018. Available online: https://tuprints.ulb.tu-darmstadt.de/8301/7/Diss%20Rita%20Schulze%2015-12-2018.pdf (accessed on 4 November 2022).
4. Dong, S.; Li, W.; Chen, H.; Han, R. The status of Chinese permanent magnet industry and R&D activities. *AIP Adv.* **2017**, *7*, 1–17. [CrossRef]
5. Dmitrievskii, V.; Prakht, V.; Kazakbaev, V. Synchronous reluctance generator with ferrite magnets for wind turbine. *J. Phys. Conf. Ser.* **2018**, *1102*, 012041. [CrossRef]
6. De Lima, W.I.B. *Rare Earth Industry*; Elsevier: Amsterdam, The Netherlands, 2015.
7. Dmitrievskii, V.; Prakht, V.; Kazakbaev, V.; Anuchin, A. Comparison of Interior Permanent Magnet and Synchronous Homopolar Motors for a Mining Dump Truck Traction Drive Operated in Wide Constant Power Speed Range. *Mathematics* **2022**, *10*, 1581. [CrossRef]

8. Papini, F.; Osama, M. Electromagnetic Design of an Interior Permanent Magnet Motor for Vehicle Traction. In Proceedings of the 2018 XIII International Conference on Electrical Machines (ICEM), Alexandroupoli, Greece, 3–6 September 2018; pp. 205–211. [CrossRef]
9. Dianov, A. Optimized Field-Weakening Strategy for Control of PM Synchronous Motors. In Proceedings of the 2022 29th International Workshop on Electric Drives: Advances in Power Electronics for Electric Drives (IWED), Moscow, Russia, 26–29 January 2022; pp. 1–6. [CrossRef]
10. The First-Ever BMW iX3, PressClub Global, Article. Available online: https://www.press.bmwgroup.com/global/article/detail/T0310696EN/the-first-ever-bmw-ix3?language=enhttps://www.netcarshow.com/bmw/2021-ix3 (accessed on 4 November 2022).
11. Feustel, S.; Huebner, B.; Loos, D.; Merwerth, J.; Tremaudant, Y.; Vollmer, K. Rotor for Separately Excited Inner Rotor Synchronous Machine, Inner Rotor Synchronous Machine, Motor Vehicle and Method. U.S. Patent US20210006105A1, 7 January 2021. Available online: https://patents.google.com/patent/US20210006105A1 (accessed on 4 November 2022).
12. Raia, M.R.; Ruba, M.; Martis, C.; Husar, C.; Sirbu, G.M. Battery electric vehicle (BEV) powertrain modelling and testing for real-time control prototyping platform integration. In Proceedings of the 2021 23rd European Conference on Power Electronics and Applications (EPE'21 ECCE Europe), Ghent, Belgium, 6–10 September 2021; pp. 1–10. Available online: https://ieeexplore.ieee.org/document/9570616 (accessed on 4 November 2022).
13. Chu, W.Q.; Zhu, Z.Q.; Zhang, J.; Ge, X.; Liu, X.; Stone, D.; Foster, M. Comparison of electrically excited and interior permanent magnet machines for hybrid electric vehicle application. In Proceedings of the 2014 17th International Conference on Electrical Machines and Systems (ICEMS), Hangzhou, China, 22–25 October 2014; pp. 401–407. [CrossRef]
14. Illiano, E. Design of a Highly Efficient Brushless Current Excited Synchronous Motor for Automotive Purposes. Ph.D. Thesis, ETH-Zürich, Zürich, Switzerland, 2014. [CrossRef]
15. Orlova, S.; Pugachov, V.; Levin, N. Hybrid Excitation of the Axial Inductor Machine. *Latv. J. Phys. Tech. Sci.* **2012**, *49*, 35–41. [CrossRef]
16. Bindu, G.; Basheer, J.; Venugopal, A. Analysis and control of rotor eccentricity in a train-lighting alternator. In Proceedings of the 2017 IEEE International Conference on Power, Control, Signals and Instrumentation Engineering (ICPCSI), Chennai, India, 21–22 September 2017; pp. 2021–2025. [CrossRef]
17. Lorilla, L.; Keim, T.; Lang, J.; Perreault, D. Topologies for future automotive generators. Part I. Modeling and analytics. In Proceedings of 2005 IEEE Vehicle Power and Propulsion Conference, Chicago, IL, USA, 7 September 2005; pp. 74–85. [CrossRef]
18. Bianchini, C.; Immovilli, F.; Bellini, A.; Lorenzani, E.; Concari, C.; Scolari, M. Homopolar generators: An overview. In Proceedings of the 2011 IEEE Energy Conversion Congress and Exposition, Phoenix, AZ, USA, 17–22 September 2011; pp. 1523–1527. [CrossRef]
19. Dmitrievskii, V.; Prakht, V.; Anuchin, A.; Kazakbaev, V. Traction Synchronous Homopolar Motor: Simplified Computation Technique and Experimental Validation. *IEEE Access* **2020**, *8*, 185112–185120. [CrossRef]
20. Lashkevich, M.; Anuchin, A.; Aliamkin, D.; Briz, F. Control strategy for synchronous homopolar motor in traction applications. In Proceedings of the 43rd Annual Conference of the IEEE Industrial Electronics Society (IECON), Beijing, China, 29 October–1 November 2017; pp. 6607–6611. [CrossRef]
21. Dmitrievskii, V.; Prakht, V.; Anuchin, A.; Kazakbaev, V. Design Optimization of a Traction Synchronous Homopolar Motor. *Mathematics* **2021**, *9*, 1352. [CrossRef]
22. Prakht, V.; Dmitrievskii, V.; Anuchin, A.; Kazakbaev, V. Inverter Volt-Ampere Capacity Reduction by Optimization of the Traction Synchronous Homopolar Motor. *Mathematics* **2021**, *9*, 2859. [CrossRef]
23. Prakht, V.; Dmitrievskii, V.; Kazakbaev, V.; Anuchin, A. Comparative Study of Electrically Excited Conventional and Homopolar Synchronous Motors for the Traction Drive of a Mining Dump Truck Operating in a Wide Speed Range in Field-Weakening Region. *Mathematics* **2022**, *10*, 3364. [CrossRef]
24. Cupertino, F.; Pellegrino, G.; Gerada, C. Design of synchronous reluctance machines with multiobjective optimization algorithms. *IEEE Trans. Ind. Appl.* **2014**, *50*, 3617–3627. [CrossRef]
25. Krasopoulos, C.T.; Beniakar, M.E.; Kladas, A.G. Robust Optimization of High-Speed PM Motor Design. *IEEE Trans. Magn.* **2017**, *53*, 1–4. [CrossRef]
26. Fatemi, A.; Ionel, D.; Popescu, M.; Chong, Y.; Demerdash, N. Design Optimization of a High Torque Density Spoke-Type PM Motor for a Formula E Race Drive Cycle. *IEEE Trans. Ind. Appl.* **2018**, *54*, 4343–4354. [CrossRef]
27. Montonen, J.; Nerg, J.; Gulec, M.; Pyrhönen, J. A New Traction Motor System with Integrated-Gear: A Solution for Off-Road Machinery. *IEEE Access* **2019**, *7*, 113740–113750. [CrossRef]
28. Guo, S.; Zhao, H.; Wang, Y.; Yin, X.; Qi, H.; Li, P.; Lin, Z. A Design Technique of Traction Motor for Efficiency Improvement Based on Multiobjective Optimization. *World Electr. Veh. J.* **2021**, *12*, 260. [CrossRef]
29. Lin, F.; Li, X.; Zhao, Y.; Yang, Z. Control Strategies with Dynamic Threshold Adjustment for Supercapacitor Energy Storage System Considering the Train and Substation Characteristics in Urban Rail Transit. *Energies* **2016**, *9*, 257. [CrossRef]
30. Nelder, J.; Mead, R. A Simplex Method for Function Minimization. *Comput. J.* **1965**, *7*, 308–313. [CrossRef]
31. IEC. *Specifications for Particular Types of Winding Wires—Part 0-2: General Requirements—Enamelled Rectangular Copper Wire*; IEC 60317-0-2:2020; IEC: Geneva, Switzerland, 2020. Available online: https://webstore.iec.ch/publication/63495 (accessed on 1 November 2022).

designs

MDPI

Article

On the Usefulness of Employing ANSYS Motor-CAD Software in Designing Permanent Magnet Synchronous Machines

Florina-Ambrozia Coteț, Iulia Văscan and Loránd Szabó *

Department of Electrical Machines and Drives, Technical University of Cluj-Napoca, 400114 Cluj-Napoca, Romania
* Correspondence: lorand.szabo@emd.utcluj.ro

Abstract: This paper deals with the simulation of a permanent magnet synchronous machine (PMSM) to be developed for driving a scroll compressor of a vehicle air conditioning system. The simulations were needed for the verification of the pre-sizing and for the electromagnetic behavior analysis of the PMSM. These were performed by using the *ANSYS Motor-CAD* software. All the obtained results emphasized that the pre-sizing of the PMSM was performed correctly, and the designed electrical machine can fulfill all the prescribed requirements. To highlight the superiority of the chosen surface-mounted PMSM topology, a comparative study was performed for four PMSMs having the same stator but different rotor topologies having the same quantity of permanent magnets. The study revealed that the PMSM with permanent magnets on its rotor surface is the best-fitting variant for the foreseen application. By intensively using the *ANSYS Motor-CAD* software, the authors appreciated that this product is a very useful tool in the hands of the designers not only in evaluating the machine's computed geometry (by checking its correctness and feasibility), but also in its in-depth electromagnetic analysis and in calculating its parameters of interest.

Keywords: electrical machine design; permanent magnet synchronous machine; design software; ANSYS Motor-CAD program; finite element method; numerical electromagnetic field computation

Citation: Coteț, F.-A.; Văscan, I.; Szabó, L. On the Usefulness of Employing ANSYS Motor-CAD Software in Designing Permanent Magnet Synchronous Machines. *Designs* **2023**, *7*, 7. https://doi.org/10.3390/designs7010007

Academic Editors: Wenbin Yu and José António Correia

Received: 6 December 2022
Revised: 14 December 2022
Accepted: 30 December 2022
Published: 5 January 2023

1. Introduction

Even though electrical machines are regarded as mature products, their design is a complex, time-consuming, and demanding engineering task as it requires profound multidisciplinary knowledge of electrical engineering (electrical machines and drives, electromagnetic field theory, electrical materials), mathematics (numeric methods), mechanical engineering, fluid mechanics and thermodynamics, acoustics, manufacturing technology, technical drawing, computer sciences, etc. [1]. Additionally, the electrical machine designers must be knowledgeable not only about the given specifications (which refers to machine ratings, customer performance expectations, etc.), but also of the existing standards, labor and materials costs, manufacturing requirements and restrictions, etc.

All electrical machine designers are facing a great variety of issues, some of which may not have a single answer, but rather several. The computations needed to be performed during the sizing of the electrical machines are complex and almost repetitive due to the frequent recalculations because of the non-fulfillment of the imposed conditions. Therefore, it is mandatory to apply some computer programs to perform the needed computations. Moreover, nowadays, the sizing of an electrical machine must be completed by adequate optimization to achieve the best structure within certain constraints [2,3].

Nowadays, it is not sufficient that the developed electrical machine works well. Global attention has been focused on the rising cost of energy over the past years, and it appears that this trend will continue. As a result, it is now more crucial than ever to design electrical machines with very high efficiency. Their use can significantly reduce world energy consumption [4]. In the post-pandemic period, during wartime, all the industries are facing massive unprecedented supply chain disruptions [5]. While the supply chain management strategies are re-examined, the material savings and recycling possibilities must also be the focus of the designers of any products [6].

Electrical machine designers have at their disposal numerous software tools to ease and improve their challenging engineering work. Most of them are using professional finite element analysis (FEA) software products for numeric electromagnetic field computations [7,8], such as *JMAG-Designer* [9,10], *ANSYS Maxwell* [11,12], *Simcenter Magnet* [13,14], *COMSOL Multiphysics* [15,16], *Flux 2D* [17–19] or its three-dimensional variant *Flux 3D* [20,21]. All such programs can perform precise computations for obtaining the magnetic field distribution within the electric machine, the generated forces, magnetic, material-related, and dynamical parameters, etc. They have versatile graphical user interfaces (GUIs) helping the engineers in adapting their models to the specific needs of the design. The keys to obtaining very precise results are the applied advanced adaptive mesh generators [22,23] and the solvers based on the most effective, time-saving, and accurate numerical methods [24,25]. Their embedded scripting tools enable the designer to simulate a great diversity of electrical machine topologies by only changing some basic parameters of the script and thus totally controlling the preprocessing phase of the simulations.

All commercial FEA software is interoperable, enabling the use of different external tools offered by other programs. This is performed by flexible communication with the other software products. The link of the FEA programs to *Simulink* must be primarily mentioned, which permits the easy development of the electrical machines' control system and thus their precise dynamic analysis [26]. These programs can export and import graphical files of a great variety of types, too.

There are specialized software solutions on the market that not only carry out a general FEA, but also greatly aid electrical machine designers by providing unique features essential in sizing and analyzing such devices. For example, *Simcenter SPEED* provides a variety of machine templates and libraries to assist users in defining the geometry, winding, drive, and control strategy of an electrical machine under development to gain information from simulations about its current, torque, speed, losses, etc. [27]. Another similar software product, *Simcenter Motorsolve*, has an easy-to-use template-based GUI for defining a great diversity of electrical machine topologies, which can be analyzed by applying both equivalent magnetic circuit-based calculations and FEA [14,28].

A special place in the market of electrical machine design and analysis software is occupied by *ANSYS Motor-CAD*. This useful program can perform in-depth multi-physical analysis of a great diversity of electrical machines [1,29,30]. The users have at their disposal four distinctive modules [31]. The main component is the electromagnetic one, which is designed for the electromagnetic performance assessment of the electrical machines by using precise analytical methods and FEA [7,8]. By applying the thermal unit, a 3D lumped thermal circuit can be built up automatically, enabling the user to determine the steady state and transient thermal characteristics of the machine taken into the study [32–37]. The third module, named *Lab*, is a special toolbox for combined electromagnetic and thermal simulations, which can ease the modeling and optimization tasks of the designers [38–41]. The mechanical module of the *ANSYS Motor-CAD* software uses the effective built-in FEA solver to calculate miscellaneous mechanical properties. This component can be very useful in the electromagnetic Noise, Vibration, and Harshness (NVH) analysis to be performed [42,43].

Our studies were focused on the development of an electrical machine for the compressor of the heating, ventilation, and air conditioning (HVAC) system of a car. Nowadays, especially in the summer, air conditioning is a vital necessity because it is the only way to circulate and cool the air inside any vehicle. Furthermore, these devices also clean and dry the indoor air [44,45]. Considering the main advantages of permanent magnet synchronous machines (PMSMs), like high efficiency, great specific power and torque, and good behavior at high speeds, such a machine was proposed for the given specific automotive application.

The paper begins with a brief survey of the software tools used by electrical machine designers. The focus is set on the *ANSYS Motor-CAD* software, since in the paper, the authors intend to share their experience in using this product in designing PMSMs. Next, the scope of designing the certain PMSM and the listing of its established design specifications follow. In this part of the paper, the results of the common pre-sizing of the machine are also detailed. The most consistent part of the paper is dealing with the surface-mounted PMSM model construction in *ANSYS Motor-CAD* and with detailing the obtained simulation results. By using the gained experience in using this software product, a comparative study on diverse PMSM topologies was performed. Four PMSMs having the same stator and permanent magnet quantities but different rotor structures were compared from diverse points of view [46–48]. The main aim of this study was to find out whether the other three PMSMs exceed the performances of the initial variant having surface-mounted permanent magnets. The study evidenced both the superiority of the surface PMSM variant and the practicality of applying *ANSYS Motor-CAD* software in designing and analyzing electrical machines.

2. The PMSM to Be Studied

2.1. Automotive Applications and the Electrical Machine Type Selection

A scroll compressor, similar to that given in Figure 1, was proposed for a car HVAC system [49,50]. They have the advantage of containing only two essential pieces (a fixed and an orbiting scroll) and thus can operate more quietly and smoothly than the reciprocating compressors [51]. Moreover, they are more reliable and energy-efficient and less prone to mechanical failure due to their smaller number of parts.

(a)

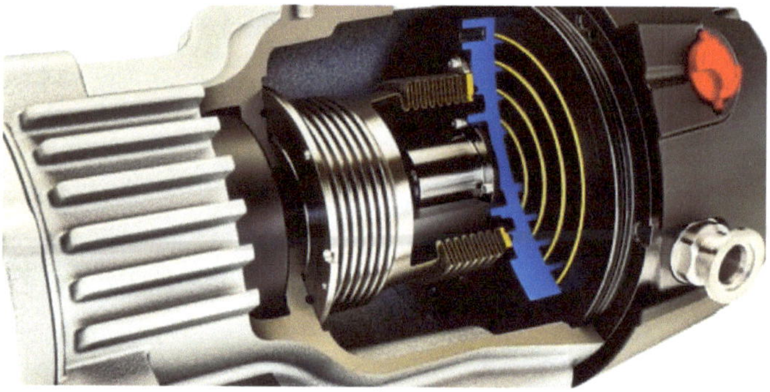
(b)

Figure 1. The scroll-type compressor. (**a**) The two co-wound spiral-shaped scrolls; (**b**) internal view of the compressor. Courtesy of the Vacuum Science World [52].

After a careful analysis, the PMSM was selected to propel the scroll compressor [53]. The main reason was that both efficiency and power density is quite high for these electrical machines. Additionally, they require little maintenance effort and produce little heat. Due to all of these factors, they are among the most significant industrial competitors of the very extensively used induction machines [54].

PMSMs are widely used in the automotive industry [55]. In 2021, 90% of the newly sold full and hybrid vehicles were powered by PMSMs [56]. Other significant automotive application areas for PMSMs include electric vehicle auxiliaries, like power brakes and steering.

Their global market share was USD 13.45 billion in 2021, and by 2028 it is anticipated to surpass USD 22.6 billion. This should result in a notable (7.57%) increase in the compound annual growth rate (CAGR) over this time span [57]. Given that the cost of the rare earth metals required to produce permanent magnets has more than doubled year over year, this gain in PMSM market share is particularly impressive and can be explained only by the outstanding properties of the PMSMs [56].

All of this emphasizes the significance of research to enhance PMSMs through the use of cutting-edge design tools, new materials, and novel manufacturing procedures [58].

2.2. The Design Specifications of the PMSM to Be Developed

The PMSMs to be studied here and designed to drive an automotive scroll compressor must have the mechanical characteristic provided in Figure 2.

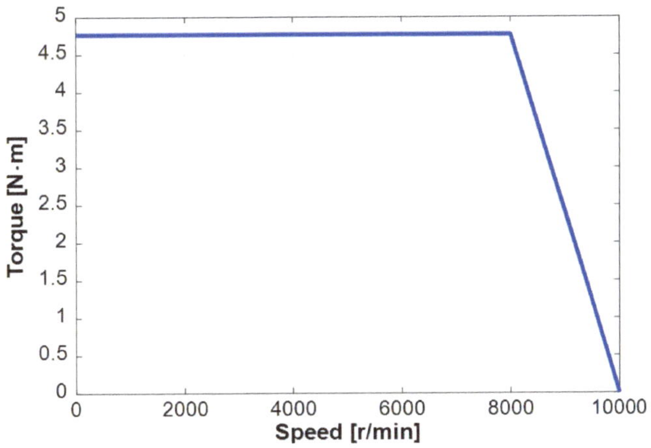

Figure 2. The mechanical characteristic of the PMSM to be designed.

The main input design data for the PMSM were established upon the torque and speed development of the given scroll compressor. During the design procedure, the following starting data were used:

- number of phases: 3;
- number of pole pairs: 3;
- DC bus voltage: 72 V;
- rated power: 4 kW;
- rated speed: 8000 r/min;
- rated torque: 4.77 N·m.

2.3. Pre-sizing the PMSM

The first step in designing the PMSM was the pre-sizing. This consisted of computing the initial geometry and winding of the PMSM by using simple analytical equations and some general rules [59–61]. During this design stage, the iron core and permanent magnet volumes, the number of slots, the main sizes of the stator and rotor, the winding diagram and the number of turns, the materials to be used, etc., were estimated by the well-known relationships among the main machine sizes and other parameters [62–65].

The main dimensions of the pre-sized three-phase, 18-slots surface PMSM are given in Figure 3.

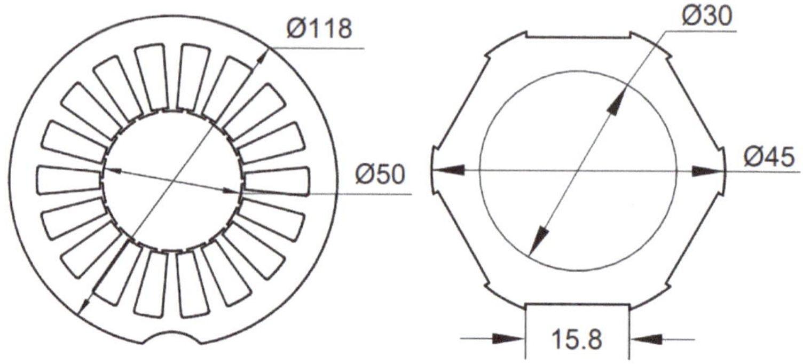

Figure 3. The main sizes (in mm) of the stator and rotor iron core laminations.

The cross-section of the rotor with the permanent magnets with its main sizes can be seen in Figure 4.

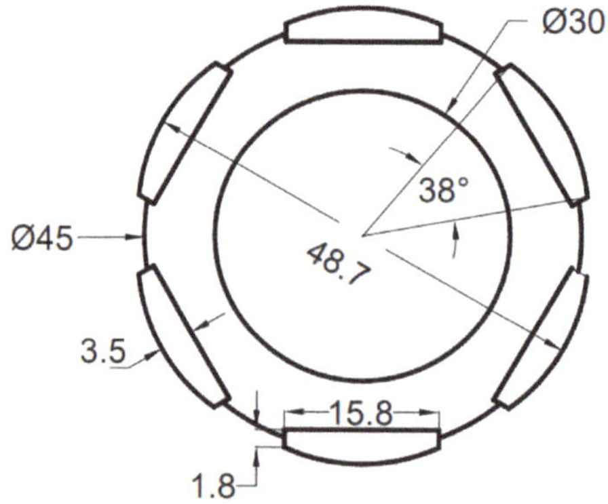

Figure 4. Cross-section of the rotor with the permanent magnets on it (all sizes are in mm).

In the 18 stator slots, a three-phase, double-layer, distributed winding was proposed to be placed. As can be seen in Figure 5, the pitch of the winding is three slots.

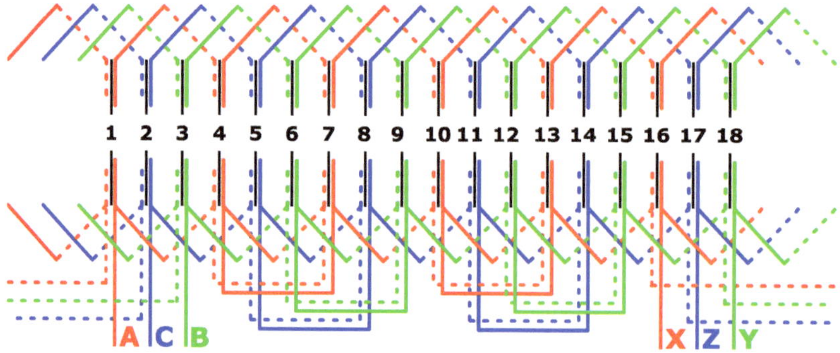

Figure 5. The winding diagram of the PMSM.

Upon these initial data, the model of the PMSM was built up in *ANSYS Motor-CAD* for evaluating the computed geometry and the selected winding, and for checking their correctness and feasibility.

3. The ANSYS Motor-CAD Model of the PMSM

By using the *ANSYS Motor-CAD* program, the building up of the surface-mounted PMSM model was very easy, as it needed only a few simple well-assisted steps to be performed.

To investigate the electromagnetic behavior of this PMSM, the electromagnetic module (named *E-Magnetic*) of the software was used. This enables the calculation of the electromagnetic performances by using both precise analytical and FEA methods [7,8].

The program offers to the designers eight predefined electromagnetic templates to be used for the study of the so-called *Brushless Permanent-Magnet Motors (BPMs)* category.

The first step in the model implementation was the definition of the stator and rotor iron cores. As can be seen in Figure 6, the sizes of the transversal and axial cross-sections of the PMSM had to be set from the *Geometry* menu of the program.

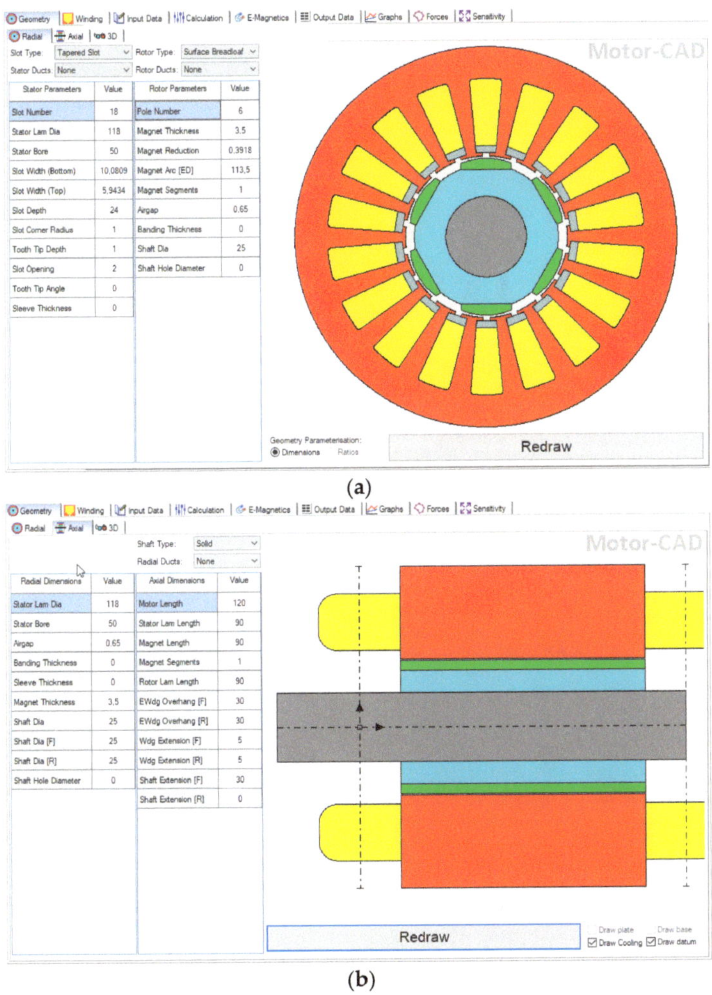

Figure 6. The panels of the *ANSYS Motor-CAD* program for setting the geometrical data of the stator and rotor iron cores. (**a**) Radial view; (**b**) axial view.

In the text boxes of the two panels, the main sizes of the iron cores (exterior and interior stator diameter, airgap length, shaft diameter, active length of the machine, and details about the stator teeth and slots) must be established.

Taking advantage of the possibility to export the cross-section of the iron cores in *dxf*-type CAD files, the resulting and the initial (obtained upon the pre-sizing of the machine) drawings were overlayed (see Figure 7). As can be seen, the designed and modeled structures are very close, so it can be stated that the iron core structure of the surface PMSM in the study was precisely modeled in *ANSYS Motor-CAD*.

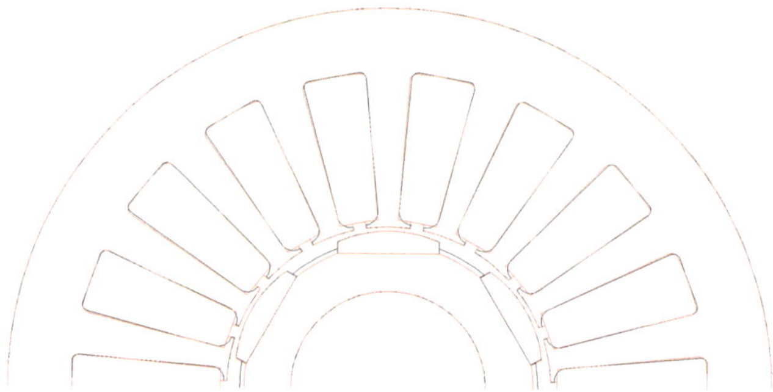

Figure 7. The overlayed drawings of the designed (black lines) and modelled (lines drawn in red) iron cores of the PMSM (showing only half of the motor cross-section).

The winding diagram of the PMSM given in Figure 5 can also be very easily implemented in *ANSYS Motor-CAD* from its *Winding* menu. The lap type of the winding with a central path had to be selected. By specifying only five winding parameters (the numbers of the phases, turns, parallel paths, layers, and the throw length in slot steps), its model could be simply completed.

In the *Pattern* panel of the *Winding* menu, two types of winding diagrams automatically generated by the program can be visualized (see Figures 8 and 9). By using these, the modeled winding diagram can be compared easily with that previously set up by the machine designer.

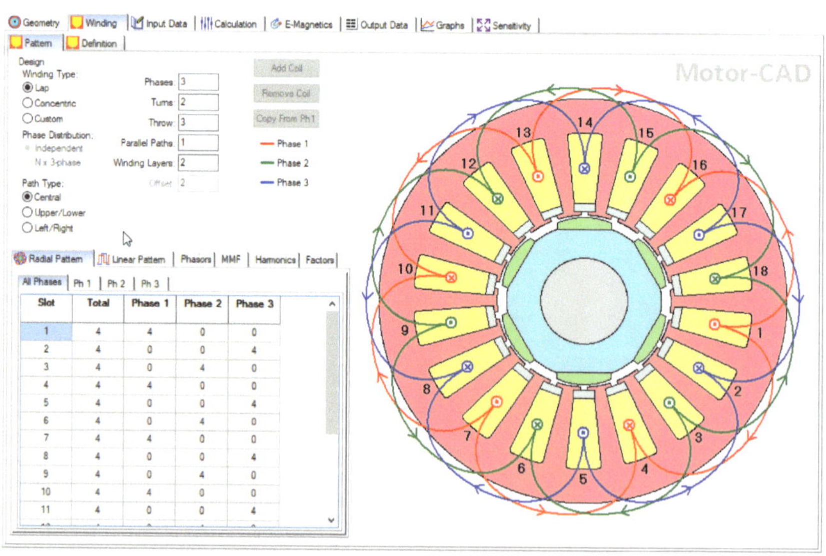

Figure 8. The radial pattern of the PMSM winding diagrams implemented in *ANSYS Motor-CAD*.

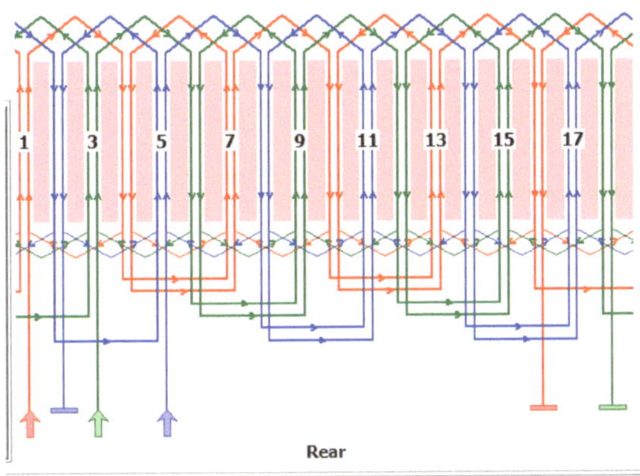

Figure 9. The linear pattern of the PMSM winding diagram as provided by the *ANSYS Motor-CAD* software.

In the same panel, the phasor diagram, the MMF distribution in the airgap, and its harmonic content can be also visualized and checked.

The *Definition* panel of the same *Winding* menu offers a specific and very useful feature of the *ANSYS Motor-CAD* software product. As can be seen in Figure 10, the filling of the slot is visualized upon the imposed number of turns, sizes of the conductors and insulators, etc. Additionally, a great variety of winding parameters are shown in this window, such as different areas, fill factors, etc.

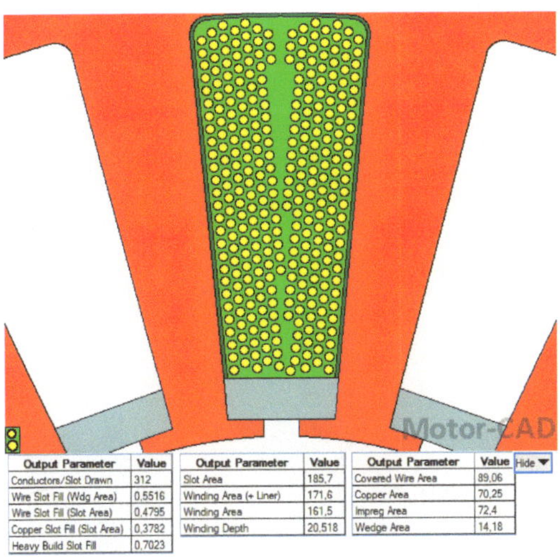

Output Parameter	Value
Conductors/Slot Drawn	312
Wire Slot Fill (Wdg Area)	0,5516
Wire Slot Fill (Slot Area)	0,4795
Copper Slot Fill (Slot Area)	0,3782
Heavy Build Slot Fill	0,7023

Output Parameter	Value
Slot Area	185,7
Winding Area (+ Liner)	171,6
Winding Area	161,5
Winding Depth	20,518

Output Parameter	Value
Covered Wire Area	89,06
Copper Area	70,25
Impreg Area	72,4
Wedge Area	14,18

Hide

Figure 10. The *Winding-Definition* panel showing the way the conductors of the windings are placed in the stator slots and listing numerous parameters of the winding.

The *Conductors/Slot Drawn* output parameter value should be colored in red to indicate an error if the designed winding does not fit into the slot area with the specified winding settings at the required fill factor.

In the next step of the modeling, the materials used in the construction of the surface PMSM had to be established. For this purpose, *ANSYS Motor-CAD* offers a huge database of the usual materials of electrical machines.

Fortunately, all the materials selected during the pre-sizing of the PMSM could be found in the provided database: *M470-50A*-type non-oriented 0.5 mm thick electric steel for the iron cores [66], pure copper for the windings, and the *N35UH*-type NdFeB permanent magnet [67], as can be seen in the material selection panel provided in Figure 11. In *ANSYS Motor-CAD*, there exists the possibility for the users to create their own databases for those materials which had not been included in the program database.

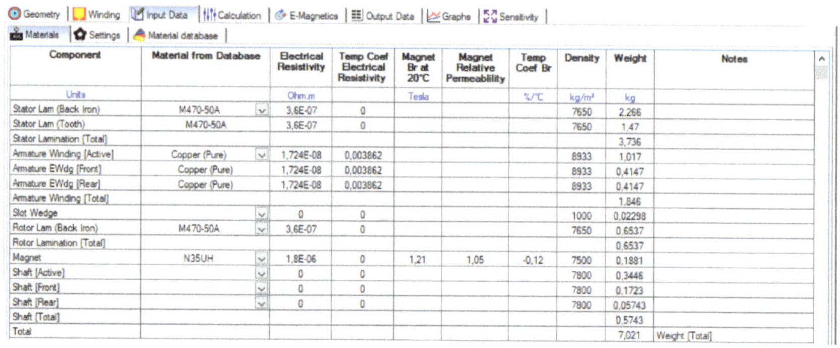

Component	Material from Database	Electrical Resistivity	Temp Coef Electrical Resistivity	Magnet Br at 20°C	Magnet Relative Permeability	Temp Coef Br	Density	Weight	Notes
Units		Ohm.m		Tesla		%/°C	kg/m³	kg	
Stator Lam (Back Iron)	M470-50A	3,6E-07	0				7650	2,266	
Stator Lam (Tooth)	M470-50A	3,6E-07	0				7650	1,47	
Stator Lamination [Total]								3,736	
Armature Winding [Active]	Copper (Pure)	1,724E-08	0,003862				8933	1,017	
Armature EWdg [Front]	Copper (Pure)	1,724E-08	0,003862				8933	0,4147	
Armature EWdg [Rear]	Copper (Pure)	1,724E-08	0,003862				8933	0,4147	
Armature Winding [Total]								1,846	
Slot Wedge		0	0				1000	0,02298	
Rotor Lam (Back Iron)	M470-50A	3,6E-07	0				7650	0,6537	
Rotor Lamination [Total]								0,6537	
Magnet	N35UH	1,8E-06	0	1,21	1,05	-0,12	7500	0,1881	
Shaft [Active]		0	0				7800	0,3446	
Shaft [Front]		0	0				7800	0,1723	
Shaft [Rear]		0	0				7800	0,05743	
Shaft [Total]								0,5743	
Total								7,021	Weight [Total]

Figure 11. The *Materials* panel of the *Input Data* menu used for selecting the materials for the different components of the modeled PMSM.

As can be seen in Figure 11, from this panel, information on the main characteristics of the used materials can be found. Important data regarding the masses of the PMSM components and the total machine mass can also be obtained from here.

With this last step, the model development in *ANSYS Motor-CAD* was practically finished. Next, the simulation conditions must be set in the *Calculation* menu of the program shown in Figure 12.

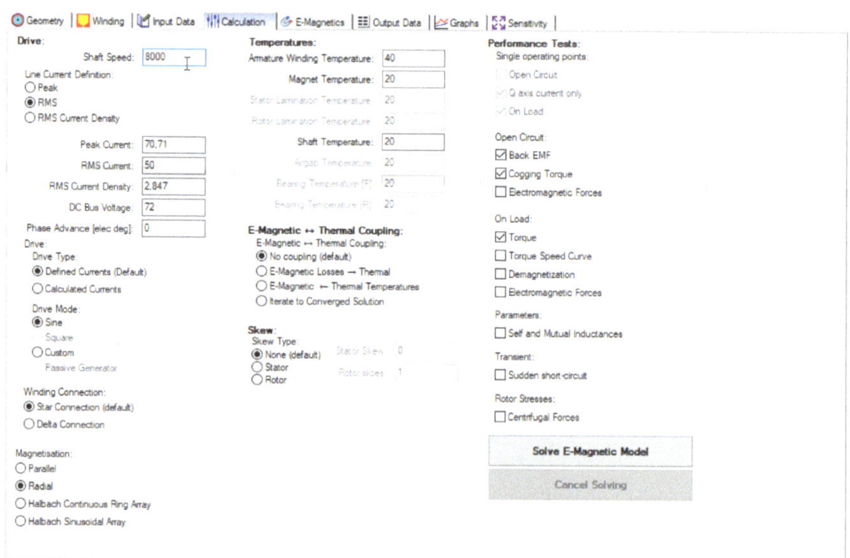

Figure 12. The *Calculation* menu where all the simulation conditions must be established.

In this panel, the drive-related data must be set, such as the imposed speed of the machine, the RMS value of the supply currents, and the DC bus voltage. Additionally, here must be selected the type of drive, the connection of the stator windings, and the magnetization direction of the permanent magnets.

By using the selection possibilities in this panel, the users can choose which performance tests besides the ones mandatorily computed have to be performed by the program.

ANSYS Motor-CAD can deliver the variations of the back EMF, cogging torque, and electromagnetic forces (both in open-circuit and on-load conditions). There is the possibility to obtain the torque versus speed (mechanical) characteristics of the analyzed electrical machines, too.

The simulations can be started from this panel by pressing the *Solve E-Magnetic Model* button.

4. Simulation Results

All the simulations to be next presented were performed at the rated conditions of the surface PMSM:

- 8000 r/min;
- 50 A RMS currents;
- 72 V DC bus voltage.

The *ANSYS Motor-CAD* software can perform simulations both on the entire machine structure or only on a part of it, considering the symmetry conditions. Despite that the FEA is perfectly performed at the default settings of the options concerning the solver, there are numerous possibilities of settings targeting the mesh fineness, the number of points in which the quantities of interest are computed, etc.

Upon the performed FEA, a great variety of results were obtained. Here only those of major interest for the study to be performed will be detailed.

The basic results are gained upon the accomplished FEA of the modeled electrical machine and can be visualized in the *E-Magnetics* panel of the program. For example, the considered regions together with the automatically generated mesh can be pictured, as can be seen in Figure 13.

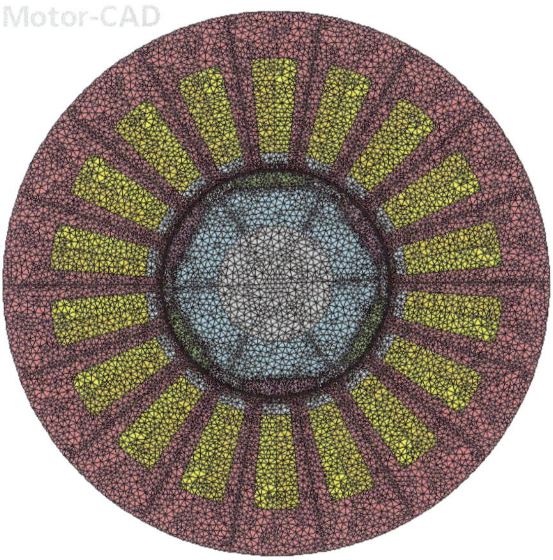

Figure 13. The different regions of the analyzed PMSM and the mesh generated by the *ANSYS Motor-CAD* program.

In the same panel, the color map of four different quantities (flux density, vector potential, current density, and eddy current density) together with the magnetic flux lines can be visualized. In Figure 14, the color maps of the flux density and the flux lines for three distinct positions during a complete 360-electrical-degree revolution of the rotor are provided.

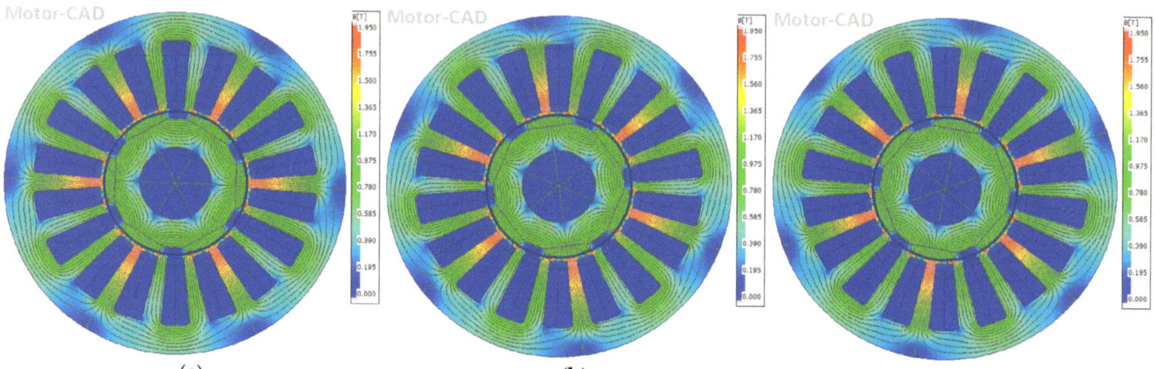

(a) (b) (c)

Figure 14. The color map of the magnetic flux density and the plotted flux lines at three different mechanical angle displacements of the rotor. (**a**) 0°; (**b**) 10°; (**c**) 20°.

The *ANSYS Motor-CAD* software can generate an animation by putting together slide-by-slide the results obtained at different rotor positions and can save this in an animated *gif*-type file.

A second possibility to analyze the obtained results is the evaluation of a great number of numeric results by accessing the *Output Data* panel. Here the following results categories (arranged in panes) can be evidenced:

- *Drive*: data concerning the back EMF, supply voltages and currents, *d-q* axes inductances and currents, machine constants, short circuit parameters, etc.;
- *E-Magnetics*: the main parameters of the analyzed electrical machines are provided here (see Figure 15), such as the electromagnetic power, torque, cogging torque, power factor, system efficiency, load angle, cogging period, base frequency, etc.;
- *Phasor Diagram*;
- *Equivalent Circuit*: shows all the parameters of the most complete equivalent circuit;
- *Flux densities* in different critical parts of the machine;
- *Losses* in all the parts of the machine;
- *Winding*: provides the parameters and current densities of all the conductors, phase resistance, slot filling factors, copper volume, etc.;
- *Miscellaneous*: data about the FEA mesh generated by the program;
- *Materials*: the properties and masses of all the used materials.

As can be seen in Figure 15 and Table 1, the designed surface-mounted PMSM achieved all its imposed rated values (see Section 2.2).

Table 1. The imposed rated values of the PMSM and the obtained ones via simulations.

Item	Rated Value	Resulted from Simulation
DC bus voltage	72 V	72 V (imposed)
Rated power	4 kW	4.0596 kW
Rated speed	8000 r/min	8000 r/min (imposed)
Rated average torque	4.77 N·m	4.82 N·m

The obtained power factor and efficiency of the PMSM are 0.975 and 95.36%, respectively (see Figure 15). Both are excellent values for such a machine having the given power. The 70.71 A phase current is under the 120 A maximum value of the power converter supplying the PMSM.

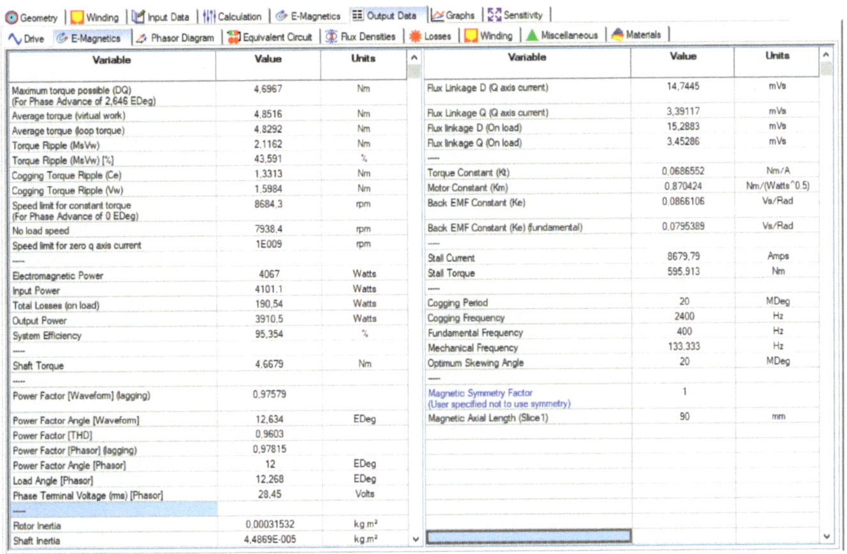

Variable	Value	Units
Maximum torque possible (DQ) (For Phase Advance of 2,646 EDeg)	4,6967	Nm
Average torque (virtual work)	4,8516	Nm
Average torque (loop torque)	4,8292	Nm
Torque Ripple (MsVw)	2,1162	Nm
Torque Ripple (MsVw) [%]	43,591	%
Cogging Torque Ripple (Ce)	1,3313	Nm
Cogging Torque Ripple (Vw)	1,5984	Nm
Speed limit for constant torque (For Phase Advance of 0 EDeg)	8684,3	rpm
No load speed	7938,4	rpm
Speed limit for zero q axis current	1E009	rpm

Electromagnetic Power	4067	Watts
Input Power	4101,1	Watts
Total Losses (on load)	190,54	Watts
Output Power	3910,5	Watts
System Efficiency	95,354	%

Shaft Torque	4,6679	Nm

Power Factor [Waveform] (lagging)	0,97579	
Power Factor Angle [Waveform]	12,634	EDeg
Power Factor [THD]	0,9603	
Power Factor [Phasor] (lagging)	0,97815	
Power Factor Angle [Phasor]	12	EDeg
Load Angle [Phasor]	12,268	EDeg
Phase Terminal Voltage (rms) [Phasor]	28,45	Volts

Rotor Inertia	0,00031532	kg.m²
Shaft Inertia	4,4869E-005	kg.m²

Variable	Value	Units
Flux Linkage D (Q axis current)	14,7445	mVs
Flux Linkage Q (Q axis current)	3,39117	mVs
Flux linkage D (On load)	15,2883	mVs
Flux linkage Q (On load)	3,45286	mVs

Torque Constant (Kt)	0,0686552	Nm/A
Motor Constant (Km)	0,870424	Nm/(Watts^0.5)
Back EMF Constant (Ke)	0,0866106	Vs/Rad
Back EMF Constant (Ke) (fundamental)	0,0795389	Vs/Rad

Stall Current	8679,79	Amps
Stall Torque	595,913	Nm

Cogging Period	20	MDeg
Cogging Frequency	2400	Hz
Fundamental Frequency	400	Hz
Mechanical Frequency	133,333	Hz
Optimum Skewing Angle	20	MDeg

Magnetic Symmetry Factor (User specified not to use symmetry)	1	
Magnetic Axial Length (Slice 1)	90	mm

Figure 15. The numerical simulation results provided in the *E-Magnetics* pane of the *ANSYS Motor-CAD* software.

The third category of results is displayed graphically. A great variety of plots can be visualized from the *Graphs* panel of the *ANSYS Motor-CAD* software. The most eloquent ones are given in Figures 16 and 17, where the phase currents, terminal voltages, back EMFs, electromagnetic torque, cogging torque (computed via two different methods), and airgap flux density are plotted versus the rotor position.

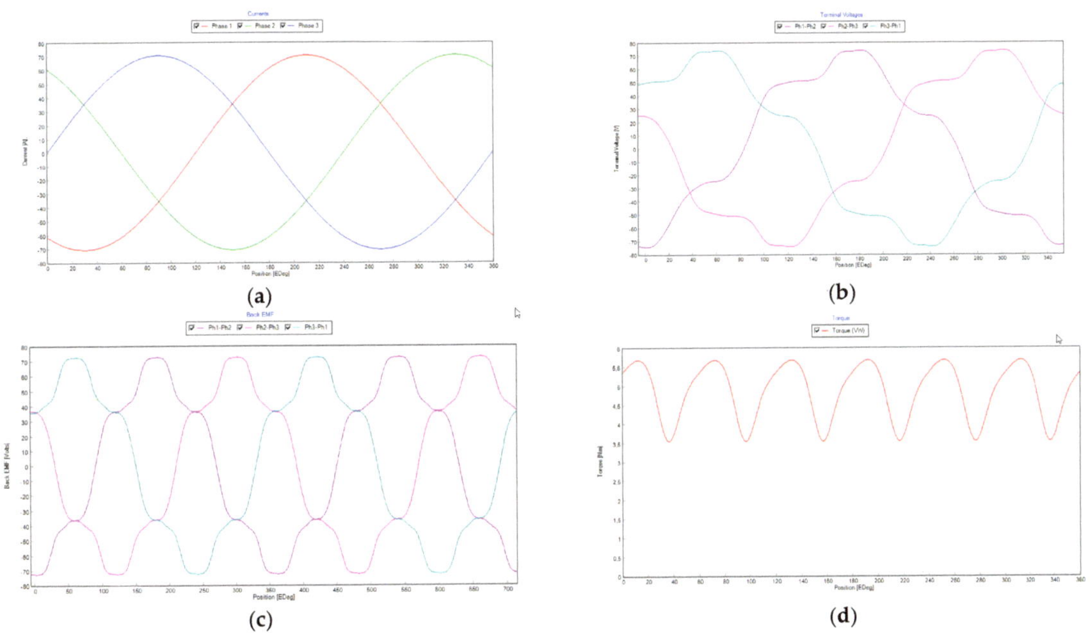

Figure 16. The variation against rotor angular position of the quantities of most interest. (**a**) Phase currents; (**b**) terminal voltages; (**c**) back EMF; (**d**) developed torque.

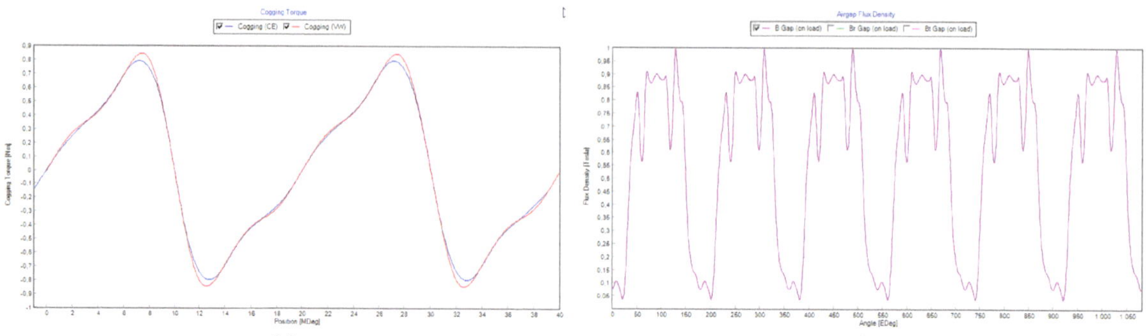

(a) **(b)**

Figure 17. The waveforms of the quantities of most interest. (**a**) Cogging torque; (**b**) airgap flux density.

As the phase currents were imposed during the simulations, they have a perfect sinusoidal shape, and their peak value is 70.71 A, as it was forced (see this in Figure 12). The maximum values of the terminal voltages do not exceed the 72 V DC bus voltage. The back EMF produced in the generator regime is not perfectly sinusoidal, as could be expected. Its harmonic content can be extracted from the *Harmonics* window of the *Graphs* panel (see Figure 18), where the great 5th harmonic content can be easily observed. The generated torque has a significant ripple (equal to 43.5% upon the result given in Figure 15). The cogging torque is small upon the result obtained via both computation methods, based on the co-energy (CE) and virtual works (VWs).

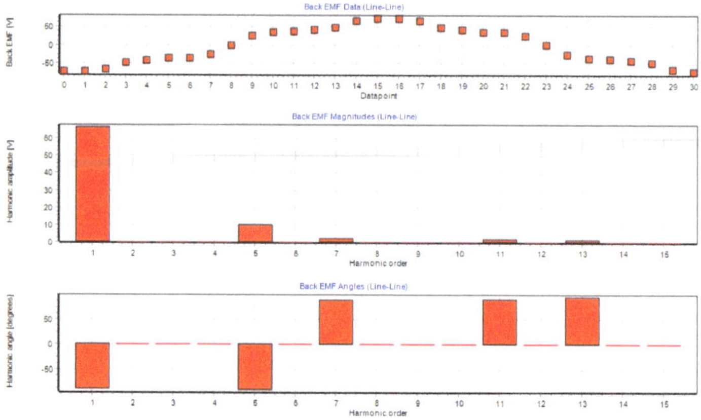

Figure 18. The harmonic content of the back EMF waveform obtained by using *ANSYS Motor-CAD* software.

There are some possibilities to manipulate the graphical results provided by the *ANSYS Motor-CAD* program from its *Graph Editor*. Various features of the plots can be set from here, and the final figure can be exported both in bitmap and metafile formats. The plotted numerical values arranged in tables can be visualized and exported in diverse formats for future processing in other programs.

The authors preferred to export the results obtained through simulations to *MATLAB* via the *XML* file format and to take advantage of the numerous graphical processing possibilities existing in this very popular computing platform [68].

5. Comparison of Four PMSM Types

By using the advanced simulation features of the *ANSYS Motor-CAD* software, the comparison of four typical PMSM topologies (with surface mount, interior, spoke-type, and V-shaped permanent magnets) was performed. All of them have the same imposed rated values (given in Section 2.2), stator (concerning both their iron core and windings), and airgap. The rotors (shown in Figure 19) have the same permanent magnet volume.

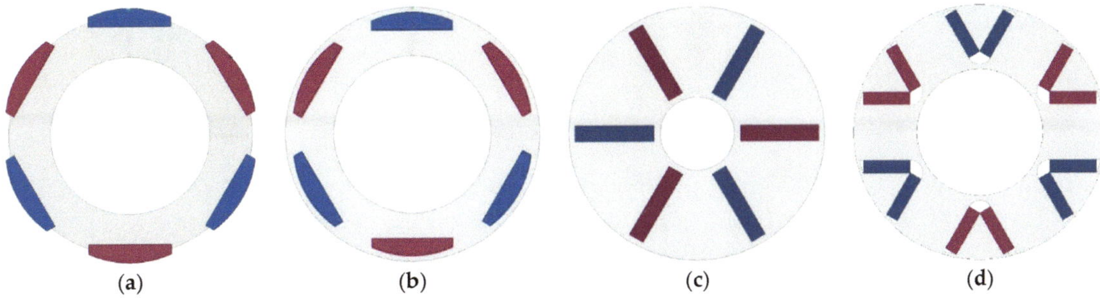

Figure 19. The four considered PMSM rotor topologies with different placements of the permanent magnets. (**a**) Surface mount; (**b**) interior; (**c**) spoke-type; (**d**) V-shaped.

5.1. Building up the Models of the Alternative PMSMs in ANSYS Motor-CAD

As in the *ANSYS Motor-CAD* software the size settings and the permanent magnet mass and volume computations can be performed very easily, the construction of the last three PMSM models was a very simple engineering task. The applied settings of the radial cross-section of the three PMSMs and the obtained structures are given in Figures 20 and 21. The settings of the axial cross-sections of the four PMSMs are identical.

(a) Slot Type: Tapered Slot; Stator Ducts: None; Rotor Type: Embedded Breadlc; Rotor Ducts: None

Stator Parameters	Value
Slot Number	18
Stator Lam Dia	118
Stator Bore	50
Slot Width (Bottom)	10,0809
Slot Width (Top)	5,9434
Slot Depth	24
Slot Corner Radius	1
Tooth Tip Depth	1
Slot Opening	2
Tooth Tip Angle	0
Sleeve Thickness	0

Rotor Parameters	Value
Pole Number	6
Notch Depth	0
Magnet Thickness	3,43
Magnet Embed Depth	0,5
Magnet Arc [ED]	113,4
Magnet Segments	1
Airgap	0,65
Banding Thickness	0
Shaft Dia	25
Shaft Hole Diameter	0

(b) Slot Type: Tapered Slot; Stator Ducts: None; Rotor Type: Spoke; Rotor Ducts: None

Stator Parameters	Value
Slot Number	18
Stator Lam Dia	118
Stator Bore	50
Slot Width (Bottom)	10,0809
Slot Width (Top)	5,9434
Slot Depth	24
Slot Corner Radius	1
Tooth Tip Depth	1
Slot Opening	2
Tooth Tip Angle	0
Sleeve Thickness	0

Rotor Parameters	Value
Pole Number	6
Rotor Lamination Inset	0
Magnet Thickness	3,25
Magnet Depth	14,3
Magnet Opening	0,1
Magnet Inset	1,2
Magnet Segments	1
Airgap	0,65
Banding Thickness	0
Shaft Dia	15
Shaft Hole Diameter	0

(c) Slot Type: Tapered Slot; Stator Ducts: None; Rotor Type: Interior V (simple); Rotor Ducts: None

Stator Parameters	Value
Slot Number	18
Stator Lam Dia	118
Stator Bore	50
Slot Width (Bottom)	10,0809
Slot Width (Top)	5,9434
Slot Depth	24
Slot Corner Radius	1
Tooth Tip Depth	1
Slot Opening	2
Tooth Tip Angle	0
Sleeve Thickness	0

Rotor Parameters	Value
Pole Number	6
Notch Depth	0
Magnet Layers	1
L1 Magnet Thickness	2,5
L1 Magnet V Width	9,29
L1 Magnet Shift	0
L1 Magnet Bar Width	9,29
L1 Bridge Thickness	0,7
L1 Pole V Angle	60
L1 Outer Extension	0,1
L1 Inner Extension	0,1
L1 Magnet Post	0,5
L1 Magnet Segments	1
L1 Mag Gap Inner	0
L1 Mag Gap Outer	0
L1 Layer Offset Angle	0
Airgap	0,65
Banding Thickness	0
Shaft Dia	25
Shaft Hole Diameter	0

Figure 20. The settings of the radial cross-section of the considered PMSMs with the different alternative permanent magnet arrangements. (**a**) Interior; (**b**) spoke-type; (**c**) V-shaped.

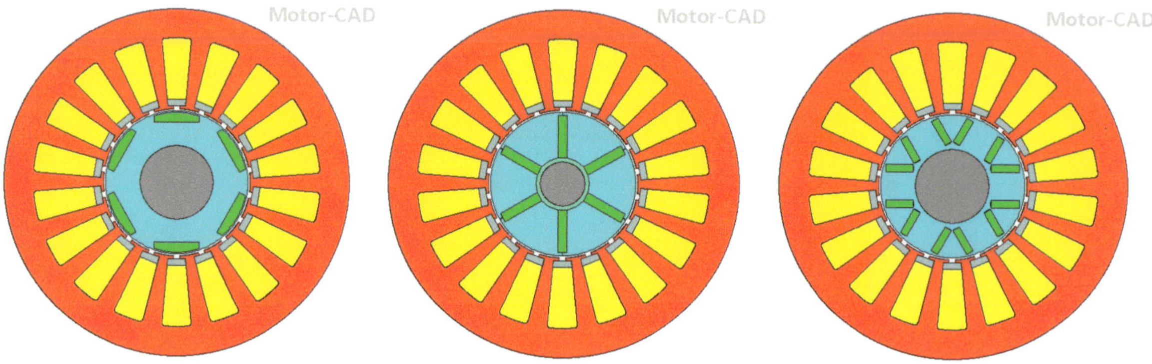

(a) (b) (c)

Figure 21. The structures generated by *ANSYS Motor-CAD* of the alternative considered PMSMs having diverse permanent magnet placements. (**a**) Interior; (**b**) spoke-type; (**c**) V-shaped.

5.2. The Comparison of the Obtained Results

The simulation in *ANSYS Motor-CAD* of the four PMSMs was performed at identical settings (see them at the beginning of Section 4).

The extraction of the multitude of results after the performed simulations enabled the comparison of the four PMSM topologies from different points of view. First, the total masses and that of each main component can be compared. In Table 2 the masses of the main machine components (stator and rotor iron cores, permanent magnets, windings, and the shaft) are given. By adding these, the total masses of the PMSMs without the housings and machine auxiliaries can be obtained.

Table 2. The masses of the four PMSM components and their total masses.

Mass [kg]	Permanent Magnet Topology			
	Surface	Interior	Spoke-Type	V-Shape
Stator core	3.736	3.736	3.736	3.736
Rotor core	0.6537	0.7301	0.9116	0.7274
Permanent magnets	0.1881	0.1881	0.1882	0.1881
Windings	1.846	1.846	1.846	1.846
Shaft	0.5743	0.5743	0.5743	0.5743
Total mass	6.9981	7.0745	7.2561	7.0718

Since the stators of the four PMSMs are identical, and the permanent magnets are of the same volume, the rotor iron core mass makes the difference. In any case, the four total masses are very close, and the least is that of the surface-mounted PMSM.

In the above table and in all which follows, the best values are highlighted in green, while the worst are in red.

The assessment of the main output characteristics is essential in such a comparative study. Therefore, eight key features of the PMSMs in the study (output power, mean torque, torque ripple, peak-to-peak value of the cogging torque, efficiency, power factor, power density, and torque density) were included in Table 3.

As can be seen, the surface-mounted PMSM has the best characteristics in almost all the considered categories, while that having V-shape permanent magnets has the worst ones. It can be concluded that under the given conditions (identical stators, external and internal dimensions of the rotor iron cores, and permanent magnet masses), the topologies considered to be compared with the surface permanent magnet variant cannot reach the imposed rated speed (8,000 r/min), nor the rated output power (4 kW), nor the rated torque (4.77 N·m).

Table 3. The main characteristics of the four PMSMs.

Characteristics	Permanent Magnet Topology			
	Surface	**Interior**	**Spoke-Type**	**V-Shape**
Output power [W]	4067	3516.6	3520.9	3129.4
Mean torque [N·m]	4.851	4.192	4.193	3.731
Torque ripple [%]	43.87	83.5	56.16	97.78
Cogging torque peak-to-peak [N·m]	1.58	2.56	1.28	4.16
Efficiency [%]	95.354	92.912	94.081	92.289
Power factor [–]	0.975	0.851	0.89	0.853
Power density [W/kg]	581.15	497.08	485.23	442.52
Torque density [N·m/kg]	0.69	0.59	0.58	0.52

Next, the variation against the angular displacement of some basic characteristics of the four considered PMSM topologies will be studied.

In Figure 22, the variation of the generated torque in all four cases is provided. For a better comparison of the PMSMs, the main torque features (the minimum, maximum, and mean values, and the torque ripple) are given in Table 4.

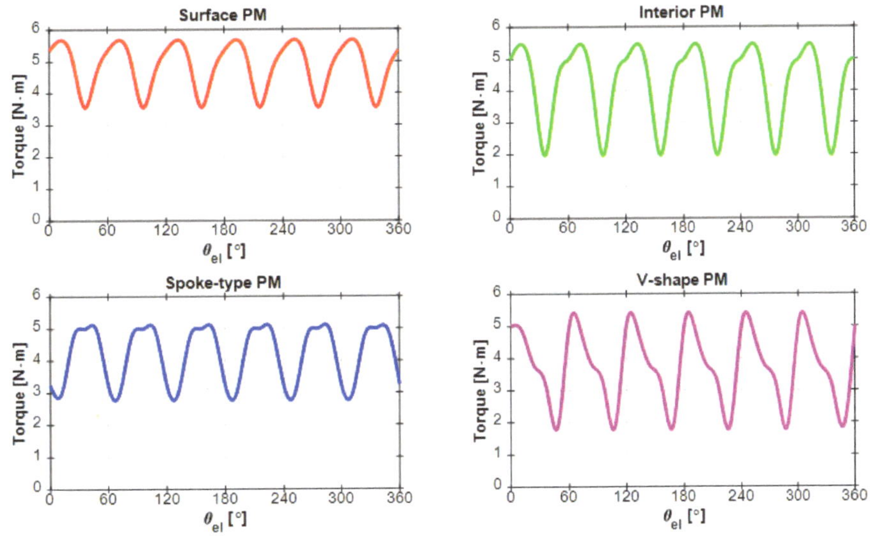

Figure 22. The torque versus angular displacement plots for the four PMSMs.

Table 4. The main torque features of the considered PMSMs.

Torque Features	Permanent Magnet Topology			
	Surface	**Interior**	**Spoke-Type**	**V-Shape**
T_{min} [N·m]	3.55	1.97	2.76	1.77
T_{max} [N·m]	5.68	5.47	5.12	5.41
T_{mean} [N·m]	4.85	4.19	4.19	3.73
T_{ripple} [%]	43.87	83.5	56.16	97.78

As is evident, the surface-mounted PMSM has the best torque development capability, due to the very close placement of the permanent magnets to the airgap of the machine. Meanwhile, the PMSM having V-shaped permanent magnets is the worst topology, also from this point of view.

In the case of all permanent magnet machines, the cogging torque is also a significant characteristic. Therefore, the variations of this physical quantity versus the angular displacement are shown in Figure 23.

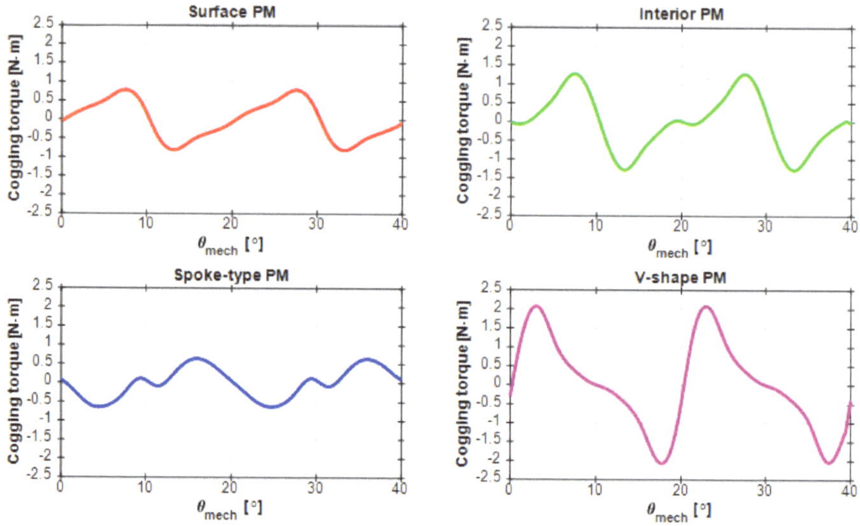

Figure 23. The cogging torque plotted against the angular displacement for the four PMSMs.

The peak-to-peak values of the cogging torque of the four PMSMs are significant comparative issues (consider Table 3). As can be seen, the lowest cogging torque is developed by the spoke-type PMSM because in this topology the permanent magnets are deeply buried into the rotor iron core far from the rotor external surface and airgap. In the case of the V-type PMSM, the permanent magnets strongly concentrate the magnetic flux into the airgap, and consequentially also cause great cogging torques.

As is evidenced, again the surface-mounted PMSM has very good behavior based on this consideration, as well.

PMSMs are often used also in generator regime; therefore, the shapes of the produced EMF (given in Figure 24) are important issues in a such comparative study.

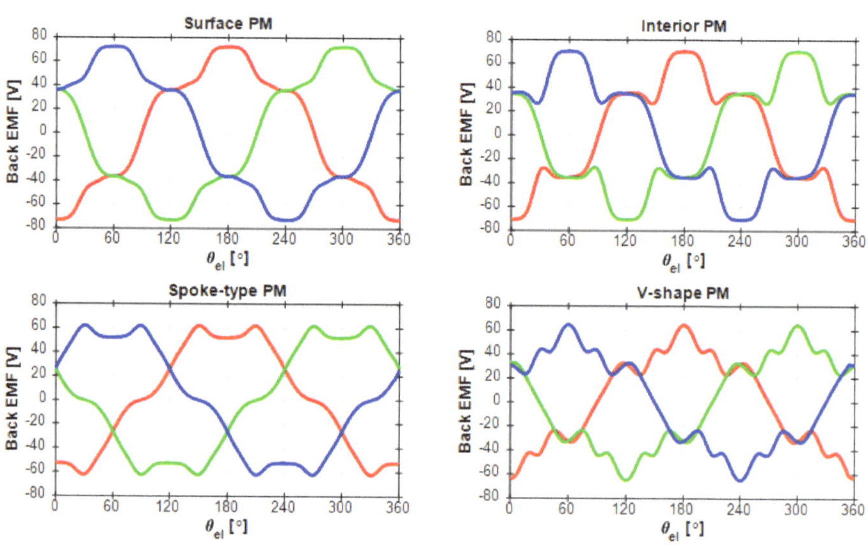

Figure 24. The back EMF generated in the case of the four PMSMs.

As could be expected and can be observed in Figure 24, the generated waveforms are far from the ideal sinusoidal shapes. Hence, their harmonic content is of major concern.

These characteristics are again easy to extract from the results provided by the *ANSYS Motor-CAD* software. The results of the harmonic analysis were plotted by using the graphical potential of the *MATLAB* program as in the previous cases. The obtained results are given in Figure 25. In such cases, the total harmonic distortion (THD) of the waveforms is also of important relevance; therefore, their values were written on each plot from the figure.

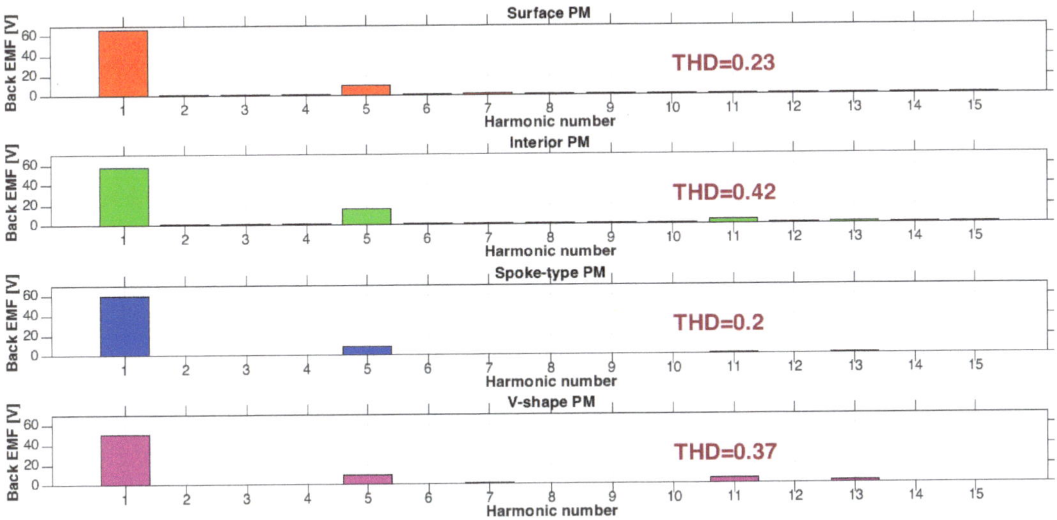

Figure 25. The harmonic contents of the back EMF waveforms given in Figure 24.

The PMSM having spoke-type permanent magnets has the lowest harmonic content, but its THD value (0.2) is only a bit smaller than the THD obtained for the surface PMSM. These results were expectable, since it was proved in numerous papers that these two types of PMSM can generate the closest EMF waveform to the perfect sinusoidal ones [69].

6. Discussion

PMSMs are essential components of modern vehicles, being used both for traction purposes in electrical and hybrid cars and for actuating several car auxiliaries. Therefore, their design is an important issue and requires advanced software tools. Such a program, used by the authors in designing and analyzing PMSMs, is the *ANSYS Motor-CAD* software.

The simulations performed by using it enabled the authors to check the pre-sizing of the designed PMSM and to verify if all the imposed designed data were fulfilled. By applying the multiple PMSM structure templates offered by the *ANSYS Motor-CAD* program, it was possible to easily perform an in-detail comparison of four rotor structures for the identical stator of the PMSM and to define the best structure among them.

ANSYS Motor-CAD can also be useful in the future for improving the structure of the proposed PMSM by assisting in an effective optimization process.

7. Conclusions

By using the *ANSYS Motor-CAD* program in our concrete design work, we concluded that it was an effective tool in our hands. By using the built-in templates, the models of all the PMSMs taken into the study were easily achieved. Hence, the consistency of the pre-computed sizes of the machines was easily checked, and the CAD drawings of the main components could be exported for further studies or manufacturing.

The program has a huge database of materials, where the designers probably can find exactly the materials proposed during the pre-sizing of the machine. If not, they can create their own material databases even for very specific materials.

The designed winding diagram could be implemented in *ANSYS Motor-CAD* straightforwardly by setting only a few parameters. The program can display diverse diagrams of the modeled winding, also including the ideal displacement of the wires in the slots. This is one of the most spectacular features of the program since these computations are complicated due to the irregular shape of the slots. By the graphical illustration of the slot with the inserted wires and insulators, the users can easily decide on the correctness of the design and can have information about the actual filling factor of the winding.

The drive systems of the simulated electrical machines can be easily configured by some simple selections, avoiding the use of complicated-to-build external circuits with adequate commands.

By using both analytical computations and FEA, *ANSYS Motor-CAD* can deliver a great variety of results (both in graphical forms and as numerical values): the magnetic flux distribution in the machine, the flux densities in diverse areas, the masses of all distinct components, etc. Numerous output parameters (such as developed torque, cogging torque, power, efficiency, power factor, etc.) can be easily visualized and thus checked by the designing engineers.

If needed, the correction of the design can be made rapidly until all the required machine parameters are achieved.

During working with this software product, we also found some limitations of it. As it enables only a template-based model construction, electrical machines with special topologies (for example, those having modular construction or transverse flux machines), not included in the offered templates, cannot be studied. Despite the valuable simulation results obtained by using the default FEA settings, such a program should enable more possibilities for setting the mesh and the used solver.

In overall conclusion concerning the used software tool: *ANSYS Motor-CAD* was very worthwhile in our design work. By applying it, we could rapidly check our designs and perform needed changes to improve them.

It can be also stated that all the simulated results showed that the surface-mounted PMSM intended to drive the scroll-type compressor of a car HVAC system meets all the set design expectations. Moreover, the comparison study's findings all underscored the advantages of this PMSM variant over the other three topologies that were taken into consideration.

Author Contributions: Conceptualization, L.S.; PMSM pre-sizing, I.V.; PMSM modeling, F.-A.C. All authors have read and agreed to the published version of the manuscript.

Funding: This research received no external funding.

Institutional Review Board Statement: Not applicable.

Informed Consent Statement: Not applicable.

Data Availability Statement: Not applicable.

Conflicts of Interest: The authors declare no conflict of interest.

References

1. Rosu, M.; Zhou, P.; Lin, D.; Ionel, D.M.; Popescu, M.; Blaabjerg, F.; Rallabandi, V.; Staton, D. *Multiphysics Simulation by Design for Electrical Machines, Power Electronics and Drives*; John Wiley & Sons: Hoboken, NJ, USA, 2017.
2. Lei, G.; Zhu, J.; Guo, Y.; Liu, C.; Ma, B. A review of design optimization methods for electrical machines. *Energies* **2017**, *10*, 1962. [CrossRef]
3. Liu, X.; Lin, Q.; Fu, W. Optimal design of permanent magnet arrangement in synchronous motors. *Energies* **2017**, *10*, 1700. [CrossRef]
4. Cepoi, R.D.; Jaşcău, F.F.; Szabó, L. Current trends in energy efficient electrical machines. *J. Electr. Electron. Eng.* **2017**, *10*, 13–18.
5. Allam, Z.; Bibri, S.E.; Sharpe, S.A. The Rising Impacts of the COVID-19 Pandemic and the Russia–Ukraine War: Energy Transition, Climate Justice, Global Inequality, and Supply Chain Disruption. *Resources* **2022**, *11*, 99. [CrossRef]
6. Bobba, S.; Carrara, S.; Huisman, J.; Mathieux, F.; Pavel, C. *Critical Raw Materials for Strategic Technologies and Sectors in the EU. A Foresight Study*; European Union: Brussels, Belgium, 2020.

7. Hameyer, K.; Belmans, R. *Numerical Modelling and Design of Electrical Machines and Devices*; WIT Press: Ashurst Lodge, UK, 1999.
8. Salon, S.J. *Finite Element Analysis of Electrical Machines*; Kluwer Academic Publishers: Boston, MA, USA, 1995.
9. JMAG-Designer. *User's Manual Solver*, 19th ed.; JSOL Corporation: Tokyo, Japan, 2020.
10. Jun, S.-B.; Kim, C.-H.; Cha, J.; Lee, J.H.; Kim, Y.-J.; Jung, S.-Y. A novel method for establishing an efficiency map of IPMSMs for EV propulsion based on the finite-element method and a neural network. *Electronics* **2021**, *10*, 1049. [CrossRef]
11. Gecer, B.; Tosun, O.; Apaydin, H.; Serteller, N.F.O. Comparative Analysis of SRM, BLDC and Induction Motor Using ANSYS/Maxwell. In Proceedings of the International Conference on Electrical, Computer, Communications and Mechatronics Engineering (ICECCME '2021), Mauritius, 7–8 October 2021. [CrossRef]
12. Aishwarya, M.; Brisilla, R. Design of Energy-Efficient Induction motor using ANSYS software. *Results Eng.* **2022**, *16*, 100616. [CrossRef]
13. Varvolik, V.; Prystupa, D.; Buticchi, G.; Peresada, S.; Galea, M.; Bozhko, S. Co-simulation analysis for performance prediction of synchronous reluctance drives. *Electronics* **2021**, *10*, 2154. [CrossRef]
14. Dobzhanskyi, O.; Grebenikov, V.; Gouws, R.; Gamaliia, R.; Hossain, E. Comparative thermal and demagnetization analysis of the PM machines with neodymium and ferrite magnets. *Energies* **2022**, *15*, 4484. [CrossRef]
15. Falkowski, K.; Kurnyta-Mazurek, P.; Szolc, T.; Henzel, M. Radial magnetic bearings for rotor–shaft support in electric jet engine. *Energies* **2022**, *15*, 3339. [CrossRef]
16. Lukočius, R.; Vilkauskas, A.; Marčiulionis, P.; Grigaliūnas, V.; Nakutis, Ž.; Deltuva, R. An analysis of axial magnetic coupling force and torque dependencies on its structure parameters using a 3D FEM. *Appl. Sci.* **2022**, *12*, 6546. [CrossRef]
17. Uberti, F.; Frosini, L.; Szabó, L. A new design procedure for rotor laminations of synchronous reluctance machines with fluid shaped barriers. *Electronics* **2022**, *11*, 134. [CrossRef]
18. Ullah, A.; Reja, M.I.; Mirza, F.; Ahmed, H. Design of permanent magnet synchronous machines of direct-driven generator with Emetor and FLUX 2D. In Proceedings of the IEEE Business, Engineering & Industrial Applications Colloquium (BEIAC '2012), Kuala Lumpur, Malaysia, 7–8 April 2012; pp. 74–79. [CrossRef]
19. Someşan, L.; Pădurariu, E.; Viorel, I.A.; Szabó, L. Design of a permanent magnet flux-switching machine. In Proceedings of the 9th International Conference ELEKTRO '2012, Žilina–Rajecké Teplice, Slovakia, 21–22 May 2012; pp. 256–259. [CrossRef]
20. Hernández, J.A.D.; Carralero, N.D.; Vázquez, E.G. A 3-D Simulation of a Single-Sided Linear Induction Motor with Transverse and Longitudinal Magnetic Flux. *Appl. Sci.* **2020**, *10*, 7004. [CrossRef]
21. Islam, M.S.; Agoro, S.; Chattopadhyay, R.; Husain, I. Heavy rare earth free high power density traction machine for electric vehicles. In Proceedings of the IEEE International Electric Machines & Drives Conference (IEMDC '2021), Hartford, CT, USA, 17–20 May 2021. [CrossRef]
22. Paiva, L.T.; Fontes, F.A.C.C. Optimal control algorithms with adaptive time-mesh refinement for kite power systems. *Energies* **2018**, *11*, 475. [CrossRef]
23. Lo, S. Finite element mesh generation and adaptive meshing. *Prog. Struct. Eng. Mater.* **2002**, *4*, 381–399. [CrossRef]
24. Okereke, M.; Keates, S. *Finite Element Applications. A Practical Guide to the FEM Process*; Springer International Publishing AG: Cham, Switzerland, 2018.
25. Li, H. The finite element method. In *Graded Finite Element Methods for Elliptic Problems in Nonsmooth Domains. Surveys and Tutorials in the Applied Mathematical Sciences*; Springer: Cham, Switzerland, 2022; Volume 10, pp. 1–12.
26. Mersha, T.K.; Du, C. Co-simulation and modeling of PMSM based on ANSYS software and Simulink for EVs. *World Electr. Veh. J.* **2021**, *13*, 4. [CrossRef]
27. Filipenko, M.; Biser, S.; Boll, M.; Corduan, M.; Noe, M.; Rostek, P. Comparative analysis and optimization of technical and weight parameters of turbo-electric propulsion systems. *Aerospace* **2020**, *7*, 107. [CrossRef]
28. Ibrahim, I.; Silva, R.; Lowther, D.A. Application of surrogate models to the multiphysics sizing of permanent magnet synchronous motors. *IEEE Trans. Magn.* **2022**, *58*, 7401604. [CrossRef]
29. Boldea, I.; Tutelea, L.N. *Electric Machines: Steady State, Transients, and Design with MATLAB®*; CRC Press: Boca Raton, FL, USA, 2010.
30. Motor-CAD Software. Available online: https://www.motor-design.com/motor-cad/ (accessed on 15 November 2022).
31. Staton, D.; Goss, J. *Open Source Electric Motor Models for Commercial EV & Hybrid Traction Motors*; Motor Design Ltd.: Ellesmere, UK, 2017.
32. Qi, J.; Hua, W.; Zhang, H. Thermal analysis of modular-spoke-type permanent-magnet machines based on thermal network and FEA method. *IEEE Trans. Magn.* **2019**, *55*, 8104105. [CrossRef]
33. Staton, D. Servo motor size reduction—Need for thermal CAD. In Proceedings of the Drives & Controls Conference (ExCeL'2001), London, UK, 13–15 March 2001; pp. 13–15.
34. Staton, D.; Hawkins, D.; Popescu, M. Motor-CAD software for thermal analysis of electrical motors–Links to electromagnetic and drive simulation models. In Proceedings of the Coil Winding, Insulation & Electrical Manufacturing Exhibitions Conference (CWIEME '2010), Berlin, Germany, 5 January 2010.
35. Jie, D. Thermal performance analysis of motor based on Motor-CAD. In Proceedings of the 3rd International Conference on Advances in Materials, Machinery, Electronics, AMME, 2019, Wuhan, China, 19–20 January 2019; AIP Conference Proceedings. Volume 2073, p. 5090703.

36. Deaconu, D.I.; Ghiță, C.; Chirilă, A.I.; Năvrăpescu, V.; Popescu, M. Thermal study of induction machine using Motor-CAD. In Proceedings of the 3rd International Symposium on Electrical and Electronics Engineering (ISEEE '2010), Galați, Romania, 16–18 September 2010; pp. 23–27. [CrossRef]
37. Staton, D.; Boglietti, A.; Cavagnino, A. Solving the more difficult aspects of electric motor thermal analysis. In Proceedings of the IEEE International Electric Machines and Drives Conference (IEMDC '03), Madison, WI, USA, 1–4 June 2003; pp. 747–755. [CrossRef]
38. Bramerdorfer, G.; Tapia, J.A.; Pyrhönen, J.J.; Cavagnino, A. Modern electrical machine design optimization: Techniques, trends, and best practices. *IEEE Trans. Ind. Electron.* **2018**, *65*, 7672–7684. [CrossRef]
39. Boglietti, A.; Cavagnino, A.; Staton, D.; Shanel, M.; Mueller, M.; Mejuto, C. Evolution and modern approaches for thermal analysis of electrical machines. *IEEE Trans. Ind. Electron.* **2009**, *56*, 871–882. [CrossRef]
40. Rivière, N.; Villani, M.; Popescu, M. Optimisation of a high speed copper rotor induction motor for a traction application. In Proceedings of the 45th Annual Conference of the IEEE Industrial Electronics Society (IECON '2019), Lisbon, Portugal, 14–17 October 2019; pp. 2720–2725. [CrossRef]
41. Salah, A.A.; Guo, Y.; Dorrell, D.G. Predicting the behavior of induction machine using motor-CAD and MATLAB packages. In Proceedings of the 12th International Conference on Compatibility, Power Electronics and Power Engineering (CPE-POWERENG '2018), Doha, Qatar, 10–12 April 2018. [CrossRef]
42. Lyshevski, S.E. *Electromechanical Systems, Electric Machines, and Applied Mechatronics*; CRC Press: Boca Raton, FL, USA, 2018.
43. Matković, K.; Splechtna, R.; Gračanin, D.; Todorović, G.; Goja, S.; Bedić, B.; Hauser, H. Getting insight into noise, vibration, and harshness simulation data. In Proceedings of the Winter Simulation Conference (WSC '2021), Phoenix, AZ, USA, 18–24 July 2021. [CrossRef]
44. Daly, S. *Automotive Air Conditioning and Climate Control Systems*; Elsevier: Amsterdam, The Netherlands, 2006.
45. Wagner, J.; VanGelder, K. *Automotive Heating, Ventilation, and Air Conditioning*; Jones & Bartlett Learning: Burlington, VT, USA, 2019.
46. Li, W.; He, Y. Parameterization and performance of permanent magnet synchronous motor for vehicle based on Motor-CAD and OptiSLang. *Wirel. Commun. Mob. Comput.* **2022**, *2022*, 1–12. [CrossRef]
47. Li, X.X.; Liu, Z.S. Optimization design and analysis of permanent magnet synchronous generator based on Motor-CAD. *J. Phys. Conf. Ser.* **2022**, *2345*, 012009. [CrossRef]
48. Chin, Y.; Soulard, J. A permanent magnet synchronous motor for traction applications of electric vehicles. In Proceedings of the International Electric Machines and Drives Conference (IEMDC '2003), Madison, WI, USA, 1–4 June 2003; pp. 1035–1041. [CrossRef]
49. Hirano, T.; Matsumura, N.; Takeda, K. Development of high efficiency scroll compressors, for air conditioners. In Proceedings of the 1988 International Compressor Engineering Conference, West Lafayette, IN, USA, 18–21 July 1988; p. 603.
50. Aurich, J.; Baumgart, R. Comparison and evaluation of different A/C compressor concepts for electric vehicles. In Proceedings of the 24th International Compressor Engineering Conference, West Lafayette, IN, USA, 9–12 July 2018; p. 2608.
51. Cardone, M.; Gargiulo, B. Numerical simulation and experimental validation of an oil free scroll compressor. *Energies* **2020**, *13*, 5863. [CrossRef]
52. Everything You Need to Know About Scroll Pumps. Available online: https://www.vacuumscienceworld.com/blog/everything-you-need-to-know-about-scroll-pumps (accessed on 1 December 2022).
53. Cosman, S.-I.; Văscan, I.; Bilațiu, C.-A.; Tintelecan, A.; Marțiș, C.S.; Zorlescu, B. Design and analysis of a heating, ventilation and air conditioning system for electric vehicles. In Proceedings of the International Symposium on Power Electronics, Electrical Drives, Automation and Motion (SPEEDAM '2020), Sorrento, Italy, 24–26 June 2020; pp. 305–310. [CrossRef]
54. Fodorean, D.; Popa, D.; Minciunescu, P.; Irimia, C.; Szabó, L. Study of a high-speed motorization for electric vehicle based on PMSM, IM and VRSM. In Proceedings of the 21st International Conference on Electrical Machines (ICEM '2014), Berlin, Germany, 2–5 September 2014; pp. 2577–2582. [CrossRef]
55. Goli, C.S.; Manjrekar, M.; Essakiappan, S.; Sahu, P.; Shah, N. Landscaping and review of traction motors for electric vehicle applications. In Proceedings of the Transportation Electrification Conference & Expo (ITEC '2021), Chicago, IL, USA, 21–25 June 2021; pp. 162–168.
56. Adamas Intelligence. *Global PMSM Market Share Continues to Rise Despite Soaring Rare Earth Prices*; Adamas Intelligence: Toronto, Canada, 2021.
57. Naphade, A. *Permanent Magnet Synchronous Motor (PMSM) Market Report, History and Forecast Breakdown Data by Types and Application till 2028*; Skyline Market Research LLP: Kalyani Nagar, India, 2022.
58. Văscan, I.; Szabó, L. Improvement trends in the development of permanent magnet synchronous machines for automotive applications. In Proceedings of the 23rd International Carpathian Control Conference (ICCC '2022), Sinaia, Romania, 29 May–1 June 2022; pp. 398–403. [CrossRef]
59. Pyrhonen, J.; Jokinen, T.; Hrabovcova, V. *Design of Rotating Electrical Machines*; John Wiley & Sons: Chichester, UK, 2013.
60. Candelo-Zuluaga, C.; Espinosa, A.G.; Riba, J.-R.; Blanch, P.T. PMSM design for achieving a target torque-speed-efficiency map. *IEEE Trans. Veh. Technol.* **2020**, *69*, 14448–14457. [CrossRef]
61. Verbelen, F.; Abdallh, A.; Vansompel, H.; Stockman, K.; Sergeant, P. Sizing methodology based on scaling laws for a permanent magnet electrical variable transmission. *IEEE Trans. Ind. Electron.* **2019**, *67*, 1739–1749. [CrossRef]

62. Rechkemmer, S.K.; Izumi, R.; Zhang, W.; Sawodny, O. Estimation of permanent magnet synchronous machine performance for pre-design using a reluctance network. *IFAC-Pap.* **2018**, *51*, 75–80. [CrossRef]
63. Shen, Q.; Zhou, Z.; Li, S.; Liao, X.; Wang, T.; He, X.; Zhang, J. Design and analysis of the high-speed permanent magnet motors: A review on the state of the art. *Machines* **2022**, *10*, 549. [CrossRef]
64. Lipo, T.A. *Introduction to AC Machine Design*; John Wiley & Sons: Hoboken, NJ, USA, 2017.
65. Kou, S.; Kou, Z.; Wu, J.; Wang, Y. Modeling and simulation of a novel low-speed high-torque permanent magnet synchronous motor with asymmetric stator slots. *Machines* **2022**, *10*, 1143. [CrossRef]
66. *Electrical Steel Sheets—JFE G-CORE, JFE N-CORE*; JFE Steel Corporation: Tokyo, Japan, 2016.
67. *N35UH Sintered Neodymium-Iron-Boron Magnets*; Arnold Magnetic Technologies Corp.: Rochester, NY, USA, 2014.
68. Majumdar, N.; Banerjee, S. *MATLAB Graphics and Data Visualization Cookbook*; Packt Publishing: Birmingham, UK, 2012.
69. Chen, A.; Nilssen, R.; Nysveen, A. Harmonic analysis and comparison of the back EMFs of four permanent magnet machine with different winding arrangements. In Proceedings of the 2008 International Conference on Electrical Machines and Systems (ICEMS '2008), Wuhan, China, 17–20 October 2008; pp. 3043–3048.

energies

Article

Nonlinear Predictive Control of Interior Permanent Magnet Synchronous Machine with Extra Current Constraint

Mengmeng Tian [1,*], Hailiang Cai [1], Wenliang Zhao [2] and Jie Ren [2]

[1] School of Electrical Engineering, University of Jinan, Jinan 250024, China
[2] School of Electrical Engineering, Shandong University, Jinan 250061, China
* Correspondence: cse_tianmm@ujn.edu.cn

Abstract: The interior permanent magnet synchronous machine (IPMSM) has been widely used in industrial applications due to its several favorable advantages. To further improve the machine performance, an improved nonlinear predictive controller for the IPMSM is proposed. In this paper, the maximum torque per ampere control law is firstly transformed to a linear function, according to the first−order Taylor expansion, and integrated with the control strategy. On this basis, an improved predictive control method is formulated by designing an optimized cost function through the input−output feedback linearization. Then the integral action is introduced to eliminate the influence of the load mutation and improve the steady−state control precision of the system. The stability of the control method is ensured by compelling the outputs to track the desired references without steady−state error. Finally, the simulation was established to verify the effective of the improved control method. Simulation results showed that the machine can reach the given reference speed without steady−state error within a short process, which means the machine has excellent dynamic and static performances. Furthermore, the machine has higher torque−to−current ratio by making full use of the reluctance torque. The simulation results verify the effectiveness of the improved control strategy.

Keywords: input−output feedback linearization; interior permanent magnet synchronous machine; nonlinear predictive control; speed control

Citation: Tian, M.; Cai, H.; Zhao, W.; Ren, J. Nonlinear Predictive Control of Interior Permanent Magnet Synchronous Machine with Extra Current Constraint. *Energies* **2023**, *16*, 716. https://doi.org/10.3390/en16020716

Academic Editors: Loránd Szabó and Feng Chai

Received: 7 December 2022
Revised: 30 December 2022
Accepted: 4 January 2023
Published: 7 January 2023

1. Introduction

Currently, the permanent magnet synchronous machine (PMSM) is widely used in industrial applications due to its favorable advantages, such as high power density, high efficiency and good controllable performance [1–3]. On some occasions, the PMSM is forced to track the desired speed, whose performance is influenced by the control policy. To increase the machine performance, many control methods have been proposed, including proportional control [4,5], sliding mode control (SMC) [6,7] and predictive control [8,9].

PI control is a classic control strategy providing such advantages as good robustness and high steady−state control precision, whereas its dynamic performance is influenced by the controller parameters. The PMSM is a complex nonlinear system operating in a wide speed range with inevitable disturbances and parameter uncertainties. Hence, a PI control with fixed regulation parameters cannot achieve a satisfying dynamic performance during the entire running process. The SMC has been widely used in some practical applications because of its high response speed and powerful ability to eliminate some disturbances. In [7], a compound control method using improved non−singular fast terminal SMC and disturbance observer compensation techniques was developed. In [10], the nonlinear control strategy for PMSM was proposed by adopting a fast terminal SMC based on the finite time sliding mode observer. In [11], an improved SMC was proposed to promote the drive performance of PMSM. Comprehensive simulation verified that the proposed controller was strongly robust to acute variations of load and machine parameters. In [12], a terminal SMC based on nonlinear disturbance observer is proposed to realize the speed and the current tracking control for the PMSM drive system. However, it is difficult for the SMC to overcome the influence of certain parameter uncertainties not satisfying the matching condition.

The model predictive control (MPC) is an optimization–based control strategy with strong robustness. It can predict future output through the input–output feedback linearization. On this basis, a cost function to evaluate the difference between the predicted output and the trajectory to be tracked is designed, whose optimal solution serves as the input of the system. The PMSM is a nonlinear multivariable system with fast dynamic, and it is sensitive to the disturbances and parameter mismatches. Hence, the discrete time linear MPC in [13] suffers from several limitations, such as the unavoidable steady–state error and high computational effort.

In order to improve the control performance, some improved MPC methods have been put forward [14–16]. In [17], a cascade MPC structure including the inner current loop and outer speed loop was developed. Typically, the speed loop MPC is embedded with different disturbance frequency modes to reduce the effect of periodic disturbance. In [18], the model of predictive direct current control based on an optimized cost function is proposed, which can achieve fast dynamic in transient state and reduce the converter losses in steady state. In [19], a cascade–free modulated predictive direct speed control scheme for PMSM drives is presented to improve the machine steady–state performance. In [20], a novel MPC mechanism referred as the Runge–Kutta MPC has been applied for speed control of a commercial PMSM. In [21], an improved MPC method is incorporated into the control design of the speed loop of PMSM, in which a compensated scheme with an extended sliding mode observer is added. In [22], a current MPC is designed for a PMSM where the speed of the motor can be regulated precisely.

The disturbance observer has been proven to be effective in estimating the bounded disturbance [23], and some papers have combined the MPC with the disturbance observer to suppress the effects of external perturbation. In [24], a composite control method combining the deadbeat predictive control and stator current observer is designed, which can reduce the control error caused by model parameter mistake and one–step delay. Another effective way to improve the disturbance attenuation is to bring in the integral action to the MPC [25]. In [26], a robust nonlinear predictive control method is proposed, which introduces the integral item by designing an optimized cost function. The simulation results verify the high performance of the controller in the presence of disturbance and mismatched parameters.

It should be noticed that the mentioned control strategies apply to the surface–mounted PMSM (SPMSM). The d-axis and q-axis inductance of SPMSM are equal, and hence, these controllers can achieve the maximum torque per ampere ratio by impelling the d-axis current to zero. As for the IPMSM, the d-axis and q-axis inductance are not equal. Hence, these predictive controllers cannot make full use of the reluctance torque caused by the interaction of the unequal d-axis and q-axis inductance. In addition, it is necessary to improve the predictive controllers to improve the torque–to–current ratio.

To improve the torque–to–current ratio of IPMSM, a nonlinear predictive control strategy with an improved cost function is proposed in this paper. By integrating the maximum torque per ampere (MTPA) control law with the predictive control strategy, the controller can make full use of the reluctance torque of IPMSM. The controller is verified to have excellent static and dynamic performance. The main content is organized as follows: In Section 2, the mathematical model of the IPMSM is presented. On this basis, the MTPA control law is expounded and transformed to a linear constraint for the armature current. Combined with the mentioned current constraint, an optimized predictive controller applied to the IPMSM is proposed in Section 3. In order to eliminate the steady–state error caused by the load mutation, the integral term is also brought in by designing a revised cost function. Section 4 gives the simulation results of the controller. In addition, the corresponding conclusion is drawn in Section 5.

2. Linear Constraint of the Armature Current

The mathematical model of the IPMSM can be depicted in *dq* coordinate system by

$$u_{\mathrm{d}} = L_{\mathrm{d}}\dot{i}_{\mathrm{d}} - p\omega L_{\mathrm{q}} i_{\mathrm{q}} + R i_{\mathrm{d}} \tag{1}$$

$$u_{\mathrm{q}} = L_{\mathrm{q}}\dot{i}_{\mathrm{q}} + p\omega L_{\mathrm{d}} i_{\mathrm{d}} + p\omega\psi_{\mathrm{f}} + R i_{\mathrm{q}} \tag{2}$$

$$J\dot{\omega} = p\left[\psi_{\mathrm{f}} i_{\mathrm{q}} + (L_{\mathrm{d}} - L_{\mathrm{q}}) i_{\mathrm{d}} i_{\mathrm{q}}\right] - B\omega - T_{\mathrm{L}} \tag{3}$$

where u_{d}, u_{q} are the *d*-axis and *q*-axis components of the armature voltage; i_{d}, i_{q} are the *d*-axis and *q*-axis components of the armature current; R, L_{d} and L_{q} are the phase resistance, the *d*-axis and *q*-axis inductances, respectively; p is the number of pole pairs; ω is the rotor angular speed; ψ_{f} represents the permanent magnet flux; J, B and T_{L} are the moment of inertia, the coefficient of friction and the load torque, respectively. The specific parameters of the machine studied in this paper are listed in Table 1.

Table 1. Main parameters of the machine.

Parameters	Values
Rated power	11 kW
d-axis inductances	18.88 mH
q-axis inductances	30.56 mH
Phase resistance	0.32 Ω
PM flux	0.317 Wb
Winding turns per slot	18
Number of pole pairs	3
Number of slots	36
Parallel branches per phase	1

For the IPMSM, there exists the reluctance torque component if i_{d} is not zero. So, it is not the best choice to adopt the common current controller which forces the *d*-axis current to zero. To get the maximum output torque under the minimum current, the maximum torque per ampere (MTPA) control is adopted in this paper.

The equations of the MTPA can be expressed as

$$\begin{cases} \min i_{\mathrm{s}} = \sqrt{i_{\mathrm{d}}^2 + i_{\mathrm{q}}^2} \\ s.t. \quad T_{\mathrm{e}} = p\left[\psi_{\mathrm{f}} i_{\mathrm{q}} + (L_{\mathrm{d}} - L_{\mathrm{q}}) i_{\mathrm{d}} i_{\mathrm{q}}\right] \end{cases} \tag{4}$$

where i_{s} and T_{e} are the armature current and the electromagnetic torque, respectively. The Lagrange multiplier is adopted to solve this equation, which is expressed as

$$H(i_{\mathrm{d}}, i_{\mathrm{q}}, \lambda) = i_{\mathrm{d}}^2 + i_{\mathrm{q}}^2 + \lambda\left\{T_{\mathrm{e}} - p\left[\psi_{\mathrm{f}} i_{\mathrm{q}} + (L_{\mathrm{d}} - L_{\mathrm{q}}) i_{\mathrm{d}} i_{\mathrm{q}}\right]\right\} \tag{5}$$

where λ is the Lagrange multiplier. Then, the partial derivatives of $H(i_{\mathrm{d}}, i_{\mathrm{q}}, \lambda)$ with respect to i_{d}, i_{q} and λ are

$$\begin{cases} \frac{\partial H}{\partial i_{\mathrm{d}}} = 2i_{\mathrm{d}} - \lambda p (L_{\mathrm{d}} - L_{\mathrm{q}}) i_{\mathrm{q}} \\ \frac{\partial H}{\partial i_{\mathrm{q}}} = 2i_{\mathrm{q}} - \lambda p\left[\psi_{\mathrm{f}} + (L_{\mathrm{d}} - L_{\mathrm{q}}) i_{\mathrm{d}}\right] \\ \frac{\partial H}{\partial \lambda} = T_{\mathrm{e}} - p\left[\psi_{\mathrm{f}} i_{\mathrm{q}} + (L_{\mathrm{d}} - L_{\mathrm{q}}) i_{\mathrm{d}} i_{\mathrm{q}}\right] \end{cases} \tag{6}$$

Set the partial derivative equations equal to zero; it yields that

$$i_{\mathrm{q}} = \sqrt{i_{\mathrm{d}}^2 + \psi_{\mathrm{f}} i_{\mathrm{d}} / (L_{\mathrm{d}} - L_{\mathrm{q}})} \tag{7}$$

$$p^2 (L_{\mathrm{d}} - L_{\mathrm{q}})^2 i_{\mathrm{q}}^4 + p\psi_{\mathrm{f}} T_{\mathrm{e}} i_{\mathrm{q}} - T_{\mathrm{e}}^2 = 0 \tag{8}$$

Equation (8) is a complex quartic equation about i_{q}, whose solution can be obtained by the numerical method. It should be noticed that the value of i_{q} is positive, whereas i_{d} is negative for the condition that $L_d < L_q$. The nonlinear curves of i_d and i_q varying with T_e are given in Figure 1.

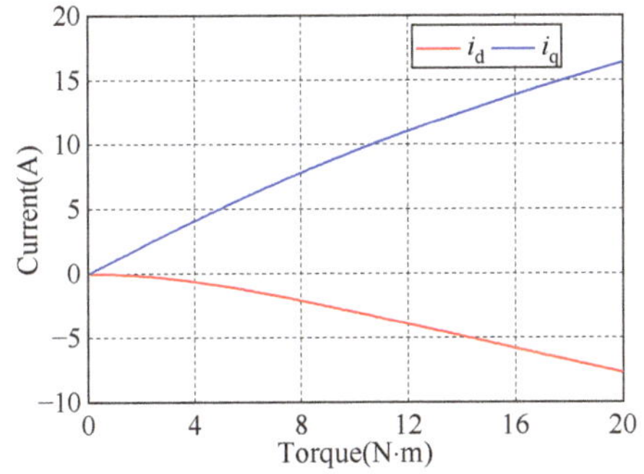

Figure 1. Curves of i_d and i_q varying with the electromagnetic torque.

In the steady state, the derivative of ω with respect to time is zero. So, the Equation (3) can be transformed into

$$T_e = B\omega + T_L \tag{9}$$

The electromagnetic torque is approximately equal to the load torque. Under this condition, it can obtain the relational between the steady−state armature current and the load torque through polynomial curve fitting method.

$$\begin{cases} i_{ds\_r} = d_1 T_L^3 + d_2 T_L^2 + d_3 T_L + d_4 \\ i_{qs\_r} = q_1 T_L^2 + q_2 T_L + q_3 \end{cases} \tag{10}$$

where i_{ds_r}, i_{qs_r} are the estimated values of i_{ds}, i_{qs}, respectively. The values of the coefficients in Equation (10) are $d_1 = 7.463 \times 10^{-4}$, $d_2 = -0.031$, $d_3 = -0.071$, $d_4 = 0.084$, and $q_1 = -0.013$, $q_2 = 1.076$, $q_3 = 0.012$. Figure 2 exhibits the comparison between the numerical and curve fitting solutions of i_{ds} and i_{qs}, respectively. The results achieved by the two methods agree well with each other, which verifies the high precision of the curve fitting method.

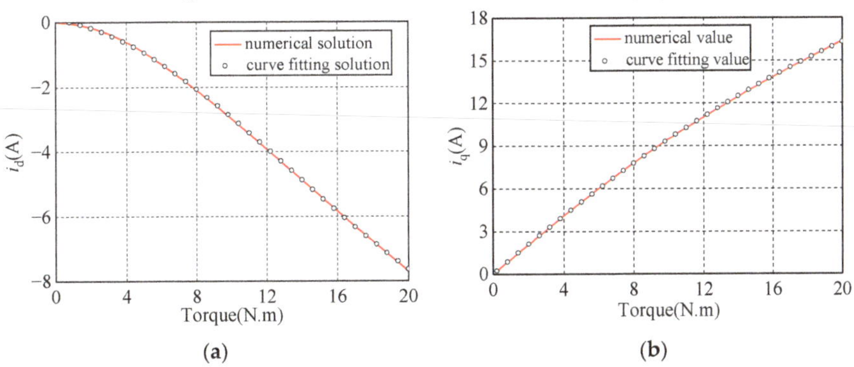

Figure 2. The numerical and curve fitting solutions of (**a**) i_d; (**b**) i_q.

The PM machine mainly operates in steady state with the required speed. So, it is reasonable to give priority to the correlation between the *d*-axis and *q*-axis steady−state current. It can assume that the *d*-axis and *q*-axis armature current are i_{ds} and i_{qs}, respectively, when the machine runs steadily with a given load torque. Then, Equation (10) can be transformed into a piecewise function based on the first−order Taylor expansion, which is diagrammed in Figure 3. For a certain interval, the governing function about i_d and i_q can be expressed as

$$i_q + \mu i_d + \varsigma = 0 \tag{11}$$

where μ and ς are two coefficients determined by the load torque.

$$\begin{cases} \mu = -\dot{i}_{q|i_d=i_{ds}} \\ \varsigma = -i_{qs} + i_{ds}\dot{i}_{q|i_d=i_{ds}} \end{cases} \tag{12}$$

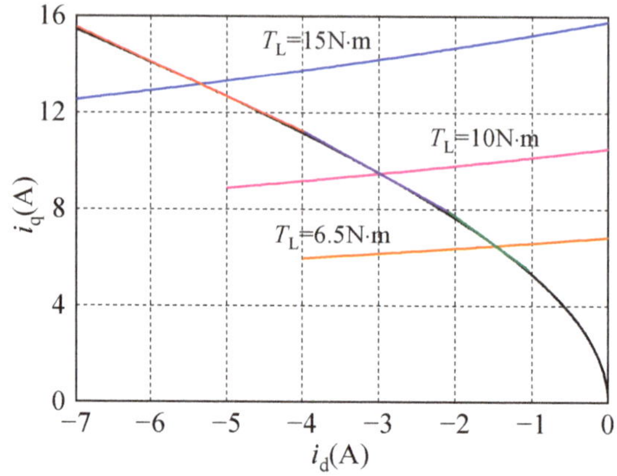

Figure 3. The first-order Taylor expansion at the certain point.

3. Proposed Controller for the IPMSM

3.1. Design of the Controller

It can adopt Equation (11) as a linear constraint to the *d*-axis and *q*-axis armature current of the PM machine. Refer to [26]; the mathematical model of the IPMSM can be expressed as Equation (13). Herein, the constraint of (11) instead of $i_d = 0$ is adopted to determine the output parameters.

$$\begin{cases} \dot{x}(t) = f(x) + g_1 u(x) + g_2 b(t) \\ y(t) = \begin{bmatrix} h_1(x) \\ h_2(x) \end{bmatrix} = \begin{bmatrix} i_q + \mu i_d + \varsigma \\ \omega \end{bmatrix} \end{cases} \tag{13}$$

where $x = [i_d \; i_q \; \omega]^T$ are the state variables, $u = [u_d \; u_q]^T$ are the inputs, $y = [h_1(x) \; h_2(x)]^T$ are the outputs, and b denotes the disturbance caused by the load torque. The reference value of $h_1(x)$ is 0, and the reference value of $h_2(x)$ is the reference speed.

$$f(x) = \begin{bmatrix} f_1(x) \\ f_2(x) \\ f_3(x) \end{bmatrix} = \begin{bmatrix} \frac{p\omega L_q i_q}{L_d} - \frac{R i_d}{L_d} \\ -\frac{p\omega L_d i_d}{L_q} - \frac{p\omega \psi_f}{L_q} - \frac{R i_q}{L_q} \\ \frac{p[\psi_f i_q + (L_d - L_q) i_d i_q]}{J} - \frac{B\omega}{J} \end{bmatrix} \tag{14}$$

$$g_1 = \begin{bmatrix} 1/L_d & 0 \\ 0 & 1/L_q \\ 0 & 0 \end{bmatrix} \quad g_2 = \begin{bmatrix} 0 \\ 0 \\ -1/J \end{bmatrix} \tag{15}$$

According to Equation (13), the following equations are derived based on the standard Lie derivative.

$$L_{g_1} h_1(x) = \begin{bmatrix} \mu/L_d & 1/L_q \end{bmatrix} \tag{16}$$

$$L_{g_1}h_2(x) = [\,0 \quad 0\,] \tag{17}$$

$$L_{g_1}L_f h_2(x) = \begin{bmatrix} \dfrac{p(L_\mathrm{d} - L_\mathrm{q})i_\mathrm{q}}{JL_\mathrm{d}} & \dfrac{p\psi_\mathrm{f} + p(L_\mathrm{d} - L_\mathrm{q})i_\mathrm{d}}{JL_\mathrm{q}} \end{bmatrix} \tag{18}$$

It is observed that $L_{g1}h_1(x)$ and $L_{g1}L_f h_2(x)$ are the non−singular matrixes. In addition, the elements of $L_{g1}h_2(x)$ are equal to zero. Hence, the relative degrees of the outputs y_1 and y_2 are $\rho_1 = 1$ and $\rho_2 = 2$, respectively. The total relative degree of the nonlinear system is $\rho = \rho_1 + \rho_2 = 3$, which is equal to the number of state variables. Therefore, the nonlinear system can be transformed into a linear model through the input−output feedback linearization. Based on the standard Lie derivative, the ρ_i−time derivatives of the outputs y_i with respect to time are given by

$$\begin{cases} \dot{y}_1(t) = L_f h_1(x) + L_{g_1}h_1(x)u(t) \\ \dot{y}_2(t) = L_f h_2(x) \\ \ddot{y}_2(t) = L_f^2 h_2(x) + L_{g_1}L_f h_2(x)u(t) \end{cases} \tag{19}$$

The outputs of the machine are forced by the controller to track the reference trajectory in the predictive period. For this purpose, a cost function should be firstly designed, whose optimum solutions serve as the inputs to the system. As stated in [25], the integral action can eliminate the steady error caused by the disturbance and parameter uncertainty. Herein, the integral term is introduced to the controller by defining a generalized cost function.

$$\min M = \frac{1}{2}\int_0^{T_1} m_1(t+\tau)^2 d\tau + \frac{1}{2}\int_0^{T_2} m_2(t+\tau)^2 d\tau \tag{20}$$

where T_1 and T_2 are the prediction horizons for the current loop and speed loop, respectively. The response speed of the current loop is much faster than the speed loop, so T_1 can be set to a positive number smaller than T_2.

$$m_1(t+\tau) = c_1 e_{\mathrm{I}1}(t) + c_3\tau e_1(t) + \frac{\tau^2}{2}\dot{e}_1(t) \tag{21}$$

$$m_2(t+\tau) = c_2 e_{\mathrm{I}2}(t) + c_4\tau e_2(t) + \frac{\tau^2}{2}\dot{e}_2(t) + \frac{\tau^3}{6}\ddot{e}_2(t) \tag{22}$$

$$e_\mathrm{i}(t) = y_\mathrm{i}(t) - y_{\mathrm{i\_r}}(t) \quad e_{\mathrm{Ii}}(t) = \int_0^t e_\mathrm{i}(\gamma)d\gamma \quad i = 1,2 \tag{23}$$

where c_1, c_2, c_3 and c_4 are the coefficients to be designed, and $y_{\mathrm{i\_r}}$ is the desired reference.

Substituting Equation (19) into Equation (20), the cost function given in Equation (20) transforms to

$$M = \frac{1}{2}E^\mathrm{T}\overline{\eta}E \tag{24}$$

where

$$E = W + \begin{bmatrix} 0_{5\times 2} \\ H \end{bmatrix} u(t) \tag{25}$$

$$W = \begin{bmatrix} e_{\mathrm{I}1} & e_1 & e_{\mathrm{I}2} & e_2 & \dot{e}_2 & L_f h_1(x) - \dot{y}_{\mathrm{r}1} & L_f^2 h_2(x) - \ddot{y}_{\mathrm{r}2} \end{bmatrix}^\mathrm{T} \tag{26}$$

$$H = \begin{bmatrix} L_{g_1}h_1(x) \\ L_{g_1}L_f h_2(x) \end{bmatrix} \tag{27}$$

$$\overline{\eta} = \int_0^{T_1} \eta_1(\tau)^{\mathrm{T}}\eta_1(\tau)d\tau + \int_0^{T_2} \eta_2(\tau)^{\mathrm{T}}\eta_2(\tau)d\tau = \begin{bmatrix} \overline{\eta}_{1(5\times5)} & \overline{\eta}_{2(5\times2)} \\ \overline{\eta}_{3(2\times5)} & \overline{\eta}_{4(2\times2)} \end{bmatrix} \tag{28}$$

$$\eta = \begin{bmatrix} \eta_1(\tau) \\ \eta_2(\tau) \end{bmatrix} = \begin{bmatrix} c_1 & c_3\tau & 0 & 0 & 0 & \tau^2/2 & 0 \\ 0 & 0 & c_2 & c_4\tau & \tau^2/2 & 0 & \tau^3/6 \end{bmatrix} \tag{29}$$

Taking partial derivative of M with respect to u gives

$$\frac{\partial M}{\partial u} = H^{\mathrm{T}}[\overline{\eta}_3 \quad \overline{\eta}_4]W + H^{\mathrm{T}}\overline{\eta}_4 H u(t) \tag{30}$$

The solutions of the equation $\partial M/\partial u = 0$ are the optimal values such that the cost function is minimized. By solving the function $\partial M/\partial u = 0$, the input voltage of the machine can be determined by

$$u(t) = -H^{-1}\left[\overline{\eta}_4^{-1}\overline{\eta}_3 \quad I_{(2\times2)}\right]W \tag{31}$$

3.2. Constraints of the Controller

The capacity of the inverter is limited. During the transient process, the armature voltage and current may exceed its limit values. Hence, the necessary constraints should be added to the control system. The upper and lower bound limits are chosen based on the maximum output voltage and maximum output current of the inverter. By this method, the machine voltage and current are within the required range, which avoids the damage of excess current to the inverter.

$$\begin{cases} u_{\mathrm{d\_min}} \le u_{\mathrm{d}} \le u_{\mathrm{d\_max}} \\ u_{\mathrm{q\_min}} \le u_{\mathrm{q}} \le u_{\mathrm{q\_max}} \end{cases} \tag{32}$$

$$\begin{cases} i_{\mathrm{d\_min}} \le i_{\mathrm{d}} \le i_{\mathrm{d\_max}} \\ i_{\mathrm{q\_min}} \le i_{\mathrm{q}} \le i_{\mathrm{q\_max}} \end{cases} \tag{33}$$

As the armature current i_{d} and i_{q} are the state variables and cannot be constrained directly, the current constraints can be achieved by setting proper limitations to the input voltage. The Lie derivatives of the armature current along the field vector g_1 are given by

$$\begin{cases} L_{g_1} i_{\mathrm{d}} = [1/L_{\mathrm{d}} \quad 0] \\ L_{g_1} i_{\mathrm{q}} = [0 \quad 1/L_{\mathrm{d}}] \end{cases} \tag{34}$$

Equation (34) implies that the relative degrees of the armature current i_{d} and i_{q} to the input variables are $\rho_3 = 1$ and $\rho_4 = 1$, respectively. So, the predicted current can be expanded into the first−order Taylor series as

$$\begin{cases} i_{\mathrm{d}}(t+\tau) = i_{\mathrm{d}}(t) + \tau f_1(x) + \tau u_{\mathrm{d}}(t)/L_{\mathrm{d}} \\ i_{\mathrm{q}}(t+\tau) = i_{\mathrm{q}}(t) + \tau f_2(x) + \tau u_{\mathrm{q}}(t)/L_{\mathrm{q}} \end{cases} \tag{35}$$

The predicted current $i_{\mathrm{d}}(t+\tau)$ and $i_{\mathrm{q}}(t+\tau)$ satisfy the inequality constraints given in Equation (33). Substituting Equation (33) into Equation (35) yields

$$L_{\mathrm{d}}[(i_{\mathrm{d\_min}} - i_{\mathrm{d}})/\tau - f_1(x)] \le u_{\mathrm{d}} \le L_{\mathrm{d}}[(i_{\mathrm{d\_max}} - i_{\mathrm{d}})/\tau - f_1(x)] \tag{36}$$

$$L_{\mathrm{q}}[(i_{\mathrm{q\_min}} - i_{\mathrm{q}})/\tau - f_2(x)] \le u_{\mathrm{q}} \le L_{\mathrm{q}}[(i_{\mathrm{q\_max}} - i_{\mathrm{q}})/\tau - f_2(x)] \tag{37}$$

Equation (32) together with Equations (36) and (37) constitute the constraints to the proposed controller.

3.3. Stability Analysis

Assume that the disturbance $b(t)$ is bounded. As for the nonlinear system for Equation (13), it can be concluded that the closed loop system under the proposed control law is asymptotically stable. In addition, the outputs will track the desired references with the error approaching zero asymptotically. The proof process is summarized briefly in the following.

$$\lim_{t\to\infty} e_i(t) = 0 \quad i = 1,2 \tag{38}$$

The first order derivative of $h_1(x)$ along the vector g_2 is given by

$$L_{g_2}h_1(x) = 0 \tag{39}$$

Combing Equation (31) with Equation (39), the following function about the error e_1 is derived.

$$\dot{\kappa}_1(t) = A_1\kappa_1(t) \tag{40}$$

where

$$\kappa_1(t) = [e_{I1}(t) \quad e_1(t)]^{\mathrm{T}} \tag{41}$$

$$A_1 = \begin{bmatrix} 0 & 1 \\ -c_1/T_1^2 & -c_3/T_1 \end{bmatrix} \tag{42}$$

The real parts of the eigenvalues of A_1 are negative, indicating that the current loop under the control law (31) is Hurwitz asymptotically stable. On this basis, the error e_1 will converge to zero if $t\to\infty$ deduced from the Barbalat lemma [27].

Taking the disturbance into consideration, the first and second order Lie derivatives of y_2 are given by

$$\begin{cases} \dot{y}_2(t) = L_f h_2(x) + L_{g_2}h_2(x)b(t) \\ \ddot{y}_2(t) = L_f^2 h_2(x) + L_{g_1}L_f h_2(x)u(t) + L_{g_2}L_f h_2(x)b(t) \end{cases} \tag{43}$$

Substituting Equation (31) into Equation (43), we can get the governing equation of the error e_2 as

$$\dot{\kappa}_2(t) = A_2\kappa_2(t) + B(x)b(t) \tag{44}$$

where

$$\kappa_2(t) = \begin{bmatrix} e_{I2}(t) & e_2(t) & \dot{e}_2(t) \end{bmatrix}^{\mathrm{T}} \tag{45}$$

$$A_2 = \begin{bmatrix} 0 & 1 & 0 \\ 0 & 0 & 1 \\ -c_2/T_2^3 & -c_4/T_2^2 & -3.5/T_2 \end{bmatrix} \tag{46}$$

$$B(x) = [0 \quad 0 \quad \varpi(x)]^{\mathrm{T}} \tag{47}$$

$$\varpi(x) = -L_{g_2}L_f h_2(x) - \frac{3.5}{T_2}L_{g_2}h_2(x) = -\frac{B}{J^2} + \frac{3.5}{JT_2} \tag{48}$$

A_2 is a Hurwitz matrix, so there exists a symmetric positive matrixes P, such that

$$A_2^{\mathrm{T}}P + PA_2 = -I \tag{49}$$

Then, establish a Lyapunov function candidate as

$$V_2(\kappa_2) = \frac{1}{2}\kappa_2^{\mathrm{T}}P\kappa_2 \tag{50}$$

The derivative of V_2 with respect to time is expressed as

$$\dot{V}_2(\kappa_2) = -\frac{1}{2}\kappa_2^{\mathrm{T}}\kappa_2 + \frac{1}{2}\kappa_2^{\mathrm{T}}PBb + \frac{1}{2}b^{\mathrm{T}}B^{\mathrm{T}}P\kappa_2 \le -\frac{1}{2}\|\kappa_2\|^2 + \xi\|\kappa_2\|\|B\|\|b\| \tag{51}$$

where $\xi = \|P\|$ is a positive value.

Under the condition that $b(t)$ and $B(x)$ are bounded, it can be deduced that the derivative of V_2 with respect to time is negative if the integral of e_2 is sufficiently large, which shows that the speed loop is asymptotically stable despite the existence of disturbance. Similarly, the output error e_2 satisfies $e_2(t) = 0$ when $t \to \infty$.

4. Simulation Analysis

The block diagram of the IPMSM predictive control system is shown in Figure 4. To verify the effectiveness of the proposed control method (strategy 1), a comparison with the robust nonlinear predictive controller [26] (strategy 2) is carried out.

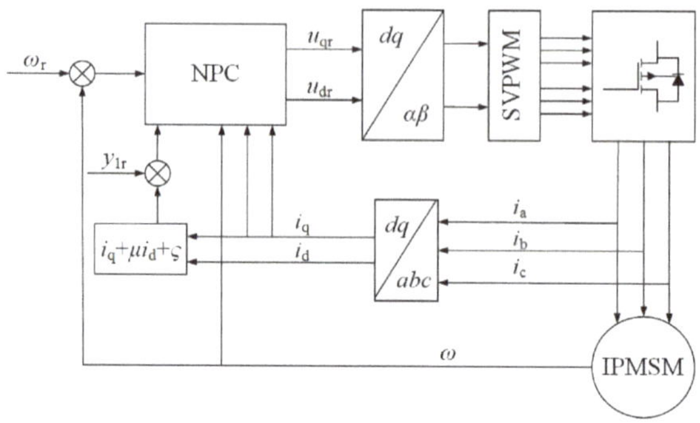

Figure 4. The block diagram of the predictive control system.

The prediction horizon of the current loop T_1 is set to 1ms, whereas that of the speed loop T_2 is 5 ms. The sampling time of the system is 0.01ms. It should be noticed that the integral gain c_1 and c_2 would influence the machine performance. The bigger integral gain will improve the response speed but may cause the undesired overshoot. However, an unfavorable decrease in the integral gain would enlarge the setting time of the system.

During the simulation process, the load torque is set to 5 N·m and the speed reference is 100 rad/s. Figure 5 exhibits the speed response of the IPMSM with two control methods. It can be observed from Figure 5a that the machine can reach the reference speed within a very short period, indicating that the system has excellent dynamic and static performances with strategy 1. Control strategy 2 has similar response speed during the start−up process. However, the phase current would exceed its limited value for this condition. Figure 6 shows the d-axis and q-axis components of the armature current with control strategy 1 and strategy 2. It can be seen that the d-axis current under strategy 2 maintains to zero. By comparison, the d-axis current is not zero but a negative value when strategy 1 is employed.

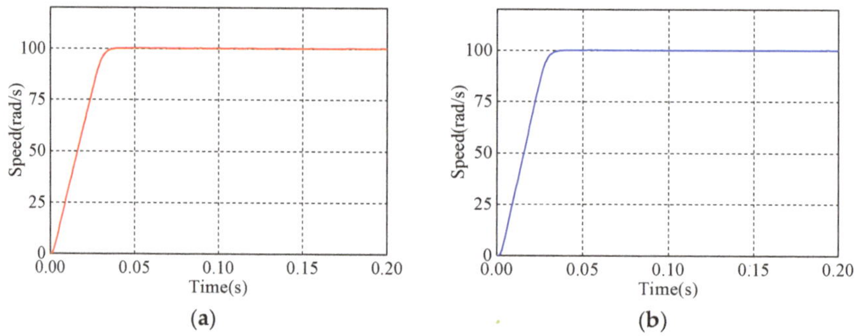

Figure 5. The speed response of the machine with (**a**) control strategy 1; (**b**) control strategy 2.

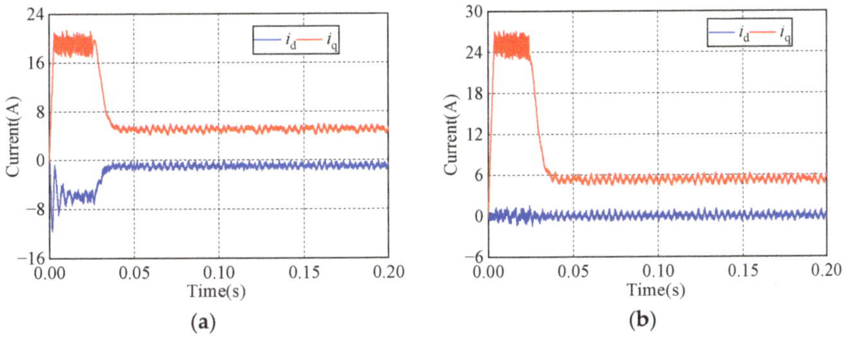

Figure 6. The *d*-axis and *q*-axis current with (**a**) control strategy 1; (**b**) control strategy 2.

The three phase currents with control strategy 1 and strategy 2 are compared in Figure 7. As stated in [26], strategy 2 is an unconstrained controller, and the speed response of this control strategy is related to the gradient of the speed trajectory. When the speed trajectory is changing quickly, the armature current would exceed its limit value during the dynamic process, which is reflected clearly in Figure 7b. So, the speed trajectory must be chosen to keep the armature current within its threshold, and the integral gain needs to be adjusted simultaneously. However, the change of integral gain would affect the machine performance. In addition, it is difficult to obtain the appropriate value of these parameters. By comparison, the strategy 1 proposed in this paper solves this problem by adding the voltage and current constraints.

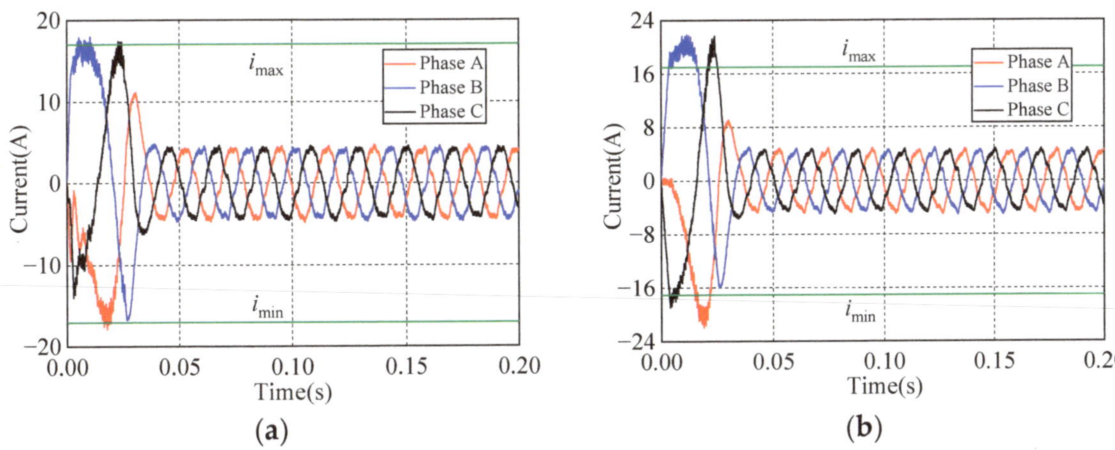

Figure 7. The phase current with (**a**) control strategy 1; (**b**) control strategy 2.

In addition, Figure 8 gives the speed response and armature current curves when the gradient of the speed trajectory is slow with strategy 2. As shown in Figure 8b, the armature current is limited within the threshold value. However, it takes more time for the machine to reach the steady state, which has an adverse influence on the machine dynamic performance. By contrast, the machine with strategy 1 could approach the steady state within 0.05 s, indicating that a better dynamic performance is achieved with the proposed control method.

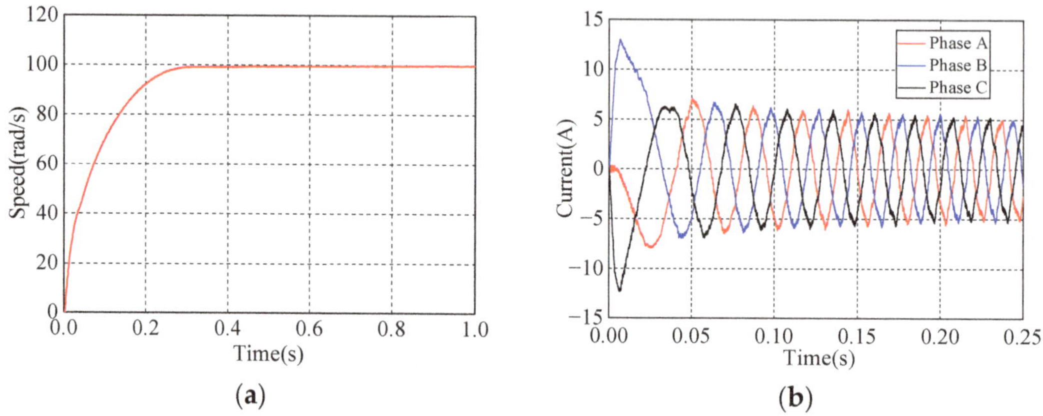

Figure 8. The (**a**) speed response; and (**b**) phase current with control strategy 2 of slow speed trajectory.

For further verification, the load torque of the machine is set to 5 N·m originally, but stepped up to 20 N·m at 1 s. Figure 9 gives the speed response and the phase current curves of the machine, respectively. It can be seen that it takes about 1 s for the steady−state error to maintain to zero as the load torque is increased suddenly up to 400%. Figure 10 shows the d-axis and q-axis components of the current corresponding to the proposed controller. In the steady state, as clearly shown in Figure 10, the d-axis current maintains to about −8A, whereas the q-axis current turns into 16A, approximately. In addition, the two components of the armature current i_d and i_q satisfy the linear constraint given in Equation (9). Figure 11 compares the A phase armature current when the machine operates with the reference speed. The machine is controlled by two different control strategies. On the premise of obtaining the same electromagnetic torque, it is obvious that the steady−state armature current under strategy 1 is smaller than that of strategy 2. It indicates that the proposed controller has a higher torque−to−current ratio than the other control strategy.

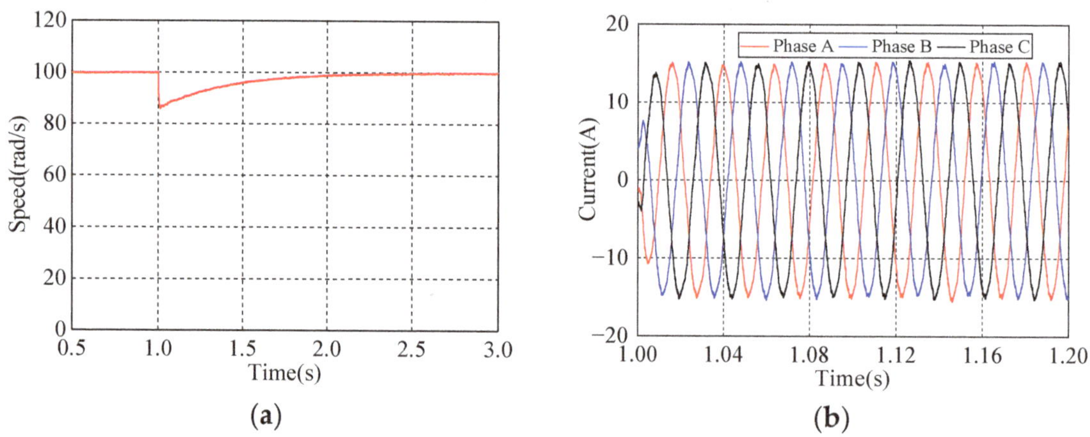

Figure 9. The (**a**) speed response; and (**b**) phase current with control strategy 1 under load mutation.

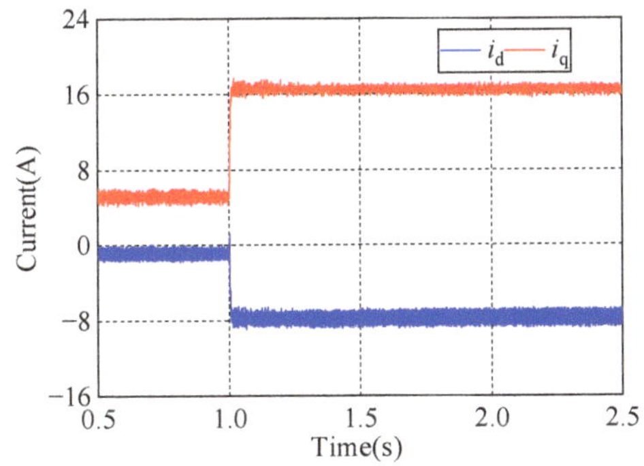

Figure 10. The d-axis and q-axis current with control strategy 1.

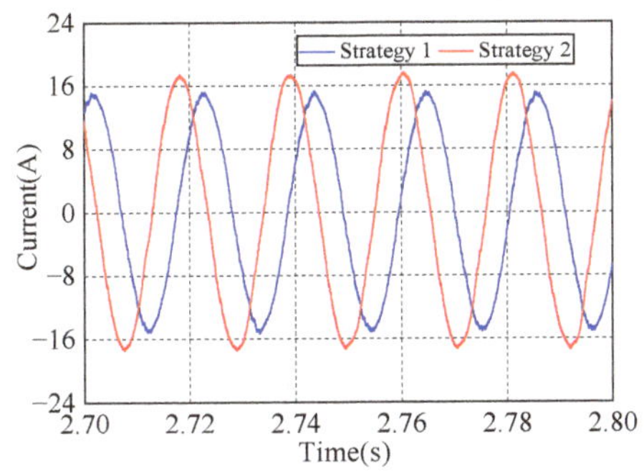

Figure 11. The A phase current with the two different controllers.

Then, the speed response of the machine for step speed change and rapid speed change conditions are researched. The reference speed is set to 100 rad/s originally, but stepped up to 110 rad/s and 200 rad/s at 1 s, respectively. Figure 12 gives the speed response of the machine. As shown in Figure 12a, the machine can reach the reference speed within 0.02 s without overshoot. For the rapid speed change condition, the machine speed would approach the reference speed rapidly with a little overshoot. Then, the static speed error is eliminated within 0.1 s. Simulation results indicate that the machine has excellent dynamic performances for speed change condition.

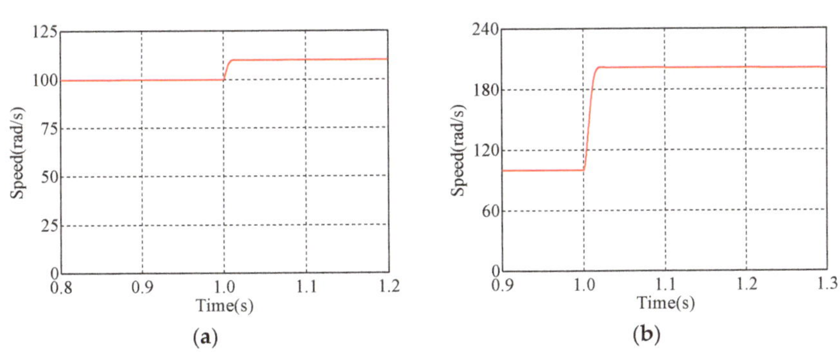

Figure 12. The speed response for (**a**) step speed change condition; and (**b**) rapid speed change condition.

The load torque is set to rated load and the speed reference is rated speed. Figure 13a exhibits the speed response of the IPMSM with the proposed control strategy, and Figure 13b gives the phase current when the machine operates with rated speed and full load. As can be seen in Figure 13, the machine can operate with full load at rated speed. Furthermore, it can be observed from Figure 13a that the machine can reach the rated speed within 1.5 s. The response time is short. Moreover, the machine can reach the rated speed without overshoot, which indicates that the machine has excellent dynamic performances with the proposed control strategy.

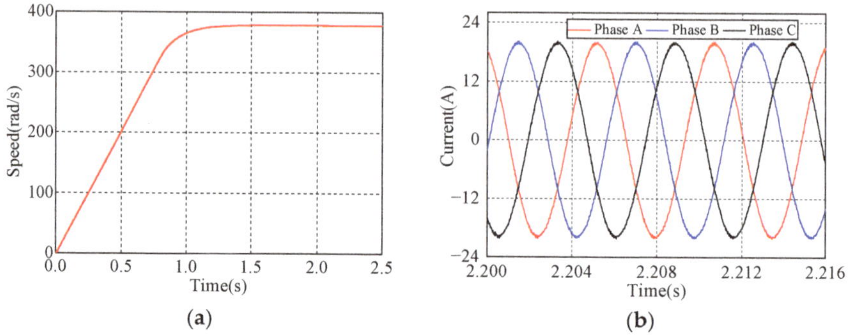

Figure 13. The (**a**) speed response; and (**b**) phase current with proposed control strategy.

Simulation results indicate that: (1) the machine with the proposed controller can reach the specified speed rapidly during the start−up process, and the phase current is within its limited value; (2) the machine can eliminate the disturbance resulting from load change within a short period; (3) the machine can rapidly reach the reference speed for speed change conditions; (4) the machine has a higher torque−to−current ratio. Hence, the machine has excellent dynamic and static performances with the proposed control strategy, which verifies the effectiveness of the control strategy.

5. Conclusions

An improved nonlinear predictive control method for the IPMSM was presented in this paper. To make full use of the reluctance torque of the machine, the d-axis and q-axis components of the armature current were compelled to follow the linear constraint derived by the maximum torque per ampere control rule and first−order Taylor expansion during the running process. On this basis, the predictive controller was established by defining a generalized cost function whose solutions served as the inputs to the system, and the integral term was also introduced to avoid the effect of the external perturbations. The stability of the proposed control method was proved by the Lyapunov function and the Barbalat lemma. The simulation was established to verify the effectiveness of the proposed control method. Simulation results showed that the system has excellent dynamic and static performances. The d-axis and q-axis components of the armature current satisfied the given linear constraint, indicating that the proposed controller has a higher torque−to−current ratio than the one that forced the d-axis current to zero.

Author Contributions: Conceptualization, M.T.; Data curation, M.T. and H.C.; Formal analysis, M.T. and W.Z.; Investigation, M.T. and J.R.; Methodology, M.T.; Project administration, M.T.; Supervision, M.T.; Validation, M.T.; Visualization, W.Z.; Writing—original draft, M.T.; Writing—review and editing, H.C., W.Z. and J.R. All authors have read and agreed to the published version of the manuscript.

Funding: This research was funded by the Shandong Provincial Natural Science Foundation, China, grant number ZR2021QE057.

Data Availability Statement: Not applicable.

Conflicts of Interest: The authors declare no conflict of interest.

References

1. Wu, L.; Yin, H.; Wang, D.; Fang, Y. A nonlinear subdomain and magnetic circuit hybrid model for open−circuit field prediction in surface−mounted PM machines. *IEEE Trans. Energy Convers.* **2019**, *34*, 1485–1495. [CrossRef]
2. Hajdinjak, M.; Miljavec, D. Analytical calculation of the magnetic field distribution in slotless brushless machines with U−shaped interior permanent magnets. *IEEE Trans. Ind. Electron.* **2020**, *67*, 6721–6731. [CrossRef]
3. Peng, B.; Wang, X.; Zhao, W.; Ren, J. Study on shaft voltage in fractional slot permanent magnet machine with different pole and slot number combinations. *IEEE Trans. Magn.* **2019**, *55*, 8102305. [CrossRef]
4. Riaz, S.; Lin, H.; Anwar, M.B.; Ali, H. Design of PD−type second−order ILC law for PMSM servo position control. *J. Phys. Conf. Ser.* **2020**, *1707*, 012002. [CrossRef]
5. Liu, G.; Chen, B.; Wang, K.; Song, X. Selective current harmonic suppression for high−speed PMSM based on high−precision harmonic detection method. *IEEE Trans. Ind. Informat.* **2019**, *15*, 3457–3468. [CrossRef]
6. Junejo, A.K.; Xu, W.; Mu, C.; Ismail, M.M.; Liu, Y. Adaptive speed control of PMSM drive system based a new sliding−mode reaching law. *IEEE Trans. Power Electron.* **2020**, *35*, 12110–12121. [CrossRef]
7. Xu, B.; Zhang, L.; Ji, W. Improved non−singular fast terminal sliding mode control with disturbance observer for PMSM drives. *IEEE Trans. Transp. Electrif.* **2021**, *7*, 2753–2762. [CrossRef]
8. Nawaz, M.K.; Dou, M.; Riaz, S.; Sardar, M.U. Current control of permanent magnet synchronous motors using improved model predictive control. *Math. Probl. Eng.* **2022**, *2022*, 1736931.
9. Chen, W.-H.; Ballance, D.J.; Gawthrop, P.J. Optimal control of nonlinear systems: A predictive control approach. *Automatica* **2003**, *39*, 633–641. [CrossRef]
10. Xu, W.; Junejo, A.K.; Tang, Y.; Shahab, M.; Habib, H.U.R.; Liu, Y.; Huang, S. Composite speed control of PMSM drive system based on finite time sliding mode observer. *IEEE Access* **2021**, *9*, 151803–151813. [CrossRef]
11. Jiang, Y.; Xu, W.; Mu, C.; Liu, Y. Improved deadbeat predictive current control combined sliding mode strategy for PMSM drive system. *IEEE Trans. Veh. Technol.* **2018**, *67*, 251–263. [CrossRef]
12. Liu, X.; Yu, H.; Yu, J.; Zhao, L. Combined speed and current terminal sliding mode control with nonlinear disturbance observer for PMSM drive. *IEEE Access* **2018**, *6*, 29594–29601. [CrossRef]
13. Bolognani, S.; Peretti, L.; Zigliotto, M. Design and implementation of model predictive control for electrical motor drives. *IEEE Trans. Ind. Electron.* **2009**, *56*, 1925–1936. [CrossRef]
14. Sun, C.; Sun, D.; Chen, W.; Nian, H. Improved model predictive control with new cost function for hybrid-inverter open-winding PMSM system based on energy storage model. *IEEE Trans. Power Electron.* **2021**, *36*, 10705–10715. [CrossRef]
15. Xue, C.; Zhou, D.; Li, Y. Finite-control-set model predictive control for three-level NPC inverter-fed PMSM Drives with LC filter. *IEEE Trans. Ind. Electron.* **2021**, *68*, 11980–11991. [CrossRef]
16. Abu−Ali, M.; Berkel, F.; Manderla, M.; Reimann, S.; Kennel, R.; Abdelrahem, M. Deep learning-based long-horizon MPC: Robust, high performing, and computationally efficient control for PMSM drives. *IEEE Trans. Power Electron.* **2022**, *37*, 12486–12501.
17. Wang, L.; Chai, S.; Rogers, E.; Freeman, C.T. Multivariable repetitive -predictive controllers using frequency decomposition. *IEEE Trans. Control Syst. Technol.* **2012**, *20*, 1597–1604. [CrossRef]
18. Preindl, M.; Schaltz, E.; Thogersen, P. Switching frequency reduction using model predictive direct current control for high power voltage source inverters. *IEEE Trans. Ind. Electron.* **2011**, *58*, 2826–2835. [CrossRef]
19. Zheng, C.; Yang, J.; Gong, Z.; Xiao, Z.; Dong, X. Cascade-free modulated predictive direct speed control of PMSM drives. *Energies* **2022**, *15*, 7200. [CrossRef]
20. Akpunar, A.; Iplikci, S. Runge-Kutta model predictive speed control for permanent magnet synchronous motors. *Energies* **2020**, *13*, 1216. [CrossRef]
21. Shao, M.; Deng, Y.; Li, H.; Liu, J.; Fei, Q. Sliding mode observer-based parameter identification and disturbance compensation for optimizing the mode predictive control of PMSM. *Energies* **2019**, *12*, 1857. [CrossRef]
22. Tang, M.; Zhuang, S. On speed control of a permanent magnet synchronous motor with current predictive compensation. *Energies* **2019**, *12*, 65. [CrossRef]
23. Yang, J.; Zheng, W.X.; Li, S.; Wu, B.; Cheng, M. Design of a prediction-accuracy-enhanced continuous-time MPC for disturbed systems via a disturbance observer. *IEEE Trans. Ind. Electron.* **2015**, *62*, 5807–5815. [CrossRef]
24. Zhang, X.; Hou, B.; Mei, Y. Deadbeat predictive current control of permanent-magnet synchronous motors with stator current and disturbance observer. *IEEE Trans. Power Electron.* **2017**, *32*, 3818–3834. [CrossRef]
25. Hedjar, R.; Toumi, R.; Boucher, P.; Dumur, D. Finite horizon nonlinear predictive control by the Taylor approximation: Application to robot tracking trajectory. *Int. J. Appl. Math. Comput. Sci.* **2005**, *15*, 527–540.
26. Errouissi, R.; Ouhrouche, M.; Chen, W.-H.; Trzynadlowski, A.M. Robust nonlinear predictive controller for permanent-magnet synchronous motors with an optimized cost function. *IEEE Trans. Ind. Electron.* **2012**, *59*, 2849–2858. [CrossRef]
27. Tao, G. A simple alternative to the Barbalat lemma. *IEEE Trans. Autom. Contr.* **1997**, *42*, 698. [CrossRef]

vehicles

MDPI

Article

Technical and Economic Analysis to Select Suitable Design Parameters of an E-Machine for Electric Commercial Vehicles

Achim Kampker, Heiner Heimes, Benjamin Dorn, Florian Brans, José Guillermo Dorantes Gómez * and Chetan Sharma

Chair of Production Engineering of E-Mobility Components (PEM), RWTH Aachen University, 52074 Aachen, Germany
* Correspondence: j.dorantes@pem.rwth-aachen.de

Abstract: In the European Union (EU), road transport contributes a major proportion of the total greenhouse gas (GHG) emissions, of which a significant amount is caused by heavy-duty commercial vehicles (CV). The increasing number of emission regulations and penalties by the EU have forced commercial vehicle manufacturers to investigate powertrain technologies other than conventional internal combustion engines (ICE). Since vehicle economics plays an important role in purchase decisions and the powertrain of a battery electric vehicle (BEV) contributes to about 8–20% of the total vehicle cost and the electric machine (EM) alone contributes to 33–43% of the drivetrain cost, it is necessary to analyze suitable EM topologies for the powertrain. In this paper, the authors aim to analyze the technical and cost aspects of an EM for electric commercial vehicles (ECV). Based on prior research and literature on this subject, an appropriate methodology for selecting suitable geometrical parameters of an e-machine for the use case of a heavy-duty vehicle is developed using MATLAB and Simulink tools. Then, for the economic analysis of the e-machine, reference ones are used, and their design parameters and cost structures are utilized to develop a cost function. Different use cases are evaluated according to the vehicle's application. The results for a use case are compared by varying the design parameters to find the most cost-effective EM. Later, an analysis is performed on other decisive factors for EM selection. This highlights the importance of collaborative consideration of technological as well as the economic aspects of EMs for different use cases in ECVs. The method developed in this work contributes to understand the economic aspect of EMs as well as considering their performance factors. State-of-the-art methods and research are used to develop a novel methodology that helps with the selection of the initial geometry of the electric motor during the design process, which can serve to aid future designers and converters of electric heavy-duty vehicles.

Keywords: design parameters; PMSM; e-machine cost; e-machine geometry; electric commercial vehicles

Citation: Kampker, A.; Heimes, H.; Dorn, B.; Brans, F.; Dorantes Gómez, J.G.; Sharma, C. Technical and Economic Analysis to Select Suitable Design Parameters of an E-Machine for Electric Commercial Vehicles. *Vehicles* **2023**, *5*, 75–93. https://doi.org/10.3390/vehicles5010005

Academic Editors: Loránd Szabó, Feng Chai and Chen Lv

Received: 24 October 2022
Revised: 10 December 2022
Accepted: 3 January 2023
Published: 10 January 2023

1. Introduction

As the demand for vehicles continues to grow, the roads are overflowing with new vehicles at an extremely fast rate. For the same reason, the rate of pollution is increased swiftly [1]. Although, due to the development of technology, IC engines are becoming less common and the amount of pollution has lessened, even from the existing data of various studies, IC engines are still the primary source of pollution after electricity [2].

Electrification is being adopted into the heavier vehicle forms apart from passenger cars, and it is driven by various factors. Some of the motivation factors for this research are summarized here:

- Depleting fossil fuels availability [3].
- The recent evolution in battery technology is making electric commercial vehicles technically and commercially feasible [4]. Additionally, the price tag for electric vehicles is expected to fall continuously courtesy of falling battery costs [5,6].

- There is an increase in the availability and capacity of public charging infrastructure [6].
- There are similarities between the powertrain architecture, structure, and components among light-, medium-, and heavy-duty vehicles as compared to those of much more technologically mature personal electric vehicles. Very recently, battery, pantograph, or fuel cell technology has matured enough to turn long-haul heavy electric trucks into a viable alternative to standard ICE trucks [7].
- Half of all trucks operating in the EU travel distances of under 300 km. These journeys can now be completed by electric trucks thanks to the new models currently on the market with a range of about 300 km (enough to make nine out of ten journeys). The range of existing electric trucks is expected to rapidly increase to 500 km, with the trips covering about two-thirds of the driven kilometers and 19 out of 20 trips [8].
- Further modelling and analyses are needed in the area of electrification to perfect the drive cycles of electric vehicles [9,10].

EMs are an integral part of either the full-time or part-time propulsion of the electric vehicle. With battery prices being expected to drop significantly in the next decade [11], it is extremely important to address other components as well. The objective of this research paper is to evaluate what type of EM topology is suitable for electric commercial vehicles without compromising on their performance and costs concerning the suitable EM topology/design.

2. State-of-the-Art Technology

The current literature offers methods to design e-machines using various parameters and calculations, however, there are no studies wherein we can gain an initial understanding of the cost of the e-machine that is being designed.

In an electric powertrain, electric energy is converted into mechanical energy by the electric machine (EM), and then, it is transmitted to the drive shaft, which is mechanically coupled to a mechanical reduction, such as a gearbox or a differential, or it is sent straight to the wheels [12]. A range of diverse machine types or topologies has resulted from the development of EM for various application use cases. In general, electric vehicles can use any known EM topology.

EMs supply power and torque to the wheels for propulsion in electric vehicles. Compared to an internal combustion engine, the efficiency of energy conversion by an EM is higher. With improved torque characteristics at lower speeds and instantaneous power ratings of two or three times the motor's rated power, an EM provides high torque and a large power density. During the braking of the vehicle, the EM turns in the opposite direction and converts mechanical energy into electrical energy, which is known as regenerative braking [13].

The selection of EM that drives the electric vehicles is a critical step that demands careful consideration. The automotive industry is still on the lookout for the best EM drive for EVs given the broad variety of topologies and individual EM designs available and the difficulties to characterize or standardize the requirements of the electric powertrain without over-dimensioned components. It is thus crucial to understand the requirements of the powertrain, and particularly, the EM.

The most basic demands for EMs are as follows [14]:

- A high-power density.
- High torque at low speeds during the beginning of the journey and ascending and high power at high velocities.
- An extensive speed range through a constant power area.
- A low level of operational noise.
- High efficiency throughout a wide speed range with steady torque and power levels.
- Excellent regenerative braking efficiency.
- A small size, a low weight, and a lower moment of inertia.
- Excellent reliability and resilience for a variety of vehicle operating environments.
- Cost efficiency.

Every EM has different highest efficiency operating points, which are dependent on a variety of design characteristics. For example, permanent magnet synchronous machines (PMSM) are preferable for applications near the rated speed and higher torque ones, whereas induction machines (IM) are better at higher speeds [15]. In comparison to passenger electric cars, the EM design for heavy-duty trucks must be changed due to the required high start-up torque but low speed requirements in the main operation points.

To start the analysis, the four most relevant EM topologies for the mobility applications of EMs were categorized, and these are the asynchronous motor (ASM)/induction motor (IM), the permanent magnet synchronous motor (PMSM), the externally excited synchronous motor (EESM), and the switched reluctance motor (SRM), which were analyzed to select the topology that best suits an EV and an electric commercial vehicle (eCV) application. Several performance characteristics such as the power density, efficiency, cost, noise levels, and control systems of the different electric machine types were examined, and a conclusion was made. The regenerative braking was assumed to be the same for all of the EMs. The results of this comparison are in Table 1.

Table 1. Rating of different e-machine types [2,16,17].

Characteristic	ASM/IM	PMSM	EESM	SRM
Power density	-	++	+	-
Efficiency	–	++	+	-
Cost	++	-	+	+
Noise level	++	++	-	-
Control systems	++	+	+	-

Advantage +; Disadvantage -.

Long-haul commercial vehicles must meet three vital criteria: low operation costs, high payloads, and range. A PMSM is usually favored for high-torque and low-speed applications due to the high efficiency areas of a PMSM that account for less power consumption and thus, the increased range of the vehicle compared to reduced efficiency and range due to the large magnetizing currents incurred in an IM and the consequentially high excitation losses in this type of machine.

Therefore, based on the comparison table and the critical performance requirements of electric commercial vehicles, the PMSM was selected to perform the analysis.

2.1. Selection of PMSM Topology

A PMSM runs as a synchronous machine when the air gap flux and stator windings are distributed in a sinusoidal pattern. PMSM's rare earth magnet material helps to boost the flux density in the air gap, the power density, and the torque-to-inertia ratio, allowing it to run at a wide constant power speed range [13].

The surface-mounted permanent magnet synchronous machine (SPMSM), in which the permanent magnets are attached to the rotor's surface, and the internal permanent magnet synchronous machine (IPMSM), in which the magnets are inserted into the rotor, are the two types of PMSM rotors [18].

Because of its simple production method, the SPMSM is quite affordable, and it also demonstrates high torque, although it experiences mechanical instability [19]. Eddy currents, on the other hand, can heat the magnets and cause irreversible demagnetization. Furthermore, to prevent the surface magnets from becoming detached at high rotational speeds, they must be bandaged to the rotor's surface.

Since the magnets are attached to the rotor, the centrifugal forces are absorbed by the superimposing laminated core. This problem can be overcome by utilizing IPMSM technology. It also protects magnets from mechanical damage and eddy currents due to the laminated core design [20]. Therefore, IPMSM was selected as the suitable topology.

2.2. Terms and Definitions

- Stack length: The length of the lamination stack that gives rigidity to the stator is referred to as the EM stack length. The winding end-turn lengths, which are not torque producing, are not included in the stack length.
- Rotor diameter: The rotor diameter of an EM is defined as the circle described by the moving rotor.
- Stator Diameter: The stator diameter of an EM is defined as the circle, which is described by the stationary stator.
- Shape Ratio (SR): The shape ratio of a PMSM is defined as the ratio of the stack length to the rotor diameter [21].

$$\text{Shape ratio, SR} = \frac{L_{stk}}{D_r} \quad (1)$$

- Split ratio (α): The split ratio of a PMSM is defined as the ratio of the rotor diameter to the stator diameter [21].

$$\text{Split ratio, } \alpha = \frac{D_r}{D_s} \quad (2)$$

3. Methodology

The methodology was developed to keep the solution as simple as possible, while considering critical design parameters and technical and economic aspects. Both quantitative and qualitative analyses have been performed for the selection of a suitable EM for the defined use cases. An overview of the followed steps is presented below in Figure 1.

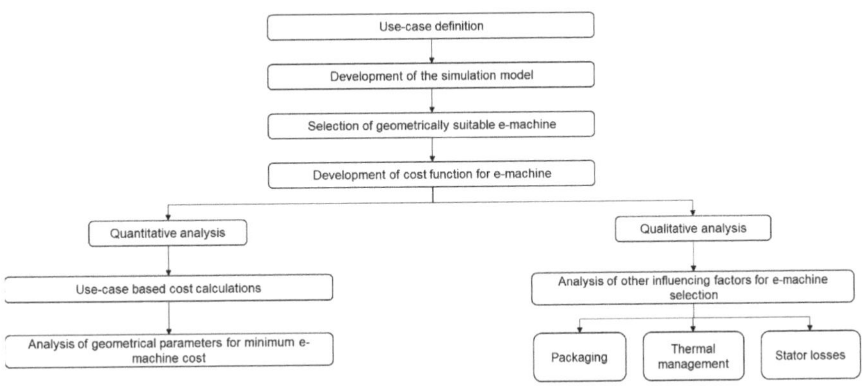

Figure 1. Overview of steps followed for methodology development.

3.1. Use Case Definition

Three different gross vehicle weights, three different values of shape ratios, and one value of split ratio were selected to define the use cases for the battery electric commercial vehicles. The WLTC drive cycles were modified, the maximum speed was limited to 90 km/h, and they were named WLTC1_mod, WLTC2_mod, and WLTC3_mod. The use cases are summarized in Table 2.

Table 2. Overview of all the use cases evaluated based on the weight of the vehicle, drive cycle, shape, and split ratio.

Weight of the Vehicle (GVW)	Drive Cycle	Shape Ratio (SR)	Split Ratio (α)
7.5 t	WLTC1_mod	0.5–2.0	0.62
	WLTC2_mod		
	WLTC3_mod		
18 t	WLTC1_mod	0.5–2.0	0.62
	WLTC2_mod		
	WLTC3_mod		
40 t	WLTC1_mod	0.5–2.0	0.62
	WLTC2_mod		
	WLTC3_mod		

The vehicle and powertrain parameters used in the research were selected from a previous study by the Chair of Production Engineering of E-Mobility Components (PEM) at RWTH Aachen. The overview of these parameters can be seen in Table 3.

Table 3. Overview of the vehicle and powertrain parameters used in the evaluation of the 7.5 t, 18 t, and 40 t vehicles.

	7.5 t	**18 t**	**40 t**	
Parameter	**Value**	**Value**	**Value**	**Unit**
Configuration of the vehicle	4 × 2	6 × 4	6 × 4	
Truck mass, fully loaded	7500	14,500	35,000	kg
Load of the vehicle	3500	8785	27,600	kg
Unloaded truck mass	4000	5715	7400	kg
Rolling resistance	0.01	0.01	0.01	
Frontal area	7.4	8.0	10.0	m^2
Air drag coefficient	0.6	0.6	0.6	
Dynamic wheel radius	0.37	0.5	0.45	m
Differential ratio	5.57	4.6	4.55	
Differential efficiency	0.98	1.0	0.98	
Gear ratio	2.5	2.5	2.5	
Gear efficiency	0.9	0.9	0.9	

3.2. Simulation Model Development

As can be seen in Figure 2, the complete layout of the simulation model consists of various sub-systems or blocks, namely, the driving cycle, vehicle, electric powertrain, BEV battery, fuel cell, and H_2 tank.

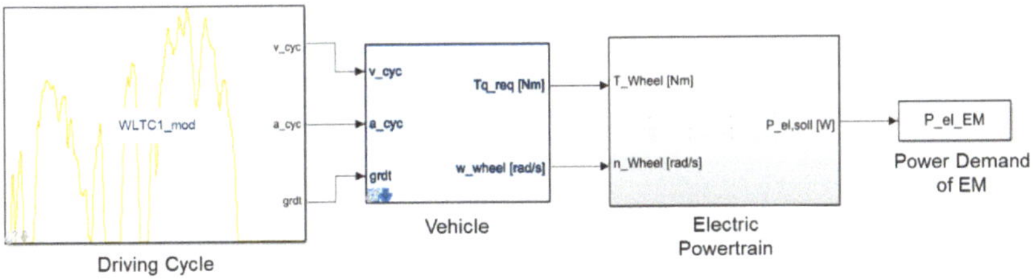

Figure 2. Overview of the simulation model.

All of the simulation blocks, apart from electric powertrain, were developed at PEM during previous studies and for the current project.

The basic process of this block is to use the torque and speed at the wheel of the vehicle as the inputs and perform calculations at the differential and transmission block, as illustrated in the process flow in Figure 3, while considering the final gear ratio and the transmission gear ratio along with the efficiencies of the respective gears. The torque and speed values after the transmission block were fed into the e-Machine block as inputs, thereby enabling the calculation of its power, torque, energy, and speed demands. This block selects the efficiencies at different operating points according to the efficiency map for a PMSM. The process to select the efficiency map is defined in the next subsection.

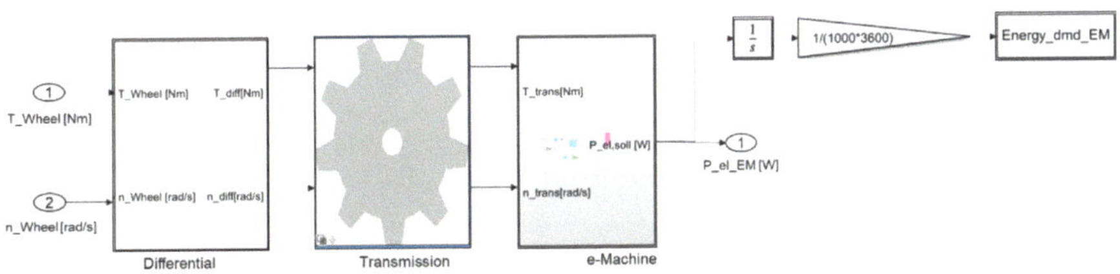

Figure 3. Inside view of the electric powertrain block.

3.2.1. Selection of Efficiency Maps

The efficiency maps that we used in our research (refer to Section 3.2.2) are available as open source for study and research purposes [22].

Additionally, these already generated efficiency maps correspond to the data in Tables 4 and 5. These efficiency maps were created by a research team from TUM Munich using their in-house developed tool, "Modell für den Entwurf und die Analyse einer PMSM oder ASM (MEAPA)" [20].

There are two steps for the selection of an EM efficiency map. The details of the steps used in the selection of efficiency maps are elaborated below. Steps 1 and 2, which are written below, provide a background of how the studies by the authors Wolff, Kalt et al. [20] and Kalt, Erhard et al. [22], respectively, were conducted and explain the underlying assumptions, parameters, and equations, etc., which have been used during their calculations for efficiency maps generation.

Table 4. Summary of heavy-duty trucks of various OEMs and their corresponding powertrain topology parameters considered [22].

S.No.	Concept	Topology	Gears	No. of Machines	Machine Type	Rated Power (kW)	Rated Rot. Velocity (rpm)	Max. EM Torque (Nm)
1	Axletech EPS785	Central motor	1	1	PMSM	350	-	3500
2	DAF CF Electric	Central motor	1	1	PMSM	210	8837	2000
3	Tesla Semi	distributed	1	4	PMSM	223	2525	380
4	E-Force EF18 SZM	Dual central motor	3	2	PMSM	150	-	2025
5	Nikola Tre	Dual central motor, eAxle	1	2	PMSM	240		900

Note: Rated rotational velocity is calculated using Equations (3) and (4).

Table 5. Assumptions of input parameters for the machine design tool MEAPA [22].

Parameter	Symbol	Value	Unit
Voltage (rated)	U	800	V
Number of pole pairs	P	4	-
Rot. Velocity (max)	n_{max}	1.75 n_N	rpm
Power factor	Cos	1	-
Number of phases	M	3	-
Magnet arrangements	-	Internal, embedded	-
Circuit wiring	-	Star	-
Wire type	-	Round wire	-
Cooling type	-	Liquid	-
Iron material	-	VACOFLUX50	-
Conductor material	-	Copper	-
Winding type	-	Single layer, integral slot	-

➢ Step 1:

In the first step, some existing EMs in heavy-duty trucks (40 t GVW) in the production or prototype stages are summarized [22]. Aside from the publicly known vehicle parameters, assumptions were made about the gearbox design and the electrical machine's subsequent operating speed. Machine efficiency maps are created based on the vehicle specifications.

Table 4 summarizes the powertrain architecture parameters of some on-the-market heavy-duty electric trucks and their corresponding electric powertrains. The topology, number of gears and e-machines, as well as the e-machine type, rated power, and torque, were obtained from fitting sources, whilst the rotational velocity of the machine identified as "n_{85}", which was rated for a maximum vehicle speed of 85 km/h, is determined using Equations (3) and (4) [22].

$$n_{85,\ rated} = \frac{85\frac{\mathrm{km}}{\mathrm{h}} * \frac{60}{3.6}}{2 * \pi * r_{dyn}} * i_{gear} \tag{3}$$

$$i_{gear} = \frac{35,000\ \text{Nm}}{(No.\,of\ machines) * (\text{max}torque)} \quad (4)$$

r_{dyn} = dynamic rolling radius of tires; i_{gear} = gear ratio.

Wolff, Kalt et al. considered the starter torque that is required for 40-tonne heavy-duty vehicles as 35,000 Nm as this enables typical road acceleration (>0.646 m/s^2) and sufficient torque reserves for starting on roads with gradients of 15–17% [22].

➢ Step 2: MEAPA Tool

In the second step, based on the input parameters from the above step, with the help of MATLAB, efficiency maps are generated using the MEAPA (Modell für den Entwurf und die Analyse einer PMSM oder ASM) tool developed by TUM, Munich.

In the initial step of the design model, the primary dimensions are determined based on the input parameters. The stator and rotor layouts can then be designed. The design parameters are then sent to the analysis model, which calculates the currents and voltages for the operating points within the valid operating range using a motor and generator model. Finally, the losses are calculated, and the efficiency diagram is created [20].

Assumptions used in the tool for creating efficiency maps:

- Because the number of pole pairs for the specified trucks is unknown, a value of 4 was used because lower pole pair numbers are often used for low-speed and high-torque applications [22].
- An IPMSM was selected with an embedded magnet assembly [22].

Apart from the number of pole pairs, several assumptions are made for the input parameters for the e-machine design tool MEAPA. These assumptions are stated below in Table 5.

3.2.2. Selected Efficiency Maps

A pre-processing approach was followed to select the efficiency maps for the research that was conducted. The demand torque and speed for selected use cases were derived from the simulation model that is explained in the previous section of the chapter. Two IPMSM efficiency maps were selected based on the torque and speed demands derived from the use cases considered for two vehicle categories; there was one efficiency map for 7.5 t vehicles and another one for the 18 t and 40 t vehicles [22]. The selected efficiency maps can cover the demand range of the use cases based on the GVW, drive cycles, and vehicle dynamics defined in Section 3.1. The efficiency map selected for the 7.5 t vehicle is for serial no. 3 and the one for the 18 t and 40 t vehicles are for serial no. 2, which can be seen in Table 4. The summary of the selected efficiency maps is given in Table 6.

Table 6. Summary of the selected efficiency maps [22].

Weight of the Vehicle (GVW)	Efficiency Map Characteristics	Value	Unit
7.5 t	Max. rated torque	380.88	Nm
	Max. rated speed	17,900	rpm
18 t and 40 t	Max. rated torque	2018	Nm
	Max. rated speed	15,465	rpm

3.3. *Selection of a Geometrically Suitable E-Machine*

In the design of a PMSM, the variables for sizing include the shape ratio (SR), the split ratio (α), the torque per rotor volume (TRV), and the torque density (TD), and these variables are important for determining the mechanical and electrical characteristics of the motor.

The rotor diameter, stack length, and stator diameter of the motor can be determined using these independent variables. According to the research conducted by Kim, H.-J. et al., the TD can be represented by Equation (5) [21]:

$$TD = \frac{T_{max}}{\frac{\pi}{4} * (D_s)^2 * L_{stk}} \quad (5)$$

To determine the rotor diameter, stator diameter, and stack length values of the e-machine, some assumptions about the values were made using the reviewed literature based on an experiment conducted on an IPMSM.

Based on Equations (1), (2) and (5), the following relationships were derived:

$$D_s = \sqrt[3]{\frac{T_{max}}{\frac{\pi}{4} * TD * SR * \alpha}} \quad (6)$$

$$D_r = \alpha * D_s \quad (7)$$

$$L_{stk} = SR * D_r \quad (8)$$

D_s = stator outer diameter (mm); D_r = rotor outer diameter (mm); L_{stk} = stack length (mm); T_{max} = max. torque required for application.

The assumed values of the parameters based on the literature are as follows:

Shape ratio (SR) = 0.5–2.0 [21].

Torque density (TD) = (7832 [21] + 28,860 [23])/2 = 18,346 Nm/m^3.

The shape ratio (SR) is commonly defined as being between 0.5 and 2 [21]. The value of the torque density (TD) is considered as an average of two values found in different works of literature, whereby 7832 Nm/m^3 is the value from the work of Kim and their team [21] and 28,860 Nm/m^3 is the typical value of torque density of a PMSM [23]. As an EM with a very small power rating was considered in the work of Kim and their team [21], another value of torque density was considered to have tangible results for the analysis in the research paper.

3.4. Development of the Cost Function for E-Machine

The cost function for the EM cost is based on a cost study of a PMSM which was conducted under the Lebenszykluskostenreduktion im elektrischen Verteilerverkehr (LiVe) project at PEM. The study is based on market data as well as assumptions. Though the detailed results of this study are not relevant to this research paper, the price components of the PMSM are. Three different reference e-machines, EM 1, EM 2, and EM 3, their geometrical parameters, and their price components were considered for this research.

Using the values of the stator diameter, rotor diameter, and stack length along with their corresponding costs, a cost function was developed to calculate the cost of the PMSM, which is directly dependent on the maximum value of torque (the max-load torque required for the application), which can be derived from the simulation model.

A relationship has been established between the geometric parameters, D_s, D_r, L_{stk}, T_{max}, α, and TD, as shown is Equations (6)–(8).

Once the geometric parameters D_s, D_r, and L_{stk} were calculated, a system of linear equations was developed using the three reference e-machines, EM 1, EM 2, and EM 3. The values of the geometric parameters for these machines along with their total cost form the components of linear equations. MATLAB was used to solve these linear equations to find the total cost of the EM.

In the next step, the component costs of the core components of the EM, i.e., the magnets, copper, shaft, rotor lamination stack, and stator lamination stack, were also considered during the analysis. The function is defined based on the geometry of the EM and various component costs from EM 1, EM 2, and EM 3. These functions were then implemented into the overall EM cost function script which was created with the MATLAB tool.

The assumed values of the parameters based on the reference e-machines are as follows:

- All of the parameters for EM 1, EM 2, and EM 3 were assumed to be constant, apart from the costs and geometry, while we were calculating the cost function.
- The split ratio for the use case analysis was considered as the average value of split ratios for the three reference e-machines. Split ratio (α) = (0.619 + 0.620 + 0.618)/3 = 0.62.
- T_{max} = Max_Torque (max. torque (Nm) required for applications; this was obtained from the simulation model).
- SR = 0.5 to 2.0.
- $\alpha = 0.62$.
- TD = 18,346.
- $D_s = 1000 * ((T_{max}/(0.25 * pi * TD * SR * \alpha))\ \hat{}\ (1/3))$.
- $D_r = \alpha * D_s$.
- $L_{stk} = SR * D_r$.

3.5. *Overview of the Selected Parameters and Assumptions*

The assumptions considered during this research are summarized below in Table 7. The assumptions have been already used, and their backgrounds are defined in the previous sub-sections under the methodology section.

Table 7. Overview of the assumptions and considered parameter values.

Parameter	Value	Comments
Type of vehicle	BEV	As defined in Section 3.1
GVW (t)	7.5, 18 and 40	
Drive Cycles	WLTC1_mod, WLTC2_mod, and WLTC3_mod	
Gear efficiency	0.9	
Differential efficiency	0.98	
Vehicle & powertrain parameters	various	
Shape ratio (SR)	0.5–2.0	As defined in Section 3.3
Split ratio (α)	0.62	
Torque density (Nm/m^3)	18,346	
Rated Voltage (V)	800	As defined in Section 3.2.1
Number of pole pairs	4	
Max. rotational speed (rpm)	1.75 n_N	
Power factor	1	
Number of phases	3	
Magnet arrangements	Internal, embedded	
Circuit wiring	Star	
Wire type	Round wire	
Cooling type	Liquid	
Iron material	VACOFLUX50	
Conductor material	Copper	
Winding type	Single layer, integral slot	

Table 7. *Cont.*

Parameter	Value	Comments
Efficiency map: 7.5 t	Max. rated torque (Nm)—380.88	As defined in Section 3.2.2
	Max. rated speed (rpm)—17,900	
Efficiency map: 18 t and 40 t	Max. rated torque (Nm)—2018	
	Max. rated speed (rpm)—15,465	
Gear reduction	Single speed gear reduction	-

4. Results and Analysis

In this section, the results of various use cases are shown. Both the quantitative and qualitative analyses were performed to research design parameters for the most cost-efficient e-machine and the effect of other decisive factors on the selection of an e-machine.

4.1. Quantitative Analysis: Results for 7.5 t Vehicle

Table 8 displays the consolidated results of the simulation with varying shape ratios. The effects of varying shape ratios on the geometry and cost of the e-machine are displayed. It can be inferred that the cost of the e-machine, stator OD and rotor OD decreases along with an increasing SR while the stack length increases.

Table 8. Overview of the behavior of e-machine cost based on different combinations of geometrical parameters split ratio (α) and shape ratio (SR) for 7.5 t vehicle and drive cycle WLTC1_mod.

Weight of the Vehicle	Drive Cycle	Split Ratio (α)	Shape Ratio (SR)	Stator OD (mm)	Rotor OD (mm)	Stack Length (mm)	E-Machine Cost (€)
7.5 t	WLTC1_mod	0.62	0.5	361.97	224.42	112.21	748.89
			0.6	340.63	211.19	126.71	753.61
			0.7	323.57	200.61	140.43	762.29
			0.8	309.48	191.88	153.50	773.52
			0.9	297.57	184.49	166.04	786.43
			1.0	287.30	178.12	178.12	800.52
			1.1	278.31	172.55	189.81	815.42
			1.2	270.36	167.62	201.15	830.91
			1.3	263.24	163.21	212.17	846.80
			1.4	256.82	159.23	222.92	862.99
			1.5	250.98	155.61	233.41	879.39
			1.6	245.64	152.29	243.67	895.92
			1.7	240.72	149.25	253.72	912.54
			1.8	236.18	146.43	263.57	929.20
			1.9	231.96	143.82	273.25	945.89
			2.0	228.03	141.38	282.75	962.58

Here are the values of the material cost distribution of the core components of the e-machine for the first and last sub-use cases, as shown above in Table 8.

Figures 4 and 5 demonstrate the material cost distribution of the core components both as a function of percentage as well as the cost (EUR/unit). It can be observed from the figures that while the rotor OD and stator OD decreased, the stack length increased with an increase in the values of the shape ratio (SR). This contributes to the increased magnet costs from 95 (EUR/unit) to approx. 240 (EUR/unit) and to the percentage increase in the magnets cost contribution from 53% to 60%.

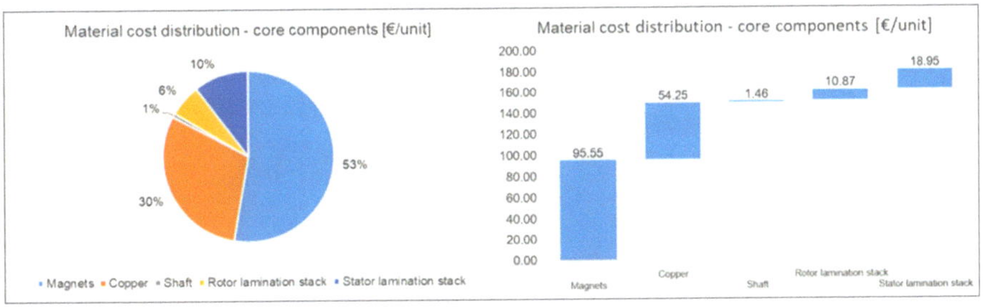

Figure 4. Material cost distribution of core components for drive cycle: WLTC1_mod, SR = 0.5, and α = 0.62.

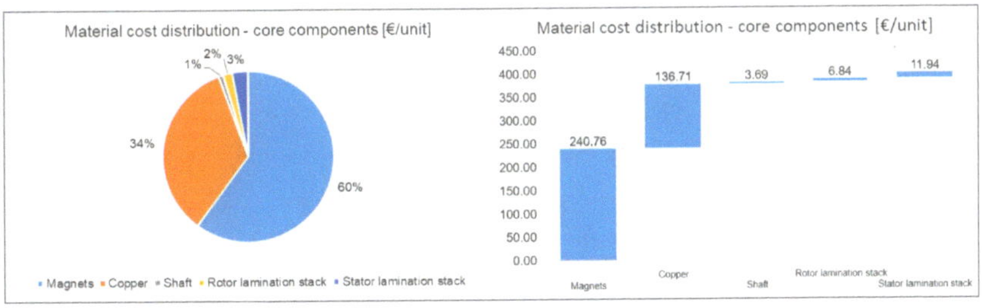

Figure 5. Material cost distribution of core components for drive cycle: WLTC1_SeLv, SR = 2.0, and α = 0.62.

Summary of Use Cases

Similarly, each use case defined in Section 3.1 was analyzed with a total of 144 sub-use cases for all of the vehicles, 7.5 t, 18 t and 40 t along with the selected drive cycles, and out of them, the following most cost-efficient combinations have been observed. The rest of the use cases have not been displayed due to repetitiveness in the pattern of the results, as similar kinds of results were observed for the three vehicles.

The E-machine cost behavior analysis demonstrates increasing SR values and a fixed α value for 7.5 t vehicle and WLTC1_mod drive cycle.

As can be inferred from Figure 6, the most cost-efficient e-machine is at the very beginning of the plot, and this is true for all of the different use cases that have been simulated.

It can be observed in Tables 8–11 that the cost of the e-machine increases as the SR increases. This is because the stack length of the e-machine increases, which results in a significant increase in the core component costs. Thus, α = 0.62 and SR = 0.5 are the values of the design parameters for the most cost-efficient e-machine for every use case that was considered.

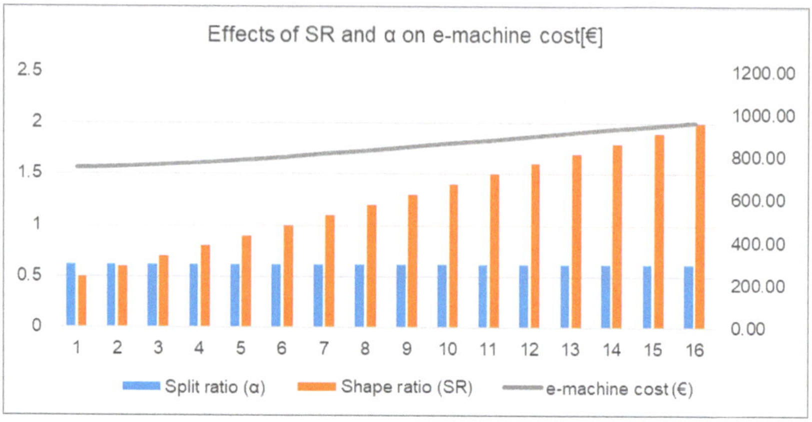

Figure 6. Cost-efficient e-machine geometry and costs for a 7.5 t vehicle.

Table 9. Overview of the most cost-effective e-machines for different use cases in the 7.5 t vehicle category.

Gross Vehicle Weight (t)	Drive Cycle	Split Ratio (α)	Shape Ratio (SR)	Stator OD (mm)	Rotor OD (mm)	Stack Length (mm)	Cost of the Core Components (€)					E-Machine Cost (€)
							Magnets	Copper	Shaft	Rotor Lamination Stack	Stator Lamination Stack	
7.5	WLTC1_mod	0.62	0.5	361.97	224.42	112.21	95.55	54.25	1.46	10.87	18.95	748.89
	WLTC2_mod		0.5	388.98	241.17	120.58	102.68	58.30	1.57	11.68	20.37	804.77
	WLTC3_mod		0.5	416.27	258.09	129.04	109.88	62.39	1.68	12.50	21.79	861.24

Table 10. Overview of the most cost-effective e-machines for different use cases in the 18 t vehicle category.

Gross Vehicle Weight (t)	Drive Cycle	Split Ratio (α)	Shape Ratio (SR)	Stator OD (mm)	Rotor OD (mm)	Stack Length (mm)	Cost of the Core Components (€)					E-Machine Cost (€)
							Magnets	Copper	Shaft	Rotor Lamination Stack	Stator Lamination Stack	
18	WLTC1_mod	0.62	0.5	381.55	236.56	118.28	100.71	57.19	1.54	11.45	19.98	789.40
	WLTC2_mod		0.5	395.65	245.30	122.65	104.44	59.30	1.60	11.88	20.71	818.57
	WLTC3_mod		0.5	541.31	335.61	167.81	142.88	81.13	2.19	16.25	28.34	1111.93

Table 11. Overview of the most cost-effective e-machines for different use cases in the 40 t vehicle category.

Gross Vehicle Weight (t)	Drive Cycle	Split Ratio (α)	Shape Ratio (SR)	Stator OD (mm)	Rotor OD (mm)	Stack Length (mm)	Cost of the Core Components (€)					E-Machine Cost (€)
							Magnets	Copper	Shaft	Rotor Lamination Stack	Stator Lamination Stack	
40	WLTC1_mod	0.62	0.5	445.46	276.18	138.09	117.58	66.76	1.80	13.37	23.32	921.62
	WLTC2_mod		0.5	519.95	322.37	161.18	137.24	77.93	2.10	15.61	27.22	1075.72
	WLTC3_mod		0.5	718.21	445.29	222.64	189.58	107.64	2.90	21.56	37.60	1485.91

4.2. Qualitative Analysis: Other Influencing Factors for E-Machine Selection

As observed, the values of the stator OD and rotor OD usually decrease with the increasing value of SR, and the stack length increases with an increasing SR. As per the analysis performed by Kim, H.-J. et al., the efficiency of the analyzed e-machine is 94.8% and 95.7% at $\alpha = 0.48$ and $\alpha = 0.70$, respectively. It is also stated that the efficiency tends to increase as the α increases, but it decreases as the α exceeds by about 0.7. Therefore, we can assume that during our analysis at $\alpha = 0.62$, the efficiency of the e-machine lies between 94.8% and 95.7%.

The geometries of the e-machine are referred to as 'e-machine A', 'e-machine B', and 'e-machine C', hereafter.

As illustrated in Figure 7, when one is considering the vehicle's powertrain/e-machine packaging requirements, either of these geometries might be more suitable compared to the other. Therefore, the values of design parameters α and SR were considered with other vehicle requirements.

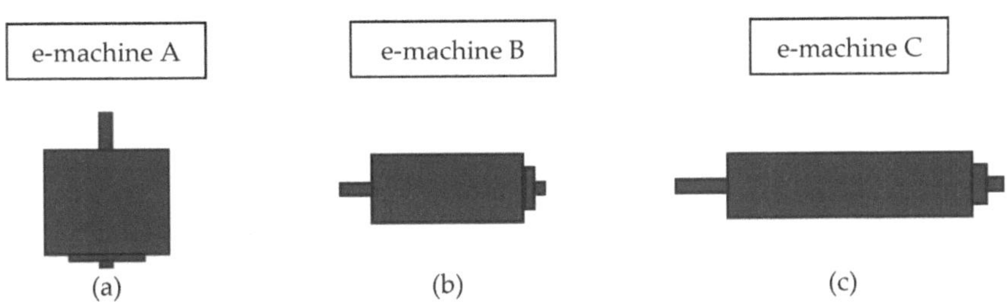

Figure 7. Illustration of geometry of an e-machine. (**a**) represents a shape ratio (SR) = 0.5 and (**b**) represents a shape ratio (SR) = 1.2 and (**c**) represents a shape ratio (SR) = 2.0.

4.2.1. Packaging Analysis

In the case of the in-wheel e-machines, the geometry of the e-machine with SR = 0.5 (e-machine A) might be more suitable compared to that of an e-machine with SR = 2.0 (e-machine C). This is illustrated below in Figure 8a.

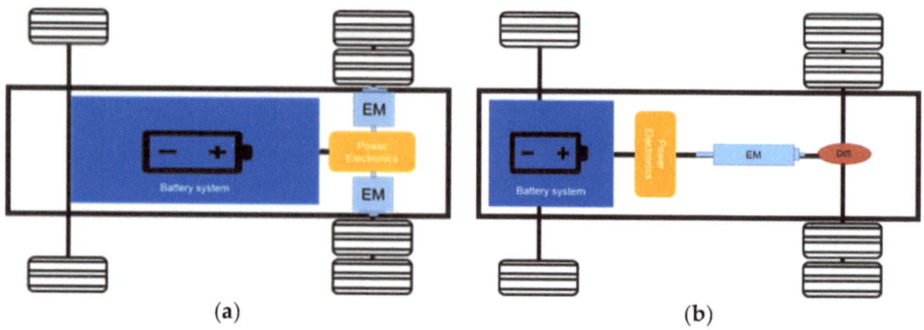

Figure 8. Illustration of suitable geometry for (**a**) an in-wheel e-machine and (**b**) central arrangement.

As the 'e-machine C′' has a longer stack length, it will not leave enough room in the frame to accommodate two e-machines as it accommodates two e-machines with geometry 1. The 'e-machine B′' geometry could be suitable for this use case depending upon the chassis structure of the vehicle.

The battery layout and capacity play a vital role in deciding on an electric commercial vehicle, particularly for long-haul applications.

Though, due to the advancements in battery cell technology, smaller batteries are becoming more efficient, still, for heavier segment electric vehicles, a large battery is considered to be essential to meet the application requests. This factor also inclines towards the e-machine with a larger stator OD and smaller stack length (e-machine A). Nevertheless, e-machine C is suitable for a central arrangement as illustrated in Figure 8b.

4.2.2. Thermal Management Analysis

It is possible to improve the overall efficiency of an electric motor by adequately cooling the copper windings. Additionally, maintaining a high level of cooling allows the motor's power output to be enhanced. As a result, it can be concluded that adequately cooling the copper windings of electric motors is critical to increasing the power density of electric machines [24].

To remove heat energy from within, most of the e-machines available today use the conventional jacket cooling method. A coolant travels through channels, running through the e-machine casing, and it removes the thermal energy from the perimeter of the motor in this operation. The heat created by components such as motor windings must pass through the stator into the motor casing, and then, into the cooling fluid to be evacuated for this cooling process to work. There is a significant thermal resistance that prevents the created thermal energy from being dissipated because there is such a long distance to travel and so many contacts to conduct through.

Figure 9 depicts the alternative paths for heat to go to the extraction points, which are represented by the blue sections of the diagram.

The following equation can be used to compute the thermal resistance of individual components in the thermal network.

$$R = \frac{l}{\kappa\, A} \tag{9}$$

The length of the conduction path, l, the thermal conductivity of the medium, κ, and the cross-sectional area of the conduction path, A, can be used to calculate the thermal resistance, R.

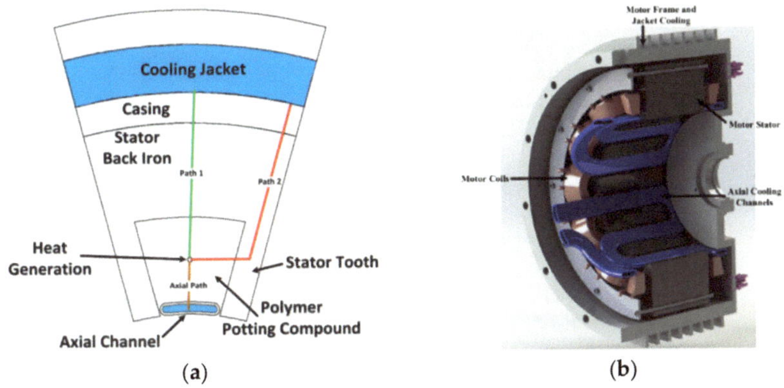

Figure 9. (**a**) Conduction paths and methods for thermal management: the cooling jacket and axial channel and (**b**) a schematic view of e-machine cross-section with conduction paths [24].

The most reasonable strategy to improve the thermal energy removal efficiency is to lower the thermal resistance encountered by the energy generated. As inferred from Equation (9), this can be accomplished in several ways, including reducing the conduction length, increasing the conduction area, or boosting the thermal conductivity of the heat-transfer medium. For different methods, the effects of e-machine geometry will be different, and these are explained below.

We assume that the thermal conductivity of the medium and the cross-sectional area of the conduction path remains constant.

- Cooling jacket method: As shown in Figure 9a, the e-machine with a shorter conduction path will offer lesser thermal resistance. This can be achieved by having an 'e-machine C' with a smaller stator diameter.
- Axial channel method: In this method, the heat transfer takes place along the length of the e-machine, as shown in Figure 9a,b. Therefore, an 'e-machine A' with a smaller stack length will offer less thermal resistance.

The 'e-machine B' will fall in between 'e-machine A' and 'e-machine C' for both of the methods.

4.2.3. Stator Loss Analysis

The major losses in a PMSM are the stator losses. According to Yu, Y. et al. [25], the components of the stator loss, i.e., hysteresis loss, eddy current loss, and stray loss, can be expressed as Equation (10):

$$W_{Fe} = k_h\, B_m^2\, f k_f + \sum\nolimits_{\Delta T}\left[\sigma\frac{d^2}{12}\left(\frac{\Delta B}{\Delta T}\right)^2 + k_e\left(\frac{\Delta B}{\Delta T}\right)^{1.5}\right]\Delta T f k_f \tag{10}$$

where B_m is the max. the magnetic field strength in one electric period, ΔB is an increment of the magnetic field intensity with one step, ΔT is the time step, k_h is hysteresis loss factor, k_f is coefficient of lamination, k_e is stray loss, σ is the conductivity of steel sheet, and d is the length of stator iron core.

If the parameters such as Bm, ΔB, the hysteresis loss factor, the coefficient of lamination, and the conductivity of the sheet are assumed to be constant over the time step, ΔT, it can be interpreted that the stator loss is directly proportional to the length of the stator iron core. Therefore, a larger stack length might result in larger stator losses, thereby reducing the efficiency of the e-machine. Hence, the geometry of 'e-machine A' might be more suitable in this case, however, bench test results will provide sufficient data to analyze the impact of other factors as well, apart from the length of the stator iron core.

5. Discussion

As learned during the research, $\alpha = 0.62$ and SR = 0.5 are the values of the split and shape ratios, respectively, for the most cost-efficient EM for any use case. However, as it was observed, the values of the stator and rotor OD usually decrease with the increasing value of SR, and the stack length increases with an increasing SR.

A qualitative analysis of the packaging requirements, thermal management, and stator losses was performed to analyze what geometry would be most suitable in different conditions.

Limitations and Future Research

Furthermore, an experimental increase in the efficiency of the EM was observed in the work of Kim and their team [21] when the α increased to 0.70, and then, the efficiency fell. Thus, an experimental setup might offer a more accurate scenario.

Another point could be to include multiple reference e-machines which will broaden the scope of the design and cost function while utilizing several values of design parameters along with the associated component costs. Additionally, a similar methodology can be utilized by designing another type of e-machine, such as an EESM or an IM, etc., and then, deriving the costs.

The factors such as the operating conditions cost over the lifetime and the ease of manufacturing cost-effective EM components could be analyzed.

Apart from this, the study could collaborate with several test bench analyses, such as a reliability analysis of the selected e-machine geometry, load, and no-load tests, as well as finite element analysis (FEA), which would enable one to comment on the experimental values of the EM parameters, such as the efficiency at operating points of the use case, reliability over defined time and distance, etc.

6. Conclusions

There was a gap in the available literature, whereby the initial understanding of the cost of the e-machine was under development, and it was not available in other studies. This work highlights a methodology that was developed to help understand the economic aspect as well as the performance factors. A novel methodology is presented in this work to help select the initial geometry of the e-machine during the design process, a cost estimation based on the geometry of the electric motor, and an estimate of the effect of design parameters on the cost of the electric motor and its components.

The most cost-efficient EM might not be suitable for all of the use cases in a real-life scenario, therefore, other methods such as a qualitative analysis, a reliability analysis, and the operating conditions over a lifetime, etc., should be considered to obtain the best trade-off between the cost and the performance. The developed methodology that is presented in this work is helpful for:

- Selecting the initial geometry of the electric motor during the design process.
- Carrying out a cost estimation based on the geometry of the electric motor.
- Estimating the effect of the design parameters on the cost of the electric motor and its components.

Moreover, the developed methodology can be used for other electric powertrain topologies as well, regardless of the energy source.

Author Contributions: Conceptualization, J.G.D.G. and C.S.; methodology, J.G.D.G., C.S. and F.B.; software, C.S.; validation, J.G.D.G. and C.S.; formal analysis, C.S.; investigation, J.G.D.G. and C.S.; resources, A.K., H.H. and B.D.; data curation, C.S.; writing—original draft preparation, C.S.; writing—review and editing, J.G.D.G. and F.B.; visualization, J.G.D.G. and C.S.; supervision, F.B. and B.D.; project administration, H.H., B.D. and A.K.; funding acquisition, A.K. and H.H. All authors have read and agreed to the published version of the manuscript.

Funding: This research was funded by the German Federal Ministry for the Environment, Nature Conservation, Nuclear Safety and Consumer Protection (BMUV), grant number 16EM4016-1. The project name was "LiVePLuS".

Data Availability Statement: The data that support the findings of this study are available from the corresponding author, J.G.D.G., upon reasonable request.

Conflicts of Interest: The authors declare no conflict of interest.

References

1. Fries, A.M.; Kerler, A.M.; Rohr, A.S.; Schickram, A.B.S.; Sinning, A.M.; Lienkamp, A.B.M.; Kochhan, A.B.R.; Fuchs, A.S.; Reuter, A.B.; Burda, A.P.; et al. An Overview of Costs for Vehicle Components, Fuels, Greenhouse Gas Emissions and Total Cost of Ownership Update 2017. *Res. Gate* **2017**, *4*.
2. Sharma, S.; Kumar, V. Optimized Motor Selection for Various Hybrid and Electric Vehicles. In Proceedings of the SAE Technical Paper Series 8th SAEINDIA International Mobility Conference & Exposition and Commercial Vehicle Engineering Congress 2013 (SIMCOMVEC), Chennai, India, 4 December 2013; SAE International400 Commonwealth Drive: Warrendale, PA, USA, 2013.
3. Kampker, A.; Pandey, R.; Gomez, J.G.D.; Wessel, S.; Treichel, P.-E.; Malatyali, I. Cost Optimal Design Strategy of Electric Drivetrains for Medium Heavy-Duty Vehicles Based on Product Development and Production Costs. In Proceedings of the 2019 9th International Electric Drives Production Conference (EDPC), Esslingen, Germany, 3–4 December 2019; pp. 1–8, ISBN 978-1-7281-4319-4.
4. Liimatainen, H.; van Vliet, O.; Aplyn, D. The potential of electric trucks—An international commodity-level analysis. *Appl. Energy* **2019**, *236*, 804–814. [CrossRef]
5. Unterlohner, F. *Comparison of Hydrogen and Battery Electric Trucks*; Transport & Environment: Brussels, Belgium, 2020.
6. Mathieu, L.; Cornelis, S.; Nix, J.; Bannon, E. *Recharge EU Trucks: Time to Act!—A Roadmap for Electric Truck Charging Infrastructure Deployment*; Transport & Environment: Brussels, Belgium, 2020.
7. Kampker, A.; Krciskother, K.; Buning, M.K.; Gomez, J.G.D. Technological and Total Cost of Ownership Analysis of Electric Powertrain Concepts for Long-Haul Transport in Comparison to Traditional Powertrain Concepts. In Proceedings of the 8th International Electric Drives Production Conference, Schweinfurt, Germany, 4–5 December 2018; IEEE: Piscataway, NJ, USA, 2018.
8. Mathieu, L. *Unlocking Electric Trucking in EU: Recharging in Cities*; Transport & Environment: Brussels, Belgium, 2020.
9. Ekberg, K.; Eriksson, L. Electrification of a Heavy-Duty CI Truck—Comparison of Electric Turbocharger and Crank Shaft Motor. *Energies* **2021**, *14*, 1402. [CrossRef]
10. Dollinger, M.; Fischerauer, G. Model-Based Range Prediction for Electric Cars and Trucks under Real-World Conditions. *Energies* **2021**, *14*, 5804. [CrossRef]
11. Sanguesa, J.; Torres-Sanz, V.; Garrido, P.; Martinez, F.; Marquez-Barja, J. A Review on Electric Vehicles: Technologies and Challenges. *Smart Cities* **2021**, *4*, 372–404. [CrossRef]
12. Fischer, R. *Elektrotechnik Für Maschinenbauer—Sowie für Studierende der Versorgungstechnik, des Wirtschaftsingenieurwesens und Anderer Technischer Fachbereiche*; Springer: Berlin/Heidelberg, Germany, 2016.
13. Karki, A.; Phuyal, S.; Tuladhar, D.; Basnet, S.; Shrestha, B.P. Status of Pure Electric Vehicle Power Train Technology and Future Prospects. *ASI* **2020**, *3*, 35. [CrossRef]
14. Zeraoulia, M.; Benbouzid, M.; Diallo, D. Electric Motor Drive Selection Issues for HEV Propulsion Systems: A Comparative Study. In Proceedings of the 2005 IEEE Vehicle Power and Propulsion Conference, Chicago, IL, USA, 7–9 September 2005; pp. 280–287, ISBN 0-7803-9280-9.
15. Xue, X.D.; Cheng, K.W.E.; Cheung, N.C. Selection of Electric Motor Drives for Electric Vehicles. In Proceedings of the 2008 Australasian Universities Power Engineering Conference, Sydney, NSW, Australia, 14–17 December 2008.
16. de Santiago, J.; Bernhoff, H.; Ekergård, B.; Eriksson, S.; Ferhatovic, S.; Waters, R.; Leijon, M. Electrical Motor Drivelines in Commercial All-Electric Vehicles: A Review. *IEEE Trans. Veh. Technol.* **2012**, *61*, 475–484. [CrossRef]
17. Fang, Q.; Scharfenstein, D.; Weiss, C.; Müller, C.; Schwarzer, U. *Motor Handbook*; Elsevier: Amsterdam, The Netherlands, 2019.
18. Scharfenstein, D.; Weiss, C.; Qi, F.; Müller, C.; Schwarzer, U. *Handbook of Electric Machines*; Infineon AG: Munich, Germany, 2019.
19. Lee, D.-K.; Ro, J.-S. Analysis and Design of a High-Performance Traction Motor for Heavy-Duty Vehicles. *Energies* **2020**, *13*, 3150. [CrossRef]
20. Kalt, S.; Erhard, J.; Lienkamp, M. Electric Machine Design Tool for Permanent Magnet Synchronous Machines and Induction Machines. *Machines* **2020**, *8*, 15. [CrossRef]
21. Kim, H.-J.; Jeong, J.-S.; Yoon, M.-H.; Moon, J.-W.; Hong, J.-P. Simple Size Determination of Permanent-Magnet Synchronous Machines. *IEEE Trans. Ind. Electron.* **2017**, *64*, 7972–7983. [CrossRef]
22. Wolff, S.; Kalt, S.; Bstieler, M.; Lienkamp, M. Influence of Powertrain Topology and Electric Machine Design on Efficiency of Battery Electric Trucks—A Simulative Case-Study. *Energies* **2021**, *14*, 328. [CrossRef]
23. Ehsani, M.; Gao, Y.; Gay, S. Characterization of electric motor drives for traction applications. In Proceedings of the 29th Annual Conference of the IEEE Industrial Electronics Society, Roanoke, VA, USA, 2–6 November 2003.

24. Rhebergen, C. The Importance of Electric Motor Thermal Management and the Role of Polymer Composites in Axial Cooling. Master thesis, McMaster University, Hamilton, ON, Canada, 2015.
25. Yu, Y.; Liang, D.; Liang, Z.; Ze, Q. Calculation for stator loss of high-speed permanent magnet synchronous machine in torque-speed envelope and restraint approach for circulating current in windings. *Trans. Electr. Mach. Syst.* **2018**, 2, 211–219. [CrossRef]

MDPI AG
Grosspeteranlage 5
4052 Basel
Switzerland
Tel.: +41 61 683 77 34

MDPI Books Editorial Office
E-mail: books@mdpi.com
www.mdpi.com/books

www.ingramcontent.com/pod-product-compliance
Lightning Source LLC
Chambersburg PA
CBHW041712270926
51850CB00027B/3410
9783725823079